Sustainability Solutions

An Interdisciplinary and Transdisciplinary Nexus

Sustainability Solutions addresses a wide range of global challenges with applied solutions in the interdisciplinary and transdisciplinary sustainability nexus. Titles in the series present a strong sustainable development focus, and topics which connect environment, economic and social applications are encouraged. Topics could include environmental impacts, vulnerability, risk assessment and reduction, human-environmental systems (e.g. food, water, health) and innovative solutions to climate change. Volumes in the series present novel solutions to societal challenges using innovative application, for example demonstrating a solutions-based approach using machine learning, artificial intelligence, decision support systems, advance geospatial technology. Case studies and shorter report volumes which present knowledge transfer for policymaking and governance are also encouraged.

Keywords:

Interdisciplinary environmental solutions, transdisciplinary knowledge transfer, geospatial technologies, natural resource management, decision support system, artificial intelligence, deep learning, smart technology, environmental science, economic, social and spatial, human-environment interactions, sustainable development

Debojyoti Moulick · Shuvasish Choudhury ·
Jayanta Kumar Biswas · Meththika Vithanage
Editors

Biochar: Achieving Agro-Environmental Sustainability

Springer

Editors
Debojyoti Moulick
Department of Environmental Science
University of Kalyani
Kalyani University, West Bengal, India

Jayanta Kumar Biswas
Department of Ecological Studies
and International Centre for Ecological
Engineering
University of Kalyani
Kalyani, West Bengal, India

Shuvasish Choudhury
Department of Life Science
and Bioinformatics
Assam (Central) University
Silchar, Assam, India

Meththika Vithanage
Department of Geochemistry
Geological Survey of Denmark
and Greenland
Copenhagen, Denmark

ISSN 3059-278X ISSN 3059-2798 (electronic)
Sustainability Solutions
ISBN 978-3-032-19854-9 ISBN 978-3-032-19855-6 (eBook)
https://doi.org/10.1007/978-3-032-19855-6

This Springer imprint is published by the registered company Springer Nature Switzerland AG
The registered company address is: Gewerbestrasse 11, 6330 Cham, Switzerland

Preface

Owing to fast industrialization and urbanization, numerous pollutants are coming up at a distressing rate, which leads to impending damage to agro-ecosystems. In the present situation, searching for mitigation strategies to combat the increasing risk of pollutants (conventional and emerging) is one of the major challenges for the global scientific community. Moreover, an upsurge in the human population is also posing an additional obstacle to achieving the SDGs. This edited volume, entitled *Biochar: Achieving Agro-Environmental Sustainability*, provides a wide spectrum of potential substitutes for the remediation of pollutants in the form of biochar prepared from diverse sources. Additionally, this book offers an inclusive discussion on the potentials and challenges of biochar production, optimization, artificial intelligence (AI), and machine learning (ML) based inputs on increasing biochar's efficacy in the cleaning up of the contamination from agro-environments. In addition, the book also discusses several eco-friendly, inexpensive, and aesthetically acceptable biochar-centered decontamination technologies. The book is an outstanding effort to bring up biochar-based expertise for sustainable environmental decontamination in solitary bind. In this volume, a total of nineteen chapters have been incorporated that comprehensively address diverse issues and evaluate the prospects and potentials of biochar as a probable yet sustainable solution with marginal effort.

Chapter 1 describes the ameliorating potentials of biochar application in crop(s) cultivated in salinity-stressed agro-ecosystem with special reference to nutrient dynamics. In Chap. 2, the authors focus on the consequences of biochar supplementation in restoring soil fertility, overall soil health, along with Carbon Sequestration, aiming for enhanced agro-ecosystem resilience to stressor(s). Chapter 3 explores a greener and sustainable solution using biochar to enhance soil fertility and minimize yield gaps. Chapter 4 deals with the consequences of biochar application on soil microbiome, evolving better climate-resilient agriculture and soil ecological refurbishment through variation in microbial diversity and metabolic pathways. In Chap. 5, the authors have comprehensively discussed the issue pertaining to the extent of application of biochar and its effectiveness in minimizing soil arsenic contamination, and the long-term stability of success is the central theme of this chapter. In Chap. 6, factors affecting biochar production, its characteristics, and effectiveness that in turn affect the ameliorating efficacy of biochar against numerous agro-environmental odds have been addressed by using AI/ML-based approaches in this chapter. In Chap. 7, the authors have addressed the issue of how effective biochar application is in agro-ecosystems, especially in GHG emission minimization, and sustainable biomass utilization has been assessed in this chapter. Chapter 8 focuses on several issues regarding the effectiveness of biochar application in ameliorating inorganic contaminants like cadmium, chromium, and arsenic content in agro-ecosystems. In Chap. 9, the authors have presented a comprehensive outline on the effective application of biochar in agriculture, especially in controlling emerging contaminants in a sustainable way. In Chap. 10, the authors discussed the potential of biochar application in regenerative agriculture, focusing on sustainable development goals (SDGs). The content of Chap. 11 deals with the modification of biochar properties with nanotechnology-based approaches, aiming for larger efficacy towards sustainable environmental management. In Chap. 12, the authors present a detailed overview of biochar's production methods, aiming for greater soil fertility, nutrient retention, reducing nutrient leaching, increasing microbial activity and diversity, and augmenting the structure of soil and water-holding capability. Chapter 13 presents a holistic assessment of various feedstock-derived biochar for sustainable wastewater treatment processes from a circular economic perspective. In Chap. 14, the authors provide a critical insight into biochar application in agro-ecosystems with special emphasis on NPK dynamics. Apart from this, existing knowledge gaps and future perspectives towards maximizing biochar's role in the quest for sustainable food production. In Chap. 15, an analysis of global research output regarding algal biochar has been presented to explain the evolving trends and research themes thereby recognizing the developing frontiers and knowledge gaps are highlighted in this chapter. In Chap. 16, the authors have assessed the potential and challenges associated with sustainable heavy metal and organic contaminant remediation by the application of biochar in the near future. In Chap. 17, an effort has been made to highlight the importance of cattle manure biochar in soil management, underlining its potential to drive a climate-resilient agro-environmental system while justifying the environmental footprint of livestock farming. Chapter 18 explores the implications of AI-driven models to improve the efficiency and accuracy of the biochar synthesis process. Advanced ML algorithms, like ANNs, SVMs, RF, and deep learning frameworks, have confirmed exceptional precision in forecasting biochar yield, surface area, porosity, and adsorption efficiency. In Chap. 19, the authors highlight the recent developments, discourses, and challenges associated with data heterogeneity and model generalization, and summarize future directions for AI-assisted biochar manufacturing for greater efficacy in dealing with challenges to achieve sustainable agriculture.

This edited volume will be extremely useful for a variety of reader, including agronomists, agro-chemists, nanotechnologists, soil scientists, biotechnologists, environmentalists, microbiologists, environmental biotechnologists, policy makers, start-up entrepreneurs, and those who are involved in clean, eco-friendly technologies (NGOs).

We are thankful to all the contributors who have contributed in this volume and to the Springer team for their considerable teamwork for printing up this volume.

Kolkata, India	Dr. Debojyoti Moulick
Silchar, India	Dr. Shuvasish Choudhury
Kalyani, India	Prof. Jayanta Kumar Biswas
Colombo, Sri Lanka	Prof. Meththika Vithanage

Contents

About the Editors

Dr. Debojyoti Moulick worked as a Post-Doctoral Fellow at Assam University, Silchar, India. In 2018 he bagged the prestigious ***Research Associateship (DBT-RA)*** award from the *Indian Institute of Science (IISc), Bangalore and Department of Biotechnology, Govt. of India*. Recently in 2025, Dr. Moulick have secured his place among ***world's top 2% most influential scientists list***, across all fields prepared by *Stanford University-USA and Elsevier*, based on bibliometric data from the *Scopus* database. He is a graduate in Microbiology and Post-graduate in Biotechnology. He was awarded with M.Phil. and Ph.D. degrees in Environmental Science by *University of Kalyani*, West Bengal, India. Dr. Moulick is also recipient of *Senior Research Fellowship (SRF)* from the Ministry of Environment and Forest for supporting his Ph.D. research. Beside these, he was also awarded *Senior Research Fellowship (SRF)* by the Indian Council for Agricultural Research (ICAR), Government of India. He carried out his PhD work under the supervision of Prof. S. C. Santra, a well-recognized Environmental Biotechnologist and a global expert of arsenic toxicity. Before joining to M.Phil., he worked on *gas-hydrates* and *Geochemical exploration in Geochemistry Division of KDMIPE of ONGC* in Dehradun, India under the supervision of Dr. A. K. Jain. Dr. Moulick has gathered extensive experience in biomonitoring, seed priming technology and phytotoxicity, grain quality analysis etc. in soil—water—crop agroecosystem. In 2018, Dr. Moulick's research work have been recognized by *Department of Science and Technology* (Government of India) in

Vigyan Prasar (India Science Wire) and several media hoses as a breakthrough in proposing an alternative, cheap and farmer-friendly mitigation option for minimizing Arsenic-induced phytotoxicity and transmission through rice plant and grain respectively. Apart from these, five edited volumes have published since 2022. Moreover, from February 2017 onwards he has reviewed 340+ papers for various international publication houses, edited book (CRC, Press), empanelled as an *editorial board member* {**BMC** ***Plant Biology*** (*Springer Nature*), Integrated Environmental Assessment and Management (***Oxford University Press***)}. As of September 2025, Dr. Moulick's publications are cited over 2300+ times as per Google Scholar with an h-index of 30 and i10 index of 59.

Dr. Shuvasish Choudhury is working as an Associate Professor of Plant Science at Assam University, Silchar, India. He has been ranked ***world's top 2% most influential scientists list*** for 2023–2024 by Stanford University and Elsevier. He completed his Ph.D. in plant science from the Assam University, Silchar, India and thereafter did his post-doctoral research from Oklahoma State University, USA. His research interest involves the understanding molecular physiology and metabolomics of abiotic stress responses in plants, nutrient management for abiotic stress management and exploring sustainable means such as biochar and compost for mitigation of heavy metals, metalloids and soil acidity of agro-ecosystems. He has published several research papers and reviews in various international journals, as well as several book chapters in the areas of abiotic stress. He has also edited six books published by leading international and national publishers. He has been awarded the Young Botanist Award by the Indian Botanical Society in 2010 and also received the DST (SERB)—Fast Track Young Scientist Award, DBT Overseas Associateship by Government of India. He has been actively engaged in editorial and review work for leading international journals in the area of plant abiotic stress.

Prof. Jayanta Kumar Biswas ranked among world's top 2% scientist list. At present he is working as the Professor and HOD of Department of Ecological Studies, and International Centre for Ecological Engineering, University of Kalyani, Kalyani 741235, West Bengal, India. He obtained M.Sc. in Zoology, M.Phil. in Ecology and Ph.D. for his work on ecotechnological management of aquatic systems. His spectrum of multidisciplinary research interest and expertise span the following areas: Soil and water contamination and remediation (toxic metals, metalloids and emerging contaminants); Ecological engineering and ecotechnology; Ecotoxicology; Environmental microbiology; Nanobiotechnology. A consistent rank holder all through his academic career Professor Biswas received many scholarships, awards and fellowships including Fellow, National Institute of Ecology (FNIE); Fellow, West Bengal Academy of Science and Technology (FAScT); Fellow, National Environmentalists Association (FNEA); Fellow, Zoological Society of India (FZSI); Fellow, Zoological Society, Kolkata (FZS); Senior Scientist Award and Gold Medal (ZSI); Best Research Paper Award (AESA); Outstanding Reviewer Award (Chemosphere (Elsevier) and Environmental Geochemistry and Health (Springer); Senior Research Fellow (ICAR); National Merit Scholarship (MHRD, GoI), etc. He is credited with publishing 7 books, 35 book chapters and 150 original research papers in reputed international journals, with about 8000+ citations, H-index: 49, i10-Index: 167, and cumulative IF > 750. He is serving as associate editor, advisory/editorial board member of several international journals of repute, namely *Science of the Total Environment* (Elsevier); *Environmental Chemistry Letters* (Springer); *Environmental Technology and Innovation* (Elsevier); *Ambio* (Springer); *iScience* (Cell Press: Elsevier); *Heliyon* (Cell Press/Elsevier); *Environmental Geochemistry and Health* (Springer); *Environmental Management* (Springer); *Ecotoxicology* (Springer); *Archives of Environmental Contamination and Toxicology* (Springer); *Current Pollution Reports* (Springer); *Environmental Chemistry and Ecotoxicology* (Elsevier and CSPM); etc. He visited several countries with government travel fellowships, to present research papers and chair scientific sessions in International conferences.

Prof. Meththika Suharshini Vithanage is working as the Professor/Director, Ecosphere Resilience Research Centre, Faculty of Applied Sciences, University of Sri Jayewardenepura, Sri Lanka. She is also having the distinction of being a Highly Cited Researcher as per the Clarivate Inc. Prof. Vithanage's contribution to science has been recognized by The World Academy of Sciences (TWAS) presenting the Fayzah M. Al-Kharafi award in 2020. She is a Fellow of the National Academy of Sciences, Sri Lanka and Institute of Chemistry, Ceylon. She was selected as one of the Early Career Women Scientists by the Organization for Women Scientists in Developing Countries, Italy. Additionally, she also received the Best Graduate Researcher Award for Natural Hazards by the American Geophysical Union in 2010. Her outstanding performance in research publications was recognized by the Presidential Awards, Sri Lanka for consecutive 8 years from 2011. Furthermore, she was listed as Top 2% of the most cited scientists across various disciplines globally from 2017 onwards. She holds a Sri Lankan patent for synthesis of organo-stabilized nano particulate zero valent iron (17,302) and two more patents for biochar in South Korea. She has co-edited 4 books with Elsevier Inc. in 2019 and 2020. She is a prominent scientist in the field of environmental science with a citation record is over to 34,000+ with an H index of 96+.

Chapter 1
Role of Biochar for the Improvement of Nutrient Dynamics in Crops Under Coastal Saline Ecosystem

Saikat Dey, Anannya Dhar, Sourav Basak, Suman Dutta, Sourav Garai, and Sukamal Sarkar

Abstract Biochar is a carbon-rich porous material produced through the thermal decomposition of organic biomass, including agricultural residues, woody biomass, and other biodegradable materials, via a process known as pyrolysis. This thermochemical conversion involves complex mechanisms that modify the structural and chemical properties of the biomass, resulting in a highly porous and stable carbonaceous product. Due to its multifunctional applications in degraded soil management, environmental remediation, and sustainable agriculture, biochar is often called "wonder char." The nutrient composition of biochar is influenced by the type of pyrolysis, temperature, and feedstock used. Plant-based biochar contains a higher carbon content but lower nutrient concentrations compared to manure-based biochar. Additionally, slow pyrolysis generally produces a higher biochar yield with elevated carbon content and greater concentrations of essential plant nutrients than fast pyrolysis. Applying biochar in salt-affected soil significantly alters soil's biophysical and chemical properties, including pH, cation exchange capacity, bulk density, porosity, moisture retention capacity, and microbial activity. These modifications are crucial in regulating soil nutrient dynamics in saline soil. Biochar acts as a direct source of essential plant nutrients, enhances nutrient bioavailability by modulating adsorption-desorption equilibria, and improves soil physicochemical properties, thereby facilitating nutrient retention and cycling processes. This chapter highlights how biochar-induced alterations in soil properties and microbial communities contribute to the stabilization and mobilization of soil nutrients, ultimately enhancing crop growth, soil fertility, and overall sustainability of the coastal saline ecosystem.

S. Dey · A. Dhar · S. Basak · S. Garai · S. Sarkar (✉)
Division of Agronomy, School of Agriculture and Rural Development, Ramakrishna Mission Vivekananda Educational and Research Institute, Narendrapur, Kolkata, India
e-mail: sukamal.sarkar@gm.rkmvu.ac.in

S. Dutta
Department of Genetics and Plant Breeding, School of Agriculture and Rural Development, Ramakrishna Mission Vivekananda Educational and Research Institute, Narendrapur, Kolkata, India

D. Moulick et al. (eds.), *Biochar: Achieving Agro-Environmental Sustainability*, Sustainability Solutions, https://doi.org/10.1007/978-3-032-19855-6_1

Keywords Biochar · Pyrolysis · Nutrient use efficiency · Soil properties

1.1 Introduction

Saline soil refers to the condition where the soil solution contains a higher concentration of soluble salts (EC more than 4 dS/m) beyond the threshold levels for most of the cultivated crops, which has a negative impact on the growth and development of the cultivated plant (Hailu and Mehari 2021; Soni et al. 2023). The presence of soil salinity is usually indicated by changes in soil structure, with a white crust of salt forming on the surface and scattered patches of salinity appearing throughout the profile (Eswar et al. 2021). Climate change affects the intensity, frequency, and duration of extreme climatic events, including flooding, moisture stress (Sarkar et al. 2024a, b). Therefore, sea levels are rising, and the intrusion of saline water into the cropland enhances the soil salinity, affects groundwater storage, increases salt concentration of freshwater, and changes nutrient fluxes, resulting in plants often being compromised with agricultural produce (Dey et al. 2025). Salt-affected soil is categorized into three such as saline, saline-alkaline, and alkaline soil. Saline soil contains high levels of soluble salts (Cl^- and SO_4^{2-} of Na and Mg), a pH less than 8.0, and an electrical conductivity (EC) is more than 4 dS/m (Mahajan et al. 2020). While in alkaline soil, Na is the predominant cation along with carbonate and bicarbonate anions, and saline-alkaline soil shows the characteristics of both saline and alkaline soil (Kumar et al. 2020).

Soil salinity is one of the critical factors that hinder crop production. It creates a significant menace to global food and nutritional security. Saline soil is mainly found in arid and semi-arid regions in Asia, Australia, South America, and Africa, encompassing approximately 1 billion ha of land (Babar et al. 2025). It is projected to increase in the upcoming years because of improper management and climate shifts (Irfan et al. 2023). Salinity imparts a wide range of negative impacts on soil, which encompasses disrupted soil physical characteristics, compromised nutrient equilibrium, osmotic stress, deficiency, and toxicity of mineral nutrients, leading to reduced crop yield and lower economic outcomes (Safdar et al. 2019). It is one of the crucial factors that exhibit poor physical features of the soil, including dispersed soil aggregates and improper drainage, which can cause waterlogging. At the same time, higher concentrations of Na^+ and Cl^- are entered through the roots of the plant and create an ionic imbalance in the cytoplasm, promoting hindrance against K^+ uptake and other macro/micro nutrients (Megha and Shankhdhar 2022). The optimum crop yield and productivity are always compromised by higher levels of salt concentration in soil, as it interferes with major plant biochemical and metabolic functions that inhibit growth, and even crops may die (Ali et al. 2017). Saline soil not only limits agricultural production, but it also affects the soil health and disrupts the ecological balance of the agroecosystems (Abbas et al. 2022). Therefore, to alleviate the excessive soluble salts or exchangeable Na^+ in soil, and sustain the crop yield, ensure essential nutrient uptake, and impose a favorable condition for plant growth, appropriate soil management is essential through organic amendments (Wang et al. 2023).

The application of biochar in coastal saline soil would be a significant approach for the reclamation of salt-affected soil, which can positively improve soil properties, more precisely, the physicochemical features (Singh et al. 2016; Fan et al. 2016). Biochar has multiple essential roles, including the rectification of degraded soil, the remediation of contaminated soil, and subsequently, increasing crop yield and productivity employing direct and indirect positive impact on all properties and nutrient dynamics of soil (Hossain et al. 2020). Biochar can affect soil nutrients in many possible ways, including acting as a direct source of essential macro and micronutrients, and helps in storing nutrients on soil colloids, resulting in the bioavailability of minerals and altering soil properties as a soil conditioner (Li et al. 2017a; Lusiba et al. 2017). According to the study conducted by Purakayastha et al. (2019), biochar can provide essential nutrients like N, P, K, Ca other micronutrients to the plants that are derived from the feedstocks used. Generally, biochar contains a low amount of N and S compared to other nutrients because most of them are lost through gaseous emission during pyrolysis (Leng et al. 2020; Mohanty et al. 2013). The application of biochar in soil can potentially decrease the nutrient losses through better adsorption capacity with soil, though it is pH-dependent (Lehmann and Joseph 2015; Lehmann et al. 2021). Since this chapter determines the influence of biochar on soil fertility, nutrient cycling, and availability to the plants, especially on coastal saline soil, which could help to promote sustainable ecosystem services and efficient soil management through optimum crop production.

1.2 Soil Salinity and Its Characteristics

Soil salinity refers to the excessive *accumulation* of soluble salts in the soil and water beyond a critical threshold, which adversely impacts plant-water relations and overall crop performance. This condition is particularly widespread in arid and semi-arid regions, where high evapotranspiration rates, limited precipitation, and reliance on irrigation exacerbate salt buildup in the soil profile (Hassani et al. 2020; Hopmans et al. 2021). Irrigation without proper drainage leads to salt retention in the root zone, as water is removed through evapotranspiration while salts remain and gradually accumulate at or near the surface (Liu et al. 2023). As a result, approximately one-third of the world's irrigated land is affected by salinity, making it a serious constraint in global crop production (Eswar et al. 2021).

Saline soils are characterized by a high concentration of soluble salts, with an electrical conductivity (EC) greater than 4 dS/m, exchangeable sodium percentage (ESP) below 15, and pH less than 8.5 (Minhas 2021). These soils contain abundant neutral salts, mainly chlorides and sulfates of sodium, calcium, and magnesium, but do not have excessive sodium bound to soil particles. As a result, their physical structure generally remains intact. However, the elevated salt content reduces the availability of water to plants by increasing osmotic pressure, leading to reduced germination, stunted growth, and yield loss in salt-sensitive crops. While slightly saline soils may only affect a few species, moderate to high salinity can severely limit the productivity of most crops unless salt-tolerant varieties are cultivated.

Soil salinization occurs through two main pathways: primary and secondary salinization. Primary salinization results from natural processes such as the weathering of mineral-rich parent rocks, deposition of salts from saline groundwater, and proximity to salt-affected landscapes like salt pans, coastal marshes, and saline lakes. In coastal regions, sea-level rise has also intensified saltwater intrusion into freshwater systems, further contributing to natural salinity (Kaushal et al. 2021). In contrast, secondary salinization arises from human-induced factors, particularly improper irrigation practices using low-quality water, excessive application of salt-based fertilizers, land-use changes such as deforestation, and inadequate drainage (Ondrasek and Rengel 2021). These activities alter the soil-water balance and facilitate the accumulation of salts in the root zone, especially in arid and semi-arid climates where evapotranspiration exceeds precipitation. Over time, these processes degrade soil structure, reduce permeability, and impair nutrient cycling, severely limiting crop productivity and land sustainability.

The global extent of salinity has reached critical levels, with over 1 billion ha of land affected, including more than 20% of the world's irrigated areas (Hopmans et al. 2021). Countries like India, China, the United States, Pakistan, Sudan, and Turkey account for the majority of salt-affected lands. Climate change is intensifying the issue by altering hydrological cycles, increasing the frequency of extreme events like floods and droughts, and shifting the balance between freshwater and saline water in coastal zones. If these trends continue without effective intervention, projections suggest that more than half of irrigated lands could become salinized within the next few decades. The economic implications are equally alarming, with potential annual losses in agricultural productivity estimated to exceed \$27 billion globally (Wang et al. 2021a). Thus, combating salinization is not only vital for soil health but also crucial for ensuring food security, environmental resilience, and the livelihoods of millions dependent on agriculture.

Soil salinity significantly alters soil properties by increasing osmotic pressure, which limits water availability to both plants and soil microbes, even when moisture is present (Rengasamy 2016; Maathuis et al. 2014). High salt concentrations restrict water and nutrient uptake, while excess sodium disrupts soil structure by causing clay dispersion, reducing infiltration, aeration, and hydraulic conductivity. The collapse of soil aggregates due to sodium-induced clay dispersion not only impairs physical properties but also restricts the movement and retention of nutrients like nitrogen, potassium, and phosphorus, thereby aggravating nutrient leaching and deficiency in plants. These changes degrade soil stability and lower its water-holding capacity. Salinity also reduces soil organic matter and cation exchange capacity (CEC), negatively affecting soil fertility (Gonçalo Filho et al. 2019). The microbial population, vital for nutrient cycling and organic matter decomposition, is highly sensitive to salt stress. High salt concentrations cause osmotic imbalance in microbial cells, inhibiting growth and activity, especially among fungi and other sensitive species (Grossiord et al. 2020; Mahmood et al. 2021). The reduction in microbial diversity, especially beneficial decomposers and nitrogen-fixing bacteria, leads to slower organic matter turnover, diminishing mineralization of key nutrients like nitrogen, phosphorus, and sulfur—ultimately exacerbating soil nutrient limitations (Farooqi et al. 2023). As a result, overall soil health and productivity decline under prolonged saline conditions.

1.3 Effect of Soil Salinity on Plant Growth

Salinity in soil adversely influences plant growth by inducing osmotic stress, toxic ion accumulation, nutrient disruption, and oxidative injury. Salinity hampers the uptake and translocation of essential macronutrients such as nitrogen, potassium, and phosphorus, while also limiting micronutrient availability like zinc and iron, due to ionic antagonism and altered root permeability (Chhabra 2022). Elevated concentrations of salts like Na^+, Cl^-, and SO_4^{2-} reduce the soil water potential, impairing water uptake and leading to physiological drought. Excessive salts also disrupt membrane stability and enzymatic functions, causing ionic toxicity and oxidative stress. Salinity also disturbs hormonal balance by increasing abscisic acid (ABA) levels and reducing auxin and cytokinin activity, which disrupts cell division and elongation, compounding growth suppression (Choudhary et al. 2021). These conditions hinder seed germination, reduce vegetative growth, limit root and shoot elongation, and ultimately suppress reproductive development and yield (Naz et al. 2021). Additionally, salinity interferes with nutrient uptake, particularly phosphorus due to precipitation with calcium, further exacerbating nutrient deficiencies. As a result, plants grown under saline conditions often display stunted growth, leaf necrosis, reduced photosynthesis, and poor productivity (Meng et al. 2018).

1.4 Biochar as a Sustainable Soil Ameliorant

Biochar, has gained global attention as an effective soil ameliorant due to its remarkable ability to rejuvenate degraded soils, enhance fertility, boost crop productivity, sequester carbon, and reduce greenhouse gas emissions (Zhao et al. 2024). When used as a remediation tool, biochar can immobilize organic pollutants, reducing their availability to plants. Biochar supports overall soil health by improving both structural and biochemical functions (Koné and Galiegue 2023; Rafiq et al. 2016). Biochar is produced through pyrolysis, a process where biomass is heated in an oxygen-limited or anoxic environment, typically at temperatures ranging from 300 to 1000 °C (Lehmann et al. 2006; Pandey et al. 2020). The quality and properties of the resulting biochar depend on factors such as the feedstock material, pyrolysis temperature, processing conditions, and residence time. Consequently, different types of biochar affect soil in different ways, and no single type of biochar is universally effective for all soil types (Van Zwieten et al. 2010; Jiao et al. 2021).

Biochar derived from lignocellulosic biomass (such as woody plants) tends to have higher fixed carbon content, greater porosity, a larger surface area, and higher ash content (Osman et al. 2023). In contrast, non-woody biomass sources like crop residues and manures produce biochar with lower carbon stability, less porosity, and reduced surface area, but higher nutrient content and pH, making it beneficial for many soil types (Dahlawi et al. 2018; Gholizadeh et al. 2024). Additionally, the temperature of pyrolysis significantly influences the nature of the biochar. Higher pyrolysis temperatures result in lower volatile matter, increased pore space, and higher ash content (Sigua et al. 2015, 2016). This also leads to greater carbon recalcitrance and aromaticity, allowing the biochar to persist in soil for years to even millennia. On the other hand, low-temperature pyrolysis results in less compositional change and a more labile biochar. The impact of biochar on soil health depends on various factors, including the soil's organic matter content, texture, structure, and mineral composition. Research has shown that applying biochar to degraded or weathered soils often leads to improvements in fertility and crop yield, more so than in already nutrient-rich, well-structured soils.

Biochar improves the physical structure, chemical properties, and microbial activity of soil, enhancing nutrient cycling and water and mineral uptake (Zheng et al. 2018; Anand et al. 2023). In salt-affected soils, where high concentrations of sodium chloride (Cl^-) and sodium sulfate (SO_4^{2-}) degrade soil structure and reduce organic matter, biochar has been particularly beneficial (Shwe et al. 2024). Its application increases cation exchange capacity (CEC), boosts soil organic carbon, and enhances the availability of essential cations like calcium, magnesium, potassium, manganese, and zinc.

The larger surface area of biochar supports better cation exchange, allowing excess sodium (Na^+) to be replaced by beneficial cations. Negatively charged soil colloids preferentially attract divalent cations (Ca^{2+}, Mg^{2+}) over monovalent ones like sodium, leading to a decrease in sodium levels in the soil solution (Jemal and Yakob 2021; Nair et al. 2023). Additionally, improved porosity and stabilized soil structure enhance water retention, facilitating the leaching of replaced sodium ions. As a carbon-rich and highly recalcitrant material, biochar promotes long-term stability of organic compounds in the soil far better than other organic amendments, which tend to decompose more quickly. The increased carbon content also makes the soil a more hospitable environment for beneficial microorganisms, which helps mitigate the adverse effects of salinity.

1.5 Genesis and Diversity of Biochar

Biochar has gained renewed significance in modern organic farming, though its roots trace back to the Amazon's fertile "terra preta" soils. These dark, nutrient-rich soils, found as deep as two meters, have sustained Amazonian agriculture for hundreds of years (Neves et al. 2004). The existence of terra preta indicates intentional human creation, demonstrating early knowledge of soil enhancement. Historical evidence suggests that biochar production through biomass carbonization dates back to ancient human civilizations (Panwar et al. 2019). By the mid-nineteenth century, the wood distillation industry thrived, but its prominence declined between 1920 and 1950 with the rise of petroleum (Panwar et al. 2019). The 1970s oil crisis renewed interest in alternative fuels, leading to the development of advanced pyrolysis reactors for bio-oil extraction (Bridgwater et al. 1999). Research into biomass pyrolysis expanded from the 1970s through the 1990s, refining the understanding of its chemical processes (Evans and Milne 1987; Piskorz et al. 1989; Bridgwater and Peacocke 1994). By the 1980s and 1990s, commercial pyrolysis methods—such as fast, flash, vacuum, and ablative pyrolysis—emerged, enabling large-scale biochar and bio-oil production (Yang et al. 1995; Roy et al. 1997; Czernik et al. 2002). Recent advancements propose bio-oil refineries capable of extracting valuable byproducts from biomass (Helle et al. 2007; Mahfud et al. 2007; Jones et al. 2009). For soil enhancement, slow pyrolysis is preferred due to its higher biochar yield (Prins et al. 2006). This method involves heating biomass between 300 and 600 °C over an extended period (Onay and Kockar 2003).

Biochar, a carbon-rich material produced through thermal decomposition of organic matter, exhibits varied properties and applications depending on production methods and feedstock (Liu et al. 2015). It is primarily generated through pyrolysis, with slow pyrolysis (300–700 °C for several hours) being most common for agricultural use due to its high biochar yield (35–50%) and stable carbon structure, while fast pyrolysis (500–1000 °C for seconds) favors bio-oil production with lower biochar output (10–20%), and gasification (700–1200 °C) mainly produces syngas with minimal biochar (5–10%) (Panwar et al. 2019). The characteristics of biochar vary significantly based on feedstock type, with woody biomass yielding high-carbon but low-nutrient biochar, and manure-based biochar containing more nitrogen and phosphorus (Ighalo et al. 2025). Pyrolysis temperature also plays a crucial role, as higher temperatures (> 600 °C) create more porous and stable biochar ideal for long-term carbon sequestration, whereas lower temperatures (300–500 °C) retain more organic compounds beneficial for soil fertility (Panwar et al. 2019; Ighalo et al. 2025). Additionally, physical and chemical modifications, such as acid/alkali treatments or mineral doping, can further enhance properties of biochar for specific uses (Kumar et al. 2022). This diversity enables tailored applications, including soil amendment for agriculture, pollutant adsorption for environmental remediation, solid fuel for energy production, and waste management solutions, making biochar a versatile tool in sustainable development. As research progresses, optimizing production techniques will continue to expand the potential of biochar across various sectors.

1.6 Mechanisms of Biochar in Improving Soil Properties in Saline Soil

Biochar is known to have significant importance on soil physical, chemical, biological, and microbial properties. The mechanisms of biochar in improving soil properties are described as follows.

1.6.1 Soil Physical Properties

The application of biochar markedly alters the physical properties of saline soils, resulting in improvements in bulk density, particle density, and porosity (Table 1.1). Specifically, biochar contributes to a reduction in bulk density, which is advantageous as it indicates decreased soil compaction, thereby facilitating enhanced root penetration and nutrient uptake by plants. This decrease in bulk density is also associated with modifications in pore-size distribution and improved soil water movement and retention capacity. Additionally, biochar has been shown to reduce particle density in saline soils, further contributing to overall improvements in soil structure. The increment of soil porosity increases the retention of soil moisture and soil nutrients. In a pot study conducted by Liu et al. (2024), it was observed that the application of biochar has enhanced the soil physical properties by increasing soil porosity and moisture holding capacity and reducing the bulk density of salt-affected soil. This may be attributed to low bulk density and large specific surface area of biochar, which reduces overall soil bulk density and improves pore structure under saline conditions. Enhanced soil porosity and water-holding capacity promote salt leaching, thereby lowering soil salinity. Application of biochar derived from residues of cotton, groundnut, and sawdust decreased the bulk density of the experimental soil (Pandian et al. 2016). It further increased the volumetric moisture content and reduced the micro aggregate and increased the macro aggregate, making the soil conditions better for plant growth (Yue et al. 2023). High Na^+ levels in saline soil disperse clay particles and break down aggregates. Biochar, rich in organic functional groups and divalent cations, helps rebind these particles, promoting aggregation. Its hydrophilic surface and porous structure improve water retention, increasing soil volumetric water content. This aligns with reduced bulk density, formation of larger pores, and enhanced water-holding capacity.

1.6.2 Soil Chemical Properties

Biochar has shown potential in ameliorating salt-affected soils by influencing soil electrical conductivity (EC) and enhancing salt leaching (Table 1.2). When biochar with a lower EC than the surrounding soil is applied, it can promote the movement of

Table 1.1 Effect of biochar on physical properties of saline soil

Crop	Soil salinity level	Dose of biochar	Effect on bulk density	Effect on particle density	Effect on porosity	Effect on water holding capacity	Effect on soil aggregates	References
Tomato	2400 µS/cm	2 t/ha	↓0.6 g/cc	↓1.57 g/cc	↑18.32%	–	–	Zonayet et al. (2023)
Lemon balm	0.60% ($NaCl + Na_2SO_4$)	150 g in 750 g of soil	↓16.67%	–	↑4.59%	↑18.93%	–	Liu et al. (2024)
Maize	1955 µS/cm	175 t/hm^2	↓14.18%	–	–	↑11.99%	↓17.38%	Yue et al. (2023)

Increment (↑) or reduction (↓) of the values are from the control treatment, i.e., no application of biochar

salts away from the root zone, thereby improving plant growth conditions (Fig. 1.1). This process also supports improved soil structure, as reduced EC is often associated with enhanced hydraulic conductivity and porosity. However, if the applied biochar has a higher EC than the native soil, it may contribute additional salts, potentially worsening salinity issues. Therefore, the effectiveness of biochar in salt management largely depends on its inherent properties, particularly its EC relative to that of the soil. The application of biochar to saline soils has been shown to reduce the exchangeable sodium percentage (ESP), which in turn contributes to a decrease in soil pH. Additionally, the initial pH of the biochar itself plays a significant role in influencing the pH of salt-affected soils.

Biochar, derived from the pyrolysis of plant-based materials, is a beneficial soil amendment for salt-affected soils due to its high nutrient content (Table 1.3). The mechanism behind nutrient availability due to biochar includes supplying nutrients, retaining them to reduce leaching and enhance availability, and altering soil properties that affect nutrient cycling and interactions (Hossain et al. 2020). Its application can improve the nutrient status of saline soils by supplying essential elements like phosphorus, potassium, calcium, magnesium and organic carbon, which are crucial for plant growth. Additionally, biochar enhances the activity of P-solubilizing bacteria in the rhizosphere, further increasing phosphorus availability. As a rich source of stable carbon, biochar supports microbial activity and contributes to long-term carbon storage in the soil. Different biochar types increased CEC in saline soil due to their oxidized acidic functional groups and high surface area, which enhances negative surface charges. Applying biochar to saline soil significantly increased concentrations of available nitrate and total nitrogen, likely due to enhanced nitrification activity. Biochar also improved phosphorus availability, consistent with studies attributing this to the release of phosphorus from biochar's inherently high phosphorus content. Additionally, available potassium levels increased, owing to potassium richness and porous structure of biochar, which supports potassium-solubilizing

Table 1.2 Effect of biochar on chemical properties of saline soil

Crop	Soil salinity level	Dose of biochar	Effect on soil pH	Effect on soil EC	Effect on soil ESP	Effect on organic carbon	Effect on CEC	References
Maize	1955 μS/cm	75 t/hm^2	↓0.08	↓44 μS/cm	↓9.64%	↑21.53 g/kg	↑1.27 cmol/kg	Yue et al. (2023)
Argula	4.12 dS/m	5% sorghum stalk biochar	–	↓3.6 μS/cm	–	–	↑1.86 cmol/kg	ELsaman et al. (2025)
Tomato	2400 μS/cm	2 t/ha	↓1.33	↓2700 μS/cm	–	↑0.87%	–	Zonayet et al. (2023)
Wheat	4 mS/cm	5 g/kg	–	↓124 μS/cm	–	↑0.56%	–	Ababsa et al. (2023)
Rice	1200 μS/cm	40 g/barrel	↑0.2	↓11.5%	–	–	–	Zhang et al. (2024)

Increment (↑) or reduction (↓) of the values are from the control treatment, i.e., no application of biochar

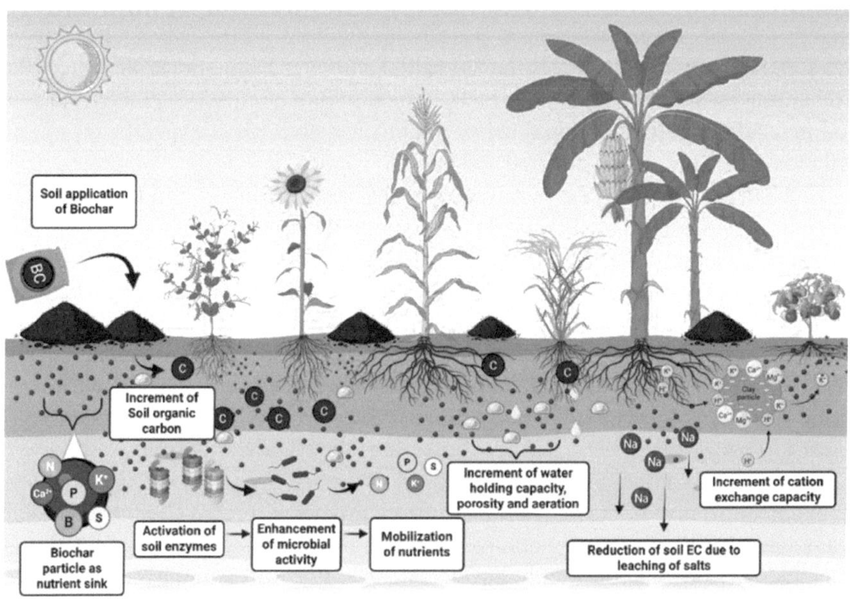

Fig. 1.1 Schematic diagram of the impact of biochar on soil properties and its associated chemistry. Image is created by Biorender.com; License Number: NW28JU76B2

microorganisms. At higher application rates, biochar further enhanced the nutrient-holding capacity, reducing nutrient leaching and improving nutrient availability for plant uptake.

1.6.3 Soil Biological Properties

The incorporation of biochar into saline soils brings about notable enhancements in soil biological functions by creating conditions that support diverse and active microbial communities. Soil containing a good number of beneficial microbes is essential for optimal plant growth and development (Sarkar et al. 2024a). Through its porous structure and chemical properties, biochar offers refuge and nutrients that promote the growth and resilience of soil microorganisms, which are typically hindered by high salinity levels. This improved microbial habitat encourages the presence of beneficial microbes, including salt-tolerant bacterial groups such as Bacteroidetes and Verrucomicrobia, while diminishing the impact of salinity-sensitive species. In tandem, biochar applications stimulate key enzymatic activities, including those involved in nutrient mineralization and organic matter decomposition, thereby reinforcing essential processes like nitrogen cycling and phosphorus availability. Addition of biochar is reported to increase the soil respiration (Moradi et al. 2019) and activity of soil enzymes like catalase, urease, sucrase, alkaline phosphatase, and (Formate dehydrogenase) FDH (Manasa et al. 2020) (Table 1.4). These biological advancements are strongly associated with the capacity of biochar to boost soil organic carbon, improve nutrient retention, and buffer soil pH, factors that collectively create a more favorable soil environment. The high stability of biochar plays a crucial role in its long-term persistence and slow breakdown in the soil, even under microbial influence. This stability supports greater microbial diversity, abundance, and activity, particularly in soils enriched with carbon. In an experiment conducted by Zhang et al. (2024), it was observed that the application of biochar derived from residues of maize, wheat, and groundnut modulated the soil enzyme activity of saline soil at a depth of 0–10 cm. The urease and alkaline phosphatase activity were increased by 194.8–744.6% and 28.2–64.8% respectively, due to biochar treatment as compared to the control. However, the opposite was observed with catalase activity in soil; with an increase in biochar dose, the enzyme's activity decreased. This improvement is mainly due to biochar enhancing soil properties, such as increasing available potassium and organic matter, which supports microbial activity. Its porous structure offers a suitable habitat for microbes that produce enzymes, boosting their activity and aiding nutrient cycling.

Table 1.3 Effect of biochar on nutrient content of saline soil

Crop	Soil salinity	Dose of biochar	Macronutrient					Micronutrient	References
			Effect on nitrogen	Effect on phosphorus	Effect on potassium	Effect of calcium	Effect of magnesium	Effect of Cl^-	
Maize	1955 μS/cm	175 t/hm^2	↑7.12 mg/kg	↑5.93 mg/kg	↑847.48 mg/kg	↑0.11 cmol/kg (Ca^{2+})	↑0.56 cmol/kg (Mg^{2+})	↑Cl^- 0.08 cmol/kg	Yue et al. (2023)
Argula (*Eruca sativa*)	4.12 dS/m	5% sorghum stalk biochar	↑34.81 mg/kg	↑12.7 mg/kg	↑34.5 mg/kg	↑Ca^{2+} 1.42 mmol/kg	↑Mg^{2+} 6.68 mmol/kg	↑Cl^- 18.45 mmol/kg	ELsaman et al. (2025)
Tomato	2400 μS/cm	2 t/ha	↑0.99%	↑10.82 μg/g	–				Zonayet et al. (2023)

Increment (↑) or reduction (↓) of the values are from the control treatment, i.e., no application of biochar

Table 1.4 Effect of biochar on soil enzymes of saline soil

Crop	Soil salinity level	Dose of biochar	Urease (%)	Alkaline phosphatase (%)	Catalase (%)	Sucrase (%)	References
Rice	1200 µS/cm	0.5–6 kg/m^2	↑194.8–744.6	↑28.2–64.8	–	–	Zhang et al. (2024)
Lemon balm	0.60% ($NaCl + Na_2SO_4$)	150 g in 750 g of soil	↑52.23	↑211.76	↑6.90	↑61.21	Liu et al. (2024)

Increment (↑) or reduction (↓) of the values are from the control treatment, i.e., no application of biochar

1.7 Biochar on Nutrient Dynamics of Saline Soil

1.7.1 Direct Nutrient Contributions from Biochar

Biochar produced under specific pyrolytic conditions contains significant amounts of nutrients such as nitrogen (N), phosphorus (P), and potassium (K), and other trace elements (Table 1.5). Even when the intention is not to use biochar as a direct fertilizer, the slow-release nature of nutrient retention by biochar can make a critical difference in increasing plant nutrient uptake (Hossain et al. 2020). Research has indicated that available N, P, and K levels in the soil are significantly increased with biochar addition compared to control treatments, contributing to higher nutrient use efficiency in crops. Among the essential nutrients, N is the most limiting factor for plant growth, development, and ultimate economic product (Cui et al. 2025). Biochar can be a potential source of nitrogen for soil and plants. It may provide some sort of organic N (water-soluble, hydrolyzed, and non-hydrolyzable N) and inorganic N content, including ammonium-N and nitrate-N (Liu et al. 2019a, b). However, most of the elemental N may be lost as gaseous emission during the pyrolysis under high temperature (Leng et al. 2020). Generally, the amount of N in biochar largely depends on the nature and composition of the feedstock. Biochar produced from crop residue, like wheat straw, contains 13.8 g/kg N, while relatively lower N has been found from woody biomass biochar. Sewage sludge, poultry litter, and chicken manure contain a higher amount of N than crop residue and woody biomass (Chen et al. 2018).

Unlike N, P content in biochar increased with high pyrolytic temperature; however, P content also depends on feedstock type. As per a study conducted by Xiao et al. (2018), when chicken manure was subjected to high temperature, a 54% increment in P content was observed. In case of K content, biochar produced from organic manures produces higher amounts as compared to other feedstocks. Potassium content is found to be highest with high pyrolysis temperature. Biochar can supply a considerable number of micronutrients, including Zn, Fe, Cu, and Mn. Though it also largely depends on feedstocks and pyrolysis temperature, such as animal manure contains higher Fe as compared to woody biomass and crop residues (Table 1.5). Studies

Table 1.5 Macro and micronutrient concentration of biochar made from different feedstocks

S. No.	Biochar feedstock	Pyrolytic temperature (°C)	Macronutrients (g/kg)							Micronutrients (mg/kg)				References
			N	P	K	Ca	Mg	S	Na	Zn	Fe	Mn	Cu	
Crop residue														
1	Wheat straw	300	13.8	2.6	30	6.3	4.5	–	–	47	158	106	–	Naeem et al. (2014)
		400	9.4	3.0	32	8.3	5.6	–	–	59	259	117	–	
		500	8.5	3.4	36	8.7	6.9	–	–	70	422	163	–	
2		300		1.98	10.17	3.53	4.74	4.05	–	43.76	89.60	61.50	–	Aon et al. (2023)
3		300	0.2	0.27	2.5	–	–	–	–	–	–	–	–	Beheshti et al. (2017)
4	Rice straw	300	11.5	1.1	36	9.1	8.1	–	–	67	195	396	–	Naeem et al. (2014)
5		400	9.8	1.3	41	9.8	9.6	–	–	89	521	649	–	
6		500	8.5	1.4	48	13.3	11.3			98	521	649		
7		500	9.32	–	28.47	5.38	2.71		0.63					
8		550–650	0.64	0.09	28.2	–	–	2.4	–	–	–	–	–	Si et al. (2018)
9	Soybean	500	–	0.06	37.8	15.7	11.7	11	–	28	699	58	–	Enders et al. (2012)
10	Corn stover	300	1.16	0.14	7.1	6.5	5.9	0.7	–	132	963	142	–	
11	Corn stover	600	1.01	0.18	24.6	9.4	8.6	0.8	–		1362	226		Enders et al. (2012)
Woody biomass														
13	Sugar maple	450	0.32	0.02	3.2	5.4	0.6	–	–	23.90	49.70	368	5.01	Noyce et al. (2017)
14	*Eucalyptus* sp.	350	–	0.005	5.1	5.4	0.4	–	–	–	500	–	–	Butnan et al. (2015)

(continued)

Table 1.5 (continued)

S. No.	Biochar feedstock	Pyrolytic temperature (°C)	Macronutrients (g/kg)							Micronutrients (mg/kg)				References
			N	P	K	Ca	Mg	S	Na	Zn	Fe	Mn	Cu	
15	Bamboo	600	9.84	0.11	27.8	–	–	4.6	–	–	–	–	–	Lu et al. (2018)
Industrial and municipal waste														
16	Sugarcane bagasse	450–500	0.67	0.07	–	–	–	–	–	–	–	–	–	Gerdeli-dani and Hosseini (2018)
17	Sewage sludge	500	2.7	1.70	5.2	65.7	6.4	–	–	1520	22,100	450	380	Zhao et al. (2018)
18	Sewage sludge	500	6.8	0.11	–	–	–	–	–	–	–	–	–	Gonzaga et al. (2019)
19	Orchard pruning	500	0.91	0.23	13.9	25	28.9	0.05	–	0.01	0.033	0.008	0.009	Baronti et al. (2014)
Organic manure														
20	Poultry litter	350	4.45	2.08	48.5	26.6	9.4	6.1	–	712	13,200	640	213	Cantrell et al. (2012)
21		700	2.07	3.12	74	40	14.5	6.3	–	1010	18,900	948	310	Cantrell et al. (2012)
23	Cow manure	300	0.71	0.19	2.6	–	–	–	–					Beheshti et al. (2017)
24	Pig manure	350	3.54	3.89	35.3	28.9	21.3	4.0	–	–	–	–	–	Subedi et al. (2016)

considered that biochar acts as a promising source of nutrients; however, the concentration may be lower due to gaseous emission during pyrolysis (Chen et al. 2018; Noyce et al. 2017).

1.7.2 Biochar for Enhanced Nutrient Cycling

Microbial communities play a central role in nutrient cycling, helping to transform and mobilize nutrients into forms available for plant uptake. Biochar provides an ideal habitat for soil microbes due to its porous structure and high surface area. These physical characteristics of biochar create microsites that protect microorganisms from environmental stress, particularly in saline conditions (Wong and Ogbonnaya 2021). Several studies have reported that biochar amendments improve the structure and functioning of soil microbial communities, promoting beneficial bacteria that are essential for nutrient cycling (Wang et al. 2021b; Manasa et al. 2020). The addition of biochar has been shown to increase the abundance of phosphorus-solubilizing bacteria (Bai et al. 2024). These microbial groups are instrumental in converting insoluble phosphates into soluble forms that can be absorbed by plant roots. In saline soils, where nutrient availability is often a limiting factor, enhancing microbial activity directly contributes to improved nutrient use efficiency.

In saline soils, processes such as denitrification can lead to significant nitrogen losses in the form of nitrous oxide (N_2O), a potent greenhouse gas. Biochar has been found to influence the abundance of microbial genes involved in nitrogen cycling. Specifically, it can reduce the copy numbers of genes such as *NIRS* and *NIRK* while increasing *NOSZ*, which are associated with the reduction of N_2O emissions (Li et al. 2025; Xiao et al. 2019).

The chemical properties of saline soils are often compromised by an excess of sodium ions (Na^+), which can displace beneficial nutrients like potassium (K^+), calcium (Ca^{2+}), and magnesium (Mg^{2+}) (Sarkar et al. 2024a; Rai et al. 2021). Biochar application is known to increase the cation exchange capacity (CEC) of soils, allowing them to hold onto essential cations and reduce their leaching losses (Domingues et al. 2020). An enhanced CEC helps maintain a balanced nutrient profile in the soil, enabling plants to access nutrients more efficiently even in challenging saline environments. Through ion exchange and adsorption processes, biochar can directly reduce soil salinity (Lee et al. 2022). The negative surface charges on biochar particles attract and bind positively charged ions from the soil solution, including both harmful sodium and beneficial nutrient ions (Tan et al. 2020). This dual action not only lowers the salinity levels but also enhances the ability of soil to retain nutrients, effectively creating a reservoir from which plants can draw essential minerals as needed.

Saline soils often suffer not only from nutrient imbalances but also from poor water retention. Under conditions of high salt concentration, the osmotic potential of the soil solution is altered, limiting water availability to plants. Biochar application has been documented to improve soil water holding capacity due to its high specific surface area and porous nature (Atkinson 2018; Suliman et al. 2017). Better water retention means that plants receive a more consistent and moderate supply of water, even under saline conditions, which in turn facilitates the uptake of soluble nutrients. Improved water dynamics in the soil also affect hydraulic conductivity. With reduced bulk density and increased microporosity, the movement of water through the soil improves, thereby increasing the leaching of potentially harmful salts and reducing the risk of salt accumulation on the surface. These effects support more balanced nutrient distribution and retention within the root zone and enhance overall plant health and productive capacity.

Consequently, the use of biochar not only benefits plant nutrition by retaining nitrogen within the soil but also contributes to a reduction in environmentally detrimental emissions. The integration of biochar into soil systems also supports a broader range of biochemical reactions that underlie nutrient cycling. Improved microbial diversity and activity accelerate processes such as the decomposition of organic matter and the transformation of nutrients between different chemical forms. This enhanced cycling facilitates a more constant and regulated supply of nutrients, effectively matching plant demand throughout the growing season. Moreover, the role of biochar in ameliorating salinity stress creates a more favourable environment for microbial activity, reinforcing a positive feedback loop in nutrient dynamics.

1.7.3 Impact of Biochar on Nutrient Use Efficiency in Saline Soils

Although the addition of biochar and enhancement of Nutrient Use Efficiency (NUE) are highly location-specific and depend on soil physicochemical factors and the type of feedstocks, it has been widely found that biochar has a positive impact on enhancing NUE in saline soils (Table 1.2). The details of nutrient-specific responses to the applied biochar are described as follows.

1.7.3.1 Nitrogen Dynamics

Nitrogen is a critical macronutrient, and its dynamics are significantly influenced by biochar in saline environments (Clough et al. 2013). Biochar increases nitrogen retention in soil by reducing leaching and gaseous losses (Li et al. 2017b). Its porous structure and high cation exchange capacity allow it to adsorb ammonium (NH^{4+}). Studies showed significant reductions in cumulative leaching of ammonia nitrogen (e.g., 80.16% reduction with 2% biochar at 400 °C pyrolysis) (Zhao et al. 2022; Clough et al. 2013). While biochar generally has a favourable adsorption effect on ammonium, its capacity to retain nitrate can be negligible in some cases. However, other research indicates that biochar can positively affect nitrate retention by indirectly adsorbing it and increasing soil water holding capacity, thereby blocking nitrate leaching (Atkinson 2018; Haider et al. 2017). Biochar alters the structure and abundance of nitrogen-related microbial communities, influencing nitrification and mineralization processes. The influence on gross mineralization rates and heterotrophic nitrification can have an overall positive effect on soil nitrogen retention. It has been found that biochar can also reduce ammonia (NH_3) volatilization from salt-affected soils (Liu et al. 2021), with low-pH biochar being more effective in sandy soils.

1.7.3.2 Phosphorus Dynamics

Phosphorus availability is often limited in saline soils, and biochar plays a crucial role in enhancing it in saline conditions (Xu et al. 2016). Biochar itself can be a source of phosphorus, particularly when produced from animal bones or manure (Hussain et al. 2021). It also improves the nutrient retention capacity of soil, reducing P leaching. Biochar enhances P availability by increasing labile P fractions (e.g., resin-P, Olsen-P) and promoting the conversion of less available P forms into plant-usable forms (Cao et al. 2021). The acidification effect of biochar and its ability to chelate metal cations (like Fe, Al, Ca) weaken their binding to phosphate, thereby releasing phosphate ions and increasing P solubility (Dai et al. 2017). Biochar can also enhance the activity of phosphatases, enzymes secreted by microorganisms that mineralize organic P compounds into inorganic P. Organic acids associated with biochar can compete with phosphate ions for adsorption sites on soil colloids, reducing P fixation. Biochar also promotes the growth and activity of phosphorus-solubilizing bacteria (PSB) (Bai et al. 2024). These microorganisms secrete organic acids and enzymes (phosphatases) that convert insoluble P into plant-available forms. Biochar acts as a carrier, protecting PSB and enhancing their survival and functional activity in the soil. While high soil pH under saline-alkali soils, P availability may be decreased by promoting P precipitation with Ca, the inherent capacity of biochar to increase P availability and promote P-solubilizing bacteria can counteract this.

1.7.3.3 Potassium Dynamics

Potassium is an essential osmoprotectant for plants, and its uptake is often impaired by salinity. Biochar application significantly reduces the uptake of toxic Na^+ ions by plants, which in turn increases the absorption of beneficial nutrients, particularly K^+ (Farhangi-Abriz and Torabian 2018). This leads to an improved K^+/Na^+ ratio in plant tissues, a key indicator of higher salt tolerance and better plant function under saline conditions. Biochar can directly contribute K^+ to the soil, especially biochar from certain feedstocks (Wang et al. 2018). It also enhances CEC, which helps retain K^+ in the rooting zone, preventing its leaching. By mitigating osmotic stress and improving water uptake, biochar indirectly supports K^+ uptake, as K^+ plays a vital role in maintaining plant water relations and turgor.

1.7.3.4 Micronutrient Availability

Biochar application to saline soils has been shown to improve the uptake of essential nutrients, including micronutrients like iron (Fe), zinc (Zn), and manganese (Mn) (Moradi et al. 2019). This is attributed to improved soil physical properties, increased CEC, and enhanced microbial activity. However, a rise in soil pH, a common effect of biochar, can sometimes immobilize certain micronutrients (e.g., Cu, Fe, Zn), potentially reducing their uptake into the edible portions of food crops. This highlights the need for site-specific considerations when applying biochar, as the optimal range for nutrient concentration may shift under saline conditions.

1.8 Challenges of Biochar Use in Agriculture

The utilization of biochar in agriculture presents several scientific and practical challenges that need careful consideration. One of the primary concerns arises from the inherent heterogeneity of biochar, which varies significantly depending on the feedstock type, pyrolysis conditions, and post-treatment processes (Xu et al. 2021). As a result, biochar cannot be uniformly characterized or applied; rather, each biochar type must be individually assessed for its suitability and effectiveness under specific agronomic conditions (Joseph et al. 2021). Low-quality feedstocks and inconsistent pyrolysis parameters often lead to the production of substandard biochar, limiting its agronomic efficacy. Even with optimal formulation and precise application, the anticipated benefits of biochar, such as enhanced nutrient retention, soil fertility, or yield improvement, may not manifest immediately and may require several cropping cycles to be realized (Burrell et al. 2016; Bai et al. 2019).

There are also concerns regarding the long-term impacts of biochar application. For instance, continuous application may lead to a cumulative rise in soil pH and thermal properties, potentially altering soil chemistry and nutrient availability (Clough et al. 2013; Glaser and Lehr 2019). Adverse effects on the soil rhizosphere have also been reported, including disruptions in microbial community structure, suppression of beneficial plant-microbe interactions, and interference with root exudate dynamics (Ren et al. 2018; Pei et al. 2020). Moreover, biochar may adsorb agrochemicals such as herbicides and nutrients, rendering them less bioavailable and thereby reducing the effectiveness of crop protection and fertilization strategies (Yadav et al. 2023). Furthermore, depending on the source material, biochar may contain harmful contaminants such as heavy metals or polycyclic aromatic hydrocarbons (PAHs), which could accumulate in the soil and pose risks to plant, microbial, and human health (Pereira Lopes and Astruc 2021). In addition, structural changes such as increased bulk density and reduced porosity can affect soil aeration and water dynamics, potentially offsetting some of biochar's intended soil amelioration benefits. Lastly, the net carbon footprint of biochar production and application, particularly in comparison to alternative organic amendments like compost, remains inadequately quantified (Mohammadi 2021). Comprehensive life cycle assessments are required to accurately evaluate its sustainability credentials and justify its widespread agricultural use.

1.9 Future Scope of Research

The multifaceted applications of biochar highlight its significant potential in fostering a sustainable and circular economy by converting agricultural and organic wastes into high-value soil amendments. Biochar contributes to nutrient cycling, carbon sequestration, and improved soil health, supporting regenerative and climate-resilient agricultural practices. Future developments should focus on scaling up biochar utilization through innovative and cost-effective technologies, including its conversion into secondary carbon products, syngas for fuel synthesis, and biodiesel for energy use (Kurniawan et al. 2022). Biochar has also been identified as a key tool in mitigating climate change, with the potential to sequester carbon and reduce up to one-eighth of global anthropogenic GHG emissions (Woolf et al. 2010), thus contributing to climate neutrality targets by 2050.

The quality of biochar is the most important factor for varied soil genesis. Soil reacts differently with biochar derived from several feedstocks and different pyrolysis conditions. Due to the wide variation in biochar properties, it is challenging to define a single overarching benefit. Each type of biochar should be considered a distinct material, and its performance must be evaluated based on the specific intended application (Joseph et al. 2021; Xu et al. 2021).

Even when the quality of biochar is properly selected and application is carefully performed, biochar may not yield desired results, especially in the first few years or even longer (Burrell et al. 2016; Bai et al. 2019). The actual carbon footprint of biochar, compared to other sources of carbon such as compost, remains to be properly studied and outlined. Therefore, these issues need to be systematically researched and documented for the application of biochar extensively in the future for agricultural and environmental uses. Moreover, enrichment of biochar would be beneficial for the specified soil type, such as gypsum-enriched biochar can be a potential ameliorant for alkali soil, where excessive Na will be replaced by Ca, resulting in improved soil structure and enhanced nutrient use efficiency (Fig. 1.2) (Katuwal et al. 2022). Likewise, biochar enrichment with several organic materials and its standardization with an appropriate procedure is required to improve the quality of the produced biochar. The necessity for crop-specific and soil-specific biochar arises from the diverse physical, chemical, and biological characteristics of soils and crops. Different soils exhibit variations in texture, pH, nutrient retention, and water-holding capacity, which influence effectiveness of biochar. Advancing biochar application requires long-term, region-specific field trials across diverse cropping systems to evaluate stability and efficacy. Integrating approaches from molecular biology, soil ecology, and materials science can elucidate the mechanisms by which biochar-based fertilizers influence microbial communities, enzyme activities, and nutrient cycling. Furthermore, efforts should be directed towards the development of

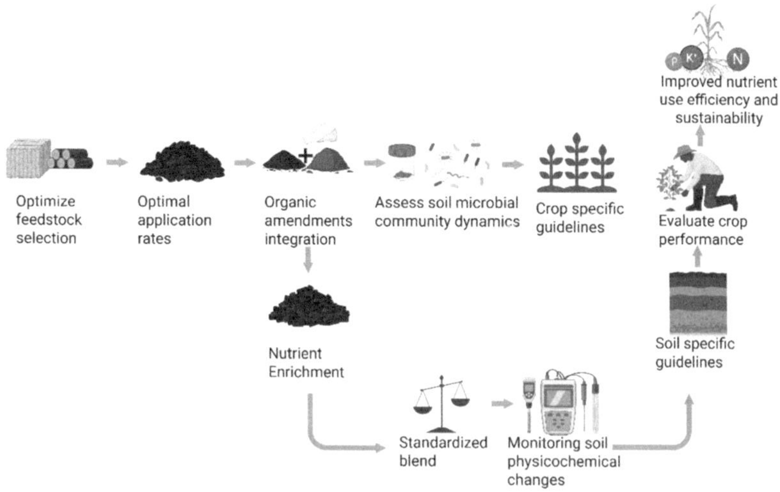

Fig. 1.2 Future scope of research for soil and crop-specific biochar enrichment. Image is created by Biorender.com; License Number: MI28JU6VRU

green, low-cost, and efficient production technologies, along with the establishment of standardized quality control measures and application guidelines to facilitate the adoption of biochar in precision agriculture and sustainable nutrient management systems.

1.10 Conclusion

Salinity is found to be a crucial factor that hinders plant growth by altering soil physicochemical and biological properties, resulting in conditions unfavorable for plant growth and development. Soil salinity alters nutrient availability and disrupts soil fertility. Excess salts, mainly sodium (Na^+) and chloride (Cl^-), compete with essential nutrients like potassium, calcium, and magnesium, reducing their uptake by plants. Increased salinity often raises soil pH, decreasing the availability of phosphorus and micronutrients such as iron and zinc. High sodium levels degrade soil structure, limit water movement, and restrict root growth, while also suppressing microbial activity and nutrient cycling. These combined effects lead to nutrient deficiencies, imbalances, and reduced crop productivity. Therefore, the reclamation of unproductive and unfertile soil becomes imperative for optimum crop production. The usefulness of biochar as a soil amendment has gained enough attention due to its efficacy in addressing salt-affected soil. Research shows that biochar improves soil physical, chemical, and biological properties, facilitating salt leaching from the root zone and improving pH, water retention, and infiltration. Biochar is a potential source of macro-, secondary, and micronutrients for plants and influences soil nutrient dynamics by altering pH and cation exchange capacity (CEC). Its nutrient contribution depends on feedstock type, pyrolysis conditions, application rate, and soil characteristics. Biochar from animal manures and wastes generally contain more nitrogen (N), phosphorus (P), potassium (K), and micronutrients than those derived from crop residues or woody materials. Higher pyrolysis temperatures enhance most nutrient availability, except nitrogen and sulfur, which are lost through volatilization. Biochar also reduces nutrient leaching, particularly phosphorus and potassium, and improves overall nutrient use efficiency in plants.

References

Abbas A, Huang P, Hussain S, Shen F, Wang H, Du D (2022) Mild evidence for local adaptation of *Solidago canadensis* under different salinity, drought, and abscisic acid conditions. Pol J Environ Stud 31(4)

Ali S, Rizwan M, Qayyum MF, Ok YS, Ibrahim M, Riaz M et al (2017) Biochar soil amendment on alleviation of drought and salt stress in plants: a critical review. Environ Sci Pollut Res 24(14):12700–12712

Anand A, Gautam S, Ram LC (2023) Feedstock and pyrolysis conditions affect suitability of biochar for various sustainable energy and environmental applications. J Anal Appl Pyrolysis 170:105881

Aon M, Aslam Z, Hussain S, Bashir MA, Shaaban M, Masood S et al (2023) Wheat straw biochar produced at a low temperature enhanced maize growth and yield by influencing soil properties of typic calciargid. Sustainability 15(12):9488

Atkinson CJ (2018) How good is the evidence that soil-applied biochar improves water-holding capacity? Soil Use Manag 34(2):177–186

Babar S, Baloch A, Qasim M, Wang J, Wang X, Iqbal R et al (2025) Unearthing the potential of aluminum-modified biochar for saline-alkali soil rejuvenation and microbial diversity enhancement. Pedosphere

Bai N, Zhang H, Li S, Zheng X, Zhang J, Zhang H et al (2019) Long-term effects of straw and straw-derived biochar on soil aggregation and fungal community in a rice–wheat rotation system. PeerJ 6:e6171. https://doi.org/10.7717/peerj.6171

Bai K, Wang W, Zhang J, Yao P, Cai C, Xie Z et al (2024) Effects of phosphorus-solubilizing bacteria and biochar application on phosphorus availability and tomato growth under phosphorus stress. BMC Biol 22(1):211

Bridgwater AV, Peacocke GVC (1994) Engineering developments in fast pyrolysis for bio-oils. In: Proceedings biomass pyrolysis oil properties and combustion meeting, Sept 1994. NREL, pp 110–127

Bridgwater AV, Meier D, Radlein D (1999) An overview of fast pyrolysis of biomass. Org Geochem 30(12):1479–1493

Burrell LD, Zehetner F, Rampazzo N, Wimmer B, Soja G (2016) Long-term effects of biochar on soil physical properties. Geoderma 282:96–102. https://doi.org/10.1016/j.geoderma.2016.07.019

Butnan S, Deenik JL, Toomsan B, Antal MJ, Vityakon P (2015) Biochar characteristics and application rates affecting corn growth and properties of soils contrasting in texture and mineralogy. Geoderma 237:105–116

Cao D, Lan Y, Sun Q, Yang X, Chen W, Meng J et al (2021) Maize straw and its biochar affect phosphorus distribution in soil aggregates and are beneficial for improving phosphorus availability along the soil profile. Eur J Soil Sci 72(5):2165–2179
Chen H, Ma J, Wei J, Gong X, Yu X, Guo H, Zhao Y (2018) Biochar increases plant growth and alters microbial communities via regulating the moisture and temperature of green roof substrates. Sci Total Environ 635:333–342
Chhabra R (2022) Nutrient management in salt-affected soils. In: Salt-affected soils and marginal waters: global perspectives and sustainable management. Springer International Publishing, Cham, pp 349–429
Choudhary P, Pramitha L, Rana S, Verma S, Aggarwal PK, Muthamilarasan M (2021) Hormonal crosstalk in regulating salinity stress tolerance in graminaceous crops. Physiol Plant 173(4):1587–1596
Clough T, Condron L, Kammann C, Müller C (2013) A review of biochar and soil nitrogen dynamics. Agronomy 3:275–293. https://doi.org/10.3390/agronomy3020275
Cui X, Li X, Wang J, Wang X, Yu F, Yang G et al (2025) A review on the production of nutrient-enriched biochar: insights from the evolution of nitrogen, phosphorus, and potassium. Crit Rev Environ Sci Technol 55(12):885–903
Czernik S, French R, Feik C, Chornet E (2002) Hydrogen by catalytic steam reforming of liquid byproducts from biomass thermoconversion processes. Ind Eng Chem Res 41(17):4209–4215
Dahlawi S, Naeem A, Rengel Z, Naidu R (2018) Biochar application for the remediation of salt-affected soils: challenges and opportunities. Sci Total Environ 625:320–335
Dai Z, Zhang X, Tang C, Muhammad N, Wu J, Brookes PC, Xu J (2017) Potential role of biochars in decreasing soil acidification—a critical review. Sci Total Environ 581:601–611
Dey S, Sarkar S, Dhar A, Brahmachari K, Ghosh A, Goswami R, Mainuddin M (2025) Potato cultivation under zero tillage and straw mulching: option for land and cropping system intensification for Indian Sundarbans. Land 14(3):563
Domingues RR, Sánchez-Monedero MA, Spokas KA, Melo LC, Trugilho PF, Valenciano MN, Silva CA (2020) Enhancing cation exchange capacity of weathered soils using biochar: feedstock, pyrolysis conditions and addition rate. Agronomy 10(6):824
Enders A, Hanley K, Whitman T, Joseph S, Lehmann J (2012) Characterization of biochars to evaluate recalcitrance and agronomic performance. Bioresour Technol 114:644–653
Eswar D, Karuppusamy R, Chellamuthu S (2021) Drivers of soil salinity and their correlation with climate change. Curr Opin Environ Sustain 50:310–318
Evans RJ, Milne TA (1987) Molecular characterization of the pyrolysis of biomass. Energy Fuels 1(2):123–137
Fan Y, Ge T, Zheng Y, Li H, Cheng F (2016) Use of mixed solid waste as a soil amendment for saline-sodic soil remediation and oat seedling growth improvement. Environ Sci Pollut Res 23(21):21407–21415
Farhangi-Abriz S, Torabian S (2018) Effect of biochar on growth and ion contents of bean plant under saline condition. Environ Sci Pollut Res 25:11556–11564
Farooqi ZUR, Qadir AA, Alserae H, Raza A, Mohy-Ud-Din W (2023) Organic amendment–mediated reclamation and build-up of soil microbial diversity in salt-affected soils: fostering soil biota for shaping rhizosphere to enhance soil health and crop productivity. Environ Sci Pollut Res 30(51):109889–109920
Gholizadeh M, Meca S, Zhang S, Clarens F, Hu X (2024) Understanding the dependence of biochar properties on different types of biomass. Waste Manag 182:142–163
Glaser B, Lehr V-I (2019) Biochar effects on phosphorus availability in agricultural soils: a meta-analysis. Sci Rep 9:9338. https://doi.org/10.1038/s41598-019-45693-z
Gonçalo Filho F, da Silva Dias N, Suddarth SRP, Ferreira JF, Anderson RG, dos Santos Fernandes C et al (2019) Reclaiming tropical saline-sodic soils with gypsum and cow manure. Water 12(1):57
Grossiord C, Buckley TN, Cernusak LA, Novick KA, Poulter B, Siegwolf RT et al (2020) Plant responses to rising vapor pressure deficit. New Phytol 226(6):1550–1566

Haider G, Steffens D, Moser G, Müller C, Kammann CI (2017) Biochar reduced nitrate leaching and improved soil moisture content without yield improvements in a four-year field study. Agric Ecosyst Environ 237:80–94

Hailu B, Mehari H (2021) Impacts of soil salinity/sodicity on soil-water relations and plant growth in dry land areas: a review. J Nat Sci Res 12(3):1–10

Hassani A, Azapagic A, Shokri N (2020) Predicting long-term dynamics of soil salinity and sodicity on a global scale. Proc Natl Acad Sci 117(52):33017–33027

Helle S, Bennett NM, Lau K, Matsui JH, Duff SJ (2007) A kinetic model for production of glucose by hydrolysis of levoglucosan and cellobiosan from pyrolysis oil. Carbohydr Res 342(16):2365–2370

Hopmans JW, Qureshi AS, Kisekka I, Munns R, Grattan SR, Rengasamy P et al (2021) Critical knowledge gaps and research priorities in global soil salinity. Adv Agron 169:1–191

Hossain MZ, Bahar MM, Sarkar B, Donne SW, Ok YS, Palansooriya KN et al (2020) Biochar and its importance on nutrient dynamics in soil and plant. Biochar 2(4):379–420

Hussain A, Nazir A, Shafiq M (2021) Potential application of biochar composite derived from rice straw and animal bones to improve plant growth. Sustainability 13(19):11104

Ighalo JO, Ohoro CR, Ojukwu VE, Oniye M, Shaikh WA, Biswas JK, Seth CS, Mohan GBM, Chandran SA, Rangabhashiyam S (2025) Biochar for ameliorating soil fertility and microbial diversity: from production to action of the black gold. Iscience 28(1)

Irfan M, Aslam H, Maqsood A, Tazeen SK, Mahmood F, Shahid M (2023) Changes in plant microbiome in response to abiotic stress. In: Plant microbiome for plant productivity and sustainable agriculture. Springer Nature Singapore, Singapore, pp 99–119

Jemal K, Yakob A (2021) Role of biochar on the amelioration of soil acidity. Agrotechnology 10:212

Jiao Y, Li D, Wang M, Gong T, Sun M, Yang T (2021) A scientometric review of biochar preparation research from 2006 to 2019. Biochar 3(3):283–298

Jones SB, Valkenburt C, Walton CW, Elliott DC, Holladay JE, Stevens DJ, Kinchin C, Czernik S (2009) Production of gasoline and diesel from biomass via fast pyrolysis, hydrotreating and hydrocracking: a design case (No. PNNL-18284 rev. 1). Pacific Northwest National Lab. (PNNL), Richland, WA (United States)

Joseph S, Cowie AL, Van Zwieten L, Bolan N, Budai A, Buss W et al (2021) How biochar works, and when it doesn't: a review of mechanisms controlling soil and plant responses to biochar. GCB Bioenergy 13:1731–1764. https://doi.org/10.1111/gcbb.12885

Katuwal S, Ashworth AJ, Rafsan NAS, Kolar P (2022) Characterization of poultry litter biochar and activated biochar as a soil amendment for valorization. Biomass 2(4)

Kaushal SS, Likens GE, Pace ML, Reimer JE, Maas CM, Galella JG et al (2021) Freshwater salinization syndrome: from emerging global problem to managing risks. Biogeochemistry 154(2):255–292

Koné S, Galiegue X (2023) Potential development of biochar in Africa as an adaptation strategy to climate change impact on agriculture. Environ Manag 72(6):1189–1203

Kumar P, Tiwari P, Biswas A, Acharya T (2020) Geophysical and hydrogeological investigation for the saline water invasion in the coastal aquifers of West Bengal, India: a critical insight in the coastal saline clay–sand sediment system. Environ Monit Assess 192(9):562

Kumar A, Bhattacharya T, Shaikh WA, Chakraborty S, Sarkar D, Biswas JK (2022) Biochar modification methods for augmenting sorption of contaminants. Curr Pollut Rep 8(4):519–555

Kurniawan TA, Othman MHD, Singh D, Avtar R, Goh HH, Setiadi T, Lo WH (2022) Technological solutions for long-term management of partially used nuclear fuel: a critical review. Ann Nucl Energy 166:108736. https://doi.org/10.1016/j.anucene.2021.108736

Lee X, Yang F, Xing Y, Huang Y, Xu L, Liu Z et al (2022) Use of biochar to manage soil salts and water: effects and mechanisms. Catena 211:106018

Lehmann J, Joseph S (2015) Biochar for environmental management: an introduction. In: Biochar for environmental management. Routledge, pp 1–13

Lehmann J, Gaunt J, Rondon M (2006) Bio-char sequestration in terrestrial ecosystems—a review. Mitig Adapt Strateg Glob Change 11(2):403–427
Lehmann J, Cowie A, Masiello CA, Kammann C, Woolf D, Amonette JE et al (2021) Biochar in climate change mitigation. Nat Geosci 14(12):883–892
Leng L, Xu SY, Liu RF, Yu T, Zhuo XM, Leng SQ et al (2020) Nitrogen-containing functional groups of biochar: an overview
Li H, Dong X, da Silva EB, de Oliveira LM, Chen Y, Ma LQ (2017a) Mechanisms of metal sorption by biochars: biochar characteristics and modifications. Chemosphere 178:466–478
Li Y, Huang L, Zhang H, Wang M, Liang Z (2017b) Assessment of ammonia volatilization losses and nitrogen utilization during the rice growing season in alkaline salt-affected soils. Sustainability 9(1):132
Li Z, Yang Y, Wang X, Qi Y, Yang X (2025) Linking N_2O emissions and nosZ gene abundance: a meta-analysis of organic carbon amendments in agricultural soils. Plant Soil 1–15
Liu WJ, Jiang H, Yu HQ (2015) Development of biochar-based functional materials: toward a sustainable platform carbon material. Chem Rev 115(22):12251–12285
Liu X, Liao J, Song H, Yang Y, Guan C, Zhang Z (2019a) A biochar-based route for environmentally friendly controlled release of nitrogen: urea-loaded biochar and bentonite composite. Sci Rep 9(1):9548
Liu Q, Liu B, Zhang Y, Hu T, Lin Z, Liu G et al (2019b) Biochar application as a tool to decrease soil nitrogen losses (NH_3 volatilization, N_2O emissions, and N leaching) from croplands: options and mitigation strength in a global perspective. Glob Change Biol 25(6):2077–2093
Liu S, Li Y, Xu J, Ma W, Liu B, Wang H et al (2021) Biochar partially offset the increased ammonia volatilization from salt-affected soil. Arch Agron Soil Sci 67(9):1202–1216
Liu J, Huang Q, Li Z, Liu N, Li J, Huang G (2023) Effect of autumn irrigation on salt leaching under subsurface drainage in an arid irrigation district. Water 15(12):2296
Lusiba S, Odhiambo J, Ogola J (2017) Effect of biochar and phosphorus fertilizer application on soil fertility: soil physical and chemical properties. Arch Agron Soil Sci 63(4):477–490
Maathuis FJ, Ahmad I, Patishtan J (2014) Regulation of Na^+ fluxes in plants. Front Plant Sci 5:467
Mahajan G, Das B, Morajkar S, Desai A, Murgaokar D, Kulkarni R et al (2020) Soil quality assessment of coastal salt-affected acid soils of India. Environ Sci Pollut Res 27(21):26221–26238
Mahfud FH, Ghijsen F, Heeres HJ (2007) Hydrogenation of fast pyrolyis oil and model compounds in a two-phase aqueous organic system using homogeneous ruthenium catalysts. J Mol Catal A Chem 264(1–2):227–236
Mahmood MZ, Bibi S, Shahzad M, Fakhar A, Rafique M, Qayyum A (2021) Mechanisms of microbes to combat salinity in soil by producing secondary metabolites. Arab J Geosci 15:45
Manasa MRK, Katukuri NR, Nair SSD, Haojie Y, Yang Z, bo Guo, R. (2020) Role of biochar and organic substrates in enhancing the functional characteristics and microbial community in a saline soil. J Environ Manag 269:110737
Megha S, Shankhdhar D (2022) Effect of different levels of salinity on morphological characteristics of some rice genotypes. Pha Innov J 11:189–193
Meng X, Zhou J, Sui N (2018) Mechanisms of salt tolerance in halophytes: current understanding and recent advances. Open Life Sci 13:149–154
Minhas P (2021) Diagnostic criteria for salt-affected soils. Sustain Agric 26
Mohammadi A (2021) Overview of the benefits and challenges associated with pelletizing biochar. Processes 9(9):1591
Mohanty P, Nanda S, Pant KK, Naik S, Kozinski JA, Dalai AK (2013) Evaluation of the physio-chemical development of biochars obtained from pyrolysis of wheat straw, timothy grass and pinewood: effects of heating rate. J Anal Appl Pyrolysis 104:485–493
Moradi S, Rasouli-Sadaghiani MH, Sepehr E, Khodaverdiloo H, Barin M (2019) Soil nutrients status affected by simple and enriched biochar application under salinity conditions. Environ Monit Assess 191:1–13

Naeem MA, Khalid M, Arshad M, Ahmad R (2014) Yield and nutrient composition of biochar produced from different feedstocks at varying pyrolytic temperatures. Pak J Agric Sci 51(1)
Nair RR, Kißling PA, Marchanka A, Lecinski J, Turcios AE, Shamsuyeva M et al (2023) Biochar synthesis from mineral and ash-rich waste biomass, part 2: characterization of biochar and co-pyrolysis mechanism for carbon sequestration. Sustain Environ Res 33(1):14
Naz T, Iqbal MM, Tahir M, Hassan MM, Rehmani MIA, Zafar MI, Ghafoor U, Qazi MA, El Sabagh A, Sakran MI (2021) Foliar application of potassium mitigates salinity stress conditions in spinach (*Spinacia oleracea* L.) through reducing NaCl toxicity and enhancing the activity of antioxidant enzymes. Horticulturae 7:566
Neves EG, Petersen JB, Bartone RN, Heckenberger MJ (2004) The timing of terra preta formation in the Central Amazon: archaeological data from three sites. In: Amazonian dark earths: explorations in space and time. Springer Berlin Heidelberg, Berlin, Heidelberg, pp 125–134
Noyce GL, Jones T, Fulthorpe R, Basiliko N (2017) Phosphorus uptake and availability and short-term seedling growth in three Ontario soils amended with ash and biochar. Can J Soil Sci 97(4):678–691
Onay O, Kockar OM (2003) Slow, fast and flash pyrolysis of rapeseed. Renew Energy 28(15):2417–2433
Ondrasek G, Rengel Z (2021) Environmental salinization processes: detection, implications & solutions. Sci Total Environ 754:142432
Osman AI, Farghali M, Ihara I, Elgarahy AM, Ayyad A, Mehta N et al (2023) Materials, fuels, upgrading, economy, and life cycle assessment of the pyrolysis of algal and lignocellulosic biomass: a review. Environ Chem Lett 21(3):1419–1476
Pandey D, Daverey A, Arunachalam K (2020) Biochar: production, properties and emerging role as a support for enzyme immobilization. J Clean Prod 255:120267
Pandian K, Subramaniayan P, Gnasekaran P, Chitraputhirapillai S (2016) Effect of biochar amendment on soil physical, chemical and biological properties and groundnut yield in rainfed Alfisol of semi-arid tropics. Arch Agron Soil Sci 62(9):1293–1310
Panwar NL, Pawar A, Salvi BL (2019) Comprehensive review on production and utilization of biochar. SN Appl Sci 1(2):168
Pei J, Li J, Fang C, Zhao J, Nie M, Wu J (2020) Different responses of root exudates to biochar application under elevated CO_2. Agric Ecosyst Environ 301:107061. https://doi.org/10.1016/j.agee.2020.107061
Pereira Lopes R, Astruc D (2021) Biochar as a support for nanocatalysts and other reagents: recent advances and applications. Coord Chem Rev 426:1–116. https://doi.org/10.1016/j.ccr.2020.213585
Piskorz J, Radlein DSA, Scott DS, Czernik S (1989) Pretreatment of wood and cellulose for production of sugars by fast pyrolysis. J Anal Appl Pyrolysis 16(2):127–142
Prins MJ, Ptasinski KJ, Janssen FJ (2006) Torrefaction of wood: part 1. Weight loss kinetics. J Anal Appl Pyrolysis 77(1):28–34
Purakayastha TJ, Bera T, Bhaduri D, Sarkar B, Mandal S, Wade P et al (2019) A review on biochar modulated soil condition improvements and nutrient dynamics concerning crop yields: pathways to climate change mitigation and global food security. Chemosphere 227:345–365
Rafiq MK, Bachmann RT, Rafiq MT, Shang Z, Joseph S, Long R (2016) Influence of pyrolysis temperature on physico-chemical properties of corn stover (*Zea mays* L.) biochar and feasibility for carbon capture and energy balance. PLoS One 11(6):e0156894
Rai AK, Basak N, Sundha P (2021) Chemistry of salt-affected soils. In: Managing salt-affected soils for sustainable agriculture, pp 128–148
Ren X, Wang F, Cao F, Guo J, Sun H (2018) Desorption of atrazine in biochar-amended soils: effects of root exudates and the aging interactions between biochar and soil. Chemosphere 212:687–693. https://doi.org/10.1016/j.chemosphere.2018.08.124
Rengasamy P (2016) Soil chemistry factors confounding crop salinity tolerance—a review. Agronomy 6:53

Roy C, Yang J, Blanchette D, Korving L, De Caumia B (1997) Development of a novel vacuum pyrolysis reactor with improved heat transfer potential: vacuum pyrolysis—improved heat transfer. In: Developments in thermochemical biomass conversion: volume 1/volume 2. Springer Netherlands, Dordrecht, pp 351–367

Safdar H, Amin A, Shafiq Y, Ali A, Yasin R, Shoukat A et al (2019) A review: impact of salinity on plant growth. Nat Sci 17(1):34–40

Sarkar S, Brahmachari K, Gaydon DS, Dhar A, Dey S, Mainuddin M (2024a) Options for intensification of cropping system in coastal saline ecosystem: inclusion of grain legumes in rice-based cropping system. Soil Syst 8(3):90

Sarkar S, Gaydon DS, Dey S, Chaki AK, Brahmachari K, Dhar A et al (2024b) Evaluation of the APSIM model in rice-lentil cropping system in a complex coastal saline environment. J Indian Soc Coast Agric Res 42(1):53–72

Shwe KK, Chaopayao N, Yampracha S (2024) Mitigation of soil salinity by addition of different rice straw biochar doses in salt-affected acid soil. Environ Res Commun 6(9):095028

Sigua GC, Stone KC, Hunt PG, Cantrell KB, Novak JM (2015) Increasing biomass of winter wheat using sorghum biochars. Agron Sustain Dev 35(2):739–748

Sigua GC, Novak JM, Watts DW, Johnson MG, Spokas K (2016) Efficacies of designer biochars in improving biomass and nutrient uptake of winter wheat grown in a hard setting subsoil layer. Chemosphere 142:176–183

Singh K, Pandey VC, Singh B, Patra DD, Singh RP (2016) Effect of fly ash on crop yield and physico-chemical, microbial and enzyme activities of sodic soils. Environ Eng Manag J (EEMJ) 11

Soni PG, Basak N, Rai AK, Sundha P, Chandra P, Yadav RK (2023) Occurrence of salinity and drought stresses: status, impact, and management. In: Salinity and drought tolerance in plants: physiological perspectives. Springer Nature Singapore, Singapore, pp 1–28

Suliman W, Harsh JB, Abu-Lail NI, Fortuna AM, Dallmeyer I, Garcia-Pérez M (2017) The role of biochar porosity and surface functionality in augmenting hydrologic properties of a sandy soil. Sci Total Environ 574:139–147

Tan Z, Yuan S, Hong M, Zhang L, Huang Q (2020) Mechanism of negative surface charge formation on biochar and its effect on the fixation of soil Cd. J Hazard Mater 384:121370

Van Zwieten LLVZ, Kimber S, Morris S, Chan KY, Downie A, Rust J et al (2010) Effects of biochar from slow pyrolysis of papermill waste on agronomic performance and soil fertility. Plant Soil 327(1):235–246

Wang L, Xue C, Nie X, Liu Y, Chen F (2018) Effects of biochar application on soil potassium dynamics and crop uptake. J Plant Nutr Soil Sci 181(5):635–643

Wang F, Yang S, Wei Y, Shi Q, Ding J (2021a) Characterizing soil salinity at multiple depth using electromagnetic induction and remote sensing data with random forests: a case study in Tarim River Basin of southern Xinjiang, China. Sci Total Environment 754:142030

Wang H, Zhang K, Gan L, Liu J, Mei G (2021b) Expansive soil-biochar-root-water-bacteria interaction: investigation on crack development, water management and plant growth in green infrastructure. Int J Damage Mech 30(4):595–617

Wang Y, Lin Q, Liu Z, Liu K, Wang X, Shang J (2023) Salt-affected marginal lands: a solution for biochar production. Biochar 5(1):21

Wong JW, Ogbonnaya UO (2021) Biochar porosity: a nature-based dependent parameter to deliver microorganisms to soils for land restoration. Environ Sci Pollut Res 28:46894–46909

Woolf D, James E, Amonette F, Street-Perrott A, Lehmann J, Stephen J (2010) Sustainable biochar to mitigate global climate change. Nat Commun 1:1–9. https://doi.org/10.1038/ncomms1053

Xiao R, Wang JJ, Gaston LA, Zhou B, Park JH, Li R et al (2018) Biochar produced from mineral salt-impregnated chicken manure: fertility properties and potential for carbon sequestration. Waste Manag 78:802–810

Xiao Z, Rasmann S, Yue L, Lian F, Zou H, Wang Z (2019) The effect of biochar amendment on N-cycling genes in soils: a meta-analysis. Sci Total Environ 696:133984

Xu G, Zhang Y, Sun J, Shao H (2016) Negative interactive effects between biochar and phosphorus fertilization on phosphorus availability and plant yield in saline sodic soil. Sci Total Environ 568:910–915

Xu H, Cai A, Wu D, Liang G, Xiao J, Xu M et al (2021) Effects of biochar application on crop productivity, soil carbon sequestration, and global warming potential controlled by biochar C:N ratio and soil pH: a global meta-analysis. Soil Tillage Res 213:105125. https://doi.org/10.1016/j.still.2021.105125

Yadav SPS, Bhandari S, Bhatta D, Poudel A, Bhattarai S, Yadav P et al (2023) Biochar application: a sustainable approach to improve soil health. J Agric Food Res 11:100498

Yang J, Tanguy PA, Roy C (1995) Heat transfer, mass transfer and kinetics study of the vacuum pyrolysis of a large used tire particle. Chem Eng Sci 50(12):1909–1922

Yue Y, Lin Q, Li G, Zhao X, Chen H (2023) Biochar amends saline soil and enhances maize growth: three-year field experiment findings. Agronomy 13(4):1111

Zhao P, Wang S, Liu D, Li H, Han S, Li M (2022) Study on influence mechanism of biochar on soil nitrogen conversion. Environ Pollut Bioavail 34(1):419–432

Zhao W, Wang S, Liu L, Xiao J, Wang S, Tang L, Chen G (2024) Effect of biochar amendment on saline-alkaline soil amelioration and plant growth: a literature review. Chin J Soil Sci 55(2):551–561

Zheng H, Wang X, Chen L, Wang Z, Xia Y, Zhang Y et al (2018) Enhanced growth of halophyte plants in biochar-amended coastal soil: roles of nutrient availability and rhizosphere microbial modulation. Plant Cell Environ 41(3):517–532

Chapter 2
Biochar as a Catalyst for Agroecosystem Resilience: Integrating Soil Health, Water Remediation and Carbon Sequestration

P. Praveenkumar, Rajashree Khatua, and S. Sachin

Abstract Agroecosystems worldwide face mounting challenges, including soil degradation, water contamination, and the impacts of climate change, which threaten both food security and environmental sustainability. Biochar, a carbon-rich material produced through the pyrolysis of biomass, offers a promising solution to enhance agroecosystem resilience by addressing these interconnected issues. This chapter examines the multifaceted role of biochar in improving soil health, purifying water, sequestering carbon, and promoting sustainable agricultural practices. Biochar enhances soil fertility by increasing nutrient retention, improving soil structure, and promoting microbial activity, with studies showing yield increases of up to 20% in nutrient-poor soils. Its high adsorption capacity enables effective water remediation by removing nitrates, heavy metals, and organic pollutants from agricultural runoff, reducing eutrophication and improving water quality. For instance, the application of biochar in drainage systems has decreased nitrate leaching by 50%. Additionally, biochar's stable carbon structure facilitates long-term carbon sequestration, potentially offsetting 1–2 gigatons of carbon annually while reducing soil greenhouse gas emissions, such as methane and nitrous oxide, by up to 34% and 50%, respectively. Despite its potential, challenges like production variability, cost, and long-term environmental impacts necessitate further research. This chapter synthesises current knowledge, highlights practical applications, and proposes future directions, including the integration of artificial intelligence to optimise biochar use. By leveraging the properties of biochar, agroecosystems can achieve greater

P. Praveenkumar (✉)
Centre for Climate Change and Disaster Management, Anna University, Chennai, India
e-mail: praveen.climate@gmail.com

R. Khatua
Department of Agricultural Meteorology, College of Agriculture, Odisha University of Agriculture and Technology, Bhubaneswar, India

S. Sachin
Tamil Nadu Agricultural University, Coimbatore, India

D. Moulick et al. (eds.), *Biochar: Achieving Agro-Environmental Sustainability*, Sustainability Solutions, https://doi.org/10.1007/978-3-032-19855-6_2

resilience, ensuring sustainable food production and environmental health for future generations.

Keywords Biochar · Agroecosystem resilience · Soil health · Water remediation · Carbon sequestration · Nutrient retention

2.1 Introduction

Agroecosystems, critical for feeding a global population expected to reach 9.7 billion by 2050, face escalating threats that jeopardize their productivity and sustainability. Soil degradation, water contamination, and climate change impacts form a triad of challenges that reduce agricultural yields, degrade environmental quality, and threaten food security. These interconnected issues demand innovative, sustainable solutions to restore resilience in agroecosystems. Biochar, a carbon-rich material produced via biomass pyrolysis, offers a promising approach by enhancing soil health, remediating water, and mitigating climate change. This introductory section outlines the scope of these vulnerabilities and emphasizes the urgent need for resilient agroecosystems, laying the foundation for exploring biochar's transformative potential. Soil degradation affects approximately 33% of global arable land, driven by erosion, nutrient depletion, and loss of organic matter. These processes reduce soil fertility, leading to yield declines of up to 10% annually in heavily degraded regions. For example, intensive South Asian agriculture has diminished soil organic carbon, increasing dependence on chemical fertilizers and compromising long-term productivity. Healthy soils are vital for nutrient cycling, water retention, and supporting microbial ecosystems, yet ongoing degradation undermines these functions, posing a significant barrier to food security. Water contamination from agricultural runoff exacerbates challenges in agroecosystems. Runoff containing nitrates, pesticides, and heavy metals accounts for 50% of surface water pollution in farming regions, contributing to eutrophication and biodiversity loss. In the European Union, 20% of groundwater bodies exceed safe nitrate levels for drinking water, while in India, pesticide residues in waterways pose a threat to both human and ecological health. Such contamination restricts irrigation potential and disrupts the balance of agroecosystems, amplifying environmental and economic consequences (Fig. 2.1).

Climate change intensifies these pressures through rising temperatures, erratic precipitation, and extreme weather events. Projections indicate that climate impacts could reduce global crop yields by 10–20% by 2050, with tropical regions, such as sub-Saharan Africa and South Asia, facing the greatest risks. For instance, prolonged droughts and monsoon variability have disrupted food production, endangering livelihoods. Additionally, agricultural soils emit substantial amounts of greenhouse gases, including methane and nitrous oxide, which account for approximately 14% of global emissions. These climate-driven challenges necessitate adaptive strategies to stabilize agroecosystems and curb environmental harm. The convergence of these issues underscores the urgent need for resilient agroecosystems that can sustain food

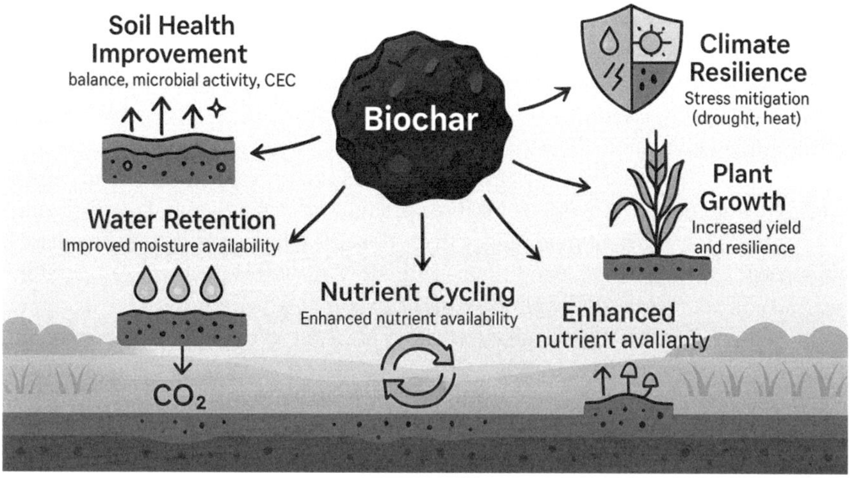

Fig. 2.1 Role of biochar in enhancing agro-ecosystem resilience

production under stress, recover from disturbances, and adapt to changing environmental conditions. Resilience is essential for ensuring food security, particularly in vulnerable regions, and for advancing global sustainability goals, such as those in the United Nations' Sustainable Development Goals (SDGs). For example, resilient systems can mitigate drought through enhanced soil water retention or reduce pollution by filtering runoff. However, conventional practices such as monoculture farming or excessive fertilizer use often exacerbate vulnerabilities, underscoring the need for sustainable interventions.

Biochar emerges as a versatile solution to address these multifaceted challenges. Its capacity to improve nutrient retention, adsorb water contaminants, and sequester carbon positions it as a tool for building agroecosystem resilience. This chapter synthesises the applications of biochar, ranging from enhancing soil fertility to mitigating greenhouse gas emissions while addressing implementation challenges and exploring future innovations, including artificial intelligence-driven optimisation. By leveraging the potential of biochar, researchers, policymakers, and farmers can foster sustainable agricultural systems that secure food production and promote environmental health for future generations.

2.2 Biochar: Properties and Production

Biochar, a carbon-rich material derived from the thermal decomposition of biomass, is a versatile amendment for enhancing the resilience of agroecosystems. Its unique properties enable improvements in soil health, water remediation, and carbon sequestration, addressing key challenges in sustainable agriculture. This section defines biochar, outlines its production through biomass pyrolysis, and highlights its key characteristics relevant to agroecosystem applications, providing a foundation for understanding its multifaceted benefits. Biochar is produced via pyrolysis, which heats biomass in a low-oxygen environment at temperatures typically between 300 and 700 °C (Lehmann and Joseph 2015). Feedstocks include agricultural residues (e.g., rice husks, corn stover), forestry byproducts (e.g., wood chips), and organic waste (e.g., manure). Pyrolysis yields three products: biochar (solid), bio-oil (liquid), and syngas (gas). Biochar yield ranges from 20 to 50%, depending on the feedstock, temperature, and pyrolysis conditions, with slow pyrolysis at lower temperatures maximising output (Woolf et al. 2010). For example, slow pyrolysis of rice straw at 400 °C yields up to 40% biochar, which is ideal for agricultural use. Production methods vary from small-scale kilns in regions like India to industrial reactors for large-scale operations. However, variability in feedstock and pyrolysis conditions affects biochar quality, necessitating standardized protocols to ensure consistency for agroecosystem applications.

Biochar's effectiveness in agroecosystems derives from its physical, chemical, and biological properties, which enhance soil fertility, remediate water, and sequester carbon. Its high surface area (up to 500 m^2/g) and porous structure enable strong adsorption of nutrients and contaminants (Ahmad et al. 2014). This reduces nutrient leaching in soils, with studies reporting a 20–30% increase in nitrogen retention in sandy soils amended with biochar (Jeffery et al. 2011). It also supports water remediation by adsorbing nitrates, heavy metals (e.g., lead, cadmium), and organic pollutants from runoff, reducing nitrate leaching by up to 50% in drainage systems. Due to its aromatic carbon structure, biochar's chemical stability ensures long-term carbon sequestration, with a residence time in soils exceeding 1000 years. If applied globally, this stability could offset 1–2 gigatons of carbon annually, aiding climate change mitigation. Biochar's alkaline nature (pH 7–10) corrects soil acidity, improving microbial activity and crop growth conditions, with pH increases of 0.5–1 unit observed in acidic tropical soils. Additionally, biochar fosters microbial activity by providing a habitat for beneficial microbes, enhances soil enzyme activity by 10–20%, and supports nutrient cycling. These properties collectively make biochar a powerful tool for agroecosystems, though its performance varies with soil type, climate, and application rate.

Biochar's production and properties position it as a sustainable solution for agroecosystem challenges. Understanding how feedstock choices and pyrolysis conditions influence its performance is critical for optimizing its application. Subsequent sections explore biochar's practical applications in enhancing soil health, remediation of water, and carbon sequestration, alongside challenges and innovations, such as artificial intelligence-driven optimisation, to maximise its impact in sustainable agriculture.

2.3 Enhancing Soil Health with Biochar

Soil health is fundamental to the resilience of agroecosystems, supporting food security and sustainable agriculture. Biochar, a carbon-rich material produced through biomass pyrolysis, significantly enhances soil health by improving fertility, nutrient retention, soil structure, and microbial activity. Its application is particularly effective in nutrient-poor soils, boosting crop yields and mitigating environmental degradation. This section examines the mechanisms by which biochar enhances soil health and reviews evidence of its impact on agricultural productivity, underscoring its role in sustainable agroecosystems.

2.3.1 Improving Soil Fertility and Nutrient Retention

Biochar's high surface area (up to 500 m^2/g) and porous structure enable it to adsorb nutrients, reducing leaching and enhancing soil fertility. By binding cations such as ammonium and potassium, biochar increases nutrient availability, especially in sandy or degraded soils. Research indicates that biochar application at 10–20 t/ha reduces nitrogen leaching by 20–30% in tropical soils, thereby improving fertiliser efficiency. Its alkaline nature (pH 7–10) corrects soil acidity, raising the pH by 0.5–1 unit in acidic soils, which enhances nutrient uptake for crops such as maize and rice. In South Asia, where soil organic carbon depletion is widespread, biochar from rice husks increases available phosphorus by 15%, supporting sustainable nutrient management (Jeffery et al. 2011).

2.3.2 Enhancing Soil Structure and Microbial Activity

Biochar improves soil physical properties, including porosity, water-holding capacity, and aggregate stability, which are essential for root growth and water retention. Its porous structure increases soil aeration and reduces bulk density by 10–15%, aiding root penetration in compacted soils (Lehmann and Joseph 2015). Biochar also enhances water retention by up to 20% in sandy soils, thereby mitigating drought stress in regions such as sub-Saharan Africa. Biologically, biochar serves as a habitat for beneficial microbes, increasing microbial biomass by 10–20% and enhancing enzyme activity associated with nutrient cycling. For example, biochar-amended soils exhibit higher populations of nitrogen-fixing bacteria, thereby improving fertility without the need for synthetic inputs. These structural and microbial improvements foster a resilient soil environment, supporting long-term productivity. Biochar's impact on crop productivity is most pronounced in nutrient-poor, degraded soils, which address multiple constraints. A meta-analysis found that biochar application increases crop yields by 10–20% on average in tropical soils, with gains up to 25% in acidic, sandy soils. In Kenya, maize yields rose by 23% with 15 t/ha biochar from wood residues due to improved nutrient retention and water availability. Similarly, rice yields in nutrient-depleted soils in India increased by 18% with the use of rice straw biochar, thereby reducing fertiliser needs. However, yield responses vary with soil type, biochar feedstock, and application rate, with limited benefits in fertile soils. These findings highlight the potential of biochar to rehabilitate degraded lands, thereby enhancing food security in vulnerable regions (Fig. 2.2).

Biochar's capacity to enhance soil health is vital for sustainable agriculture. Improving fertility, structure, and microbial activity addresses soil degradation while boosting yields in nutrient-poor environments. Subsequent sections explore biochar's roles in water remediation and carbon sequestration alongside challenges and innovations, such as the use of artificial intelligence for optimising soil-specific biochar use to advance agroecosystem resilience.

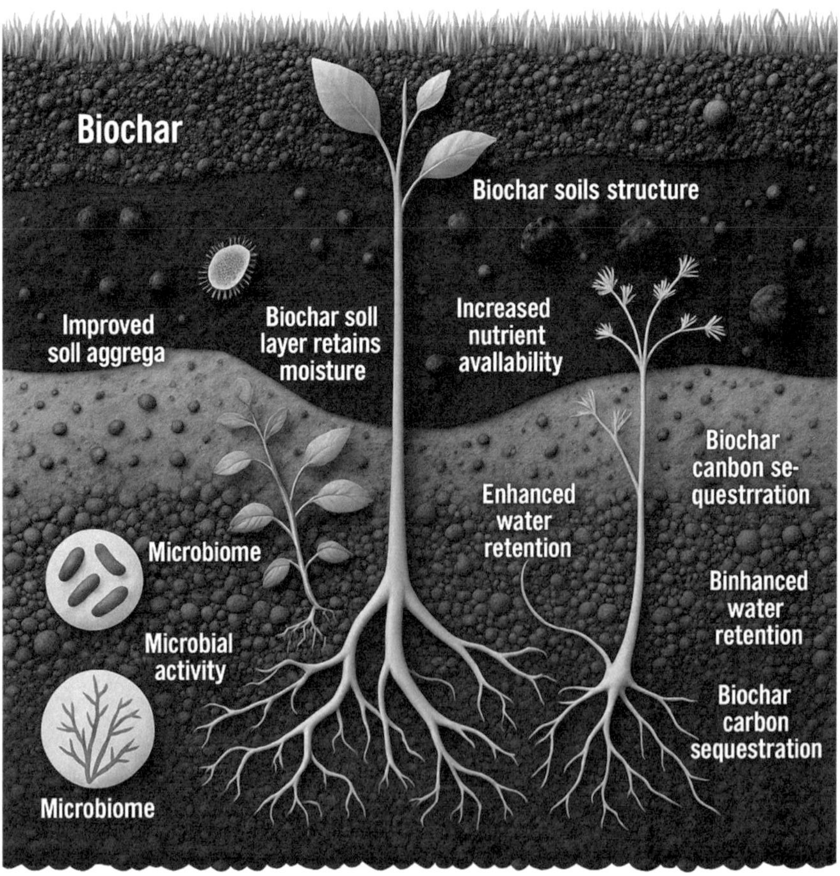

Fig. 2.2 Impact of biochar on soil health and plant growth

2.4 Biochar in Water Remediation

Water contamination from agricultural activities poses a significant threat to the sustainability of agroecosystems, contributing to pollution and ecosystem degradation. Biochar, a carbon-rich material produced through the pyrolysis of biomass, offers a promising solution for water remediation due to its exceptional adsorption properties. Biochar mitigates agricultural runoff and eutrophication by removing nitrates, heavy metals, and organic pollutants from water, enhancing water quality and agroecosystem resilience. This section examines the adsorption capacity of biochar, its applications in reducing agricultural runoff, and case studies that demonstrate its effectiveness in reducing nitrate leaching, highlighting its potential as a sustainable tool for water management.

2.4.1 Adsorption Capacity for Nitrates, Heavy Metals, and Organic Pollutants

Biochar's high surface area (up to 500 m^2/g) and porous structure enable it to act as a powerful adsorbent for a wide range of contaminants. For nitrates, biochar's negatively charged surfaces bind ammonium and nitrate ions, reducing their mobility in water. Studies show that biochar can adsorb up to 20 mg/g of nitrate, depending on feedstock and pyrolysis conditions (Mohan et al. 2014). For heavy metals, such as lead, cadmium, and copper, biochar's functional groups (e.g., carboxyl and hydroxyl) facilitate ion exchange and complexation, achieving removal efficiencies of 70–90% in contaminated water. Biochar also effectively adsorbs organic pollutants, including pesticides and pharmaceuticals, with removal rates of 60–80% for compounds such as atrazine. These adsorption capabilities make biochar a versatile tool for addressing diverse water contaminants in agroecosystems (Fig. 2.3).

2.4.2 Applications in Reducing Agricultural Runoff and Eutrophication

Agricultural runoff, a primary source of water pollution, carries nitrates, pesticides, and sediments into waterways, contributing to eutrophication and biodiversity loss. Biochar's integration into soil amendments, filter systems, and drainage ditches reduces runoff by adsorbing contaminants before they reach water bodies. For instance, biochar-amended soils retain 20–30% more nitrogen, decreasing nitrate runoff from fertilized fields. In drainage systems, biochar filters remove up to 50% of nitrates and heavy metals, mitigating eutrophication in lakes and rivers. In regions like India, where pesticide runoff threatens groundwater, biochar from rice straw has been used to reduce contaminant loads by 40%, improving irrigation water

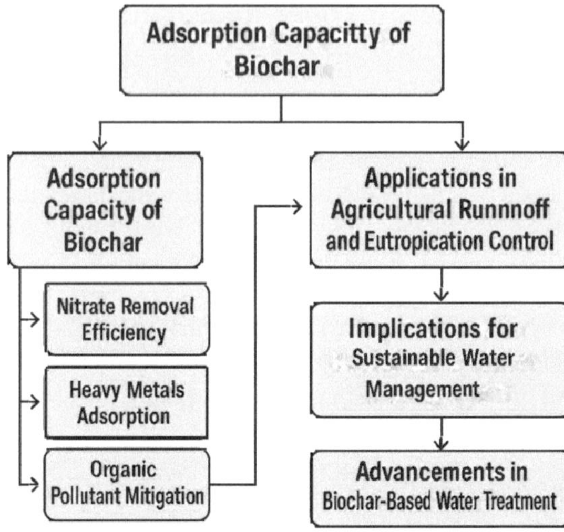

Fig. 2.3 Role of biochar in water remediation for sustainable agroecosystems

quality. These applications demonstrate biochar's role in sustainable water management, protecting agroecosystems from pollution-driven degradation. In a field trial in Iowa, USA, wood-based biochar applied at 10 t/ha in maize fields reduced nitrate leaching by 50% over a two-year period, attributed to enhanced soil retention and adsorption. In China, rice husk biochar incorporated into paddy soils decreased nitrate runoff by 45%, lowering eutrophication risks in nearby waterways. In India, a study using sugarcane bagasse biochar in vegetable farms reported a 40% reduction in nitrate leaching, improving groundwater quality. These cases highlight the efficacy of biochar, although outcomes vary depending on the type of biochar, soil conditions, and application methods, necessitating tailored approaches for optimal results. Biochar's adsorption capacity and practical applications make it a critical tool for water remediation in agroecosystems. Reducing runoff and eutrophication supports sustainable water management and environmental health. Subsequent sections explore biochar's roles in soil health and carbon sequestration, alongside challenges and innovations, such as the use of artificial intelligence for optimising biochar-based filtration to enhance agroecosystem resilience further.

2.5 Carbon Sequestration and Greenhouse Gas Mitigation

Agriculture contributes significantly to global greenhouse gas emissions, accounting for approximately 14% of the total, primarily through the release of soil-derived methane and nitrous oxide (IPCC 2021). Biochar, a stable, carbon-rich material produced through the pyrolysis of biomass, offers a dual benefit in mitigating climate change by sequestering carbon in the long term and reducing soil emissions of greenhouse gases. Its integration into agroecosystems supports sustainable practices that address global warming while enhancing environmental resilience. This section examines the role of biochar in long-term carbon storage, its potential for offsetting global carbon emissions, and its effectiveness in reducing soil emissions of methane and nitrous oxide, highlighting its contribution to climate-smart agriculture. Biochar's Role in Long-Term Carbon Storage: Biochar's chemical stability, derived from its aromatic carbon structure, enables it to persist in soils for centuries to millennia, making it an effective tool for carbon sequestration. Unlike organic matter, which decomposes rapidly, biochar resists microbial degradation, with a mean residence time exceeding 1000 years. When applied to soils, biochar locks carbon that would otherwise be released as carbon dioxide during biomass decomposition. For instance, biochar from wood residues can sequester 50–60% of the carbon content of the original biomass. In tropical soils, such as those in South Asia, the application of biochar at 10 t/ha can store 2–3 t/ha of carbon annually, thereby contributing to long-term climate mitigation (Fig. 2.4).

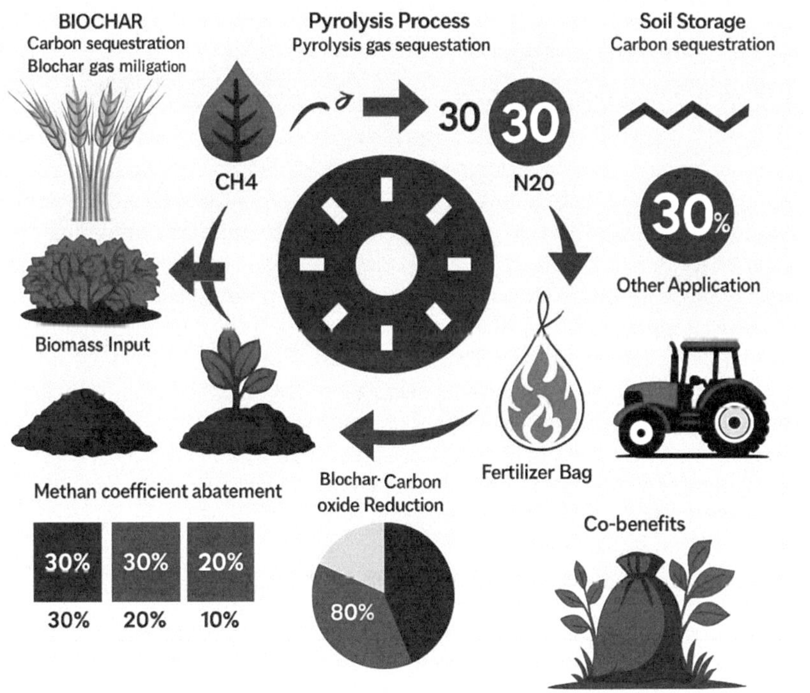

Fig. 2.4 Biochar production and application pathway for carbon sequestration and greenhouse gas mitigation

2.5.1 Potential for Offsetting Global Carbon Emissions

Biochar's scalability offers substantial potential to offset global carbon emissions. International adoption of biochar at rates of 10–50 t/ha across agricultural lands could sequester 1–2 gigatons of carbon annually, equivalent to 10–20% of current fossil fuel emissions. In regions like India, where agricultural residues are abundant, converting rice straw into biochar could sequester 0.5 gigatons of carbon yearly if implemented widely. Biochar's carbon sequestration potential is enhanced by its co-benefits, including improved soil fertility and water retention, which incentivise its adoption. However, achieving this potential requires overcoming barriers such as production costs and logistical constraints, particularly in resource-constrained regions.

2.5.2 Reduction of Soil Methane and Nitrous Oxide Emissions

Biochar mitigates soil greenhouse gas emissions by altering microbial processes and soil chemistry. In wetland rice systems, biochar reduces methane emissions by 20–34% by improving soil aeration and suppressing methanogenic bacteria. For instance, rice husk biochar applied at 15 t/ha in Chinese paddy fields decreased methane emissions by 30% while maintaining yields. Similarly, biochar reduces nitrous oxide emissions by 30–50% by enhancing soil nitrogen retention and inhibiting denitrification processes (Cayuela et al. 2014). In a Brazilian study, sugarcane bagasse biochar reduced nitrous oxide emissions by 40% in fertilised soils, thereby mitigating the climate impact of nitrogen fertilisers. These reductions are most pronounced in acidic, waterlogged, or heavily fertilised soils, making biochar a valuable tool for mitigating emissions in emission-intensive agroecosystems.Biochar's capacity for carbon sequestration and greenhouse gas mitigation positions it as a cornerstone of climate-smart agriculture. Storing long-term carbon and reducing soil emissions supports agroecosystem resilience and global climate goals. Subsequent sections explore biochar's applications in soil health and water remediation alongside challenges and innovations, such as the use of artificial intelligence to optimise biochar's climate benefits to advance sustainable agriculture further.

2.6 Challenges and Limitations of Biochar Use

Biochar promises to enhance agroecosystem resilience, yet several challenges hinder its widespread adoption. Variability in production and quality, high costs and scalability constraints, and uncertainties surrounding long-term environmental impacts limit its practical application. Addressing these barriers is essential to fully harnessing biochar's potential in sustainable agriculture. This section explores these challenges, their implications for agroecosystem management, and critical research gaps that must be addressed to advance biochar's role in resilient farming systems.

2.6.1 Variability in Biochar Production and Quality

The properties of biochar, including porosity, pH, and adsorption capacity, are highly dependent on the type of feedstock and pyrolysis conditions, resulting in inconsistent performance across different agroecosystems. For example, biochar from corn stover pyrolyzed at 350 °C may have a surface area of 200 m^2/g, while bamboo biochar at 600 °C can reach 450 m^2/g, affecting its nutrient retention or contaminant adsorption capabilities. In smallholder farms in India, locally produced biochar from mixed residues often yields unpredictable results, with crop yield improvements ranging from 5 to 15% due to non-standardised production methods. This variability complicates the development of universal application guidelines, as biochar is effective in one soil type (e.g., sandy) but may be less so in another (e.g., clay). The absence of global standards for biochar production remains a significant hurdle, requiring coordinated efforts to ensure consistent quality for agroecosystem applications.

2.6.2 Cost Considerations and Scalability

Biochar production entails significant costs, including feedstock acquisition, pyrolysis infrastructure, and labour, which limit its accessibility, particularly for smallholder farmers. In developing regions, the costs of small-scale biochar production range from \$100 to \$300 per ton, translating to \$1000–\$6000/ha at typical application rates of 10–20 t/ha (Shackley et al. 2011). Industrial-scale systems, while more efficient, demand substantial upfront investment, often exceeding \$1 million for large reactors. In sub-Saharan Africa, competition for biomass as fuel or fodder further restricts feedstock availability, hampering scalability. Although co-products like bio-oil and syngas can offset costs, their commercial markets are still in their infancy, particularly in low-income regions. These economic barriers necessitate innovative financing models and low-cost production technologies to make biochar viable for widespread use in agroecosystems.

2.6.3 Long-Term Environmental Impacts and Research Gaps

Although biochar is valued for its stability, its long-term environmental impacts remain unclear, posing potential risks to agroecosystems. Some biochars, particularly from sewage sludge or treated wood, may release contaminants like polycyclic aromatic hydrocarbons (PAHs), which could accumulate in soils or leach groundwater. High biochar application rates (> 50 t/ha) have also been linked to reduced soil microbial diversity in certain tropical soils, raising concerns about the health of ecosystems. Furthermore, the long-term effects of biochar on carbon cycling and soil fertility in diverse climates, such as arid or saline environments, remain understudied. The abstract's claims of reduced greenhouse gas emissions require validation across varied agroecosystems, as some studies report negligible effects in alkaline soils. Long-term field trials spanning decades are necessary to assess the environmental safety and efficacy of biochar, thereby addressing critical knowledge gaps and ensuring its sustainable application. These challenges highlight the need for targeted research and policy interventions to optimise the use of biochar in agroecosystems. Standardizing production, reducing costs, and evaluating long-term impacts are pivotal to scaling its benefits. Subsequent sections explore practical applications, case studies, and innovations, such as the use of artificial intelligence for customising biochar applications, to overcome these limitations and advance sustainable agriculture.

2.7 Practical Applications and Case Studies

Biochar's versatility as a soil amendment and water remediation tool has led to its adoption in diverse agroecosystems, offering practical solutions to enhance agricultural productivity and environmental sustainability. Real-world applications demonstrate its efficacy in improving soil fertility, reducing water pollution, and seamlessly integrating into existing farming and drainage systems. This section presents case studies of biochar use in agriculture and water management, highlighting its integration into conventional practices and its potential to support resilient agroecosystems in varied regional contexts. Biochar has been successfully applied to enhance crop yields and remediate water in multiple regions, particularly in degraded or resource-constrained agroecosystems. In Sri Lanka, coconut shell biochar applied at 15 t/ha in tea plantations increased yields by 20% over three years by improving soil water retention and nutrient availability. In Indonesia, the use of rice husk biochar in paddy fields reduced irrigation needs by 15% and increased rice yields by 18%, thereby mitigating drought stress in rainfed systems. For water management, a pilot project in Australia employed wood-based biochar in constructed wetlands to treat agricultural runoff, removing 60% of nitrates and 70% of phosphorus, thus reducing eutrophication in downstream waterways. In Bangladesh, biochar made from jute sticks was applied to vegetable farms, resulting in a 50% decrease in pesticide runoff and an improvement in groundwater quality. These examples illustrate biochar's role in enhancing agricultural productivity and protecting water resources, particularly in tropical and subtropical agroecosystems. Biochar's practical application is enhanced by its compatibility with existing agricultural and water management infrastructure, minimizing the need for systemic overhauls. In farming systems, biochar is often mixed with compost or fertilisers to optimise nutrient delivery. This approach is exemplified in Vietnam, where maize farmers combined bamboo biochar with manure at a rate of 10 t/ha, resulting in a 22% increase in yields while reducing fertiliser use by 30% (Doan et al. 2015). In India, smallholder farmers incorporate rice straw biochar into furrow irrigation systems, which enhances soil moisture retention and reduces runoff by 25%. Biochar is integrated into filter strips or bioretention cells for drainage systems to capture contaminants. A Swedish study demonstrated that birch biochar filters installed in field drainage ditches reduced nitrate leaching by 45% and heavy metal loads by 60%, protecting local streams. In China, biochar-amended buffer strips along rice paddies decreased sediment and nutrient runoff by 40%, integrating effectively with traditional flood irrigation. These integrations leverage existing infrastructure, making biochar a cost-effective and scalable solution for agroecosystem management.

2.8 Future Directions and Innovations

Biochar's potential to enhance agroecosystem resilience is well established, yet its full realisation requires innovative approaches to overcome current limitations and scale its adoption. Advances in artificial intelligence (AI), targeted research to improve efficacy and sustainability, and strategic policies for widespread use are critical to integrating biochar into global agricultural systems. This section examines the role of AI in optimising biochar applications, identifies key research needs, and proposes strategies to promote the adoption of biochar in agroecosystems, thereby paving the way for sustainable food production and environmental health.

2.8.1 Role of Artificial Intelligence in Optimising Biochar Applications

AI technologies, including machine learning and predictive modelling, offer transformative opportunities to tailor biochar applications to specific agroecosystems. AI can analyse complex datasets—encompassing soil types, climate conditions, feedstock properties, and pyrolysis parameters—to predict the performance of biochar in enhancing soil health, water remediation, and carbon sequestration. For instance, machine learning models have been utilised to optimise biochar production, resulting in a 15% increase in adsorption capacity for heavy metals by adjusting the pyrolysis temperature (Zhu et al. 2022). In India, AI-driven decision-support systems are being piloted to recommend biochar application rates for smallholder farms, resulting in a 10% improvement in yield consistency in nutrient-poor soils. AI can dynamically adjust biochar strategies by integrating real-time data from sensors that monitor soil moisture and nutrient levels, thereby enhancing efficiency and reducing costs. These advancements address the abstract's call for AI-driven optimisation, ensuring that the benefits of biochar are maximised across diverse agricultural contexts.

2.8.2 Research Needs for Enhancing Efficacy and Sustainability

To improve the efficacy and sustainability of biochar, research must address several critical gaps. First, long-term field studies (10–50 years) are necessary to evaluate biochar's stability and environmental impacts in diverse climates, particularly in arid or saline soils where data are scarce. Second, research should focus on standardising biochar production to minimise variability, thereby ensuring consistent quality for applications such as nitrate adsorption or carbon sequestration. Third, the sustainability of feedstock sourcing requires investigation, as biomass competition in regions like sub-Saharan Africa limits scalability. Developing biochar from waste materials, such as sewage sludge or invasive plants, could enhance sustainability while reducing costs. Finally, the abstract's claim of greenhouse gas reductions (e.g., 34% methane and 50% nitrous oxide) needs validation across diverse agroecosystems to confirm scalability (Biederman and Harpole 2013). These research priorities will enhance the role of biochar in sustainable agriculture.

2.8.3 Strategies for Widespread Adoption in Agroecosystems

Promoting the adoption of biochar requires integrated approaches that address economic, technical, and policy barriers. Financial incentives, such as subsidies or carbon credit schemes, can help offset the high initial costs of biochar ($100–300/t), making it more accessible to smallholder farmers in regions like South Asia. Training programs and extension services can educate farmers on the benefits and application methods of biochar, as demonstrated in Vietnam, where farmer workshops increased adoption by 25%. Policy frameworks, such as integrating biochar into national agricultural plans, can drive large-scale implementation. For example, Australia's biochar initiatives have boosted its use on 10% of farms. Partnerships between researchers, industry, and governments can develop low-cost pyrolysis technologies, thereby enhancing scalability in biomass-rich regions such as India. Combined with AI-driven optimisation, these strategies can effectively integrate biochar into global agroecosystems, thereby supporting food security and sustainability. By leveraging AI, addressing research gaps, and implementing strategic policies, biochar can transition from a promising tool to a cornerstone of resilient agroecosystems. The concluding section synthesises the multifaceted benefits of biochar and envisions its role in sustainable agriculture.

2.9 Conclusion

Biochar is a transformative tool for addressing the interconnected challenges of soil degradation, water contamination, and climate change, paving the way for resilient agroecosystems that ensure food security and environmental sustainability. This chapter synthesises the multifaceted benefits of biochar, ranging from enhancing soil health to remediating water and mitigating greenhouse gas emissions, while acknowledging the challenges and future innovations. As global agriculture faces increasing pressures, biochar offers a pathway to sustainable practices that balance productivity with ecological health, envisioning a future where agroecosystems thrive amidst environmental uncertainties. Biochar's contributions to agroecosystems are comprehensive. Agriculture rejuvenates nutrient-poor soils, boosting crop yields by up to 20% in degraded lands through improved fertility and enhanced water retention, as observed in regions such as South Asia and sub-Saharan Africa. Its ability to adsorb contaminants, reducing nitrate leaching by up to 50% and heavy metal loads in runoff, safeguards water quality, protecting ecosystems from eutrophication and pollution. For climate mitigation, biochar's stable carbon structure sequesters significant carbon, potentially offsetting gigatons of emissions annually, cutting soil methane and nitrous oxide emissions by 34% and 50%, respectively. These benefits are amplified when biochar is integrated into existing farming and drainage systems, as demonstrated by case studies in Indonesia, Australia, and India, where it enhances productivity without disrupting traditional practices despite challenges such as production variability, high costs, and long-term environmental uncertainties, innovations like artificial intelligence-driven optimization promise to tailor biochar applications, ensuring consistency and affordability.

The vision for sustainable agriculture and environmental health hinges on scaling the adoption of biochar through strategic policies and research. By leveraging abundant agricultural residues, such as rice straw in India and maize stover in Africa, biochar production can become a model for a circular economy, transforming waste into a valuable resource. Financial incentives, farmer training, and low-cost pyrolysis technologies can make biochar accessible to smallholder farmers, who manage a significant portion of the world's arable land. Collaborative efforts among researchers, policymakers, and communities can standardise production, ensuring the efficacy of biochar across diverse soils and climates. Moreover, integrating biochar into national climate and agricultural plans can align it with global sustainability goals, thereby fostering agroecosystems that are more resilient to drought, pollution, and climate variability. Biochar is not a panacea but a critical component of a broader strategy to build resilient agroecosystems. Its ability to address multiple environmental challenges simultaneously positions it as a cornerstone of sustainable agriculture. As research advances and adoption grows, biochar can empower farmers to produce more with less, safeguard water resources, and contribute to a stable climate. This chapter envisions a future where biochar, combined with innovative technologies and inclusive policies, transforms agroecosystems into models of resilience, ensuring food security and environmental health for future generations.

References

Ahmad M, Rajapaksha AU, Lim JE, Zhang M, Bolan N, Mohan D et al (2014) Biochar as a sorbent for contaminant management in soil and water: a review. Chemosphere 99:19–33

Biederman LA, Harpole WS (2013) Biochar and its effects on plant productivity and nutrient cycling: a meta-analysis. GCB Bioenergy 5:202–214

Cayuela ML, van Zwieten L, Singh BP, Jeffery S, Roig A, Sánchez-Monedero MA (2014) Biochar's role in mitigating soil nitrous oxide emissions: a review and meta-analysis. Agric Ecosyst Environ 191:5–16

Doan TT, Henry-des-Tureaux T, Rumpel C, Janeau JL, Jouquet P (2015) Impact of compost, vermicompost and biochar on soil fertility, maize yield and soil erosion in Northern Vietnam: a three-year mesocosm experiment. Sci Total Environ 514:147–154

Intergovernmental Panel on Climate Change (IPCC) (2021) Climate change 2021: the physical science basis. Cambridge University Press, Cambridge

Jeffery S, Verheijen FGA, van der Velde M, Bastos AC (2011) A quantitative review of the effects of biochar application to soils on crop productivity using meta-analysis. Agric Ecosyst Environ 144:175–187

Lehmann J, Joseph S (eds) (2015) Biochar for environmental management: science, technology and implementation, 2nd edn. Routledge, London

Mohan D, Sarswat A, Ok YS, Pittman CU Jr (2014) Organic and inorganic contaminants removal from water with biochar, a renewable, low-cost and sustainable adsorbent—a critical review. Bioresour Technol 160:191–202

Shackley S, Carter S, Knowles T, Middelink E, Haefele S, Sohi S et al (2011) Sustainable gasification–biochar systems: a case study of rice-husk gasification in Cambodia, part I: context, chemical properties, environmental and health and safety issues. Energy Policy 39:563–574

Woolf D, Amonette JE, Street-Perrott FA, Lehmann J, Joseph S (2010) Sustainable biochar to mitigate global climate change. Nat Commun 1:56

Zhu X, Chen B, Zhu L, Xing B (2022) Machine learning for optimizing biochar properties for environmental applications. Environ Sci Technol 56:11342–11352

Chapter 3
Biochar for Improving Soil Quality and Carbon Cycling

Md. Zubair Al Islam, Annika Jahan Aonti, Sabikunnahar, Md. Mostofa Kamal Sohel, and Akbar Hossain

Abstract With the growing global food gap and increasing climate change concerns, there is a need to identify green solutions to increase soil fertility and productivity while improving soil biochemical quality and lowering the environmental effects of agriculture. Biochar is a solid, stable, carbon-rich, and highly aromatic material made from plant feedstock with a rich pore structure and a high specific surface area, making it a cost-effective carbonaceous resource. As biochar is composed of carbon, hydrogen and oxygen, it may remain in the environment for a long time and retain its carbon content in the soil for 100 years, which is critical for soil health. Biochar has the ability to capture estimated amounts of CO_2 ranging from 1 to 35 gigatons ($GtCO_2$) and 78 to 477 $GtCO_2$ per century. It also helps mitigate climate change by sequestering carbon in the soil for a longer period of time, consequently reducing GHG emissions, and can also be employed to establish carbon credits. As an amendment to soil, it enhances the physical and biochemical characteristics of the soil, as well as its fertility and productivity, especially in the long run. It also increases soil aggregation, water retention, pH, and microbial activity, all of which contribute to the overall improvement in soil quality and may eventually lessen the need for chemical fertilizers. The distinctive qualities, broad applicability, affordability, and reliable development prospects of biochar show how much potential it has for achieving sustainable development goals (SDGs).

Keywords Biochar · Soil quality · Carbon cycle · Cost-effectiveness · Environmentally friendly

Md. Zubair Al Islam
Soil Science Division, Bangladesh Institute of Nuclear Agriculture, Mymensingh, Bangladesh

A. J. Aonti · A. Hossain (✉)
Soil Science Division, Bangladesh Wheat and Maize Research Institute, Dinajpur, Bangladesh
e-mail: akbarhossainwrc@gmail.com

Sabikunnahar
Agronomy Division, Bangladesh Agricultural Research Institute, Gazipur, Bangladesh

Md. Mostofa Kamal Sohel
Plant Pathology Division, Bangladesh Wheat and Maize Research Institute, Nashipur, Dinajpur, Bangladesh

D. Moulick et al. (eds.), *Biochar: Achieving Agro-Environmental Sustainability*, Sustainability Solutions, https://doi.org/10.1007/978-3-032-19855-6_3

3.1 Introduction

Biochar, which is a carbon-rich outcome of pyrolysis, has gained prominence as an efficient amendment for enhancing agricultural sustainability. This material positively influences soil properties, enhances nutrient retention, and improves water-holding capacity, which are critical factors for sustainable farming practices (Mensah and Frimpong 2018; Raj et al. 2023). The introduction of biochar into agricultural soils not only sequesters carbon but also reduces greenhouse gas emissions, making it a dual solution for ecological and agricultural challenges (Carvalho et al. 2020; Zheng et al. 2012).

The varying impacts of biochar are closely linked to its feedstock and pyrolysis conditions, which determine its physical and chemical properties. Research shows that different types of biochar can enhance soil characteristics in diverse climates and soil types (Khan et al. 2024; Kuryntseva et al. 2023). For example, studies have shown that biochar can lead to significant improvements in soil fertility and crop yields in low-fertility, degraded soils, whereas the benefits may be less pronounced in already fertile soils (Blanco-Canqui 2020). Biochar has a relatively high cation exchange capacity (CEC) and porosity, which contributes significantly to its effectiveness as a soil amendment, facilitating better nutrient uptake by plants and fostering robust microbial communities (Mensah and Frimpong 2018; Raj et al. 2023; Adirianto and Bachtiar 2023).

The historical development of biochar is closely linked to ancient agricultural practices, particularly the use of charcoal in the Amazon, where indigenous peoples created fertile "Terra Preta" soils. These soils are characterized by high levels of organic matter and nutrients and have sustained agricultural productivity for centuries (Manyà 2012; Vithanage et al. 2013). The incorporation and mixing of charcoal into soil to enhance health and state dates back thousands of years and has significant implications for contemporary agriculture (Hood-Nowotny et al. 2018).

The scientific understanding of biochar has evolved over time. The deliberate pyrolysis of biomass to produce biochar has been documented for over 150 years in Western agriculture, with its use noted in animal husbandry and as a soil amendment (Schmidt et al. 2021). In these contexts, biochar has been recognized as being capable of improving soil characteristics, such as the water holding capacity and availability of plant nutrients, thus increasing crop production (Kalina et al. 2022; Nair et al. 2017).

Biochar possesses a range of key properties that render it a valuable amendment in agricultural practices and environmental management. Understanding these properties is essential for maximizing their benefits for soil health, carbon sequestration, and overall agricultural sustainability.

One of the primary attributes of biochar is its elemental composition, which influences its effectiveness as a soil amendment. Biochar is typically made of carbon (C), hydrogen (H), oxygen (O), and nitrogen (N), with the proportions of these elements relying on the biomass used and the degree of pyrolysis (Wijitkosum and Jiwnok 2019; Crombie and Mašek 2014). The C content is especially critical because it is related to the firmness of biochar, whereas the H and O contents impact its reactivity and nutrient exchange capabilities. The hydrogen–carbon ratio (H/C) and oxygen–carbon ratio (O/C) further influence the stability of biochar in soil and its interactions with soil nutrients (Wijitkosum and Jiwnok 2019; Crombie and Mašek 2014).

The use of biochar as a soil amendment is recognized for its significant benefits on soil health and stability, with numerous studies highlighting various mechanisms through which it improves soil quality and crop yields. Biochar enhances the physical properties of the soil, including increasing its water retention capacity and improving aeration, which is particularly beneficial for light-textured soils (Rizhiya et al. 2015; Mensah and Frimpong 2018). For example, Luan et al. demonstrated that biochar significantly enhances soil fertility and facilitates better growth of crops such as Chinese cabbage by improving soil aggregation and water retention (Luan et al. 2023). This claim is corroborated by researchers, as they reported that biochar can increase the availability of nutrients over time; thus, nutrients become more readily available to plants (Bista et al. 2019; Crane-Droesch et al. 2013).

Furthermore, biochar has an important role in improving soil chemical properties. Studies indicate that biochar can improve the CEC of soils, thereby allowing better nutrient retention and availability. This effect is further augmented when biochar is used together with organic and inorganic fertilizers, resulting in increased nutrient absorption by crops (Liu et al. 2022). The alkalinity of biochar also helps increase the soil pH, which can further benefit crop growth by increasing nutrient availability (Laird et al. 2010; Biederman and Harpole 2012).

Owing to its multidimensional benefits in improving soil health, improving crop productivity, and contributing to the mitigation of climate change, biochar is increasingly recognized as a promising tool for sustainable agriculture. The incorporation of biochar into soils positively affects their physical, chemical, and biological properties, which, in turn, fosters a more sustainable agricultural ecosystem.

3.2 Concepts, Importance, Properties and Benefits of Biochar for Agricultural Sustainability

3.2.1 Concepts of Biochar

Biochar is a carbon-rich material produced through the pyrolysis of biomass, which is a thermochemical process in which organic materials are decomposed at elevated temperatures in an oxygen-limited environment. This process transforms biomass into a stable solid, making it ideal for long-term carbon storage in soil. Biochar is characterized by its porous structure and high surface area, which contributes to its ability to improve soil properties and enhance plant growth.

3.2.2 Importance of Biochar

Biochar has a surface area characterized by < 2 nm, 2–50 nm mesopores along with > 50 nm macropores (Jatav et al. 2020). Owing to its unique physicochemical qualities, such as high surface area, porous structure, and various functional groups, biochar is an important tool for eliminating contaminants from wastewater and soil. Furthermore, biochar has shown efficiency in combating emerging contaminants such as microplastics (Kumar et al. 2023). The process by which biochar is synthesized via pyrolysis and then its structure is altered to remove aquatic pollutants is clearly described in Fig. 3.1, and Table 3.1 shows the different types of biochar used for the removal of organic pollutants.

Biochar has gained significant attention as a compelling tool for environmental sustainability, agriculture, and carbon sequestration. This material is generated primarily through the pyrolysis of organic waste, resulting in a charcoal-like substance that has unique properties that are beneficial for improving soil health.

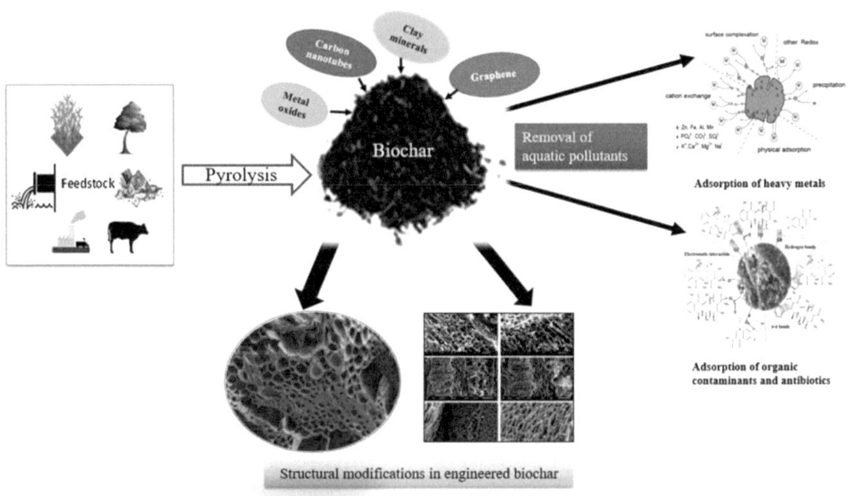

Fig. 3.1 Biochar implications for the removal of pollutants from soil and water (Murtaza et al. 2022)

Table 3.1 Different biochar types for the removal of organic pollutants

Biochar type	Organic pollutants	Procedures	References
Rice straw	Diazinon	Hydrogen bonding with polar group	Taha et al. (2014)
Maize	Perfluorooctane sulfonate	Hydrophobic adsorption	Chen et al. (2011)
Dairy manure	Atrazine	Partitioning	Cao et al. (2009)
Poultry litter	Ethinylestradiol	Pore filling	Sun et al. (2011)

One of the foremost advantages of biochar is its ability to enhance soil structure and fertility. Studies have demonstrated that biochar application leads to improved soil bulk density, increased porosity, and increased water retention capabilities, which are essential for fostering plant growth in various agricultural contexts, particularly in regions with tropical climates (Laghari et al. 2016; Guo 2020). For example, agricultural experiments have shown significant improvements in soil nutrient retention and availability, facilitating a healthier ecosystem and ultimately contributing to increased crop yields (Laghari et al. 2016; Jindo et al. 2020a, b). The incorporation of biochar enriches soils with critical nutrients such as phosphorus, potassium, calcium, and magnesium, supporting the overall nutrient cycling necessary for sustainable farming practices (Latawiec et al. 2019; Ghosh and Maiti 2022).

3.2.3 Classification of Biochar

3.2.3.1 Feedstock Methods

Lignocellulosic Biochar

Lignocellulosic waste biochar, which is produced from biomass rich in cellulose, hemicellulose, and lignin, has attracted considerable interest as a sustainable material for various applications, particularly in soil enhancement and pollutant remediation. The dominant feedstocks for lignocellulosic biochar typically derive from agricultural residues, forestry byproducts, and municipal waste, which are characterized by their high carbon content and potential to improve soil properties (Bednik et al. 2022; López-Cano et al. 2018).

The production of biochar from lignocellulosic waste predominantly occurs through pyrolysis—a thermochemical process that converts organic material into a stable form, increasing its carbon content and structural integrity. During pyrolysis, the decomposition of lignocellulose leads to the volatilization of water and gases, resulting in biochars whose physicochemical properties vary depending on the feedstock and pyrolysis conditions employed (Chemerys and Baltrėnaitė 2018; Mota-Panizio et al. 2023; Novak et al. 2013). For example, biochars produced from agricultural residues such as wheat straw or sugarcane bagasse present higher nutrient concentrations and cation exchange capacities (CECs) than traditional wood biochars do, significantly increasing their agronomic value (López-Cano et al. 2018; Mukome et al. 2013).

Animal-Derived Biochars

Animal-derived biochars are characterized by their organic compound composition and nutrient profiles, which often differ significantly from those of plant-derived biochars. The pyrolysis of animal waste typically results in biochar that retains essential nutrients, particularly nitrogen, phosphorus, and potassium, which are beneficial for soil enrichment (Domingues et al. 2017; Lin et al. 2022). Studies indicate that animal manure-based biochars tend to have a lower carbon concentration than their plant counterparts do, especially as the pyrolysis temperature increases. For example, when the pyrolysis temperature of poultry litter biochar was raised from 350 to 700 °C, a considerable reduction in the carbon concentration (approximately 6%) was observed (Domingues et al. 2017).

Municipal Waste Biochar

Waste biomass, including municipal solid waste (MSW), has emerged as a crucial feedstock for biochar production. Studies have demonstrated that biochar can be derived from diverse sources, such as paper sludge and agricultural residues, which can significantly contribute to waste management strategies and environmental sustainability (Yaro et al. 2023; Ippolito et al. 2020). The use of municipal and industrial waste as feedstock helps lower the operational costs associated with waste disposal while ensuring that carbon is preserved in a stable form, potentially sequestering it in the soil for centuries (Ippolito et al. 2020; Zhu et al. 2022). Furthermore, using waste for biochar production aligns with circular economy principles, where the material cycle is closed by reusing waste (Ana et al. 2024).

3.2.3.2 Pyrolysis

The classification of biochar on the basis of pyrolysis parameters is vital for understanding its properties, applications, and potential benefits in environmental contexts. Pyrolysis, defined as the thermal degradation of biomass in an oxygen-free environment, can be executed under various conditions, resulting in diverse biochar characteristics. The primary classifications of pyrolysis include slow, fast, and intermediate pyrolysis, which are distinguished on the basis of the temperature and heating rates employed during the process (Aboughaly and Fattah 2023; Ronsse et al. 2012; Roberts et al. 2015).

Slow pyrolysis typically occurs at temperatures between 300 and 600 °C and is aimed primarily at maximizing biochar yield while also producing syngas and bio-oil as byproducts. This method is often associated with the production of biochars that exhibit enhanced stability and a relatively high carbon content, which are vital for soil amendment applications (Ronsse et al. 2012; Roberts et al. 2015). Fast pyrolysis, on the other hand, generally occurs at relatively high temperatures (approximately 400–500 °C) and is focused on maximizing liquid bio-oil yields, resulting in biochars that may have different physicochemical properties than those resulting from slow pyrolysis (Cordeiro et al. 2024; Brown et al. 2010).

3.2.3.3 Other Thermochemical Processes

Gasification

Biochar production through gasification involves a thermochemical conversion process in which biomass is partially oxidized at high temperatures (typically between 700 and 1000 °C) in an environment with limited oxygen. This method serves as a waste management strategy and facilitates the production of syngas, which can be utilized for energy generation (Lugato et al. 2012). The resulting biochar can be characterized by its unique physicochemical properties and potentially beneficial applications, especially in agriculture and environmental remediation.

Hydrothermal Carbonization (HTC)

Hydrothermal carbonization (HTC) represents an emerging thermochemical technology for converting wet biomass into biochar under mild temperature and pressure conditions, typically between 180 and 250 °C. This method diverges from traditional carbonization processes by the use of water as a solvent, which facilitates the breakdown of lignocellulosic materials and their transformation into stable carbonaceous substrates. The resulting biochar, often termed "hydrochar," possesses unique properties that render it valuable in agricultural and environmental applications (Köves et al. 2024).

Torrefaction

Biochar is a carbon-rich byproduct derived from the thermochemical conversion of biomass and is characterized by its production methods in oxygen-limited environments. Among the various techniques used for biochar production, torrefaction is gaining recognition as an effective method for improving the handling and energy properties of biomass. The torrefaction process generally involves heating biomass to temperatures between 200 and 300 °C, leading to the removal of moisture and some volatile compounds, thus increasing the energy density and hydrophobic characteristics of the biomass material (Pawar and Panwar 2020; Ayadi et al. 2020).

Unlike pyrolysis and gasification, which occur at relatively high temperatures and yield various outputs, including gases and oils, torrefaction primarily focuses on producing a solid biochar product while further concentrating the carbon content. Compared with raw biomass, the resulting biochar has improved quality, making it suitable for energy applications and soil enhancement because of its increased stability and reduced ash content (Osman et al. 2023). This process not only increases the calorific value of the biomass but also allows for easier transportation and storage, thereby adding economic value to otherwise low-cost agricultural residue (Jia et al. 2021).

3.2.3.4 Physical and Chemical Properties

Carbon Content

Biochar classification on the basis of carbon content can be divided into two primary categories: high-carbon biochar and low-carbon biochar. This differentiation is essential, as it influences biochar properties, stability, and potential applications in fields such as agriculture, environmental remediation, and carbon sequestration.

High-Carbon Biochar

High-carbon biochar typically has a carbon content exceeding 70% and is often derived from biomass subjected to high-temperature pyrolysis (typically above 500 °C). This category includes biochars produced from materials with high lignin content, such as pine needles, which contribute to more stable carbon structures (Nain 2023). High-carbon biochar is notable for its stability and durability in soils, providing long-term carbon sequestration potential. The formation of fused aromatic structures during pyrolysis contributes to the longevity of high-carbon biochars in soil systems, as these structures are less susceptible to microbial decay than lower-carbon biochars are (Woo 2013; Azzi et al. 2023).

Low-Carbon Biochar

The production of low-carbon biochar often occurs at lower pyrolysis temperatures (below 500 °C), resulting in a higher volatile content and a less stable carbon structure than those of high-carbon biochar produced at higher temperatures. The decreased stability makes low-carbon biochar more susceptible to degradation when applied to soil, which can lead to faster carbon emissions into the atmosphere unless managed properly (Woolf et al. 2010). However, despite its lower carbon retention capabilities, low-carbon biochar still has significant agronomic value, assisting in soil amendment and nutrient retention (Harvey et al. 2012; Wyzińska et al. 2023).

3.2.3.5 pH-Based Methods

Acidic Biochar

Acidic biochar refers to a specific type of biochar characterized by a relatively low pH, typically below seven, which can be beneficial in certain soil management contexts. The production of biochar involves the thermal decomposition of biomass in low-oxygen environments, a process known as pyrolysis, which can significantly alter its chemical properties, including pH. Research has shown that the acidic properties of biochar can be influenced by various factors, such as the source of biomass and the conditions employed during pyrolysis.

For example, Zhang et al. demonstrated the varied effects of acidic biochar compared with those of alkaline biochar in saline–alkali soils, highlighting that acidic biochar can alleviate nitrogen limitation under specific soil conditions while impacting microbial nutrient limitation differently from its alkaline counterparts (Zhang et al. 2025). Furthermore, Wang et al. supported this perspective by describing the preparation of acidic phosphorus-rich biochar from halophyte species, which positively affects phosphorus availability in alkaline soil environments, revealing how acidic biochar can enhance nutrient management in soils requiring specific amendments (Wang et al. 2022).

Alkaline Biochar

Alkaline biochar is a type of biochar with a pH level typically greater than 7, often exceeding pH 8. This alkalinity can arise from various factors, including the materials used in its production and the pyrolysis conditions. Alkaline biochar has gained attention for its potential agronomic benefits, particularly in enhancing soil properties and fertility, especially in acidic or nutrient-poor soils.

Research has demonstrated that the alkalinity of biochar can significantly influence soil conditions. Alazzaz et al. reported that the liming effect of alkaline biochar can improve nutrient availability in acidic soils, suggesting its role in correcting soil pH imbalances to promote plant growth (Alazzaz et al. 2020). Additionally, biochar produced at high pyrolysis temperatures is typically alkaline, thereby increasing soil carbon sequestration and improving nutrient retention, ultimately supporting a healthier soil ecosystem (Liu et al. 2014; Al-Wabel et al. 2017).

3.2.3.6 Surface Area and Porosity

High-Surface-Area Biochar

High surface area biochar is a form of biochar characterized by its elevated specific surface area, which typically develops through pyrolysis processes at high temperatures. The specific surface area of biochar is fundamentally important, as it influences various physical and chemical interactions within the soil, including adsorption, nutrient retention, and provision of habitats for microorganisms.

One of the primary determinants of the specific surface area of biochar is the pyrolysis temperature. Higher-temperature pyrolysis commonly yields biochar with a greater surface area due to the thermal decomposition of organic materials, leading to a more porous structure (Zhang et al. 2015; Kloss et al. 2012). Studies indicate that biochars produced at temperatures exceeding 500 °C have significantly greater surface areas than those produced at lower temperatures, primarily due to increased aromatization and loss of volatile compounds (Kloss et al. 2012; Jindo et al. 2014). These structural characteristics contribute to enhanced adsorption capacities, allowing biochar to effectively sequester contaminants and nutrients in the soil (Hossain et al. 2020).

Low Surface Area Biochar

The low surface area of biochars can be attributed to several factors, predominantly the feedstock used and the pyrolysis conditions. For example, biochars produced from feedstocks with high ash contents tend to have reduced porosity. Research by Fan et al. indicates that a high ash content may block pores, thereby restricting accessibility for gases adsorbed within them (Fan et al. 2017). Similarly, the study by Jindo et al. supports this observation, indicating that the pyrolysis temperature and type of feedstock play critical roles in determining the physical characteristics, including surface area (Jindo et al. 2014).

Specifically, they noted that higher pyrolysis temperatures often result in biochars with larger surface areas and more developed porous structures.

3.2.3.7 Ash Content

Biochar can be classified into three categories on the basis of its ash content: low ash content (< 5% wt), medium ash content (5–10% wt), and high ash content (> 10% wt) (Mogaji et al. 2020). This classification is crucial because ash content significantly influences the physical and chemical properties of biochar and its subsequent applications in agriculture and environmental management.

Biochars with low ash contents typically result from feedstocks that are low in inorganic minerals, such as hardwoods and certain agricultural residues. These biochars are often characterized by relatively high carbon contents and surface areas, making them relatively effective for soil amendment and carbon sequestration (Domingues et al. 2017); conversely, biochars with high ash contents, such as those derived from biomass rich in minerals—such as agricultural residues or certain grasses—are rich in nutrients and facilitate soil fertility but may possess relatively low carbon sequestration potential (Domingues et al. 2017; Sun et al. 2017).

3.2.4 Benefits and Prospects of Biochar

3.2.4.1 Effects of Biochar Amendment on SOC

Plants take up CO_2 through photosynthesis. Half of this CO_2 is then exhaled back into the atmosphere, while the other half is used for plant growth, retained inside the plants, and then released as plant waste back into the soil. After being mineralized and broken down, these leftovers release CO_2 back into the environment without reducing carbon. Furthermore, greenhouse gas emissions from fossil fuels can be decreased by using bioenergy generated by biomass pyrolysis. This whole process is known as "carbon negativity," which emphasizes how applying biochar can help the soil retain carbon. The response of SOC to biochar amendment in croplands is influenced by a number of variables, including soil characteristics, pyrolysis settings, and biochar feedstocks. Biochar has been shown to enhance the biological, physical, and chemical characteristics of soil, increasing crop yields (Bo et al. 2023).

3.2.4.2 Impact on Soil Biological Properties

Biochar addition has the potential to significantly alter the biological characteristics of soil, hence altering the variety and activity of soil microbes and fauna (Zhang et al. 2017). Rhizosphere and soil ecosystem functions are affected differently by the characteristics of biochar and the associated changes in soil properties (Hussain et al. 2017). Microbial activity can be stimulated by biochar, increasing the amount of soil biota (Lehmann et al. 2011). Microbes can use resources more because of the physical characteristics of biochar, such as its specific surface area (SSA) and porosity, as well as its ability to absorb nutrients and organic matter surrounding its particles. The shape of biochar can protect soil microorganisms (0.3–3.0 μm for bacteria, 2–80 μm for fungi, and 7–30 μm for protozoa) from larger predators such as microarthropods (Gul et al. 2015).

3.2.4.3 Impact on Microbial Communities

The growth of microorganisms involved in the cycling of nitrogen, phosphorus, and carbon is impacted by biochar, as is the case in the soil food web (Kolton et al. 2016). For example, one incubation study revealed that adding 10% w/w green waste biochar that had been pyrolyzed at 700 °C to a sandy loam increased the number of *Pseudomonas stutzeri* and *Pedobacter saltans* bacteria. According to Harter et al. (2014), these bacterial species may be engaged in N_2O reduction by increasing nosZ gene activity. In nutrient-poor soil, in contrast to nutrient-rich soil, the addition of biochar appears to have a greater effect on microbial diversity (Wang et al. 2016). The negative charges on the surface of biochar can increase microbial and enzymatic activity and improve the ability of the soil to buffer the pH in acidic soils. For example, in acidic sandy loam soil, adding 20% and 40% wheat straw biochar with a pH (H_2O) of 10.4 increased the activity of alkaline phosphatases by 67% and 73%, respectively (Chen et al. 2011).

3.2.4.4 Adsorption of Heavy Metals

Through soil food chain transmission, heavy metal and metalloid contamination of agricultural soils, including arsenic (As), cadmium (Cd), lead (Pb), and zinc (Zn), which are commonly recognized as heavy metal(loid)s, can pose health hazards (Bian et al. 2014). According to Khalid et al. (2017), heavy metal(loid)s can linger in soil for a considerable amount of time after being introduced because they are extremely resistant to degradation caused by either biological or chemical processes. Biochar has drawn much attention for its ability to immobilize heavy metals (loids) in soil and has attractive qualities for remediating soil contamination (Inyang et al. 2010). These components are immobilized on biochar, which lowers the possibility of pollution from runoff and leachate and may have positive environmental effects. These components are immobilized on biochar, which lowers the possibility of pollution from runoff and leachate and may have positive environmental effects.

To reduce metal mobility, bioavailability, and bioaccessibility, biochar was found to be an immobilizing agent. According to Ahmad et al. (2014), biochars made at a comparatively high pyrolysis temperature are often better at adsorbing pollutants. Similarly, three years of field research have been conducted in which biochar produced by slowly pyrolyzing wheat straw at temperatures ranging from 350 to 550 °C was added to a rice paddy (Bian et al. 2014). They discovered that adding biochar to the soil considerably reduced the amount of extractable Cd and Pb in the soil, as well as the concentrations of Cd and Pb in the tissues of rice plants and roots, respectively. An increase in soil pH and organic carbon was linked to this effect.

3.2.4.5 Adsorption of Pesticides

There are concerns about the possible impacts of the widespread use of pesticides in agricultural soils on the environment and public health. An immobilization method for remediating pesticide-contaminated soils and thus reducing diffuse herbicide pollution has been proposed: the addition of biochar (Ouyang et al. 2016). In fact, charcoal addition can cause hydrophobic organic molecules, such as insecticides, to sorb quickly and strongly (Khorram et al. 2016). Thus, pesticide leaching and bioavailability are impacted by the adsorption and desorption of biochar. For example, in homogenized soil columns, the application of pine chip biochar decreased cumulative atrazine leaching by 52% (Delwiche et al. 2014).

3.2.4.6 Mitigation of Greenhouse Gas Emissions from Agricultural Soil

Biochar affects soil non-CO_2 greenhouse gas (GHG) emissions (N_2O and CH_4) because of its effects on the physical, chemical, and biological characteristics of the soil. The very powerful greenhouse gas N_2O has the potential to cause 265 times as much global warming as CO_2 over a 100-year period (IPCC 2015). N_2O emissions from soil are caused primarily by three nitrogen cycle processes: nitrate ammonification, denitrification, and microbial nitrification (Baggs 2011). A shift in soil microbial abundance is one of the major biotic processes that has been found to be responsible for reducing N_2O emissions following soil biochar amendment. Additionally, scientists have concluded that biochar may influence soil CH_4 fluxes via CH_4 sorption to the surface. High-pH and high-surface-area biochar might be more suitable for reducing CH_4 emissions from soil (Jeffery et al. 2016).

While biochar offers wide-ranging benefits, ongoing research is needed to optimize its production, understand its long-term environmental impacts, and address potential risks such as the release of heavy metals or organic pollutants. Overall, biochar represents a promising tool for sustainable agriculture, environmental remediation, and climate change mitigation (Wang and Wang 2019).

3.3 Role of Biochar in Carbon Sequestration

By enhancing the resilience of soil carbon stocks, carbon sequestration through biochar represents a pivotal strategy for mitigating climate change. Biochar, a carbon-rich product formed through pyrolysis, exhibits notable longevity in the soil, with an average residence time ranging up to 1000 years (Hua et al. 2013; Minamino et al. 2019). This durability is crucial, as it allows for the stabilization of carbon, thereby preventing carbon dioxide release (CO_2) back into the environment. Biochar has the capacity to sequester carbon, which is largely attributed to its composition and structure, which includes a high level of recalcitrant carbon that is resistant to microbial degradation (Crombie and Mašek 2014; Nan et al. 2018).

3.4 The Carbon Sequestration Mechanisms of Biochar

3.4.1 Long-Term Carbon Stability

The pyrolysis of biomass produces biochar, which has emerged as a potential tool for long-term carbon sequestration in soil. Its stability as a carbon repository is influenced by several mechanisms, including its chemical structure, physical properties, and interactions with soil components. Understanding these mechanisms is crucial in evaluating the effectiveness of biochar in mitigating climate change through reduced atmospheric CO_2 levels.

One significant mechanism contributing to the durability of carbon in biochar is its chemical recalcitrance. Biochar is characterized by a complex aromatic structure that resists microbial decomposition much more than does conventional organic matter. Studies indicate that most of the carbon derived from biochar remains in the soil for extended periods, potentially for centuries or even millennia (Shin et al. 2016; Anand et al. 2022). For example, biochar created through pyrolysis at 400 °C has been noted to persist in soils for up to 2000 years (Anand et al. 2022). This high resistance to microbial degradation is pivotal for creating a stable carbon sink in agricultural soils.

3.4.2 Physical Protection of Organic Matter

Biochar has drawn attention in terms of carbon sequestration due to its potential mechanisms in protecting organic matter and enhancing its stability within soil environments. The physical protection of organic matter through biochar occurs through several pivotal mechanisms, primarily through its porous structure, its chemical interactions with soil components, and the enhancement of aggregate stability.

One of the most significant attributes of biochar is its porous structure, which facilitates the physical retention of organic matter. The high surface area of biochar, coupled with its microporous morphology, effectively traps organic particles and prevents their decomposition through microbial action. This physical encapsulation not only retains carbon within the soil matrix but also contributes to the initiation and establishment of firm soil aggregates (Akhmetzyanova et al. 2022). The pore spaces in biochar can provide refugia for soil microorganisms and organic molecules, thus indirectly promoting the retention of soil organic carbon (SOC) (Jusoh et al. 2021).

3.4.3 Greenhouse Gas Emission Reduction

Biochar is increasingly recognized as a strategic approach for mitigating emissions from crop fields while increasing carbon sequestration in soils. Its efficacy as a negative emission technology is supported by several mechanisms that mitigate the emission of potent greenhouse gases, specifically methane (CH_4) and nitrous oxide (N_2O). Additionally, biochar contributes to overall soil health and agricultural productivity, which play significant roles in long-term climate change mitigation.

One primary mechanism by which biochar affects GHG emissions is its impact on methane and nitrous oxide dynamics in soil. Studies indicate that biochar application can significantly reduce N_2O emissions, particularly in agricultural systems where nitrogen fertilizers are utilized (Woolf et al. 2010). This reduction is attributed to the ability of biochar to increase soil aeration and improve microbial processes that mitigate N_2O production. However, biochar may also increase CH_4 emissions under certain conditions, such as anaerobic environments such as paddy rice fields; nevertheless, the overall net reduction in total GHG emissions is typically observed when biochar is used.

3.4.4 Improved Soil Conditions

Owing to its unique physicochemical properties, biochar has emerged as an important strategy for improving soil conditions and sustainable sequestration of soil carbon, which contributes to prolonged carbon storage in soil ecosystems. Through the pyrolysis process, biochar is created from biomass, which leads to the formation of stable carbon substances that are highly resistant to microbial degradation. This stability is a critical factor in its ability to act as an agent for trapping carbon, which can last in soils for significantly longer periods than unpyrolyzed organic substances (Wang et al. 2015).

One key mechanism by which biochar facilitates carbon sequestration is by increasing the total carbon (TC) pool in the soil. Research indicates that the incorporation of biochar into different kinds of soil, including sandy and loamy soils, results in an evident increase in organic carbon content, especially in fine-textured soils that are more susceptible to carbon loss (El-Naggar et al. 2018; Liu et al. 2019; Biederman and Harpole 2012). The active surface area and porosity of biochar are crucial, as they allow it to interact effectively with soil particles, stabilizing organic carbon through physical and chemical bonding (Ghorbani et al. 2023). This phenomenon enhances soil structure and fertility and encourages the establishment of soil aggregates, which further capture carbon (Ghorbani et al. 2023; Keskinen et al. 2019).

3.4.5 *Adsorption and Immobilization*

One of the fundamental processes by which biochar aids in the sequestration of soil carbon is through its high adsorption capacity for nutrients and contaminants, which can lead to increased soil carbon content over time. The structure of biochar, characterized by its high porosity and surface area, allows for substantial adsorption of cationic species such as ammonium nitrogen (NH_4^+-N). Research indicates that certain types of biochar exhibit remarkable NH_4^+-N adsorption capacities, which can reach as high as 909 mg kg^{-1} in some formulations (Wang et al. 2017). This adsorption not only enhances nitrogen retention in soils, thereby reducing nitrogen losses and associated gas emissions, particularly N_2O but also promotes the immobilization of carbon through increased microbial activity and enhanced organic matter stabilization (Shin et al. 2018; Jindo et al. 2020a, b).

3.5 Factors Influencing Effective Soil Carbon Sequestration by Biochar

The efficiency of biochar in the sequestration process is influenced by various factors, including its chemical composition, stability, interaction with soil matrices, and specific soil environment in which it is applied. One major factor is the inherent stability of the biochar itself. The carbon contained in biochar is often highly recalcitrant, meaning that it resists degradation over time. Studies indicate that the carbon released from biochar remains within soil for a long period of time, often persisting for centuries (Wang et al. 2015, 2020; Liu et al. 2019). This stability is largely a function of the degree of pyrolysis, including temperature and duration, which affect its chemical structure and thereby its environmental resilience (Biederman and Harpole 2012; Mohan et al. 2018).

Other factors that influence biochar effectiveness in terms of carbon sequestration include biomass and the method used to generate the biochar. The type of biomass from which biochar is derived has a significant effect on its carbon content, stability, and overall effectiveness as a climate change mitigation strategy. Higher pyrolysis temperatures enhance the stability and recalcitrance of biochar, regardless of the feedstock, by promoting the formation of aromatic carbon structures and facilitating carbon sequestration (Conz et al. 2017).

3.6 Biochar–Soil–Plant Interactions

Biochar also enhances crop growth and productivity, increases natural N overload in legumes, and facilitates carbon insulation (Murtaza et al. 2023).

The pH and N, P, K, Ca, Mg, and OC contents of the soil were markedly affected by the biochar. The optimal soil pH can prompt some nutrients to take on other forms, making it easier for roots to adsorb specific nutrients. Compared with the control treatment, the biochar treatment significantly increased the cation exchange capacity by 4–30 (Laird et al. 2010). Nevertheless, the increase in interchangeable cations in the amended soil indicated better nutrient retention and soil fertility. The use of biochar can increase the WHC (22) (Peake et al. 2014). Eventually, the increased bulk viscosity, water-holding capacity, and aggregation capability of the soil may increase water and nutrient retention, directly perfecting the soil structure (Lu et al. 2014). In addition to increasing the likelihood of base achromatism (6.4–26) and increasing hydraulic conductivity (16.7–33.1 cm h^{-1}), biochar treatment might increase total C (2.27–2.78), total N (0.24–0.25), and available P (15.7–15.8 mg kg^{-1}) and decrease the soil corrosion rate from 1458 to 532 g m^{-2} h^{-1} (Jones et al. 2012; Jien and Wang 2013). The betterment of the soil profile may contribute, directly or laterally, to increased nutrient vacuity and decreased nutrient filtering, all of which are considered pathways for advanced soil fertility.

The effectiveness of biochar as a soil correction in various soil types is widely believed. It enhances the fertility and physicochemical characteristics of soil, which promotes plant growth. Low-rate biochar promoted the growth of the seedlings and promoted their germination. A high biochar application rate can be dangerous, yet a low biochar application rate can be stimulating. Kochanek et al. (2016) reported that biochars derived from organic fly karrikins can promote seed and seedling growth. The type of soil, biochar, and biochar treatment temperature affect how important or how little plant growth occurs. For example, Kwapinski et al. (2010) reported that applying biochar made from tableware lawn fields heated to 400 °C inhibited the development of sludge. Nonetheless, the biochar produced at 600 °C encouraged development. The growth of rice, wheat, and cowpea was greatly increased by treating the soil with biochar at an operation rate of 68 t ha^{-1} (Vaccari et al. 2011; Solaiman et al. 2012), and the operation rates and biochar types considerably impacted the growth of clover, mug sap, and wheat in a laboratory trial.

Biochar operation appears to be a promising approach for mitigating heavy substance impurities in plants, enhancing disease resistance against stress conditions and increasing disease severity against environmental stress, which may lead to advanced agrarian productivity and guarding our plant community.

3.7 Formulation, Production, Morphology, Biochemistry, Storage and Incorporation Methods of Biochar

3.7.1 Formulation

Biochar formulation is the process of customizing the physical, chemical, and biological characteristics of biochar through deliberate choices of raw material, pyrolysis parameters, and postprocessing and enrichment techniques. This enables the production of biochar with specific functions such as nutrient delivery, soil conditioning, pollutant removal, or microbial support. Proper formulation ensures that biochar is not a generic product but rather a targeted tool for sustainable agriculture and environmental management.

3.7.1.1 Components of the Biochar Formulation

Feedstock Selection

Feedstock selection is critical in determining the base properties of biochar. The chemical composition (C:N ratio, lignin, cellulose, mineral content) varies significantly among different biomass types (Fig. 3.2).

Pyrolysis Process Parameters

Pyrolysis parameters control the decomposition chemistry and structural development of biochar. The key variables include:

(1) Temperature: This factor influences biochar stability and porosity. High-temperature biochar (> 500 °C) is more aromatic and durable.
(2) Heating rate: Effects on volatile release and char structure.
(3) Residence time: Longer times promote greater carbonization.
(4) Atmosphere: Inert gases such as N_2 prevent combustion and favour specific chemical structures.

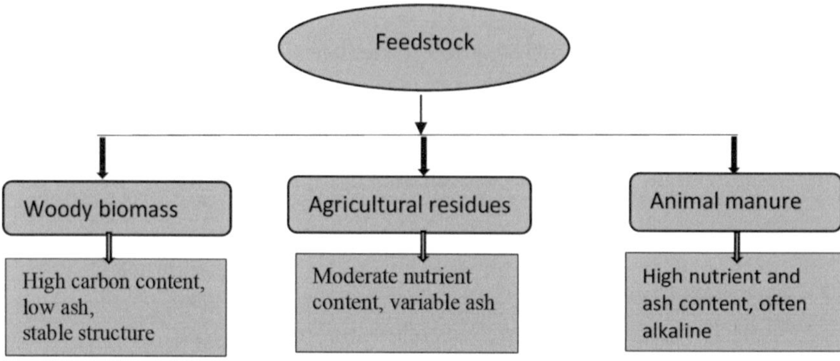

Fig. 3.2 Flowchart showing the influence of feedstock type on final biochar properties

Postpyrolysis Treatments

Posttreatment steps improve functionality and performance. Common methods include the following:

(1) Acid/base washing to remove tar and ash (increasing purity).
(2) Steam or chemical activation (H_3PO_4, KOH) to increase the surface area and porosity.
(3) Sieving for particle size uniformity.
(4) Aging or oxidation to improve compatibility with soil biology.

Enrichment and Functionalization

Formulations often include enrichment with nutrients, microorganisms, or nanomaterials to enhance biochar function. Examples:

- Nutrient-enriched biochar using urea, phosphate, or compost leachates.
- Microbial inoculants (e.g., PGPR, Trichoderma) added for plant growth promotion.
- Coated or composite biochar used for controlled nutrient release or pollutant adsorption

Particle Size and Physical Form

To suit application needs, biochar can be formulated into different physical forms:

- Powdered for seed coating or mixing in compost.
- Granular or pelletized for ease of application in fields.
- Activated forms for adsorption in water filtration or air purification.

Table 3.2 Application-specific formulation strategies

Application	Formulation goal	Required biochar traits
Soil amendment	Improve fertility and structure	Moderate pH, high CEC, porous, nutrient-enriched
Heavy metal remediation	Adsorb and immobilize contaminants	High surface area, functional groups (OH, –COOH)
Composting	Reduce GHG, enhance compost quality	Low ash, high microbial compatibility
Water filtration	Remove organic/inorganic pollutants	Microporous, high adsorption capacity
Carbon sequestration	Long-term C stabilization in soil	High aromaticity, thermal stability (pyrolysis > 500 °C)

Smaller particles offer greater surface areas but may pose dust hazards; larger particles are more stable in soil but react more slowly (Table 3.2).

Biochar formulation is a critical innovation in modern soil science and environmental engineering. By tailoring the feedstock, pyrolysis conditions, and post-treatment steps, biochar products with precise, application-specific properties can be created. As a result, formulated biochar has become a highly versatile tool for addressing challenges in terms of soil fertility, water quality, carbon management, and sustainable agriculture.

Production

The processes used to produce biochar vary, and each has a unique impact on the characteristics of this carbon-rich substance. An overview of popular biochar manufacturing techniques is provided in this discussion, highlighting the critical role that feedstock selection and processing conditions play in determining the properties of biochar.

Pyrolysis

Pyrolysis, which involves the heat breakdown of organic compounds without oxygen, is a fundamental step in the synthesis of biochar. The three stages of this procedure are chilling, pyrolysis, and drying. The characteristics of biochar are strongly influenced by the temperature and duration of each phase. Biochar produced by high-temperature pyrolysis typically has a relatively high carbon content and is relatively stable. The choice of feedstock, whether agricultural residues, wood, or manure, profoundly influences the physicochemical attributes of the resulting biochar. Figure 3.3 shows the pyrolysis process.

Gasification

Gasification represents an alternative thermochemical process wherein organic materials are converted into synthesis gas (syngas) under controlled conditions. The remaining solid residue is biochar, and syngas can be used to generate electricity. A wide variety of feedstocks, including woody biomass and agricultural leftovers,

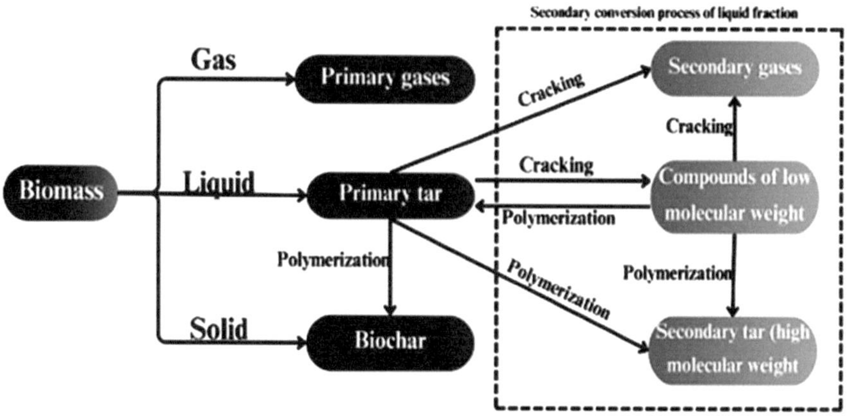

Fig. 3.3 Pyrolysis process and production of its derivatives (adapted)

can be used with gasification. The gaseous environment in this process influences biochar properties, with lower oxygen levels typically yielding biochar with enhanced stability.

Combustion

Combustion: Combustion-based reactors are mainly made to produce heat, which is frequently utilized for secondary heat applications such as drying grain or curing lumber in addition to steam turbine electricity generation. The process continues because the biomass is heated in the presence of oxygen, and the resulting gases are burned close to the biomass.

Even if the term "combustion" seems incompatible with the process of producing biochar, most of the biochar generated in North America may be the result of combustion as a technology. Importantly, not all combustion technologies provide 100% total combustion, which produces ash as a solid product. In fact, the contrary is more accurate because most combustion technologies do not produce complete combustion. It is possible to remove biochar from the system when there is oxygen present but not enough for full combustion. In a boiler, combustion typically produces 1.5–2% biochar.

Morphology and Biochemistry

The diverse qualities of biochar are crucial in determining how well it works in a range of settings, from environmental remediation to agriculture. This section scrutinizes the physical, chemical, and biological attributes of biochar, elucidating the pivotal roles of surface area, porosity, and functional groups in shaping its behaviour across different environments. The biochar properties and their implications are described below.

Physical Properties

Surface Area: One important factor affecting the reactivity and adsorption capability of biochar is its surface area. It offers additional locations for nutrient adsorption and chemical reactions. The process of manufacturing, particularly high-temperature pyrolysis, frequently results in the production of biochar with a relatively large surface area, increasing its potential for use in environmental and soil improvement.

Porosity: The distribution of pore sizes in biochar determines its porosity, which influences microbial colonization, aeration, and water retention in soils. Micropores contribute to the preservation of nutrients, whereas macropores help transport water. For biochar to effectively improve the soil structure, these pore diameters must be balanced. A macroporous structure is often caused by a relatively high lignin concentration, whereas a microporous structure is produced by a relatively high cellulose content.

Density: Biochar density influences nutrient retention capacity and durability in soil. Low-density biochar may breakdown more rapidly, releasing nutrients, whereas high-density biochar may persist longer in the soil. Finding an optimal density is essential for achieving the desired agronomic and environmental outcomes. Wijitkosum and Jiwnok (2019) used scanning electron microscopy (Fig. 3.4) to analyse the morphology of three biochar products and reported that the biomass type influenced the morphology under identical pyrolysis conditions.

Chemical Properties

Like biomass, biochar can be viewed as consisting of both organic and inorganic components, such as volatile matter, carbon, oxygen, hydrogen and ash. Heterogeneous organic matter with different thermal stabilities makes up the organic fraction; components that degrade at high temperatures are referred to as volatile matter (VM), whereas those that remain stable at high temperatures are known as fixed carbon (FC). However, both the FC and VM are solids at room temperature. Ash plus a tiny quantity of remaining moisture make up the inorganic part of biochar (Wood et al. 2023).

Functional Groups

Compared with their biomass antecedents, biochar has a substantially lower number of functional groups. Through chemical changes, these compounds are connected to those present in biomass. Aryl, hydroxyl, carbonyl, and many other functionalities based on carbon and oxygen predominate. A range of functionalities based on nitrogen, phosphorus, and sulfur may also be present, depending on the initial biomass composition; however, their concentrations are much lower than those of oxygen-based groups (Wood et al. 2023).

Numerous functional groups may be present in the biomass, depending on where it comes from, but many of them become unstable during pyrolysis. This suggests that each functional group degrades throughout a wide temperature range as opposed to at a specific value. This results from the impact of a group's surrounding chemical

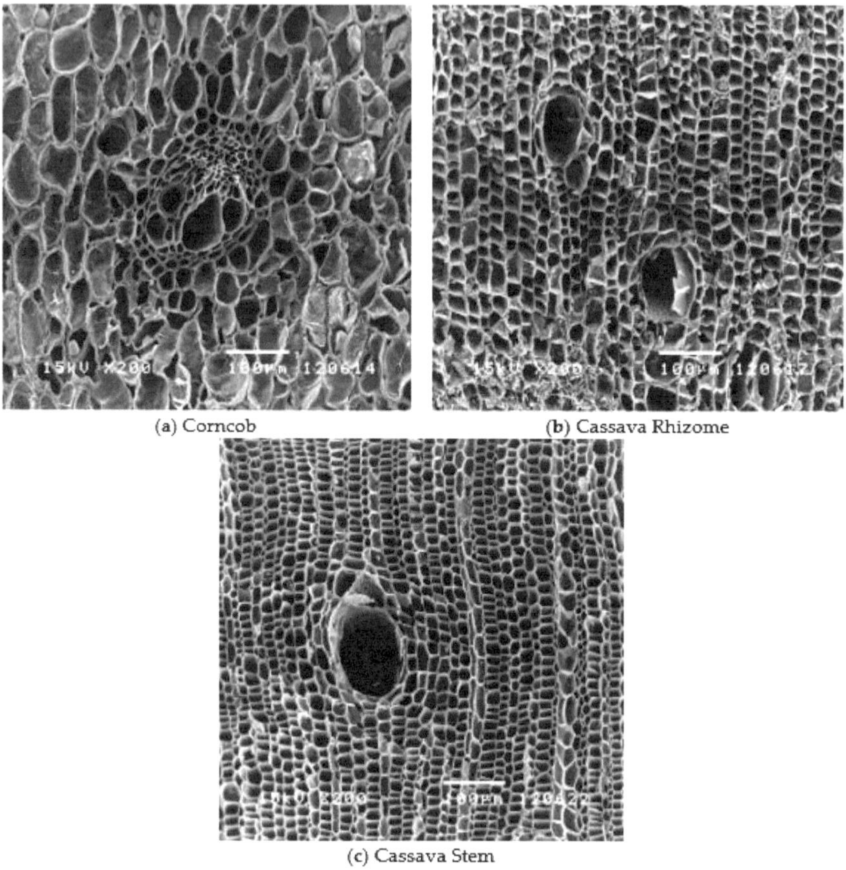

Fig. 3.4 Electron microscopy (SEM) images showing differences in the morphology of biochar produced from different biomass feedstocksWijitkosum and Jiwnok 2019)

environment. This impact is especially significant when biochar is produced since a variety of factors influence the formation of functional groups during pyrolysis (Wood et al. 2023).

Biological Properties

Microbial Activity: Biochar can alter soil microbial communities by providing a home for beneficial microbes. According to Lehmann et al. (2011), its porous structure can act as a haven for microorganisms, creating an environment that is favourable for soil health and nutrient cycling.

Plant Growth Promotion: Biochar's ability to increase nutrient availability, water retention, and microbial activity in the rhizosphere is associated with its beneficial effects on plant growth. According to Agegnehu et al. (2016), the presence of biochar can improve plant nutrient intake and root development.

Storage Method

Proper storage of biochar is essential to preserve its physical and chemical integrity, maintain its effectiveness for soil and environmental applications, and minimize losses due to degradation, contamination, or exposure to moisture. While the production of biochar has received considerable research attention, the importance of post-production storage methods is often overlooked. This chapter provides a comprehensive overview of biochar storage considerations, technologies, and best practices on the basis of recent scientific findings.

Principles of Biochar Storage

The main goals of biochar storage are to prevent physical disintegration, chemical degradation, and microbial colonization before its intended use. Several factors influence storage stability, including the following:

- **Moisture**: Biochar is highly hygroscopic and absorbs moisture from the air, potentially leaching nutrients or reducing the energy content if stored for fuel use.
- **Oxygen and Microbial Exposure** can lead to surface oxidation and microbial colonization, altering surface chemistry and reducing the number of reactive sites.
- **Temperature**: High storage temperatures may accelerate chemical changes and off-gases.
- **Contamination**: Exposure to dust, soil, or other organics can compromise purity.

Biochar Storage Methods

- **Bulk Storage in Sheds or Silos**

Dry biochar is often stored in enclosed silos or sheds that protect against rain and humidity. Ventilation is essential to prevent moisture accumulation, and fire prevention measures should be in place because of the potential for self-heating in large piles.

- **Bagged Storage**

For small to medium quantities, biochar is commonly stored in polypropylene or polyethylene bags. These bags should be UV resistant and sealed to prevent moisture ingress. Stacking on pallets in dry warehouses is recommended.

- **Covered Outdoor Storage**

In some rural settings, biochar is stored outdoors under tarpaulins or plastic covers. This method is low-cost but poses risks of water infiltration and wind dispersion. Biochar should be stored above ground on impermeable surfaces if this method is used.

- **Inert Atmosphere or Sealed Containers**

For research or high-value applications, biochar can be stored in sealed containers under inert gases (e.g., nitrogen) to prevent oxidation. This approach helps maintain surface chemistry and is particularly useful for nanostructured or chemically modified biochar.

- **Blended Storage with Compost or Soil**

Some storage systems mix biochar with compost or soil. While this may reduce dust and ease its application, it can alter the original properties of biochar through microbial activity and nutrient leaching during storage.

Safety and Handling Considerations

Biochar dust is a fine particulate matter that may pose respiratory risks. During storage and handling:

- Use dust masks and PPE during loading/unloading.
- Avoid sparks and open flames near dry biochar storage areas.
- Regularly monitor the temperature in bulk piles to prevent self-ignition.
- Store biochar away from chemical fertilizers or volatile materials.

The effective storage of biochar is vital for maintaining its quality and ensuring its safety. The selection of the appropriate storage method depends on the scale of operation, intended application, and environmental conditions. Research is needed to further understand the long-term effects of storage on the functional properties and field performance of biochar.

Incorporation Method

Since the majority of nutrient cycling and plant uptake occurs in the root zone, the addition of biochar close to the soil surface is recommended to increase its fertility. Applying biochar in layers under the root zone may help some systems, such as when landscaping to sequester carbon or when using biochar to control moisture. Several techniques of biochar incorporation are described below:

(1) **Surface Application and Mechanical Incorporation**

- **Surface Broadcasting**

Surface applications involve spreading biochar uniformly over the soil surface without immediate mixing. This method is straightforward and inexpensive and is often used in no-till or pasture systems. However, owing to the low bulk density and high surface area of biochar, it is susceptible to wind erosion and runoff losses. Furthermore, biochar applied in this way may remain ineffective in the short term due to limited interactions with plant roots and soil microbes.

To mitigate these limitations, surface-applied biochar is often subjected to irrigation or rainfall to assist its movement into the soil or in combination with organic amendments such as compost.

- **Mechanical Incorporation (Tillage)**

In conventional agricultural systems, biochar is commonly incorporated into the soil through tillage—using disc ploughs, rotavators, or harrows. This ensures:

Improved physical contact with soil particles
Reduced surface losses
Increased microbial colonization
Enhanced nutrient retention.

The tillage depth (typically 10–20 cm) depends on the crop root zone and soil type. While effective for distribution, this method can disturb natural soil structure and microbial habitats, especially in conservation agriculture systems.

(2) **Subsurface Placement Techniques**

- **Deep Banding**

Deep banding involves placing biochar in narrow, concentrated bands within the root zone—usually at depths of 15–30 cm. It is especially effective in row crops such as maize or wheat. This method allows the following:

Efficient use of biochar by positioning it closer to roots
Enhanced nutrient uptake
Reduced volatilization and leaching
Precision equipment such as strip–till planters or deep banders is often used, which may be cost-prohibitive in resource-poor settings.

- **Trenching and Pit Application**

In perennial cropping systems (e.g., orchards), biochar is placed in trenches or planting pits. Biochar is mixed with compost or soil and filled back into the trench/pit around the root zone. The benefits include:

Localized improvement of the root zone soil
Enhanced water retention and microbial activity
Reduced nutrient leaching in deep soils
This method is labor intensive but particularly suitable for high-value crops or degraded soils.

(3) **Combined Application with Organic Amendments**

- **Compost–Biochar Blending**

Blending biochar with compost before application is increasingly favoured. Biochar acts as a **carrier** for microbes, enhances compost maturity, and reduces odor and greenhouse gas emissions during composting. When applied, this blend:

- Provides a balanced nutrient supply
- Improving microbial colonization
- Mitigation of nitrogen immobilization
- Increasing carbon stability.

The application methods include broadcasting followed by incorporation or direct pit/trench filling, depending on the cropping system.

- **Coapplication with Farmyard Manure or Crop Residues**

In smallholder systems, biochar is often applied together with locally available organic materials such as cattle dung, poultry litter, or crop residues. This increases soil microbial activity, buffering capacity, and long-term fertility. However, nutrient synchronization (time of availability) is critical.

3.8 Negative Impacts of Biochar

Biochar is widely promoted for its environmental and agricultural benefits, but research highlights several potential negative impacts that must be considered. Biochar can introduce harmful substances, alter soil properties, and negatively affect soil organisms and environmental quality, depending on its composition, production, and application.

3.8.1 *Environmental and Soil Risks*

- Release of Harmful Compounds: Biochar can contain and release toxic substances such as heavy metals, polycyclic aromatic hydrocarbons (PAHs), polychlorinated dibenzodioxins, and dibenzofurans, which may harm soil, water, and air quality and pose risks to human health (Tan et al. 2021).
- Soil Property Alterations: Application can increase the soil pH, especially in alkaline soils, leading to nutrient precipitation, reduced nutrient availability, and decreased soil fertility. High doses in clay soils may reduce the water content, whereas surface application in sandy soils can increase erosion and particulate emissions (Chang et al. 2019).
- Suppression of Crop Productivity: In some cases, biochar reduces plant nutrient uptake and soil carbon mineralization, potentially suppressing crop yields, particularly in fertile soils (Sarraf et al. 2023).

3.8.2 *Effects on the Soil Biota and Agrochemicals*

- Toxicity to Soil Organisms: Biochar may negatively affect the soil biota, resulting in reduced reproduction, growth, and DNA integrity in earthworms, and shifts in soil microbial communities, such as changes in the fungi-to-bacteria ratio (Al-Sadi et al. 2017).

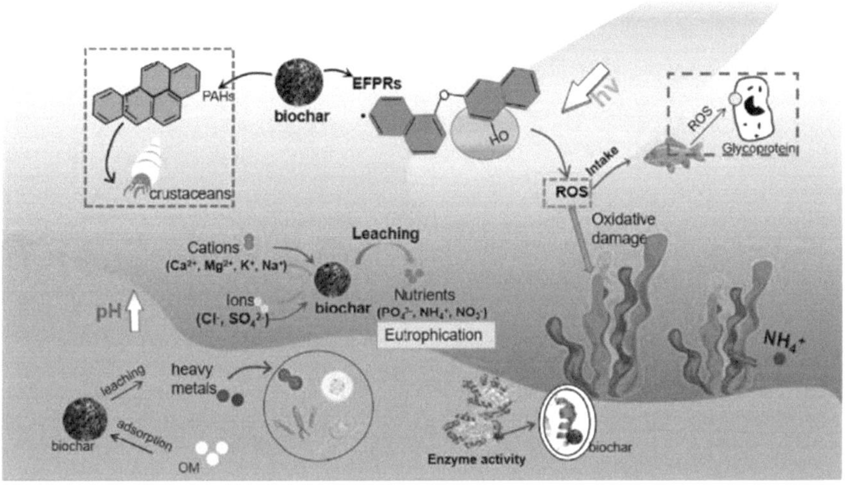

Fig. 3.5 Possible adverse environmental effects of applying biochar (Pan et al. 2023)

- Agrochemical Interactions: Biochar can retain pesticides and herbicides, reducing their effectiveness and complicating weed control (Datta et al. 2021).

Figure 3.5 shows an overview of the negative effects of biochar in the ecosystem.

3.9 Conclusion

Biochar can contribute to the sustainable maintenance of soil health. Crop production depends on the stability of biochar in the soil since it fixes, maintains, and enhances soil quality properties. It is widely recognized as a promising soil amendment for improving soil quality and enhancing carbon cycling. Its application can increase soil fertility, promote carbon sequestration, and help mitigate climate change. Its benefits are most pronounced when it is tailored to specific soil and environmental conditions.

References

Aboughaly M, Fattah I (2023) Production of biochar from biomass pyrolysis for removal of PFAS from wastewater and biosolids: a critical review. https://doi.org/10.20944/preprints202304.0309.v1

Adirianto B, Bachtiar T (2023) Effect of biochar in soil on microbial diversity: a meta-analysis. IOP Conf Ser Earth Environ Sci 1263(1):012047. https://doi.org/10.1088/1755-1315/1263/1/012047

Ahmad M, Rajapaksha AU, Lim JE, Zhang M, Bolan N, Mohan D et al (2014) Biochar as a sorbent for contaminant management in soil and water: a review

Akhmetzyanova L, Kuritsin I, Zinnatov M, Gordeev A, Kamalova A, Karamova K (2022) The potential of using corn straw biochar to compensate the loss of available soil organic matter. E3S Web Conf 341:01002. https://doi.org/10.1051/e3sconf/202234101002

Alazzaz A, Usman A, Ahmad M, Elfaki J, Sallam A, Ibrahim H, Al-Wabel M (2020) Potential short-term negative versus positive effects of olive mill-derived biochar on nutrient availability in a calcareous loamy sandy soil. https://doi.org/10.1101/2020.04.23.057695

Al-Sadi A, Ok Y, Alghamdi S, Farooq M, Hussain M, Ammara U, Nawaz A, Solaiman Z, Siddique K (2017) Biochar for crop production: potential benefits and risks. J Soils Sediments 17:685–716. https://doi.org/10.1007/s11368-016-1360-2

Al-Wabel M, Hussain Q, Usman A, Ahmad M, Abduljabbar A, Sallam A et al (2017) Impact of biochar properties on soil conditions and agricultural sustainability: a review. Land Degrad Dev 29(7):2124–2161. https://doi.org/10.1002/ldr.2829

Ana L, Oliveira J, Gonçalves G, Tusset Â, Lenzi G (2024) Circular economics in agricultural waste biomass management. Biomass 4(2):543–554. https://doi.org/10.3390/biomass4020029

Anand A, Kumar R, Kumar V, Kaushal P (2022) Assessment of sugarcane residues derived biochar for carbon sequestration in the soil in India. https://doi.org/10.46855/energy-proceedings-10045

Ayadi R, Koubaa A, Braghiroli F, Migneault S, Wang H, Bradaï C (2020) Effect of the pyro-gasification temperature of wood on the physical and mechanical properties of biochar-polymer biocomposites. Materials 13(6):1327. https://doi.org/10.3390/ma13061327

Azzi E, Sundberg C, Söderqvist H, Källgren T, Cederlund H, Li H (2023) Guidelines for estimation of biochar durability: background report. https://doi.org/10.54612/a.lkbuavb9qc

Baggs EM (2011) Soil microbial sources of nitrous oxide: recent advances in knowledge, emerging challenges and future direction. Curr Opin Environ Sustain 3:321–327. https://doi.org/10.1016/j.cosust.2011.08.011

Bednik M, Medyńska-Juraszek A, Ćwieląg-Piasecka I (2022) Effect of six different feedstocks on biochar's properties and expected stability. Agronomy 12(7):1525. https://doi.org/10.3390/agronomy12071525

Bian R, Joseph S, Cui L, Pan G, Li L, Liu X et al (2014) A three-year experiment confirmed the continuous immobilization of cadmium and lead in contaminated paddy fields with biochar amendment. J Hazard Mater 272:121–128. https://doi.org/10.1016/j.jhazmat.2014.03.017

Biederman L, Harpole W (2012) Biochar and its effects on plant productivity and nutrient cycling: a meta-analysis. GCB Bioenergy 5(2):202–214. https://doi.org/10.1111/gcbb.12037

Bista P, Ghimire R, Machado S, Pritchett L (2019) Biochar effects on soil properties and wheat biomass vary with fertility management. Agronomy 9(10):623. https://doi.org/10.3390/agronomy9100623

Blanco-Canqui H (2020) Does biochar improve all soil ecosystem services? GCB Bioenergy 13(2):291–304. https://doi.org/10.1111/gcbb.12783

Bo X, Zhang Z, Wang J, Guo S, Li Z, Lin H, Zou J (2023) Benefits and limitations of biochar for climate-smart agriculture: a review and case study from China. Biochar 5(1):77

Brown T, Wright M, Brown R (2010) Estimating profitability of two biochar production scenarios: slow pyrolysis vs fast pyrolysis. Biofuels Bioprod Biorefin 5(1):54–68. https://doi.org/10.1002/bbb.254

Cao X, Ma L, Gao B, Harris W (2009) Dairy-manure derived biochar effectively sorbs lead and atrazine. Environ Sci Technol 43(9):3285–3291

Carvalho M, Moraes M, Cerri C, Cherubin M (2020) Biochar amendment enhances water retention in a tropical sandy soil. Agriculture 10(3):62. https://doi.org/10.3390/agriculture10030062

Chang S, Shaheen S, Tsang D, Sarkar B, Rinklebe J, El-Naggar A, El-Naggar A, Ok Y (2019) Biochar composition-dependent impacts on soil nutrient release, carbon mineralization, and potential environmental risk: a review. J Environ Manage 241:458–467. https://doi.org/10.1016/j.jenvman.2019.02.044

Chemerys V, Baltrėnaitė E (2018) A review of lignocellulosic biochar modification towards enhanced biochar selectivity and adsorption capacity of potentially toxic elements. Ukr J Ecol 8(1):21–32. https://doi.org/10.15421/2018_183

Chen X, Xia X, Wang X, Qiao J, Chen H (2011) A comparative study on sorption of perfluorooctane sulfonate (PFOS) by chars, ash and carbon nanotubes. Chemosphere 83(10):1313–1319

Conz RF, Abbruzzini TF, de Andrade CA, Milori DM, Cerri CE (2017) Effect of pyrolysis temperature and feedstock type on agricultural properties and stability of biochars. Agric Sci 8(09):914. https://doi.org/10.4236/as.2017.89067

Cordeiro E, Arenas Calle L, Woolf D, Sherpa S, Poonia S, Kritee K et al (2024) The fate of rice crop residues and context-dependent greenhouse gas emissions: model-based insights from eastern India. J Clean Prod 435:140240. https://doi.org/10.1016/j.jclepro.2023.140240

Crane-Droesch A, Abiven S, Jeffery S, Torn M (2013) Heterogeneous global crop yield response to biochar: a meta-regression analysis. Environ Res Lett 8(4):044049. https://doi.org/10.1088/1748-9326/8/4/044049

Crombie K, Mašek O (2014) Pyrolysis biochar systems, balance between bioenergy and carbon sequestration. GCB Bioenergy 7(2):349–361. https://doi.org/10.1111/gcbb.12137

Datta R, Ahmed N, Kučerík J, Mravcová L, Šudoma M, Danish S, Bielská L, Holátko J, Brtnický M, Radziemska M, Hammerschmiedt T, Pecina V, Kintl A, Gusiatin Z, Fahad S (2021) A critical review of the possible adverse effects of biochar in the soil environment. Sci Total Environ 796:148756. https://doi.org/10.1016/j.scitotenv.2021.148756

Delwiche KB, Lehmann J, Walter MT (2014) Atrazine leaching from biochar-amended soils. Chemosphere 95:346–352. https://doi.org/10.1016/j.chemosphere.2013.09.043

Domingues R, Trugilho P, Silva C, Melo I, Melo L, Magriotis Z et al (2017) Properties of biochar derived from wood and high-nutrient biomasses with the aim of agronomic and environmental benefits. PLoS ONE 12(5):e0176884. https://doi.org/10.1371/journal.pone.0176884

El-Naggar A, Awad Y, Tang X, Liu C, Niazi N, Jien S, Lee S (2018) Biochar influences soil carbon pools and facilitates interactions with soil: a field investigation. Land Degrad Dev 29(7):2162–2171. https://doi.org/10.1002/ldr.2896

Fan F, Zheng Y, Huang Y, Lu Y, Wang J, Chen B et al (2017) Preparation and characterization of biochars from waste *Camellia oleifera* shells by different thermochemical processes. Energy Fuels 31(8):8146–8151. https://doi.org/10.1021/acs.energyfuels.7b00269

Ghorbani M, Neugschwandtner R, Soja G, Konvalina P, Kopecký M (2023) Carbon fixation and soil aggregation affected by biochar oxidized with hydrogen peroxide: considering the efficiency of pyrolysis temperature. Sustainability 15(9):7158. https://doi.org/10.3390/su15097158

Ghosh D, Maiti S (2022) Invasive weed-based biochar facilitated the restoration of coal mine degraded land by modulating the enzyme activity and carbon sequestration. Restor Ecol 31(3). https://doi.org/10.1111/rec.13744

Gul S, Whalen JK, Thomas BW, Sachdeva V, Deng H (2015) Physico-chemical properties and microbial responses in biochar-amended soils: mechanisms and future directions. Agr Ecosyst Environ 206:46–59. https://doi.org/10.1016/j.agee.2015.03.015

Guo M (2020) The 3r principles for applying biochar to improve soil health. Soil Syst 4(1):9. https://doi.org/10.3390/soilsystems4010009

Harter J, Krause H-M, Schuettler S, Ruser R, Fromme M, Scholten T et al (2014) Linking N_2O emissions from biochar-amended soil to the structure and function of the N-cycling microbial community. ISME J 8:660–674

Harvey O, Kuo L, Zimmerman A, Louchouarn P, Amonette J, Herbert B (2012) An index-based approach to assessing recalcitrance and soil carbon sequestration potential of engineered black carbons (biochars). Environ Sci Technol 46(3):1415–1421. https://doi.org/10.1021/es2040398

Hood-Nowotny R, Watzinger A, Wawra A, Soja G (2018) The impact of biochar incorporation on inorganic nitrogen fertilizer plant uptake; an opportunity for carbon sequestration in temperate agriculture. Geosciences 8(11):420. https://doi.org/10.3390/geosciences8110420

Hossain M, Bahar M, Sarkar B, Donne S, Ok Y, Palansooriya K et al (2020) Biochar and its importance in nutrient dynamics in soil and plants. Biochar 2(4):379–420. https://doi.org/10.1007/s42773-020-00065-z

Hua L, Lu Z, Ma H, Jin S (2013) Effect of biochar on carbon dioxide release, organic carbon accumulation, and aggregation of soil. Environ Prog Sustain Energy 33(3):941–946. https://doi.org/10.1002/ep.11867

Hussain M, Farooq M, Nawaz A, Al-Sadi AM, Solaiman ZM, Alghamdi SS et al (2017) Biochar for crop production: potential benefits and risks. J Soils Sediments 17:685–716. https://doi.org/10.1007/s11368-016-1360-2

Inyang M, Gao B, Pullammanappallil P, Ding W, Zimmerman AR (2010) Biochar from anaerobically digested sugarcane bagasse. Bioresour Technol 101:8868–8872. https://doi.org/10.1016/j.biortech.2010.06.088

IPCC (2015) Agriculture, forestry and other land use (AFOLU). In: Climate change 2014: mitigation of climate change: working group III contribution to the IPCC fifth assessment report. Cambridge University Press, Cambridge, pp 811–922. https://doi.org/10.1017/CBO9781107415416.017

Ippolito J, Cui L, Kammann C, Wrage N, Estavillo J, Fuertes-Mendizábal T et al (2020) Feedstock choice, pyrolysis temperature and type influence biochar characteristics: a comprehensive meta-data analysis review. Biochar 2(4):421–438. https://doi.org/10.1007/s42773-020-00067-x

Jatav HS, Singh SK, Jatav SS, Rajput VD, Parihar M, Mahawer SK, Singhal RK (2020) Importance of biochar in agriculture and its consequence. In: Applications of biochar for environmental safety, p 109

Jeffery S, Verheijen FGA, Kammann C, Abalos D (2016) Biochar effects on methane emissions from soils: a meta-analysis. Soil Biol Biochem 101:251–258. https://doi.org/10.1016/j.soilbio.2016.07.021

Jia H, Ben H, Wu F (2021) Effect of biochar prepared from food waste through different thermal treatment processes on crop growth. Processes 9(2):276. https://doi.org/10.3390/pr9020276

Jien SH, Wang CS (2013) Effects of biochar on soil properties and erosion potential in a highly weathered soil. Catena 110:225–233

Jindo K, Mizumoto H, Sawada Y, Sánchez-Monedero M, Sonoki T (2014) Physical and chemical characterization of biochars derived from different agricultural residues. Biogeosciences 11(23):6613–6621. https://doi.org/10.5194/bg-11-6613-2014

Jindo K, Audette Y, Higashikawa F, Silva C, Akashi K, Mastrolonardo G, Mondini C (2020a) Role of biochar in promoting circular economy in the agriculture sector. Part 1: a review of the biochar roles in soil N, P and K cycles. Chem Biol Technol Agric 7(1). https://doi.org/10.1186/s40538-020-00182-8

Jindo K, Sánchez-Monedero M, Mastrolonardo G, Audette Y, Higashikawa F, Silva C, Mondini C (2020b) Role of biochar in promoting circular economy in the agriculture sector. Part 2: a review of the biochar roles in growing media, composting and as soil amendment. Chem Biol Technol Agric 7(1). https://doi.org/10.1186/s40538-020-00179-3

Jones DL, Rousk J, Edwards-Jones G, DeLuca TH, Murphy DV (2012) Biochar-mediated changes in soil quality and plant growth in a three year field trial. Soil Biol Biochem 45:113–124

Jusoh M, Xin L, Ywih C, Abdullah P, Radzi N, Abidin M, Muttalib M (2021) Effect of wood vinegar and rice husk biochar on soil properties and growth performances of immature kenaf (*Hibiscus cannabinus*) planted on bris soil. J Trop Resour Sustain Sci (JTRSS) 9(1):48–57. https://doi.org/10.47253/jtrss.v9i1.709

Kalina M, Sovová Š, Švec J, Trudičová M, Hajzler J, Kubíková L, Enev V (2022) The effect of pyrolysis temperature and the source biomass on the properties of biochar produced for the agronomical applications as the soil conditioner. Materials 15(24):8855. https://doi.org/10.3390/ma15248855

Keskinen R, Hyväluoma J, Sohlo L, Help H, Rasa K (2019) Fertilizer and soil conditioner value of broiler manure biochars. Biochar 1(3):259–270. https://doi.org/10.1007/s42773-019-00020-7

Khan S, Irshad S, Mehmood K, Hasnain Z, Nawaz M, Rais A et al (2024) Biochar production and characteristics, its impacts on soil health, crop production, and yield enhancement: a review. Plants 13(2):166. https://doi.org/10.3390/plants13020166

Khorram MS, Zhang Q, Lin D, Zheng Y, Fang H, Yu Y (2016) Biochar: a review of its impact on pesticide behavior in soil environments and its potential applications. J Environ Sci 44:269–279. https://doi.org/10.1016/j.jes.2015.12.027

Kloss S, Zehetner F, Dellantonio A, Hamid R, Ottner F, Liedtke V et al (2012) Characterization of slow pyrolysis biochars: effects of feedstocks and pyrolysis temperature on biochar properties. J Environ Qual 41(4):990–1000. https://doi.org/10.2134/jeq2011.0070

Kochanek J, Long RL, Lisle AT, Flematti GR (2016) Karrikins identified in biochars indicate post-fire chemical cues can influence community diversity and plant development. PLoS ONE 11(8):e0161234

Kolton M, Graber ER, Tsehansky L, Elad Y, Cytryn E (2016) Biochar-stimulated plant performance is strongly linked to microbial diversity and metabolic potential in the rhizosphere. New Phytol 213:1393–1404. https://doi.org/10.1111/nph.14253

Köves M, Madár V, Ringer M, Kocsis T (2024) Overview of traditional and contemporary industrial production technologies for biochar along with quality standardization methods. Land 13(9):1388. https://doi.org/10.3390/land13091388

Kumar R, Liu C, Ha GS, Kim KH, Chakrabortty S, Tripathy SK, Jeon BH (2023) A novel membrane-integrated sustainable technology for downstream recovery of molybdenum from industrial wastewater. Resour Conserv Recycl 196:107035

Kuryntseva P, Karamova K, Galitskaya P, Selivanovskaya S, Evtugyn G (2023) Biochar functions in soil depending on feedstock and pyrolyzation properties with particular emphasis on biological properties. Agriculture 13(10):2003. https://doi.org/10.3390/agriculture13102003

Kwapinski W, Byrne CM, Kryachko E, Wolfram P, Adley C, Leahy JJ, Hayes MH (2010) Biochar from biomass and waste. Waste Biomass Valorization 1:177–189

Laghari M, Naidu R, Xiao B, Hu Z, Mirjat M, Hu M, Fazal S (2016) Recent developments in biochar as an effective tool for agricultural soil management: a review. J Sci Food Agric 96(15):4840–4849. https://doi.org/10.1002/jsfa.7753

Laird D, Fleming P, Davis D, Horton R, Wang B, Karlen D (2010) Impact of biochar amendments on the quality of a typical midwestern agricultural soil. Geoderma 158(3–4):443–449. https://doi.org/10.1016/j.geoderma.2010.05.013

Latawiec, A., Strassburg, B., Junqueira, A., Araújo, E., Moraes, L., Pinto, H., & Hale, S. (2019). Biochar amendment improves degraded pasturelands in Brazil: environmental and cost–benefit analysis. Scientific Reports, 9(1). https://doi.org/10.1038/s41598-019-47647-x

Lehmann J, Rillig MC, Thies J, Masiello CA, Hockaday WC, Crowley D (2011) Biochar effects on soil biota—a review. Soil Biol Biochem 43:1812–1836. https://doi.org/10.1016/j.soilbio.2011.04.022

Lin L, Luo J, Wang Q, Jia C, Zhang S, Wei X, Zhu X (2022) Copyrolysis of recycled plastics and biomass reduces biochar bioavailable silicon production and cadmium phytotoxicity. ACS ES&T Eng 2(7):1356–1364. https://doi.org/10.1021/acsestengg.1c00478

Liu G, Zheng H, Wang Z (2014) Analysis of material properties with biochar improve Indian mustard (*Brassica juncea*) growth in acidic soil in northern China. Appl Mech Mater 540:239–242. https://doi.org/10.4028/www.scientific.net/amm.540.239

Liu Z, Zhu M, Wang J, Liu X, Guo W, Zheng J, Pan G (2019) The responses of soil organic carbon mineralization and microbial communities to fresh and aged biochar soil amendments. GCB Bioenergy 11(12):1408–1420. https://doi.org/10.1111/gcbb.12644

Liu M, Cholidah L, Ma S, Ma Q, Song W, Shen M, Wang L (2022) Biochar combined with organic and inorganic fertilizers promoted the rapeseed growth and improved the soil quality in purple soil. https://doi.org/10.21203/rs.3.rs-1883568/v1

López-Cano I, Cayuela M, Mondini C, Takaya C, Ross A, Sánchez-Monedero M (2018) Suitability of different agricultural and urban organic wastes as feedstocks for the production of biochar—part 1: physicochemical characterization. Sustainability 10(7):2265. https://doi.org/10.3390/su10072265

Lu SG, Sun FF, Zong YT (2014) Effect of rice husk biochar and coal fly ash on some physical properties of expansive clayey soil (Vertisol). Catena 114:37–44

Luan J, Fu Y, Tang W, Yang F, Li X, Yu Z (2023) Impact of interaction between biochar and soil microorganisms on growth of Chinese cabbage by increasing soil fertility. Appl Sci 13(23):12545. https://doi.org/10.3390/app132312545

Lugato E, Vaccari F, Genesio L, Baronti S, Pozzi A, Rack M, Miglietta F (2012) An energy-biochar chain involving biomass gasification and rice cultivation in northern Italy. GCB Bioenergy 5(2):192–201. https://doi.org/10.1111/gcbb.12028

Manyà J (2012) Pyrolysis for biochar purposes: a review to establish current knowledge gaps and research needs. Environ Sci Technol 46(15):7939–7954. https://doi.org/10.1021/es301029g

Mensah A, Frimpong K (2018) Biochar and/or compost applications improve soil properties, growth, and yield of maize grown in acidic rainforest and coastal savannah soils in Ghana. Int J Agron 2018:1–8. https://doi.org/10.1155/2018/6837404

Minamino Y, Fujitake N, Suzuki T, Yoshitake S, Koizumi H, Tomotsune M (2019) Effect of biochar addition on leaf-litter decomposition at soil surface during three years in a warm-temperate secondary deciduous forest, Japan. Sci Rep 9(1). https://doi.org/10.1038/s41598-019-53615-2

Mogaji T, Moses E, Idowu E, Jen T (2020) Thermal degradation conditions effects on selected biomass wastes and characterization of their produced biochar. J Energy Res Rev 46–59. https://doi.org/10.9734/jenrr/2020/v4i330131

Mohan D, Kumar A, Sarswat A, Patel M, Singh P, Pittman C (2018) Biochar production and applications in soil fertility and carbon sequestration—a sustainable solution to crop-residue burning in India. RSC Adv 8(1):508–520. https://doi.org/10.1039/c7ra10353k

Mota-Panizio R, Assis A, Calado L, Nobre C, Longo A, Silveira J, Brito P (2023) Production, characterization, and activation of biochars from a mixture of waste insulation electric cables (WIEC) and waste lignocellulosic biomass (WLB). C J Carbon Res 9(2):49. https://doi.org/10.3390/c9020049

Mukome F, Zhang X, Silva L, Six J, Parikh S (2013) Use of chemical and physical characteristics to investigate trends in biochar feedstocks. J Agric Food Chem 61(9):2196–2204. https://doi.org/10.1021/jf3049142

Murtaza G, Ahmed Z, Dai DQ, Iqbal R, Bawazeer S, Usman M, Ali I (2022) A review of mechanism and adsorption capacities of biochar-based engineered composites for removing aquatic pollutants from contaminated water. Front Environ Sci 10:1035865

Murtaza G, Ahmed Z, Eldin SM, Ali B, Bawazeer S, Usman M et al (2023) Biochar-soil-plant interactions: a cross talk for sustainable agriculture under changing climate. Front Environ Sci 11:1059449

Nain P (2023) Characterization of pine needle and pine needle biochar: a potential soil amendment for sustainable forest waste management. Ann Plant Soil Res 25(2):211–220. https://doi.org/10.47815/apsr.2023.10259

Nair V, Nair P, Dari B, Freitas A, Chatterjee N, Pinheiro F (2017) Biochar in the agroecosystem–climate-change–sustainability nexus. Front Plant Sci 8. https://doi.org/10.3389/fpls.2017.02051

Nan H, Yang F, Zhao L, Mašek O, Cao X, Xiao Z (2018) Interaction of inherent minerals with carbon during biomass pyrolysis weakens biochar carbon sequestration potential. ACS Sustain Chem Eng 7(1):1591–1599. https://doi.org/10.1021/acssuschemeng.8b05364

Novak J, Cantrell K, Watts D, Busscher W, Johnson M (2013) Designing relevant biochars as soil amendments using lignocellulosic-based and manure-based feedstocks. J Soils Sediments 14(2):330–343. https://doi.org/10.1007/s11368-013-0680-8

Osman A, Lai Z, Farghali M, Yiin C, Elgarahy A, Hammad A, Yap P (2023) Optimizing biomass pathways to bioenergy and biochar application in electricity generation, biodiesel production,

and biohydrogen production. Environ Chem Lett 21(5):2639–2705. https://doi.org/10.1007/s10311-023-01613-2
Ouyang W, Zhao X, Tysklind M, Hao F (2016) Typical agricultural diffuse herbicide sorption with agricultural waste-derived biochars amended soil of high organic matter content. Water Res 92:156–163. https://doi.org/10.1016/j.watres.2016.01.055
Pan Y, Chen H, Bolan N, Sarkar B, Wang H, Chen C (2023) Both sides of coin: benefits and potential negative consequences of biochar in sediment remediation. Rev Environ Contam Toxicol 261(1):4
Pawar A, Panwar N (2020) Experimental investigation on biochar from groundnut shell in a continuous production system. Biomass Convers Biorefin 12(4):1093–1103. https://doi.org/10.1007/s13399-020-00675-4
Peake LR, Reid BJ, Tang X (2014) Quantifying the influence of biochar on the physical and hydrological properties of dissimilar soils. Geoderma 235:182–190
Raj A, Karol A, Shaji A, Pandey R, Gupta H (2023) Biochar: a comprehensive overview of its role in soil health. Int J Environ Clim Change 13(10):1621–1628. https://doi.org/10.9734/ijecc/2023/v13i102817
Rizhiya E, Buchkina N, Mukhina I, Belinets A, Balashov E (2015) Effect of biochar on the properties of loamy sand Spodosol soil samples with different fertility levels: a laboratory experiment. Eurasian Soil Sci 48(2):192–200. https://doi.org/10.1134/s1064229314120084
Roberts D, Cole A, Paul N, Nys R (2015) Algal biochar enhances the revegetation of stockpiled mine soils with native grass. J Environ Manage 161:173–180. https://doi.org/10.1016/j.jenvman.2015.07.002
Ronsse F, Hecke S, Dickinson D, Prins W (2012) Production and characterization of slow pyrolysis biochar: influence of feedstock type and pyrolysis conditions. GCB Bioenergy 5(2):104–115. https://doi.org/10.1111/gcbb.12018
Sarraf M, Ibranshahib I, Kayabasi E, Bispo M, Iftikhar J, Chen Z, Elkhlifi Z, Ali B, Saleem M, Meili L, Hegedűsová A, Ercişli S, Ansari N (2023) Potential role of biochar on capturing soil nutrients, carbon sequestration and managing environmental challenges: a review. Sustainability. https://doi.org/10.3390/su15032527
Schmidt H, Kammann C, Hagemann N, Leifeld J, Bucheli T, Sánchez-Monedero M, Cayuela M (2021) Biochar in agriculture—a systematic review of 26 global meta-analyses. GCB Bioenergy 13(11):1708–1730. https://doi.org/10.1111/gcbb.12889
Shin J, Choi Y, Lee S (2016) Estimation of carbon sequestration and its profit analysis with different application rates of biochar during corn cultivation periods. J Korea Org Resour Recycl Assoc 24(3):83–90. https://doi.org/10.17137/korrae.2016.24.3.83
Shin J, Choi E, Jang E, Hong S, Lee S, Ravindran B (2018) Adsorption characteristics of ammonium nitrogen and plant responses to biochar pellet. Sustainability 10(5):1331. https://doi.org/10.3390/su10051331
Solaiman ZM, Murphy DV, Abbott LK (2012) Biochars influence seed germination and early growth of seedlings. Plant Soil 353:273–287
Sun K, Ro K, Guo M, Novak J, Mashayekhi H, Xing B (2011) Sorption of bisphenol A, 17α-ethinyl estradiol and phenanthrene on thermally and hydrothermally produced biochars. Bioresour Technol 102(10):5757–5763
Sun X, Shan R, Li X, Pan J, Xin L, Deng R, Song J (2017) Characterization of 60 types of Chinese biomass waste and resultant biochars in terms of their candidacy for soil application. GCB Bioenergy 9(9):1423–1435. https://doi.org/10.1111/gcbb.12435
Taha SM, Amer ME, Elmarsafy AE, Elkady MY (2014) Adsorption of 15 different pesticides on untreated and phosphoric acid treated biochar and charcoal from water. J Environ Chem Eng 2(4):2013–2025
Tan X, Ye S, Liu S, Zeng G, Song B, Qin F, Tan C, Shen M, Yang H, Xiang L (2021) Potential hazards of biochar: the negative environmental impacts of biochar applications. J Hazard Mater 420:126611. https://doi.org/10.1016/j.jhazmat.2021.126611

Vaccari FP, Baronti S, Lugato E, Genesio L, Castaldi S, Fornasier F, Miglietta F (2011) Biochar as a strategy to sequester carbon and increase yield in durum wheat. Eur J Agron 34(4):231–238
Vithanage M, Gunarathne V, Bandara T (2013) Mini-review on influence of biochar for sustainable agriculture. J Food Agric 6(1–2):1. https://doi.org/10.4038/jfa.v6i1-2.5182
Wang J, Wang S (2019) Preparation, modification and environmental application of biochar: a review. J Clean Prod. https://doi.org/10.1016/J.JCLEPRO.2019.04.282
Wang J, Xiong Z, Kuzyakov Y (2015) Biochar stability in soil: meta-analysis of decomposition and priming effects. GCB Bioenergy 8(3):512–523. https://doi.org/10.1111/gcbb.12266
Wang J, Xiong Z, Kuzyakov Y (2016) Biochar stability in soil: meta-analysis of decomposition and priming effects. GCB Bioenergy 8:512–523. https://doi.org/10.1111/gcbb.12266
Wang L, Wu J, Wang Z, Li W, Zhang K (2017) Effects of three biochars as adsorbents on soils adsorbing ammonium nitrogen in biogas slurry. J Chem 2017:1–10. https://doi.org/10.1155/2017/4627928
Wang L, O'Connor D, Rinklebe J, Ok Y, Tsang D, Shen Z, Hou D (2020) Biochar aging: mechanisms, physicochemical changes, assessment, and implications for field applications. Environ Sci Technol 54(23):14797–14814. https://doi.org/10.1021/acs.est.0c04033
Wang Y, Gu X, Huang Y, Ding Z, Chen Y, Hu X (2022) Insight into biomass feedstock on formation of biochar-bound environmentally persistent free radicals under different pyrolysis temperatures. RSC Adv 12(30):19318–19326. https://doi.org/10.1039/d2ra03052g
Wijitkosum S, Jiwnok P (2019) Elemental composition of biochar obtained from agricultural waste for soil amendment and carbon sequestration. Appl Sci 9(19):3980. https://doi.org/10.3390/app9193980
Woo S (2013) Biochar for soil carbon sequestration. Clean Technol 19(3):201–211. https://doi.org/10.7464/ksct.2013.19.3.201
Wood R, Masek O, Erastova V (2023) Biochars at the molecular level. Part 1—insights into the molecular structures within biochars. arXiv preprint arXiv:2303.09661
Woolf D, Amonette J, Street-Perrott F, Lehmann J, Joseph S (2010) Sustainable biochar to mitigate global climate change. Nat Commun 1(1). https://doi.org/10.1038/ncomms1053
Wyzińska M, Berbeć A, Grabinski J (2023) Impact of biochar dose and origin on winter wheat grain quality and quantity. Agriculture 14(1):39. https://doi.org/10.3390/agriculture14010039
Yaro N, Sutanto M, Habib N, Usman A, Kaura J, Murana A et al (2023) A comprehensive review of biochar utilization for low-carbon flexible asphalt pavements. Sustainability 15(8):6729. https://doi.org/10.3390/su15086729
Zhang J, Lü F, Zhang H, Shao L, Chen D, He P (2015) Multiscale visualization of the structural and characteristic changes of sewage sludge biochar oriented towards potential agronomic and environmental implication. Sci Rep 5(1). https://doi.org/10.1038/srep09406
Zhang C, Lin Y, Tian X, Xu Q, Chen Z, Lin W (2017) Tobacco bacterial wilt suppression with biochar soil addition associates to improved soil physiochemical properties and increased rhizosphere bacteria abundance. Appl Soil Ecol 112:90–96. https://doi.org/10.1016/j.apsoil.2016.12.005
Zhang G, Zhang L, Shi Z, Yang Y, Liu J (2025) Microbial nutrient limitation and carbon use efficiency in saline–alkali soil amended with biochar: insights from ecoenzymatic C:N:P stoichiometry. Biochar 7(1). https://doi.org/10.1007/s42773-025-00458-y
Zheng J, Stewart C, Cotrufo M (2012) Biochar and nitrogen fertilizer alters soil nitrogen dynamics and greenhouse gas fluxes from two temperate soils. J Environ Qual 41(5):1361–1370. https://doi.org/10.2134/jeq2012.0019
Zhu X, Labianca C, He M, Luo Z, Wu C, You S, Tsang D (2022) Life-cycle assessment of pyrolysis processes for sustainable production of biochar from agro-residues. Bioresour Technol 360:127601. https://doi.org/10.1016/j.biortech.2022.127601

Chapter 4
Biochar as a Modulator of Soil Microbiomes: Implications for Microbial Diversity, Biomass, and Enzymatic Functions Across Ecosystems

Jyotirmay Roy and Tirunagari Rupesh

Abstract As a carbon-rich byproduct of biomass pyrolysis, biochar has drawn a lot of interest as a sustainable soil amendment that has an immense effect on the microbial communities and functions of the soil. This chapter comprehensively examines the impact of biochar on soil microbiomes across diverse ecosystems, focusing on microbial diversity, community structure, biomass, and enzymatic activity. Evidence from recent studies indicates that biochar modulates the soil environment by altering pH, nutrient availability, porosity, and moisture retention, thereby promoting beneficial microbial taxa and enhancing environmental processes. The nature of biochar's influence is highly context-dependent, governed by its physicochemical properties (feedstock, pyrolysis temperature, surface area) and interactions with soil type, climate, and land-use practices. While most findings highlight biochar's capacity to support microbial-mediated soil health and resilience, inconsistencies across studies reveal critical knowledge gaps. Limitations include variability in biochar effects, short-term experimental designs, and insufficient understanding of biochar aging and microbe-plant interactions. Future research should aim to standardize biochar characterization, investigate long-term impacts, and explore synergistic formulations tailored to specific agroecosystems. This chapter underscores biochar's potential in advancing climate-resilient agriculture and soil ecological restoration through microbial pathway modulation.

Keywords Biochar properties · Nutrient cycling · Microbial diversity · Microbe communities · Soil health

---

J. Roy (✉) · T. Rupesh
Division of Soil Science and Agricultural Chemistry, ICAR-Indian Agricultural Research Institute, New Delhi, India
e-mail: jyotirmayroy176@gmail.com

D. Moulick et al. (eds.), *Biochar: Achieving Agro-Environmental Sustainability*, Sustainability Solutions, https://doi.org/10.1007/978-3-032-19855-6_4

4.1 Introduction

A carbon-rich substance made by pyrolyzing biomass, biochar has drawn a lot of interest as a sustainable soil supplement that may enhance soil health and enhance microbial communities across diverse ecosystems. The physicochemical characteristics of biochar (porous structure, surface chemistry, nutrient content, and stability) enable it to influence soil physicochemical parameters including pH, moisture retention, nutrient availability, and cation exchange capacity (Mukherjee et al. 2022; Anyebe et al. 2025). These changes, in turn, create favorable microhabitats that support diverse and functionally important microbial populations, thereby contributing to improved nutrient cycling, soil fertility, and ecosystem resilience (Tian et al. 2025).

The impact of biochar on soil microbiomes is highly context-dependent, varying widely across various ecosystems, soils, and environmental conditions (Xiang et al. 2023; Sani et al. 2023). Studies have documented biochar's ability to enrich bacterial and fungal diversity, promote beneficial microbial taxa, and reshape microbial community structure over time (Wolna-Maruwka et al. 2021; Hu et al. 2023). By offering both physical habitats and labile carbon sources, biochar amendments have been demonstrated to promote microbial diversity and improve nutrient cycling in karst and temperate soils. (Dai et al. 2021; Yan et al. 2021). In contrast, in saline-alkaline soils, biochar acts as a remediation agent by alleviating salinity stress and selectively enriching salt-tolerant microbial taxa that support plant growth under adverse conditions (Egamberdieva et al. 2022).

Beyond ecological diversity, biochar significantly influences soil biological characteristics that are central to soil biochemical processes. The magnitude and direction of these effects depend on biochar properties (Li et al. 2023). Low-temperature biochars generally stimulate nutrient-acquisition enzymes like dehydrogenase, urease, and alkaline phosphatase by providing labile carbon and functional groups favorable for microbial metabolism (Liao et al. 2022; Zhao et al. 2022a, b). Conversely, high-temperature biochars, with their more aromatic and inert nature, often exhibit minimal stimulation or even inhibitory effects on certain enzymes due to changes in surface chemistry and enzyme adsorption dynamics (Liu et al. 2022).

Importantly, biochar's role extends beyond direct chemical and physical interactions with enzymes to indirect modulation via shifts in microbial diversity. By altering the balance between fungi and bacteria and influencing the abundance of key functional groups, biochar affects the diversity and activity of extracellular enzymes (Ali et al. 2025). However, biochar's effects are not universally positive or uniform. Variability in feedstock, pyrolysis conditions, soil types, and environmental stressors such as freeze–thaw cycles, heavy metal contamination, and microplastic pollution contribute to complex and sometimes contradictory outcomes (Yang et al. 2023).

This chapter delves into the multifaceted and dynamic impacts of biochar on soil microbial communities across a diverse array of ecosystems, including agricultural lands, forest soils, grasslands, and contaminated or degraded environments. This chapter synthesizes a growing body of recent research that illustrates both the potential and the limitations of biochar as a tool for promoting microbial-mediated soil functions and fostering ecosystem sustainability.

Central to the discussion is an exploration of the underlying mechanisms through which biochar exerts its influence-ranging from its inherent physicochemical characteristics. These properties can profoundly affect microbial habitats, energy availability, and microbe-microbe interactions, thereby shaping both microbial diversity and function. Moreover, the chapter examines how biochar interacts with different soil types, climates, land management practices, and environmental stressors such as heavy metal contamination, drought, and salinity. Special attention is given to the modulation of soil enzymatic activities, which serve as sensitive indicators of soil biochemical processes, nutrient availability, and microbial functional potential. By influencing key enzymes involved in carbon, nitrogen, phosphorus, and sulfur cycling, biochar can indirectly regulate plant productivity and soil fertility.

In addition, this chapter addresses emerging challenges and knowledge gaps, including variability in biochar effects. It also considers the temporal dynamics of biochar's influence and the potential for synergistic or antagonistic interactions with other soil amendments. By integrating mechanistic insights with ecosystem-level outcomes, this chapter aims to provide a comprehensive and nuanced understanding of how biochar amendments can be strategically optimized to support biologically active, resilient, and productive soils.

4.2 Impact of Biochar on Soil Microbiomes Across Different Ecosystems

Biochar has shown promise as a soil supplement to enhance microbial populations and soil health. Its effects on soil microbiomes vary across ecosystems, soil types, and environmental conditions, reflecting complex interactions between biochar properties, soil physicochemical parameters, and native microbial ecology (Table 4.1).

4.2.1 Karst and Temperate Soils

Yan et al. (2021) conducted a 24-month field study in karst soils investigating biochar-based fertilizers combined with compost and mineral fertilizers. Biochar treatments significantly enhanced both bacterial and fungal diversity, mainly due to porous structure which provides microhabitats, its nutrient content from pyrolyzed biomass. The combined biochar-compost-fertilizer treatments (BMF and B4MF) showed the greatest microbial diversity and soil nutrient enrichment. Dominant bacterial phyla included Actinobacteria, which increased notably in BMF and B4MF treatments due

Table 4.1 Impact of biochar amendment on soil microbial communities across different regions, feedstocks, and application conditions

Location	Biochar feedstock	Pyrolysis temperature	Application rate	Soil type	Impact on microbial community	Possible mechanism	References
Changsha County, Hunan Province, China	–	–	24 and 48 t ha^{-1}	Hydragric Anthrosol	The bacterial 16S rRNA gene abundance increased by 4.2–101.8% with 24–48 t ha^{-1} biochar compared to control	Degradation of aged biochar likely boosted bacterial and fungal abundance by supplying more substrates	Wang et al. (2021)
Taihu Lake region, Yangtze River Delta, China	Rice straw	600 °C	20 and 40 t ha^{-1}	Hydragric Anthrosol	Biochar had no significant effect on bacterial abundance but reduced fungal Shannon diversity compared to control	Biochar altered the soil microenvironment, increasing bacterial community differences, reducing uniformity, and lowering diversity	Jiang et al. (2021)
Henan, China	Maize straw	450 °C	3, 6 and 12 t ha^{-1}	Calcaric Aqui-alluvic Primisol	Biochar reduced 16S rRNA gene copies by 9.25–35.77% but had no effect on bacterial Shannon index	Biochar adsorbed low molecular organic C, reducing its availability and thus bacterial abundance	Zheng et al. (2022)
Zhejiang, China	Rice straw and canola stalk	350 and 550 °C	21 and 15.5 g pot^{-1}	Acidic Ultisol	Fungal Shannon diversity increased with rice straw biochar but decreased with canola stalk biochar	Soil pH increase favored bacterial growth over fungi by influencing their relative abundance	Yang et al. (2022)
Wuhan, Hubei Province, China	Rice straw	500 °C	2% (w/w)	Saline-alkali soil	Biochar application increased Actinobacteriota, Bdellovibrionota, and Gemmatimonadota abundance	Biochar created a favorable habitat for soil bacteria and microbial communities	Wang et al. (2024)

(continued)

Table 4.1 (continued)

Location	Biochar feedstock	Pyrolysis temperature	Application rate	Soil type	Impact on microbial community	Possible mechanism	References
Qitai Wheat Test Station, Xinjiang, China	Corn straw	450 °C	20 t ha^{-1}	Sandy loam soil	Biochar with nitrogen fertilizer increased average well color development in BIOLOG ECO plates compared to control	Biochar promoted microbial growth directly by improving conditions and indirectly by enhancing soil aeration, moisture, and chemistry	Yang et al. (2025)
Ghana, West Africa	Corn cobs	500–550 °C	15 and 30 t ha^{-1}	Haplic Acrisols	Biochar increased arbuscular mycorrhiza by 1.8–4.5 folds and fungal abundance by 3.2–5.4 folds compared to control	Increased MBC and MBN boosted microbial growth and activity	Amoakwah et al. (2022)
Lucknow, Uttar Pradesh, India	Sewage sludge	300, 500, and 600 °C	25%, 50%, 75% (w/w)	Typic haplusepts	Biochar amendment showed a microbial count of 2.7×10^5 to 4.4×10^5 CFU g^{-1} in agricultural soil	–	Ahmad et al. (2022)
Tianjin, China	Wheat straw and rice husk	300, 500, and 700 °C	4 mg g^{-1}	–	Biochar addition significantly reduced the bacterial Chao1 index from 18,599.34 to 14,356.02	–	Sun et al. (2022)
Xuchang City, Henan Province, China	Peanut shells	450 °C	3.75, 7.5, 11.25, and 15 kg ha^{-1}	Cinnamon	Bacterial abundance was 69.61 and 58.18% higher than the control	Higher soil TOC boosted bacterial abundance by providing essential organic carbon nutrition	Wang et al. (2022)
Zhaoqing City, Guangdong Province, China	Maize straw	300, 450, and 600 °C	1 and 3% (w/w)	Anthrosols	While 3% biochar decreased bacterial richness and diversity, 1% biochar boosted bacterial and fungal richness and diversity	Excessive biochar could be toxic to soil microbes due to its inherent inhibitory concentration and possible toxic substances	Han et al. (2024)

(continued)

Table 4.1 (continued)

Location	Biochar feedstock	Pyrolysis temperature	Application rate	Soil type	Impact on microbial community	Possible mechanism	References
Wenling, Zhejiang, China	Swine manure	700 °C	5% (w/w)	–	Biochar amendment upregulated 45 genera utilizing 13C-rhizodeposits, mainly from *Actinobacteria* and *Proteobacteria*. *Actinobacteria* genera with high abundance included *Micrococcaceae*, *Oryzihumus*, *Nocardioides*, and *Methylophilus*	Upregulated genera after biochar amendment likely prefer recalcitrant organic C like aromatic compounds	Fu et al. (2022)
Jhansi, Bundelkhand region, Uttar Pradesh, India	*Parthenium hysterophorus* L. and *Lantana camara* L.	400 °C	2.5, 5, and 10 t ha^{-1}	Typic haplusepts	The 10 t ha^{-1} biochar with 75% RDF showed the highest bacterial (7.33 × 10^8 CFU g^{-1}) and fungal (1.33 × 10^5 CFU g^{-1}) populations	–	Choudhary et al. (2023)
Harbin, China	Corn stalk	450 °C	0.3, 0.6, and 1.2%	Mollisol	The Shannon and Inverse Simpson indices were considerably reduced by biochar treatments at 0.3 and 1.2%	–	He et al. (2021)
Wenling region, Zhejiang Province, China	Maize straw	500 and 600 °C	20 g kg^{-1}	Alfisol	Biochar increased Alphaproteobacteria, Gammaproteobacteria, Gemmatimonadetes, and Bacteroidia, but decreased Bacilli, Acidobacteriia, Thermoleophilia, Clostridia, and Acidimicrobiia	Increased bacterial diversity results from diverse DOM molecules, boosting Gammaproteobacteria, Betaproteobacteria, and Flavobacteria, which selectively decompose specific DOM components	Ling et al. (2022)

to their ability to degrade complex carbon compounds in biochar, and Proteobacteria, which were abundant across biochar treatments given their role in nitrogen fixation and phosphorus solubilization. This study highlights biochar's role as a habitat provider and nutrient modulator, enhancing microbial-mediated nutrient cycling in karst soils (Dai et al. 2021).

4.2.2 Saline-Alkaline Soils: Biochar as a Remediation Agent and Microbial Enhancer

In saline-alkaline soils, Wang et al. (2024) showed that rice straw biochar at 2% application effectively reduced soil pH, electrical conductivity (EC), and soluble salts, thereby alleviating salinity stress. High salinity typically suppresses microbial richness and enzyme activity, but biochar amendment boosted microbial diversity and extracellular enzymes related to nitrogen and phosphorus cycling by 32.79% and 34.25%, respectively. Biochar also increased salt-tolerant taxa such as *Actinobacteriota*, *Pseudomonas*, and *Bacillus*, which promote plant growth under abiotic stress. This demonstrates biochar's dual role in improving soil physicochemical properties and fostering beneficial microbial communities that enhance soil resilience in harsh environments (An et al. 2023).

4.2.3 Paddy Fields and Double Rice Cropping Systems

Wang et al. (2021) examined biochar's impact three to four years after a single application in double rice cropping systems. Straw-derived biochar (24 and 48 t ha^{-1}) led to an increase in *Acidobacteria*-a group known for thriving under low nutrient conditions and promoting total organic carbon (TOC) accumulation through reduced mineralization. Conversely, populations of *Proteobacteria* and gram-positive bacteria such as *Actinobacteria* and *Firmicutes*, which utilize labile biochar-derived compounds, decreased as those easily degradable substrates were depleted. Fungal community shifts were observed with increased abundance of *Zygomycota* and decreased *Basidiomycota*, possibly due to niche competition and organic matter enrichment. Genera like *Mortierella*, known for saprotrophic decomposition and phosphorus acquisition support to plants, were also enriched. These findings indicate biochar's capacity to restructure soil microbiomes over time, favoring oligotrophic and recalcitrant carbon decomposers, thus affecting nutrient cycling dynamics and organic matter stabilization (Huang et al. 2024).

4.2.4 Heavy Metal Contaminated Soils

Mei et al. (2022) explored the effect of composite biochar with Bacillus cereus RC-1 on cadmium (Cd) immobilization and microbial diversity. The combined treatment (BC + M) maintained higher bacterial diversity compared to biochar or bacteria alone, suggesting biochar's role as a carrier enhancing bacterial colonization and activity. The dominant phyla included *Proteobacteria*, *Actinobacteria*, and *Firmicutes*, with a notable decrease in *Actinobacteria* relative abundance following Cd immobilization, reflecting microbial stress responses. Genera such as *Bacillus*, *Streptomyces*, and *Arthrobacter*, known metal reducers, thrived particularly with biochar-assisted inoculation. This shows biochar's potential to mitigate heavy metal toxicity by stabilizing contaminants and supporting functional microbial consortia that can promote bioremediation (Wu et al. 2024).

4.2.5 Tropical Sandy Loam Soils

Amoakwah et al. (2022) conducted field experiments in tropical ecosystems applying corn cob biochar at different rates and combined with phosphate fertilizer. Biochar treatments increased total phospholipid fatty acid (PLFA) content, reflecting higher microbial biomass due to increased labile carbon and nutrient availability. Higher pyrolysis temperatures raised biochar pH and enhanced porous surface area, which influenced microbial community composition. Interestingly, biochar with higher pore volume (from higher pyrolysis temperature) was less favorable for fungal hyphae, reducing fungal abundance. The fungal:bacteria ratio-a proxy for carbon sequestration potential-was higher in biochar treatments, indicating enhanced soil carbon stability. This underscores how biochar production parameters modulate microbial responses and ecosystem carbon dynamics (Bai et al. 2024).

4.2.6 Field Soils: Dose-Dependent Effects on Microbial Diversity and Bacterial Community Composition

Zhang and Shen (2022) showed the impact of varying rates of biochar application (0, 10, and 20 t m^{-3}) on soil bacterial communities under field conditions. Their study revealed a dose-dependent increase in bacterial community diversity, with both the Shannon and Chao1 indices showing significant enhancement at higher biochar application rates. This suggests that biochar provides a more conducive environment for microbial colonization and proliferation. A notable shift was observed in the composition of the bacterial community, with enrichment of dominant phyla such as *Proteobacteria* and *Actinobacteria*. *Proteobacteria* are indicative of an active and functionally diverse microbial population. The increase in *Actinobacteria*, a copiotrophic (eutrophic) group that thrives in nutrient-rich environments and is capable of rapidly utilizing available carbon sources, pointed toward an improvement in soil nutritional status as a result of biochar application. This shift not only reflects the enhanced carbon availability and habitat quality but also the stimulation of microbial taxa involved in key soil biochemical processes. Biochar amendments can significantly reshape soil microbial community composition and increase biodiversity (Dai et al. 2021). By enriching beneficial microbial populations and altering the physicochemical environment, biochar enhances soil ecological functions. This microbial restructuring is vital for sustainable soil management, particularly in degraded or intensively farmed soils where biological function needs to be restored or enhanced.

4.2.7 Freeze–Thaw Affected Mollisols: Biochar and Aged Biochar Effects on Pollutant Degradation and Microbial Functional Genes

Wang et al. (2025) studied the effects of fresh and freeze–thaw aged corn straw biochar (BC and FBC) on microbial communities in DEHP-contaminated Mollisols. Both types of biochar significantly enhanced microbial richness, as reflected by increases in Chao 1 and Shannon diversity indices, and substantially altered the microbial community structure compared to the unamended control. Notably, the relative abundance of *Proteobacteria* increased by 18.35% with BC and 12.58% with FBC. This phylum includes many functionally important genera, such as *Rhodococcus*, *Arthrobacter*, *Pseudomonas*, and *Lysobacter*, which were particularly enriched in biochar-treated soils. These genera are well-known for their capabilities in degrading xenobiotic compounds and participating in nitrogen cycling processes. The enrichment of such functional microbes suggests that biochar, whether fresh or aged, creates a more favorable microhabitat for pollutant-degrading and nutrient-cycling bacteria. This could be attributed to biochar's high porosity, surface area, and potential to adsorb toxic pollutants, thereby reducing stress on microbial populations. Additionally, freeze–thaw aging may further alter the biochar's surface chemistry, potentially enhancing its interactions with both soil particles and microbial biofilms. Overall, the study underscores biochar's potential as a multifunctional amendment in contaminated and climatically stressed soils. By improving microbial resilience and ecological function, particularly under freeze–thaw stress and organic pollutant pressure, biochar contributes to enhanced bioremediation and long-term soil health restoration (Shi et al. 2024).

4.2.8 *Microplastic (MP)-Contaminated Soils Under Variable Moisture Regimes*

The effects of biochar on MP-contaminated soils were examined by Dissanayake et al. (2022) at two moisture levels (30 and 70% water holding capacity, WHC). Microbial colonization was facilitated by the increased surface area and ash concentration of biochar generated at higher temperatures (700 °C). Under 30% WHC, biochar increased bacterial richness indices (Chao1, ACE), especially for OSR 700 biochar. However, bacterial diversity (Shannon index) decreased significantly in some biochar treatments, possibly due to altered soil pH and EC. *Proteobacteria*, *Bacteroidetes*, *Actinobacteria*, and *Acidobacteria* were the most prevalent bacterial phyla; changes were influenced by the kind of biochar and the moisture regime.

Notably, *Proteobacteria* dominated in both moisture conditions and are known for broad metabolic capabilities important for C, N, and S cycling as well as plastic degradation. Actinobacteria, enriched under drought conditions, have polyethylene degradation capabilities. This study highlights the complex and moisture-dependent impacts of biochar on microbial communities in contaminated soils, indicating the need for further field-level research incorporating plant interactions and environmental variability (Bogati et al. 2025).

4.3 Impact of Biochar on Soil Microbial Biomass and Underlying Mechanisms

The decomposition of organic matter, nitrogen cycling, and general soil health are all significantly influenced by soil microbial biomass. The ability of biochar, a carbon-rich byproduct of pyrolyzing organic material, to increase soil microbial activity has attracted a lot of interest. Depending on its feedstock, pyrolysis conditions, and application rate, biochar can modify soil physicochemical properties and microenvironments, thereby influencing microbial dynamics. Numerous studies across diverse agroecological regions have explored how different biochars impact microbial biomass and associated soil functions. Zhang et al. (2014) studied that the application of crushed corncob biochar (360 °C, 9.0 t ha^{-1} yr^{-1}) significantly enhanced microbial biomass carbon (MBC) by 118–763% at 0–5 cm and 5–10 cm soil depths compared to control. Though no mechanism was explicitly mentioned, the increase likely resulted from improved carbon availability and microbial habitat formation. Adebajo et al. (2022) demonstrated that rice husk biochar (350 °C, 7.5 t ha^{-1}) increased microbial nitrogen (MBN), carbon (MBC), and phosphorus (MBP) by 9.75%, 22.94%, and 12.01%, respectively. This enhancement was attributed to improved nitrogen and phosphorus dynamics driven by biochar amendment. Hale et al. (2021) found that yellow pine biochar (350 °C, 11.2 and 22.4 t ha^{-1}) had an insignificant effect on microbial biomass and nutrient availability. The biochar failed to increase inorganic N, P, K, or PLFA biomass, highlighting that not all biochars elicit positive microbial responses, especially depending on feedstock and soil type.

Wang et al. (2023) reported that cotton straw biochar pyrolyzed at a high temperature (850 °C) and applied at 5–20 t ha^{-1} yr^{-1} led to MBC increases of 6.26–27.09% and MBN increases of 19.45–37.80%. The improvement was linked to increased soil nutrients (TOC, TN, NO3-N, $NH4^{+}$-N), reduced pH, and the provision of ample carbon and nitrogen to soil microbes. Amoakwah et al. (2022) observed that corn cob biochar (500–550 °C, 15 and 30 t ha^{-1}) raised MBC by 4.5 and 8.3 times and MBN by 2.7 times compared to control. Although no mechanism was specified, the results suggest enhanced nutrient availability and habitat formation. Wang et al. (2022) showed that peanut shell biochar (450 °C, 3.75–15 kg ha^{-1}) increased MBC by 38.69–40.90% and MBN by 32.14–42.66%. The improvement was attributed to the biochar's porous structure and carbon content, which promoted microbial colonization. According to Choudhary et al. (2023), biochar made from *Lantana camara* and *Parthenium hysterophorus* (400 °C, 2.5–10 t ha^{-1}) raised soil microbial biomass carbon (SMBC) by about 61%. The biochar's improved microbial habitats were the cause of the effect.

Ali et al. (2022) reported that cassava straw biochar (300–500 °C, 10–30 t ha^{-1}) increased SMBC by 126–134%. The underlying mechanism was attributed to the biochar's alkalinity, which supported microbial activity and population growth, particularly in acidic soils. Aziz et al. (2024) showed that poultry litter biochar (400 °C, 10 and 20 t ha^{-1}) increased MBC by 21% at the higher rate. This enhancement was associated with the supply of organic substrates that boosted microbial enzymatic activity. Overall, biochar can positively affect microbial biomass by improving nutrient availability, altering soil pH, supplying labile carbon, and creating favorable physical conditions for microbial colonization. However, its effectiveness is dependent on biochar properties and soil context.

The effect of biochar on soil microbial biomass is multifaceted and largely influenced by its physicochemical properties, feedstock origin, pyrolysis temperature, and application rate. Most studies indicate a stimulatory effect on microbial biomass due to enhanced pH balance in acidic soils, and improved habitat conditions. However, as shown in some cases, such as with yellow pine biochar, not all materials uniformly enhance microbial activity. Therefore, tailored biochar formulations considering soil type and ecological conditions are essential for optimizing microbial benefits and achieving sustainable soil health outcomes.

4.4 Biochar Properties Influencing Soil Microbial Biomass

Because of its ability to boost microbial activity, improve nitrogen cycling, and improve soil health, biochar has drawn more and more attention as a soil supplement. Among its most studied effects are its impacts on soil enzymatic functions and microbial biomass, which are critical indicators of soil fertility and ecological function. However, the extent of these benefits varies considerably depending on several biochar characteristics. Understanding the interactions between these variables and their influence on microbial and enzymatic dynamics is essential for optimizing biochar use in sustainable land management.

4.4.1 Feedstock Type and Its Influence

The kind of biomass used to make biochar has a significant influence on how it affects microbial biomass and soil enzymatic activity. MBC, MBN, and MBP have continuously improved when biochars made from nutrient-rich agricultural wastes including crushed corncob, rice husk, cotton straw, and cassava straw are used. Zhang et al. (2014) reported a substantial increase in MBC (up to 763%) with crushed corncob biochar, while Adebajo et al. (2022) observed increases in MBN, MBC, and MBP using rice husk biochar. These outcomes suggest that the intrinsic nutrient composition and decomposition characteristics of the feedstock contribute to microbial proliferation and enhanced enzymatic functioning.

4.4.2 Pyrolysis Temperature and Biochar Reactivity

The temperature at which biochar is pyrolyzed significantly alters its physical and chemical properties. Low- to moderate-temperature biochars (300–500 °C) retain more labile carbon and functional groups that support microbial growth and enzyme production. For example, Ali et al. (2022) used cassava straw biochar produced between 300 and 500 °C and noted an increase of 126–134% in soil microbial biomass carbon. Conversely, high-temperature biochar (850 °C from cotton straw used by Wang et al. 2023) led to moderate increases in microbial biomass, potentially due to its more recalcitrant nature and reduced labile organic content. Thus, low-temperature biochars are generally more conducive to enzyme activity due to greater availability of accessible carbon substrates.

4.4.3 Application Rate and Dose-Dependent Responses

Biochar application rate also influences its effectiveness in enhancing enzymatic activity. A dose-dependent positive response is evident across multiple studies. For instance, Amoakwah et al. (2022) found that increasing the application of corn cob biochar from 15 to 30 t ha^{-1} increased MBC by 4.5–8.3 times, and MBN by 2.7 times. Similarly, in Guangxi, cassava straw biochar at 10, 20, and 30 t ha^{-1} led to consistent increases in microbial biomass, indicating that higher biochar doses can more effectively modify the soil microenvironment and stimulate microbial and enzymatic functions, especially in degraded or acidic soils.

4.4.4 Biochar Surface Structure and Porosity

Biochar's porous structure provides a large surface area that supports microbial colonization and enzyme stabilization. Wang et al. (2022) emphasized that peanut shell biochar with appropriate porosity and surface carbon promoted MBC and MBN by over 40%. The physical protection and microhabitats provided by the porous structure likely shield soil enzymes from denaturation and enhance their persistence and activity. Similarly, Choudhary et al. (2023) observed a 61% increase in SMBC with weed biochar, attributing the boost to improved microbial habitat availability.

4.4.5 Biochar Alkalinity and Soil pH Modulation

The pH-buffering capacity of biochar is especially critical in acidic soils, where microbial activity and enzymatic processes are often inhibited. Ali et al. (2022) reported that cassava straw biochar increased SMBC and SMBN due to its alkaline nature, which neutralized soil acidity and created a favorable environment for microbes. Similar trends were observed in saline-alkali soils (Wang et al. 2023), where high-temperature cotton straw biochar improved MBC and MBN, likely by adjusting pH and supplying essential nutrients.

4.4.6 Nutrient Content and Labile Carbon Supply

Biochar's contribution of nutrients and labile carbon strongly influences microbial metabolism. The poultry litter biochar applied in Alfisols (Aziz et al. 2024) led to a 21% increase in MBC, which was attributed to the supply of organic substrates that enhanced enzymatic activity. Similarly, cotton straw and peanut shell biochars supplied essential C and N sources, stimulating microbial biomass growth. These results support the hypothesis that biochar acts as a slow-release nutrient source, boosting microbial-driven processes and enzyme-mediated nutrient cycling.

4.4.7 Soil Type Compatibility and Background Fertility

Lastly, the interaction between biochar and specific soil types determines the extent of its influence on enzymatic activity. In highly weathered or degraded soils like Ultisols and Acrisols, biochar has shown pronounced benefits. In contrast, biochar application to already fertile or well-buffered soils (Hale et al. 2021) showed insignificant effects, likely due to the absence of nutrient or microbial limitations in the unamended plots.

Therefore, the initial soil condition plays a major role in determining the effectiveness of biochar in enhancing enzymatic activities (Table 4.2).

4.5 Effect of Biochar on Soil Enzymatic Activities in Relation to Different Enzymes and Factors Affecting It

Soil enzymes are critical mediators of biogeochemical cycles and serve as sensitive indicators of soil health. Numerous studies have demonstrated that biochar influences soil enzymatic activities, but the extent and direction of these effects vary with biochar type, enzyme group, soil characteristics, and environmental conditions. A global meta-analysis conducted by Liao et al. (2022) revealed that biochar significantly enhances the activity of key nutrient-acquisition enzymes, such as dehydrogenase (DHA), alkaline phosphatase (AlkP), protease (PRO), urease (URE), leucine aminopeptidase (LAP), and invertase (InV). In contrast, enzymes like β-cellobiohydrolase (CBH), xylanase (XYL), and N-acetylglucosaminidase (NAG) showed negligible responses, while acid phosphatase (AcP) activity decreased by approximately 10%. Importantly, the pyrolysis temperature of biochar emerged as a determining factor, with low-temperature biochars (< 500 °C) increasing the activities of N- and P-acquisition enzymes and DHA by 23%, 12%, and 22%, respectively, whereas high-temperature biochars had little to no effect.

4.5.1 *Dehydrogenase*

Dehydrogenase activity was the only enzyme that consistently responded favorably to the administration of charcoal. According to Garbuz et al. (2022), applying 1% w/w of biochar generated from willow chips at 350 °C enhanced DHA activity in both Sil-andic Andosol and Dystric Cambisol soils. Increased root biomass and improved microbial biomass carbon (MBC) were strongly linked to this rise. Similar findings were made by Amoakwah et al. (2022), who discovered that adding 30 Mg ha^{-1} of corn cob biochar considerably increased DHA activity in tropical soils. This was mostly because the soil had more organic matter and moisture. Liao et al. (2022) found that applying low-temperature biochar increased DHA by an average of 22% worldwide.

Table 4.2 Effects of biochar application on microbial biomass in different soil types across global locations

Location	Biochar feedstock	Pyrolysis temperature	Application rate	Soil type	Impact on microbial biomass	Possible mechanism	References
Huantai County, Shandong Province, China	Crushed corncob	360 °C	9.0 t ha^{-1} yr^{-1}	Sandy loam Fluvic Cambisol	At 9.0 t ha^{-1} yr^{-1}, MBC increased 118–763% at 0–5 and 5–10 cm versus control	–	Zhang et al. (2014)
Abeokuta, Ogun State, South western Nigeria	Rice husk	350 °C	7.5 t ha^{-1}	–	MBN, MBC, and MBP increased by 9.75%, 22.94%, and 12.01%, respectively, compared to control	Biochar amendment positively influenced nitrogen and phosphorus	Adebajo et al. (2022)
Riverside, California, United States	Yellow pine	350 °C	11.2 and 22.4 t ha^{-1}	Hanford fine sandy loam	When compared to unaltered plots, biochar showed no discernible impact on microbial biomass or nutrient availability	Inorganic N, P, K, and PLFA biomass were not increased by biochar derived from oak and pine	Hale et al. (2021)
Binzhou, Shandong province, China	Cotton straw	850 °C	5, 10, and 20 t ha^{-1} yr^{-1}	Saline-alkali soil	In the 0–20 cm layer, MBC increased 6.26–27.09%, and MBN rose 19.45–37.80%	Biochar increased soil nutrients (TOC, TN, NO^{3}–N, NH^{4+}-N), lowered pH, and boosted MBC and MBN by supplying microbes with ample C and N	Wang et al. (2023)

(continued)

Table 4.2 (continued)

Location	Biochar feedstock	Pyrolysis temperature	Application rate	Soil type	Impact on microbial biomass	Possible mechanism	References
Ghana, West Africa	Corn cobs	500–550 °C	15 and 30 t ha^{-1}	Haplic acrisols	MBC increased by 4.5 and 8.3 times at 15 and 30 t ha^{-1}, respectively, while MBN increased by 2.7 times at 30 t ha^{-1} compared to control	–	Amoakwah et al. (2022)
Xuchang City, Henan Province, China	Peanut shells	450 °C	3.75, 7.5, 11.25 and 15 kg ha^{-1}	Cinnamon	Biochar increased MBC by 38.69–40.90% and MBN by 32.14–42.66% compared to control	The carbon content and surface structure of biochar encourage the growth of soil microorganisms	Wang et al. (2022)
Jhansi, Bundelkhand region, Uttar Pradesh, India	*Parthenium hysterophorus* L. and *Lantana camara* L	400 °C	2.5, 5, and 10 t ha^{-1}	Typic haplusepts	Biochar treatment increased SMBC by ~ 61% over control	Biochar enhanced SMBC by providing more microbial habitats	Choudhary et al. (2023)

(continued)

Table 4.2 (continued)

Location	Biochar feedstock	Pyrolysis temperature	Application rate	Soil type	Impact on microbial biomass	Possible mechanism	References
Guangxi University, China	Cassava straw	300–500 °C	10, 20, and 30 t ha^{-1}	Ultisols	SMBC increased by an average of 126–134% in biochar-treated soil versus control	SMBN and SMBC increased likely due to biochar's alkalinity, which improves microbial activity and populations in acidic soils	Ali et al. (2022)
Koont, Rawalpindi district, Pothwar plateau, northern Punjab	Poultry litter	400 °C	10 and 20 t ha^{-1}	Udic Haplustoll Alfsols	Maximum MBC at 20 t ha^{-1} biochar was 21% higher than control	Biochar supplied organic substrates that enhanced enzymatic activity	Aziz et al. (2024)

4.5.2 Urease

The addition of biochar also significantly increased the activity of urease, an important enzyme in nitrogen cycle. In contrast to control and lime-treated soils, urease activity rose in Cambisols treated with biochar (Garbuz et al. 2022). However, this impact was not present in Andosols without effluent. Yao et al. (2021) showed that urease activity increased by 40.00–53.33% in the first year and by 47.06–58.82% in the second year across biochar treatments ranging from 33.75 to 101.25 t ha^{-1} in severely saline-sodic paddy soils. In the study by Amoakwah et al. (2022), urease activity showed a strong correlation ($r = 0.93$) with MBC, suggesting microbial biomass as a significant source of this enzyme. Zeng et al. (2024) explored the adsorption mechanisms of enzymes onto biochar and showed that urease, being the largest enzyme tested (640 kDa) with a high isoelectric point (~5.0), strongly adsorbed onto wheat straw biochar due to favorable electrostatic interactions.

4.5.3 Phosphatase (Acid and Alkaline)

Phosphatases responded variably depending on enzyme type and biochar characteristics. Garbuz et al. (2022) found an increase in alkaline phosphatase activity in Cambisols following biochar addition, attributed to increased P availability. Liao et al. (2022) similarly observed enhanced AlkP activity but noted a reduction in acid phosphatase under biochar treatment. Yao et al. (2021) recorded a 17.54–21.13% increase in AlkP activity with increasing rates of biochar in saline-sodic paddy fields. Wojewódzki et al. (2022) emphasized the role of biochar feedstock in determining enzyme response: mellow compost and oak leaf biochars increased AlP activity, while biochars derived from pine and sycamore sawdust decreased it. These variations were linked to changes in enzyme conformation upon adsorption to biochar surfaces with different pore structures and surface chemistries.

4.5.4 Cellulase and Sucrase

Enzymes involved in carbon cycling, such as cellulase and sucrase, also responded positively in several cases. Garbuz et al. (2022) found elevated cellulase activity in Cambisols treated with biochar, which was closely related to increased root biomass and necromass input. In another study, Yao et al. (2021) demonstrated that sucrase activity improved significantly with biochar addition and increased over time, showing the highest enhancement (110.34% over control) in the second year. These increases reflect improved substrate availability and enhanced microbial activity due to changes in soil structure and C input.

4.5.5 *Nitrate Reductase*

Nitrate reductase (NR) activity, which is sensitive to pH and nitrate availability, also increased with biochar treatment. Garbuz et al. (2022) reported that NR activity was highest in Andosols treated with lime but was also significantly higher in biochar-treated soils compared to the control. The biochar-induced pH increase and higher nitrate availability, particularly in soils under effluent irrigation or high-legume pastures, were likely drivers. Moreover, the high water retention and micropore structure of Andosols promoted anaerobic microsites, further enhancing nitrate-reducing conditions.

4.5.6 *Other Enzymes*

Other enzymes, such as β-glucosidase (BG), NAG, LAP, and acid phosphatase (AP), showed complex responses to biochar and compost. Zhao et al. (2022a, b) found that biochar alone often inhibited NAG and LAP but biochar-compost blends significantly boosted their activity (NAG: + 65–90%, LAP: + 12–29%). The AP activity was also most improved under biochar-compost treatments (170–296% increase), suggesting synergistic effects of combined amendments. In contrast, Baldoni et al. (2023) observed negligible changes in soil enzyme activity following the application of 60 Mg ha^{-1} of wood gasification biochar under Mediterranean conditions, highlighting the variability in biochar response across climates and soil types.

Enzyme adsorption and activity are influenced not only by biochar's chemical properties but also by its physical characteristics. Zeng et al. (2024) showed that enzyme adsorption was governed by the size of the enzyme, surface area of biochar, and electrostatic interactions between enzyme and biochar surfaces. Smaller enzymes like α-amylase accessed more surface area and showed stronger adsorption to high-SSA biochars, while larger enzymes like urease were limited by pore size restrictions. The chemical oxidation state of biochar also modulated enzyme binding affinity; more oxidized (negatively charged) surfaces led to weaker enzyme adsorption due to electrostatic repulsion.

4.6 Biochar Properties Influencing Soil Enzymatic Activities

The influence of biochar on soil enzymatic activities is governed by a complex interplay of its physicochemical properties, which include feedstock type, pyrolysis temperature, surface area and porosity, functional group chemistry, and the content of labile organic matter. These properties affect not only the availability of substrates for microbial activity but also the interaction between biochar and extracellular enzymes, which in turn modulate soil biochemical processes.

4.6.1 Feedstock and Pyrolysis Temperature

Biochar feedstock origin and pyrolysis temperature significantly determine its nutrient content, surface chemistry, and structural attributes. Liao et al. (2022) demonstrated that low-temperature biochars (< 500 °C) enhance the activity of nutrient-acquisition enzymes (dehydrogenase, alkaline phosphatase, urease) more than high-temperature biochars. This is attributed to the higher content of labile carbon and oxygen-containing functional groups in low-temperature biochars, which provide better substrates for microbial metabolism. In contrast, high-temperature biochars tend to be more aromatic and inert, resulting in negligible stimulation of enzyme activities. Similarly, biochars derived from different feedstocks exhibit distinct influences on enzymatic dynamics. Wojewódzki et al. (2022) reported that biochars from mellow compost and oak leaves increased alkaline phosphatase activity, while those from pine and sycamore sawdust reduced it, indicating that the biochemical nature of the original biomass affects the biochar's interaction with soil enzymes.

4.6.2 Surface Area, Porosity, and Pore Size Distribution

The porosity and specific surface area (SSA) of biochar are critical for enzyme adsorption and microbial colonization. Zeng et al. (2024) showed that smaller enzymes like α-amylase were strongly adsorbed onto high-SSA biochars, especially those with accessible mesopores, whereas larger enzymes like urease exhibited limited adsorption due to pore size constraints. They further found that N2-SSA was a better predictor of enzyme adsorption than CO_2-SSA, as micropores (< 1.5 nm) captured by CO_2 adsorption are generally too small for enzyme entry. This implies that the effectiveness of enzyme–biochar interactions is size-dependent and varies with the accessibility of biochar pores to specific enzymes.

4.6.3 Surface Chemistry and Charge Characteristics

Electrostatic interactions between enzyme and biochar surfaces are key drivers of enzyme adsorption. Zeng et al. (2024) reported that enzymes with higher isoelectric points, such as urease (IEP ~ 5.0), were more strongly adsorbed onto negatively charged biochar surfaces due to favorable electrostatic attraction. The presence of oxygen-containing functional groups (e.g., carboxyl and phenolic groups) imparts negative charges to biochar, particularly after oxidation, which may lead to electrostatic repulsion for enzymes with lower IEPs. Thus, the surface functionalization and oxidation state of biochar regulate the extent and strength of enzyme binding, which can enhance or inhibit enzymatic activity depending on the enzyme's charge properties.

4.6.4 Labile Carbon Content and Biochar-C/N Ratio

Biochars with higher labile carbon fractions and favorable C/N ratios can stimulate microbial growth and enzyme activity. Garbuz et al. (2022) observed that willow-chip-derived biochar at 350 °C increased cellulase and dehydrogenase activity in Dystric Cambisols, likely due to the presence of labile carbon that supported microbial proliferation and positive priming effects. Their biochar also had a high Cox/Corg and H/Corg ratio, suggesting a considerable amount of labile C, which enhanced C mineralization and soil MBC. Similarly, Amoakwah et al. (2022) linked the increase in urease and DHA activity to improved soil moisture and SOM content, suggesting that biochar not only serves as a habitat but also as an energy source for microbial metabolism.

4.6.5 Biochar Interaction with Soil Type and Moisture Regimes

Soil type modulates the effect of biochar on enzyme activity. For instance, Garbuz et al. (2022) showed that biochar application enhanced urease activity in Cambisols but had no effect in Andosols unless effluents were present. Andosols, with high microporosity and water retention, create anaerobic microsites conducive for nitrate-reductase activity under udic conditions. This highlights that biochar effects are not universal but interact strongly with soil texture, mineralogy, and hydrological behavior.

4.6.6 Enzyme Conformation and Activity Modulation via Adsorption

Biochar can alter enzyme conformation upon adsorption, thereby affecting their catalytic efficiency. Wojewódzki et al. (2022) emphasized that biochar surface interaction may rotate enzyme active sites or alter their configuration, leading to inhibition or enhancement. The reduction in acid phosphatase activity observed with certain biochars was attributed to such conformational changes. Zhao et al. (2022a, b) further confirmed this by showing that biochar alone inhibited β-glucosidase, NAG, and LAP, whereas compost or biochar-compost treatments alleviated these effects, possibly by modifying the adsorption environment or providing additional substrates.

4.6.7 Biochar-Mediated Shifts in Microbial Community Composition

Biochar influences soil enzyme activity not only through direct mechanisms, such as providing protective surfaces for enzyme adsorption or altering pH and nutrient availability, but also indirectly through reshaping the soil microbial community structure. Garbuz et al. (2022) highlighted this indirect pathway by demonstrating significant shifts in the fungi-to-bacteria ratio and changes in the relative abundance of specific microbial taxa following biochar amendment. These shifts have profound implications for the soil biochemical environment, as fungi and bacteria differ in their ecological strategies, substrate preferences, and enzyme production capacities. Fungal-dominated communities, for instance, are often associated with greater production of ligninolytic and cellulolytic enzymes, which are crucial for the degradation of complex organic matter such as cellulose and lignin. Bacteria, on the other hand, typically respond rapidly to labile carbon inputs and contribute enzymes like urease, phosphatase, and β-glucosidase, which are involved in nutrient mineralization and short-term cycling. Thus, by altering the relative abundance of these groups, biochar indirectly influences the quantity, type, and efficiency of extracellular enzymes released into the soil.

The study also found that biochar-induced increases in root biomass and root-derived carbon inputs—a result of improved soil structure, nutrient retention, and moisture availability—further stimulated microbial activity and enzymatic function. In particular, elevated levels of cellulase activity in biochar-treated soils were closely linked to increased root exudation and turnover. This reinforces the tripartite interaction among plants, microbes, and biochar, wherein enhanced plant growth leads to greater carbon flow into the rhizosphere, stimulating microbial growth and enzyme production. Such positive feedback loops are essential for maintaining a dynamic and efficient soil organic matter turnover system. Moreover, biochar can influence the spatial organization and microhabitat heterogeneity of soil, creating protected niches where microbial consortia can thrive. These niches may harbor specialized enzyme-producing microbes that are otherwise suppressed in less structured soils. Biochar's surface may also facilitate enzyme–substrate proximity, improving catalytic efficiency by retaining both enzymes and their target organic compounds within close spatial range.

4.6.8 *Salinity and Soil Stress Alleviation by Biochar*

Biochar has shown encouraging promise in improving soil biochemical activity and resilience in stress-prone soils, especially saline-sodic paddy fields. A two-year field study by Yao et al. (2021) examined how increasing rates of biochar application altered the activity of soil enzymes in paddy soils affected by salinity. According to their findings, biochar dramatically increased the activity of several important enzymes in a dose-dependent and time-progressive manner, including catalase, urease, sucrase, and alkaline phosphatase. These enzymes are very useful for evaluating soil functional restoration because they are important markers of microbial oxidative stress response (catalase), nitrogen mineralization (urease), carbon cycling (sucrase), and phosphorus availability (alkaline phosphatase).

The application of biochar resulted in several simultaneous improvements in soil conditions, which were responsible for the observed increase in enzymatic activity. In order to produce a more conducive microenvironment for microbial communities and enzyme stabilization, biochar first increased the aggregate stability, porosity, and water retention of the saline-sodic soils. Second, by lowering electrical conductivity (EC), raising the availability of calcium and other exchangeable cations that aid in the removal of salt from soil exchange sites, and boosting the sodium adsorption ratio (SAR), biochar helped to alleviate the effects of salinity stress. These changes lowered the osmotic and ionic stress on microbes, enabling them to maintain metabolic activity under otherwise inhibitory conditions. Additionally, biochar's surface functional groups and porous architecture may have played a direct role in adsorbing enzyme inhibitors and protecting extracellular enzymes from denaturation or degradation, thus prolonging their functional lifespan in the soil matrix. The cumulative nature of the enzyme activity improvements observed over the two-year period also suggests a progressive microbial adaptation and possible development of a more functionally diverse and stable soil microbial community.

4.6.9 *Inconsistencies and Context-Specific Outcomes*

Despite the generally positive trend observed across many studies, there remains a subset of research reporting negligible, inconsistent, or even negative effects of biochar on soil enzymatic activity. Baldoni et al. (2023) found no significant enhancement in key enzyme activities following the application of wood gasification biochar at a high rate (60 Mg ha^{-1}) under Mediterranean field conditions. This outcome contrasts with many studies showing biochar-induced stimulation of enzyme-mediated nutrient cycling, highlighting the context-dependent nature of biochar efficacy. Several factors may explain such limited responses. First, gasification biochar is typically produced at very high temperatures (>700 °C), resulting in a highly recalcitrant, carbonized structure with low levels of labile organic matter and minimal surface functional groups. This reduces its potential to serve as a substrate or habitat for soil microbes and enzymes. Unlike lower-temperature pyrolysis biochars that retain more volatile and bioavailable organic compounds, gasified biochars may be less biologically active and therefore have limited ability to stimulate microbial metabolism or enzyme production.

Second, the prevailing soil conditions can influence how biochar interacts with native soil enzymes. In Mediterranean soils, which often face seasonal extremes in temperature and moisture, environmental stresses such as repeated drying-rewetting cycles may further complicate biochar-soil interactions. These cycles can cause microbial stress, enzyme denaturation, or fluctuating substrate availability, potentially overriding any beneficial effects introduced by biochar application. Moreover, the timing of sampling post-biochar application is critical. Some effects on enzyme activities may take time to manifest or may be transient, peaking shortly after application and declining thereafter. A lack of temporal resolution in field studies may therefore miss the windows where biochar effects are most pronounced. These findings emphasize the need to tailor biochar applications to specific site conditions, target outcomes, and biochar characteristics. Simply applying biochar, especially in high doses or of a single type, without regard to its compatibility with the local soil and climate may lead to ineffective or even counterproductive results. It also underscores the importance of biochar pre-assessment, including characterization of its nutrient content, surface chemistry, and reactivity, as well as soil diagnostics to evaluate site-specific constraints or opportunities (Table 4.3).

4.7 Limitations and Future Scope

Despite the increasing evidence supporting biochar's benefits for soil microbial diversity, biomass, and enzymatic activity, several limitations hinder its broad application. One major challenge is the variability in biochar properties, which depend heavily on feedstock types, pyrolysis temperatures, and production methods. This variability often results in inconsistent effects on soil microbiomes and biochemical processes, making it difficult to develop universally applicable guidelines for biochar use across different soils and ecosystems (Brtnicky et al. 2021). Moreover, the impact of biochar is highly context-dependent. Soil characteristics such as texture, pH, moisture regime, and native microbial communities significantly influence the outcomes of biochar amendments (Xiang et al. 2023). Environmental stressors like salinity, heavy metal contamination, freeze–thaw cycles, and microplastic pollution further complicate biochar's interactions with soil microbes and enzymes. Consequently, biochar effects observed in one ecosystem may not be directly transferable to others, necessitating site-specific investigations (Vijay et al. 2021).

Another limitation is the scarcity of long-term field studies. Most research has been conducted over short to medium durations and often under controlled conditions, which may not fully capture biochar's ageing, persistence, and cumulative effects on microbial communities and enzymatic functions (Joseph et al. 2021). Additionally, the intricate interactions among biochar, plants, and soil microbiomes remain

Table 4.3 Global impacts of biochar application on soil enzyme activities: effects of feedstock, pyrolysis conditions, and soil type

Location	Biochar feedstock	Pyrolysis temperature	Application rate	Soil	Impact on enzyme activity	Possible mechanism	References
Karaj, Alborz province, Iran	Raw maize stalks	200, 400, and 600 °C	5 and 10 g kg−1	Typic Haplocalcid	Biochar-amended soils had FDA values that were considerably higher than control at 5 g kg^{-1} (28–207%) and 10 g kg^{-1} (44–280%) rates	Biochar boosted microbial activity in soils low in organic matter and substrates	Khadem et al. (2021)
Xiangfang district, Harbin city, Heilongjia ng Province, China	Soybean straw	500 °C	20 t ha^{-1}	Degraded black soil	Biochar treatment increased soil sucrase activity by 2.5–10.3% over control	Soil sucrase hydrolyzed sucrose, enhancing soil nutrient availability	Zhang et al. (2023)
Wuhan, Hubei Province, China	Rice straw	500 °C	2% (w/w)	Saline-alkali soil	Biochar treatment increased β-1,4-glucosidase, β-D-cellobiosidase, and β-xylosidase by 18.49%, 15.00%, and 18.04%, respectively, over control	Biochar's high porosity, surface area, and adsorption enhanced salt leaching and Na + adsorption	Wang et al. (2024)
Ghana, West Africa	Corn cobs	500–550 °C	15 and 30 t ha^{-1}	Haplic Acrisols	Biochar at 30 t ha^{-1} increased urease activity 1.5-fold; dehydrogenase activity rose 1.5- and 3.2-fold at 15 and 30 t ha^{-1}, respectively, versus control	Biochar enhanced N-related enzyme activity and significantly increased dehydrogenase activity, likely due to higher soil water and SOM content	Amoakwah et al. (2022)

(continued)

Table 4.3 (continued)

Location	Biochar feedstock	Pyrolysis temperature	Application rate	Soil	Impact on enzyme activity	Possible mechanism	References
Xuchang City, Henan Province, China	Peanut shells	450 °C	3.75, 7.5, 11.25 and 15 kg ha^{-1}	Cinnamon	The application of biochar exhibited the highest urease activity, which was 18.06% higher than the control. Invertase activity increased by 23.40%	Biochar raised soil organic and active carbon, key drivers of enzyme activity	Wang et al. (2022)
Jiangnan Town, Zhejiang Province, China	Rice straw	350, 500, and 650 °C	1% (w/w)	–	Catalase activity declined with rising pyrolysis temperature	High-temperature biochar has greater surface area and adsorption, causing non-selective enzyme and substrate adsorption that reduces enzyme activity	Zhao et al. (2023)
Jhansi, Bundelkhand region, Uttar Pradesh, India	*Parthenium hysterophorus* L. and *Lantana camara* L.	400 °C	2.5, 5, and 10 t ha^{-1}	Typic haplusepts	Biochar application at 2.5–10 t ha^{-1} with 75% RDF reduced acid phosphatase activity by 21.80–35.64%	Biochar application increased alkaline phosphatase activity by promoting P uptake, root growth into biochar pores, enzyme synthesis, and soil pH changes	Choudhary et al. (2023)
Hangzhou City, China	Swine manure	350 and 500 °C	0.5, 1, and 2% (w/w)	Laterite	Sucrase activity was highest at 2.0% biochar compared to 0.5%, 1%, and control	Biochar raised soil carbon, improving the C/N ratio and enhancing microbial and enzyme activities	Jiang et al. (2021)

(continued)

Table 4.3 (continued)

Location	Biochar feedstock	Pyrolysis temperature	Application rate	Soil	Impact on enzyme activity	Possible mechanism	References
Jincheng Township, Lin'an, Hangzhou City, Zhejiang Province, China	Corn straw	450–500 °C	8.49 t ha^{-1}	Ferrasol	In particular, the activities of β-glucosidase (BG) were significantly reduced by 49%, β-D-cellobiosidase (CB) by 60%, β-xylosidase (XYL) by 25%, phenol oxidase (PHOX) by 40%, peroxidase (PERO) by 31%, and GMEA by 36% when biochar and manure were applied together	Biochar reduced enzyme and microbial activities but kept microbial biomass steady, indicating higher microbial metabolic efficiency	Wu et al. (2021)
Shaoguan City, Guangdon g Province, China	Rice straw	700 °C	5% (w/w)	–	Biochar addition enhanced sucrase activity compared to control	BC increased soil alkalinity, inhibiting sucrase activity, which peaks in acidic conditions	Xu et al. (2022)
Zhejiang province, Eastern China	Rice straw and canola stalk	350 and 550 °C	1% (w/w)	Ultisol	Straw and biochar significantly increased ACP and ALP, respectively; biochar led to lower ACP but higher ALP than straw	Straw boosted soil P by enhancing microbial activity, while biochar's porous structure provides microbes with water and nutrient-rich habitat	Yang and Lu (2022)
Koont, Rawalpind i district, Pothwar plateau, northern Punjab	Poultry litter	400 °C	10 and 20 t ha^{-1}	Udic Haplustoll Alfsols	In comparison to control, 20 t/ha of biochar enhanced the activities of phosphatase, protease, and β-glucosidase by 23%, 69%, and 86%, respectively	Biochar supplied organic substrates that enhanced enzymatic activity	Aziz et al. (2024)

insufficiently understood, limiting our knowledge of how biochar influences plant growth through microbially mediated nutrient cycling. Furthermore, while biochar can adsorb extracellular enzymes, this interaction may alter enzyme conformation and activity, sometimes leading to unexpected inhibition rather than stimulation of enzyme functions. The molecular mechanisms underlying these effects are still poorly elucidated and represent an important area for future research (Zhou et al. 2021).

To address these challenges, future studies should focus on standardizing biochar production and characterization to enable comparability and reproducibility. Long-term, ecosystem-specific field trials across diverse soils and climatic conditions are essential to better understand biochar's real-world impacts, especially under environmental stresses. Deeper mechanistic investigations into biochar's effects on microbial functional genes, enzyme adsorption, and conformational changes will clarify how biochar influences soil biochemical processes at the molecular level. Moreover, research on the interactions among plants, microbes, and biochar can provide valuable insights into enhancing nutrient use efficiency and plant productivity. Finally, integrating biochar with organic amendments and microbial inoculants, and assessing the environmental and economic feasibility of biochar use at scale will be crucial for its widespread adoption. By addressing these gaps, we can optimize biochar applications tailored to specific agroecosystems, thereby promoting soil health, ecosystem resilience, and sustainable food production.

4.8 Conclusion

Biochar has become a very versatile and multipurpose soil supplement. A more varied, active, and robust soil microbiome is promoted by its porosity structure and surface chemistry, which provide ideal microhabitats for microbial colonization and biofilm development. The application of biochar has been shown in numerous studies conducted in a variety of agroecosystems and soil types to considerably enhance soil microbial diversity, MBC, MBN, and soil enzymatic activities, including those involved in the cycling of carbon, nitrogen, and phosphorus. However, a complex web of interrelated processes shapes the overall effect of biochar on soil microbial populations and their functional capacity. These include the kind of feedstock (woody, herbaceous, or manure-based), the temperature and length of the pyrolysis, the size of the particles, the rate of application, and the natural characteristics of the soil, such as its texture, pH, and organic matter content.

This variability highlights the need for site-specific and goal-oriented biochar management strategies. For instance, biochar designed for acidic, nutrient-depleted soils may differ significantly in composition and function from that intended for use in saline or contaminated soils. The development of customized biochar formulations, potentially combined with other organic or microbial amendments (e.g., compost, biofertilizers), could enhance functional outcomes and minimize unintended consequences. Despite promising short-term findings, significant knowledge gaps remain.

The long-term ecological impacts of repeated or high-dose biochar application, the aging and weathering processes of biochar in situ, and the tripartite interactions among biochar, soil microbes, and plant roots are still poorly understood. Additionally, inconsistencies in biochar characterization and reporting standards make it difficult to compare results across studies and to develop universal guidelines for its use.

Moving forward, future research must prioritize mechanistic and molecular-level investigations to unravel the causal pathways by which biochar influences microbial functions. The standardization of biochar characterization protocols, coupled with long-term, multi-site field trials under realistic management scenarios, is crucial to translate experimental successes into scalable agricultural practices. Integrating biochar research with systems approaches, such as meta-omics, isotope tracing, and ecological modeling, will help bridge knowledge gaps and enhance our predictive capabilities. In conclusion, the strategic and science-based application of biochar represents a powerful tool in climate-smart and regenerative agriculture. By promoting microbial-mediated processes critical for soil fertility, carbon sequestration, pollutant degradation, and overall ecosystem resilience, biochar can play a vital role in addressing global challenges. Its success, however, will depend on tailored application strategies grounded in a deep understanding of soil–microbe–biochar–environment interactions.

References

Adebajo SO, Oluwatobi F, Akintokun PO, Ojo AE, Akintokun AK, Gbodope IS (2022) Impacts of rice-husk biochar on soil microbial biomass and agronomic performances of tomato (*Solanum lycopersicum* L.). Sci. Rep. 12(1):1787

Ahmad A, Chowdhary P, Khan N, Chaurasia D, Varjani S, Pandey A, Chaturvedi P (2022) Effect of sewage sludge biochar on the soil nutrient, microbial abundance, and plant biomass: a sustainable approach towards mitigation of solid waste. Chemosphere 287:132112

Ali A, Jabeen N, Chachar Z, Chachar S, Ahmed S, Ahmed N, Laghari AA, Sahito ZA, Farruhbek R, Yang Z (2025) The role of biochar in enhancing soil health and interactions with rhizosphere properties and enzyme activities in organic fertilizer substitution. Front Plant Sci 16:1595208

Ali I, Yuan P, Ullah S, Iqbal A, Zhao Q, Liang H, Khan A, Imran, Zhang H, Wu X, Wei S (2022) Biochar amendment and nitrogen fertilizer contribute to the changes in soil properties and microbial communities in a paddy field. Front Microbiol 13:834751

Amoakwah E, Arthur E, Frimpong KA, Lorenz N, Rahman MA, Nziguheba G, Islam KR (2022) Biochar amendment impacts on microbial community structures and biological and enzyme activities in a weathered tropical sandy loam. Appl Soil Ecol 172:104364

An X, Liu Q, Pan F, Yao Y, Luo X, Chen C, Liu T, Zou L, Wang W, Wang J, Liu X (2023) Research advances in the impacts of biochar on the physicochemical properties and microbial communities of saline soils. Sustainability 15(19):14439

Anyebe O, Sadiq FK, Manono BO, Matsika TA (2025) Biochar characteristics and application: effects on soil ecosystem services and nutrient dynamics for enhanced crop yields. Nitrogen 6(2):31

Aziz MA, Khan KS, Khalid R, Shabaan M, Alghamdi AG, Alasmary Z, Majrashi MA (2024) Integrated application of biochar and chemical fertilizers improves wheat (*Triticum aestivum*) productivity by enhancing soil microbial activities. Plant Soil 502(1):433–448

Bai J, Huang Y, Bai Y, Chen D, Haider S, Song J, Moreira BRDA, Ren G, Yang G, Feng Y, Wang X (2024) Impact of straw-biochar amendments on microbial activity and soil carbon dynamics in wheat-maize system. Soil Tillage Res 244:106284
Baldoni N, Francioni M, Trozzo L, Toderi M, Fornasier F, D'Ottavio P, Corti G, Cocco S (2023) Effect of wood gasification biochar on soil physicochemical properties and enzyme activities, and on crop yield in a wheat-production system with sub-alkaline soil. Biomass Bioenerg 176:106914
Bogati KA, Sewerniak P, Walczak M (2025) Unraveling the effect of soil moisture on microbial diversity and enzymatic activity in agricultural soils. Microorganisms 13(6):1245
Brtnicky M, Datta R, Holatko J, Bielska L, Gusiatin ZM, Kucerik J, Hammerschmiedt T, Danish S, Radziemska M, Mravcova L, Fahad S (2021) A critical review of the possible adverse effects of biochar in the soil environment. Sci Total Environ 796:148756
Choudhary P, Prasad M, Choudhary M, Kumar A, Kumar S, Srinivasan R, Mahawer SK (2023) Exploring invasive weed biochar as soil amendment: a study on fodder oats productivity and soil biological properties. Environ Res 216:114527
Dai Z, Xiong X, Zhu H, Xu H, Leng P, Li J, Tang C, Xu J (2021) Association of biochar properties with changes in soil bacterial, fungal and fauna communities and nutrient cycling processes. Biochar 3:239–254
Dissanayake PD, Palansooriya KN, Sang MK, Oh DX, Park J, Hwang SY, Igalavithana AD, Gu C, Ok YS (2022) Combined effect of biochar and soil moisture on soil chemical properties and microbial community composition in microplastic-contaminated agricultural soil. Soil Use Manag 38(3):1446–1458
Egamberdieva D, Alaylar B, Kistaubayeva A, Wirth S, Bellingrath-Kimura SD (2022) Biochar for improving soil biological properties and mitigating salt stress in plants on salt-affected soils. Commun Soil Sci Plant Anal 53(2):140–152
Fu Y, Luo Y, Auwal M, Singh BP, Van Zwieten L, Xu J (2022) Biochar accelerates soil organic carbon mineralization via rhizodeposit-activated Actinobacteria. Biol Fertile Soils 58(5):565–577
Garbuz S, Mackay A, Camps-Arbestain M, DeVantier B, Minor M (2022) Biochar increases soil enzyme activities in two contrasting pastoral soils under different grazing management. Crop Pasture Sci
Hale L, Curtis D, Azeem M, Montgomery J, Crowley DE, McGiffen ME Jr (2021) Influence of compost and biochar on soil biological properties under turfgrass supplied deficit irrigation. Appl Soil Ecol 168:104134
Han L, Lu C, Chen L, Wang F, Chen QA, Gao K, Yu Y, Xu C (2024) Carbon sequestration potential of biochar in soil from the perspective of organic carbon structural modification. Appl Soil Ecol 198:105389
He X, Xie H, Gao D, Khashi U, Rahman M, Zhou X, Wu F (2021) Biochar and intercropping with potato–onion enhanced the growth and yield advantages of tomato by regulating the soil properties, nutrient uptake, and soil microbial community. Front Microbiol 12:695447
Hu W, Zhang Y, Xiangmin R, Fei J, Peng J, Luo G (2023) Coupling amendment of biochar and organic fertilizers increases maize yield and phosphorus uptake by regulating soil phosphatase activity and phosphorus-acquiring microbiota. Agric Ecosyst Environ 355:108582
Huang M, Hu T, Wang J, Ding Y, Köster K, Sun L (2024) Effects of biochar on soil carbon pool stability in the Dahurian larch (*Larix gmelinii*) forest are regulated by the dominant soil microbial ecological strategy. Sci Total Environ 175725
Jiang Z, Yang S, Pang Q, Xu Y, Chen X, Sun X, Qi S, Yu W (2021) Biochar improved soil health and mitigated greenhouse gas emission from controlled irrigation paddy field: insights into microbial diversity. J Clean Prod 318:128595
Joseph S, Cowie AL, Van Zwieten L, Bolan N, Budai A, Buss W, Cayuela ML, Graber ER, Ippolito JA, Kuzyakov Y, Luo Y (2021) How biochar works, and when it doesn't: a review of mechanisms controlling soil and plant responses to biochar. Gcb Bioenergy 13(11):1731–1764
Khadem A, Raiesi F, Besharati H, Khalaj MA (2021) The effects of biochar on soil nutrients status, microbial activity and carbon sequestration potential in two calcareous soils. Biochar 3(1):105–116

Li L, Long A, Fossum B, Kaiser M (2023) Effects of pyrolysis temperature and feedstock type on biochar characteristics pertinent to soil carbon and soil health: a meta-analysis. Soil Use Manag 39(1):43–52

Liao X, Kang H, Haidar G, Wang W, Malghani S (2022) The impact of biochar on the activities of soil nutrients acquisition enzymes is potentially controlled by the pyrolysis temperature: a meta-analysis. Geoderma 411:115692

Ling L, Luo Y, Jiang B, Lv J, Meng C, Liao Y, Reid BJ, Ding F, Lu Z, Kuzyakov Y, Xu J (2022) Biochar induces mineralization of soil recalcitrant components by activation of biochar responsive bacteria groups. Soil Biol Biochem 172:108778

Liu J, Pan Z, Sun K, Chen Y, Yang Y, Gao B, Xing B (2022) The preferential preservation of both different minerals and polyethylene microplastics on aromatic or aliphatic carbon fractions within low or high pyrolysis temperature biochar under mineralization. Renew Sustain Energy Rev 170:112963

Mei C, Wang H, Cai K, Xiao R, Xu M, Li Z, Zhang Z, Cui J, Huang F (2022) Characterization of soil microbial community activity and structure for reducing available Cd by rice straw biochar and Bacillus cereus RC-1. Sci Total Environ 839:156202

Mukherjee A, Patra BR, Podder J, Dalai AK (2022) Synthesis of biochar from lignocellulosic biomass for diverse industrial applications and energy harvesting: effects of pyrolysis conditions on the physicochemical properties of biochar. Front Mater 9:870184

Sani MNH, Amin M, Siddique AB, Nasif SO, Ghaley BB, Ge L, Wang F, Yong JWH (2023) Waste-derived nanobiochar: a new avenue towards sustainable agriculture, environment, and circular bioeconomy. Sci Total Environ 905:166881

Shi G, Li H, Fu Q, Li T, Hou R, Chen Q, Xue P (2024) Effects of biochar and compost on the abundant and rare microbial communities assembly and multifunctionality in pesticide-contaminated soil under freeze-thaw cycles. Environ Pollut 362:125003

Sun Y, Lyu H, Cheng Z, Wang Y, Tang J (2022) Insight into the mechanisms of ball-milled biochar addition on soil tetracycline degradation enhancement: physicochemical properties and microbial community structure. Chemosphere 291:132691

Tian L, Wang Y, Jin D, Zhou Y, Mukhamed B, Liu D, Feng B (2025) The application of biochar and organic fertilizer substitution regulates the diversities of habitat specialist bacterial communities within soil aggregates in proso millet farmland. Biochar 7(1):6

Vijay V, Shreedhar S, Adlak K, Payyanad S, Sreedharan V, Gopi G, Sophia van der Voort T, Malarvizhi P, Yi S, Gebert J, Aravind PV (2021) Review of large-scale biochar field-trials for soil amendment and the observed influences on crop yield variations. Front Energy Res 9:710766

Wang C, Chen D, Shen J, Yuan Q, Fan F, Wei W, Li Y, Wu J (2021) Biochar alters soil microbial communities and potential functions 3–4 years after amendment in a double rice cropping system. Agric Ecosyst Environ 311:107291

Wang H, Zhang R, Zhao Y, Shi H, Liu G (2022) Effect of biochar on rhizosphere soil microbial diversity and metabolism in tobacco-growing soil. Ecologies 3(4):539–556

Wang S, Gao P, Zhang Q, Shi Y, Guo X, Lv Q, Wu W, Zhang X, Li M, Meng Q (2023) Biochar improves soil quality and wheat yield in saline-alkali soils beyond organic fertilizer in a 3-year field trial. Environ Sci Pollut Res 30(7):19097–19110

Wang X, Riaz M, Babar S, Eldesouki Z, Liu B, Xia H, Li Y, Wang J, Xia X, Jiang C (2024) Alterations in the composition and metabolite profiles of the saline-alkali soil microbial community through biochar application. J Environ Manage 352:120033

Wang L, Zhang W, Yao J, Qi Z, Liu Y, Li Z, Qu J, Ma Y, Zhang Y (2025) Effect of freeze-thaw cycle aging on biochar application in DEHP contaminated Mollisols: insights from soil properties and microbial community. Ecotoxicol Environ Saf 294:118082

Wojewódzki P, Lemanowicz J, Debska B, Haddad SA (2022) Soil enzyme activity response under the amendment of different types of biochar. Agronomy 12(3):569

Wolna-Maruwka A, Piechota T, Niewiadomska A, Kamiński A, Kayzer D, Grzyb A, Pilarska AA (2021) The effect of biochar-based organic amendments on the structure of soil bacterial community and yield of maize (*Zea mays* L.). Agronomy 11(7):1286

Wu Q, Lian R, Bai M, Bao J, Liu Y, Li S, Liang C, Qin H, Chen J, Xu Q (2021) Biochar co-application mitigated the stimulation of organic amendments on soil respiration by decreasing microbial activities in an infertile soil. Biol Fertil Soils 57(6):793–807

Wu J, Fu X, Zhao L, Lv J, Lv S, Shang J, Lv J, Du S, Guo H, Ma F (2024) Biochar as a partner of plants and beneficial microorganisms to assist in-situ bioremediation of heavy metal contaminated soil. Sci Total Environ 171442

Xiang Y, Liu Y, Niazi NK, Bolan N, Zhao L, Zhang S, Xue J, Yao B, Li Y (2023) Biochar addition increased soil bacterial diversity and richness: large-scale evidence of field experiments. Sci Total Environ 893:164961

Xu M, Dai W, Zhao Z, Zheng J, Huang F, Mei C, Huang S, Liu C, Wang P, Xiao R (2022) Effect of rice straw biochar on three different levels of Cd-contaminated soils: Cd availability, soil properties, and microbial communities. Chemosphere 301:134551

Yan T, Xue J, Zhou Z, Wu Y (2021) Biochar-based fertilizer amendments improve the soil microbial community structure in a karst mountainous area. Sci Total Environ 794:148757

Yang C, Lu S (2022) Straw and straw biochar differently affect phosphorus availability, enzyme activity and microbial functional genes in an Ultisol. Sci Total Environ 805:150325

Yang C, Liu J, Ying H, Lu S (2022) Soil pore structure changes induced by biochar affect microbial diversity and community structure in an Ultisol. Soil Tillage Res 224:105505

Yang X, Hou R, Fu Q, Li T, Wang J, Su Z, Shen W, Zhou W, Wang Y (2023) Effect of freeze-thaw cycles and biochar coupling on the soil water-soil environment, nitrogen adsorption and N_2O emissions in seasonally frozen regions. Sci Total Environ 893:164845

Yang W, Zhang L, Wang Z, Zhang J, Li P, Su L (2025) Effects of biochar and nitrogen fertilizer on microbial communities, CO_2 emissions, and organic carbon content in soil. Sci Rep 15(1):9789

Yao T, Zhang W, Gulaqa A, Cui Y, Zhou Y, Weng W, Wang X, Liu Q, Jin F (2021) Effects of peanut shell biochar on soil nutrients, soil enzyme activity, and rice yield in heavily saline-sodic paddy field. J Soil Sci Plant Nutr 21:655–664

Zeng L, Zimmerman AR, Huang R (2024) Adsorption of extracellular enzymes by biochar: impacts of enzyme and biochar properties. Geoderma 451:117082

Zhang J, Shen JL (2022) Effects of biochar on soil microbial diversity and community structure in clay soil. Ann Microbiol 72(1):35

Zhang QZ, Dijkstra FA, Liu XR, Wang YD, Huang J, Lu N (2014) Effects of biochar on soil microbial biomass after four years of consecutive application in the north China plain. PLoS ONE 9(7):e102062

Zhang C, Zhao X, Liang A, Li Y, Song Q, Li X, Li D, Hou N (2023) Insight into the soil aggregate-mediated restoration mechanism of degraded black soil via biochar addition: emphasizing the driving role of core microbial communities and nutrient cycling. Environ Res 228:115895

Zhao K, Wang N, Jiang S, Li F, Luo S, Chen A, Li H, Lin X, Zhang J, Zhang L, Huang H (2022a) Potential implications of biochar and compost on the stoichiometry-based assessments of soil enzyme activity in heavy metal-polluted soils. Carbon Research 1(1):29

Zhao Y, Li X, Li Y, Bao H, Xing J, Zhu Y, Nan J, Xu G (2022b) Biochar acts as an emerging soil amendment and its potential ecological risks: a review. Energies 16(1):410

Zhao Y, Lu Y, Zhuang H, Shan S (2023) In-situ retention of nitrogen, phosphorus in agricultural drainage and soil nutrients by biochar at different temperatures and the effects on soil microbial response. Sci Total Environ 904:166292

Zheng H, Liu D, Liao X, Miao Y, Li Y, Li J, Yuan J, Chen Z, Ding W (2022) Field-aged biochar enhances soil organic carbon by increasing recalcitrant organic carbon fractions and making microbial communities more conducive to carbon sequestration. Agric Ecosyst Environ 340:108177

Zhou X, Zhu Y, Niu Q, Zeng G, Lai C, Liu S, Huang D, Qin L, Liu X, Li B, Yi H (2021) New notion of biochar: a review on the mechanism of biochar applications in advanced oxidation processes. Chem Eng J 416:129027

Chapter 5
Biochar for the Remediation of Arsenic-Contaminated Agroecosystems: Mechanisms, Applications, and Future Directions

Somrita Nag, Ajay Kumar Mishra, Malay Kumar Bhowmick, Dipti Grover, Piyush Kumar Maurya, Dev Narayan Yadav, Ruby Pandey, Sarita Mishra, and Anthony Fulford

Abstract Arsenic contamination in agroecosystems, particularly in South and Southeast Asia, where groundwater irrigation is prevalent, poses significant risks to food safety, soil health, and crop growth. In this chapter, we assess the use of biochar as a disaster mitigation strategy for managing arsenic risk in contaminated agricultural soils. Biochar is a high-carbon material produced through the pyrolysis of biomass. This chapter begins by exploring the chemistry and environmental cycling of arsenic in soil–plant systems, with particular attention to its speciation, movement, and mechanism of uptake by crop plants. The chemical and physical properties of biochar, along with its various chemical functional groups, cation exchange capacity, and redox behavior, enable its effectiveness in arsenic remediation. Key mechanisms, including surface complexation, precipitation, and adsorption, are discussed in detail. Other sections examine the effects of biochar on soil microbial communities, the role of biochar in reducing arsenic accumulation in crops, and engineered biochars (such as iron-impregnated biochars) that could be effectively utilized in remediation processes. Relevant case studies on vegetable production in arid ecosystems and rice systems from the Bengal Basin exemplify the practical application of biochar. Despite the potential of biochar for addressing contaminants such as arsenic, significant challenges such as long-term stability need to be addressed.

S. Nag · A. K. Mishra (✉) · P. K. Maurya · D. N. Yadav · R. Pandey · A. Fulford
International Rice Research Institute South Asia Regional Centre (ISARC), Varanasi, Uttar Pradesh, India
e-mail: a.k.mishra@cgiar.org

M. K. Bhowmick
Rice Research Station, Chinsurah, West Bengal, India

D. Grover
Institute of Environmental Studies, Kurukshetra University, Kurukshetra 136119, Haryana, India

S. Mishra
Faculty of Science, Sri Sri University, Cuttack, Odisha, India

D. Moulick et al. (eds.), *Biochar: Achieving Agro-Environmental Sustainability*, Sustainability Solutions, https://doi.org/10.1007/978-3-032-19855-6_5

Keywords Agroecosystems · Arsenic remediation · Engineered biochar · Soil contamination

5.1 Introduction

Arsenic (As), a naturally occurring metalloid and significant environmental contaminant, poses risks to food safety, public health, and soil productivity. While trace amounts of As can naturally be found in various geological formations, agricultural lands are becoming increasingly contaminated with As through both natural and anthropogenic mobilization processes. Chronic exposure to As from irrigation water, soil, and plants, especially rice, has been linked to cancer, neurological diseases, cardiovascular issues, and other chronic illnesses. Arsenic exists in both organic and inorganic forms, whereas many other heavy metals do not. In agroecosystems, inorganic As [As(III) and As(V)] is often more prevalent and more toxic than organic forms. Arsenic contamination is particularly impactful in agriculture, as it easily mobilizes in wet soils and is bioavailable to plants. Researchers have identified As contamination in soils and crop plants due to industrial pollution, application of As-containing pesticides, and irrigation of crops with As-contaminated groundwater. Arsenic poses a global threat by hindering plant growth upon entering into the soil–plant system and subsequently the food chain (Rai et al. 2023; Genchi et al. 2022; Rahaman et al. 2022; Sharma et al. 2021; Shahid et al. 2021; Rehman et al. 2021; Mishra et al. 2023; Pandey et al. 2023).

There are numerous pathways for As to enter into the agroecosystems. The primary pathway for As in South and Southeast Asia is through irrigation with As-contaminated groundwater. In these regions, excessive groundwater withdrawal releases arsenic from naturally occurring aquifers into irrigation water, which then transports As into surface soils.

Plants absorb inorganic As species, such as arsenate [As(V)] and arsenite [As(III)], primarily under anaerobic (flooded) conditions when As(III) is more bioavailable. In addition to irrigation pathways, agricultural fields experience long-term residual contact from the historical use of As-based pesticides and herbicides, such as lead arsenate and monosodium methanearsonate (MSMA). Arsenic loading to agricultural soils, particularly in peri-urban areas, is also linked to mining, industrial effluents, and atmospheric deposition. Upon sorption to soils, As speciation, mobility, and its presence in crops depend on interactions with minerals, organic matter, and microbial communities (Singh et al. 2023; Srivastava et al. 2021; Hussain et al. 2021; Sharma et al. 2022; Rehman et al. 2021; Alam et al. 2021).

Arsenic contamination affects more than 140 million people globally, with agriculture being perhaps the most vulnerable sector. The problem is especially acute in South and Southeast Asia, including the Indo-Gangetic Plains of India, Bangladesh, Nepal, and the rice-growing countries of Vietnam and Cambodia. Rice is grown in flooded soils of these regions, where As is more mobile (in flooded soils) and thus represents a major pathway to As exposure. For example, more than 60% of the groundwater used for irrigation in the Bengal Basin has As concentrations of above the WHO's recommended limit (10 μg/L). Thus, rice represents a significant exposure pathway for As that accumulates into the grains of rice, thereby comprehensively impacting the physical safety of food. Moreover, As not only affects humans but also decreases crop yields, disrupts soil microbial activity, and reduces soil biological activity for fertility. Chronic exposure to As contamination threatens food security and the sustainability of agriculture and food systems in resource-limited contexts (Bhowmik et al. 2022). Common remediation techniques such as soil excavation, chemical washing, or phytoremediation have high cost, low effectiveness, or potential negative environmental impacts, and have, therefore, been impractical on a broad-scale agricultural landscape. Consequently, there is an urgent need for approaches that are affordable, sustainable, and environmentally sound to mitigate As contamination in agroecosystems. In this context, biochar, a carbon-rich material derived from the pyrolysis of organic biomass, has emerged as a viable option. Owing to its high surface area, porosity, and functional groups, biochar can absorb and immobilize As in soils, reducing its bioavailability to plants. Additionally, biochar can enhance microbial activity, improve pH buffering, and enhance the soil physical structure to create potential synergistic effects for long-term enhancement of soil health overall. Taken together, these agricultural properties demonstrate that biochar is a flexible agronomic approach for As remediation (Liu et al. 2015; Vithanage et al. 2017; Beiyuan et al. 2023; Palansooriya et al. 2019; Mishra et al. 2022a, b; Chaudhary et al. 2023).

This chapter highlights biochar as a potential bioremediation resource for As-contaminated agroecosystems. This study provides a comprehensive discussion on the chemistry of As in soil–plant systems and on the function of biochar in immobilizing the effects of As and biochar on crop and soil health. Particular attention has been given to case studies of As-affected areas, especially in South Asia. The present chapter is intended to illustrate the environmental dynamics of As in agroecosystems; highlight the characteristics of biochar that regulates As remediation; assess biochar–soil–plant interactions in polluted systems; and promote operational challenges and future research agendas to improve biochar use. Given the numerous opportunities for As contamination, the chapter aims to educate researchers, practitioners, and policymakers in the use of biochar to promote sustainable agriculture and restoration (Fig. 5.1).

5.2 Chemistry and Behaviour of Arsenic in Soil–Plant Systems

In terms of its redox sensitivity, its many chemical forms, and its tight complications with soil constituents, As behaves distinctly from other heavy metals in soil. Understanding As speciation, mobility, and uptake mechanisms is essential for developing

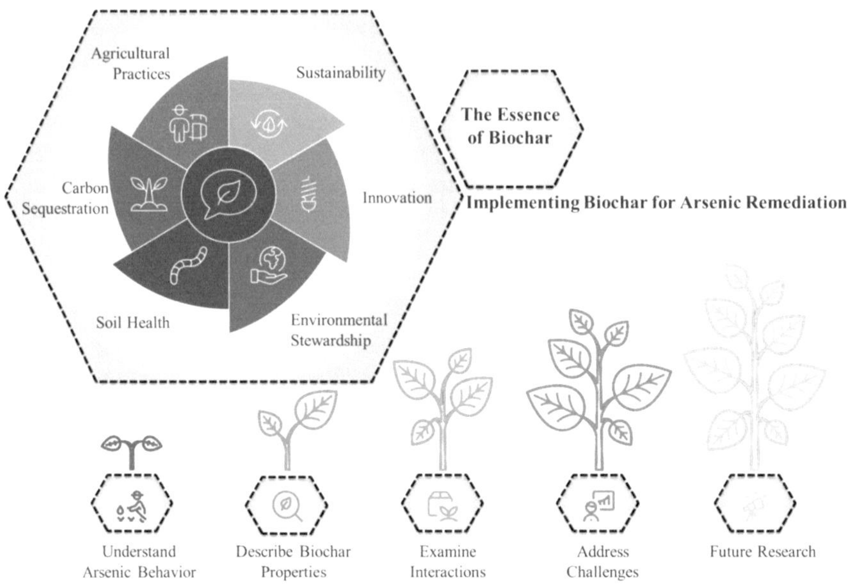

Fig. 5.1 Multifunctional role of biochar in sustainable agriculture and As mitigation

effective remediation strategies and for ensuring food safety and agricultural viability (Khalid et al. 2017; Khan et al. 2021).

5.2.1 Forms of Arsenic in Soil

Both inorganic and organic forms of As are present in soils, but inorganic forms are more dominant and more toxic. Inorganic forms of As fall into two groups:

- As(V): This form is energetically more favourable under oxidized (aerobic) conditions, and chemically similar to phosphate.
- As(III): This form is energetically more favourable under reduced (anaerobic or waterlogged) conditions, and more mobile and more toxic than As(V).

In contrast to inorganic As, organic As is less common in soils. The most common forms include monomethylarsonic acid (MMA) and dimethylarsinic acid (DMA), which are usually produced by microbial methylation or human inputs such as herbicides. Compared with inorganic forms, organic forms of As are generally less toxic and bioavailable; however, both can be converted to more toxic forms depending on the conditions. The speciation of As impacts its mobility, adsorption characteristics, and bioavailability. For example, As(III) has a lower binding affinity for iron (Fe) and aluminum (Al) oxides in soil than As(V); therefore, As(III) has a greater chance of leaching and transpiration (plant uptake) (Smith et al. 1998; Lim et al. 2014; Luo et al. 2020; Hemmat-Jou et al. 2024; Di et al. 2019).

5.2.2 *Soil Factors Affecting Arsenic Mobility and Bioavailability*

The mobility and availability of As in plants are controlled by following soil parameters:

- **Redox Potential (Eh)**: It is an important factor in determining the dominance of As(V) or As(III). Anaerobic or water-logged soils promote the reduction of As(V) to As(III), increasing both As(III) solubility and bioavailability.
- **Soil pH**: As binds more tightly to soil particles at lower pH values (<6), and desorption becomes more effective at higher pH values, particularly for As(V). Mobility is better under alkaline conditions.
- **Organic Matter**: Organic matter can either make As more mobile or less mobile. Dissolved organic carbon (DOC) may enhance As movement by forming complexes with As, or it may stimulate microbial activity that alters the form of As or immobilizes it.
- **Iron and Manganese Oxides**: These oxides serve as As sorbents. Fe oxides strongly retain As(V) in oxidized soils. If iron or manganese are reduced, then the As may be released into the soil solution when these oxidic forms dissolve.
- **Phosphate Concentration**: Phosphate and As(V) compete for adsorption sites and plant uptake. The precipitation of As(V) can occur with the application of large amounts of phosphatic fertilizers, which further increase the bioavailability of As(V).
- **Microbial Activity**: Arsenic speciation and distribution are affected by methylation processes and redox transformations [i.e., As(III) oxidizers; As(V) reducers] mediated by soil microbes.

Particularly in rice-based systems, the interaction of these variables produces a dynamic soil environment where As availability can fluctuate both seasonally and spatially (Yao et al. 2021; Zhang et al. 2014, 2007; Campbell and Nordstrom 2014).

5.2.3 *Arsenic Uptake Mechanisms in Plants*

Plants take up As through both specific and nonspecific transport mechanisms, depending on the form of As present. For example:

- **As(V) Uptake**: As(V) mimics phosphate because As(V) has a similar structure and is processed by phosphate transporters (PHTs). Once inside a plant, As(V) reductases (ACR2s) can convert As(V) to As(III), which can then be either effluxed or sequestered in vacuoles.

- **As(III) Uptake**: As(III) passively crosses biological membranes through aquaporin channels, especially aquaporin channels containing nodulin 26-like intrinsic proteins (NIPs). This phenomenon is important in flooded soils with relatively high concentrations of As(III), where As uptake is preferentially more effective.
- **Organic As (DMA) Uptake**: Generally, this process is very limited, but some aquaglyceroporins and other transporters can take up DMA.

With respect to the translocation of As from roots to shoots (i.e., the concentration of As in the edible parts of crops and some plants shows different rates of translocation concerning the As concentration in their roots). The ability of As(III) to readily translocate into the rice paddy grains rice or the edible portion of the crop highlights the potential for human exposure to As through food. The ability of several crops to translocate and concentrate As in edible parts of the crop is limited; in effect, the As content in their roots, as opposed to that in the leaves or edible parts of the crop, is concentrated (Zvobgo et al. 2018; Nahar et al. 2012; Pommerrenig et al. 2015; Guenther 2002; Mukhopadhyay et al. 2014).

5.2.4 Arsenic Phytotoxicity and Impact on Crop Yield

At relatively low concentrations, As is phytotoxic and affects various physiological and biochemical functions.

- **Inhibition of Photosynthesis**: Arsenic affects chloroplasts and inhibits the synthesis of chlorophyll, which in turn decreases photosynthetic efficiency.
- **Inhibition of Enzyme Activity**: The pyruvate dehydrogenase (PDH) complex enzyme inhibits metabolic activity by binding to the sulfhydryl (–SH) groups on proteins/enzymes.
- Impact on lipid peroxidation: It causes oxidative stress via reactive oxygen species (ROS), which damage DNA, degrade cell membranes, and affect lipid peroxidation.
- Disruption of nutrient uptake and metabolism: It is caused by disrupting phosphorus metabolism and competing with phosphates that limits important nutrients such as Mn, Zn, and Fe.

Signs of As toxicity include decreased seed germination, chlorosis, leaf necrosis, stunted growth, and decreased grain yield, among others. Leafy vegetables, rice, and wheat seem to be particularly affected, despite reductions in yield observed at available As concentrations lower than global soil safety levels (Gusman et al. 2013; Mishra and Dubey 2006; Sharma 2012; Howladar et al. 2019).

5.2.5 Arsenic Transfer in the Food Chain

The transfer of As from the soil-to-plant-to-food-chain presents a serious health risk for humans:

- **Edible Parts**: Arsenic can accumulate at very high levels in grains, especially in the bran layer of raw rice. Depending upon the condition of the soil, it can also bioaccumulate in root vegetables (e.g., carrots) and leafy vegetables (e.g., spinach and coriander).
- **Animal Pathways**: Arsenic does not biomagnify very well. However, As-contaminated feed and forage can bioaccumulate in dairy products and solid meat.
- **Human Exposure**: Chronic intake of As can lead to adverse human health effects, including skin lesions; cancers of the skin, lung, and bladder; and developmental effects in children resulting from the consumption of rice, vegetables, and drinking water.

Transfer along the food chain is modified by plant species, agronomic practices, and the chemical speciation of As. Therefore, limiting the bioavailability of As in the root zone is important for limiting the exposure pathway.

The behaviour of As in soil–plant systems is determined by a complex interplay of chemical forms, soil characteristics, and plant physiology (Fig. 5.2). Because As is highly mobile under specific conditions, and some crops, such as rice, can accumulate As in edible plant tissue, it is a priority contaminant in agricultural systems. Solutions should decrease As bioavailability, disrupt uptake pathways, and improve soil health. Biochar provides an effective way to reach these goals in the context of a diverse set of remediation treatments, as we outline in the next sections (Bundschuh et al. 2012; Bhattacharya et al. 2012).

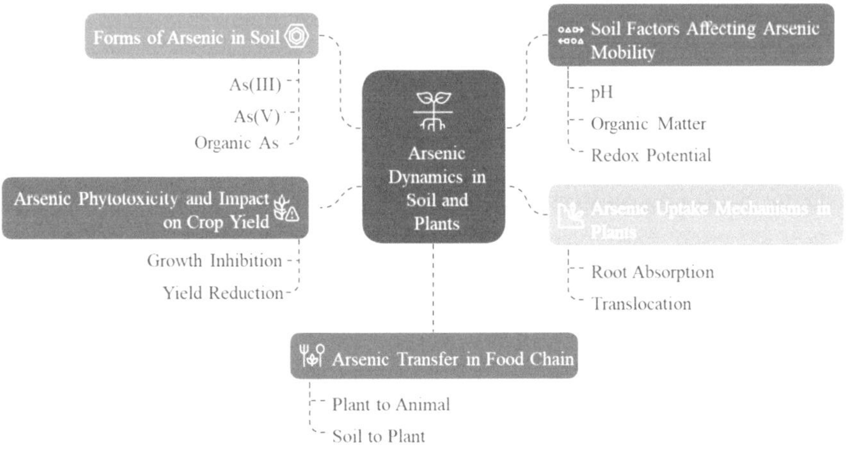

Fig. 5.2 Framework for understanding arsenic behavior in agricultural soils

5.3 Biochar: Properties and Mechanisms

Biochar is increasingly recognized as a sustainable material for use in environmental remediation and soil amendment. Owing to its unique physicochemical properties, biochar not only improves overall soil quality but also immobilizes heavy metals and metalloids such as As. This chapter reviews what biochar is and how its properties, production and biochar-specific mechanisms make it a potential remediation agent for As-contaminated agroecosystems.

5.3.1 Biochar: Production Processes and Feedstocks

Pyrolysis is the thermal decomposition of organic biomass under low or no oxygen, and biochar is a porous and carbon-dense end product in this process of decomposition.

Biochar has many structural and chemical properties that are related to its reactivity in soils, depending on the conditions under which it is made (Zhao et al. 2013; Mašek et al. 2016).

Some important pyrolysis parameters are as follows:

- The temperature generally ranges between 300 and 700 °C. A relatively high temperature generally increases the surface area and aromaticity, whereas a relatively low temperature results in the production of more biochar and a relatively high volatile content.
- The heating rate and residence time can both influence and alter the porosity, surface functionality, and ash content.
- Different feedstock types can also affect the quality of biochar:
- Biochars created from lignocellulosic residues, such as wood chips, sawdust, and crop straw, tend to be very porous and carbon stable.
- Biochars produced from manure and sewage sludge contain many nutrients but usually contain more ash and possible contaminants.
- In developing countries, agricultural waste (e.g., rice husk, sugarcane bagasse, coconut shell) is used for biochar production because it is abundant and inexpensive.
- Maintaining the right mix between feedstock chemistry and production variables is necessary for modifying biochar for specific remediation purposes, such as As immobilization.

5.3.2 Physicochemical Properties Relevant to Remediation

The successful use of biochar in environmental applications relies on its highly variable and modifiable physicochemical properties, including the following:

- **High Porosity and Surface Area**: It improves the potential for microbial colonization and pollutant adsorption.
- **Ability to Buffer pH**: As many biochars are alkaline, biochar can increase the pH of acidic soils and impact As speciation, favouring reduced mobility.
- **Cation Exchange Capacity (CEC)**: It enhances the interaction between contaminants and nutrients.
- **Electrical Conductivity (EC)**: It influences ion exchange and redox reactions, and varies with feedstock and pyrolysis temperature.
- **Ash Content**: It may decrease the total carbon content, but influences nutrient contents (K, Ca and Mg).

These properties contribute to the dual advantages of biochar in contaminated agroecosystems by reducing As mobility and increasing soil fertility (Zhang et al. 2018; Xie et al. 2015; Munera-Echeverri et al. 2018; Gabhi et al. 2024).

5.3.3 Surface Chemistry and Functional Groups

Many oxygen-containing functional groups present in large quantities on the surface of biochars are important for the adsorption of contaminants. The functional groups include standard carboxyl (–COOH), hydroxyl (–OH), carbonyl (C=O), and phenolic groups (–PhOH). The distribution and concentration of functional groups depend on both the type of feedstock and the pyrolysis temperature. Compared with biochars produced at low temperatures, high temperature biochars have a more aromatic character and lower polar character, which increases the structural stability of the biochar but diminishes the surface reactivity.

Functional groups may act as nucleation sites for the precipitation of minerals, as well as inner- or outer-sphere complexes with As species. In addition, redox-active functional groups may also indirectly influence As speciation by mediating oxidation/reduction reactions such as the oxidization of As(III) to the less toxic As(V) (Dong et al. 2024; Campbell and Nordstrom 2014).

5.3.4 Sorption Mechanisms: Adsorption, Precipitation, and Complexation

Biochar remediation of As is controlled by several sorption mechanisms that can work at the same time:

a. **Adsorption**

   - **Electrostatics**: Biochar surfaces can adsorb neutral As(III) or anionic As(V), depending on the relative surface charge.

- **Ligand Exchange**: As(V) can swap out surface hydroxyl or phosphate surface sites associated with the metal oxides of biochar surfaces.

b. **Precipitation**

- Arsenic can precipitate as a soluble, calcium arsenate or ferric arsenate in high-calcium or iron biochars. This will be more pronounced with manures or biochar sourced from manures engineered with iron or manganese.

c. **Complexation of Surfaces**

- To avoid desorption, arsenic can complex with metal oxides or reactive functional groups found in the biochar matrix, becoming an inner-sphere complex.

d. **Coprecipitation**

Arsenic can coprecipitate with Fe(III) hydroxides or phosphate in iron systems, particularly when the pH is neutral to alkaline.

e. **Redox**

Conversely, aged biochar can, under reducing conditions (e.g., flooded soils), sometimes reduce As(V) to As(III) depending on the microbial and chemical system. Some biochars promote the oxidation of As(III) to As(V), which is a stronger adsorbate to surfaces and less mobile.

These processes make biochar an efficient immobilizing agent by lowering the solubility of As, limiting plant uptake, and preventing leaching into groundwater (Strawn 2021; Ford and Sparks 2001; Benis et al. 2020).

5.3.5 Ageing and Stability of Biochar in Soils

Upon their addition to soils, biochars become aged physically, chemically, and biologically, with consequent physical and chemical changes in the biochars as well as any ongoing efficacy.

- **Oxidation of Surfaces**: Over time, this enhances the sorptive capacity of the biochar, as new functional groups are formed.
- **Microbial Colonization**: By colonizing biochar, microbes change the reactivity of the biochar and, in some cases, can promote the transformation of As through microbial metabolism.
- **Interaction with Soil Minerals**: Over time, biochar particles may interact with (mature or newly formed) iron oxides, clay minerals, or organomineral complexes, leading to new reactivity.

Because of surface oxidation and the formation of secondary minerals, aged biochar generally has improved As sorption potential. However, arsenic capitalized by biochar may be remobilized under certain conditions (e.g., increased DOC or

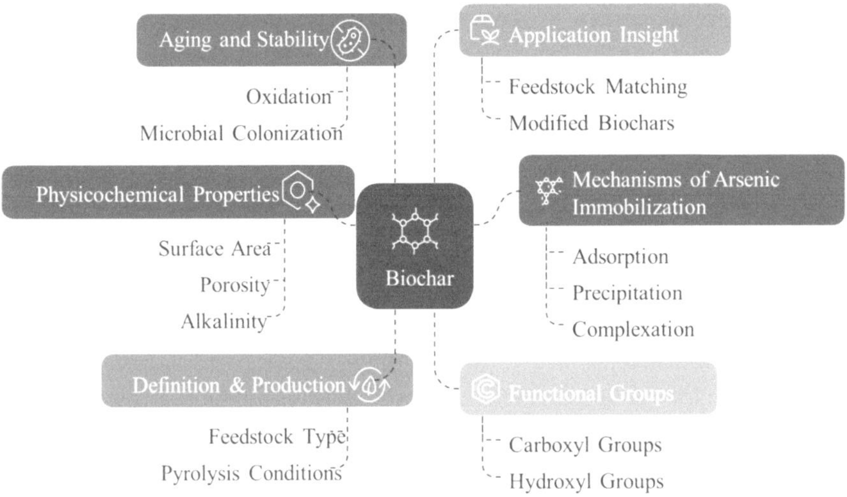

Fig. 5.3 Biochar functional attributes supporting sustainable soil remediation

reducing conditions), especially if the iron or manganese associated with the biochar undergoes microbially induced reduction.

Given the estimated mean residence times of decades to centuries of biochar, it is also remarkably recalcitrant in terms of carbon stability. Hence, it is a carbon sequestration mechanism under climate-smart agriculture and acts as a remediation strategy.

Biochar is a practical tool for the remediation of arsenic because of its unique physical structure, surface chemistry, and engaging interactions in soil environments. The proper feedstock, pyrolysis parameters, and, in some cases, posttreatment modifications (*e.g.*, Fe impregnation) further optimize the mechanisms that immobilize As adsorption, precipitation, and complexation. An understanding of these mechanisms (Fig. 5.3) is essential for developing efficient and application-ready biochar solutions to reduce As risk in agroecosystems (Wang et al. 2020; Yang et al. 2021a, b).

5.4 Biochar in Arsenic Remediation

Given its ability to reduce crop uptake, change microbial pathways, and reduce As mobilization, biochar is being used in As-contaminated agroecosystems. This section examines practical aspects of field and experimental research results and explores the functions of biochar for remediation (Sun et al. 2022; Beiyuan et al. 2023).

5.4.1 Immobilization of Arsenic in Soils

The immobilization of As is one of the key remediation mechanisms of biochar. Immobilization occurs when biochar reduces the solubility and mobility of As through various chemical and physical mechanisms.

- **Adsorption and Complexation**: If biochar has sufficient metal oxides or oxygenated functional groups, it will most likely adsorb As at its surface through ligand exchange and surface complexation.
- **pH Buffering**: Many types of biochar increase the soil pH, especially in acidic soils. This has the potential to reduce As solubility, but it may also impact speciation. For example, As(III) will be less mobile.
- **Fe and Mn Interactions**: Arsenic can coprecipitate with Fe or Mn hydroxides or can form stable complexes via inner-sphere complexes with biochar containing, or coapplied with, Fe or Mn oxides.
- **Mineralization**: In calcareous soils or with biochar from manure, arsenic can precipitate as insoluble calcium arsenate.
- **Immobilization**: It is a critical first step in stabilizing contaminated soils and reducing the likelihood of leaching to groundwater and bioaccumulation in plants.

5.4.2 Interactions with Soil Microorganisms and Rhizospheres

The rhizosphere, the soil affected by plant roots, is a particularly active site of chemical and biological activity and therefore an active site of As transformation due to high rates of microbial and biochemical activity. Biochar has multiple important effects on this microecosystem:

- **Microbial Diversity**: Biochar enhances microbial diversity, stabilizes microbial populations, and provides microhabitats for the germination of beneficial microorganisms. Microorganisms involved in the transformation of As, such as bacteria that reduce As(V) or bacteria that oxidize As(III), may benefit from the application of biochar.
- **Redox Modulation**: Biochar can indirectly influence the release of As by partially moderating the redox potential and slowing the reductive dissolution of Fe oxides in flooded or anaerobic soils.
- **Microbial Enzymatic Activity**: Biochar-amended soils may promote the methylation of As via microbial enzymes, although this can also promote the production of more mobile organic As species.
- **Root Exudate Modulation**: The volume and type of root exudates influence As speciation and bioavailability in the rhizosphere, and biochar can influence their volume and type.

Through these mechanisms, biochar can change the composition of microbial communities to increase the abundance of functional groups that directly promote the immobilization or detoxification of As (Fitz and Wenzel 2002).

5.4.3 *Reduction in Arsenic Uptake in Plants*

Biochar plays an essential role in decreasing As uptake by crops, especially staple crops such as wheat, spinach, and rice:

- **Reduced Root Zone Availability**: Sorption and precipitation in the root zone decrease the amount of plant-available As in the primary mode.
- **Inhibition of Root Transport**: Research has shown that biochar decreases the expression of phosphate transporters that participate in As(V) uptake.
- **Healthy Soil**: Biochar improves nutrient balance (e.g., P, Fe, Zn), fosters healthy plant growth, and decreases As in tissues.
- **Induced Defence in Plants**: Biochar may induce antioxidant enzymes that make plants more tolerant to oxidative stress from As.

Multiple field trials in rice have reported a decrease in grain As concentrations, ranging from 30 to 60% with biochar, demonstrating that biochar promotes a safer food supply (Zama et al. 2018; Khan et al. 2020).

5.4.4 *Biochar Modifications*

By improving the sorption capacity and redox reactivity, engineered or modified biochar exhibits better performance than conventional biochar. Some major modifications include the following:

- **Iron-Impregnated Biochar (Fe-Biochar)**: By impregnating biochar with iron salts (e.g., $FeCl_3$, $FeSO_4$) or zero-valent iron (ZVI), strong surfaces that bind arsenic are created. Fe-oxides form on biochar, and they enhance inner-sphere complexation with As(V) and As(III).
- **Manganese-Modified Biochar**. It enhances redox reactivity and promotes oxidation of As(III) to less mobile As(V).
- **Nanobiochar**: By increasing the surface area and reactivity, nanoscale biochar has a relatively high As absorption potential, but it deserves caution because it can leach into the environment.
- **Phosphate-Loaded Biochar**: Phosphate–arsenate competition can decrease the uptake of As(V) by roots by competing for adsorption sites.

Although the performance observations for these engineered variants are promising, there are doubts about scalability, production costs, and regulatory approval for agricultural use (Lee and Shin 2021; Hu et al. 2015; Wang et al. 2016; Parlayıcı et al. 2024).

5.4.5 Case Studies from Laboratory and Field Trials

5.4.5.1 Rice Fields in Bengal Basin

The Bengal Basin (including parts of Bangladesh and eastern India, such as West Bengal) has gained a global reputation as an As contamination hotspot due to widespread groundwater irrigation. The effectiveness of biochar in these systems has been evaluated via a number of pot experiments and field studies.

- **Improved Rice Yield**: Application of rice husk biochar at 10–20 t/ha results in a significant decrease in As uptake in rice grains, accompanied by increased biomass and yield.
- **Reduced Soil Arsenic Solubility**: Fe-rich biochars produced from chicken litter or coconut shell can result in a 30–50% reduction in water-soluble As fractions.
- **Synergistic Solutions**: Addition of compost and iron sulfate in combination with biochar can significantly reduce As availability and improve soil microbial health.

Overall, biochar use in Bengal Basin rice fields has considerable potential for improving food safety and rehabilitating soils (Sarkar et al. 2022; Leksungnoen et al. 2019; Yang et al. 2021a, b; Mensah et al. 2022).

5.4.5.2 Vegetable Systems in Arid Lands

Vegetable production in As-affected soils in arid and semiarid regions (such as Rajasthan and Central Iran) is further compounded by low organic matter, alkaline pH, and limited water resources; hence, the many functions of biochar are particularly advantageous.

- **Better Water Retention**: Biochar from crop residues or woody biomass increases soil water retention (which is critical in arid regions).
- **Less Arsenic Accumulation**: Depending on the feedstock source and application rate, biochar can reduce the As-levels in edible crops such as spinach, okra, and carrots by up to 40%.
- **Nutrient Interactivity**: As a result of positive interactivity in the availability of micronutrients (i.e., Fe and Zn), arsenic toxicity may be reduced indirectly through competitive uptake.
- **Sustainable Management**: Numerous experiments have demonstrated how biochar continues to exhibit sorption capabilities and beneficial functions in soils during fluctuations in moisture and temperature, which are intended to represent extreme climatic conditions.

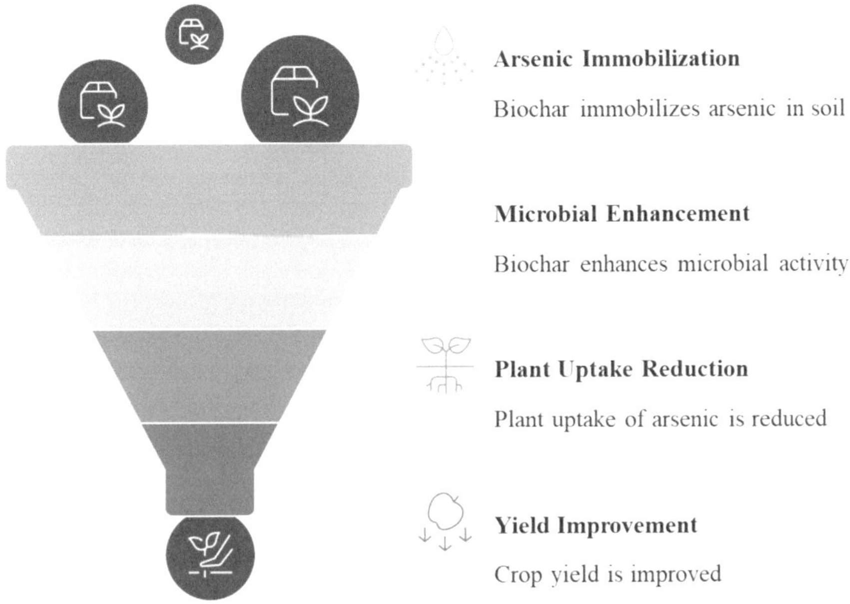

Fig. 5.4 Cascading benefits of biochar in arsenic-contaminated agroecosystems

These results indicate the potentials and usefulness of biochar in a number of cropping systems and in various environmental contexts.

The physical, chemical, and biological characteristics of biochar make it a viable remediation option for arsenic. Biochar can be a powerful, flexible, and clean-up option for contaminated agricultural systems by immobilizing only As, supporting good microbes, or stopping plant uptake (Fig. 5.4). While field case studies and engineered biochars demonstrate the value of biochar, future studies need to focus on cost, local adaptations of biochar, and connections with holistic land management approaches (Li et al. 2021; Ray and Bharti 2023; Natasha et al. 2022; Acharya et al. 2024).

5.5 Biochar–Soil–Plant Interactions

When applied, biochar initiates a series of physical, chemical, and biological changes in the soil–plant continuum. In the case of As-stressed systems, these changes can increase crop productivity, food safety, chemical and microbial processes, and increase soil fertility. Therefore, potential trade-offs should also be materialized to avoid 'surprises' (Murtaza et al. 2023).

5.5.1 Impact on Soil pH, CEC and Nutrient Availability

The physicochemical status of soil is drastically affected by biochar, which may have a direct or indirect impact on plant health and As mobility (Antonangelo et al. 2024; Li et al. 2023; Darma et al. 2021).

- **Buffering Soil pH**: Most biochars, particularly wood and manure biochars, have an alkaline pH (8–10). This increase in pH affects sorption to Fe/Al oxides, as well as the solubility of As, particularly As(V), when applied to acidic soil. This buffering pH creates a more favourable environment for root growth and beneficial microorganisms.
- **Cation Exchange Capacity (CEC)**: The porous structure and surface functional groups in biochar increase the CEC of the soil. Consequently, this encourages the retention of important nutrients in the soil, such as K^+, Ca^{2+}, Mg^{2+}, and NH_4^+. This improves competition for phosphorus and As(V) availability for uptake, thereby promoting plant health.
- **Nutrient Retention and Cycling**: Biochar can provide macro- and micronutrients (i.e., P, K, Zn, and Fe) depending on the feedstock. In addition, biochar contributes to increased nutrient-use efficiency, which is attributed to the large surface area of biochar, which absorbs nutrients and releases them more slowly. The increased availability of Fe and Zn in soils contaminated with As possibly enables competition that prevents plants from taking up additional As. The synergistic improvement in soil chemistry suggests that the benefits to each type of soil improvement may provide long-term reclamation of fertility, another important cobenefit in degraded agroecosystems.

5.5.2 Effects on Microbial Community and Enzyme Activity

Soil microorganisms significantly influence organic matter decomposition, nutrient cycling, and contaminant transformation. Through the provision of a habitat and an electron mediator, biochar modifies the ecology of soil microorganisms.

- **Microbial Biomass and Diversity**: Biochar provides carbon substrates and microniches that increase microbial diversity and biomass. This phenomenon is especially important in soils that are deficient or under stress, where microbial communities are suppressed.
- **Functional Group Shift**: Biochar supports communities that participate in the beneficial transformations of nitrogen and phosphorus (i.e., fixators or solubilizers) and in the oxidation/reduction of As. There may be (for example) more As(III)-oxidizing bacteria that convert more mobile As(III) into the less harmful As(V).
- **Soil Enzyme Activities**: Once biochar is applied to the soil, the activity of soil enzymes such as urease, phosphatase, and dehydrogenase improves, which further

indicates improved microbial metabolic function, along with the promotion of soil respiration and nutrient cycling.

As we can see, changes in the microbial community in soils contaminated with As may help detoxify, regenerate the ecological balance, and immobilize As.

5.5.3 *Crop Performance and Food Safety Improvements*

Continuous application of biochar improves food safety, crop growth, and yield, especially in the context of As stress.

- **Increased Growth of Plants**: Enhanced root aeration, nutrient supply, and water holding capacity provided by biochar increase biomass production and stress tolerance. This results in increased tiller production, canopy coverage, and grain or leaf yield in plants such as rice and leafy vegetables.
- **Decreased As Accumulation**: Biochar nearly eliminates As in food-value plants by ensuring the misalignment of uptake pathways and reducing As availability in the root zone. Field tests revealed a reduction in As in the grains of rice of as much as 60%, whereas spinach, mustard greens, and carrots also presented similar trends.
- **Vital Nutritional Quality**: Nutrient-dense crops grown in contaminated soil (i.e., higher nutrient density of zinc, iron, and magnesium) can impact human health in places where diets are deficient in such micronutrients, especially in As-affected zones.
- **Stress Tolerance**: Biochar reduces oxidative damage due to As toxicity as a result of antioxidant defence mechanisms in plants.

These benefits highlight how continuously using biochar enhances agronomic productivity and food safety in line with sustainable intensification objectives (Kumar et al. 2024; Natasha et al. 2022; Moulick et al. 2021; Khan et al. 2022).

5.5.4 *Trade-Offs and Risks (Potential for Other Contaminant Mobilization)*

Although biochar has several benefits, it has several limitations and considerations that must be managed depending on the biochar type and soil.

- **Contaminant Mobilization**: If biochar alters redox conditions or competes for binding sites, it may even mobilize additional contaminants [e.g., lead (Pb), cadmium (Cd), or nitrate]. In situations with many contaminants, this can be problematic.

- **Excessive Application**: Excess application of biochar can increase the pH high enough to prevent crops from taking up certain micronutrients (e.g., Zn or Mn).
- **Feedstock Risks**: If biochar derived from sewage, sludge or industrial wastes has not been adequately characterized, it may contain nonorganic pollutants and toxic metals, which could risk soil and food systems over time.
- **Temporal Dynamics**: The effectiveness of biochar in immobilizing As is subject to age, microbial activity, and organic matter turnover, and may very rarely cause delayed remobilization of bound As.
- **Practical and Economic Constraints**: Despite the known benefits, costs, availability, and inadequate technical guidelines (especially in areas with limited resources) of biochar remain barriers to its field-scale production and use.

To maximize the benefits and minimize the undesirable effects of biochar, site-specific risk–benefit evaluations and regulations are important. By understanding the mechanisms underlying the ability of biochar to remediate As in contaminated agroecosystems, biochar may be beneficial for removing contaminants through interactions with soil and plants. The synergy among plants and biochar, which leads to increased resilience, improved microbial health, and improved soil nutrient availability, may increase agricultural production while lowering exposure to contaminants in food sources (Fig. 5.5). Long-term sustainability relies on understanding these contingent effects and managing them. The impact and success of biochar across different agricultural systems depend on selection, application rates, and integration into broader soil management protocols (El-Naggar et al. 2021; Younas et al. 2024; Rangabhashiyam et al. 2022; Heinrich et al. 2023; Luo et al. 2023).

5.6 Challenges and Limitations

Although biochar is an attractive option for remediation in As-contaminated agroecosystems, limitations in implementation, science, and policy have hindered its adoption. It is highly important to understand these limitations, focus on future research, inform stakeholder decisions, and design scalable interventions (Singh et al. 2023; Sun et al. 2022).

5.6.1 Variability in Biochar Properties

The greatest disadvantage of biochar is its heterogeneity. The properties of biochar vary greatly on the basis of the following:

- Feedstock type (wood, manure, crop residue, etc.);
- Pyrolysis conditions (temperature, heating rate, and retention time);
- Posttreatment options (chemical activation, grinding, washing).

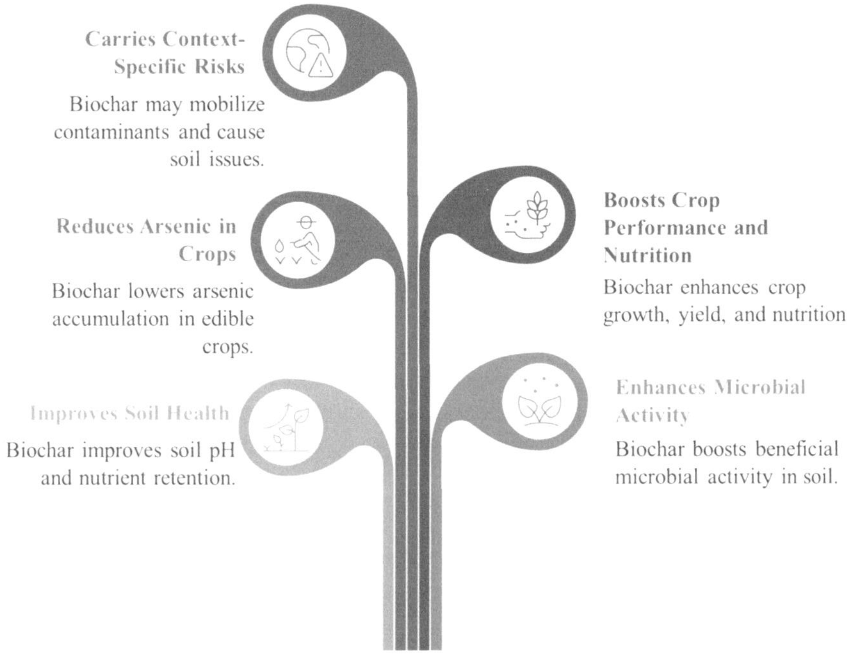

Fig. 5.5 Multifunctional impacts of biochar in arsenic-contaminated agroecosystems

All the variability noted will shape important characteristics, such as pH, porosity, surface area, functional groups, mineral content and contaminant sorption capacity. Therefore, even if biochar is applied to similar soils, the biochar produced under it will have very different impacts because of these variations. This may be important in a case such as As remediation, where the interactions with metals and surface loading capacity and chemistry differ. Ensuring the reliability of performance across applications will require the standardization of production processes and an improved framework for biochar characterization.

5.6.2 *Long-Term Efficacy and Field Variability*

Long-term field performance remains unpredictable, even with multiple lab studies demonstrating that biochar passively immobilizes arsenic.

- Climate, crop type, irrigation application, and soil variability can all affect the effective longevity of biochar.
- Under saturated conditions, such as puddled transplanted rice, redox changes can not only remobilize any immobilized As but also dissolve the iron oxide coatings

on the biochar surface, along with a portion of the biochar itself, thus reducing its effectiveness.
- Physical, chemical, and microbiological aging processes result in a change in the structure and reactivity of biochar and can result in a decrease in its sorption capacity.

Additionally, there is a significant uncertainty with regard to biochar behaviour over decadal timescales because very few studies have followed the persistence of As immobilization beyond a two-to three-year period. As a result, uncertainty renders biochar a substantially less appealing option for risk-averse farmers and for legislators looking for long-term remediation options.

5.6.3 Potential for as Remobilization

Biochar can immobilize phosphorus; however, there are several routes whereby phosphorus can be remobilized when biochar is added to a soil.

- The microbial activity processes via anaerobic environments, such as wet soils, may degrade the Fe(III) oxides that adhere to the biochar, thereby potentially remobilizing the associated phosphorus back to the soil solution.
- Microbial methylation of phosphorus from organic matter may result in biochar-associated phosphorus becoming increasingly mobile organic phosphorus species via the dissolved organic carbon (DOC) matrices of biochar.
- DOC, as with phosphorus, can also change the relative bonds formed between phosphorus and biochar surfaces to eventually reduce sorption in soils due to desorption properties in response to changes in pH and ionic strength, as well as the presence of competing anions in the soil solution that also bind to biochar surfaces.

These findings indicate that localized site assessments need to be completed and that other amendments should be applied to biochar, which can mitigate the risk to soil phosphorus stability.

5.6.4 Economic and Logistical Barriers to Adoption

In terms of environmental issues, the use of biochar has considerable potential. Farmers struggle to adopt biochar because of the wide variety of costs and logistical challenges, including the following:

- **Higher Production Cost**: Producing high-quality biochar uses specialized tools, inputs of energy, and feedstock handling that can increase production costs.
- **Transportation and Bulk Application**: Biochar involves both a confinement of mass with weight and a constraint of mass with size when it scales to larger

quantities, which translates into an increased cost of consistent transport and application in large quantities.

- **Inadequate Availability**: Supply chain limitations can be a result of insufficient localized biochar production in As-affected areas.
- **Gaps in Farmer Knowledge and Training**: Many farmers are simply unaware of how to use biochar or lack technical information on when, how, and how to combine it with other agronomic practices.

There are still few examples of biochar being used at scale on farms rather than valuable research plots, pilot types of projects without direct subsidies, extension, or business models in place.

5.6.5 Regulatory and Policy Gaps

Currently, many countries have limited or no regulatory frameworks for controlling and monitoring biochar application in agriculture (Meyer et al. 2017; Vereš et al. 2014; Hammond et al. 2016).

- No international standard exists for the safety, quality, or classification of biochar, leading to variations in the products on the market.
- In cases where biochars are potentially hazardous due to contamination from industrial wastes or sewage sludges, there are inconsistent regulatory bodies and expectations.
- The institutional uptake of biochar is unclear, as few national policies acknowledge biochar as a legitimate soil amendment or remediation strategy.

Moreover, until now, financial returns for producers have not been robust because the value of biochar's contribution to climate change mitigation through carbon sequestration has not yet been widely included in carbon markets or incentive programs.

To support responsible scaling, governments and international organizations should develop certification standards, performance objectives, and financial incentives that recognize the multifaceted agronomic and environmental benefits of biochar.

Biochar holds significant potential for reducing agricultural arsenic risk. However, the wide-scale adoption of biochar is hindered by systemic challenges such as regulatory issues, inconsistent product quality, uncertain long-term impacts, and As remobilization factors (Fig. 5.6). Resolving these impediments will require broad, multidisciplinary engagement concentrating on robust field evaluations, technology standardization, economic modelling, and regulatory support to position biochar as a trusted component of sustainable land remediation.

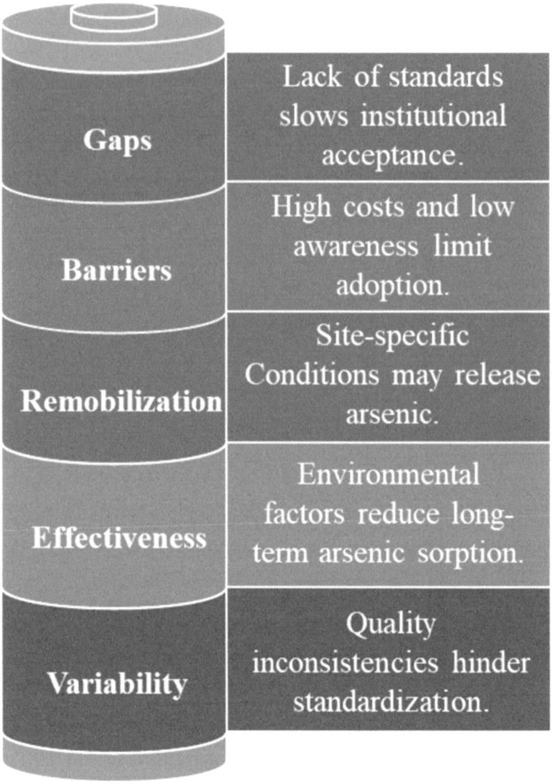

Fig. 5.6 Key challenges limiting the widespread adoption of biochar for arsenic remediation

5.7 Prospects and Research Needs

Biochar research has progressed rapidly, from experimental studies to application-oriented approaches involving environmental remediation. As the need to manage As-contaminated soils continues to grow, especially in sensitive areas, so does the need for more intelligent, customized, and scalable biochar products. This section identifies important areas for future directions for biochar to enable even greater effectiveness, safety, and integration for As remediation and sustainable agriculture (Alchouron et al. 2022; Kumar et al. 2025).

5.7.1 Advanced Functional Biochar and Nanobiochar

To improve As immobilization, next-generation biochars are being developed for the purpose of enhancing specific sorption and redox properties.

- Directed research involves the use of metal-doped biochars (Fe-, Mn-, or Al-impregnated biochars) with improved affinities for arsenic via surface precipitation and complexation.
- In water treatment applications, magnetic biochar with magnetite or zero-valent iron can facilitate recovery for later application and may exploit the dynamic absorption of contaminants into contaminated soil.
- Nanobiochars that are made via direct pyrolysis nanoscale or very fine grinding have high surface-to-volume ratios and may interact quickly with pollutants, requiring careful evaluation of ecotoxicity, mobility, and long-term fate.

In addition, research on the structure–function relationship and synthesis optimization of biochar are warranted to develop bespoke biochar tailored to specific soil types and contamination characteristics.

5.7.2 Integration with Phytoremediation and Microbial Tools

Biochar and biological remediation complements, including microbes and plants, are being more frequently adopted in synergistic approaches.

- **Biochar + Phytoremediation**: Biochar can limit leaching and maintain root health, as some hyperaccumulator plants can take As from the soil. Using both methods together could decrease crop uptake and As mobility in the food web and increase phytoremediation rates.
- **Biochar + Microbial Consortia**: Biochar can be added to microbial consortia of native or engineered microbes, which can reductively or oxidatively transform As, including an As(III)-oxidizing bacterium. Biochar allows for electron transfer, provides a microbial habitat, and improves microbial activity and stability.
- **Rhizosphere Engineering**: Engineering nutrient uptake may be feasible for remediating plants through favourable root–microbe interactions (e.g., arbuscular mycorrhizal fungi), combining with biochar and reducing As uptake into shoots.

To consider the complicated interactions in real-life agricultural settings, these systems should be incorporated with multiscale experimental design and modelling (Xiang et al. 2022).

5.7.3 Modelling and Monitoring Tools

Sophisticated monitoring and modelling frameworks (Shahid et al. 2025; Khalid et al. 2023; Long et al. 2021) are necessary to predict outcomes and improve interventions effectively. This is due to the inherent variability and complexity of biochar–soilarsenic interactions.

- Arsenic speciation, sorption, and precipitation under varying soil pH and redox conditions influenced by biochar can be simulated via the geochemical models PHREEQC or Visual MINTEQ.
- Soil–plant–atmosphere models can be updated to include biochar parameters (e.g., CEC, porosity, and As sorption coefficients) to predict crop uptake, yield, and environmental fates of biochar-amended soils.
- Real-time (in situ) monitoring of biochar performance in various environments can be performed through sensor-based field tracking systems to monitor the soil pH, redox potential (Eh), and arsenic concentration.
- Remote sensing and GIS mapping may be used to identify areas at high risk of As contamination and the landscape-scale effects of biochar interventions.

Investing in open-access tools/decision support tools will help to strategically target and manage biochar remediation initiatives.

5.7.4 Biochar Certification and Guidelines

To ensure proper and safe use, biochar cannot be used without strict quality assurance and regulatory requirements.

- Biochar standards (i.e., IBI and EBC protocols) should provide reference values for assessments of carbon stability, ash content, pH, and heavy metal concentrations, but As-specific sorption parameters must also be included and present, largely absent (Lin et al. 2025; Verheijen et al. 2019).
- Agencies such as national agricultural departments, environmental departments, and research entities should establish biochar classification systems on the basis of feedstock-specific pyrolysis parameters and intended purpose (e.g., soil amendment *vs.* contaminant remediation).
- Biochars produced from domestic sewage sludge or industrial feedstocks should have safety protocols and risk assessment mechanisms.
- The development of "fit-for-purpose" certifications or labels will assist consumers in following biochars with consistent performance assessments.

These frameworks should also assist in the legal use of biochar under public soil health stewardship programs, extension services, and subsidization programs (Park and Lee 2023).

5.7.5 Biochar Upscaling in Climate-Smart Agriculture

The benefits of biochar for resilience, adaptation, and mitigation fit well with the concepts of climate-smart agriculture (CSA):

- **Carbon Sequestration**: Because biochar is a climate mitigation tool, it can be linked to carbon markets or offset credits that lock away carbon in stable forms for decades.
- **Water Retention and Drought Resilience**: Biochar increases soil moisture retention and decreases irrigation requirements in arid areas.
- **Stability of Yield When Cultivated Is Under Stress**: Biochar serves as a buffer against heavy metals, acidity, and salinity stress, which are expected to intensify due to climate change.
- **Circular Economy**: By creating value in farm waste through the production of biochar from agricultural residues, environmental pollutants are reduced, and communities gain access to environmental remediation equipment.

It will be necessary to include biochar in national policies such as Bangladesh's mitigation for arsenic, India's Soil Health Card program, the Sustainable Development Goals (SDG) in Africa for CSA, and other opportunities for scaling biochar in CSA (Milton et al. 2012; Reddy 2019; Newell et al. 2019; Mrabet and Moussadek 2022).

Innovation, integration, and institutional support represent the future of biochar use in As remediation technologies. The possible biochar uses are massive, ranging from state-of-the-art materials, such as Fe-enhanced biochars and nanobiochars, to collaboration with plant–microbe systems and CSA technologies. It is crucial to combine biochar use with technologies, policies, and predictability that resonate with farmers if biochar is to be widely adopted for a long-term frame (Fig. 5.7). Overall, better interdisciplinary research and stakeholder interactions are needed for the agricultural and environmental potential of biochar to be actualized (Ngwoke et al. 2025).

5.8 Conclusion

The mining of arsenic in agroecosystems creates an ongoing hazard to soil health, human health, and food safety, especially in circumstances with widespread groundwater recharge for irrigation and little enforcement of controls. This chapter highlights the complex chemistry of As in soil–plant systems, the mechanism of crop uptake, and the consequent risks to public health and agricultural sustainability. Owing to its diverse biological and physicochemical traits, biochar often emerges as a potential solution. The scientific records from laboratory and field-based studies show its ability to immobilize arsenic through adsorption, precipitation, and surface complexation. In addition, biochar in even degraded and/or As-susceptible soils improves crop performance, reduces As bioavailability, enhances beneficial microbial activities, and improves soil properties.

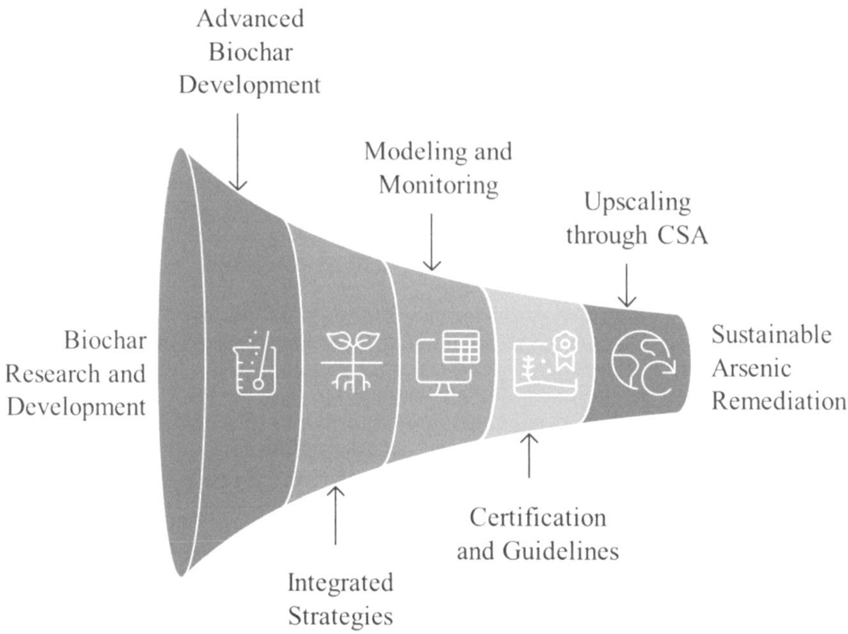

Fig. 5.7 Strategic roadmap toward sustainable arsenic remediation using biochar

Role of Biochar in Sustainable Arsenic Management

Owing to the advantages and length of time biochar is effective, biochar is uniquely positioned to provide overall and longer-term As management. Together with short-term effects such as reduced mobility of As and improved retention of soil nutrients, biochar provides long-term co-benefits such as carbon sequestration and reduced risks of soil degradation. When tailored to local conditions using appropriate feedstocks, combined with pyrolysis methods and the potential for co-applications using iron or other amendments, biochar can be a viable and sustainable solution for low-input farming systems at scale. Its ability to mitigate plant uptake, stimulate microbially mediated detoxification of As, and regenerate soil ecosystem functions supports its use as a remediation and regeneration agent.

Interdisciplinary and Policy-Driven Approaches

Although biochar has potential for widespread use, additional effort is needed across different sectors to achieve that potential.

- The continuation of scientific research is necessary to establish the risks of As remobilization or unintentional contaminant release, develop biochar properties, and measure long-term performance.
- Agricultural extension services need to establish application guidelines for a certain region and its area, train farmers, and educate farmers.

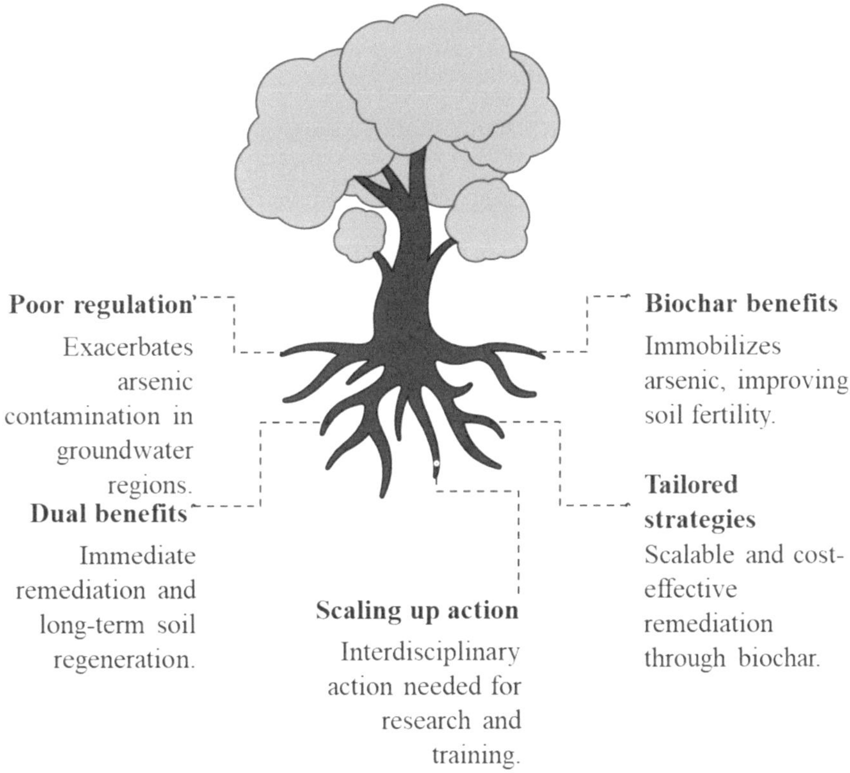

Fig. 5.8 Enabling conditions and constraints for sustainable biochar deployment

- Critical environmental issues such as national soil health, food safety, and climate resilience programs should integrate biochar, create quality standards, and promote incentives for local production.

To maximize the use of biochar, we need to collaborate across various disciplines, including soil science, agronomy, environmental engineering, microbiology, and policy (Fig. 5.8). We need collaboration to scale from disparate pilot projects to achieve meaningful landscape-scale impacts on arsenic-laden regions.

References

Acharya BS, Dodla S, Wang JJ, Pavuluri K, Darapuneni M, Dattamudi S, Maharjan B, Kharel G (2024) Biochar impacts on soil water dynamics: knowns, unknowns, and research directions. Biochar 6(1):34

Alam MF, Villholth KG, Podgorski J (2021) Human arsenic exposure risk via crop consumption and global trade from groundwater-irrigated areas. Environ Res Lett 16(12):124013

Alchouron J, Fuentes ALB, Musser A, Vega AS, Mohan D, Pittman Jr CU, Mlsna TE, Navarathna C (2022) Arsenic removal from household drinking water by biochar and biochar composites: a focus on scale-up. In: Sustainable biochar for water and wastewater treatment. Elsevier, pp 277–320

Antonangelo JA, Culman S, Zhang H (2024) Comparative analysis and prediction of cation exchange capacity via summation: influence of biochar type and nutrient ratios. Front Soil Sci 4:1371777

Beiyuan J, Qin Y, Huang Q, Wang J, Sarkar B, Bolan N, Wu X, Xu W, Liu J, Chen X, Xu S (2023) Modified biochar for arsenic immobilization in soil: a critical review. Rev Environ Contam Toxicol 261(1):20

Benis KZ, Damuchali AM, Soltan J, McPhedran KN (2020) Treatment of aqueous arsenic—a review of biochar modification methods. Sci Total Environ 739:139750

Bhattacharya S, Gupta K, Debnath S, Ghosh UC, Chattopadhyay D, Mukhopadhyay A (2012) Arsenic bioaccumulation in rice and edible plants and subsequent transmission through food chain in Bengal basin: a review of the perspectives for environmental health. Toxicol Environ Chem 94(3):429–441

Bhowmik T, Sarkar S, Bhattacharya A, Mukherjee A (2022) A review of arsenic mitigation strategies in community water supplies with insights from South Asia: options, opportunities and constraints. Environ Sci Water Res Technol 8(11):2491–2520

Bundschuh J, Nath B, Bhattacharya P, Liu CW, Armienta MA, López MVM, Lopez DL, Jean JS, Cornejo L, Macedo LFL, Tenuta Filho A (2012) Arsenic in the human food chain: the Latin American perspective. Sci Total Environ 429:92–106

Campbell KM, Nordstrom DK (2014) Arsenic speciation and sorption in natural environments. Rev Mineral Geochem 79(1):185–216

Chaudhary A, Venkatramanan V, Mishra AK, Sharma S (2023) Agronomic and environmental determinants of direct seeded rice in South Asia. Circ Econ Sustain 3(1):253–290. https://doi.org/10.1007/s43615-022-00173-x

Darma A, Yang J, Bloem E, Możdżen K, Zandi P (2021) Arsenic biotransformation and mobilization: the role of bacterial strains and other environmental variables. Environ Sci Pollut Res 1–25

Di X, Beesley L, Zhang Z, Zhi S, Jia Y, Ding Y (2019) Microbial arsenic methylation in soil and uptake and metabolism of methylated arsenic in plants: a review. Int J Environ Res Public Health 16(24):5012

Dong W, Xing J, Chen Q, Huang Y, Wu M, Yi P, Pan B, Xing B (2024) Hydrogen bonds between the oxygen-containing functional groups of biochar and organic contaminants significantly enhance sorption affinity. Chem Eng J 499:156654

El-Naggar A, Chang SX, Cai Y, Lee YH, Wang J, Wang SL, Ryu C, Rinklebe J, Ok YS (2021) Mechanistic insights into the (im) mobilization of arsenic, cadmium, lead, and zinc in a multicontaminated soil treated with different biochars. Environ Int 156:106638

Fitz WJ, Wenzel WW (2002) Arsenic transformations in the soil–rhizosphere–plant system: fundamentals and potential application to phytoremediation. J Biotechnol 99(3):259–278

Ford RG, Scheinost AC, Sparks DL (2001) Frontiers in metal sorption/precipitation mechanisms on soil mineral surfaces. Adv Agron 74:41–62

Gabhi R, Tan K, Feng T, Kirk DW, Giorcelli M, Tagliaferro A, Jia CQ (2024) Intrinsic electrical conductivity of monolithic biochar. Biomass Bioenerg 181:107051

Genchi G, Lauria G, Catalano A, Carocci A, Sinicropi MS (2022) Arsenic: a review on a great health issue worldwide. Appl Sci 12(12):6184

Guenther JF (2002) Aquaporins and aquaglyceroporins of legume nodules: structure, function, and regulation. The University of Tennessee

Gusman GS, Oliveira JA, Farnese FS, Cambraia J (2013) Arsenate and arsenite: the toxic effects on photosynthesis and growth of lettuce plants. Acta Physiol Plant 35:1201–1209

Hammond J, Schmidt HP, Van Scholl L, Ruysschaert G, Nelissen V, Ibarrola R, O'Toole A, Shackley S, Van Laer T (2016) The legality of biochar use: regulatory requirements and risk assessment. In: Biochar in European soils and agriculture. Routledge, pp 227–252

Heinrich T, Park H, Orozco R, Ding Z, Álvarez-López V, Mosquera-Losada MR, Steinbeis L, Hoffmann T (2023) Biochar production from late-harvest grass—challenges and potential for farm-scale implementation. Sustain Prod Consump 37:256–267

Hemmat-Jou MH, Liu S, Liang Y, Chen G, Fang L, Li F (2024) Microbial arsenic methylation in soil–water systems and its environmental significance. Sci Total Environ 173873

Howladar MM, Uddin MJ, Islam MM, Parveen Z, Rahman MK (2019) Effects of arsenic and phosphorus on the growth and nutrient concentration in rice plant. J Biodivers Conserv Bioresour Manage 5(1):31–38

Hu X, Ding Z, Zimmerman AR, Wang S, Gao B (2015) Batch and column sorption of arsenic onto iron-impregnated biochar synthesized through hydrolysis. Water Res 68:206–216

Hussain MM, Bibi I, Niazi NK, Shahid M, Iqbal J, Shakoor MB, Ahmad A, Shah NS, Bhattacharya P, Mao K, Bundschuh J (2021) Arsenic biogeochemical cycling in paddy soil–rice system: interaction with various factors, amendments and mineral nutrients. Sci Total Environ 773:145040

Khalid S, Shahid M, Alothman ZA, Al-Kahtani AA, Murtaza B, Dumat C (2023) Predicting chemical speciation of metals in soil using Visual Minteq. Soil Ecol Lett 5(3):220162

Khalid S, Shahid M, Niazi NK, Rafiq M, Bakhat HF, Imran M, Abbas T, Bibi I, Dumat C (2017) Arsenic behaviour in soil–plant system: biogeochemical reactions and chemical speciation influences. In: Enhancing cleanup of environmental pollutants, Vol 2: Nonbiological approaches, pp 97–140

Khan AZ, Ding X, Khan S, Ayaz T, Fidel R, Khan MA (2020) Biochar efficacy for reducing heavy metals uptake by Cilantro (*Coriandrum sativum*) and spinach (*Spinaccia oleracea*) to minimize human health risk. Chemosphere 244:125543

Khan I, Awan SA, Rizwan M, Ali S, Zhang X, Huang L (2021) Arsenic behavior in soil–plant system and its detoxification mechanisms in plants: a review. Environ Pollut 286:117389

Khan MI, Ahmad MF, Ahmad I, Ashfaq F, Wahab S, Alsayegh AA, Kumar S, Hakeem KR (2022) Arsenic exposure through dietary intake and associated health hazards in the Middle East. Nutrients 14(10):2136

Kumar A, Bhattacharya T, Shaikh WA, Roy A (2024) Sustainable soil management under drought stress through biochar application: immobilizing arsenic, ameliorating soil quality, and augmenting plant growth. Environ Res 259:119531

Kumar R, Lamba J, Adhikari S, Kesara N, Torbert HA (2025) Cotransport and retention of arsenic and pine wood biochar in saturated porous media: fate and interaction mechanisms under solution chemistry and practical feasibility. ACS ES&T Water

Lee HS, Shin HS (2021) Competitive adsorption of heavy metals onto modified biochars: comparison of biochar properties and modification methods. J Environ Manage 299:113651

Leksungnoen P, Wisawapipat W, Ketrot D, Aramrak S, Nookabkaew S, Rangkadilok N, Satayavivad J (2019) Biochar and ash derived from silicon-rich rice husk decrease inorganic arsenic species in rice grain. Sci Total Environ 684:360–370

Li L, Zhang YJ, Novak A, Yang Y, Wang J (2021) Role of biochar in improving sandy soil water retention and resilience to drought. Water 13(4):407

Li Q, Wang L, Fu Y, Lin D, Hou M, Li X, Hu D, Wang Z (2023) Transformation of soil organic matter subjected to environmental disturbance and preservation of organic matter bound to soil minerals: a review. J Soils Sediments 23(3):1485–1500

Lim KT, Shukor MY, Wasoh H (2014) Physical, chemical, and biological methods for the removal of arsenic compounds. BioMed Res Int 503784

Lin G, Wang Y, Wu X, Meng J, Ok YS, Wang CH (2025) Enhancing agricultural productivity with biochar: evaluating feedstock and quality standards. Bioresour Technol Rep 102059

Liu WJ, Jiang H, Yu HQ (2015) Development of biochar-based functional materials: toward a sustainable platform carbon material. Chem Rev 115(22):12251–12285

Long X, Liu F, Zhou X, Pi J, Yin W, Li F, Huang S, Ma F (2021) Estimation of spatial distribution and health risk by arsenic and heavy metals in shallow groundwater around Dongting lake plain using GIS mapping. Chemosphere 269:128698

Luo L, Wang J, Lv J, Liu Z, Sun T, Yang Y, Zhu YG (2023) Carbon sequestration strategies in soil using biochar: advances, challenges, and opportunities. Environ Sci Technol 57(31):11357–11372

Luo T, Huang Z, Li X, Zhang Y (2020) Anaerobic microbe mediated arsenic reduction and redistribution in coastal wetland soil. Sci Total Environ 727:138630

Mašek O, Ronsse F, Dickinson D (2016) Biochar production and feedstock. In: Biochar in European soils and agriculture, pp 17–40

Mensah AK, Marschner B, Shaheen SM, Rinklebe J (2022) Biochar, compost, iron oxide, manure, and inorganic fertilizer affect bioavailability of arsenic and improve soil quality of an abandoned arsenic-contaminated gold mine spoil. Ecotoxicol Environ Saf 234:113358

Meyer S, Genesio L, Vogel I, Schmidt HP, Soja G, Someus E, Shackley S, Verheijen FG, Glaser B (2017) Biochar standardization and legislation harmonization. J Environ Eng Landsc Manag 25(2):175–191

Milton AH, Hore SK, Hossain MZ, Rahman M (2012) Bangladesh arsenic mitigation programs: lessons from the past. Emerg Health Threats J 5(1):7269

Mishra AK, Das R, George Kerry R, Biswal B, Sinha T, Sharma S, Arora P, Kumar M (2023) Promising management strategies to improve crop sustainability and to amend soil salinity. Front Environ Sci 10:962581

Mishra AK, Shinjo H, Jat HS, Jat ML, Jat RK, Funakawa S, Sutaliya JM (2022a) Farmers' perspectives as determinants for adoption of conservation agriculture practices in the Indo-Gangetic plains of India. Resour Conserv Recycl Adv 15:200105

Mishra AK, Sinha DD, Grover D, Roohi, Mishra S, Tyagi R, Sheoran HS, Sharma S (2022b) Regenerative agriculture as climate smart solution to improve soil health and crop productivity thereby catalysing farmers' livelihood and sustainability. In: Towards sustainable natural resources, vol 1, pp 295–309

Mishra S, Dubey RS (2006) Inhibition of ribonuclease and protease activities in arsenic exposed rice seedlings: role of proline as enzyme protectant. J Plant Physiol 163(9):927–936

Moulick D, Samanta S, Sarkar S, Mukherjee A, Pattnaik BK, Saha S, Awasthi JP, Bhowmick S, Ghosh D, Samal AC, Mahanta S (2021) Arsenic contamination, impact and mitigation strategies in rice agro-environment: an inclusive insight. Sci Total Environ 800:149477

Mrabet R, Moussadek R (2022) Development of climate smart agriculture in Africa. In: Conservation agriculture in Africa: climate smart agricultural development, pp 17–65. CABI, GB

Mukhopadhyay R, Bhattacharjee H, Rosen BP (2014) Aquaglyceroporins: generalized metalloid channels. Biochim Biophys Acta (BBA) General Subjects 1840(5):1583–1591

Munera-Echeverri JL, Martinsen V, Strand LT, Zivanovic V, Cornelissen G, Mulder J (2018) Cation exchange capacity of biochar: an urgent method modification. Sci Total Environ 642:190–197

Murtaza G, Ahmed Z, Eldin SM, Ali B, Bawazeer S, Usman M, Iqbal R, Neupane D, Ullah A, Khan A, Hassan MU (2023) Biochar–Soil–Plant interactions: a cross talk for sustainable agriculture under changing climate. Front Environ Sci 11:1059449

Nahar N, Rahman A, Moś M, Warzecha T, Algerin M, Ghosh S, Johnson-Brousseau S, Mandal A (2012) *In silico* and *in vivo* studies of an *Arabidopsis thaliana* gene, ACR2, putatively involved in arsenic accumulation in plants. J Mol Model 18:4249–4262

Natasha N, Shahid M, Khalid S, Bibi I, Naeem MA, Niazi NK, Tack FM, Ippolito JA, Rinklebe J (2022) Influence of biochar on trace element uptake, toxicity and detoxification in plants and associated health risks: a critical review. Crit Rev Environ Sci Technol 52(16):2803–2843

Newell P, Taylor O, Naess LO, Thompson J, Mahmoud H, Ndaki P, Rurangwa R, Teshome A (2019) Climate smart agriculture? Governing the sustainable development goals in Sub-Saharan Africa. Front Sustain Food Syst 3:55

Ngwoke UN, Oduware J, Yusuf H, Omowonuola AA, Ero-Omoghe DE, Onaiwu GE, Ikhajiagbe B (2025) Overcoming production and environmental challenges in Sub-Saharan Africa's rice (*Oryza sativa*) sector: a sustainable approach. Trop J Chem 1(2):1–13

Palansooriya KN, Wong JTF, Hashimoto Y, Huang L, Rinklebe J, Chang SX, Bolan N, Wang H, Ok YS (2019) Response of microbial communities to biochar-amended soils: a critical review. Biochar 1:3–22

Pandey A, Singh SK, Sharma S, Mishra AK, Jatav SS, Patra A, Bahuguna A, Mukharjee S, Yadav B, Pankaj B (2023) Effect of different arsenic and biochar levels on soil microbial population and enzymatic activity. Int J Plant Soil Sci 35(16):443–451

Park S, Lee JY (2023) Carbon removal role and certification standards for biochar. In: Proceedings of the annual conference of Japan society of material cycles and waste management, The 34th annual conference of Japan Society of Material Cycles and Waste Management. Japan Society of Material Cycles and Waste Management, p 297

Parlayıcı Ş, Bahadir M, Pehlivan E (2024) Nanoporous carbonaceous materials (biochar and activated carbon): recent progress and potential applications for arsenic removal. J Dispers Sci Technol 1–22

Pommerrenig B, Diehn TA, Bienert GP (2015) Metalloido-porins: essentiality of Nodulin 26-like intrinsic proteins in metalloid transport. Plant Sci 238:212–227

Rahaman MS, Mise N, Ichihara S (2022) Arsenic contamination in food chain in Bangladesh: a review on health hazards, socioeconomic impacts and implications. Hygiene Environ Health Adv 2:100004

Rai PK, Sonne C, Kim KH (2023) Heavy metals and arsenic stress in food crops: elucidating antioxidative defense mechanisms in hyperaccumulators for food security, agricultural sustainability, and human health. Sci Total Environ 874:162327

Rangabhashiyam S, dos Santos Lins PV, de Magalhães Oliveira LM, Sepulveda P, Ighalo JO, Rajapaksha AU, Meili L (2022) Sewage sludge-derived biochar for the adsorptive removal of wastewater pollutants: a critical review. Environ Pollut 293:118581

Ray PK, Bharti P (2023) Biochar: a quality enhancer for fruit crops. International Year of Millets 77

Reddy AA (2019) The soil health card Scheme in India: lessons learned and challenges for replication in other developing countries. J Nat Resour Policy Res 9(2):124–156

Rehman MU, Khan R, Khan A, Qamar W, Arafah A, Ahmad A, Ahmad A, Akhter R, Rinklebe J, Ahmad P (2021) Fate of arsenic in living systems: Implications for sustainable and safe food chains. J Hazard Mater 417:126050

Sarkar A, Paul B, Darbha GK (2022) The groundwater arsenic contamination in the Bengal Basin—a review in brief. Chemosphere 299:134369

Shahid M, Khalid S, AL-Othman Z, Al-Kahtani AA, Dumat C, Murtaza B (2025) A Comparative study on chemical speciation of trace metals as a function of solution pH: a multivariate analysis and simulation approach. Commun Soil Sci Plant Anal 1–13

Shahid M, Khalid S, Niazi NK, Murtaza B, Ahmad N, Farooq A, Zakir A, Imran M, Abbas G (2021) Health risks of arsenic buildup in soil and food crops after wastewater irrigation. Sci Total Environ 772:145266

Sharma I (2012) Arsenic induced oxidative stress in plants. Biologia 67(3):447–453

Sharma R, Nanda R, Bhagat N (2022) Background level, occurrence, speciation, bioavailability, uptake, detoxification mechanisms and management of arsenic polluted soil. In: Appraisal of metal (loids) in the ecosystem. Elsevier, pp 221–254

Sharma S, Kaur I, Nagpal AK (2021) Contamination of rice crop with potentially toxic elements and associated human health risks—a review. Environ Sci Pollut Res 28:12282–12299

Singh S, Rajput VD, Upadhyay SK, Minkina T (2023) Arsenic contamination in rice agroecosystems: mitigation strategies for safer crop production. J Plant Growth Regul 42(10):6413–6424

Smith E, Naidu R, Alston AM (1998) Arsenic in the soil environment: a review. Adv Agron 64(149195):60504

Srivastava S, Pathak S, Ponsin M, Hensawang S, Chanpiwat P, Yoeurn C, Phan K (2021) Sustainable solutions to arsenic accumulation in rice grown in south and south–east Asia. Crop Pasture Sci 73(2):149–159

Strawn DG (2021) Sorption mechanisms of chemicals in soils. Soil Syst 5(1):13

Sun Y, Wang T, Bai L, Han C, Sun X (2022) Application of biochar-based materials for remediation of arsenic contaminated soil and water: preparation, modification, and mechanisms. J Environ Chem Eng 10(5):108292

Vereš J, Koloničnýa J, Ochodeka T (2014) Biochar status under international law and regulatory issues for the practical application. Chem Eng 37:799–804

Verheijen FG, Bastos AC, Schmidt HP, Jeffery S (2019) Biochar and certification 1. In: Sustainability certification schemes in the agricultural and natural resource sectors. Routledge, pp 113–136

Vithanage M, Herath I, Joseph S, Bundschuh J, Bolan N, Ok YS, Kirkham MB, Rinklebe J (2017) Interaction of arsenic with biochar in soil and water: a critical review. Carbon 113:219–230

Wang L, O'Connor D, Rinklebe J, Ok YS, Tsang DC, Shen Z, Hou D (2020) Biochar aging: mechanisms, physicochemical changes, assessment, and implications for field applications. Environ Sci Technol 54(23):14797–14814

Wang S, Gao B, Li Y (2016) Enhanced arsenic removal by biochar modified with nickel (Ni) and manganese (Mn) oxyhydroxides. J Ind Eng Chem 37:361–365

Xiang L, Harindintwali JD, Wang F, Redmile-Gordon M, Chang SX, Fu Y, He C, Muhoza B, Brahushi F, Bolan N, Jiang X (2022) Integrating biochar, bacteria, and plants for sustainable remediation of soils contaminated with organic pollutants. Environ Sci Technol 56(23):16546–16566

Xie T, Reddy KR, Wang C, Yargicoglu E, Spokas K (2015) Characteristics and applications of biochar for environmental remediation: a review. Crit Rev Environ Sci Technol 45(9):939–969

Yang F, Xu Z, Huang Y, Tsang DC, Ok YS, Zhao L, Qiu H, Xu X, Cao X (2021) Stabilization of dissolvable biochar by soil minerals: release reduction and organo-mineral complexes formation. J Hazard Mater 412i, 125213

Yang X, Pan H, Shaheen SM, Wang H, Rinklebe J (2021b) Immobilization of cadmium and lead using phosphorus-rich animal-derived and iron-modified plant-derived biochars under dynamic redox conditions in a paddy soil. Environ Int 156:106628

Yao BM, Chen P, Zhang HM, Sun GX (2021) A predictive model for arsenic accumulation in rice grains based on bioavailable arsenic and soil characteristics. J Hazard Mater 412:125131

Younas Z, Rahman U, Ikram M, Raja NI (2024) Biochar and micronutrient availability: problem and future prospects. In: Biochar-assisted remediation of contaminated soils under changing climate. Elsevier, pp 101–127

Zama EF, Reid BJ, Sun GX, Yuan HY, Li XM, Zhu YG (2018) Silicon (Si) biochar for the mitigation of arsenic (As) bioaccumulation in spinach (*Spinacia oleracea*) and improvement in the plant growth. J Clean Prod 189:386–395

Zhang C, Liu L, Zhao M, Rong H, Xu Y (2018) The environmental characteristics and applications of biochar. Environ Sci Pollut Res 25:21525–21534

Zhang G, Liu F, Liu H, Qu J, Liu R (2014) Respective role of Fe and Mn oxide contents for arsenic sorption in iron and manganese binary oxide: an X-ray absorption spectroscopy investigation. Environ Sci Technol 48(17):10316–10322

Zhang G, Qu J, Liu H, Liu R, Wu R (2007) Preparation and evaluation of a novel Fe–Mn binary oxide adsorbent for effective arsenite removal. Water Res 41(9):1921–1928

Zhao L, Cao X, Mašek O, Zimmerman A (2013) Heterogeneity of biochar properties as a function of feedstock sources and production temperatures. J Hazard Mater 256:1–9

Zvobgo G, LwalabaWaLwalaba J, Sagonda T, Mapodzeke JM, Muhammad N, Shamsi IH, Zhang G (2018) Phosphate alleviates arsenate toxicity by altering the expression of phosphate transporters in the tolerant barley genotypes. Ecotoxicol Environ Saf 147:832–839

Chapter 6
Application of Artificial Intelligence and Machine Learning in Biochar Optimization

Mohammad Vaheed, Chandra Bhushan, Sunil Kumar Verma, Sameer Shrivastava, Shaik Mohammed Shahid, Srishti Singh, Sidra Qidwai, Ashish Kumar Verma, Adesh, and Lakey Radha

Abstract Artificial intelligence (AI) and machine learning (ML) offer transformative approaches for optimizing biochar production and application, overcoming challenges posed by the inherent variability in feedstock composition and pyrolysis conditions. Traditional optimization methods like Response Surface Methodology (RSM) struggle with the complex, nonlinear relationships governing biochar yield and its physicochemical properties (e.g., surface area, porosity, pH, adsorption capacity). AI/ML techniques, including Artificial Neural Networks (ANN), Support Vector Machines (SVM), Random Forests (RF), and Adaptive Neuro-Fuzzy Inference Systems (ANFIS), effectively model these intricate interactions. ML algorithms leverage extensive datasets to predict biochar characteristics (yield, fixed carbon, specific surface area) and performance (e.g., heavy metal adsorption efficiency) based on inputs like feedstock type, pyrolysis temperature, residence time and heating rate. Techniques like Principal Component Analysis (PCA) identify key influential variables. Hybrid models, integrating ANN with metaheuristic algorithms includes Genetic Algorithms (GA) or Particle Swarm Optimization (PSO), further enhance prediction accuracy and enable multi-objective optimization (e.g., maximizing yield while enhancing adsorption). Gradient Boosting Regression (GBR) and Gaussian

M. Vaheed (✉) · C. Bhushan · S. K. Verma · S. Shrivastava · L. Radha
Department of Agronomy, Institute of Agricultural Sciences, Banaras Hindu University, Varanasi, Uttar Pradesh, India
e-mail: vaheedias@bhu.ac.in

S. M. Shahid
Department of Agronomy, ICAR-IARI-Central Institute for Cotton Research, Nagpur, Nagpur, Maharashtra, India

S. Singh · S. Qidwai · A. K. Verma
Department of Agronomy, Acharya Narendra Deva University of Agriculture and Technology Kumarganj, Ayodhya (U.P.), Faizabad, India

Adesh
Department of Agronomy, Chandra Shekhar Azad University of Agriculture and Technology Kanpur, Kanpur, India

D. Moulick et al. (eds.), *Biochar: Achieving Agro-Environmental Sustainability*, Sustainability Solutions, https://doi.org/10.1007/978-3-032-19855-6_6

Process Regression (GPR) provide robust predictions with uncertainty quantification, even for limited datasets. These data-driven approaches significantly reduce experimental burden and cost compared to traditional trial-and-error methods. They facilitate the tailored design of biochar with properties optimized for specific applications, such as soil amendment, carbon sequestration, or pollutant remediation. The integration of AI/ML represents a critical advancement towards intelligent, efficient and scalable biochar technology for environmental sustainability.

Keywords Artificial intelligence · Artificial neural network · Biochar · Feedstock · Pyrolysis

6.1 Introduction

Biochar, a carbon-rich material derived from the thermal conversion of biomass in an oxygen-limited conditions, has evolved as a sustainable and versatile solution in addressing critical environmental and agricultural challenges. Derived from feedstocks viz., agricultural residues, forestry waste and organic matter, biochar is primarily generated through processes including pyrolysis, hydrothermal carbonization, gasification and torrefaction, with pyrolysis being the most widely utilized method (Manikandan et al. 2023). Its unique physicochemical properties, such as high surface area, porosity and stability, make biochar a valuable resource for applications in soil enhancement, carbon sequestration, pollutant remediation and various industrial processes (Ukoba and Jen 2022; Huang et al. 2021). The growing interest in biochar stems from its potential to mitigate climate change, improve soil health and contribute to circular economy principles by utilizing organic waste for high-value products.

The role of biochar in soil amendment is particularly significant, as its application improves soil physical, chemical and biological properties by enhancing nutrient retention, increasing water-holding capacity, improve cation exchange capacity (CEC) and fostering microbial activity. Furthermore, biochar‘s ability to adsorb contaminants like heavy metals, pesticides and pharmaceutical residues underscores its utility in pollution remediation efforts (Jeyasubramanian et al. 2021; Lakshmi et al. 2021). Additionally, biochar‘s potential contribution to carbon sequestration is evident in its capacity to store carbon in soils for extended periods, thereby reducing atmospheric CO_2 levels and mitigating greenhouse gas emissions (Wang and Yao 2023). Studies have shown that biochar application can also suppress the emission of potent greenhouse gases eg., methane (CH_4) and nitrous oxide (N_2O) by altering soil microbial processes (Ruan et al. 2019). However, despite its promising applications, biochar production and utilization are hindered by several challenges. Variability in feedstock composition, differences in pyrolysis conditions and environmental influences contribute to heterogeneity in biochar properties, affecting its performance in diverse applications (Liao and Yao 2021). The optimization of biochar for specific uses necessitating advanced analytical and optimization approaches of its physicochemical characteristics and their relationship with production parameters.

The advent of artificial intelligence (AI) and machine learning (ML) offers a transformative approach to addressing the challenges associated with biochar optimization. AI-driven predictive modelling enables the assessment of complex, nonlinear interactions between biochar properties and production parameters, facilitating process optimization and quality control (Chen et al. 2023). Machine learning algorithms like artificial neural networks (ANN), support vector machines (SVM) and random forests (RF) have been employed to predict biochar yield, physicochemical attributes and adsorption efficiencies with high accuracy. Moreover, ML techniques have been instrumental in optimizing biochar-based pollutant remediation by predicting the adsorption behaviour of various contaminants, including heavy metals, pesticides and pharmaceuticals (Manatura et al. 2023; Yuan et al. 2022). By leveraging large datasets and advanced computational models, AI facilitates the identification of optimal biochar synthesis conditions, enhances material characterization and improves the efficiency of biochar applications in diverse environmental contexts (Goswami et al. 2022).

Deep learning, a subset of ML, offers further advancements by enabling highly complex pattern recognition and predictive analytics in biochar research. Convolutional Neural Networks (CNNs) and Recurrent Neural Networks (RNNs) have shown promise in modeling biochar characteristics, predicting its adsorption performance and optimizing process conditions with minimal experimental input (Ascher et al. 2022). The integration of AI and ML with biochar research not only accelerates data analysis but also contributes to the development of intelligent biochar systems tailored for specific environmental and agricultural applications. Furthermore, the application of explainable AI (XAI) methods, such as Shapley Additive Explanations (SHAP), enhances the interpretability of ML models, allowing researchers to understand key variables influencing biochar performance.

Despite these advancements, there remains a need for comprehensive studies that systematically explore the intersection of biochar science and AI-driven methodologies. A holistic framework that integrates machine-learning techniques with biochar production, characterization and application can pave the way for innovative solutions in sustainable agriculture, waste management and environmental remediation (Chen et al. 2023). This chapter aims to provide a detailed overview of current research trends in biochar applications and the role of artificial intelligence in optimizing biochar technology, offering insights into the future prospects and challenges associated with biochar-based solutions for environmental sustainability.

6.2 Need for Optimization in Biochar Production

Optimization of biochar production processes is crucial to:

- Maximize yield and carbon content.
- Improve physicochemical properties like surface area, porosity and pH.
- Tailor biochar characteristics to specific applications (e.g., soil amendment, adsorbent, catalyst support).
- Reduce environmental impact and energy consumption.

6.3 Biochar Production and Optimization Parameters

Biochar, a carbon-rich byproduct of biomass pyrolysis under oxygen-limited conditions, plays a crucial role in soil amendment, carbon sequestration and pollutant removal. Its production and optimization depend on factors e.g., feedstock selection, pyrolysis conditions and physicochemical properties, all of which influence its yield and functionality (Gupta et al. 2023). Several research efforts on optimizing biochar process parameters are compiled in Table 6.1. Different biomass sources, including agricultural residues, wood waste and organic waste, affect biochar's carbon retention and porosity, with lignin-rich feedstocks producing more stable biochar (Tomczyk et al. 2020). Pyrolysis conditions, particularly temperature, residence time and heating rate, determine biochar's structural and adsorption characteristics, where low-temperature pyrolysis (300–400 °C) yields more biochar, whereas high-temperature pyrolysis (600–900 °C) enhances porosity and adsorption properties (Tripathi et al. 2016). Physicochemical traits like surface area, porosity and functional groups influence biochar's effectiveness in pollutant removal and soil improvement, with functional groups like hydroxyl (–OH) and carboxyl (–COOH) enhancing adsorption (Cho et al. 2024). Optimization techniques, including Design of Experiments (DOE) and Response Surface Methodology (RSM), are commonly used to refine biochar production parameters, while emerging machine learning (ML) models, includes Random Forest (RF) and Artificial Neural Networks (ANN), provide predictive insights for process optimization (Leng et al. 2022). An integrated approach considering feedstock properties, pyrolysis parameters and advanced optimization techniques is essential for producing high-quality biochar with applications in sustainable agriculture, carbon sequestration and environmental remediation (Gupta et al. 2023). The parameters of the reaction through the pyrolytic are essentially involved in the formation of biochar. The properties of biochar are essentially determined by the feedstocks, the temperature, the particle size and the heating rate. These elements directly affect the quality of biochar (Leng and Huang 2018). Meanwhile, the yield of biochar is influenced by the type and conditions of pyrolysis. A thorough understanding of biochar property analysis is critical for selecting biochar usage. Various biomass sources, including natural, agricultural residues, wood, solid waste, etc., can produce biochar (Senthil and Lee 2021). Biochar is produced in greater quantities from solid and animal waste than other material resources, including wood biomass, crop residues, etc.

6.3.1 Feedstock's Selection

Biomass is a solid material made from natural substances that come from both living and non-living things (Amalina et al. 2022). It is mainly of two types: woody and non-woody. Woody biomass includes things like tree trunks and leftover parts from forests (Leng et al. 2021; Yang et al. 2016). This type has low moisture and waste, is not very reactive, is dense, and gives a lot of energy when used (Danish and Ahmed 2018). On the other hand, non-woody biomass such as animal dung and waste from farms or factories has more moisture and debris, reacts more easily, is less dense, and gives less energy (Tomczyk et al. 2020). One important feature of biomass is how much moisture it contains. Moisture can be in the form of water, steam, or absorbed in the material. If the moisture is too high, it reduces the amount of char produced and takes more energy to reach the right temperature during heating (Nidheesh et al. 2021; Oliveira et al. 2017). So, using dry biomass is better for making biochar, as it saves energy and time.

Biomass, a composite, solid, biological, widely available, organic, or inorganic material generated from recently living organisms or non-living organisms, has been primarily utilized as a raw material for biochar preparation via pyrolysis. Nonwoody biomass primarily includes agricultural, industrial, urban and animal wastes, whereas woody biomass contains residues generated from trees and forests. Nonwoody biomass possesses higher ash and moisture content, lower bulk density, low calorific value and much less void age than woody biomass (Vassilev et al. 2012).

Biomass with high moisture content is not suitable for biochar production because it needs more energy and time during pyrolysis compared to dry biomass. When moist materials like sewage sludge are heated, they release a lot of steam, which reduces the amount of biochar produced through steam-related reactions (Xiong et al. 2013). Also, if the raw material has different levels of moisture, it can change the physical and chemical properties of the biochar made. The moisture in the material affects the surface structure of the biochar. When moisture is low, the surface shows more clear polyaromatic or graphitic patterns (Darmstadt et al. 2000)

Biomass is mainly made up of lignin, hemicellulose, and cellulose. Biomass with high amounts of lignin is good for making char, while cellulose helps form tar, which is a mix of liquids and solids like ketones, aldehydes, and organic compounds. However, as pyrolysis continues, the amount of char from lignin and cellulose starts to reduce because particles stick together. Lignin-rich biomass is more heat-resistant and gives better mass and energy output during biochar production (Ma et al. 2019).

Fracture of relatively weak bonds and subsequent condensation into a solid structure are the major processes behind biochar preparation from lignin rich biomass. Coniferous lignin rich biomass is thermally more stable and produces higher biochar yield compared to deciduous lignin containing biomass due to the difference in

Table 6.1 Summary of research on optimization of biochar production parameters

Biochar type	Process parameters	Optimization method	Optimized values	Key findings	References
Parthenium	T, t	Surface plot	T: 300–350 °C, t: 30–45 min	SOMYI ↓ with ↑ T, moderate t better	Kumar et al. (2013)
Eichhornia	T, t	Surface plot	T: 300–350 °C, t: 30–45 min	Similar to above	Masto et al. (2013)
Marine brown algae	CD, T, Q_h	RSM (BBD)	CD: 38.78 mA/cm^2, T: 584.1 °C, Q_h: 6.91 °C/min	Q_h most influential for phosphate adsorption	Jung et al. (2016)
Oil palm	P, T, N_2 flow	RSM (CCD)	H_2: P = 510 W, T = 700 °C, N_2 = 1200 cm^3/min	N_2 > P > T for H_2 yield; T > P > N_2 for biochar yield	Hossain et al. (2017)
Sagwan sawdust	T, h, QN	RSM (BBD)	T: 639.5 °C, h: 8 cm, QN: 181.6 ml/min	↑ T and h ↑ bio-oil yield; ↑ T and QN ↓ biochar yield	Gupta and Mondal (2019)
Pomegranate peel	T, t, particle size	RSM (CCD)	T: 300 °C, t: 20 min, size: 3 mm	T is most critical for yield	Siddiqui et al. (2019)
Rice husk	T, t, Q_h	RSM (CCD)	T: 432°C, t: 40 min, Q_h: 4 °C/min	↓ T, t, Q_h ↑ energy yield; strong model fit	Yadav et al. (2019)
Eichhornia stem	T, t, Q_h	RSM (BBD)	T: 393°C, t: 2.42 h, Q_h: 15.56 °C/min	T > t > Q_h for Cd adsorption	Zhou et al. (2019)
Cabbage waste	T, sample size, sample mass	RSM (CCD)	T: 360 °C, size: 0.9 mm, mass: 45 g	T > size > mass; low T gives better properties	Pradhan et al. (2020)
Rice husk	T, t, Q_h	Taguchi	T: 300 °C, t: 5400 s, Q_h: 20 °C/min	T most influential for HHV and FC	Vieira et al., 2020
Water hyacinth	T, t, Q_h	RSM (BBD)	T: 433 °C, t: 2.65 h, Q_h: 19.96 °C/min	T > Q_h > t for Pb^{2+} adsorption	Zhou et al. (2020)

(continued)

Table 6.1 (continued)

Biochar type	Process parameters	Optimization method	Optimized values	Key findings	References
Arecanut husk	T, t	Surface plot	T: 250 °C, t: 60 min	Moderate T and t favoured for SOMYI	Gujre et al. (2022)
Peanut shell	T, Q_h, QN	RSM (BBD)	T: 400 °C, Q_h: 20 °C/min, QN: 100 ml/min	↑ QN and T ↓ biochar yield	Kumar et al. (2021)
Banana peel waste	T, t, Q_h	RSM (CCD)	T: 356.1 °C, t: 180 min, Q_h: 14.7°C/min	T ↓ yield and O/C ratio	Te et al. (2021)
Maize-cob biomass waste	T, t, PS	RSM (BBD)	T: 463.7 °C, t: 15.3 min, PS: 3.8 mm	T, t, PS ↑ yield and HHV	Adelawon et al. (2022)
Algal biomass	T, t, QN	RSM (BBD)	T: 450 °C, t: 30 min, QN: 0.3 l/min	T, t, QN ↓ biochar yield	Jeeru et al. (2023)
Date stones, SCG, CM	FFC, T, Q_h	RSM (BBD)	T: 350°C, Q_h: 7 °C/min, FFC: 14–22%	T, FFC ↓, Q_h ↑ yield; quadratic model significant	Vieira et al. (2020)

Note T = Temperature; t = resident time; ↓ = decreasing; ↑ = increasing; Qh = heating rate (°C/min); QN=nitrogen flow rate; CD = current density (mA/cm^2); P = power input (Watt) in microwave-assisted pyrolysis; PS = particle size (mm); FFC = fixed feedstock composition; RSM = response surface methodology; BBD = Box-Benken design (a type of RSM); HHV = higher heat value; CEC = cation exchange capacity; ECE = electrical conductivity equivalent; ζ potential = Zeta potential (indicator of surface charge), ANOVA = Analysis of Variance; DMRT = Duncan's multiple range test (a statistical test for significance); R^2 = Coefficient of determination (measures model fit in regression); F-value = F-statistic from ANOVA indicating overall model significance; *p*-value = probability value (used to assess significance in statistical testing, usually $p < 0.05$); Pb^{2+}, Cd^{2+} = Lead and Cadmium ions (tested for adsorption/removal efficiency); SCG = spent coffee grounds; CM = cow manure; DS = date stones

degradation kinetics. Coconut husk had more content lignin (28.2%) compared to coconut frond feedstock (21.5%), which makes coconut husk more favourable for char generation during slow pyrolysis (Shariff et al. 2016). Apart from yield, the high proportion of lignin in plant biomass also helps to increase the carbon as well as ash content of produced biochar (Wang et al. 2015a, b). Different inorganic constituents also affect the produced biochar yield by delivering a higher ash content and slightly volatile matter content. Silica (SiO2) and other diverse inorganic constituents influence biochar production yield from sewage sludge (Zielinska et al. 2015).

The ash content and the carbon-to-nitrogen (C/N) ratio of biochar mostly depend on the type of raw biomass used (Tomczyk et al. 2020). Woody biomass produces biochar with less ash compared to non-woody biomass, making it easier to transport and better for mixing into soil, as it doesn't blow away easily in the wind. On the other hand, non-woody biochar contains salts like sodium and potassium chlorides and carbonates, which raise its pH. This higher pH can help improve acidic soils (Mukome et al. 2013). During pyrolysis, certain chemicals like levoglucosan and its by-products (such as levo glucosenone, furfural, and 5-methyl furfural) are formed, which add oxygen-rich groups to the biochar surface and increase its pH.

The ability of biochar to hold and exchange nutrients known as Cation Exchange Capacity (CEC) varies with the type of biomass used. For example, biochar made from crop straw has a CEC of about 222.32, which is much higher than that of hardwood biochar, which has a CEC of 111.36 (Yang et al. 2015). The presence of alkali and alkaline earth metals in the biomass helps form oxygen-rich surface groups like carboxylic, phenolic, and lactonic groups. These surface features improve the CEC of biochar (Cely et al. 2015a, b; Tag et al. 2016).

6.3.2 Lignin-Rich Biomass

Lignin-rich biomass significantly influences biochar properties, particularly by yielding higher biochar content with increased carbon stability. Lignin, being a complex aromatic polymer, decomposes at higher temperatures compared to cellulose and hemicellulose, leading to a more thermally stable biochar. During pyrolysis, the structural rigidity and resistance of lignin contribute to reduced volatile losses, resulting in a greater proportion of fixed carbon in the final biochar product. Studies indicate that lignin-rich residues, such as digestate and olive stones, produce biochar with lower volatile matter content and higher fixed carbon percentages, making it more suitable for long-term carbon sequestration and pollutant adsorption. The decomposition mechanism of lignin involves the breaking of weaker bonds at lower temperatures, like hydrogen and C–OH bonds, followed by the cleavage of stronger β-O-4 linkages at higher temperatures, leading to the formation of thermally stable char. As a result, biochar derived from lignin-rich biomass exhibits enhanced porosity and a higher aromatic carbon content, which improves its adsorption capacity for environmental applications. Additionally, the high carbon retention in lignin-derived biochar contributes to its suitability for soil amendment and carbon sequestration, as it resists microbial degradation more effectively than cellulose- and hemicellulose-derived biochars. Studies have demonstrated that lignin-rich biochar, especially from biomass sources such as kraft lignin and hydrolysis lignin, yields up to 45.9% char at 800 °C, significantly higher than cellulose and hemicellulose, which produce only 19.1% and 23.5%, respectively. This higher yield and carbon stability make lignin-rich biochar an ideal candidate for applications in wastewater treatment, soil fertility enhancement and carbon capture (Gul et al. 2021; Wan et al. 2022 and Kopp Alves et al. 2024).

6.3.3 Ash Content

High ash content in agricultural residues significantly influences the surface area and porosity of biochar by affecting its structural and physicochemical properties. Biochars derived from biomass with high ash content tend to exhibit lower specific surface areas and reduced porosity due to the presence of mineral residues that block or fill micropores. During pyrolysis, the decomposition of organic matter and the volatilization of labile components create a porous structure, but excessive ash content can lead to pore clogging, reducing the overall adsorption capacity of biochar. Additionally, ash components such as silica, calcium and magnesium can alter the biochar's surface chemistry, affecting its ability to retain nutrients and adsorb contaminants. Studies indicate that biochar produced from crop residues and manures, which typically have higher ash content compared to wood-based biochars, often exhibits lower porosity and specific surface area due to the accumulation of inorganic material in the carbon matrix. As a result, high-ash biochars may have reduced effectiveness in applications such as soil amendment and pollutant adsorption, where surface area and porosity are critical factors in determining biochar performance (Tomczyk et al. 2020; Wang et al. 2020).

6.3.4 Pyrolysis Conditions

6.3.4.1 Temperature

As explained earlier, pyrolysis is a common method used to turn biomass into biochar. This is done by heating the biomass at high temperatures in an environment with little or no oxygen. The temperature used during pyrolysis affects the physical and chemical properties of the biochar, such as its makeup, pore size, surface area and surface features (Jawad et al. 2019). At higher temperatures, more gases and vapours are released, which removes elements like carbon, hydrogen, and oxygen. This leads to a lower O/C and H/C ratio and increases the amount of stable carbon in the biochar (Han et al. 2018). This makes the biochar more stable, and this effect becomes stronger as the temperature increases.

When the pyrolysis temperature is above 500 °C, the biochar becomes more water-repellent (hydrophobic), gains a larger surface area, and develops more tiny pores. These features make it useful for removing organic pollutants (Sakhiya et al. 2020). On the other hand, if the temperature is below 500 °C, the biochar has smaller pores, a smaller surface area, and more oxygen-containing surface groups. This kind is better for removing inorganic contaminants (Senthil and Lee 2021). Higher temperatures also raise the pH of the biochar because more ash is left behind (Leng et al. 2021; Oliveira et al. 2017).

Among all the factors in the pyrolysis process, temperature plays the most important role in determining how much biochar is produced and what its qualities are. Pyrolysis usually happens in three steps:

1. **Pre-pyrolysis (Room temperature to 200 °C)**: Moisture and light gases are removed and this helps form some functional groups.
2. **Main pyrolysis (200–500 °C)**: Rapid breakdown of cellulose and hemicellulose takes place.
3. **High-temperature pyrolysis (> 500 °C):** Lignin and other materials break down, forming the final carbon-rich product.

As the pyrolysis temperature increases, the amount of biochar produced generally decreases because more liquids and gases are formed instead (Tripathi et al. 2016). This pattern is seen in different materials like maize straw, pine sawdust, and sugarcane bagasse between 200 and 700 °C, no matter the heating speed or conditions (Luo et al. 2015). Still, higher temperatures improve surface area, pH, carbon content, and volatile matter, but reduce the cation exchange capacity (CEC) and number of useful surface groups (Tomczyk et al. 2020).

The fragmentation of organic matter and generation of micropores are the significant reasons behind the increased surface area and the large number of micropores at higher temperatures of pyrolytic transformation. As the carbonization temperature rises, dissociation of aliphatic alkyls and esters group occurs and also aromatic lignin core is exposed, which helps to enhance the surface area of produced biochar (Murtaza et al. 2022a, b). Also, externally attainable surface areas are exposed at higher carbonization temperatures by eliminating the pore-blocking substance through thermal cracking. Extensive numbers of pores and greater surface area in the biochar offer potential habitats for soil microbiota, helping to form diverse microbial community structures and their proliferation inside the soil (Yu et al. 2019a, b). The pore size distribution is the predominant factor that increases the specific surface area of biochar. The enlarged porous surface is advantageous for soil nutrients, moisture retention, adsorption characteristics and enriched transfer of nutrients to the plant (Pradhan et al. 2018; Hussain et al. 2020).

In most types of biochar, the amount of oxygen (O), hydrogen (H), and nitrogen (N) goes down as the pyrolysis temperature increases. This is because higher temperatures remove water and other light substances from the biomass. As the temperature rises, the H:C and O:C ratios also decrease, showing that the biochar becomes richer in stable, ring-shaped (aromatic) carbon structures (Weber and Quicker 2018). Other ratios like H:C, (C + H):O, and (N + O):C, which show properties like how stable, polar, or chemically reactive the biochar is, are also mainly controlled by the pyrolysis temperature (Chen et al. 2018).

At very high temperatures (above 650 °C), the surface of biochar becomes water-repellent (hydrophobic) because lignin in the biomass turns into hydrophobic compounds called polycyclic aromatic hydrocarbons (PAHs). In contrast, biochar made at 500 °C or lower is more water-attracting (hydrophilic) (Ghani et al. 2013). When pyrolysis happens between 350 and 650 °C, new surface groups form like phenol, carboxyl, ether, lactone, and others through breaking and reshaping of chemical bonds (Tomczyk et al. 2020). The carbohydrates in cellulose and lignin help form these oxygen-rich surface groups. But at higher temperatures, there is less hydrogen and oxygen left in the biochar, which means fewer of these helpful surface groups are present (Ahmad et al. 2014a, b). These oxygen-rich groups help biochar attract and hold heavy metals, making it good for cleaning up polluted soil or water (Chen et al. 2018). Biochar made at lower temperatures keeps more plant-like and soft structures, showing a variety of organic features.

The cation exchange capacity (CEC) which means how well the biochar can hold and exchange nutrients is strongly affected by how many oxygen-containing groups are on its surface. Many studies have shown that as the pyrolysis temperature goes up, the CEC usually goes down. This is because high temperatures reduce oxygen-rich groups and create more stable, non-reactive carbon structures. The way these oxygen groups are spread across the surface plays a key role in determining the biochar's CEC (Banik et al. 2018a, b).

Increased pyrolysis temperature helps release a large quantity of volatile matter and volatilize the intermediate melts. The release of volatile matter from micropores at higher temperatures enhances the biochar's surface area and accessibility to ions. Usually, at a lower temperature, biochar contains carboxyl, carbonyl, hemi acetyl and hydroxyl, while biochar produced at high carbonization temperature contains pyranones, quinines and ether. The quantity of volatile matter present in the biochar majorly affects the N availability, sorption capacity and stability of the biochar. Furthermore, volatile matter in biochar also had a promising impact on plant growth, either by inhibiting root growth or by microbial decomposition of labile carbon (Shaaban et al. 2014; Tomczyk et al. 2020).

Total base cations or carbonate contents are increased at rising carbonization temperature, resulting in the biochar's increased alkaline behaviour. Uncoupling of alkaline salts from the organic matter at high pyrolysis temperature (> 300 °C) causes an increase in pH values. In contrast, organic acid and phenolic acid are produced at lower carbonization temperatures (250–300 °C), decreasing pH value (Ding et al. 2014a, b; Yu et al. 2014).

6.3.4.2 Residence Time

Residence time Extending the residence period at a low pyrolysis temperature (300 °C) resulted in a gradual reduction in biochar productivity and increased biochar pH and iodine adsorption. Nevertheless, increasing residence period at a higher pyrolysis temperature (600 °C) have a more negligible impact on biochar production or pH, while decreasing biochar iodine sorption capacity (Zhou et al. 2020). Residence time, or how long the biomass is heated during pyrolysis, is an important factor that affects both the amount and quality of the biochar produced. When the biomass is held at a certain temperature for a longer time, the biochar yield usually goes down, but the quality improves. This means the easily decomposable organic parts are reduced, hydrogen and oxygen levels drop, less volatile matter remains, and more fixed carbon is formed. The structure of the biomass also breaks down more, leading to a larger surface area and making the biochar more resistant to microbial breakdown (Leng and Huang 2018). At lower temperatures (300–400 °C), increasing the holding time makes the biochar less water-repellent (less hydrophobic). Less hydrophobic biochar is better for use in soil, as it helps improve water movement and water-holding ability. How well biochar absorbs water depends on its organic content, surface features, and pore size. Some studies also suggest that shorter residence times can lead to faster release of carbon from biochar when it is added to soil (Tripathi et al. 2016). However, when the pyrolysis temperature goes above 550 °C, the effect of residence time becomes very small, as most of the carbonization process is already complete. In general, temperature has a stronger effect than residence time, so it's hard to clearly separate the impact of residence time on biochar yield and properties (Leng and Huang 2018; Tripathi et al. 2016).

6.3.4.3 Pre-treatment

The way biomass is prepared before pyrolysis, called pre-treatment, has a big effect on the quality and features of the biochar. Pre-treatment is the first step in turning different materials into biochar. These methods can be physical, chemical, or biological. Common pre-treatment techniques include soaking the material in liquids or reducing its particle size. Breaking the biomass into smaller pieces usually increases the amount of biochar produced. For example, pine wood was soaked in a mild acid solution before pyrolysis to improve results (Yaashikaa et al. 2020a, b). Adding substances like nitrogen or metals during pre-treatment can also help improve the formation and stability of biochar. Soaking or steaming the biomass can change its chemical makeup and overall properties. Strong chemicals such as acids, bases, and oxidizers are sometimes used to create biochar with a larger surface area, special pore structures, and useful surface features. The type of biomass determines how it should be physically prepared. For example, dried sludge is often oven-dried overnight, crushed, filtered, and stored in sealed containers to control its moisture.

Biological methods, like using bacteria through anaerobic digestion or biofuel processing, can also improve biomass before pyrolysis. This type of treatment helps make biochar with better properties, including a larger surface area and stronger ability to absorb pollutants. Different feedstocks and uses of biochar require different pre-treatment methods whether physical, chemical, or biological. After the biochar is produced, it can also be improved further using post-treatment methods. These include physical or chemical changes that boost surface area, pore size, surface chemistry, and add functional materials like nanoparticles. However, current post-treatment technologies are still being developed and improved.

6.3.5 Heating Rate

The heating rate is an important factor in the pyrolysis process because it affects what kind of products are formed, how much biochar is made, and the final properties of the biochar—such as its shape, surface features, and chemical makeup. When the heating rate is low, the biomass breaks down more slowly, which helps increase the amount of solid biochar produced and reduces the chance of cracking. But when the heating rate is high, the biomass breaks down quickly into gases and liquids, which reduces the amount of solid biochar formed (Tripathi et al. 2016). Studies have shown that increasing the heating rate reduces biochar yield for different types of biomass, especially at temperatures between 400 and 600 °C. For example, Fu et al. (2012) found that rice straw biochar yield dropped by about 16% when the heating rate increased between 600 and 1000 °C. A faster heating rate also changes the surface structure of biochar, like in the case of rapeseed stems, due to the quick breakdown of organic materials. Low heating rates help keep the biomass structure more intact, while high heating rates can cause changes like swelling, melting, and breaking of cell walls (Tripathi et al. 2016). At high heating rates, small ring-like structures were found in the biochar, although some parts of the biomass stayed unchanged.

6.3.5.1 Post-processing Treatments

Post-processing treatments significantly influence biochar characteristics by altering its nutrient content, stability, surface area and suitability for various applications. Laghari et al. (2021) evaluated two primary post–processing treatments—char oxidation and char steam gasification—to enhance biochar properties for use as a soil amendment and phosphorus (P) fertilizer. Char oxidation, performed at 550 °C in limited airflow, increases the concentrations of essential nutrients such as phosphorus (P), potassium (K) and magnesium (Mg) while reducing nitrogen (N) and sulphur (S). However, this process also leads to carbon loss, reducing biochar's potential for carbon sequestration and lowers its BET surface area, thereby affecting its adsorption properties. In contrast, char steam gasification at 800 °C in the presence of superheated steam results in gasification ash with higher plant-available P, making it more effective than oxidation for phosphorus-rich biochars derived from biogas fibre (BF), cattle manure (CM) and poultry manure (PM), except for sewage sludge-derived biochar where oxidation performs better. Gasification also enhances the surface area of biochar from biogas fibre and wheat straw while significantly increasing alkalinity. Both oxidation and gasification substantially increase P concentration but at the cost of carbon loss, limiting their ability to improve soil organic matter. The choice of treatment depends on the intended use, with oxidation being preferable for improving phosphorus release in sewage sludge biochar and gasification being the better approach for increasing P bioavailability in other P-rich biochars.

6.3.5.2 Storage and Environmental Factors

Storage and environmental factors significantly affect the characteristics of biochar by influencing its stability, chemical composition and functionality. Exposure to environmental conditions such as oxygen, moisture and microbial activity leads to biochar aging, altering its physical and chemical properties over time. Oxidation processes, particularly when biochar is exposed to air and water, result in surface modification, increasing the formation of oxygen-containing functional groups such as carboxyl and hydroxyl groups, which can enhance nutrient retention but may also lead to degradation. High humidity levels promote water adsorption, affecting biochar's porosity and reducing its structural integrity. Additionally, biochar stored under humid conditions may undergo microbial colonization, leading to biochemical transformations that further modify its surface chemistry. Temperature fluctuations also play a critical role, with higher temperatures accelerating decomposition and increasing the release of volatile organic compounds. Leaching of minerals and nutrients due to precipitation can alter the elemental composition of biochar, potentially reducing its effectiveness as a soil amendment. Proper storage in dry, oxygen-limited and controlled temperature environments is essential to preserve biochar's stability and functionality for long-term applications (Yaashikaa et al. 2020a, b; Xiang et al. 2021).

6.4 Physicochemical Properties of Biochar

Biochar is a carbon-rich material produced through the pyrolysis of biomass under limited oxygen conditions and its physicochemical properties determine its effectiveness in various applications such as soil amendment, carbon sequestration and environmental remediation. The properties of biochar vary depending on factors such as feedstock type, pyrolysis temperature and post-processing treatments (Tomczyk et al. 2020).

6.4.1 Specific Surface Area and Porosity

The specific surface area and porosity of biochar are among its most critical physical characteristics, directly influencing its environmental and agricultural applications. Biochar is well-known for its inherently porous and complex structure, which is a result of the carbonization of biomass under pyrolysis. During the pyrolysis process, the dehydration and thermal degradation of biomass result in the loss of water and volatile organic compounds. This leads to the development of a network of pores, significantly enhancing the surface area of the resulting biochar. These pores are categorized into micropores (< 2 nm), mesopores (2–50 nm), and macropores (> 50 nm), which together create a highly intricate internal structure that facilitates the retention of water, nutrients and other agrochemicals (Leng et al. 2021; Yaashikaa et al. 2020a, b).

The pore distribution and surface characteristics of biochar make it particularly effective for use in soil amendment, pollutant adsorption and nutrient retention. High porosity improves the soil's physical, chemical and biological properties when biochar is applied to agricultural fields. The porosity is not only responsible for moisture and nutrient retention but also plays a significant role in enhancing microbial habitats in soil ecosystems (Leng et al. 2021). Moreover, biochar's large surface area provides abundant active sites for chemical interactions, especially with contaminants and nutrients, making it a valuable material for environmental remediation purposes (Liu et al. 2020).

The specific surface area of biochar varies depending on several factors, including the type of biomass feedstock, pyrolysis temperature, heating rate and post-processing methods such as chemical or physical activation. Smaller particle sizes of biomass tend to produce biochar with greater surface area due to more uniform thermal decomposition. Liu et al. (2020) demonstrated that reducing particle size from 2.0 to < 0.075 mm increased the biochar's surface area from 15.88 to 20.96 m^2/g. Similarly, biochar derived from lignocellulosic feedstocks, such as wood, tends to have higher surface areas compared to manure-based biochar due to lower ash content, which otherwise blocks pore formation (Leng et al. 2021).

Higher pyrolysis temperatures generally lead to an increase in surface area and microporosity. As temperature rises, more volatiles are driven off and a more developed pore network is formed. Tomczyk et al. (2020) reported that pyrolysis at elevated temperatures increased the carbonized fraction, surface area and pH of biochar while reducing its cation exchange capacity and surface functional groups. In addition to temperature, the heating rate significantly affects surface characteristics. Rapid heating tends to create biochar with larger and more defined pores. Shagali et al. (2021) noted a linear increase in specific surface area with increasing heating rate, ranging from 19.3 to 126.5 m^2/g. More specifically, a higher heating rate leads to a biochar with very defined and big pores (Jouhara et al. 2018).

The pyrolysis atmosphere also plays a crucial role in determining the surface properties of biochar. For instance, biochar produced under a carbon dioxide atmosphere exhibits higher surface area and aromaticity, whereas production under a nitrogen atmosphere results in biochar with increased alkalinity and cation exchange capacity (Aktar et al. 2020). Furthermore, post-treatments such as physical activation (e.g., exposure to steam or CO_2) or chemical activation (e.g., using acids or bases) can enhance the surface area and pore structure. These treatments further increase the biochar's reactivity and adsorption capacity, broadening its applicability in various environmental technologies (Yaashikaa et al. 2020a, b).

6.4.2 Carbon Content and Stability

Biochar has a high proportion of carbon, which contributes to its stability and resistance to microbial degradation. The carbon content increases with pyrolysis temperature as cellulose and hemicellulose decompose, leaving behind a more condensed aromatic structure. Biochars derived from lignin-rich biomass tend to have higher carbon content and stability, making them suitable for long-term carbon sequestration (Tomczyk et al. 2020). The duration of pyrolysis also is responsible for biochar properties. It is reported that longer pyrolysis times can lead to biochar with greater stability and reduced reactivity. He et al. (2021) found that oxygen content is significantly dropped with increasing pyrolysis duration from 8.9 to 7.4%, this dropping means that a longer pyrolysis duration can increase the stability of biochar.

The carbon content and stability of biochar are fundamental parameters that influence its long-term environmental and agronomic performance. Biochar is a carbon-rich material whose stability in soil is primarily governed by its chemical structure and the degree of aromaticity. During pyrolysis, biomass undergoes thermal decomposition, where higher temperatures result in the breakdown of labile components (like cellulose and hemicellulose) and the enrichment of more stable, aromatic carbon structures such as turbostratic and graphitic carbon. These stable forms are resistant to microbial degradation, contributing to biochar's longevity in the soil for hundreds to thousands of years (Tomczyk et al. 2020).

The carbon content of biochar typically increases with the pyrolysis temperature. This is because elevated temperatures drive off hydrogen and oxygen, leaving behind a more carbon-enriched matrix. Biochars produced at temperatures above 500 °C generally possess higher fixed carbon content and lower volatile matter, enhancing their resistance to oxidation and microbial breakdown (Yaashikaa et al. 2020a, b). Consequently, such biochars are more effective for carbon sequestration—a key factor in mitigating climate change.

According to Maryam Afshar (2024), biochar can trap 2.0–2.6 tons of CO_2 per ton of biochar, making it a promising negative emission technology (NET). Approximately 50% of avoided emissions from biochar systems come from this long-term carbon sequestration capability, while additional benefits arise from replacing fossil fuels and avoiding methane and nitrous oxide emissions.

The stability of biochar also varies with the feedstock used. Lignocellulosic biomass, which is rich in lignin, tends to produce biochar with more stable carbon structures compared to manure- or sludge-derived biochars, which may contain more labile compounds. The condensed aromatic structure formed from lignin pyrolysis enhances the durability of the biochar in soil (Patel and Panwar 2023).

Moreover, biochar's stability can be inferred from its resistance to microbial decomposition and its H/C and O/C atomic ratios. Low H/C and O/C ratios are indicative of high aromaticity and greater stability. A biochar with an H/C atomic ratio below 0.2 is typically considered to be highly stable and recalcitrant in soil (Tomczyk et al. 2020; Yaashikaa et al. 2020a, b).

6.4.3 pH and Alkalinity

The pH of biochar varies based on the feedstock and pyrolysis conditions. Generally, biochar has an alkaline nature, with higher pyrolysis temperatures leading to increased pH due to the formation of alkaline oxides and carbonates. High-pH biochars can help neutralize acidic soils and improve soil fertility. The pH and alkalinity of biochar are critical physicochemical parameters that influence its role in soil amendment, pollutant remediation, and nutrient management. Biochar typically exhibits an alkaline pH, usually ranging between 7.1 and 12.9, depending on the type of feedstock and the conditions of pyrolysis. The alkalinity of biochar arises from the accumulation of basic oxides, carbonates, and soluble cations like Ca^{2+}, K^+, and Mg^{2+} during thermal decomposition of biomass at elevated temperatures (Tomczyk et al. 2020).

The pyrolysis temperature has a pronounced effect on the pH of biochar. As the temperature increases, volatile acidic functional groups such as carboxylic and phenolic groups decompose, leaving behind more basic surface chemistries. This results in an increase in pH. For example, biochars produced at 300 °C from corn straw, peanut shells, and soybean straw exhibited pH values of 9.4, 8.6, and 7.7, respectively, whereas canola straw-derived biochar had a pH of 6.5 due to lower ash content and inherent feedstock differences (Tomczyk et al. 2020).

Ash content in the biochar plays a vital role in determining alkalinity. Feedstocks with higher mineral content, such as agricultural residues and animal manures, produce biochars with higher pH values. Conversely, woody biomass, which contains less ash, tends to yield biochars with lower alkalinity. This is why manure- and crop-residue-derived biochars are often more effective in neutralizing acidic soils than wood-derived counterparts (Yaashikaa et al. 2020a, b).

Furthermore, biochar pH correlates with oxygen content and functional group chemistry. Mukome et al. (2013) found that biochar pH strongly correlates with the oxygen content ($R_2 = 0.7$), suggesting that oxygen-containing groups such as chromene, c-pyrone, and quinone contribute to basicity through deprotonation mechanisms (Tomczyk et al. 2020). These oxygen-rich functional groups contribute to the biochar's buffering capacity, making it useful for ameliorating acidic soils and increasing nutrient retention.

From a methodological perspective, the pH of biochar is typically measured by mixing it with deionized water at a fixed solid-to-liquid ratio (commonly 1:10 or 1:20 w/v), shaking the mixture, and measuring the equilibrium pH with a calibrated pH meter (Ngan et al. 2019).

The impact of biochar alkalinity on soil is well recognized. When applied to acidic soils, biochar increases soil pH, reduces exchangeable aluminium and improves the availability of essential nutrients such as phosphorus and potassium. These effects are especially beneficial in tropical and subtropical regions where acidic soils limit crop productivity (Tomczyk et al. 2020; Patel and Panwar 2023).

6.4.4 Cation Exchange Capacity (CEC)

Biochar's cation exchange capacity (CEC) influences its ability to retain essential soil nutrients such as potassium, calcium and magnesium. Lower-temperature biochar tends to have higher CEC due to the presence of functional groups such as carboxyl and phenolic groups. However, high-temperature biochar, while having greater stability, exhibit lower CEC due to the loss of these functional groups. The Cation Exchange Capacity (CEC) of biochar refers to its ability to retain and exchange positively charged ions (cations) such as potassium (K^+), calcium (Ca^{2+}), magnesium (Mg^{2+}) and ammonium (NH_4^+), which are essential nutrients for plant growth. CEC is a key indicator of biochar‘s potential to enhance soil fertility and nutrient retention. The CEC of biochar is influenced by its surface functional groups, especially carboxylic, phenolic, and hydroxyl groups, which develop predominantly at lower pyrolysis temperatures and contribute negative charges to the surface (Tomczyk et al. 2020).

Low-temperature biochar (below 400 °C) tends to have higher CEC due to a greater abundance of oxygen-containing functional groups, while high-temperature biochar (above 500 °C) typically have lower CEC as these functional groups degrade with increasing pyrolysis temperature (Yaashikaa et al. 2020a, b). This trade-off between stability and functionality means that biochar optimized for nutrient exchange may be less stable, while those optimized for carbon sequestration may have lower nutrient retention capacities.

The feedstock type also plays a significant role in determining biochar CEC. Biochar made from manure and agricultural residues tend to exhibit higher CEC values than those derived from woody biomass, primarily due to higher ash content and more readily available surface functional groups (Patel and Panwar 2023). Furthermore, post-treatment processes such as acid or alkali activation can significantly enhance the CEC of biochar by introducing or preserving active surface sites (Yaashikaa et al. 2020a, b).

CEC is not only essential for nutrient retention but also enhances soil buffering capacity, reducing nutrient leaching and increasing the availability of micronutrients over time. It also supports microbial activity by improving the soil's physicochemical environment. In acidic soils, biochar with high CEC can contribute to ameliorating aluminium toxicity, thus improving root development and overall plant productivity (Tomczyk et al. 2020).

Measurement of CEC in biochar typically involves exchange with ammonium acetate or barium chloride, followed by quantification using spectroscopy or titration methods. However, due to biochar's variable surface properties, measurement protocols must be carefully selected and standardized to obtain accurate and comparable results (Yaashikaa et al. 2020a, b).

6.4.5 Ash Content and Nutrient Composition

The ash content in biochar depends on the type of feedstock used. Agricultural residues and manure-derived biochar tend to have higher ash content than wood-based biochar. The presence of mineral elements such as calcium, magnesium, potassium and phosphorus in the ash contributes to soil fertility but can also influence biochar's physical properties by reducing surface area and porosity (Tomczyk et al. 2020).

The ash content and nutrient composition of biochar are essential indicators of its quality and functionality in soil systems, and both are strongly influenced by the type of feedstock and pyrolysis conditions. Ash in biochar primarily comprises mineral constituents such as carbonates, oxides, and silicates of calcium, potassium, magnesium, and sodium, which originate from the inorganic fraction of the biomass feedstock. The ash content can significantly vary depending on whether the biomass is woody or non-woody. Non-woody biomass, such as agricultural residues and sewage sludge, generally yields biochar with a higher ash content compared to woody biomass due to its inherently higher inorganic matter content (Zielińska et al. 2015; Vassilev et al. 2012). The ash content contributes to the alkalinity of biochar, which can be beneficial for neutralizing acidic soils. This is largely due to the presence of alkaline salts such as carbonates and oxides in the ash, which increase the biochar's pH (Mukome et al. 2013). High ash content biochar are often associated with elevated pH levels and can improve the cation exchange capacity (CEC) and buffering capacity of soils, enhancing nutrient retention and availability. However, excessive ash can also dilute the organic carbon content, potentially reducing the carbon sequestration potential of the biochar (Tomczyk et al. 2020).

Nutrient composition in biochar is determined by both the elemental makeup of the biomass and the pyrolysis temperature. Elements like nitrogen (N), phosphorus (P), potassium (K), calcium (Ca), magnesium (Mg) and sulphur (S) are commonly found in biochar and can provide essential nutrients to plants. For instance, biochar derived from nutrient-rich feedstocks such as manure and sewage sludge typically contains higher concentrations of N, P, and K (Gupta et al. 2023). These nutrients exist in mineral or organic forms and their availability depends on the extent of thermal transformation during pyrolysis.

The carbonization temperature also plays a pivotal role in defining the nutrient profile. As the pyrolysis temperature increases, nitrogen and other volatile elements tend to volatilize, leading to a decrease in their concentration in the final biochar product (Tripathi et al. 2016). On the other hand, elements like potassium and calcium, being more thermally stable, tend to accumulate in the ash fraction, thereby increasing with rising pyrolysis temperatures (Ding et al. 2014a, b). Biochar produced at moderate temperatures (around 400–500 °C) often balances nutrient retention and structural integrity, making it suitable for soil amendment purposes (Tomczyk et al. 2020).

The presence of oxygenated functional groups, influenced by both ash content and nutrient composition, is critical for nutrient adsorption and release. These include carboxyl, hydroxyl, and phenolic groups that contribute to the biochar's surface reactivity and CEC. Ash components can also modify the spatial distribution of organic matter within the biochar matrix, which in turn affects the accessibility and reactivity of these functional groups (Ulusal et al. 2021).

In terms of functionality, ash content has been shown to moderately correlate with improved cation exchange capacity and particle size, both of which are positively linked with the adsorption efficiency of heavy metals and nutrient retention (Yang et al. 2023). However, an inverse relationship is generally observed between ash and carbon content, implying that a trade-off often exists between maximizing nutrient functionality and maintaining high carbon sequestration potential (Chen et al. 2018).

6.4.6 Surface Functional Groups

The presence of oxygen-containing functional groups such as carboxyl, hydroxyl and carbonyl groups in biochar influences its reactivity and adsorption properties. These groups contribute to nutrient retention and enhance biochar's interaction with soil microorganisms. The abundance of these functional groups decreases with increasing pyrolysis temperature, leading to reduced nutrient-holding capacity. The surface functional groups of biochar play a pivotal role in determining its physico-chemical behaviour, environmental interactions, and suitability for applications such as soil amendment, pollutant adsorption, and nutrient retention. These functional groups, primarily oxygenated in nature, are formed during the pyrolysis of biomass through thermal decomposition, dehydration, decarboxylation, and aromatization reactions. The type, abundance and distribution of these functional groups are significantly influenced by the feedstock composition and pyrolysis conditions, especially temperature and residence time (Tomczyk et al. 2020).

During the early stages of pyrolysis, particularly at lower temperatures (below 400 °C), biochar retains a substantial quantity of aliphatic and cellulosic structures, which are rich in labile organic functional groups such as hydroxyl (–OH), carboxyl (–COOH), carbonyl (–C=O), and ether (–C–O–C–) groups. These polar functional groups contribute to the biochar's hydrophilic character and are highly reactive, enhancing its ability to chelate metal ions and interact with nutrients and soil colloids (Ahmad et al. 2014a, b; Chen et al., 2018). As the pyrolysis temperature increases beyond 500 °C, there is a marked decrease in the abundance of these groups due to deoxygenation and aromatization processes. Consequently, high-temperature biochars are typically characterized by the presence of aromatic structures and a greater degree of graphitization, leading to a decline in surface polarity and cation exchange capacity (Banik et al. 2018a, b).

The formation and stability of surface functional groups are also contingent upon the chemical composition of the biomass. Lignin-rich feedstocks, for example, tend to generate biochar with a higher concentration of phenolic and aromatic functional groups, which exhibit strong affinity for heavy metal ions and persistent organic pollutants (Wang et al. 2015a, b). Furthermore, the presence of mineral constituents such as alkali and alkaline earth metals in the biomass can catalyse the formation of oxygenated functional groups during pyrolysis, thereby enhancing the sorption characteristics of the biochar (Cely et al. 2015a, b; Tag et al. 2016).

The diversity of surface functional groups in biochar is not only critical for its sorption capacity but also for its role in improving soil chemical properties. Carboxyl and phenolic groups contribute significantly to the cation exchange capacity (CEC), enabling the retention of essential nutrients such as ammonium, potassium, and calcium in the soil matrix (Gupta et al. 2023). At moderate pyrolysis temperatures (350–500 °C), a balance is achieved between the preservation of reactive oxygenated groups and the development of porous structures, leading to biochar with optimal properties for environmental applications (Tomczyk et al. 2020; Ding et al. 2014a, b).

Moreover, the functional groups on the surface of biochar influence its interaction with soil microbial communities. Oxygenated groups can serve as binding sites for microbial colonization and enzymatic activity, promoting nutrient cycling and organic matter decomposition (Yu et al. 2019a, b). The introduction of specific functional groups through post-pyrolysis modification, such as acid or base treatments, has also been explored to enhance the functionality of biochar for targeted applications, including adsorption of specific contaminants (Chen et al. 2018).

Quantitative and qualitative analysis of surface functional groups is typically performed using techniques such as Fourier-transform infrared spectroscopy (FTIR), X-ray photoelectron spectroscopy (XPS), and Boehm titration. These analyses reveal the relative abundance of acidic (carboxylic, phenolic, lactonic) and basic (pyrone, chromene) groups and provide insight into the chemical reactivity and sorption behavior of the biochar (Murtaza et al. 2022a, b). The surface chemistry of biochar, therefore, serves as a fundamental determinant of its environmental reactivity, with direct implications for its effectiveness as a soil amendment and pollutant adsorbent.

6.5 Traditional Method of Biochar Optimization

6.5.1 Design of Experiments (DOE)

The traditional methods of biochar optimization have predominantly relied on experimental design frameworks that systematically evaluate the influence of multiple process variables on desired outcomes such as yield, surface area and adsorption capacity. These methods provide a structured approach to biochar process development by reducing the number of experimental runs while maximizing the data output. Among these, Design of Experiments (DOE) has been widely adopted for its efficacy in optimizing pyrolysis parameters and establishing mathematical relationships between input variables and biochar properties (Gupta et al. 2023).

6.5.1.1 Full Factorial Design (FFD)

Full Factorial Design (FFD) is a foundational approach in DOE where experiments are conducted at all possible combinations of the levels of factors under study. For example, a two-level FFD involving three factors—such as temperature, residence time, and feedstock type—requires 23 = 82^3 = 823 = 8 runs. FFD offers comprehensive interaction analysis among all variables, making it ideal for understanding the synergistic or antagonistic effects of factors in the pyrolysis process. However, its main drawback is the exponential increase in the number of experiments with the addition of more variables, which can be resource-intensive (Gupta et al. 2023).

6.5.1.2 Box-Behnken Design (BBD)

The Box-Behnken Design (BBD), a type of response surface methodology (RSM), is more efficient for exploring quadratic response surfaces without requiring a full factorial experiment. It reduces the number of experimental runs compared to FFD by avoiding combinations where all factors are at their extreme levels simultaneously. BBD uses a three-level factorial scheme, placing experimental points at the midpoints of the edges of the process space and at the centre. This design is particularly useful for modelling the nonlinear relationship between process parameters and responses in biochar optimization. Studies have shown that BBD can effectively predict the biochar yield, surface area and contaminant removal efficiency, especially when multiple variables such as temperature, heating rate and residence time are simultaneously considered (Gupta et al. 2023).

6.5.1.3 Central Composite Design (CCD)

The Central Composite Design (CCD) is another RSM-based method that includes factorial or fractional factorial points, axial points (star points) and center points to fit a second-order (quadratic) model. CCD provides a rotatable design, which ensures that the prediction variance is constant at all points equidistant from the center, making it highly suitable for optimization studies. CCD has been successfully applied to optimize biochar production from various feedstocks, as demonstrated in the work of Machrouhi et al. (2019), where CCD was used to determine the optimal activation temperature, impregnation ratio, and activation time for preparing activated carbon from *Thapsia transtagana* stems. The model accurately predicted iodine number, methylene blue index and dye removal efficiencies, highlighting CCD's effectiveness in multivariate optimization (Machrouhi et al. 2019).

All these experimental designs allow the exploration of main effects, interaction effects, and quadratic effects of process variables. They provide critical insight into the interdependence of pyrolysis parameters, enabling the production of biochar with tailored properties suitable for specific applications such as soil amendment, pollutant adsorption, and carbon sequestration. While these traditional methods have proven effective, they are increasingly being supplemented or replaced by machine learning and artificial intelligence models due to their superior efficiency in handling complex, high-dimensional data (Wang et al. 2024).

6.5.1.4 Response Surface Methodology (RSM)

Response Surface Methodology (RSM) has become an indispensable statistical and mathematical tool in optimizing biochar production processes. It enables the modeling and analysis of the interactions among multiple variables with a reduced number of experimental runs, making it both time- and resource-efficient. RSM is especially beneficial in tailoring pyrolysis conditions to achieve desired biochar characteristics such as high surface area, optimized porosity, enhanced adsorption capacity, and improved carbon yield. In biochar studies, designs such as Box–Behnken Design (BBD) and Central Composite Design (CCD) are most frequently employed under the RSM framework to predict process outcomes with high precision. For instance, Ba-Abbad et al. (2022) utilized BBD to optimize the synthesis parameters of Fe_3O_4 nanoparticles, demonstrating the significance of variables like Fe^{2+}/Fe^{3+} molar ratio and pH on nanoparticle size. The application of RSM not only provided significant interaction effects through analysis of variance (ANOVA) but also successfully predicted optimal conditions, minimizing experimental efforts. Similarly, Irfan et al. (2022) employed RSM for predicting pyrolysis reaction rate constants to enhance oil yields from plastic waste, validating the statistical model's ability to estimate kinetic parameters efficiently. In another example, Zhang et al. (2020) used RSM to optimize the extraction of yellow pigment from *Gardenia jasminoides*, identifying ethanol concentration, solid–liquid ratio, extraction time, and temperature as influential factors, thus showcasing RSM's adaptability across varied biomaterial processes.

Further demonstrating the versatility of RSM, Chen et al. (2024) implemented CCD to optimize microwave-assisted pyrolysis parameters of sugarcane bagasse for activated carbon production, achieving predicted results with minimal deviations and yielding biochar with a high iodine number and adsorption capacity. Obieogu et al. (2022) compared RSM and artificial neural networks (ANN) in modelling the hydrolysis of Colocynthis vulgaris peel, finding both methods effective, though ANN showed slightly higher predictive accuracy. Nevertheless, RSM provided a statistically valid and experimentally consistent model, making it a robust tool in biomass conversion optimization. Similarly, Campos et al. (2020) used RSM in a full factorial design to evaluate enzymatic hydrolysis of pretreated sugarcane bagasse, optimizing glucose yield by examining enzyme interactions and pretreatment variables. These studies collectively emphasize that RSM is not only a powerful tool for optimization but also a reliable method for modelling and understanding the complex interactions in biochar and biomass transformation processes. Its ability to identify optimal operational windows while minimizing resource use makes it a cornerstone in advancing sustainable biochar technologies.

6.6 Artificial Intelligence and Machine Learning Based Techniques Used in Biochar Process Optimization

Artificial Intelligence (AI) and Machine Learning (ML) have emerged as transformative tools in the optimization of biochar production, addressing the inherent complexity and variability of pyrolysis processes. Unlike conventional methods such as Response Surface Methodology (RSM) or Design of Experiments (DOE), which require a substantial number of experimental trials and often fail to capture nonlinear interactions, AI and ML algorithms can model high-dimensional data efficiently (Figures 6.1, 6.2 and Table 6.2). These approaches allow researchers to predict the effects of multiple process parameters-such as feedstock type, pyrolysis temperature, heating rate and residence time-on biochar yield and properties including surface area, porosity and adsorption capacity. Among the various ML techniques employed, Artificial Neural Networks (ANN), Random Forest (RF), Support Vector Machines (SVM) and Gradient Boosting Machines (GBM) are widely used due to their robustness and predictive power. ANN models, for example, are capable of capturing complex nonlinear dependencies between inputs and outputs, while RF and SVM are known for their accuracy and resistance to overfitting in small and noisy datasets (Ukoba et al. 2022; Wang et al. 2024).

In recent years, ML models have been successfully used to predict biochar performance in diverse applications such as heavy metal adsorption, soil amendment and carbon sequestration. For instance, Random Forest models have demonstrated superior predictive accuracy over ANN in simulating the adsorption capacity of biochar for contaminants like Pb^{2+}, Zn^{2+} and Cd^{2+} in aqueous solutions, highlighting the practical utility of ML in environmental remediation (Ukoba et al. 2022).

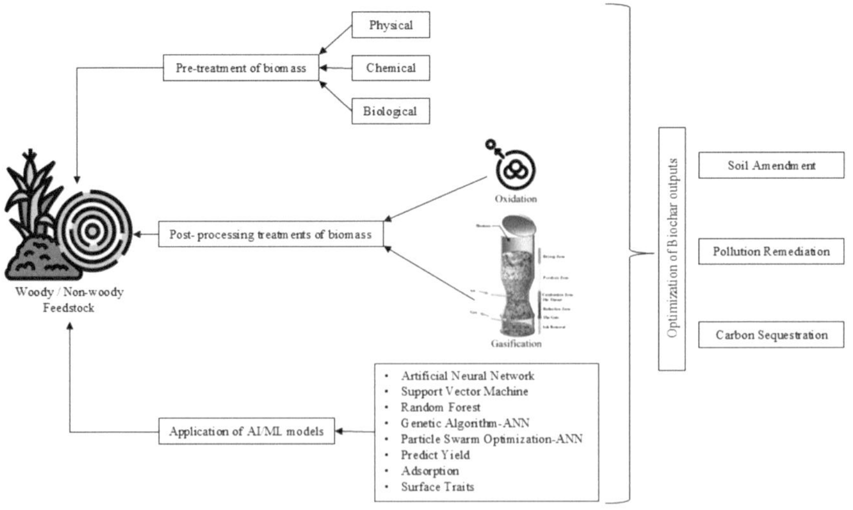

Fig. 6.1 Application of AI/ML in Biochar optimization

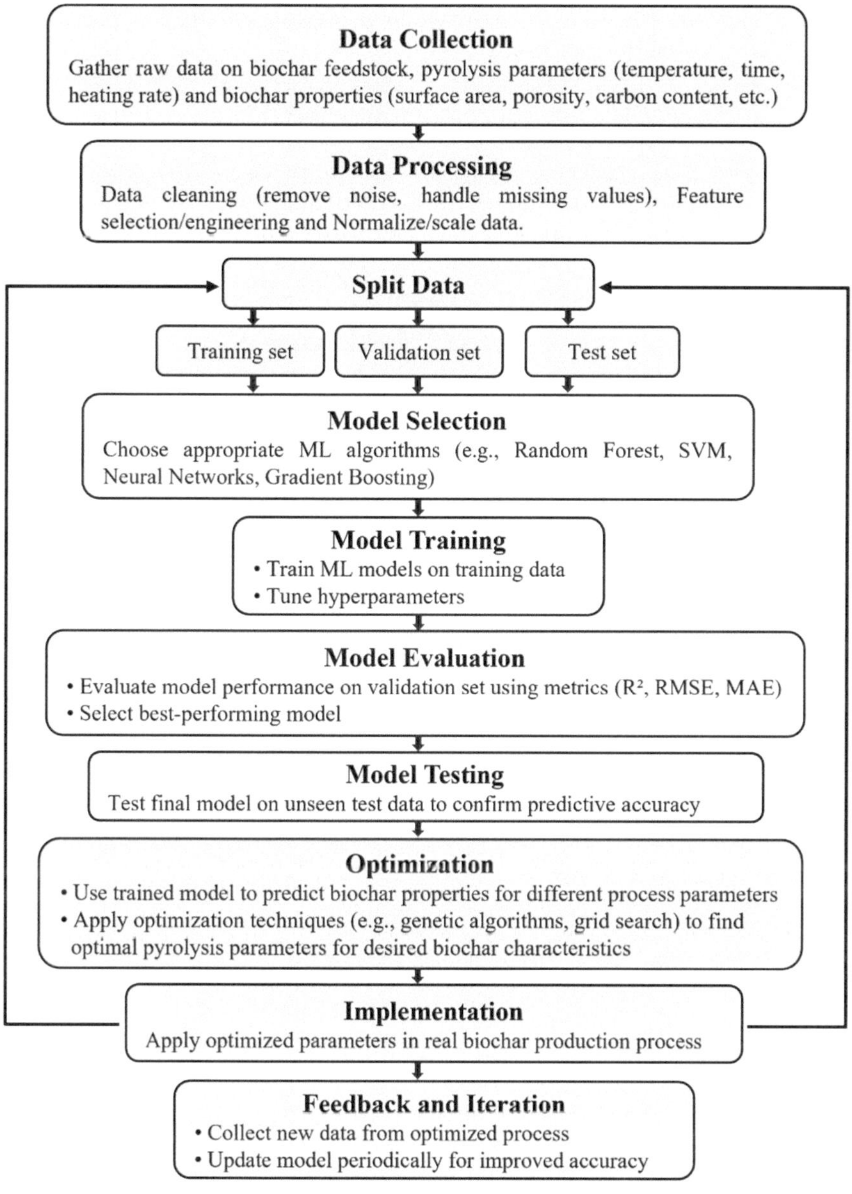

Fig. 6.2 Flowchart of machine learning application in biochar optimization

Table 6.2 Applications of Artificial Neural Network and other machine learning techniques in modeling and optimizing biochar processes

Biochar Source	ML Technique(s)	Optimized outcome/ best model performance	References
Safflower	ANN + RSM (custom)	ANN (2–5–1): R^2 = 0.999, RMSE = 0.0759 (train); R^2 = 0.922 (valid); T = 400 °C, Q_h = 10	Angın and Tiryaki (2016)
Cattle manure	LSSVM versus ANN	LSSVM better (R^2 = 0.9625, RMSE = 0.8347); mass most sensitive	Cao et al. (2016a, b)
Cotton shell	MLP-ANN	Best ANN: 3–10–3; R^2 = 0.9978; T = 450 °C, PS = 0.8 mm, QN = 1.75 m^3/h	Madhu et al. (2017)
Manure	ANFIS–PSO versus GA, ANN, LSSVM	ANFIS–PSO best: R^2 = 0.9842, RMSE = 0.2673	El Aziz et al. (2017)
Manure	ANFIS–SSO, ABC, PSO, LSSVM	ANFIS–SSO best: R^2 = 0.9804, RMSE = 0.2636, T = 500.7 °C, Q_h = 11.5	Ewees et al. (2017)
Various datasets (245)	Random forest regression (RFR)	T most influential; R^2 = 0.8548 (yield); R^2 = 0.8480 (carbon)	Zhu et al. (2019)
Manure	ANFIS–GWO versus other ANFIS hybrids	ANFIS–GWO best: R^2 = 0.980, RMSE = 0.259, AAPRE = 4.525	Ewees and Elaziz (2018)
Multiple biomass (402 sets)	ANN + meta-heuristic (GWO, PSO, GA, RA)	ANN–GWO best: R^2 = 0.93, RMSE = 1.74; T > t > Q_h most impactful	Khan et al. (2022)

(continued)

Table 6.2 (continued)

Biochar Source	ML Technique(s)	Optimized outcome/ best model performance	References
19 docs (226 sets)	MLP-ANN versus ANFIS	Single-layer ANN better: $R^2 = 0.964$ (yield), $R^2 = 0.914$ (prox. Comp.), HHV = high	Li et al. (2022)
Agri waste	ANN, SVM, GPR (Gaussian process)	ANN–LM best: $R^2 = 0.999$, RMSE = 2.473; SVM poorest: $R^2 = 0.620$	Kanthasamy et al. (2023)

Note T = temperature; t = resident time; Q_h = heating rate (°C/min); QN = nitrogen flow rate; PS = particle size (mm); m = mass of feedstock; FMC/MC = feedstock moisture content/moisture content; ANN = artificial neural network; MLP-ANN = multi-layer perception ANN; ANFIS = adaptive neuro-fuzzy inference system; PSO=Particle Swarm Optimization; SSO = social spider optimization; GA = genetic algorithm; ABC = Artificial Bee Colony algorithm; GWO = grey wolf optimizer; GOA = Grasshopper Optimization Algorithm; SCA = Sine Cosine Algorithm; SVM = support vector machine; LSSVM = leas squares support vector machine; RFR = random forest regression; GPR = Gaussian process regression; RA = Rao Algorithms; R^2 = coefficient of determination (model accuracy); RMSE = root mean square error; MAE = mean absolute error; MAPE = mean absolute percentage error; AAPRE = absolute average percentage relative error; SD = standard deviation; HHV = higher heating value; Prox. Comp. = proximate composition

ML has also been integrated with kinetic and isotherm models to improve theinterpretability and reliability of adsorption predictions. Moreover, optimization algorithms such as Extreme Gradient Boosting (XGBoost) and Least Squares Support Vector Machine (LS-SVM) have been applied to identify the optimal pyrolysis conditions for maximizing surface area and carbon content of biochar (Gupta et al. 2023). These models can guide the selection of appropriate biomass types and operational parameters, significantly reducing the number of trial-and-error experiments and associated costs. The application of AI and ML has not only improved the scalability and commercial feasibility of biochar technologies but also accelerated the development of functionalized biochars with enhanced performance in specific applications such as pollutant remediation and nutrient retention in soils (Wei Wang et al. 2024). As the biochar research domain continues to grow, the integration of AI and ML is expected to become increasingly central to the design, evaluation and deployment of advanced biochar-based solutions.

6.7 Common AI/ML Techniques Used

6.7.1 Artificial Neural Networks (ANN)

An Artificial Neural Network (ANN) is a computer-based tool that works like the human brain. It learns from data and is useful for solving complex problems where it's hard to understand the relationship between input and output using regular math or statistics. ANN is made up of three main parts: the input layer, hidden layers, and the output layer. The input layer takes in data, the hidden layers process it, and the output layer gives the final result. In each layer, the input values are multiplied by certain weights, and small constant values called biases are added.

Inside the hidden layers, special formulas called activation functions like sigmoid, tanh, ReLU (Rectified Linear Unit), and linear functions are used to process the data. Among different ANN types, the Multilayer Perceptron (MLP) is the most common for predicting and optimizing process parameters. MLP is a basic type of ANN that follows a feed-forward system, where data flows in one direction from input to output without looping back.

A simple MLP has three layers: input, hidden and output. Figure 6.1 shows a typical example of this structure. Inputs like pyrolysis temperature (T), residence time (t), heating rate (Q), and feedstock features (mass, size, moisture, ash content, and lignocellulose amount) are entered as $x_1, x_2, x_3, \ldots x_n$. The outputs, labelled $y_1, y_2, y_3, \ldots y_m$, include results such as biochar yield (Y), and its properties like stable organic matter (SOM), fixed carbon (FC), high heating value (HHV), and surface area (SSA). The computation of ANN output may be achieved by activation functions of hidden and output layers to the weighted sum of the inputs of the previous layers. The kth output of the ANN is mathematically expressed as:

$$y_k = f_o\left[\sum_{j=1}^{n_h} w_{jk} \cdot f_h\left(\sum_{i=1}^{n} w_{ij}x_i + (b_h)_j\right) + (b_o k)\right] \quad (6.1)$$

where y_k is the kth output variable; x_i is the ith input variable; f_h and f_o are the activation functions in the hidden and the output layer, respectively. i, j, and k denote the input layer neuron, hidden layer neuron, and output layer neuron, respectively, and given as $i = 1, 2, 3, \ldots, n$, $j = 1, 2, 3, \ldots, n_h$, and $k = 1, 2, 3, \ldots, m$. w_{ij} represents the connection weight between the ith input neuron and the jth hidden neuron. W_{jk} represents the connection weight between the jth hidden neuron and the kth output neuron. $(b_h)_j$ denotes the bias for the jth neuron in the hidden layer, and $(b_o)k$ is the bias for the kth neuron in the output layer. Commonly used activation functions are logistic sigmoid, hyperbolic tangent sigmoid, and linear. The mathematical expressions of activation functions are as follows:

$$\begin{aligned} f(x) &= \frac{1}{1+e^{-x}}, 0 \le f(x) \le 1 \\ f(x) &= \frac{e^{x}-e^{-x}}{e^{x}+e^{-x}}, -\infty \le f(x) \le +\infty \\ f(x) &= x, -\infty \le f(x) \le +\infty \end{aligned} \tag{6.2}$$

The error backpropagation algorithm is the most frequently used ANN model training method. However, the algorithm may get stuck in local minima and poor convergence. Therefore, a combination of Gauss-Newton and steepest descent and algorithm, known as the Levenberg–Marquardt (LM) backpropagation learning algorithm is considered as most robustly used ANN training algorithm as it is capable to exhibit a stable and fast convergence. The expression for weights updates by LM algorithm by minimizing the means squared error (MSE) between actual vs. models predicted output are expressed as follows:

$$\{w\}_{k+1} = \{w\}_k - \left(Z_k^T Z_k + \mu I\right)^{-1} (Z_k)\{e_k\} \tag{6.3}$$

Input parameter sensitivity of ANN models can be achieved by the Garson model expressed as follows:

$$S_{ik}(\%) = \frac{\sum_{j=1}^{n_h}\left\{\left(\frac{|w_{ij}|}{\sum_{i=1}^{n_h}|w_{ij}|}\right)\right\} \times |w_{jk}|}{\sum_{i=1}^{n}\left[\sum_{j=1}^{n_h}\left\{\left(\frac{|w_{ij}|}{\sum_{i=1}^{n_h}|w_{ij}|}\right)\right\} \times |w_{jk}|\right]} \times 100 \tag{6.4}$$

where, S_{ik} is the relative importance of the input variable on output variable, n and n_h are the number of input and hidden neurons, w_{ij} and w_{jk} are weights between input-hidden, hidden-output respectively.

6.7.2 *Support Vector Machines (SVM)*

Support Vector Machines (SVM), particularly their regression variant known as Support Vector Regression (SVR), are increasingly recognized as effective machine learning (ML) techniques for modelling complex and nonlinear relationships in biochar production systems. These models utilize kernel functions—such as radial basis function (RBF), polynomial, or sigmoid—to transform input data into higher-dimensional spaces, enabling the construction of optimal hyperplanes for predictive modelling (Gupta et al. 2023).

In the context of biochar optimization, SVR has been employed to model and predict biochar yield and quality parameters from pyrolysis process conditions. For instance, Cao et al. (2016a, b) applied a Least Squares Support Vector Machine (LSSVM) model with an RBF kernel to estimate the yield of biochar produced from cattle manure. The model demonstrated superior predictive capability compared to artificial neural networks (ANN), achieving a coefficient of determination (R2) of 0.9625 during training, indicating high accuracy and robustness. Through sensitivity analysis, key variables such as feedstock mass, pyrolysis temperature, and moisture content were identified as having the most significant impact on biochar yield, while heating rate and residence time were found to be less influential (Gupta et al. 2023).

SVM models have also been integrated into biochar process optimization frameworks, where they serve as surrogate models within multi-objective or single-objective optimization algorithms. These models help in identifying optimal process parameter settings that enhance biochar yield or desired physicochemical properties, such as porosity, fixed carbon content, or heavy metal adsorption efficiency (Gupta et al. 2023).

While SVM-based approaches are less commonly used in biochar research compared to ANN models, they offer several advantages, including a strong theoretical foundation, the ability to handle high-dimensional datasets, and minimal need for hyperparameter tuning. These attributes position SVM as a promising tool for future studies focused on optimizing biochar production under varying pyrolysis conditions (Gupta et al. 2023).

6.7.3 Decision Trees and Random Forests

The production and application of biochar—a carbon-rich material derived from thermochemical conversion of biomass—has gained significant attention in environmental remediation, carbon sequestration, and soil improvement. Optimizing biochar properties and application conditions is essential to maximize its effectiveness in these diverse roles. Machine learning (ML), especially ensemble learning techniques like decision trees and random forests (RF), has proven valuable in modelling and predicting biochar performance due to their interpretability, scalability, and robustness.

6.7.3.1 Overview and Suitability for Biochar Studies

Decision trees are hierarchical models that split data into branches based on feature values, ultimately forming predictive paths for regression or classification tasks. They are intuitive, accommodate nonlinear relationships, and can handle both categorical and numerical variables. However, single decision trees often suffer from high variance and overfitting.

Random forests overcome this limitation by aggregating the predictions of multiple decision trees trained on bootstrap samples of the data and considering a random subset of features at each split. This ensemble approach leads to reduced variance and improved generalization, making random forests especially suitable for modeling complex environmental systems such as biochar interactions in soil or water.

Random forests have been successfully used to predict biochar adsorption capacities for pesticides and heavy metals, pyrolysis yield outcomes, and soil remediation efficiency. For instance, in a study by Nighojkar et al. (2023), RF was used alongside CatBoost and LightGBM to identify the critical variables affecting pesticide removal from aqueous matrices. The model revealed that biochar textural properties, pesticide concentration, and dosage were the most important variables influencing adsorption efficiency.

6.7.3.2 Variable Importance in Random Forests: Insight for Biochar Optimization

A key advantage of RF is its ability to estimate the relative importance of input variables using metrics such as Mean Decrease in Accuracy (MDA) and Mean Decrease in Gini Impurity. These metrics provide insight into the relative influence of feedstock characteristics, pyrolysis conditions, and environmental parameters on biochar performance.

Genuer et al. (2010) proposed a variable selection strategy using RF scores of importance, demonstrating that RF can robustly rank variables without assuming specific model forms. This capability is highly beneficial in biochar research where underlying relationships are often nonlinear and interactions between parameters are complex.

For example, in studies related to heavy metal adsorption, features such as pyrolysis temperature, surface area, pH, and initial metal concentration were ranked by RF models. Yang et al. (2023) used Bayesian-optimized RF to identify the relative impact of environmental conditions, biochar properties, and synthesis parameters, finding that environmental variables (e.g., pH and temperature) were more influential than preparation conditions.

6.7.3.3 Regression and Classification Applications in Biochar Research

Random forests are applicable to both regression and classification problems in biochar research. In regression, they predict continuous outcomes such as adsorption efficiency, yield, or surface area. In classification, RF can distinguish between categories such as "high" versus "low" performing biochars or biochar types suited for specific contaminants. For instance, in Cho et al. (2024), RF was employed to predict the long-term lead (Pb) immobilization potential of different biochars based on feedstock and pyrolysis temperature. The study combined machine learning with small datasets using data augmentation to improve prediction robustness, illustrating RF's flexibility even with limited data. Similarly, Ang et al. (2023) used rough set machine learning to classify biochar surface properties based on biomass and pyrolysis parameters, a task also well-suited to decision tree-based methods due to their rule-based structure.

6.7.3.4 Robustness Against Overfitting

A prominent advantage of random forests is their resilience to overfitting, particularly when compared to single decision trees. This robustness arises from the aggregation of multiple trees trained on different data subsets and feature spaces, which averages out individual biases.

Wang et al. (2024) emphasized the importance of model generalizability in biochar applications and noted that ensemble models like RF are particularly reliable, even when only lab-scale data is available. In comparative studies, RF has consistently outperformed or matched the predictive accuracy of more complex algorithms such as support vector machinesor artificial neural networks, with far less risk of overfitting.

6.7.3.5 Integration with Other Techniques and Future Directions

Random forests are increasingly being integrated with Bayesian optimization, genetic algorithms, and hybrid models to further refine predictions. For example, Yang et al. (2023) employed Bayesian optimization to tune RF hyperparameters, significantly enhancing model performance ($R_2 = 0.998$) for heavy metal adsorption.

Furthermore, future research may benefit from combining mechanistic models with ML frameworks to overcome the black-box limitations of data-driven models. As suggested by Wang et al. (2024), hybrid modelling frameworks could enhance the interpretability and practical applicability of ML-based biochar optimization strategies.

6.7.4 Genetic Algorithms (GA) and Particle Swarm Optimization (PSO)

The production of biochar via thermochemical processes such as pyrolysis depends on a range of interacting parameters including biomass composition, temperature, heating rate and residence time. Given the complexity and non-linearity of these interactions, the integration of artificial neural networks (ANNs) with metaheuristic optimization techniques such as Genetic Algorithms (GA) and Particle Swarm Optimization (PSO) has emerged as a powerful approach for process modelling and optimization.

6.7.4.1 Metaheuristic Optimization in Biochar Yield Prediction

Artificial neural networks (ANNs) are widely recognized for their ability to model complex, nonlinear systems. However, their performance depends heavily on appropriate selection of input features and network architecture. To address this challenge,

Khan et al. (2022) developed and compared several ANN-based hybrid models integrated with different metaheuristic algorithms, including GA and PSO, for predicting biochar yield using biomass properties and pyrolysis parameters. Their results indicated that ANN models tuned using GA (ANN-GA) and PSO (ANN-PSO) performed exceptionally well, achieving high prediction accuracy ($R2 \approx 0.92$, RMSE < 2%) and demonstrating robust generalization capabilities. These models also required fewer input parameters and network layers compared to other algorithms, thus reducing computational cost and enhancing model interpretability.

6.7.4.2 Genetic Algorithm–Artificial Neural Network (GA–ANN)

Genetic Algorithms (GAs) simulate the process of natural selection to evolve optimal solutions. Within ANN modelling, GA is frequently employed to optimize weights, biases, number of neurons, and even feature selection. This is particularly valuable in biochar optimization due to the large parameter space and complex interactions involved. In GA–ANN models, GA uses fitness functions (e.g., root mean squared error, RMSE) to assess candidate ANN structures. The best-performing individuals are selected, crossed over, and mutated to generate improved architectures. In Khan et al. (2022), GA was effective in reducing the number of required features to seven out of eleven in their proximate analysis dataset while maintaining high model performance ($R_2 = 0.92$). This underlines the efficacy of GA in identifying optimal configurations that balance accuracy and computational efficiency.

6.7.4.3 Particle Swarm Optimization–Artificial Neural Network (PSO–ANN)

Particle Swarm Optimization (PSO) is inspired by the social behaviours of birds and fish. Each potential solution (particle) adjusts its position in the search space based on its own experience and that of neighbouring particles. In PSO–ANN frameworks, PSO is used to find the optimal weights and structure of the neural network that minimize prediction errors.

PSO–ANN models have been widely used in environmental modelling, including forbiochar. The PSO technique demonstrates strong global search capabilities and fast convergence, making it well-suited for high-dimensional optimization problems such as those encountered in pyrolysis modelling. According to Khan et al. (2022), the ANN-PSO model selected only five input features and still achieved competitive performance metrics ($R^2 \approx 0.92$, RMSE ≈ 1.82). This demonstrates that PSO can effectively reduce model complexity while preserving predictive power.

6.7.4.4 Hybrid Optimization Frameworks: GA–ANN Versus PSO–ANN

Both GA–ANN and PSO–ANN models offer distinct advantages. GA tends to explore the solution space more broadly, which may help escape local optima, while PSO generally converges more quickly. A comparative assessment in the study by Khan et al. (2022) concluded that both models delivered nearly equivalent performance in predicting biochar yield. The choice between GA and PSO may therefore depend on specific application goals—e.g., if computational efficiency is prioritized, PSO may be favoured; if feature importance and model simplification are emphasized, GA may offer superior interpretability.

In wastewater modelling, similar optimization paradigms have been adopted. Al Saleh (2021) applied ANN to predict parameters like BOD_5 and COD in wastewater treatment systems, demonstrating the flexibility and accuracy of ANN-based approaches when optimized appropriately.

6.7.5 Principal Component Analysis (PCA)

Principal Component Analysis (PCA) is a widely used statistical method for dimensionality reduction that enables researchers to analyse large, complex datasets by transforming correlated variables into a smaller set of uncorrelated variables called principal components. In biochar research, which often involves the study of numerous physical and chemical parameters influenced by feedstock characteristics, pyrolysis temperature, and environmental interactions, PCA serves as a crucial tool for identifying the most influential variables affecting biochar performance (Fig. 6.3).

Biochar's effectiveness as a soil amendment, adsorbent, or remediation agent is determined by a range of properties such as surface area, porosity, elemental composition (C, H, O, N), pH, cation exchange capacity (CEC) and thermal stability. These variables are frequently interdependent, making it difficult to draw meaningful conclusions using traditional univariate analysis. PCA addresses this by capturing

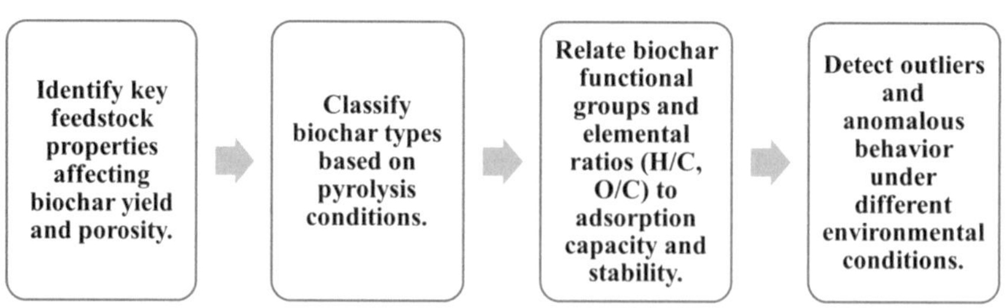

Fig. 6.3 Identifying key variables using PCA

the underlying structure of the data, summarizing the variance in a reduced number of dimensions without significant loss of information.

One of the key advantages of PCA is its ability to identify and rank the importance of variables that contribute to the variance in the dataset. This feature is particularly beneficial when comparing biochar produced from different biomass types or pyrolysis conditions. For instance, a PCA-based study by Cho et al. (2024) employed dimensionality reduction to identify the optimal biochar characteristics for lead (Pb) immobilization in contaminated soils. Their analysis revealed that biochar with high pH and exchangeable nutrient contents—especially those derived from oilseed rape at 700 °C—were most effective for heavy metal remediation. PCA helped isolate these dominant factors from a larger set of measured soil and biochar properties, enabling a clearer understanding of which properties most strongly influence performance.

Moreover, PCA is commonly used in conjunction with spectroscopic and thermal characterization methods like FTIR, XRD and TGA. In such applications, PCA assists in interpreting complex spectra by highlighting which spectral features correlate with important physicochemical properties or functional group composition. This application is essential when assessing modifications to biochar surface chemistry or when differentiating between raw and engineered biochar.

PCA also enhances the performance of predictive machine learning models by reducing input complexity. In data-driven modelling of biochar yield, adsorption efficiency, or pollutant removal, large numbers of input features can lead to overfitting and reduced generalizability. Wang et al. (2024) emphasized that PCA, when used as a preprocessing step in artificial neural networks (ANN), support vector machines (SVM), or random forest models, can reduce noise and multicollinearity, streamline model training, and improve interpretability. By selecting only the most significant components, researchers can maintain high predictive accuracy with fewer computational resources. López et al. (2022) further highlighted PCA's value in the context of genomic and high-dimensional agricultural datasets, noting its strength in capturing major sources of variation and allowing for better visualization and interpretation of complex multivariate relationships. While their application was in genomic prediction, the underlying statistical principles are equally applicable to biochar systems, where hundreds of interacting variables may influence environmental outcomes.

Despite its utility, PCA has limitations. It assumes linear relationships among variables and the resulting principal components, being combinations of original variables, may lack direct physical interpretability. Additionally, PCA does not inherently distinguish between relevant and irrelevant features in a predictive sense—it captures variance, not necessarily importance with respect to an outcome variable. Nevertheless, its simplicity, speed, and effectiveness make PCA a foundational technique in biochar research and beyond.

6.7.6 *Adaptive Neuro-Fuzzy Interface System (ANFIS)*

The ANFIS is a hybrid soft computing technique that integrates the learning capability of Artificial Neural Networks (ANN) with the approximate reasoning and decision-making framework of fuzzy logic systems. ANFIS has emerged as a robust modelling and optimization tool for nonlinear systems, particularly in biochar research, where complex interactions among multiple process variables govern outcomes such as yield, surface area and adsorption efficiency. Its ability to deal with imprecise, uncertain, or sparse data makes it well-suited for predicting and optimizing biochar properties under varying pyrolysis conditions.

6.7.6.1 Understanding ANFIS Architecture and Working Principles

ANFIS models are typically constructed using a Sugeno-type fuzzy inference system. The architecture consists of five layers: (1) fuzzification, (2) rule evaluation, (3) normalization, (4) defuzzification and (5) output summation. The neuro-adaptive learning algorithm adjusts the parameters associated with membership functions using a combination of gradient descent and least squares estimation. This structure enables ANFIS to capture both qualitative rules (from fuzzy logic) and quantitative learning (from neural networks), offering high interpretability and learning ability (Uzuner and Cekmecelioglu 2016).

6.7.6.2 Application of ANFIS in Biochar Process Optimization

Biochar production through pyrolysis involves a multitude of nonlinear variables—such as pyrolysis temperature, heating rate, residence time, particle size, and biomass composition—that interact in complex ways. Traditional statistical approaches like Response Surface Methodology (RSM) may not capture these nonlinear relationships effectively. ANFIS, in contrast, has shown superior capability in modelling these interactions and predicting biochar properties with high accuracy. Gupta et al. (2023) highlighted that ANFIS, along with other soft computing tools, is increasingly applied in biochar optimization because of its capacity to identify critical parameters while minimizing experimental effort. For instance, ANFIS can accurately predict biochar yield and adsorption performance based on limited input data, offering a reliable alternative to extensive laboratory trials (Fig. 6.4).

6.7.6.3 Comparative Performance of ANFIS and ANN in Biochar Modelling

Artificial Neural Networks (ANN) and Least Squares Support Vector Machines (LS-SVM) have both been effectively employed to predict biochar production. Among

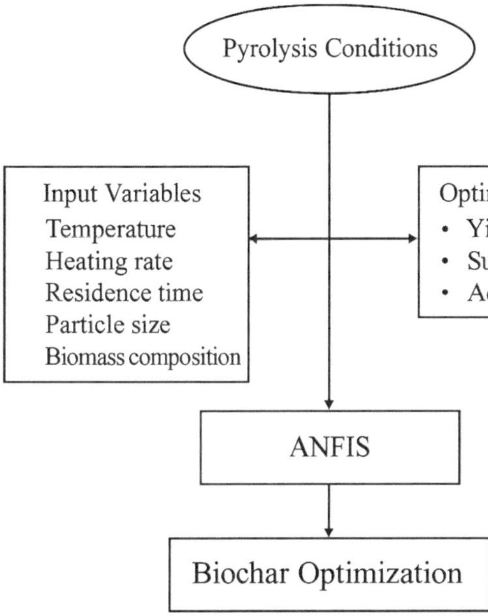

Fig. 6.4 Mechanism of adaptive Neuro-Fuzzy Inference System (ANFIS) in Biochar optimization

these, the LS-SVM model demonstrated superior prediction performance and robustness compared to the ANN model. Nonetheless, a notable limitation of the LS-SVM algorithm is its susceptibility to entrapment in local optima. To address this drawback, Ewees et al. (2017) developed an adaptive neuro-fuzzy inference system enhanced with the Social Spider Optimization algorithm (ANFIS-SSO), specifically tailored for biochar prediction. This hybrid model outperformed classical ANFIS, Particle Swarm Optimization (PSO), Artificial Bee Colony (ABC), and LS-SVM in terms of predictive accuracy.

Building on this approach, Ewees and Elaziz (2018) proposed another hybrid framework by integrating ANFIS with the Grey Wolf Optimization (GWO) algorithm. This model further mitigated the risk of local optimum convergence while enhancing prediction reliability. In the broader domain of bioprocess modelling, comparative studies between ANFIS and ANN have highlighted that, although both frameworks yield high predictive accuracy, ANFIS often offers enhanced interpretability owing to its fuzzy rule-based structure. For instance, Uzuner and Cekmecelioglu (2016) applied both ANN and ANFIS to model enzyme production and reported R_2 values exceeding 0.97 for both models. While ANN exhibited marginally higher precision, ANFIS provided deeper insight into the influence of individual input parameters. Such interpretability is particularly advantageous in biochar applications, where understanding the interactions among variables is essential for process optimization, scale-up and industrial deployment.

6.7.6.4 ANFIS for Adsorption and Remediation Modelling

Biochar is widely used in environmental applications such as heavy metal remediation, wastewater treatment, and soil amendment. The performance of biochar in these domains is largely dictated by its physicochemical properties, which in turn depend on pyrolysis conditions. Zhang et al. (2023) emphasized that machine learning models, including ANFIS, have successfully predicted adsorption behaviour and optimized conditions for pollutant removal. These models help in understanding how variations in synthesis parameters influence key outputs like surface area, functional groups, and adsorption capacity.

6.7.6.5 Integration with Hybrid Models and Future Prospects

One promising direction is the integration of ANFIS with other optimization algorithms such as Genetic Algorithm (GA), Particle Swarm Optimization (PSO), or Grey Wolf Optimizer (GWO), leading to hybrid models like GA-ANFIS or PSO-ANFIS. These models further enhance predictive accuracy and robustness in parameter tuning. Khan et al. (2022) discussed similar integrations in their biochar yield prediction models, noting that coupling ANNs with optimization algorithms outperformed standalone models in accuracy and feature reduction. Another potential application of ANFIS is in developing surrogate models for real-time process control in pilot or industrial-scale pyrolysis units. ANFIS can also be used to interpret spectral data (e.g., FTIR, TGA) for classification and quality assessment of biochar, enhancing its utility in standardization and regulatory frameworks.

6.7.7 Gaussian Process Regression (GPR)

Biochar, a carbon-rich material produced via the thermal decomposition of organic biomass under limited oxygen conditions (pyrolysis), plays a pivotal role in sustainable agriculture through its applications in soil enhancement, carbon sequestration, and pollution mitigation. The optimization of biochar production parameters (such as pyrolysis temperature, residence time and heating rate) is critical for tailoring its physicochemical properties to specific environmental or agronomic needs. Traditional design of experiment (DOE) methods such as Response Surface Methodology (RSM) and Central Composite Design (CCD) have been widely used, but recent advancements in artificial intelligence (AI) and machine learning (ML) have introduced more powerful tools, notably Gaussian Process Regression (GPR), for predictive modelling and optimization of biochar systems.

6.7.7.1 Overview of Gaussian Process Regression (GPR)

GPR is a non-parametric, Bayesian approach to regression that is particularly useful when dealing with small datasets and nonlinear relationships. It provides not only predictive estimates but also associated uncertainties, making it highly suitable for optimizing experimental processes such as biochar production, where data acquisition is often resource-intensive (Shafizadeh et al. 2022; Gupta et al. 2023). In GPR, the function space is defined by a mean function and a covariance (kernel) function, the latter capturing assumptions about the function's smoothness and variation. This characteristic enables GPR to flexibly model complex input-output relationships typical of thermochemical processes like pyrolysis.

6.7.7.2 Application of GPR in Biochar Optimization

Gupta et al. (2023) emphasize the utility of GPR in the context of process modelling and optimization in biochar production. Their review highlights that GPR, when trained on historical or experimental data, can predict biochar yield and its properties such as carbon content, surface area, and adsorption capacity based on varying input parameters. Its predictive accuracy and uncertainty quantification provide robust foundations for selecting optimal processing conditions.

Compared to conventional ML methods like Artificial Neural Networks (ANN), GPR offers a key advantage in that it does not require extensive hyperparameter tuning and is less prone to overfitting, especially in low-sample-size scenarios. This makes it particularly relevant for novel biochar feedstocks or emerging pyrolysis techniques where large datasets may not be available.

6.7.7.3 GPR Coupled with Optimization Algorithms

The performance of GPR can be further enhanced through hybrid modelling approaches that integrate it with optimization algorithms. For example, Bayesian Optimization (BO) uses GPR as a surrogate model to guide the exploration of the design space towards optimal regions with minimal experimental effort. In recent studies, GPR models have been coupled with metaheuristic algorithms (e.g., Particle Swarm Optimization or Genetic Algorithms) to simultaneously optimize multiple biochar properties, enabling multi-objective optimization (Alruqi and Sharma 2023; Tao et al. 2024).

This approach is especially beneficial for optimizing trade-offs—such as maximizing adsorption efficiency while maintaining a high biochar yield—which are common in practical applications.

6.7.7.4 Role of GPR in Sustainable and Precision Agriculture

GPR also finds utility beyond the laboratory, serving as a decision-support tool in field applications. By modelling and predicting how variations in biochar properties affect soil health, nutrient dynamics, or pollutant remediation, GPR-based frameworks can guide practitioners toward more sustainable and effective use of biochar in different agroecological contexts (Ghosh et al. 2022).

Furthermore, GPR's capacity to incorporate prior knowledge (via kernels) allows for the transfer of insights across different biomass types or pyrolysis setups, making it a powerful generalization tool in precision agriculture and waste valorisation.

6.7.8 Gradient Boosting Regression

Biochar, a carbon-rich material produced through pyrolysis of biomass under oxygen-limited conditions, is recognized for its versatile applications in agriculture, environmental remediation, and carbon sequestration. The optimization of biochar's physicochemical properties—such as specific surface area (SSA), yield, pH and adsorption capacity—requires modelling techniques capable of capturing the complex nonlinear interactions between feedstock characteristics and pyrolysis parameters. In recent years, Gradient Boosting Regression (GBR) has emerged as a powerful machine learning method for predictive modelling and optimization in biochar research.

6.7.8.1 Conceptual Framework of Gradient Boosting Regression

Gradient Boosting Regression (GBR) is an ensemble machine learning technique that builds a strong predictive model by sequentially combining multiple weak learners, typically decision trees. Each new model corrects the errors of the preceding one by minimizing a specified loss function through gradient descent in function space. This iterative improvement allows GBR to effectively model complex nonlinear relationships with high predictive accuracy (Zhou et al. 2024).

6.7.8.2 Application of GBR in Biochar Property Prediction

Several studies have successfully applied GBR to predict key biochar properties such as yield and SSA. Zhou et al. (2024) utilized GBR and other ensemble models, including XGBoost and LightGBM, to predict biochar yield and SSA from a dataset of 622 samples derived from various biomass feedstocks and pyrolysis conditions. Among the models tested, GBR demonstrated high predictive power, particularly when input parameters such as ash content, pyrolysis temperature, and residence time were included. The model achieved an R_2 of 0.92 for SSA prediction, indicating excellent performance.

Similarly, Pathy et al. (2020) employed an eXtreme Gradient Boosting (XGB) model—a more advanced form of GBR—to predict algal biochar yield. The model effectively captured the influence of parameters like H/C ratio, ash content, pyrolysis temperature, and residence time. The XGB model achieved a regression coefficient (R_2) of 0.84 for the testing dataset, underscoring the model's robustness in handling nonlinearity and feature interactions.

6.7.8.3 GBR for Heavy Metal Adsorption Optimization

Yaseen and Alhalimi (2025) extended GBR applications to predict heavy metal adsorption efficiencies of biochar based on its characteristics (e.g., CEC, surface area, pH) and environmental parameters (e.g., initial metal concentration, solution pH). Their work compared several ensemble models—Random Forest (RF), Adaboost, GBR, XGBoost, and LightGBM—on a dataset of 353 samples. GBR ranked among the top performers, particularly in modeling the influence of complex interactions among pyrolysis temperature, feedstock type, and surface functionality. The authors highlighted that chemical features like pH and concentration ratio had greater importance than physical features like SSA in determining adsorption performance.

6.7.8.4 Feature Importance and Interpretability

A major advantage of GBR and its variants lies in the interpretability of results through feature importance rankings. Zhou et al. (2024) used Shapley Additive Explanations (SHAP) to reveal that pyrolysis temperature, biomass type, and ash content were the most influential variables for yield and SSA predictions. This insight is valuable for targeted process optimization, reducing experimental burden by focusing on the most impactful parameters.

6.7.8.5 Integration with Experimental Design and Decision Support

GBR models can be integrated with experimental design strategies (e.g., Design of Experiments, DOE) and decision support systems to iteratively refine experimental protocols. Gupta et al. (2023) argue that such machine learning approaches are instrumental in modernizing biochar research by offering cost-effective, scalable, and precise tools for guiding experiments and enhancing product consistency.

References

Adelawon BO, Latinwo GK, Eboibi BE, Agbede OO, Agarry SE (2022) Comparison of the slow, fast, and flash pyrolysis of recycled maize-cob biomass waste, Box-Benhken process optimization and characterization studies for the thermal fast pyrolysis production of bio-energy. Chem Eng Commun 209(9):1246–1276. https://doi.org/10.1080/00986445.2021.1957851

Ahmad M, Lee SS, Lim JE, Lee SE, Cho JS, Moon DH, Hashimoto Y, Ok YS (2014a) Speciation and phytoavailability of lead and antimony in a small arms range soil amended with mussel shell, cow bone, and biochar: EXAFS spectroscopy and chemical extractions. Chemosphere 95:433–441. https://doi.org/10.1016/j.chemosphere.2013.09.077

Ahmad M, Rajapaksha AU, Lim JE, Zhang M, Bolan N, Mohan D, Vithanage M, Lee SS, Ok YS (2014b) Biochar as a sorbent for contaminant management in soil and water: a review. Chemosphere 99:19–33

Aktar M, Rahman MM, Yasmeen R (2020) Influence of pyrolysis atmospheres on biochar properties and carbon sequestration potential. J Anal Appl Pyrolysis 147:104797

Al Saleh HA (2021) Wastewater pollutants modeling using artificial neural networks. J Ecol Eng 22(7):35–45. https://doi.org/10.12911/22998993/138872

Alruqi M, Sharma P (2023) Biomethane production from the mixture of sugarcane vinasse, solid waste and spent tea waste: a Bayesian approach for hyperparameter optimization for Gaussian process regression. Fermentation 9(2):120. https://doi.org/10.3390/fermentation9020120

Amalina F, Abd Razak AS, Krishnan S, Zularisam AW, Nasrullah M (2022) A comprehensive assessment of the method for producing biochar, its characterization, stability, and potential applications in regenerative economic sustainability–a review. Cleaner Mater 3:100045

Ang JC, Tang JY, Chung BYH, Chong JW, Tan RR, Aviso KB, Chemmangattuvalappil NG, Thangalazhy-Gopakumar S (2023) Development of predictive model for biochar surface properties based on biomass attributes and pyrolysis conditions using rough set machine learning. Biomass Bioenergy 174:106820. https://doi.org/10.1016/j.biombioe.2023.106820

Angın D, Tiryaki AE (2016) Application of response surface methodology and artificial neural network on pyrolysis of safflower seed press cake. Energy Sources Recover Util Environ Eff 38(8):1055–1061. https://doi.org/10.1080/15567036.2013.862585

Ascher S, Watson I, You S (2022) Machine learning methods for modelling the gasification and pyrolysis of biomass and waste. Renew Sust Energ Rev 155:111902

Ba-Abbad MM, Chai PV, Benamour A, Ewis D, Mohammad AW, Mahmoudi E (2022) Optimizing and control of effective synthesize parameters for Fe_3O_4 nanoparticles using response surface methodology. Chem Pap 76:6359–6370. https://doi.org/10.1007/s11696-022-02320-y

Banik C, Lawrinenko M, Bakshi S, Laird DA (2018a) Impact of pyrolysis temperature and feedstock on surface charge and functional group chemistry of biochars. J Environ Qual 47(3):452–461. https://doi.org/10.2134/jeq2017.11.0432

Banik C, Sistani KR, McLaughlin RA (2018b) Biochar and nitrate: a review of mechanisms and applications. Agronomy 8(8):151

Campos LMA, Moura HOMA, Cruz AJG, Assumpção SMN, de Carvalho LS, Pontes LAM (2020) Response surface methodology for assessing the effects of pretreatment, feedstock, and enzyme complex association on cellulose hydrolysis. Biomass Convers Biorefinery 12:2811–2822. https://doi.org/10.1007/s13399-020-00756-4

Cao H, Xin Y, Yuan Q (2016b) Prediction of biochar yield from cattle manure pyrolysis via least squares support vector machine intelligent approach. Bioresour Technol 202:158–164. https://doi.org/10.1016/j.biortech.2015.12.024

Cao X, Ma L, Gao B, Harris W (2016a) Dairy-manure derived biochar effectively sorbs lead and atrazine. Environ Sci Technol 43(9):3285–3291

Cely P, Gascó G, Paz-Ferreiro J, Méndez A (2015a) Agronomic properties of biochars from different manure wastes. J Anal Appl Pyrolysis 111:173–182. https://doi.org/10.1016/j.jaap.2014.11.014

Cely P, Tarquis AM, Paz-Ferreiro J, Méndez A, Gascó G (2015b) Factors driving the biochar stability and effect in labile organic matter in soil. Geoderma 245–246:80–87

Chen MW, Chang MS, Mao Y, Hu S, Kung CC (2023) Machine learning in the evaluation and prediction models of biochar application: a review. Sci Prog 106(1):00368504221148842
Chen W, Meng J, Han X, Lan Y, Zhang W (2018) Past, present, and future of biochar. Biochar 1:75–87
Chen X, Pei Y, Wang X, Zhou W, Jiang L (2024) Response surface methodology—central composite design optimization sugarcane bagasse activated carbon under varying microwave-assisted pyrolysis conditions. PRO 12:497. https://doi.org/10.3390/pr12030497
Cho Y, Lim JY, Igalavithana AD, Hwang G, Sang MK, Mašek O, Ok YS (2024) AI-guided investigation of biochar's efficacy in Pb immobilization for remediation of Pb contaminated agricultural land. Appl Biol Chem 67:82. https://doi.org/10.1186/s13765-024-00933-3
Danish M, Ahmad T (2018) A review on utilization of biochar for environmental applications. Environ Technol Innovation 10:108–121
Darmstadt H, Pantea D, Sümmchen L, Roland U, Kaliaguine S, Roy C (2000) Surface and bulk chemistry of charcoal obtained by vacuum pyrolysis of bark: influence of feedstock moisture content. J Anal Appl Pyrolysis 53:1–17. https://doi.org/10.1016/S0165-2370(99)00051-0
Ding W, Dong X, Ime IM, Gao B, Ma LQ (2014a) Pyrolytic temperatures impact lead sorption mechanisms by bagasse biochars. Chemosphere 105:68–74. https://doi.org/10.1016/j.chemosphere.2013.12.042
Ding Y, Liu Y, Liu S, Li Z, Tan X, Huang X, Zeng G, Zhou L, Zheng B (2014b) Biochar to improve soil fertility: a review. Agron Sustain Dev 34:679–712. https://doi.org/10.1007/s13593-014-0234-4
El Aziz MA, Hemdan AM, Ewees AA, Elhoseny M, Shehab A, Hassanien AE, Xiong S (2017) Prediction of biochar yield using adaptive neuro-fuzzy inference system with particle swarm optimization. In: 2017 IEEE PES power Africa, IEEE. pp 115–120. https://doi.org/10.1109/PowerAfrica.2017.7991209
Ewees AA, Elaziz MA (2018) Improved adaptive neuro-fuzzy inference system using gray wolf optimization: a case study in predicting biochar yield. J Intell Syst 29(1):924–940. https://doi.org/10.1515/jisys-2017-0641
Ewees AA, Abd El Aziz M, Elhoseny M (2017) Social-spider optimization algorithm for improving ANFIS to predict biochar yield. In: In: In 2017 8th international conference on computing, communication and networking technologies (ICCCNT), IEEE, pp 1–6. https://doi.org/10.1109/ICCCNT.2017.8203950
Genuer R, Poggi JM, Tuleau-Malot C (2010) Variable selection using random forests. Pattern Recogn Lett 31(14):2225–2236. https://doi.org/10.1016/j.patrec.2010.03.014
Ghani WAWAK, Mohd A, da Silva G, Bachmann RT, Taufiq-Yap YH, Rashid U, Al-Muhtaseb AH (2013) Biochar production from waste rubber-wood-sawdust and its potential use in C sequestration: chemical and physical characterization. Ind Crop Prod 44:18–24. https://doi.org/10.1016/j.indcrop.2012.10.017
Ghosh SS, Dey S, Bhogapurapu N, Homayouni S, Bhattacharya A, McNairn H (2022) Gaussian process regression model for crop biophysical parameter retrieval from multi-polarized C-band SAR data. Remote Sens 14(4):934. https://doi.org/10.3390/rs14040934
Goswami L, Kushwaha A, Kafle SR, Kim BS (2022) Surface modification of biochar for dye removal from wastewater. Catalysts 12(8):817. https://doi.org/10.3390/catal12080817
Gujre N, Mitra S, Agnihotri R, Sharma MP, Gupta D (2022) Novel agrotechnological intervention for soil amendment through areca nut husk biochar in conjunction with vetiver grass. Chemosphere 287:132443. https://doi.org/10.1016/j.chemosphere.2021.132443
Gul E, Alrawashdeh KAB, Masek O, Skreiberg Ø, Corona A, Zampilli M, Wang L, Samaras P, Yang Q, Zhou H, Bartocci P (2021) Production and use of biochar from lignin and lignin-rich residues (such as digestate and olive stones) for wastewater treatment. J Anal Appl Pyrolysis 158:105263. https://doi.org/10.1016/j.jaap.2021.105263
Gupta D, Das A, Mitra S (2023) Role of modelling and artificial intelligence in process parameter optimization of biochar: a review. Bioresour Technol 390:129792. https://doi.org/10.1016/j.biortech.2023.129792

Gupta GK, Mondal MK (2019) Bio-energy generation from sagwan sawdust via pyrolysis: product distributions, characterizations and optimization using response surface methodology. Energy 170:423–437. https://doi.org/10.1016/j.energy.2018.12.166

Han L, Ro KS, Wang Y, Sun K, Sun H, Libra JA, Xing B (2018) Oxidation resistance of biochars as a function of feedstock and pyrolysis condition. Sci Total Environ 616–617:335–344. https://doi.org/10.1016/j.scitotenv.2017.11.014

He M, Xu Z, Sun Y, Chan PS, Lui I, Tsang DCW (2021) Critical impacts of pyrolysis conditions and activation methods on application-oriented production of wood waste-derived biochar. Bioresour Technol 341:125811. https://doi.org/10.1016/j.biortech.2021.125811

Hossain MA, Ganesan P, Jewaratnam J, Chinna K (2017) Optimization of process parameters for microwave pyrolysis of oil palm fiber (OPF) for hydrogen and biochar production. Energy Convers Manag 133:349–362. https://doi.org/10.1016/j.enconman.2016.10.046

Huang WH, Lee DJ, Huang C (2021) Modification on biochars for applications: a research update. Bioresour Technol 319:124100. https://doi.org/10.1016/j.biortech.2020.124100

Hussain R, Ravi K, Garg A (2020) Influence of biochar on the soil water retention characteristics (SWRC): potential application in geotechnical engineering structures. Soil Tillage Res 204:104713. https://doi.org/10.1016/j.still.2020.104713

Irfan M, Nabi RAU, Hussain H, Naz MY, Shukrullah S, Khawaja HA et al (2022) Response surface methodology analysis of pyrolysis reaction rate constants for predicting efficient conversion of bulk plastic waste into oil and gaseous fuels. Energies 15(24):9594. https://doi.org/10.3390/en15249594

Jawad AH, Razuan R, Appaturi JN, Wilson LD (2019) Adsorption and mechanism study for methylene blue dye removal with carbonized watermelon (*Citrullus lanatus*) rind prepared via one-step liquid phase H_2SO_4 activation. Surf Interfaces 16:76–84. https://doi.org/10.1016/j.surfin.2019.04.01

Jeeru LR, Abdul FK, Anireddy JS, Ch VP, Dhanavath KN (2023) Optimization of process parameters for conventional pyrolysis of algal biomass into bio–oil and bio–char production. Chem Eng Process 185:109311. https://doi.org/10.1016/j.cep.2023.109311

Jeyasubramanian K, Thangagiri B, Sakthivel A, Raja JD, Seenivasan S, Vallinayagam P, Madhavan D, Devi SM, Rathika B (2021) A complete review on biochar: production, property, multifaceted applications, interaction mechanism and computational approach. Fuel 292:120243. https://doi.org/10.1016/j.fuel.2021.120243

Jouhara H, Ahmad D, Van Den Boogaert I, Katsou E, Simons S, Spencer N (2018) Pyrolysis of domestic based feedstock at temperatures up to 300 °C. Therm Sci Eng Prog 5:117–143. https://doi.org/10.1016/j.tsep.2017.11.007

Jung KW, Jeong TU, Kang HJ, Chang JS, Ahn KH (2016) Preparation of modified-biochar from Laminaria japonica: simultaneous optimization of aluminum electrode-based electro-modification and pyrolysis processes and its application for phosphate removal. Bioresour Technol 214:548–557. https://doi.org/10.1016/j.biortech.2016.05.005

Kanthasamy R, Almatrafi E, Ali I, Sait HH, Zwawi M, Abnisa F, Peng LC, Ayodele BV (2023) Biochar production from valorization of agricultural wastes: data-driven modelling using machine learning algorithms. Fuel 351:128948. https://doi.org/10.1016/j.fuel.2023.128948

Khan M, Ullah Z, Mašek O, Naqvi SR, Khan MNA (2022) Artificial neural networks for the prediction of biochar yield: a comparative study of metaheuristic algorithms. Bioresour Technol 355:127215. https://doi.org/10.1016/j.biortech.2022.127215

Kopp Alves A, Hauschild T, Basegio TM, Amorim Berutti F (2024) Influence of lignin and cellulose from termite-processed biomass on biochar production and evaluation of chromium VI adsorption. Sci Rep 14(1):14937. https://doi.org/10.1038/s41598-024-53426-6

Kumar M, Rai D, Bhardwaj G, Upadhyay SN, Mishra PK (2021) Pyrolysis of peanut shell: kinetic analysis and optimization of thermal degradation process. Ind Crop Prod 174:114128. https://doi.org/10.1016/j.indcrop.2021.114128

Kumar S, Masto RE, Ram LC, Sarkar P, George J, Selvi VA (2013) Biochar preparation from Parthenium hysterophorus and its potential use in soil application. Ecol Eng 55:67–72. https://doi.org/10.1016/j.ecoleng.2013.02.011

Lakshmi D, Akhil D, Kartik A, Gopinath KP, Arun J, Bhatnagar A, Rinklebe J, Kim W, Muthusamy G (2021) Artificial intelligence (AI) applications in adsorption of heavy metals using modified biochar. Sci Total Environ 801:149623. https://doi.org/10.1016/j.scitotenv.2021.149623

Leng L, Huang H (2018) An overview of the effect of pyrolysis process parameters on biochar stability. Bioresour Technol 270:627–642. https://doi.org/10.1016/j.biortech.2018.09.030

Leng L, Huang H, Li H, Li J, Zhou W (2021) Biochar stability assessment methods: a review. Sci Total Environ 763:144204. https://doi.org/10.1016/j.scitotenv.2020.144204

Leng L, Yang L, Lei X, Zhang W, Ai Z, Yang Z, Yuan X, Peng H, Li H (2022) Machine learning predicting and engineering the yield, N content, and specific surface area of biochar derived from pyrolysis of biomass. Biochar 4:63. https://doi.org/10.1007/s42773-022-00173-3

Li Y, Gupta R, You S (2022) Machine learning assisted prediction of biochar yield and composition via pyrolysis of biomass. Bioresour Technol 359:127511. https://doi.org/10.1016/j.biortech.2022.127511

Liao M, Yao Y (2021) Applications of artificial intelligence-based modeling for bioenergy systems: a review. GCB Bioenergy 13(5):774–802. https://doi.org/10.1111/gcbb.12817

Liu Z, Quek A, Hoekman SK, Balasubramanian R (2020) Production of solid biochar fuel from waste biomass by hydrothermal carbonization. Fuel 103:943–949. https://doi.org/10.1016/j.fuel.2012.07.069

López OAM, Montesinos López A, Crossa J (2022) Multivariate statistical machine learning methods for genomic prediction. Springer. https://doi.org/10.1007/978-3-030-89010-0

Luo L, Xu C, Chen Z, Zhang S (2015) Properties of biomass-derived biochars: combined effects of operating conditions and biomass types. Bioresour Technol 192:83–89. https://doi.org/10.1016/j.biortech.2015.05.054

Ma Z, Yang Y, Wu Y, Xu J, Peng H, Liu X, Zhang W, Wang S (2019) In-depth comparison of the physicochemical characteristics of bio-char derived from biomass pseudo components: hemicellulose, cellulose, and lignin. J Anal Appl Pyrolysis 140:195–204. https://doi.org/10.1016/j.jaap.2019.03.015

Machrouhi A, Alilou H, Farnane M et al (2019) Statistical optimization of activated carbon from *Thapsia transtagana* stems and dyes removal efficiency using central composite design. J Sci Adv Mater Dev 4(4):544–553. https://doi.org/10.1016/j.jsamd.2019.10.003

Madhu P, Matheswaran MM, Periyanayagi G (2017) Optimization and characterization of bio-oil produced from cotton shell by flash pyrolysis using artificial neural network. Energy Sources A Recover Util Environ Eff 39(23):2173–2180. https://doi.org/10.1080/15567036.2017.1403508

Manatura K, Chalermsinsuwan B, Kaewtrakulchai N, Kwon EE, Chen WH (2023) Machine learning and statistical analysis for biomass torrefaction: a review. Bioresour Technol 369:128504. https://doi.org/10.1016/j.biortech.2023.128504

Manikandan S, Vickram S, Subbaiya R, Karmegam N, Chang SW, Ravindran B, Awasthi MK (2023) Comprehensive review on recent production trends and applications of biochar for greener environment. Bioresour Technol 388:129725. https://doi.org/10.1016/j.biortech.2023.129725

Masto RE, Kumar S, Rout TK, Sarkar P, George J, Ram LC (2013) Biochar from water hyacinth (*Eichornia crassipes*) and its impact on soil biological activity. Catena 111:64–71. https://doi.org/10.1016/j.catena.2013.06.025

Mukome FND, Zhang X, Silva LCR, Six J, Parikh SJ (2013) Use of chemical and physical characteristics to investigate trends in biochar feedstocks. J Agric Food Chem 61(9):2196–2204. https://doi.org/10.1021/jf3049142

Murtaza G, Ahmed Z, Usman M (2022a) Feedstock type, pyrolysis temperature and acid modification effects on physiochemical attributes of biochar and soil quality. Arab J Geosci 15(3):305. https://doi.org/10.1007/s12517-022-09539-9

Murtaza G, Mustaqeem M, Zafar M, Ali S (2022b) Role of biochar surface chemistry in contaminant adsorption and microbial interactions: a critical review. J Environ Chem Eng 10(1):106958. https://doi.org/10.1016/j.jece.2021.106958

Ngan A, Jia CQ, Tong S-T (2019) Production, characterization and alternative applications of biochar. In: Fang Z (ed) Production of materials from sustainable biomass resources. Springer, pp 345–368. https://doi.org/10.1007/978-981-13-3730-5_12

Nidheesh PV, Gopinath A, Ranjith N, Akre AP, Sreedharan V, Kumar MS (2021) Potential role of biochar in advanced oxidation processes: a sustainable approach. Chem Eng J 405:126582. https://doi.org/10.1016/j.cej.2020.126582

Nighojkar A, Pandey S, Naebe M, Kandasubramanian B, Soboyejo WW, Plappally A, Wang X (2023) Using machine learning to predict the efficiency of biochar in pesticide remediation. *Npj sustainable.* Agriculture 1:1. https://doi.org/10.1038/s44264-023-00001-1

Obieogu KN, Adeyi O, Dzarma G (2022) Response surface methodology and artificial neural network modelling and optimization of *Colocynthis vulgaris* peel sulphuric acid hydrolysis. Clean Circ Bioeconomy 2:100019. https://doi.org/10.1016/j.clcb.2022.100019

Oliveira FR, Patel AK, Jaisi DP, Adhikari S, Lu H, Khanal SK (2017) Environmental application of biochar: current status and perspectives. Bioresour Technol 246:110–122. https://doi.org/10.1016/j.biortech.2017.08.122

Patel MR, Panwar NL (2023) Biochar from agricultural crop residues: environmental, production, and life cycle assessment overview. Resour Conserv Recycl Adv 19:200173. https://doi.org/10.1016/j.rcradv.2023.200173

Pathy A, Meher S, Balasubramanian P (2020) Predicting algal biochar yield using eXtreme gradient boosting (XGB) algorithm of machine learning methods. Algal Res 50:102006. https://doi.org/10.1016/j.algal.2020.102006

Pradhan A, Chan C, Roul PK, Halbrendt J, Sipes B (2018) Potential of conservation agriculture (CA) for climate change adaptation and food security under rainfed uplands of India: a transdisciplinary approach. Agric Syst 163:27–35. https://doi.org/10.1016/j.agsy.2017.01.002

Pradhan S, Shahbaz M, Abdelaal A, Al-Ansari T, Mackey HR, McKay G (2020) Optimization of process and properties of biochar from cabbage waste by response surface methodology. Biomass Convers Biorefinery 1–13. https://doi.org/10.1007/s13399-020-01101-5

Ruan X, Sun Y, Du W, Tang Y, Liu Q, Zhang Z, Doherty W, Frost RL, Qian G, Tsang DCW (2019) Formation, characteristics, and applications of environmentally persistent free radicals in biochars: a review. Bioresour Technol 281:457–468. https://doi.org/10.1016/j.biortech.2019.02.105

Sakhiya AK, Anand A, Kaushal P (2020) Production, activation, and applications of biochar in recent times. Biochar 2(3):253–285. https://doi.org/10.1007/s42773-020-00047-1

Senthil C, Lee CW (2021) Biomass-derived biochar materials as sustainable energy sources for electrochemical energy storage devices. Renew Sust Energ Rev 137:110464. https://doi.org/10.1016/j.rser.2020.110464

Shaaban A, Se SM, Dimin MF, Juoi JM, Mohd Husin MH, Mitan NMM (2014) Influence of heating temperature and holding time on biochars derived from rubber wood sawdust via slow pyrolysis. J Anal Appl Pyrolysis 107:31–39. https://doi.org/10.1016/j.jaap.2014.01.021

Shafizadeh M, Khosravi A, Nahavandi S, Nguyen QV (2022) A review on Gaussian process regression: applications and challenges in the era of big data. Eng Appl Artif Intell 113:104995. https://doi.org/10.1016/j.engappai.2022.104995

Shagali A, Abba M, Sulaimon AA (2021) Influence of heating rate and temperature on the characteristics of biochar derived from agricultural waste. Renew Energy 172:1051–1060. https://doi.org/10.1016/j.renene.2021.03.087

Shariff A, Aziz NSM, Saleh NM, Ruzali NSI (2016) The effect of feedstock type and slow pyrolysis temperature on biochar yield from coconut wastes. Int J Chem Mol Nucl Mater Metall Eng 10(12):1335–1339

Siddiqui MTH, Nizamuddin S, Mubarak NM, Shirin K, Aijaz M, Hussain M, Baloch HA (2019) Characterization and process optimization of biochar produced using novel biomass, waste pomegranate peel: a response surface methodology approach. Waste Biomass Valoriz 10(3):521–532. https://doi.org/10.1007/s12649-017-0091-y

Tag AT, Duman G, Ucar S, Yanik J (2016) Effects of feedstock type and pyrolysis temperature on potential applications of biochar. J Anal Appl Pyrolysis 120:200–206. https://doi.org/10.1016/j.jaap.2016.05.006

Tao H, Basem A, Alizadeh A, Sharma K, Jasim DJ, Rajab H, Ahmed M, Kassim M, Singh NS, Maleki H (2024) Optimizing Gaussian process regression (GPR) hyperparameters with three metaheuristic algorithms for viscosity prediction of suspensions containing microencapsulated PCMs. Sci Rep 14:20271. https://doi.org/10.1038/s41598-024-71027-9

Te WZ, Muhanin KNM, Chu YM, Selvarajoo A, Singh A, Ahmed SF, Vo DVN, Show PL (2021) Optimization of pyrolysis parameters for production of biochar from banana peels: evaluation of biochar application on the growth of Ipomoea aquatica. Front Energy Res 8:637846. https://doi.org/10.3389/fenrg.2020.637846

Tomczyk A, Sokołowska Z, Boguta P (2020) Biochar physicochemical properties: pyrolysis temperature and feedstock kind effects. Rev Environ Sci Biotechnol 19(1):191–215. https://doi.org/10.1007/s11157-020-09523-3

Tripathi M, Sahu JN, Ganesan P (2016) Effect of process parameters on production of biochar from biomass waste through pyrolysis: a review. Renew Sust Energ Rev 55:467–481. https://doi.org/10.1016/j.rser.2015.10.122

Ukoba KM, Tiako R, Mamat R (2022) Biochar and application of machine learning: a review. Cleaner Energy Syst 3:100013. https://doi.org/10.1016/j.cles.2022.100013

Ukoba KO, Jen TC (2022) Biochar and application of machine learning: a review. In: In: Biochar-productive technologies, properties and applications. Intech Open. https://doi.org/10.5772/intechopen.1004101

Ulusal E, Karaer H, Karagöz S (2021) Characterization and evaluation of biochars produced from different biomass feedstocks. J Clean Prod 295:126477. https://doi.org/10.1016/j.jclepro.2021.126477

Uzuner S, Cekmecelioglu D (2016) Comparison of artificial neural networks (ANN) and adaptive neuro-fuzzy inference system (ANFIS) models in simulating polygalacturonase production. Bio Resour 11(4):8676–8685

Vassilev SV, Baxter D, Andersen LK, Vassileva CG, Morgan TJ (2012) An overview of the organic and inorganic phase composition of biomass. Fuel 94:1–33. https://doi.org/10.1016/j.fuel.2011.09.030

Vieira FR, Luna CMR, Arce GL,´Avila I (2020) Optimization of slow pyrolysis process parameters using a fixed bed reactor for biochar yield from rice husk. Biomass Bioenergy 132:105412. https://doi.org/10.1016/j.biombioe.2019.105412

Wan J, Liu L, Wang G, Sang L, Liang W, Zhang W, Peng C, Fu R (2022) Unveiling the mechanisms of carbon conversion and loss in biochars derived from characteristic lignocellulosic biomass. J Environ Chem Eng 10(5):108403. https://doi.org/10.1016/j.jece.2022.108403

Wang HSH, Yao Y (2023) Machine learning for sustainable development and applications of biomass and biomass-derived carbonaceous materials in water and agricultural systems: a review. Resour Conserv Recycl 190:106847. https://doi.org/10.1016/j.resconrec.2022.106847

Wang L, Wang Y, Ma F, Tankari Dan-Badjo A (2015b) Review on utilization of biochar for nitrogen retention in soil. Environ Technol Rev 4(1):27–33. https://doi.org/10.1080/21622515.2015.1028950

Wang S, Gao B, Zimmerman AR, Li Y, Ma L, Harris WG, Migliaccio KW (2015a) Physicochemical and sorptive properties of biochars derived from woody and herbaceous biomass. Chemosphere 134:257–262. https://doi.org/10.1016/j.chemosphere.2015.04.062

Wang S, Zhang H, Huang H, Xiao R, Li R, Zhang Z (2020) Influence of temperature and residence time on characteristics of biochars derived from agricultural residues: a comprehensive evaluation. Process Saf Environ Prot 139:218–229. https://doi.org/10.1016/j.psep.2020.04.043

Wang W, Chang JS, Lee DJ (2024) Machine learning applications for biochar studies: a mini-review. Bioresour Technol 394:130291. https://doi.org/10.1016/j.biortech.2023.130291

Weber K, Quicker P (2018) Properties of biochar. Fuel 217:240–261. https://doi.org/10.1016/j.fuel.2017.12.054
Xiang L, Liu S, Ye S, Yang H, Song B, Qin F, Shen M, Tan C, Zeng G, Tan X (2021) Potential hazards of biochar: the negative environmental impacts of biochar applications. J Hazard Mater 420:126611. https://doi.org/10.1016/j.jhazmat.2021.126611
Xiong S, Zhuo J, Zhang B, Yao Q (2013) Effect of moisture content on the characterization of products from the pyrolysis of sewage sludge. J Anal Appl Pyrolysis 104:632–639. https://doi.org/10.1016/j.jaap.2013.05.003
Yaashikaa PR, Kumar PS, Varjani S, Saravanan A (2020a) A critical review on the biochar production techniques, characterization, stability and applications for circular bioeconomy. Biotechnol Rep 28:e00570. https://doi.org/10.1016/j.btre.2020.e00570
Yaashikaa PR, Kumar PS, Varjani S, Saravanan A (2020b) A review on bioremediation of heavy metals using carbonaceous materials. Chemosphere 250:126278. https://doi.org/10.1016/j.chemosphere.2020.126278
Yadav K, Tyagi M, Kumari S, Jagadevan S (2019) Influence of process parameters on optimization of biochar fuel characteristics derived from rice husk: a promising alternative solid fuel. Bioenergy Res 12(4):1052–1065. https://doi.org/10.1007/s12155-019-10027-4
Yang F, Lee X, Wang B (2015) Characterization of biochars produced from seven biomasses grown in three different climate zones. Chin J Geochem 34(4):592–600. https://doi.org/10.1007/s11631-015-0072-4
Yang H, Huan B, Chen Y, Gao Y, Li J, Chen H (2016) Biomass-based pyrolytic polygeneration system for bamboo industry waste: evolution of the char structure and the pyrolysis mechanism. Energy Fuel. https://doi.org/10.1021/acs.energyfuels.6b00732
Yang H, Liu X, Liu Y, Cui J, Xiao Y (2023) Application of an optimized machine learning model based on Bayesian algorithm in predicting heavy metal adsorption by biochar. SSRN Electron J. https://doi.org/10.2139/ssrn.4438236
Yaseen ZM, Alhalimi FL (2025) Heavy metal adsorption efficiency prediction using biochar properties: a comparative analysis for ensemble machine learning models. Sci Rep 15:13434. https://doi.org/10.1038/s41598-025-96271-5
Yu H, Zhang Z, Li Z (2014) Characteristics of tar formation during cellulose, hemicellulose and lignin gasification. Fuel 118:250–256. https://doi.org/10.1016/j.fuel.2013.10.080
Yu H, Zou W, Chen J, Chen H, Yu Z, Huang J, Tang H, Wei X, Gao B (2019b) Biochar amendment improves crop production in problem soils: a review. J Environ Manag 232:8–21. https://doi.org/10.1016/j.jenvman.2018.10.117
Yu L, Yuan Y, Tang J, Yang Y, Yu H (2019a) Biochar as an electron shuttle for reductive degradation of nitrobenzene by *Pseudomonas aeruginosa* strain. Bioresour Technol 255:39–45. https://doi.org/10.1016/j.biortech.2018.01.075
Yuan X, Wang J, Deng S, Suvarna M, Wang X, Zhang W, Hamilton ST, Alahmed A, Jamal A, Park AH, Bi X, Ok YS (2022) Recent advancements in sustainable upcycling of solid waste into porous carbons for carbon dioxide capture. Renew Sust Energ Rev 162:112413. https://doi.org/10.1016/j.rser.2022.112413
Zhang J, Wu J, Yu X, Shu Y, Zhang S, Zhang Y (2020) Extraction optimization by using response surface methodology and purification of yellow pigment from *Gardenia jasminoides* var. *radicans* Makikno. Food Sci Nutr 9:822–832. https://doi.org/10.1002/fsn3.2046
Zhang W, Chen R, Li J, Huang T, Wu B, Ma J, Wen Q, Tan J, Huang W (2023) Synthesis optimization and adsorption modelling of biochar for pollutant removal via machine learning. Biochar 5(25). https://doi.org/10.1007/s42773-023-00225-x
Zhou R, Zhang M, Li J, Zhao W (2020) Optimization of preparation conditions for biochar derived from water hyacinth by using response surface methodology (RSM) and its application in Pb^{2+} removal. J Environ Chem Eng 8(5):104198. https://doi.org/10.1016/jjece.2020.104198
Zhou R, Zhang M, Zhou J, Wang J (2019) Optimization of biochar preparation from the stem of Eichhornia crassipes using response surface methodology on adsorption of Cd^{2+}. Sci Rep 9(1):1–17. https://doi.org/10.1038/s41598-019-54105-1

Zhou X, Liu X, Sun L, Jia X, Tian F, Liu Y, Wu Z (2024) Prediction of biochar yield and specific surface area based on integrated learning algorithm. Carbon Res 10(1):10. https://doi.org/10.3390/c10010010

Zhu X, Li Y, Wang X (2019) Machine learning prediction of biochar yield and carbon contents in biochar based on biomass characteristics and pyrolysis conditions. Bioresour Technol 288:121527. https://doi.org/10.1016/j.biortech.2019.121527

Zielińska A, Oleszczuk P, Charmas B (2015) Properties of sewage sludge-derived biochars: effect of feedstock and pyrolysis temperature. Appl Energy 140:196–205. https://doi.org/10.1016/j.apenergy.2014.11.067

Chapter 7
Biochar for Sustainable Air Pollution Control

Deep Chakraborty, Sonal Singh, Jayachelvi Babu, Nidhi Shukla, Krishnendu Mukhopadhyay, and Sankar Sambandam

Abstract Air quality and climate change are impacted by the discharge of hazardous compounds into the atmosphere by a variety of industries, including waste management, construction, chemical manufacture, and agriculture. Eliminating these contaminants prior to their release into the environment has emerged as a key strategy for reducing air pollution. The inexpensive cost, high porosity, big specific surface area, abundance of surface functional groups, and high removal capacity of biochar have all contributed to its growing popularity. Biochar characteristics have been influenced by factors like preparation method & feedstock types. Biochar also have ability to remove different air pollutants released from agriculture biomass (AB) burning, industrialization, transportation and other anthropogenic activities. In this article we have discussed about composition of air release from AB burning, characteristics of biochar & different methods of biochar formation. We have also explored the production of biochar as an alternative to burning biomass in agriculture for sustainable agriculture.

Keywords Pollutants · Agriculture · Waste management · Human health

7.1 Introduction

Air pollution in the Earth's atmosphere is currently most dangerous issue to address in the sustainability conservation due to the growing difficulties. Air-pollution continues to expand in tandem with agricultural productivity which affects ecological and human health (Howell et al. 2021). Despite the fact that agriculture accounts for

D. Chakraborty (✉) · J. Babu · K. Mukhopadhyay · S. Sambandam
Department of Environmental Health Engineering, Sri Ramachandra Faculty of Public Health, Sri Ramachandra Institute of Higher Education and Research (Deemed to be University), Chennai, India
e-mail: deepchakraborty@sriramachandra.edu.in

S. Singh · N. Shukla
Department of Environmental Science, Amity School of Life Science, Amity University Madhya Pradesh, Gwalior, India

D. Moulick et al. (eds.), *Biochar: Achieving Agro-Environmental Sustainability*, Sustainability Solutions, https://doi.org/10.1007/978-3-032-19855-6_7

21% of global greenhouse gas emissions (GHGs), efforts are being made to achieve zero agricultural solid waste. Using agricultural leftovers known as agriculture waste changes into Charcoal, pulp & paper, fuel in addition to using them as animal feed and to fertilize fields. Furthermore, using agricultural byproducts such as biofuels to produce renewable energy is a feasible alternative (Pattanaik et al. 2019). In addition to value addition, producing biochar from agricultural leftover waste is a practical method of reducing the bulk (Kannan et al. 2013). A byproduct of the thermochemical conversion of carbon-rich biomass material created in absence or little amount of oxygen is known as biochar (Ahmad et al. 2014). Numerous pollutants, including carbon monoxide (CO), sulfur dioxide (SO_2), nitrogen oxides (NO_X), and particulate matter (PM), are released by agriculture biomass burning and other anthropogenic sources (Zorpas et al. 2016). Large volumes of carbon dioxide (CO_2), the main greenhouse gas causing global warming, are released concurrently by these processes (Ritchie et al. 2020).

According to the United Nations Department of Economics and Social Affairs (UNDESA), there will be 9.8 billion people on this planet posing an increasing threat to the world's agricultural system by 2050. Presently, technology is crucial to farming and chemical supply in order to feed the expanding population and fulfill the demand for organic produce and grains. Some parts of the world to meet their food needs by applying more sophisticated agricultural system technologies. These farming methods are classified into following categories: agroforestry, organic farming, sustainable agriculture & agroecology (Muhie 2022). All of these innovations in agricultural techniques attempt to minimize waste and boost crop output for achieving environmentally & agriculturally sustainable practices. Biochar, one of the products of scientific study, is essential for creating agricultural and environmentally sustainable practices. In addition to being a viable soil conditioner, a type of charcoal called biochar produced by a feedstock's burning with little to no oxygen present is an effective strategy to trap carbon to counteract worldwide warming as well as climate change (Lorenz and Lal 2014). The production and use of biochar have skyrocketed in the last several years. Regarding agriculture, climate change, energy, and environmental enhancement in general, biochar has generated a lot of interest. The adsorption capacity of biochar is significantly influenced by its size structure, pore volume, and surface chemistry (Tan et al. 2015). Humans have used biochar for a variety of purposes over the years because of its inherent properties, which include the presence of functional groups on its surface, high thermal resistance, the ability to exchange cations, heat capacity, specific surface area, permeability, conductivity, high fixed carbon content, volatility, and ionic strength (Wang and Wang 2019). Biochar has been directly linked to the development of sustainable farming, carbon capture, and agronomic techniques in recent years. The growing emphasis on environmental safety regulations, greenhouse gas emissions, and food shortages is partially to blame for this (Oni et al. 2019). Biochar is unique because it is effective in eliminating both organic and inorganic pollutants and has a large surface area, a non-carbonized fraction, oxygen-containing functional groups on the surface, and a strong cation exchange capability (Tomczyk et al. 2020). Biochar, a carbonaceous byproduct of the pyrolysis process, can be used to achieve and protect soil, water, and air sustainably during extensive agricultural operations and major industrial or anthropogenic activities (Zabaniotou and Stamou 2020). Biochar is widely recognized for its ability to eliminate pollutants, make a substantial contribution to the fight against climate change, and generate bioenergy (Bis et al. 2018). Different types of respiratory diseases like asthma, lung infections, allergies, dehydration, hyperthermia, cardiopulmonary disease and numerous other airborne illnesses have all been linked to air pollution for decades (Verma and Devi 2024). In this article we have discussed about different sources of air pollution, types of air pollutants. We also discussed about biochar formation methods and its use in reducing air pollutants & characteristics of biochar for sustainable agriculture.

7.2 Composition of Air Pollution Release from Agriculture

Due to continuous increase in population India gets Ist position in term of population globally, now this situation is responsible for severe air pollution problem. Balakrishnan et al. (2019) stated that the high levels of atmospheric PM especially the finer ones PM 2.5 is responsible for the high death rates worldwide. According to Garaga et al. (2018) Indian cities have been among the top 20 most polluted in the world for a number of years, exceeding the regulatory limits for pollutants set by the central Pollution Control Board (CPCB) and the World Health Organization. A lot of air pollutants were produced by agricultural operations due to the quick expansion of modern agriculture, which is causing air pollution to become a more serious problem (Ge et al. 2021; Xu et al. 2020). The health of animals and humans of densely populated cities can be affected by atmospheric aerosols, gaseous pollutants & particulate matter released by different anthropogenic activities (Balasubramanian et al. 2021). Greenhouse gases (GHGs) releases from different sources also responsible for regional climate change (Herrero et al. 2016; Lu et al. 2021).

The air emissions of urban and rural regions may not be uniform due to the emission sources, topography, weather and ineffective use of regulatory resources (Han et al. 2020; Long et al. 2022). More research on agricultural pollution and the development of more effective mitigation strategies are required due to the rising demand for food brought on by the world's expanding population. Nine original papers and one review make up this Special Issue, which presents new research on air pollution in areas with a high agricultural concentration and only a small percentage of them took urban air pollution into account.

Various air pollutants were (Fig. 7.1) pointed out in this Special Issue, including particulate matter (PM) (Zhao et al. 2022a, 2022b), ammonia (NH_3) (Oh et al. 2021), hydrogen sulfide (H_2S) (Guo et al. 2022), odor [(Guo et al. 2022; Rzeźnik and Mielcarek-Bocheńska 2022), greenhouse gases [GHG: carbon dioxide (CO_2), methane (CH_4), nitrous oxide (N_2O)] (Gao et al. 2021; Huang et al. 2022), NO_X (Choi and Sunwoo 2022; Jiao et al. 2022), carbon monoxide (CO) (Jiao et al. 2022), volatile organic compounds (VOCs) (Choi and Sunwoo 2022; Guo et al. 2022), sulfur

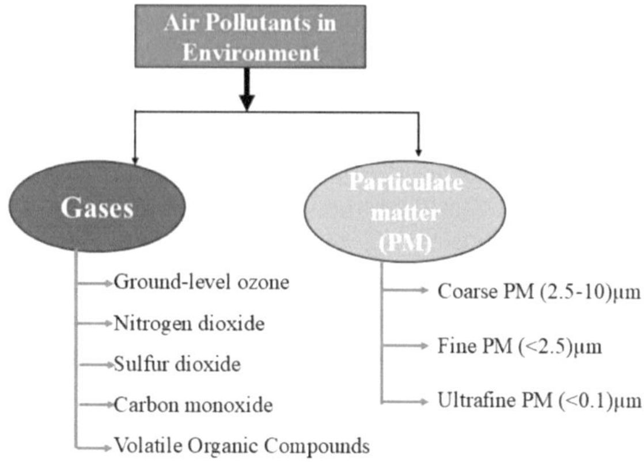

Fig. 7.1 Composition of Air pollution, Gases & particulate matter

dioxide (SO_2) (Choi and Sunwoo 2022; Jiao et al. 2022), as well as bacteria, fungi and viruses (Guo et al. 2022; Yan et al. 2021).

7.3 Major Sources of Emissions of Air Pollutants

The topic of air pollution and its consequences on Indian urban cities has been the subject of numerous research; however, the majority of these studies are city-specific and may not give a comprehensive picture of the situation (Pandey and Venkataraman 2014). According to their research, compared to other forms of transportation including trains, rivers, and airplanes, road transportation in India is responsible for over 97% of estimated emissions (Gurjar and Nagpure 2016, Gurjar et al. 2016). There are different sources which are responsible for air pollution (Fig. 7.2). These are as follows:

Industry

India has seen significant industrialization in recent decades. As a result, the majority of four big cities have seen a decline in air quality. Polluting industries are divided into seven categories by the Central Pollution Control Board (CPCB), which correspond to small and medium airways (Gurjar and Nagpure 2016, Gurjar et al. 2016). Seven of these sectors—steel, sugar, paper, cement, fertilizers, copper, and aluminium—have been designated as "critical" industries. SPM, SOX, NO_X, and CO_2 emissions are the primary pollutants (Gurjar 2021). In addition to the state-provided power, small businesses, which are less regulated than large industry, use a variety of energy sources. Crude oil, polymers, and biomass are used in some of these fuels. Current assessments of emissions inventories ignore these energy sources. Many industries in Delhi were moved from urban areas to nearby rural areas following court intervention in 2000 (Nagpure et al. 2016). The brick business, which is situated on the outskirts of Delhi, is largely to blame for the city's pollution. According to Nagpure et al. (2016),

Fig. 7.2 Air pollution releases by different industries

industry is a major contributor to the highest levels of PM2.5 in Rajkot (42%) and Pune (30%). Industry sector contributes total 52% of total air pollution release from different industries which was given in Fig. 7.3.

Transportation

In practically every city, the transportation sector is the primary cause of air pollution, although in metropolitan areas, the problem is more severe (Gurjar et al. 2004). This might be because there are more cars on the road than there are gas stations, passenger terminals for public transportation, and other pieces of infrastructure. By 2012, there were 159.5 million motorized vehicles in India, up from 0.3 million in 1951 (Gurjar et al. 2016; Gurjare et al. 2016). Delhi, Mumbai, Bengaluru, and Calcutta are among the urban cities that contribute significantly to car emissions. The primary pollutants

Fig. 7.3 Sources & Emissions of Air Pollutants from different sectors

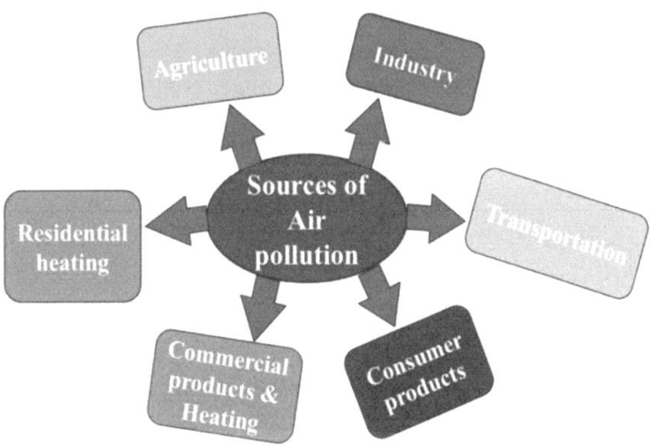

(> 80%) of vehicle emissions are carbon monoxide (CO), NO_X, and NMVOCs (Non-methane Volatile Organic Compounds) (Gurjar et al. 2004) Methane (CH4), carbon dioxide (CO_2), sulfur oxides (SO_X), and total particulate matter (TSP) are other trace emissions (Gurjar 2021). One of the main sources of air pollution in urban areas is emissions from motor vehicles. The main source of particulate matter emissions in Delhi (37%), Mumbai (30%), and Calcutta (61%), respectively, is road dust. In Bengaluru (41%), Chennai (34%), Surat (42%), and Indore (47%), road transportation is the primary source of PM2.5 (Nagpure et al. 2016). The great lack of exhaust measures, the highly diverse character of the vehicles, and poor fuel quality are some of the key causes of excessive traffic emissions in India (Gurjar 2021).

Agriculture.

Emissions from agricultural operations have the potential to harm the environment. Nitrous oxide (N_2O) and ammonia (NH_3) are the primary pollutants emitted during agricultural operations. The addition of fertilizers and other residues to agricultural soils can result in nitrogen emissions (N_2O, NO_X, and NH_3), methane emissions from enteric fermentation processes, nitrogen excretions from animal manure (CH_4, N_2O, and NH_3), and methane emissions from wetlands (Gurjar et al. 2004). The smoke produced by agricultural practices like slash-and-burn agriculture is one of the main causes of photochemical haze. Another method that releases harmful pollutants into the atmosphere is the burning of crop wastes. Delhi's surrounding towns are part of the agricultural pollution problem. In this case, the hazard of urban air pollution is exacerbated by external sources (Gurjar and Nagpure 2016; Gurjar et al. 2016).

Agricultural biomass burning

Air quality management may be a global concern since it is closely related to Sustainable Development Goal 3 (excellent health and well-being) (Verma and Devi 2025). Due to incomplete combustion, burning agricultural biomass is the second largest source of air pollutants. This has a significant negative impact on the environment, human health, and climate change affecting both regional and global atmospheric chemistry (Kang et al. 2019). According to Andreae and Merlet 2001, burning crop residue ranks as the fourth most common type of biomass burning pollution of the environment in 2021. In the past, Indian agriculture has prioritized production maximization over sustainable post-harvest methods. Farmers frequently utilize crop waste burning as a rapid and low-cost way to clear fields, particularly in the brief time between rice harvest and wheat planting. This approach, however, has negative effects on the environment and human health. A number of hazardous pollutants such as CO_2 (carbon dioxide), methane (CH_4), sulfur dioxide (SO_2), ammonia (NH_3) & VOCs (Volatile organic compounds) are released when agricultural waste, household wastes & other wastes are burned (Kumar et al. 2018). These gases are responsible for severe air pollution, global warming, and harmful health issues like cardiovascular and respiratory disorders. According to the Ministry of Agriculture and Farmers Welfare (2020), India produces between 500 and 550 million tonnes (MTs) of crop leftovers within a year, with Uttar Pradesh accounting for the largest amount at 60 million tonnes (MTs), followed by Punjab at 51 million tonnes (MTs) and Maharashtra at 46 million tonnes (MTs). Instead of Buring of crop leftovers we can use it in the production of biochar which is a sustainable agriculture method also improves soil fertility, soil nutrients availability etc. According to the Indian Ministry of New and Renewable Energy (MNRE), India generates on an average 500 million tons (Mt here after) of crop residue per year (NPCMR 2019). Table 7.1 shows different crop with their dry residue production in India adapted from Economic Survey 2020; (CRIDA 2018; Nan et al. 2020).

According to the Mor et al. 2022 carbon monoxide (CO), nitrogen oxides (NO_X), and particulate matter (PM) are the most dangerous pollutants released into the atmosphere when crop residue is burned. Crop residue burning is especially prevalent in Punjab and Haryana due to the large amounts of wheat and rice that are grown there (Kumar and Singh 2020). According to data from the Indian Council of Agricultural Research (Govindaraj et al. 2019) about 14% of Total crop residue burning in Haryana India in 2018. Chhabra & Mehta claimed that one kilogram of paddy yields 1–1.5 kg of straw. It is quite challenging to dispose of such a large amount of crop leftovers. Kaur et al. 2022, stated that, due to the lack of efficient in situ agriculture residue management system farmers face various obstacles, including the height of crop stubbles, the low nutritional value of paddy straw, the cost of

Table 7.1 India generates huge quantity of crop residue (million ton/year)

Crop	Quantity of residues generated (million ton/year)	Quantity of dry residues generated (million ton/year)	Biochar yield %
Wheat straw *(Triticum aestivum)*	170	149	25–30
Rice straw *(Oryza sativa)*	169	145	28–33
Rice husk *(O. Sativa)*	23	20	33–38
Sorghum (*Sorghum vulgare*)	24.2		25–30
Corn stalks *(Zea mays)*	27		25–30
Castor (*Ricinus communis*)	180	150	17–24
Pigeon pea *(Cajanus cajan)*	10.33		30–35
Cotton stalk (*Gossypium* sp.)	16	13	25–30
Sugarcane bagasse (*Saccharum officinarum*)	1521	133	24–28

(Economic Survey 2020) (Nan et al. 2020)

collection and transportation, the lack of crop stubble markets. For these reasons, farmers who are unaware of public health concerns view open crop residue burning as the simplest and most cost-effective method of crop residue management (Abdurrahman et al. 2020). According to Chanana et al. 2023, the toxic pollutants emitted during agriculture residue burning can cause serious health effects specially heart diseases and respiratory issues. These pollutants, which are seen in Fig. 7.1, disperse throughout the ecosystem after being released into the atmosphere, may undergo physical and chemical changes, and eventually have an adverse effect on the environment and human health. Additionally, by releasing large quantities of greenhouse gases into the atmosphere, such as carbon dioxide and methane, the practice adds to global warming (Chawala and Sandhu 2020). But it also presents serious problems including nitrogen depletion and soil deterioration (Nunes et al. 2020).

7.4 Formation of Biochar from Agriculture Crop Residue

A new word for carbonaceous materials used in carbon sequestration, soil remediation, and sustainable energy is "biochar." With the exception of fossil fuel products, biochar is defined as charcoal and carbon-rich material made by partial oxidation (pyrolysis at $\leq$700 °C without or with a limited supply of oxygen) of carbonaceous organic sources like wood and plants.

The increasing interest in turning agricultural waste into biochar is driven by the use of natural, plentiful, and renewable resources to create carbonaceous compounds using straightforward and energy-efficient processes (Kumar et al. 2020). Biochar can be used to sustainably manage 500–550 million tons of crop residue in India, lowering pollution from burning stubble and storing carbon. 74,000 tons of carbon may be conserved by converting just 1% of the agricultural trash in Northeast India (Verma and Devi 2025). According to B2B Market Report 2024, the biochar market in India is expected to grow at an annual pace of 14.1% to reach $75.7 million by 2032. Based on the degree of development, several methods have been identified for the manufacture of biochar, such as gasification, hydrothermal carbonization, torrefaction, and others; pyrolysis is the most common method (Gabhane et al. 2020).

Pyrolysis

Pyrolysis is of two type-slow and fast, with varying residence durations and heating rates. Slow pyrolysis has a low heating rate and a lengthy residence time, whereas fast pyrolysis has a high heating rate (Fig. 7.4) and a short residence time (Tripathi et al. 2016). Three steps make up the process: (a) pre-pyrolysis, which removes moisture and volatiles, at 200–300 °C; (b) main pyrolysis, which breaks down cellulose and hemicellulose at 300–600 °C; and (c) formation of carbonaceous products, which breaks down lignin and other compounds with stronger bonds at 600–900 °C (Zhao et al. 2022a, 2022b; Hassan et al. 2020).

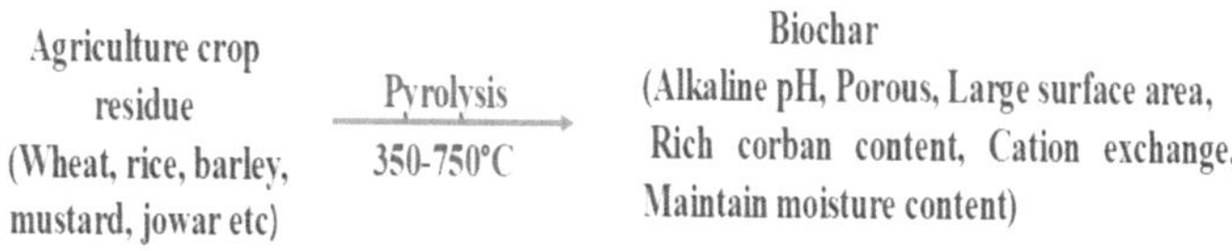

Fig. 7.4 Formation of biochar from different crop residue with its characteristics

Gasification

Gasification is a heat-based thermal reaction & with regulated steam and oxygen supply that converts biomass into energy fuel, producing charcoal as a byproduct at temperatures higher than 700 °C. By breaking down hemicellulose and producing stable torrefied wood, torrefaction—a biomass treatment without oxygen at atmospheric pressure and 230–300 °C—improves feedstock qualities for gasification and combustion (Tripathi et al. 2016).

Hydrothermal carbonization (HTC)

Hydrothermal carbonization (HTC) entails raising the temperature of wet biomass to between 220 and 240 °C, decreasing the O/C ratio, and enhancing properties such as calorific value (Sun et al. 2014; Gabhane et al. 2020).

Microwave pyrolysis

Microwave pyrolysis, an effective technique that uses dielectric heating to reduce temperatures by 200 °C without requiring direct contact between the reaction mixture and the energy source (Sakhiya et al. 2020; Gabhane et al. 2020; Zhou et al. 2021a, 2021b).

7.5 Potential Applications, Regeneration and Disposal of Biochar

Research aimed at combating air pollution is absolutely essential. Advanced nanotechnology, metal-organic frameworks, plant-based phytoremediation, and biochar are resourceful and efficient techniques that current potential solutions for enhancing ambient air quality (Sadiq et al. 2025). Biochar is an affordable and renewable resource and there are opportunities to develop and broaden the application of biochar technology in the field of air pollution control since it effectively eliminates organic pollutants, heavy metal vapours, acidic gases, nitrogen oxides, ozone, and industrial (Hama Aziz et al. 2023). In spite of the potential advantages of biochar-based air filter technology; there are a number of challenges, mainly because of current limitations and outstanding questions. To be able to fully achieve the benefits of these filters, more research needs to be conducted to overcome these limitations and address the knowledge gaps.

Biochar—Opportunities and potential applications in air pollution remediation

Biochar has diverse physical and chemical characteristics. Biochar is typically much smaller than soil particles, and its incorporation makes soil lighter, but it helps in increasing soil moisture, maintaining the pH of soil, and increasing nutrient viability with cation exchange capacity (Soni et al. 2026). The properties of biochar are a function of feedstock type and pyrolysis conditions (Iwuozor et al. 2023). The selection of feedstock is based on its local availability, which helps minimize transportation costs and reduce the carbon footprint (Chandra et al. 2018).

Biochar has numerous environmental applications, including waste treatment, carbon sequestration, water purification, and soil rejuvenation (Fig. 7.5). All of these functions contribute to ecological sustainability (Qambrani et al. 2017). Its high efficiency and stability in eliminating contaminants show its ability to act as a catalyst and an essential ingredient in enhancing soil fertility & Rejuvenation (Rajput et al. 2024).

Over the past few years, carbon dioxide emissions have increased, and they are a major cause of global warming (Kabir et al. 2023). Agriculture contributes to almost 25% of the world's greenhouse gas emissions, mainly from livestock and fertilizers. Disturbed soils emit carbon dioxide, but they can also sequester atmospheric carbon by plant growth. The growing population of the world is increasing food and land requirements, which are a challenge as well as an opportunity. Sustainable practices like cover crops and reduced tillage can improve soil storage of carbon and reduce emissions, enabling agriculture to contribute to carbon neutrality in agricultural production. Biochar, which is a carbonated by-product derived from pyrolysis of agricultural waste, has the potential to boost soil health. Its properties including slow decomposition, high adsorbing capacity, and dense nutrient load

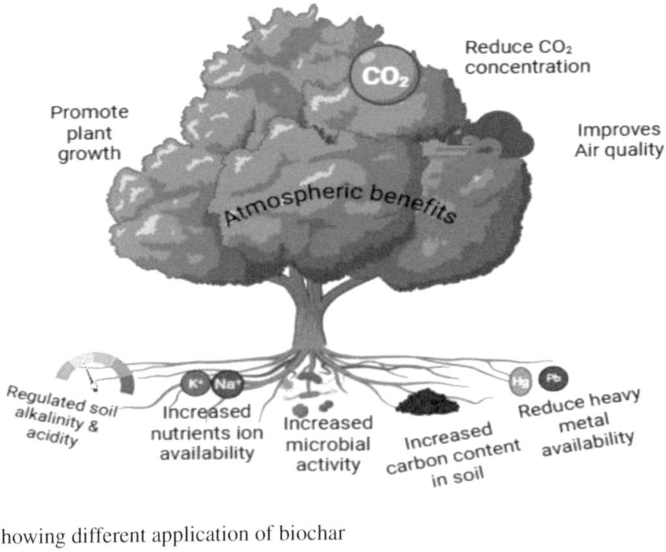

Fig. 7.5 Showing different application of biochar

increase soil fertility and water retention capacity. In addition to that, biochar stores carbon for the long term and hinders soil mineralization of carbon, which, in turn, curbs the emissions of carbon from the land. Biochar enhances the residence of green residues, enhancing organic matter, structural aspects, fertility, and microorganism levels of the soil, all in aid of optimal soil health (Bai et al. 2025). Proper treatment of industrial wastewater utilizes several methods, such as adsorption, precipitation, surface redox reactions, and catalytic degradation. Incorporating biochar into municipal wastewater processes dramatically accelerates flocculation, dewatering, and nutrient recovery and improves sludge management as well as reduces odors. Utilizing biochar in sustainable drainage systems is also beneficial in reducing the effects of stormwater by improving soil structure, enhancing erosion resistance, and efficiently removing pollutants (He et al. 2022). Biochar, a carbon-rich solid, is an effective and sustainable solution for managing organic solid waste. It also serves as a valued use in material adaptation technology. As the population continues to grow and solid waste increases, integrated technologies like biochar can aid humanity in achieving a circular economy (Sudarsan et al. 2023). Municipal solid waste (MSW) largely consists of household waste, plastics, organic matter, metals, and glass. Pyrolyzing MSW can produce biochar, which varies in physicochemical properties due to the diverse feedstocks. Research shows that municipal solid waste biochar (MSW-BC) can effectively remove volatile organic compounds (VOCs) from water. However, the high ash content common in both MSW and food waste materials, like meat and bones, makes them less suitable for creating high-specific-surface-area biochar. Despite these challenges, MSW has great potential because of its abundance and ease of handling (Wang et al. 2023). In the construction industry, cement production is a major source of these emissions. While cementitious composites can sequester CO_2 through carbonation, their efficiency is limited by the slow diffusion of CO_2. To reduce net emissions during concrete production, enhancing carbon capture through accelerated carbonation curing (ACC) and CO2 capture and storage (CCS) is effective. Biochar stands out as a green, low-cost option with strong CO2 adsorption capacity, offering significant economic benefits (Liu et al. 2022). Biochar also promotes the growth of microbes. Microbes use their cellular metabolisms to convert biodegradable, complicated hazardous chemicals into safe, acceptable end products like CO2 and H2O in soil (Singh et al. 2025).

The United Nations Charter aimed at creating a sustainable future has led to the introduction of the Sustainable Development Goals (SDGs) on a global scale. Specifically, SDG 11 focuses on developing cities and communities that promote a high quality of life while achieving net-zero discharge and self-sustainability. In support of the global community's initiatives, biochar has emerged as a viable solution due to its capability to transform waste into valuable resources (Ganesapillai et al. 2023).

7.6 Regeneration of Spent Biochars after Air Pollution Remediation

Biochar is known to be an effective adsorbent, capable of removing various pollutants. However, its saturation points limit long-term use, but regeneration can enhance its applicability. The process of biochar regeneration is essential to preserving the sustainability and efficacy of biochar as an adsorbent for the removal of contaminants from water and wastewater. Biochar's adsorption capacity is restored by a variety of regeneration techniques when it becomes saturated with impurities, allowing for reusing it and lowering operating expenses.

Regeneration Methods

Chemical regeneration

Biochar saturated with organic and inorganic contaminants is frequently regenerated using chemical agents such hydrogen peroxide, peroxymonosulfate, persulfate, and alkaline solutions. The type of pollutants adsorbed the pore structure, and the functional groups of the biochar all affect how well chemicals regenerate. Secondary pollution and the requirement for sustainable procedures, however, continue to be obstacles to widespread use (Alsawy et al. 2022).

Thermal and Oxidative Regeneration

Oxidation and ultrasonic decomposition techniques generally require significant thermal energy, classifying them as thermal decomposition methods. In contrast, desorption regeneration techniques utilize either thermal or non-thermal approaches to break the bonds between biochar and pollutants. Thermal regeneration, which is a destructive method, is only applied when the adsorbate is considered waste and recovery is not necessary. This process involves high energy consumption, often reaching temperatures of up to 800 °C, which can increase the overall cost significantly—potentially accounting for as much as 50% of the total expense of producing fresh biochar. Non-thermal methods, however, allow adsorbents to be preserved and reused across multiple cycles (Alsawy et al. 2022).

Synergistic and Advanced Regeneration Approaches

Combining chemical agents, such as H_2O_2 and peroxymonosulfate, or utilizing advanced oxidation processes can significantly enhance regeneration efficiency and pollutant mineralization. These techniques can achieve higher regeneration rates and are regarded as more sustainable and environmentally friendly (Ding et al. 2023).

Mechanisms and Efficiency

Physical and Chemical Interactions

Regeneration typically involves both the physical desorption and chemical degradation of adsorbed pollutants. This process can be influenced by factors such as temperature, pH, and the presence of metal oxides or persistent free radicals in the biochar. These components can mediate redox reactions and enhance the regeneration process (Ding et al. 2023).

Regeneration Performance

Effective regeneration can restore or even enhance the adsorption capacity of biochar over multiple cycles. For example, biochar activated with iron and biochar modified with multiple metals demonstrate high regeneration efficiencies and improved pollutant removal after several cycles (Zeng and Kan 2021).

Final disposal of biochars after uses

Final disposal options for biochar after its use typically involve safely incorporating it into the environment, with soil application being the most common method. This is regarded as a final disposal option because once biochar is applied to the soil, it becomes nearly impossible to separate it. Disposal methods can be categorized into temporary options, such as recycling or reuse, and final options like incineration, landfilling, or permanent application to the soil.

Recycling and Reuse

Biochar can be repurposed into similar or different products through a process known as cascading uses, as described in several studies. For example, biochar can first be used in water or gas filtration, then treated as an additive for animal manure, and ultimately applied to agricultural land. In urban areas, biochar may initially be used in green roof substrates, reused in other green roofs, and finally converted into landscaping soil for the horticulture sector (Ruett et al. 2024).

Incineration

Incineration of biochar releases all of its carbon back into the atmosphere, meaning it cannot be considered a long-term carbon sink. However, incineration may be a viable option in cases where biochar has been used in applications that make it unsuitable for other disposal methods, such as the treatment of highly toxic substances. Many biochar producers and users, who value the carbon removal benefits of biochar, want to ensure that their products are not incinerated (Gamaralalage et al. 2025).

Landfilling

Landfilling biochar can occur when soil products are disposed of. It is generally believed that the carbon permanence of biochar in landfills is similar to that in soil environments. However, landfilling biochar can have both positive and negative effects. On one hand, it may enhance soil conditions and promote carbon sequestration. On the other hand, it could increase landfill gas emissions and requires careful management to prevent leachate contamination (Brtnicky et al. 2021).

Final Soil Application

The application of biochar to soil is considered a final disposal method because, once applied, it is nearly impossible to separate biochar from the soil matrix (Brtnicky et al. 2021).

Dispersion in the Environment

During the use and disposal of biochar, some portion may unintentionally spread into the environment. For instance, biochar used in green roof substrates can be washed away into nearby waterways due to erosion and runoff. Similarly, biochar incorporated into asphalt may be released into the surrounding environment as the asphalt deteriorates over time (Liao et al. 2022).

7.7 Future Perspectives

Biochar has significant potential as a net-zero carbon resource and plays a critical role in a circular economy. The global market for biochar is expected to grow substantially. Its applications include improving soil health, mitigating climate change, and potentially contributing to sustainable energy storage. Future research and development will focus on optimizing production methods, characterizing biochar properties for specific applications, and ensuring its safe and environmentally responsible use, particularly in large-scale implementations.

7.8 Key Future Perspectives of Biochar

Biochar aids in achieving net-zero emissions by storing carbon in soil and potentially removing CO_2 from the atmosphere. It enhances soil health by improving structure, water retention, and nutrient availability, which can boost crop yields and reduce chemical fertilizer use. Additionally, biochar helps mitigate climate change by decreasing greenhouse gas emissions and storing carbon long-term. Researchers are also exploring its use as a renewable battery electrode material for sustainable energy storage. Biochar helps pollutant control, reduce soil and water contaminants. Waste Valorization transforms biomass waste into a valuable resource, minimizing landfill use. Circular Economy—Biochar promotes sustainable resource management.

7.9 Future Research Directions

Future research will focus on optimizing biochar production techniques, understanding its long-term effects on ecosystems and soil, and developing advanced biochar-based materials for specific applications, such as nano-biochar.

7.10 Knowledge Gaps

There are large knowledge gaps in biochar studies, most notably about its long-term implications, the processes by which it influences soil, and the best ways of making biochar with particular characteristics to suit intended purposes. Additional research is required to determine how long biochar remains in the soil, how it interacts with microbial networks, and how to reduce potential adverse effects, i.e., leaching of pollutants. Additional research and development can maximize the potential of biochar in solving key environmental and agricultural problems by bridging these knowledge gaps.

There is limited research on the potential adverse impacts of biochar on ecosystems since there are comparatively few studies that consider its potential detrimental impacts, which are usually overlooked. Yet, there is growing evidence that biochar does not always have beneficial impacts on soil and soil biota, and the impact on certain microorganisms cannot be generalized. Additionally, even though biochar is long-term stable, it also experiences chemical, physical, and biological alterations in the course of time, which makes it hard to make exact predictions about its impacts, both qualitatively and quantitatively. Hence, more research is necessary to examine the possibly negative impacts of biochar in agricultural soils (Brtnicky et al. 2021).

7.11 Constraints of Using Biochar in Air Pollutant Remediation

Though biochar promises to remediate air pollutants, various constraints discourage its general use. These include the potential of biochar to release contaminants, limited research into its efficacy in gaseous pollutants, and the effect of several factors on its performance, including pH values, competing ions, and organic matter. Also, producing and handling biochar can require intensive resources and generate concerns related to possible emissions of dust particles with associated adverse effects on human health.

References

Abdurrahman MI, Chaki S, Saini G (2020) Stubble burning: effects on health and environment, regulations and management practices. Environ Adv 2:100011

Ahmad M, Rajapaksha AU, Lim JE, Zhang M, Bolan N, Mohan D, Vithanage M, Lee SS, Ok YS (2014) Biochar as a sorbent for contaminant management in soil and water: a review. Chemosphere 99:19–33. https://doi.org/10.1016/j.chemosphere.2013.10.071

Alsawy T, Rashad E, El-Qelish M, Mohammed RH (2022) A comprehensive review on the chemical regeneration of biochar adsorbent for sustainable wastewater treatment. NPJ Clean Water 5(1). https://doi.org/10.1038/s41545-022-00172-3

Bai J, De Almeida Moreira BR, Bai Y, Nadar CG, Feng Y, Yadav S (2025) Assessing biochar's impact on greenhouse gas emissions, microbial biomass, and enzyme activities in agricultural soils through meta-analysis and machine learning. Sci Total Environ 963:178541. https://doi.org/10.1016/j.scitotenv.2025.178541

Balakrishnan K et al (2019) The impact of air pollution on deaths, disease burden, and life expectancy across the states of India: the global burden of disease study 2017 lancet planet. Health 3:e26–e39

Balasubramanian S, Domingo NGG, Hunt ND, Gittlin M, Colgan KK, Marshall JD, Robinson AL, Azevedo IML, Thakrar SK, Clark MA, Tessum CW, Adams PJ, Pandis SN, Hill JD (2021) The food we eat, the air we breathe: a review of the fine particulate matter-induced air quality health impacts of the global food system. Environ Res Lett 16(10):103004. https://doi.org/10.1088/1748-9326/ac065f

Bis Z, Kobyłecki R, Ścisłowska M, Zarzycki R (2018) Biochar—potential tool to combat climate change and drought. Ecohydrol Hydrobiol 18(4):441–453. https://doi.org/10.1016/j.ecohyd.2018.11.005

Brtnicky M, Datta R, Holatko J, Bielska L, Gusiatin ZM, Kucerik J, Hammerschmiedt T, Danish S, Radziemska M, Mravcova L, Fahad S, Pecina V (2021) A critical review of the possible adverse effects of biochar in the soil environment. Sci Total Environ 796:148756. https://doi.org/10.1016/j.scitotenv.2021.148756

Chanana I, Sharma A, Kumar P, Kumar L, Kulshreshtha S, Kumar S, Patel SKS (2023) Combustion and stubble burning: a major concern for the environment and human health. Fire 6(2):79. https://doi.org/10.3390/fire6020079

Chandra R, Castillo-Zacarias C, Delgado P, Parra-Saldívar R (2018) A biorefinery approach for dairy wastewater treatment and product recovery towards establishing a biorefinery complexity index. J Clean Prod 183:1184–1196. https://doi.org/10.1016/j.jclepro.2018.02.124

Chawala P, Sandhu HAS (2020) Stubble burns area estimation and its impact on ambient air quality of Patiala & Ludhiana district, Punjab, India. Heliyon 6(1):e03095. https://doi.org/10.1016/j.heliyon.2019.e03095

Chhabra V, Mehta CM (2019) Rice straw management for sustainable agriculture—a review. Plant Arch 19(2):47–49

Choi H, Sunwoo Y (2022) Environmental benefits of ammonia reduction in an agriculture-dominated area in South Korea. Atmos 13(3):384. https://doi.org/10.3390/atmos13030384

CRIDA (2018). https://www.icarcrida.res.in/assets/img/Technicalblletins/Biochar%20Research%20Bulletin%20March%202018.pdf

Ding H, Tong G, Sun J, Ouyang J, Zhu F, Zhou Z, Zhou N, Zhong M (2023) Regeneration of methylene blue-saturated biochar by synergistic effect of H_2O_2 desorption and peroxymonosulfate degradation. Chemosphere 316:137766. https://doi.org/10.1016/j.chemosphere.2023.137766

Economic Survey (2020). indiabudget.gov.in/budget2020–21/economicsurvey/doc/echapter_vol2.pdf

Gabhane JW, Bhange VP, Patil PD, Bankar ST, Kumar S (2020) Recent trends in biochar production methods and its application as a soil health conditioner: a review. SN Appl Sci 2(7). https://doi.org/10.1007/s42452-020-3121-5

Gamaralalage D, Rodgers S, Gill A, Meredith W, Bott T, West H, Alce J, Snape C, McKechnie J (2025) Biowaste to biochar: a techno-economic and life cycle assessment of biochar production from food-waste digestate and its agricultural field application. Biochar 7(1). https://doi.org/10.1007/s42773-025-00456-0

Ganesapillai M, Mehta R, Tiwari A, Sinha A, Bakshi HS, Chellappa V, Drewnowski J (2023) Waste to energy: a review of biochar production with emphasis on mathematical modelling and its applications. Heliyon 9(4):e14873. https://doi.org/10.1016/j.heliyon.2023.e14873

Gao Y, Liang A, Zhang Y, McLaughlin N, Zhang S, Chen X, Zheng H, Fan R (2021) Dynamics of microbial biomass, nitrogen mineralization and crop uptake in response to placement of maize residue returned to Chinese mollisols over the maize growing season. Atmos 12(9):1166. https://doi.org/10.3390/atmos12091166

Garaga R, Sahu SK, Kota SH (2018) A review of air quality modeling studies in India: local and regional scale. Curr Pollut Rep 4(2):59–73. https://doi.org/10.1007/s40726-018-0081-0
Ge P, Chen M, Cui Y, Nie D (2021) The research Progress of the influence of agricultural activities on atmospheric environment in recent ten years: a review. Atmos 12(5):635. https://doi.org/10.3390/atmos12050635
Govindaraj, G., Hiremath, J., Reddy, G.B., Siju, S.J., Yogishardhaya, R. and Prajapati, A., (2019). ICAR NIVEDI annual report 2018–19
Guo L, Zhao B, Jia Y, He F, Chen W (2022) Mitigation strategies of air pollutants for mechanical ventilated livestock and poultry housing—a review. Atmos 13(3):452. https://doi.org/10.3390/atmos13030452
Gurjar, B. R. (2021), Air pollution in India: major issues and challenges—16 sustainability, Agri, food and environmental research, 12, special issue. (ISSN: 0719–3726)
Gurjar BR, Nagpure AS (2016) Indian megacities as localities of environmental vulnerability from air quality perspective. J Smart Cities 1(1). https://doi.org/10.18063/jsc.2015.01.003
Gurjar BR, van Aardenne JA, Lelieveld J, Mohan M (2004) Emission estimates and trends (1990–2000) for megacity Delhi and implications. Atmos Environ 38(33):5663–5681. https://doi.org/10.1016/j.atmosenv.2004.05.057
Gurjar BR, Ravindra K, Nagpure AS (2016) Air pollution trends over Indian megacities and their local-to-global implications. Atmos Environ 142:475–495. https://doi.org/10.1016/j.atmosenv.2016.06.030
Hama Aziz KH, Mustafa FS, Omer KM, Hama S, Hamarawf RF, Rahman KO (2023) Heavy metal pollution in the aquatic environment: efficient and low-cost removal approaches to eliminate their toxicity: a review. RSC Adv 13(26):17595–17610. https://doi.org/10.1039/d3ra00723e
Han W, Li Z, Guo J, Su T, Chen T, Wei J, Cribb M (2020) The urban–rural heterogeneity of air pollution in 35 metropolitan regions across China. Remote Sens 12(14):2320. https://doi.org/10.3390/rs12142320
Hassan M, Liu Y, Naidu R, Parikh SJ, Du J, Qi F, Willett IR (2020) Influences of feedstock sources and pyrolysis temperature on the properties of biochar and functionality as adsorbents: a meta-analysis. Sci Total Environ 744:140714. https://doi.org/10.1016/j.scitotenv.2020.140714
He M, Xu Z, Hou D, Gao B, Cao X, Ok YS, Rinklebe J, Bolan NS, Tsang DCW (2022) Waste-derived biochar for water pollution control and sustainable development. Nat Rev Earth Environ 3(7):444–460. https://doi.org/10.1038/s43017-022-00306-8
Herrero M, Henderson B, Havlík P, Thornton PK, Conant RT, Smith P, Wirsenius S, Hristov AN, Gerber P, Gill M, Butterbach-Bahl K, Valin H, Garnett T, Stehfest E (2016) Greenhouse gas mitigation potentials in the livestock sector. Nat Clim Chang 6(5):452–461. https://doi.org/10.1038/nclimate2925
Howell N, Pimentel A, Bhattacharia S (2021) Material properties and environmental potential of developing world-derived biochar made from common crop residues. Environ Challenges 4:100137. https://doi.org/10.1016/j.envc.2021.100137
Huang D, Chen X, Zhang S, Zhang Y, Gao Y, Zhang Y, Liang A (2022) No-tillage improvement of nitrogen absorption and utilization in a Chinese mollisol using 15n-tracing method. Atmos 13(4):530. https://doi.org/10.3390/atmos13040530
Inkwood Research (2024) India biochar market [B2B Market Report]. https://www.Inkwoodhttps://www.Inkwoodresearch.com/reports/india-biochar-market
Iwuozor KO, Emenike EC, Omonayin EO, Bamigbola JO, Ojo HT, Awoyale AA, Eletta OA, Adeniyi AG (2023) Unlocking the hidden value of pods: a review of thermochemical conversion processes for biochar production. Bioresour Technol Rep 22:101488. https://doi.org/10.1016/j.biteb.2023.101488
Jiao M, Yan M, Liu Y, Mawusi SK, Shrestha P, Xue C, Song H, Wang H, Liu Z, Xu Y, Liu G (2022) Laboratory performance evaluation of novel bituminous coal pellet combustion in an automatic heating stove. Atmos 13(2):159. https://doi.org/10.3390/atmos13020159
Kabir M, Habiba UE, Khan W, Shah A, Rahim S, los Rios-Escalante PRD, Shafiq M (2023) Climate change due to increasing concentration of carbon dioxide and its impacts on environment in 21st

century; a mini review. J King Saud Univ Sci 35(5):102693. https://doi.org/10.1016/j.jksus.2023.102693
Kang M, Guo H, Wang P, Fu P, Ying Q, Liu H, Zhang H (2019) Characterization and source apportionment of marine aerosols over the East China Sea. Sci Total Environ 651: 2679–2688. https://doi.org/10.1016/j.scitotenv.2018.10.174
Kannan P, Arunachalam P, Prabukumar, Govindaraj M (2013) Biochar an alternate option for crop residues and solid waste disposal and climate change mitigation. Afr J Agric Res 8:2403–2412
Kaur, A., Kaur, P. and Kumar, P., 2022. Problems with paddy straw management technologies in South-Western Punjab
Kumar A, Saini K, Bhaskar T (2020) Hydrochar and biochar: production, physicochemical properties and techno-economic analysis. Bioresour Technol 310:123442. https://doi.org/10.1016/j.biortech.2020.123442
Kumar P, Singh RK (2020) Selection of sustainable solutions for crop residue burning: an environmental issue in northwestern states of India. Environ Dev Sustain 23(3):3696–3730. https://doi.org/10.1007/s10668-020-00741-x
Kumar P, Kumar S, Sharma P (2018) Crop residue burning in India: a review. J Environ Sci Health B 53(1):37–46
Liao W, Sifton MA, Thomas SC (2022) Biochar granulation reduces substrate erosion on green roofs. Biochar 4(1). https://doi.org/10.1007/s42773-022-00186-7
Liu J, Liu G, Zhang W, Li Z, Xing F, Tang L (2022) Application potential analysis of biochar as a carbon capture material in cementitious composites: a review. Constr Build Mater 350:128715. https://doi.org/10.1016/j.conbuildmat.2022.128715
Long F, Liu J, Zheng L (2022) The effects of public environmental concern on urban-rural environmental inequality: evidence from Chinese industrial enterprises. Sustain Cities Soc 80:103787. https://doi.org/10.1016/j.scs.2022.103787
Lorenz K, Lal R (2014) Biochar application to soil for climate change mitigation by soil organic carbon sequestration. J Plant Nutr Soil Sci 177(5):651–670. https://doi.org/10.1002/jpln.201400058
Lu X, Ye X, Zhou M, Zhao Y, Weng H, Kong H, Li K, Gao M, Zheng B, Lin J, Zhou F, Zhang Q, Wu D, Zhang L, Zhang Y (2021) The underappreciated role of agricultural soil nitrogen oxide emissions in ozone pollution regulation in North China. Nat Commun 12(1). https://doi.org/10.1038/s41467-021-25147-9
Ministry of Agriculture and Farmers Welfare. (2020). National policy for management of crop residues
Mor S, Singh T, Bishnoi NR, Bhukal S, Ravindra K (2021) Understanding seasonal variation in ambient air quality and its relationship with crop residue burning activities in an agrarian state of India. Environ Sci Pollut Res 29(3):4145–4158. https://doi.org/10.1007/s11356-021-15631-6
Muhie SH (2022) Novel approaches and practices to sustainable agriculture. J Agric Food Res 10:100446. https://doi.org/10.1016/j.jafr.2022.100446
Nagpure AS, Gurjar BR, Kumar V, Kumar P (2016) Estimation of exhaust and non-exhaust gaseous, particulate matter and air toxics emissions from on-road vehicles in Delhi. Atmos Environ 127:118–124. https://doi.org/10.1016/j.atmosenv.2015.12.026
Nan Q, Wang C, Yi Q, Zhang L, Ping F, Thies JE, Wu W (2020) Biochar amendment pyrolysed with rice straw increases rice production and mitigates methane emission over successive three years. Waste Manag 118:1–8. https://doi.org/10.1016/j.wasman.2020.08.013
NPMCR (2019). Available online: http://agricoop.nic.in/sites/default/files/NPMCR_1.pdf. Accessed on 6 March 2019
Nunes MR, Karlen DL, Veum KS, Moorman TB, Cambardella CA (2020) Biological soil health indicators respond to tillage intensity: a US meta-analysis. Geoderma 369:114335. https://doi.org/10.1016/j.geoderma.2020.114335
Oh S, Kim S-G, Lee JB, Park J, Jee J-B, Hong S-W, Kwon K-S, Song M (2021) Spatial distributions of atmospheric ammonia in a rural area in South Korea and the associated impact on a nearby urban area. Atmos 12(11):1411. https://doi.org/10.3390/atmos12111411

Oni BA, Oziegbe O, Olawole OO (2019) Significance of biochar application to the environment and economy. Ann Agric Sci 64(2):222–236. https://doi.org/10.1016/j.aoas.2019.12.006

Pandey A, Venkataraman C (2014) Estimating emissions from the Indian transport sector with on-road fleet composition and traffic volume. Atmos Environ 98:123–133

Pattanaik L, Pattnaik F, Saxena DK, Naik SN (2019) Biofuels from agricultural wastes. In: Second and third generation of feedstocks. Elsevier. Retrieved from, pp 103–142. https://doi.org/10.1016/b978-0-12-815162-4.00005-7

Qambrani NA, Rahman MM, Won S, Shim S, Ra C (2017) Biochar properties and eco-friendly applications for climate change mitigation, waste management, and wastewater treatment: a review. Renew Sust Energ Rev 79:255–273. https://doi.org/10.1016/j.rser.2017.05.057

Rajput V, Saini I, Parmar S, Pundir V, Kumar V, Kumar V, Naik B, Rustagi S (2024) Biochar production methods and their transformative potential for environmental remediation. Discov Appl Sci 6(8). https://doi.org/10.1007/s42452-024-06125-4

Ritchie, H., Roser, M., Rosado, P.J.O., (2020). CO_2 and greenhouse gas emissions

Ruett J, Abdelshafy A, Walther G (2024) Using miscanthus and biochar as sustainable substrates in horticulture: an economic and carbon footprint assessment of their primary and cascading value chains. Sustain Prod Consum 49:163–178. https://doi.org/10.1016/j.spc.2024.06.016

Rzeznik W, Mielcarek-Bochenska P (2022) Odour emissions from livestock buildings. Atmos 13:254

Sadiq FK, Sadiq AA, Matsika TA, Momoh BA (2025) Sustainable remediation of persistent organic pollutants: a review on recent innovative technologies. Curr Res Biotechnol 9:100293. https://doi.org/10.1016/j.crbiot.2025.100293

Sakhiya AK, Anand A, Kaushal P (2020) Production, activation, and applications of biochar in recent times. Biochar 2(3):253–285. https://doi.org/10.1007/s42773-020-00047-1

Singh S, Dwivedi K, Chakraborty D, Pathak S, Shukla N (2025) Toxic pollutants in environment: microbial technology in wastewater. In: Sustainable Landscape Planning and Natural Resources Management. Springer Nature Switzerland, pp 1–7. https://doi.org/10.1007/978-3-031-80845-6_1

Soni M, Kanani K, Singh S, Shukla N, Dwivedi K (2026) Biochar production and characteristics. In: Biochar as Remediation Tools for Pollutant Removal. Springer Nature Switzerland, Cham, pp 1–26. Retrieved from https://doi.org/10.1007/978-3-032-16746-0_1

Sudarsan JS, Prasanna K, Shiam Babu R, Sai Krishna VMV (2023) Biochar: a sustainable solution for organic waste management a way forward towards circular economy. In: Recent trends in solid waste management. Elsevier. Retrieved from, pp 215–230. https://doi.org/10.1016/b978-0-443-15206-1.00005-0

Sun Y, Gao B, Yao Y, Fang J, Zhang M, Zhou Y, Chen H, Yang L (2014) Effects of feedstock type, production method, and pyrolysis temperature on biochar and hydrochar properties. Chem Eng J 240:574–578. https://doi.org/10.1016/j.cej.2013.10.081

Tan X, Liu Y, Zeng G, Wang X, Hu X, Gu Y, Yang Z (2015) Application of biochar for the removal of pollutants from aqueous solutions. Chemosphere 125:70–85. https://doi.org/10.1016/j.chemosphere.2014.12.058

Tomczyk A, Sokołowska Z, Boguta P (2020) Biochar physicochemical properties: pyrolysis temperature and feedstock kind effects. Rev Environ Sci Biotechnol 19(1):191–215. https://doi.org/10.1007/s11157-020-09523-3

Tripathi M, Sahu JN, Ganesan P (2016) Effect of process parameters on production of biochar from biomass waste through pyrolysis: a review. Renew Sust Energ Rev 55:467–481. https://doi.org/10.1016/j.rser.2015.10.122

Verma N, Devi NL (2024) Correlation and concentration trends of PM10 and PM2. 5 in Bihar, India during the winter season

Verma N, Devi NL (2025) Removal of atmospheric pollutants using biochar: preparation, application, regeneration and its future research. Air Qual Atmos Health. https://doi.org/10.1007/s11869-025-01692-w

Wang J, Wang S (2019) Preparation, modification and environmental application of biochar: a review. J Clean Prod 227:1002–1022. https://doi.org/10.1016/j.jclepro.2019.04.282

Wang W, Huang J, Wu T, Ren X, Zhao X (2023) Research on the preparation of biochar from waste and its application in environmental remediation. Water 15(19):3387. https://doi.org/10.3390/w15193387

Xu J, Chen J, Zhao N, Wang G, Yu G, Li H, Huo J, Lin Y, Fu Q, Guo H, Deng C, Lee S-H, Chen J, Huang K (2020) Importance of gas-particle partitioning of ammonia in haze formation in the rural agricultural environment. Atmos Chem Phys 20(12):7259–7269. https://doi.org/10.5194/acp-20-7259-2020

Yan X, Ma J, Ren J, Cui M, Chen X, Qiu D, Lei M, Li T, Guo L, Chen C, Han Y (2021) Concentrations, size distribution, and community structure characteristics of culturable airborne antibiotic-resistant bacteria in Xinxiang, Central China. Atmos 12(8):1077. https://doi.org/10.3390/atmos12081077

Zabaniotou A, Stamou K (2020) Balancing waste and nutrient flows between urban agglomerations and rural ecosystems: biochar for improving crop growth and urban air quality in the Mediterranean region. Atmos 11(5):539. https://doi.org/10.3390/atmos11050539

Zeng S, Kan E (2021) Adsorption and regeneration on iron-activated biochar for removal of microcystin-LR. Chemosphere 273:129649. https://doi.org/10.1016/j.chemosphere.2021.129649

Zhang S, Zhou L, Jia L, Li J, Liu B, Yuan Y (2022) Numerical simulation on particulate matter emissions from a layer house during summer in Northeast China. Atmos 13(3):435. https://doi.org/10.3390/atmos13030435

Zhao Y, Xu R, Xu Z, Wang L, Wang P (2022a) Temporal and spatial patterns of biomass burning fire counts and carbon emissions in the Beijing–Tianjin–Hebei (BTH) region during 2003–2020 based on GFED4. Atmos 13(3):459. https://doi.org/10.3390/atmos13030459

Zhao Z, Wang B, Theng BKG, Lee X, Zhang X, Chen M, Xu P (2022b) Removal performance, mechanisms, and influencing factors of biochar for air pollutants: a critical review. Biochar 4(1). https://doi.org/10.1007/s42773-022-00156-z

Zhou X, Moghaddam TB, Chen M, Wu S, Zhang Y, Zhang X, Adhikari S, Zhang X (2021a) Effects of pyrolysis parameters on physicochemical properties of biochar and bio-oil and application in asphalt. Sci Total Environ 780:146448. https://doi.org/10.1016/j.scitotenv.2021.146448

Zhou Y, Qin S, Verma S, Sar T, Sarsaiya S, Ravindran B, Liu T, Sindhu R, Patel AK, Binod P, Varjani S, Rani Singhnia R, Zhang Z, Awasthi MK (2021b) Production and beneficial impact of biochar for environmental application: a comprehensive review. Bioresour Technol 337:125451. https://doi.org/10.1016/j.biortech.2021.125451

Zorpas AA, Skouroupatis A (2016) Indoor air quality evaluation of two museums in a subtropical climate conditions. Sustain Cities Soc 20:52–60. https://doi.org/10.1016/j.scs.2015.10.002

Chapter 8
Evaluation of the Efficacy of Biochar on Minimization as, Cd, Cr Content in Crops

Priti Smita Nayak, Sumit Acharya, and Subhrajyoti Chatterjee

Abstract Biochar application has emerged as a promising strategy for mitigating heavy metal contamination in agricultural soils and reducing their uptake by crops. This chapter focuses on evaluating the efficacy of biochar in minimizing the accumulation of arsenic (As), cadmium (Cd) and chromium (Cr) in various crops. A comprehensive review of recent studies reveals that biochar derived from diverse feedstocks and produced under varying pyrolysis conditions significantly influences the bioavailability of these heavy metals. The mechanisms involved include increased soil pH enhanced cation exchange capacity and the formation of stable metal complexes with biochar's functional groups. Consequently, biochar amendment leads to a substantial reduction in As, Cd and Cr concentrations in plant tissues particularly in roots and shoots. The extent of reduction varies depending on biochar type, application rate, soil characteristics and crop species. Furthermore, long-term studies indicate that biochar can provide sustained remediation effects improving soil health and crop safety. However, further research is needed to optimize biochar application strategies for different soil-crop systems and to assess the potential long-term ecological impacts.

Keywords Biochar · Arsenic · Cadmium · Chromium · Heavy metals · Crop uptake · Remediation · Soil amendment · Phytoremediation · Bioavailability

8.1 Introduction

The contamination of agricultural soils with toxic heavy metals, particularly arsenic (As), cadmium (Cd) and chromium (Cr), has emerged as a critical global environmental and public health concern. These hazardous elements accumulate in soil through both natural processes and anthropogenic activities, including industrial

P. S. Nayak · S. Acharya · S. Chatterjee (✉)
Department of Horticulture, MS Swaminathan School of Agriculture, Centurion University of Technology and Management, Paralakhemundi, Gajapati, Odisha, India
e-mail: subhrajyoti.chatterjee@cutm.ac.in

D. Moulick et al. (eds.), *Biochar: Achieving Agro-Environmental Sustainability*, Sustainability Solutions, https://doi.org/10.1007/978-3-032-19855-6_8

discharges, mining operations, excessive use of agrochemicals and irrigation with contaminated water (Sohi 2012). Once introduced into the soil ecosystem, these metals persist for extended periods, posing severe risks to crop productivity, food safety and human health. Crops grown in contaminated soils readily absorb and translocate heavy metals to their edible parts, leading to bioaccumulation in the food chain. Chronic exposure to As, Cd and Cr through dietary intake is associated with numerous health disorders, including cancer, neurological damage, kidney dysfunction and cardiovascular diseases (Ali et al. 2018). Conventional remediation techniques, such as soil excavation, chemical stabilization and phytoremediation, often face limitations due to high costs, operational complexity and potential secondary pollution. In this scenario, materials such as activated carbon, which is subjected to oxygen treatment, can be employed to improve micro-porosity and specific surface area, consequently decreasing the bioavailability of contaminants. The term "activated" describes charcoal that has undergone heating or chemical treatment to increase its surface area (Cao et al. 2011). However, 'biochar' is regarded as a relatively economical option and an efficient adsorbent than charcoal (Tan et al. 2017). 'Biochar' is a stable, porous, carbonaceous material produced by the pyrolysis of biomass in an oxygen-limited environment, engineered for the secure, long-term sequestration of carbon (Shackley et al. 2012). International Biochar Initiative (IBI) defined biochar as a solid material that results from the decomposition of biomass in an environment with limited oxygen (IBI 2025). Though biochar and charcoal are having a few similarities regarding their chemical composition but both can be differentiated by their commercial purposes. Charcoal is an organic matter used for fuel and energy, while biochar aids environmental management through carbon sequestration (Lehman and Joseph 2009).

Biochar's efficacy in heavy metal remediation stems from its unique physicochemical properties, including high surface area, porous structure, abundant functional groups (e.g., carboxyl, hydroxyl and phenolic) and variable charge characteristics (Aslam et al. 2014). These properties enable biochar to interact with heavy metals through multiple mechanisms, such as surface adsorption, ion exchange, electrostatic attraction and precipitation. For instance, biochar's alkaline nature can elevate soil pH, thereby reducing the solubility and mobility of cationic metals like Cd and Cr (Algethami et al. 2024). Additionally, the presence of oxygen-containing functional groups on biochar surfaces facilitates the complexation and chelation of metal ions, effectively sequestering them in the soil matrix. Biochar also promotes the reduction of highly toxic Cr (VI) to less mobile and less toxic Cr (III), while its affinity for oxyanions like As (V) helps mitigate arsenic uptake by crops. Beyond metal immobilization, biochar enhances soil fertility by improving water retention, nutrient cycling and microbial activity, making it a multifaceted solution for sustainable agriculture.

Despite its potential, the effectiveness of biochar in heavy metal remediation varies depending on factors such as feedstock type, pyrolysis conditions, soil properties and the specific metal involved (Githinji et al. 2014). For example, biochar derived from manure or crop residues may exhibit distinct metal-binding capacities compared to wood-based biochar due to differences in ash content and functional group composition. Similarly, soil pH, organic matter content and redox conditions influence biochar-metal interactions, necessitating tailored application strategies for optimal results (Harter et al. 2017). Recent advancements in biochar modification, such as impregnation with minerals (e.g., iron oxides) or activation with acids, have further enhanced its metal adsorption capacity, offering promising avenues for targeted remediation (Harter et al. 2017).

This chapter critically evaluates the role of biochar in minimizing As, Cd and Cr uptake by crops, synthesizing current research findings, mechanistic insights and field-scale applications. It also addresses knowledge gaps and challenges, such as the long-term stability of biochar-metal complexes and potential trade-offs between metal immobilization and nutrient availability. By integrating fundamental science with practical applications, this discussion aims to guide policymakers, agronomists and environmental practitioners in harnessing biochar's potential to safeguard agricultural systems and food security in metal—contaminated regions (Huang et al. 2019).

8.2 History of Biochar

Biochar, often regarded as a modern innovation for carbon sequestration and soil enhancement, actually has ancient roots dating back at least 2000 years (Hunt et al. 2010). The most well-documented historical use of biochar comes from the Amazonian basin, where indigenous communities created highly fertile soils known as 'Terra Preta de Índio' (Portuguese for "black earth of the Indians") (Lehman 2009; Han et al. 2020). These dark, carbon-rich soils exhibit remarkable fertility compared to the naturally nutrient-poor soils of the Amazon rainforest (Mukherjee et al. 2020).

The exact origins of Terra Preta remain debated-whether it was intentionally produced through systematic charring of organic waste (such as crop residues, animal bones and manure) or simply a by-product of ancient slash-and-char agricultural practices. However, archaeological evidence suggests that pre-Columbian civilizations actively enriched soils with charcoal, food scraps and other organic matter, leading to long-term improvements in soil structure, nutrient retention and microbial activity.

What makes Terra Preta particularly fascinating is its exceptional persistence—centuries after its creation, these soils remain highly fertile and continue to sequester carbon, unlike typical organic matter, which decomposes rapidly in tropical climates. Modern scientific analysis reveals that biochar, the carbonized material found in these soils, plays a key role in their sustained productivity. This soil exhibits exceptionally elevated concentrations of essential nutrients, including potassium, phosphorus and nitrogen (Chen et al. 2019). Biochar's porous structure enhances water retention, reduces nutrient leaching and provides a habitat for beneficial soil microbes.

The rediscovery of Terra Preta in the twentieth century sparked renewed interest in biochar as a sustainable soil amendment. Today, biochar research focuses on replicating and optimizing these ancient practices for contemporary agriculture, climate change mitigation and environmental remediation. By studying these historical soils, scientists gain insights into how biochar can be used to improve degraded lands, enhance crop yields and combat global warming through long-term carbon storage.

Thus, while biochar is now being promoted as a cutting-edge solution for sustainable farming and carbon-negative technologies, its foundations lie in the wisdom of ancient civilizations that mastered the art of turning waste into fertile, enduring soil.

8.3 Perspective of Biochar Application in Soil

Biochar, a carbon-rich material produced through the pyrolysis of organic biomass, has gained significant attention as a sustainable soil amendment with multifaceted benefits for agriculture, environmental remediation and climate change mitigation (Karhu et al. 2011). Its unique physicochemical properties such as high porosity, large surface area and abundant functional groups make it a promising tool for improving soil health, enhancing crop productivity and reducing greenhouse gas emissions (US Environmental Protection Agency 1995). The perspective of biochar application in soil encompasses several key areas, including soil fertility enhancement, carbon sequestration, contaminant immobilization and long-term sustainability in agricultural systems (Mukherjee et al. 2013).

8.3.1 Soil Fertility and Crop Productivity

Biochar enhances soil fertility by improving physical, chemical and biological properties. Its porous structure increases water retention in sandy soils while improving aeration in clayey soils, creating an optimal environment for root growth. Additionally, biochar's high cation exchange capacity (CEC) helps retain essential nutrients like nitrogen (N), phosphorus (P) and potassium (K), reducing leaching and making them more available to plants. Studies have shown that biochar-amended soils exhibit increased microbial activity, promoting nutrient cycling and organic matter decomposition. Farmers in degraded or nutrient-poor soils can get benefit from biochar application, which has been shown to boost crop yields in cereals, vegetables and legumes (Rehman et al. 2011).

8.3.2 Carbon Sequestration and Climate Mitigation

One of the most compelling advantages of biochar is its ability to sequester carbon in soils for centuries to millennia. Unlike raw organic matter, which decomposes rapidly, biochar's stable aromatic carbon structure resists microbial breakdown, effectively locking carbon away from the atmosphere. This makes biochar a powerful tool in climate-smart agriculture, helping offset agricultural greenhouse gas emissions. When combined with sustainable farming practices, biochar can contribute to carbon-negative farming systems, aligning with global climate goals such as the Paris Agreement (Sameshima et al. 2006).

8.3.3 Remediation of Contaminated Soils

Biochar has demonstrated remarkable potential in immobilizing heavy metals (e.g., cadmium, lead, arsenic) and organic pollutants (e.g., pesticides, polycyclic aromatic hydrocarbons) in contaminated soils. Its adsorption capacity, redox activity and surface functional groups help bind toxic elements, reducing their bioavailability to plants and minimizing entry into the food chain. Engineered biochars, such as those modified with iron or sulfur, further enhance contaminant retention. This application is particularly valuable in industrial and mining—affected regions where soil pollution threatens food safety and ecosystem health (Singh et al. 2019).

8.4 Preparation

8.4.1 Feedstock Selection

The first step in biochar production is choosing the right biomass feedstock. The type of biomass affects the biochar's properties, including its porosity, nutrient content and stability (Yang et al. 2019).

Common feed-stocks:

- Agricultural waste: Crop residues (rice husks, corn stalks, wheat straw), nut shells (coconut, almond) and fruit peels.
- Woody biomass: Sawdust, wood chips, branches and bark.
- Animal manure: Poultry litter, cow dung (requires careful processing due to high ash content).
- Other organic waste: Sewage sludge, food waste and algae.

8.4.2 Factors to Consider

- Moisture content: It should ideally be below 20% for efficient pyrolysis.
- Ash content: High-ash feedstocks (e.g., rice husk) may produce biochar with lower carbon stability.
- Particle size: Smaller particles (2–10 mm) ensure uniform heating.

8.4.3 Pre-treatment of Biomass

Before pyrolysis, the feedstock may require preparation to optimize biochar yield and quality (Abbruzzini et al. 2019).

Steps:

- Drying: Sun-drying or kiln may be used to reduce moisture content (< 20%).
- Chopping/grinding: It is done to reduce biomass to small, uniform pieces for consistent pyrolysis.
- Pre-carbonization (optional): Slow heating is done at low temperatures (200–300 °C) to remove volatile compounds.

8.4.4 Pyrolysis: The Core Process

Pyrolysis is the thermal decomposition of biomass in the absence of oxygen. The key parameters are temperature, heating rate and residence time.

8.4.5 Types of Pyrolysis

Type	Temperature (°C)	Heating rate	Residence time	Biochar yield	Applications
Slow pyrolysis	300–700	Slow (5–10 °C/min)	Hours to days	High (30–40%)	Soil amendment
Fast pyrolysis	400–800	Rapid (>100 °C/s)	Seconds to minutes	Low (10–20%)	Biofuel & chemicals
Gasification	700–1200	Very fast	Minutes	Very low (<10%)	Energy production

Step-by-step pyrolysis process:

- Loading the reactor: The dried biomass is placed in a pyrolysis kiln, retort or drum.
- Sealing the system: This step is followed to ensure minimal oxygen entry to prevent combustion.
- Gradual heating:
  - Slow Pyrolysis: Heating is done at 5–10 °C/min until reaching 400–600 °C.
  - Fast Pyrolysis: Rapid heating is followed to 500–800 °C in seconds.
- Maintaining the temperature: The temperature is hold for 1–4 hours (slow pyrolysis) or seconds (fast pyrolysis).

- Cooling down: After that, the biochar is allowed to cool in an oxygen-free environment to prevent oxidation (Aborisade et al. 2023).

8.4.6 Post-processing

Raw biochar may require treatment before use to enhance its properties.

8.4.7 Activation (Optional)

- Physical activation: The biochar may be exposed to steam or CO_2 at high temperatures to increase porosity
- Chemical activation: It may be treated with acids (HCl, H_3PO_4) or bases (KOH) to modify surface chemistry

8.4.8 Grinding and Sieving

The biochar is crushed into fine particles (0.5–5 mm) for better soil integration followed by sieving to remove large, uncharred fragments (Adak et al. 2024).

8.4.9 Nutrient Enrichment (Optional)

- Composting: Biochar may be mixed with compost to enhance nutrient content.
- Mineral coating: It can also be soaked in nutrient solutions (e.g., NPK fertilizers).

8.4.10 Quality Assessment

Not all biochars are equal. Key quality indicators include:

8.4.11 Physical Properties

- Surface Area: Measured via BET analysis (ideal: 100–400 m^2/g).
- Porosity: Determined by SEM microscopy.

8.4.12 *Chemical Properties*

- pH: Should match crop requirements (typically 7–9 for acidic soils).
- Cation Exchange Capacity (CEC): Indicates nutrient retention ability.
- Ash content: High ash (> 20%) may reduce stability.

8.4.13 *Safety Testing*

- Heavy metal analysis: Compliances are ensured with safety standards (e.g., IBI guidelines).
- PAH (Polycyclic Aromatic Hydrocarbons) Testing: Should be below toxic thresholds.

8.4.14 *Storage and Application*

- Storage: Biochar is kept dry to prevent moisture absorption.
- Soil application rates: Typically 5–20 tons are applied per hectare, depending on soil type.

8.5 Effect on Soil Properties

Soil salinity and acidity are critical environmental factors that significantly impact soil fertility, crop productivity and overall ecosystem health (Ameloot et al. 2013). Excessive salinity leads to osmotic stress, ion toxicity (Na^+, Cl^-) and nutrient imbalances, while acidic soils suffer from aluminum (Al^{3+}) and manganese (Mn^{2+}) toxicity, as well as phosphorus (P) fixation. These conditions reduce microbial activity, organic matter decomposition and plant nutrient uptake, ultimately diminishing agricultural yields. Biochar, a carbon-rich material produced through pyrolysis of organic biomass, has emerged as a promising soil amendment to mitigate these challenges. Its ability to enhance soil physicochemical properties particularly in saline and acidic soils makes it a valuable tool for sustainable agriculture (Ayaz et al. 2021).

8.6 Mechanisms of Biochar in Improving Saline and Acidic Soils

- **pH modulation and liming effect**

Biochar's alkalinity plays a crucial role in neutralizing acidic soils. The pH of biochar depends on pyrolysis temperature and feedstock type. High-temperature biochar (> 400 °C) tends to be more alkaline due to the decomposition of acidic functional groups (carboxyl, phenolic) and the concentration of alkaline minerals (Ca, Mg, K) in the ash fraction. When applied to acidic soils, biochar acts similarly to lime, raising soil pH and reducing Al^{3+} toxicity (Cao et al. 2019). In contrast, low-temperature biochar (< 400 °C) retains more acidic functional groups, making it less effective for pH correction but potentially useful in already alkaline soils (Shiade et al. 2023).

8.6.1 Cation Exchange Capacity (CEC) Enhancement

Biochar improves soil CEC, which is vital for nutrient retention and availability. Its porous structure and oxygen-containing surface functional groups (e.g., carboxyl, hydroxyl) provide binding sites for cations (Ca^{2+}, Mg^{2+}, K^{+}), reducing leaching losses. In saline soils, biochar's high CEC helps displace sodium (Na^{+}) ions, mitigating sodicity and improving soil structure.

8.6.2 Salt Stress Mitigation

In saline soils, biochar reduces osmotic stress by improving water retention and decreasing Na^{+} uptake by plants. Studies show that biochar-amended soils exhibit lower electrical conductivity (EC) due to Na^{+} adsorption onto biochar surfaces. Enhanced K^{+}/Na^{+} ratio, promoting better plant growth. Improved microbial activity, which aids in organic matter decomposition and nutrient cycling.

8.6.3 Nutrient Availability and Retention

Biochar enhances the availability of essential nutrients (N, P, K, Ca, Mg) in both acidic and saline soils. In acidic soils, it reduces P fixation by aluminum and iron oxides, while in saline soils, it prevents nutrient leaching by increasing soil aggregation and CEC (Gluba et al. 2021).

8.7 Sources and Toxicity of Heavy Metals (as, Cd, Cr) in Agricultural Systems

Heavy metal contamination in agricultural soils has become a pressing global issue, with arsenic (As), cadmium (Cd) and chromium (Cr) being among the most problematic contaminants due to their persistence, toxicity and potential for bioaccumulation. These metals enter agricultural systems through both natural and anthropogenic pathways. Geological weathering of parent materials represents the primary natural source, particularly for As which is commonly associated with sulfide minerals in rocks. However, human activities have dramatically accelerated metal inputs to farmlands. Industrial activities such as mining operations release substantial quantities of these metals, with As frequently contaminating soils near gold and copper mines, while Cr pollution is prevalent near chromite mines and leather tanning facilities (Hu et al. 2020). Agricultural practices themselves contribute significantly to metal accumulation, particularly through the application of phosphate fertilizers (a major source of Cd) and contaminated irrigation water. Wastewater irrigation, common in water-scarce regions, introduces multiple metals simultaneously, with studies showing that long-term use of untreated industrial wastewater can elevate soil Cd concentrations by 10–100 times background levels (Khan et al. 2020a).

The toxic effects of these metals on both crops and human health are well-documented and particularly concerning due to their ability to enter the food chain. Arsenic exists primarily as arsenate [As (V)] and arsenite [As (III)] in soils, with the latter being more toxic and mobile. In plants, As disrupts metabolic processes by competing with phosphate uptake, inhibiting root growth and inducing oxidative stress. Chronic human exposure to As through contaminated crops like rice is associated with skin lesions, cardiovascular diseases and various cancers. Cadmium, even at low concentrations, is highly toxic to plants, causing chlorosis, stunted growth and disruption of nutrient uptake by competing with essential elements like zinc and calcium. The human health impacts are particularly severe, as Cd accumulates in the kidneys over time, potentially leading to renal dysfunction and bone diseases such as itai-itai disease. Chromium presents unique challenges as its toxicity depends on oxidation state, with Cr (VI) being 100 times more toxic than Cr (III). In plants, Cr (VI) is readily taken up and causes severe oxidative damage, while in humans it is a known carcinogen associated with lung cancer when inhaled and gastrointestinal cancers when ingested (Khan et al. 2020b).

The mobility and bioavailability of these metals in soil-plant systems are influenced by various soil factors. Soil pH plays a crucial role, with lower pH increasing the solubility and plant uptake of Cd and Cr, while As shows opposite behavior, becoming more mobile at higher pH. Organic matter content affects metal dynamics through complexation and adsorption reactions, while redox conditions particularly influence As and Cr speciation. These interactions create complex scenarios where multiple metals may co-occur, potentially exhibiting synergistic or antagonistic effects on their mobility and toxicity. Understanding these relationships is essential for developing effective remediation strategies, particularly in regions where intensive agriculture, rapid industrialization and inadequate waste management have created significant metal contamination challenges. The persistence of these metals in soils and their potential for long-term ecological and health impacts underscore the urgent need for effective mitigation approaches in agricultural systems worldwide (Kumar et al. 2021).

8.8 Biochar: Production, Properties and Mechanisms of Metal Immobilization

Biochar, a carbon-rich material produced through the thermochemical conversion of biomass under oxygen-limited conditions (pyrolysis), has emerged as a promising soil amendment for heavy metal immobilization. The production process significantly influences biochar's physicochemical properties, with slow pyrolysis (temperatures ranging from 350 °C to 700 °C and heating rates of 5–10 °C/min) being the most common method for agricultural applications. Feedstock selection plays a crucial role in determining biochar characteristics, with lignocellulosic materials (wood chips, crop residues) producing biochars with high carbon content, while manure-based feedstocks yield ash-rich biochars with elevated nutrient concentrations. The pyrolysis temperature is particularly critical, as higher temperatures (500–700 °C) generally increase surface area and porosity but may reduce functional group density, whereas lower temperatures (300–500 °C) preserve oxygen-containing functional groups that are essential for metal binding (Gul et al. 2016).

The remarkable metal immobilization capacity of biochar stems from its unique properties, including high surface area (50–400 m^2/g), porous structure and abundant surface functional groups (-COOH, -OH, -C = O). These characteristics enable multiple mechanisms of interaction with heavy metals. Cationic metals like Cd^{2+} and Cr^{3+} are primarily immobilized through surface complexation with oxygen functional groups, ion exchange with inherent cations (Ca^{2+}, Mg^{2+}, K^+) and electrostatic attraction to negatively charged biochar surfaces. For anionic species such as AsO43-, biochar's metal oxide impurities (e.g., Fe3O4, Al2O3) and positively charged edges at lower pH conditions facilitate adsorption. Additionally, biochar's alkaline nature can increase soil pH, promoting the precipitation of metal carbonates and hydroxides, particularly for Cd and Cr. The carbon matrix itself provides π-electrons that can interact with metal ions through cation-π bonding, while the developed pore structure physically traps metal particles (Naveed et al. 2021).

Biochar also influences the redox transformation of metals, notably converting highly toxic and mobile Cr (VI) to less toxic and immobile Cr (III) through electron transfer reactions mediated by biochar's redox-active functional groups. Similarly, biochar can affect arsenic speciation and mobility by altering soil redox conditions and providing sorption sites. The presence of mineral components in biochar, such as phosphates, silicates and carbonates, can further enhance metal immobilization through co-precipitation reactions. For instance, phosphate-rich biochars have shown exceptional ability to precipitate Cd as stable Cd3 (PO4)2.

The long-term effectiveness of biochar in metal immobilization depends on several factors, including biochar aging in soil, which may alter its surface properties through oxidation and microbial degradation. While fresh biochar typically shows higher metal sorption capacity, aged biochar develops more oxygen-containing functional groups that can maintain or even enhance certain metal binding mechanisms. The interaction between biochar and soil organic matter also plays a significant role, as it can either facilitate additional metal complexation or compete for binding sites. Understanding these complex interactions is crucial for optimizing biochar application strategies to ensure sustainable and effective remediation of metal—contaminated agricultural soils (Hale et al. 2020).

8.9 Impact of Biochar on Arsenic (as) Immobilization in Soil and Crops

Arsenic (As) contamination in agricultural soils poses significant risks to crop safety and human health due to its high toxicity and carcinogenic potential. Arsenic contamination of drinking water and food has been identified as a significant public health concern, disproportionately impacting populations in Asian nations, with millions experiencing adverse health effects (Zhao et al. 2009; Singh et al. 2015). Biochar has emerged as a promising soil amendment for mitigating As bioavailability, primarily through mechanisms that alter its speciation, mobility and plant uptake. The efficacy of biochar in As immobilization depends on its physicochemical properties, soil characteristics and the specific form of As present (arsenite As (III) or arsenate As (V)). Biochar's high surface area, porous structure and functional groups (e.g., carboxyl, hydroxyl and phenolic) facilitate as adsorption, while its alkaline nature can influence as solubility through pH-mediated reactions. Specifically, the heightened alkalinity introduced by biochar can remobilize arsenic into solution at pH levels exceeding 8, as hydroxyl ions displace anionic arsenic species through electrostatic repulsion (Zama et al. 2022). In aerobic soils, biochar promotes the oxidation of more mobile and toxic As (III) to less bioavailable As (V), which strongly binds to iron (Fe) and aluminum (Al) oxides present in biochar or soil. For instance, Fe-modified biochar has shown exceptional As sequestration capacity, with studies reporting up to 40–60% reduction in As concentration in rice grains grown in contaminated paddies. This is attributed to the formation of stable Fe-As complexes, which limit As translocation to edible plant parts (Rafi et al. 2022).

The impact of biochar on As immobilization is also influenced by soil redox conditions and competing ions. Under flooded conditions, such as in rice fields, anaerobic microbial activity can reduce As (V) to As (III), increasing its mobility. However, biochar's role in enhancing soil aeration and promoting Fe plaque formation on rice roots can counteract this effect, reducing As uptake. Additionally, biochar's ability to alter microbial communities may further influence As speciation by affecting the activity of As-transforming bacteria. Despite these benefits, certain challenges exist, such as the competitive inhibition of As adsorption by phosphate (PO43-), which shares similar chemical behavior and binding sites with As (V). This competition can reduce biochar's effectiveness in high-phosphate soils, necessitating the use of modified biochars with higher selectivity for As. For example, biochar impregnated with magnesium (Mg) or manganese (Mn) oxides has demonstrated enhanced As adsorption capacity, even in the presence of competing anions.

Field studies have validated biochar's potential for As mitigation in various cropping systems. In Bangladesh, where groundwater As contamination is widespread, rice fields amended with rice husk biochar exhibited a 30–50% decrease in grain As levels compared to untreated controls. Similarly, in mining-affected regions, the application of wood-derived biochar reduced As uptake by leafy vegetables by up to 45%. These field trials highlight biochar's efficacy in diverse agricultural settings, underscoring its role in enhancing crop safety and mitigating human exposure to arsenic through the food chain (Leksungnoen et al. 2019). However, long-term field data remain limited and the persistence of biochar's Asimmobilizing effects requires further investigation, especially under varying climatic conditions and cropping cycles.

Future research should focus on optimizing biochar properties (e.g., feedstock selection, pyrolysis temperature and modification techniques) to enhance As retention while minimizing unintended effects on soil chemistry. Advanced characterization techniques, such as synchrotron-based X-ray absorption spectroscopy, can provide deeper insights into Asbiochar interactions at the molecular level. Furthermore, integrating biochar with other remediation strategies, such as phytoremediation or microbial inoculants, could offer synergistic benefits for As-contaminated soils. Policymakers and agricultural practitioners must also consider economic feasibility and farmer acceptance to scale up biochar applications in As-affected regions. Overall, biochar represents a sustainable and scalable solution for reducing As accumulation in crops, contributing to safer food production and improved environmental health (Chatterjee et al. 2024).

8.10 Role of Biochar in Cadmium (Cd) Mitigation

Cadmium (Cd) contamination in agricultural soils represents a significant environmental and health challenge due to its high toxicity, persistence and bioaccumulation potential in crops (Bashir et al. 2021). Cd enters soil systems primarily through phosphate fertilizers, industrial emissions and sewage sludge application, where it becomes readily available for plant uptake, particularly in acidic soils (Bashir et al. 2018). The metal accumulates in edible plant parts, posing risks to human health, including kidney damage, bone disorders and carcinogenic effects. Biochar has emerged as a promising and sustainable soil amendment for Cd immobilization, offering multiple mechanisms to reduce its bioavailability and translocation in crops. The efficacy of biochar in Cd mitigation stems from its unique physicochemical properties, including high surface area, porous structure, cation exchange capacity (CEC) and abundant functional groups such as carboxyl (-COOH), hydroxyl (-OH) and phenolic moieties. These characteristics enable biochar to interact with Cd through adsorption, precipitation, ion exchange and surface complexation, effectively reducing its mobility in soil and uptake by plants (Danso et al. 2023).

One of the primary mechanisms by which biochar mitigates Cd toxicity is through pH modulation of the soil environment. Biochar, particularly those derived from feedstocks with high ash content (e.g., manure or crop residues), often exhibits alkaline properties. When applied to acidic soils, biochar raises soil pH, which promotes the conversion of bioavailable Cd^{2+} ions into less soluble forms, such as Cd-carbonates, Cd-hydroxides or Cd-phosphates. This pH-induced immobilization significantly reduces Cd uptake by plant roots. For example, studies have shown that biochar application in Cd-contaminated rice field can decrease grain Cd concentrations by 30–60%, depending on the biochar type and application rate.

Additionally, biochar's surface functional groups play a critical role in Cd binding. Oxygen—containing groups (e.g., carboxyl and hydroxyl) form strong complexes with Cd ions through electrostatic attraction and chemisorption, effectively sequestering them on biochar surfaces.

Another key mechanism is ion exchange, where biochar's negatively charged surfaces attract and retain Cd^{2+} ions, replacing them with less toxic cations such as Ca^{2+} or Mg^{2+}. This process is particularly effective in soils with high CEC, where biochar enhances the soil's inherent capacity to retain Cd. Furthermore, biochar's porous structure provides physical adsorption sites, trapping Cd ions within its micro—and mesopores and preventing their leaching into groundwater. Recent advancements in engineered biochar, such as phosphate—or sulfur—modified variants, have further improved Cd immobilization by promoting the formation of insoluble Cd compounds (e.g., Cd3(PO4)2 or CdS). For instance, biochar enriched with hydroxyapatite has demonstrated exceptional Cd adsorption capacity, reducing plant uptake by over 70 % in contaminated soils (Harvey et al. 2011).

Field studies and meta-analyses corroborate biochar's effectiveness in Cd mitigation across diverse crops, including wheat, rice and leafy vegetables. For example, in a field trial in China, wheat grown in biochar-amended soils exhibited 50% lower Cd concentrations in grains compared to untreated plots. Similarly, biochar derived from bamboo and rice husk has shown promise in reducing Cd accumulation in spinach and lettuce. However, the long-term stability of biochar-Cd complexes and potential effects on soil microbial communities require further investigation. Despite these challenges, biochar remains a viable and eco—friendly strategy for Cd remediation, offering a dual benefit of heavy metal immobilization and soil fertility enhancement. Future research should focus on optimizing biochar properties for site-specific conditions and integrating biochar with other remediation techniques, such as phytoremediation, to maximize Cd mitigation in contaminated agroecosystems (Rahim et al. 2024).

8.11 Chromium (Cr) Remediation Using Biochar

Chromium contamination in agricultural soils, primarily from industrial effluents, leather tanning and electroplating waste, poses significant environmental and health risks due to its toxicity and persistence. Chromium exists in two stable oxidation states in soil: Cr (III), which is relatively immobile and less toxic and Cr (VI), a highly soluble and carcinogenic form that readily enters the food chain via plant uptake. Conventional remediation methods, such as chemical reduction and soil washing, are often costly and environmentally disruptive. Biochar, a porous carbonaceous material derived from biomass pyrolysis, has emerged as a sustainable and cost-effective alternative for Cr immobilization in soils. Its efficacy stems from its unique physicochemical properties, including high surface area, abundant functional groups (e.g., carboxyl, hydroxyl and phenolic) and redox-active surfaces, which collectively facilitate Cr adsorption, reduction and stabilization (Sheikh et al. 2025).

The remediation mechanism of biochar for Cr involves two primary processes: adsorption and chemical reduction. Biochar's porous structure and negatively charged surfaces provide ample sites for Cr (VI) adsorption through electrostatic attraction and surface complexation. For instance, oxygen-containing functional groups on biochar surfaces protonate under acidic conditions, forming positively charged sites that attract Cr (VI) oxyanions (e.g., CrO42—and HCrO4-). Additionally, biochar's electron-donating capacity plays a critical role in reducing Cr (VI) to Cr (III), a less toxic and immobile form. This reduction process is enhanced by biochar's inherent reducing agents, such as phenolic groups and persistent free radicals, which donate electrons to Cr (VI). Studies have shown that biochar derived from sugarcane bagasse or rice husk can achieve Cr (VI) reduction efficiencies exceeding 90% under optimal conditions. The resulting Cr (III) forms stable complexes with biochar's functional groups or precipitates as hydroxides and oxides, further minimizing its bioavailability to plants (Hu et al. 2021).

The effectiveness of biochar in Cr remediation is influenced by several factors, including feedstock type, pyrolysis temperature and soil properties. High-temperature biochar ($\geq$ 500 °C) typically exhibits greater surface area and aromaticity, enhancing Cr adsorption, while low—temperature biochar retains more functional groups, favoring Cr (VI) reduction. For example, wood-based biochar pyrolyzed at 700 °C demonstrated a Cr (VI) adsorption capacity of 45 mg/g, whereas manure-based biochar produced at 400 °C showed superior reduction potential due to its higher oxygen content. Soil pH also plays a pivotal role; acidic conditions promote Cr (VI) adsorption and reduction, whereas alkaline soils may inhibit these processes due to competition with other anions (e.g., PO43-). To augment biochar's performance, modifications such as iron (Fe) or sulfur (S) impregnation have been explored. Fe-modified biochar, for instance, leverages Fe oxides' affinity for Cr (VI), while S-modified biochar facilitates the formation of stable Cr (III)-sulfide precipitates (Kavitha et al. 2018).

Field applications of biochar for Cr remediation have yielded promising results. In a contaminated tannery site in India, amending soil with 5% (w/w) rice husk biochar reduced Cr uptake in spinach by 70% within one growing season. Similarly, a study in China reported a 60% decrease in Cr (VI) concentration in maize grains following biochar application. Despite these successes, challenges remain, such as the long-term stability of immobilized Cr and potential biochar aging effects, which may alter its remediation efficiency over time. Future research should focus on optimizing biochar properties for site-specific conditions, exploring hybrid remediation approaches (e.g., biochar combined with phytoremediation) and assessing the economic viability of large-scale applications. By addressing these gaps, biochar can be positioned as a cornerstone technology for sustainable Cr mitigation in agriculture, ensuring safer food production and ecosystem health (Lehmann et al. 2019).

8.12 Effect of Biochar for Elimination of Lead (PB) from Soil and Crops

Lead, a hazardous heavy metal, is found in a divalent state and constitutes 0.002% of the earth's crust (Arias et al. 2010). Lead (Pb) contamination in agricultural soils poses severe risks to human health and ecosystem stability, primarily due to its persistence, toxicity and bioaccumulation in the food chain (Kushwaha et al. 2018). Industrial activities, mining, leaded gasoline residues and improper waste disposal contribute to Pb pollution in soils, where it can persist for centuries. Exposure to lead via contaminated crops can adversely affect the renal, cardiovascular, bone, hematopoietic, nervous, and reproductive systems (Kumar et al. 2022). Given these risks, effective remediation strategies are essential.

Biochar, a carbon-rich material produced through pyrolysis of organic biomass, has emerged as a promising, sustainable solution for immobilizing Pb in soils and reducing its uptake by crops. Its high surface area, porous structure and abundant functional groups enable effective Pb adsorption, while its alkaline nature helps reduce Pb bioavailability. This article explores the mechanisms by which biochar mitigates Pb contamination, its effectiveness in different soil types and practical applications for safer agriculture (Lehmann et al. 2021).

8.12.1 Mechanisms of Lead Immobilization by Biochar

Biochar reduces Pb mobility and bioavailability in soils through several key mechanisms:

8.12.2 Adsorption and Surface Complexation

Biochar's porous structure and large surface area provide numerous binding sites for Pb ions. Functional groups such as carboxyl (-COOH), hydroxyl (-OH) and phenolic groups on biochar surfaces form strong complexes with Pb^{2+}, effectively immobilizing it. Additionally, the negatively charged surfaces of biochar attract and retain Pb cations through electrostatic interactions. This multifaceted mechanism of immobilization, encompassing both physical adsorption and chemical complexation, highlights biochar's potential as a robust agent for mitigating lead contamination in various environmental matrices (Cho et al. 2024).

8.12.3 Precipitation and Mineral Co-Precipitation

In alkaline conditions (often induced by high-temperature biochar), Pb can precipitate as insoluble minerals like pyromorphite [Pb5(PO4)3Cl] or lead carbonates (PbCO3). Biochar's mineral ash content (Ca, Mg, P) further facilitates these precipitation reactions, reducing Pb solubility.

8.12.4 Cation Exchange and pH Modulation

Biochar increases soil pH, which plays a critical role in Pb immobilization. Under higher pH conditions, Pb^{2+} tends to form less soluble hydroxides and carbonates. Additionally, biochar's cation exchange capacity (CEC) allows it to retain Pb ions, preventing leaching into groundwater.

8.12.5 Redox Reactions

Some biochars, particularly those rich in iron (Fe) or sulfur (S), can promote redox reactions that convert Pb into less bioavailable forms. For example, Fe-modified biochar can facilitate the reduction of $Pb^{2+}$to less mobile Pb(0) under certain conditions. Biochar offers a sustainable, cost-effective solution for Pb-contaminated soils by immobilizing Pb through adsorption, precipitation and pH modulation. Its effectiveness depends on feedstock, pyrolysis conditions and soil properties.

When applied correctly, biochar not only reduces Pb uptake in crops but also improves soil fertility and carbon sequestration. Future research should focus on optimizing biochar formulations for specific Pb-contaminated sites and integrating biochar with other remediation techniques for maximum efficiency. By adopting biochar-based strategies, we can mitigate Pb pollution, protect food safety and promote sustainable agriculture (Beesley et al. 2015).

8.13 Factors Influencing Biochar's Efficacy in Heavy Metal Remediation

The effectiveness of biochar in immobilizing heavy metals such as arsenic (As), cadmium (Cd) and chromium (Cr) in contaminated soils is governed by a complex interplay of biochar—specific properties, soil characteristics and environmental conditions. One of the primary determinants is feedstock type, as the raw material used for biochar production significantly impacts its physical and chemical attributes. For instance, biochar derived from manure or crop residues typically exhibits higher ash content, pH and cation exchange capacity (CEC) compared to wood-based biochar, making it more effective at immobilizing cationic metals like Cd through precipitation and ion exchange. Conversely, lignin-rich feedstocks (e.g., hardwood) often yield biochar with greater porosity and surface area, which enhances physical adsorption of metals. Pyrolysis conditions, particularly temperature and duration, further modify biochar's functionality. High-temperature pyrolysis (500–700 °C) generally produces biochar with increased surface area and aromaticity but fewer oxygen-containing functional groups, favoring the adsorption of non-polar contaminants. In contrast, low—temperature biochar (300–500 °C) retains more functional groups (e.g., carboxyl and hydroxyl), which are critical for complexing metals like Cr and As (Beesley et al. 2015).

Soil properties equally dictate biochar's performance. Soil pH is a pivotal factor, as it influences metal speciation and biochar's surface charge. In acidic soils, biochar's alkaline nature can raise pH, reducing the solubility of cationic metals (Cd^{2+}, Cr^{3+}) via precipitation as hydroxides or carbonates. However, this effect may be less pronounced for anionic As, whose mobility can paradoxically increase at higher pH due to desorption from iron oxides. Soil organic matter (SOM) also interacts with biochar; while SOM can compete with biochar for metal binding sites, it may also form ternary complexes with biochar and metals, enhancing immobilization. Redox conditions play a crucial role, especially for redox-sensitive metals like Cr and As. Under reducing conditions, biochar can facilitate the reduction of toxic Cr (VI) to less soluble Cr (III), whereas aerobic conditions may promote As (III) oxidation to As (V), which binds more strongly to iron oxides on biochar surfaces (US Environmental Protection Agency 1995).

Metal-specific behavior further complicates remediation efficacy. For example, Cd primarily binds to biochar via cation exchange and surface complexation, while As forms inner-sphere complexes with iron-modified biochar. Chromium, depending on its valence state, may undergo redox transformations or electrostatic interactions. Biochar modification can optimize these interactions: iron-impregnated biochar enhances As and Cr adsorption via ligand exchange, while sulfur-modified biochar improves Hg and Cd sequestration. However, aging and weathering of biochar in soil can alter its properties over time. Oxidation of biochar surfaces may introduce additional functional groups, potentially increasing metal binding capacity, but prolonged exposure can also lead to the breakdown of biochar particles and release of previously immobilized metals (Dawar et al. 2023).

Application rate and particle size are practical considerations. Higher biochar doses (e.g., 5–10% w/w) generally improve metal immobilization but may disrupt soil structure or nutrient availability. Fine biochar particles offer greater surface area but may migrate in soil, whereas coarse particles ensure longevity. Climate factors, such as rainfall and temperature, affect biochar's stability and metal retention, with heavy rainfall potentially leaching mobilized metals (Hamid et al. 2020).

In summary, biochar's efficacy in heavy metal remediation is not universal but context—dependent. Tailoring biochar production (feedstock, pyrolysis) and application (dose, modification) to specific soil-metal systems is essential. Future research should focus on long-term field studies, standardized protocols and eco-engineered biochar designs to unlock its full potential in sustainable soil remediation (Saxena et al. 2024).

8.14 Case Studies and Field Applications

The practical efficacy of biochar in mitigating arsenic (As), cadmium (Cd) and chromium (Cr) contamination in crops has been demonstrated through numerous field studies and real—world applications across diverse agroecological regions. The case studies of some important heavy metal remediation through biochar are enlisted below in the tabular form (Table 8.1).

These case studies underscore the importance of tailoring biochar interventions to site—specific conditions. Key lessons include: (1) Feedstock selection (e.g., Fe-rich biochar for As, high-pH biochar for Cd), (2) Dosage optimization (typically 1–5% w/w or 10–20 t ha^{-1}) and (3) Integration with farming practices (e.g., biochar-compost blends). Challenges such as biochar production scalability and farmer adoption barriers (e.g., cost, knowledge gaps) remain, but participatory approaches like Vietnam's farmer cooperatives producing low-cost kilns demonstrate viable solutions. Future efforts should prioritize long-term monitoring to assess biochar's aging effects and expand policy incentives for large-scale adoption in heavy—metal-affected regions (Rehman et al. 2018).

8.15 Challenges and Future Perspectives

Despite the demonstrated potential of biochar in mitigating arsenic (As), cadmium (Cd) and chromium (Cr) uptake in crops, several challenges limit its widespread adoption and effectiveness. One major challenge is the variability in biochar performance

Table 8.1 Mitigation of heavy metals using biochar in field level (case studies)

Target heavy metal	Details of study	References
As	Where groundwater arsenic contamination severely affects rice production, field trials with iron-modified biochar showed remarkable success. Farmers in arsenic-affected regions applied rice husk biochar enriched with iron oxides (Fe-BC) at 10–20 t ha^{-1}, which reduced arsenic uptake in rice grains by 40–60%. The mechanism involved the formation of stable Fe-as complexes, limiting arsenic bioavailability.	Wu et al. (2022)
Cd	A large-scale field study on cd—contaminated paddy soils revealed that bamboo-derived biochar (applied at 5% w/w) decreased cd concentrations in rice grains by 50–70%, achieving compliance with national food safety standards (0.2 mg kg^{-1}). The alkaline nature of bamboo biochar (pH ~9.5) increased soil pH, promoting cd precipitation as insoluble carbonates and hydroxides.	Tian et al. (2021)
Cr	Farmers applied sulfur-enriched biochar (S-BC) at 3% w/w, which facilitated the reduction of toxic Cr (VI) to less mobile Cr (III) while enhancing soil organic carbon. Tomato crops grown in treated soils exhibited 70% lower Cr accumulation in fruits compared to untreated plots.	Xiao et al. (2019)
Cd and as	The walnut shell biochar application (15 t ha^{-1}) not only reduced Cd and As uptake by 30–45% but also improved crop yield in lettuce by 20% due to enhanced soil fertility.	Xiao et al. (2019)
Cr	Locally produced biochar from maize Stover, when co-applied with farmyard manure (1:1 ratio), reduced Cr bioavailability by 50% and increased maize yields by 35%. This approach also addressed socioeconomic challenges by utilizing farm waste and minimizing costs	Xu et al. (2019)
Cd	Rice straw biochar applied at 2% w/w in cd-contaminated soils reduced grain cd levels by 55%, while its porous structure improved water retention during drought periods	Xu et al. (2019)

due to differences in feedstock composition, pyrolysis conditions and soil properties. For instance, biochar derived from manure may exhibit higher metal immobilization capacity than wood—based biochar due to its greater mineral content, but its long-term stability in soil remains uncertain. Additionally, biochar's efficacy is highly dependent on soil pH organic matter content and redox conditions, which influence metal speciation and bioavailability. In acidic soils, for example, biochar may initially reduce metal mobility by increasing pH, but prolonged exposure to weathering and microbial activity could alter its surface properties, potentially leading to the re-mobilization of trapped metals. Another concern is the potential trade-off between heavy metal immobilization and nutrient availability. While biochar can reduce toxic metal uptake, it may also bind essential micronutrients like zinc and copper, inadvertently affecting crop nutrition. Furthermore, the economic feasibility of large-scale biochar production and application poses a significant barrier, particularly for smallholder farmers in developing countries where metal contamination is often most severe. The energy—intensive nature of pyrolysis and the logistical challenges of transporting bulk biochar to contaminated sites further complicate its implementation (Zama et al. 2018).

Looking ahead, future research should focus on optimizing biochar properties for site-specific remediation. Advanced characterization techniques, such as synchrotron-based spectroscopy and machine learning modeling, can help predict biochar-metal interactions and guide the design of tailored biochar formulations. For example, engineered biochars modified with iron oxides, sulfur or phosphorus have shown enhanced affinity for specific metals like As and Cd, offering a promising avenue for targeted remediation. The development of low-cost, decentralized pyrolysis units could also make biochar more accessible to resource-limited farming communities. Another critical area for exploration is the long-term fate of biochar in agricultural systems, including its aging processes and interactions with soil microbiota. Understanding how biochar evolves over time will be essential for assessing its durability as a metal immobilization agent. Additionally, integrating biochar with other remediation strategies, such as phytoremediation or microbial-assisted metal stabilization, could create synergistic effects for more comprehensive soil rehabilitation (Xu et al. 2021). Policy and regulatory frameworks must also evolve to support biochar adoption, including standardized guidelines for biochar quality, application rates and environmental risk assessments. Public awareness campaigns and farmer training programs can further facilitate the transition to biochar-based soil management practices. Ultimately, addressing these challenges through interdisciplinary collaboration and innovation will be key to unlocking biochar's full potential as a sustainable tool for reducing heavy metal contamination in crops, ensuring food safety and promoting environmental resilience in metal-polluted agroecosystems (Ramtahal et al. 2019).

8.16 Conclusion

The comprehensive evaluation of biochar's efficacy in minimizing arsenic (As), cadmium (Cd) and chromium (Cr) content in crops demonstrates its significant potential as a sustainable soil amendment for heavy metal remediation. Extensive research reveals that biochar's unique physicochemical properties, including its porous structure, high surface area and abundant functional groups, enable multiple mechanisms of metal immobilization in contaminated soils. Through processes such as surface complexation, ion exchange, electrostatic attraction and precipitation, biochar effectively reduces the bioavailability and plant uptake of these toxic elements while simultaneously improving soil health and fertility. The alkaline nature of many biochars plays a crucial role in increasing soil pH, which is particularly effective in reducing the mobility of cationic metals like Cd and Cr. Furthermore, biochar's redox-active properties facilitate the transformation of highly toxic Cr (VI) to less mobile Cr (III), while its affinity for oxyanions helps mitigate arsenic translocation to edible plant parts.

Field-scale applications and case studies across various agricultural systems have consistently shown that biochar amendment can reduce heavy metal concentrations in crops by 30–70%, depending on the specific metal, biochar characteristics and soil conditions. For instance, iron-modified biochar has demonstrated remarkable efficacy in arsenic immobilization, while sulfur-enriched biochar shows particular promise for chromium remediation. The dual benefit of biochar—simultaneously addressing soil contamination while enhancing soil quality—makes it an attractive alternative to conventional remediation techniques that often merely transfer contaminants or degrade soil functionality. Moreover, the carbon-negative nature of biochar production aligns with global climate change mitigation efforts, adding an environmental dimension to its value proposition for sustainable agriculture.

However, the long-term effectiveness and practical implementation of biochar technology face several challenges that require careful consideration. The variability in biochar performance based on feedstock type, pyrolysis conditions and soil properties necessitates the development of tailored application protocols. Potential trade-offs between metal immobilization and nutrient availability must be carefully managed to avoid unintended consequences on crop nutrition. Economic barriers related to large-scale biochar production and distribution, particularly in developing regions most affected by heavy metal contamination, remain significant hurdles to widespread adoption. Future research should focus on optimizing biochar properties for specific metal-contamination scenarios, developing cost-effective production methods from local biomass waste and establishing standardized guidelines for application rates and methods.

The integration of biochar with other remediation approaches, such as phytoremediation or microbial-assisted strategies, may offer synergistic effects that enhance overall remediation efficiency. Advances in biochar modification techniques, including mineral impregnation and nanomaterial incorporation, show particular promise for improving metal sequestration capacity. Policy frameworks and economic incentives will be crucial for promoting biochar adoption at scale, particularly in regions with severe heavy metal contamination of agricultural lands. As the scientific understanding of biochar-metal interactions continues to evolve, this versatile material is poised to play an increasingly important role in ensuring food safety, protecting ecosystem health and supporting the transition to more sustainable agricultural systems worldwide. The collective evidence presented in this chapter strongly supports biochar as a viable, multifunctional solution for mitigating heavy metal contamination in crops, though its successful implementation will require continued research, technological innovation and stakeholder collaboration across the agricultural value chain.

Conflict of Interest The authors are declaring that no conflicts of interests are associated with this manuscript.

References

Abbruzzini TF, Davies CA, Toledo FH, Cerri CEP (2019) Dynamic biochar effects on nitrogen use efficiency, crop yield and soil nitrous oxide emissions during a tropical wheat-growing season. J Environ Manage 252:109638

Aborisade MA, Oba BT, Kumar A, Liu J, Chen D, Okimiji OP, Zhao L (2023) Remediation of metal toxicity and alleviation of toxic metals-induced oxidative stress in Brassica chinensis L using biochar-iron nanocomposites. Plant Soil 493:629–645

Adak E, Halder S, Koley B, Biswas S, Sengupta S, Kundu S, Sarkar T (2024) An overview of the importance of biochar in sustainable Agriculture. J Adv Biol Biotechnol 27:924–937

Algethami JS, Ibrahim M, Javed W, Alosaimi EH, Irshad MK (2024) Efficacy of Fe-BC in enhancing growth, photosynthesis, nutrition and alleviating the toxicity of Cd and Cr in rapeseed (Brassica napus L.): a tool for managing the environment and attaining sustainable agriculture. Environ Technol Innov 36:103789

Ali S, Rizwan M, Bano R, Bharwana SA, Rehman MZU, Hussain MB, Al-Wabel MI (2018) Effects of biochar on growth, photosynthesis and chromium (Cr) uptake in Brassica rapa L. Under Cr stress. Arab J Geosci 11:1–9

Ameloot N, De Neve S, Jegajeevagan K, Yildiz G, Buchan D, Funkuin YN, Sleutel S (2013) Short-term CO_2 and N_2O emissions and microbial properties of biochar amended sandy loam soils. Soil Biol Biochem 57:401–410

Arias JA, Peralta-Videa JR, Ellzey JT, Ren M, Viveros MN, Gardea-Torresdey JL (2010) Effects of glomus deserticola inoculation on prosopis: Enhancing chromium and lead uptake and translocation as confirmed by X-ray mapping, ICP-OES and TEM techniques. Environ Exp Bot 68:139–148

Aslam Z, Khalid M, Aon M (2014) Impact of biochar on soil physical properties. J Agric Sci 4:280–284

Ayaz M, Feizienė D, Tilvikienė V, Akhtar K, Stulpinaitė U (2021) Biochar role in the sustainability of agriculture and environment. Sustainability 13:1330

Bashir S, Hussain Q, Shaaban M, Hu H (2018) Efficiency and surface characterization of different plant derived biochar for cadmium (Cd) mobility, bioaccessibility and bioavailability to Chinese cabbage in highly contaminated soil. Chemosphere 211:632–639

Bashir S, Qayyum MA, Husain A, Bakhsh A, Ahmed N, Hussain MB, Elshik MS, Alwahibi MS, Almunqedhi BMA, Hussain R, Wang YF, Zhou Y, Diao ZH (2021) Efficiency of different types of biochars to mitigate Cd stress and growth of sunflower (Helianthus; L.) in wastewater irrigated agricultural soil. Saudi J Biol Sci 28:2453–2459

Beesley L, Moreno-Jimenez E, Fellet G, Melo L, Sizmur T (2015) Biochar and heavy metals. In: Lehmann J, Joseph S (eds) Biochar for environmental management—science, technology and implementation, 1st edn. Routledge, USA, pp 563–594

Cao H, Ning L, Xun M, Feng F, Li P, Yue S, Yang H (2019) Biochar can increase nitrogen use efficiency of Malus hupehensis by modulating nitrate reduction of soil and root. Appl Soil Ecol 135:25–32

Cao X, Ma L, Liang Y, Gao B, Harris W (2011) Simultaneous immobilization of lead and atrazine in contaminated soils using dairy-manure biochar. Environ Sci Technol 45:4884–4889

Chatterjee S, Nayak P, Acharya S, Paul RAI, Ocaña-Reyes JA (2024) Regenerative agriculture: inclusive approach for transforming conventional agriculture. In: Mishra AK, Sharma S, Mishra A (eds) Key drivers and indicators of soil health management: transitioning from conventional to regenerative agriculture. Singapore, Springer Nature Singapore, pp 85–93

Chen W, Meng J, Han X, Lan Y, Zhang W (2019) Past, present, and future of biochar. Biochar. 1:75–87

Cho Y, Lim JY, Igalavithana AD, Hwang G, Sang MK, Masek O, Ok YS (2024) AI-guided investigation of biochar's efficacy in Pb immobilization for remediation of Pb contaminated agricultural land. Appl Biol Chem 67:1–17

Danso OP, Acheampong A, Zhang Z, Song J, Wang Z, Dai J, Zhi T, Yin X, Zhu R (2023) The management of Cd in rice with biochar and selenium: effects, efficiency and practices. Carbon Res 2:41

Dawar K, Asif M, Irfan M, Mian IA, Khan B, Gul N, Fahad S, Jalal A, Danish S, Syed A, Elgorban AM, Eswaramoorthy R, Hussain MI (2023) Evaluating the efficacy of activated carbon in minimizing the risk of heavy metals contamination in spinach for safe consumption. ACS Omega 8:24323–24331

Githinji LJM (2014) Effect of biochar application rate on soil physical and hydraulic properties of a sandy loam. Arch Agron Soil Sci 60:457–470

Gluba Ł, Rafalska-Przysucha A, Szewczak K, Łukowski M, Szlązak R, Vitkova J, Kobyłecki R, Bis Z, Wichlinski M, Zarzycki R, Kacprzak A, Usowicz B (2021) Effect of fine size-fractionated sunflower husk biochar on water retention properties of arable sandy soil. Materials 14:1335

Gul S, Whalen JK (2016) Biochemical cycling of nitrogen and phosphorus in biochar—amended soils. Soil Biol Biochem 103:1–15

Hale SE, Nurida NL, Mulder J, Sørmo E, Silvani L, Abiven S, Joseph S, Taherymoosavi S, Cornelissen G (2020) The effect of biochar, lime and ash on maize yield in a long-term field trial in a Ultisol in the humid tropics. Sci Total Environ 719:137455

Hamid Y, Tang L, Hussain B, Usman M, Gurajala HK, Rashid MS, He Z, Yang X (2020) Efficiency of lime, biochar, Fe containing biochar and composite amendments for Cd and Pb immobilization in a co-contaminated alluvial soil. Environ Pollut 257:113609

Han C, Chen F, Lian C, Liang R, Liang W, Chen M, Luo A, Gao T (2020) Development of preparation method and application of biochar. Int J Emerg Technol Adv Eng 10:6–13

Harter J, El-Hadidi M, Huson DH, Kappler A, Behrens S (2017) Soil biochar amendment affects the diversity of nosZ transcripts: Implications for N_2O formation. Sci Rep 7:3338

Harvey OR, Herbert BE, Rhue RD, Kuo LJ (2011) Metal interactions at the biochar-water interface: energetics and structure-sorption relationships elucidated by flow adsorption microcalorimetry. Environ Sci Technol 45:5550–5556

Herath HMSK, Arbestain MC, Hedley M (2013) Effect of biochar on soil physical properties in two contrasting soils: an alfisol and an andisol. Geoderma 209:188–197

Hu F, Xu C, Ma R, Tu K, Yang J, Zhao S, Yang M, Zhang F (2021) Biochar application driven change in soil internal forces improves aggregate stability: based on a two-year field study. Geoderma. 403:115276

Hu Y, Zhang P, Yang M, Liu Y, Zhang X, Feng S, Guo D, Dang X (2020) Biochar is an effective amendment to remediate Cd-contaminated soils-a meta—analysis. J Soils Sediments 20:3884–3895

Huang Y, Wang C, Lin C, Zhang Y, Chen X, Tang L, Liu C, Chen Q, Onwuka MI, Song T (2019) Methane and nitrous oxide flux after biochar application in subtropical acidic paddy soils under tobacco-rice rotation. Sci Rep 9:17277

Hunt J, DuPonte M, Sato D, Kawabata A (2010) The basics of biochar: a natural soil amendment. Soil Crop Manag 30:1–6

International Biochar Initiative (IBI) (2025) Available online: https://biochar-international.org/characterizationstandard/. Accessed on 25th September, 2025).

Karhu K, Mattila T, Bergstrom T, Regina K (2011) Biochar addition to agriculture soil increased CH4 uptake and water holding capacity-Results from a short-term pilot field study. Agric Ecosyst Environ 140:309–313

Kavitha B, Reddy PVL, Kim B, Lee SS, Pandey SK, Kim KH (2018) Benefits and limitations of biochar amendment in agricultural soils: a review. J Environ Manage 227:146–154

Khan AZ, Ding X, Khan S, Ayaz T, Fidel R, Khan MA (2020a) Biochar efficacy for reducing heavy metals uptake by Cilantro (Coriandrum sativum) and spinach (Spinaccia oleracea) to minimize human health risk. Chemosphere. 244:125543

Khan AZ, Khan S, Khan MA, Alam M, Ayaz T (2020b) Biochar reduced the uptake of toxic heavy metals and their associated health risk via rice (Oryza sativa L.) grown in Cr-Mn mine contaminated soils. Environ Technol Innov 17:100590

Kumar A, Bhattacharya T, Shaikh WA, Roy A, Mukherjee S, Kumar M (2021) Performance evaluation of crop residue and kitchen waste-derived biochar for eco-efficient removal of arsenic from soils of the Indo-Gangetic plain: A step towards sustainable pollution management. Environ Res 200:111758

Kumar S, Rahman MA, Islam MR, Hashem MA, Rahman MM (2022) Lead and other elements-based pollution in soil, crops and water near a lead-acid battery recycling factory in Bangladesh. Chemosphere 290:133288

Kushwaha A, Hans N, Kumar S, Rani R (2018) A critical review on speciation, mobilization and toxicity of lead in soil-microbe-plant system and bioremediation strategies. Ecotoxicol Environ Saf 147:1035–1045

Lehmann J (2009) Terra preta Nova—where to from here? In: Woods WI, Teixeira WG, Lehmann J, Steiner C, WinklerPrins AMGA, Rebellato L (eds) Amazonian dark earths: Wim Sombroek's vision, vol 28. Springer, Dordrecht, Netherlands, pp 473–486

Lehmann J, Joseph S (2009) Biochar for environmental management: an introduction. In: Lehmann J, Joseph S (eds) Biochar for environmental management science and technology, vol 1, 1st edn, London, UK, pp 1–9

Lehmann J, Cowie A, Masiello CA, Kammann C, Woolf D, Amonette JE, Cayuela ML, Camps-Arbestain M, Whitman T (2021) Biochar in climate change mitigation. Nat Geosci 14:883–892

Lehmann J, da Silva Jr JP, Rondon M, Kinyangi J (2019) Biochar as an organic amendment. Science, Technology and Implementation, Biochar for Environmental Management, pp 339–362

Leksungnoen P, Wisawapipat W, Ketrot D, Aramrak S, Nookabkaew S, Rangkadilok N, Satayavivad J (2019) Biochar and ash derived from silicon-rich rice husk decrease inorganic arsenic species in rice grain. Sci Total Environ 684:360–370

Mukherjee A, Lal R (2013) Biochar impacts on soil physical properties and greenhouse gas emissions. Agronomy. 3:313–339

Mukherjee D, Chatterjee S, Mandal AR (2020) Biochar: an environmentally sustainable residue management. Agric Food: e-Newsl 2:297–303

Naveed M, Tanvir B, Xiukang W, Brtnicky M, Ditta A, Kucerik J, Subhani Z, Nazir MZ, Radziemska M, Saeed Q, Mustafa A (2021) Co-composted biochar enhances growth, physiological and phytostabilization efficiency of Brassica napus and reduces associated health risks under chromium stress. Front Plant Sci 12:775785

Rafi MN, Imran M, Nadeem HA, Abbas A, Pervaiz M, Khan WUD, Ullah S, Hussain S, Saeed Z (2022) Comparative influence of biochar and doped biochar with Si-NPs on the growth and anti-oxidant potential of Brassica Rapa L. under Cd toxicity. Silicon 14:11699–11714

Rahim HU, Mian IA, Akbar WA, Khan K (2024) Comparative efficacy of wheat-straw biochar and chicken-waste compost on cadmium contaminated soil remediation, reducing cadmium bioavailability and enhancing wheat performance under cadmium stress. J Agric Food Res 15:101005

Ramtahal G, Umaharan P, Hanuman A, Davis C, Ali L (2019) The effectiveness of soil amendments, biochar and lime, in mitigating cadmium bioaccumulation in Theobroma cacao L. Sci Total Environ 693:133563

Rehman MZU, Rizwan M, Hussain A, Saqib M, Ali S, Sohail MI, Shafiq M, Hafeez F (2018) Alleviation of cadmium (Cd) toxicity and minimizing its uptake in wheat (Triticum aestivum) by using organic carbon sources in Cd-spiked soil. Environ Pollut 241:557–565

Rehman MH, Holmes A, Saunders SJ, Islam KR (2011) Biochar workshop. Massey University, Palmerston North, New Zealand, NZ Biochar Research Centre

Sameshima-Saito R, Chiba K, Hirayama J, Itakura M, Mitsui H, Eda S, Minamisawa K (2006) Symbiotic Bradyrhizobium japonicum reduces N_2O surrounding the soybean root system via nitrous oxide reductase. Appl Environ Microbiol 72:2526–2532

Saxena A, Srivastava PC, Pathak A, Pachauri SP, Srivastava P (2024) Effect of fortified pine needle biochar on green fodder yield, nutrient uptake of oat and soil properties. Soil Adv 2:100019

Shackley S, Carter S, Knowles T, Middelink E, Haefele S, Sohi S, Cross A, Haszeldine S (2012) Sustainable gasification–biochar systems? A case-study of rice—husk gasification in Cambodia,

Part I: Context, chemical properties, environmental and health and safety issues. Energy Policy 42:49–58
Sheikh L, Naz N, Oranab S, Younis U, Alarfaj AA, Alharbi SA, Ansari MJ (2025) Minimization of cadmium toxicity and improvement in growth and biochemical attributes of spinach by using acidified biochar. Sci Rep 15:5880
Shiade SRG, Fathi A, Minkina T, Wong MH, Rajput VD (2023) Biochar application in agroecosystems: a review of potential benefits and limitations. Environ Dev Sustain 26:19231–19255
Singh SV, Chaturvedi S, Dhyani VC, Datta D (2019) Biochar: an eco-friendly residue management approach. Indian Farm 69:27–29
Singh VP, Singh S, Kumar J, Prasad SM (2015) Hydrogen sulfide alleviates toxic effects of arsenate in pea seedlings through up-regulation of the ascorbate—glutathione cycle: possible involvement of nitric oxide. J Plant Physiol 181:20–29
Sohi SP (2012) Carbon storage with benefits. Science 338:1034–1035
Tan Z, Lin C, Ji X, Rainey T (2017) Returning biochar to fields: a review. Appl Soil Ecol 116:1–11
Tian X, Wang D, Chai G, Zhang J, Zhao X (2021) Does biochar inhibit the bioavailability and bioaccumulation of as and Cd in co-contaminated soils? A meta—Analysis. Sci Total Environ 762:143117
US Environmental Protection Agency (1995) Trends in Greenhouse Gas Emissions. Inventory of U.S. Greenhouse Gas Emissions and Sinks: 1990–2012—Trends. EPA.Gov. Environmental Protection Agency, Mar 2014. Web. 5th Aug 2025.
Wu K, Wu C, Jiang X, Xue R, Pan W, Li WC, Luo X, Xue S (2022) Remediation of arsenic-contaminated paddy field by a new iron oxidizing strain (Ochrobactrum sp.) and iron-modified biochar. J Environ Sci 115:411–421
Xiao F, Cheng J, Cao W, Yang C, Chen J, Luo Z (2019) Removal of heavy metals from aqueous solution using chitosan-combined magnetic biochars. J Colloid Interface Sci 540:579–584
Xu X, Cheng K, Wu H, Sun J, Yue Q, Pan G (2019) Greenhouse gas mitigation potential in crop production with biochar soil amendment-a carbon footprint assessment for cross-site field experiments from China. Glob Change Biol Bioenerg 11:592–605
Xu Y, Song Z, Chang X, Guo Z, Gao M (2021) Effects of Fe-Mn oxide—modified biochar composite applications on phthalate esters (PAEs) accumulation in wheat grains and grain quality under PAEs-polluted brown soil. Ecotoxicol Environ Saf 208:111624
Yang X, Zhang S, Ju M, Liu L (2019) Preparation and modification of biochar materials and their application in soil remediation. Appl Sci 9:1365
Zama EF, Gang L, Tang YT, Reid BJ, Ngwabie NM, Sun GX (2022) The removal of arsenic from solution through biochar-enhanced precipitation of calcium—Arsenic derivatives. Environ Pollut 292:118241
Zama EF, Reid BJ, Sun GX, Yuan HY, Li XM, Zhu YG (2018) Silicon (Si) biochar for the mitigation of arsenic (As) bioaccumulation in spinach (Spinacia oleracea) and improvement in the plant growth. J Clean Prod 189:386–395
Zhao FJ, Ma JF, Meharg AA, Mcgrath SP (2009) Arsenic uptake and metabolism in plants. New Phytol 181:777–794

Chapter 9
Sustainable Applications of Biochar: A Comprehensive Review in the Remediation of Emerging Environmental Contaminants

Sagorika Paul, Erom Romi Singha, Bishal Kumar Singh, Mahua Das, and Anirudha Giri

Abstract The rapid rise in the global population led to a surge in various forms of inorganic and organic contaminants, such as nanometals, polycyclic aromatic hydrocarbons, microplastics etc., posing a risk to the sustainable environmental management. Biochar is a carbon-rich byproduct of oxygen limited pyrolysis at a temperature ranging from 300 to 900 °C derived from various biological resources, such as plants, foods and other animal wastes. Apart from pollution management, biochar increases soil fertility and contributes to the climate change mitigation through carbon retention. Additionally, biochar has the of removing toxins from wastewater, thus lowering the health risk and converting it into potable water. The versatility of biochar is further emphasized by its application in biofuel production and as a catalyst in the generation of various organic compounds. The versatile applications of biochar make it an appealing alternative for environmental remediation and the implementation of environmentally sustainable farming practices. This chapter explores the different techniques applied to generate biochar from a multitude of biomaterials, characterization techniques of biochar and sustainable applications of biochar in environmental remediation, agriculture and energy generation. A brief discussion of nano-biochar is also provided emphasizing its potential to revolutionize sustainable practices. Application of biochar in environmental pollution management highlights its promising role in fostering a more sustainable and greener future.

Keywords Biochar · Nano-biochar · Pyrolysis · Remediation · Sustainable development

---

S. Paul · E. R. Singha · B. K. Singh · M. Das · A. Giri (✉)
Environment and Human Toxicology Laboratory, Department of Life Science and Bioinformatics, Assam University, Silchar, India
e-mail: agiri123@yahoo.com

D. Moulick et al. (eds.), *Biochar: Achieving Agro-Environmental Sustainability*, Sustainability Solutions, https://doi.org/10.1007/978-3-032-19855-6_9

Abbreviations

BET	Brunauer-Emmett-Teller
BPA	Bisphenol A
DNA	Deoxyribonucleic acid
EDS	Energy-dispersive X-ray spectroscopy
EEC	Emerging environmental contaminant
FTIR	Fourier-transform infrared spectroscopy
H/C	Hydrogen to carbon ratio
MIP	Mercury intrusion porosimetry
NMR	Nuclear magnetic resonance
O/C	Oxygen to carbon ratio
PAH	Polycyclic aromatic hydrocarbons
PFAS	Per- and polyfluoroalkyl substance
PFOA	Perfluorooctanoic acid
PFOS	Perfluorooctanesulfonic acid
ROS	Reactive oxygen species
SEM	Scanning electron microscopy
XPS	X-ray photoelectron spectroscopy
XRD	X-ray diffraction

9.1 Introduction

The rapid industrialization and urbanization of modern society have increased the release of emerging environmental contaminants (EECs) into ecosystems, presenting complex challenges for environmental protection and public health. These contaminants, which polycyclic aromatic hydrocarbons (PAHs), microplastics, per- and polyfluoroalkyl substances (PFAS) and other chemicals are increasingly detected in soil systems, and water bodies (Richardson and Kimura 2017). Unlike conventional pollutants, many EECs are highly persistent, bioaccumulative and capable of exerting toxic effects at trace concentrations making them particularly difficult to remediate using traditional treatment methods (Li et al. 2024). Thus, the inability of existing wastewater treatment plants to remove contaminants demands innovative, cost-effective, and sustainable remediation technologies. Among these is biochar: a carbonaceous material generated by the pyrolysis of biomass. Its exceptional adsorption capacity, catalytic properties and environmental compatibility are the key characteristics for the sustainable remediation of EECs (Sarkar et al. 2024).

Biochar is defined as a stable, porous, carbonaceous material produced by the thermal decomposition of organic feedstocks in an oxygen-deficient environment, a process considered as pyrolysis (Tomczyk et al. 2020). The material is further characterized by highly aromatic carbon, high surface area, and a high presence of oxygen-containing functional groups. All of these properties give it an exceptional ability to remove contaminants and fertilize the soil (Li et al. 2023). Depending on feedstock selection and production conditions, biochar can be engineered to possess specific physicochemical properties tailored for different remediation applications, ranging from contaminant immobilization to degradation (Haider et al. 2022). Beyond environmental remediation, biochar plays an important part in mitigation of climate change by serving as a long-term carbon sink, with estimates suggesting that biochar-amended lands may hold carbon for thousands of years to millennia (Shoudho et al. 2024).

Emerging contaminants comprise an evolving and varied group of substances that must be considered because of the threats they impose on ecological and human health owing to persistence, mobility, and potential toxicity (Wang et al. 2024a, b). Emerging contaminants are any substances introduced into the environment that should not be there from human interaction, such as endocrine disruptors like bisphenol A (BPA) and phthalates, pharmaceutical residues, PFAS, pesticides, microplastics, and engineered nanomaterials (Nishmitha et al. 2025). Effluents from wastewater treatment are able to carry many contaminates of pharmaceuticals, unwanted residue from agriculture, industrial discharge, and improper disposal of consumer products into environmental matrices (Kumari et al. 2022). Disinfection by conventional water and soil treatment technologies, these contaminants generally resist treatment because of their chemical stability and are present even at very low concentrations. Generally, wastewater treatment facilities are not equipped well for the elimination of trace concentrations of EECs; hence, their constant discharging into water bodies (Richardson and Kimura 2017). This has created an immediate demand for sophisticated remediation approaches capable of targeting these recalcitrant pollutants.

The exceptional remediation capabilities of biochar stem from its distinct physical and chemical features that can be precisely tailored through feedstock selection and pyrolysis conditions. Surface area, one of the most critical parameters, typically ranges from 10–500 m^2/g but can exceed 700 m^2/g for optimized biochars, providing abundant sites for contaminant adsorption (Leng et al. 2021). Surface functional groups, including carboxyl (–COOH), hydroxyl (–OH), and phenolic moieties, enable complexation with contaminants and involve in redox reactions (Nishmitha et al. 2025). Current research data indicates that these features collaborate in order to enhance the sorption capacity of biochar. A greater surface area, in conjunction with oxygen functional groups, may enhance the adsorption capacity on organic contaminants by 3–5 times over activated carbon (Zhang et al. 2025). Biochar also catalyzes pollutant degradation via many advanced oxidation processes because of its electron-donating capability (0.5–2 mmol e^-/g) and stability of persistent free radicals (10^{16}–10^{18} spins/g) (Jiang et al. 2024a, b).

Nano-biochar represents a cutting-edge advancement in contaminant remediation, where biochar particles are engineered at the nanoscale (1–100 nm) to dramatically enhance reactivity and mobility. Synthesis methods include ball-milling of conventional biochar to reduce particle size, ultrasonic fragmentation, and controlled pyrolysis of nano-sized feedstock particles (Amusat et al. 2021). Depending on the size of the char particles, nano-biochar will be at least 2–10 times more efficient in adsorption of contaminants than the bulk biochar due to the increase in surface area available and the easy accessibility of the binding sites (Bhandari et al. 2023). The small particle size allows nano-biochar to be transported through soil pores and groundwater systems, facilitating in situ remediation of deep contamination plumes. However, similar features have raised questions regarding environmental safety as nanomaterial can be accumulated by organisms which may generate reactive oxygen species (ROS) (Singha et al. 2024). Lifecycle assessment studies indicate that while nano-biochar offers superior remediation performance, its energy-intensive production and uncertain long- term fate necessitate careful evaluation (Hassan et al. 2024).

It becomes essential to understand environment friendly approaches for contaminants remediation in order to tackle worldwide sustainability challenges. Thus, in this chapter different types of EEC and their toxicity has been mentioned. The techniques applied to generate biochar from a range of biomaterials, characterization techniques of biochar and its sustainable applications of biochar in agriculture, environmental remediation, and energy generation has been explored. It also includes the sustainable applications of biochar in mitigating EECs, focusing on its production, characterization, modification techniques, and real-world remediation potential. A brief discussion of nano-biochar is also provided emphasizing its potential to revolutionize sustainable practices. The use of biochar in environmental pollution management underscores its importance in fostering a more sustainable and eco-friendly future.

9.2 Emerging Environmental Pollutants

A multitude of the pollutants infiltrate the environment via wastewater discharge, agricultural runoff, industrial effluents, and incorrect waste disposal, posing a serious challenge to environmental safety and public health (Table 1; Fig. 1). EECs arise from diverse human activities, each contributing to environmental pollution in unique manners (Wang et al. 2024a, b). Studies on 80% of world rivers and even in drinking water at levels up to 1 μg/L reveal that these chemicals are regularly found in the environment (Ehalt Macedo et al. 2025). The environmental contaminants can be bio-accumulated and remain in nature for an extended period making them lethal for plants and animals including humans. Among the worst health hazards of EEC is linked to endocrine disruption, developmental problems in children, diabetes and infertility (Mitra et al. 2024). Remedial plans are needed to stop long-term agricultural pollution and guarantee food safety. Contaminants change the balance of soil microbiota, therefore hampering the plant growth (Chen et al. 2024a, b) and threaten the world food security (Solanki et al. 2024). Based on reduced natural nutrient cycling, it was proposed that contaminants in agricultural soils could compromise sustainable farming methods (Chen et al. 2024a, b). These toxicants have other negative environmental implications which may hampers stability of ecosystem, disrupt carbon retention and initiate the mutagenesis cascades (Yu et al. 2023).

Table 1 Different types of emerging environmental contaminants and their sources and toxicities

Serial No.	Main classes of EECs	Names (examples)	Sources	Key chemical feature(s)	References
1	Personal care products	Triclosan, parabens	Cosmetic products, soaps	Lipophilic, endocrine-disrupting	Lee-Sarwar et al. (2018)
2	Per- and polyfluoroalkyl substances	PFOA, PFOS	Firefighting foams, non-stick cookware	Highly stable, bioaccumulative	Buck et al. (2011)
3	Microplastics	Polyethylene, polypropylene, polystyrene	Plastic waste breakdown, synthetic textiles	Persistent, adsorb other pollutants	Rochman et al. (2015)
4	Endocrine disrupting compounds	BPA, phthalates	Plastics, industrial effluents	Mimic hormones	Warner and Flaws (2018)
5	Pesticides and herbicides	Glyphosate, 2,4-dichlorophenoxyacetic acid	Agricultural runoff	Persistent	Van Bruggen et al. (2018); Ma et al. (2024)
6	Nanoparticles	Silver, nickel, nanoparticles,	Industrial processes, medical applications	High reactivity, potential toxicity	Zhang et al. (2022)
7	Artificial sweeteners	Sucralose, saccharin	Wastewater from food industry	Highly water-soluble	Ghusn et al. (2023)
8	PAHs	Naphthalene, benzo[a]pyrene, anthracene	Coal burning, oil spills	Carcinogenic, hydrophobic, persistent	Abdel Shafy and Mansour (2016)

9.2.1 *Different Types of Emerging Environmental Contaminants*

EEC in the environment have received increased attention recently as they are frequently detected in various trophic levels. Triclosan, an antimicrobial chemical, often detected in the personal care products such as toothpaste and soap and enters the environment via different anthropogenic activities (Dar et al. 2022). Phthalates are also detected in a variety of personal care products, including pharmaceuticals and hair products, in addition to the plastic material industry (Wang and Quian 2021). Studies indicates that over 90% of bottled water contains microplastics, with an average 325 particles/liter (Mason et al. 2018). Microplastics provide ecological and health hazards include leakage of harmful chemicals like phthalates and persistent

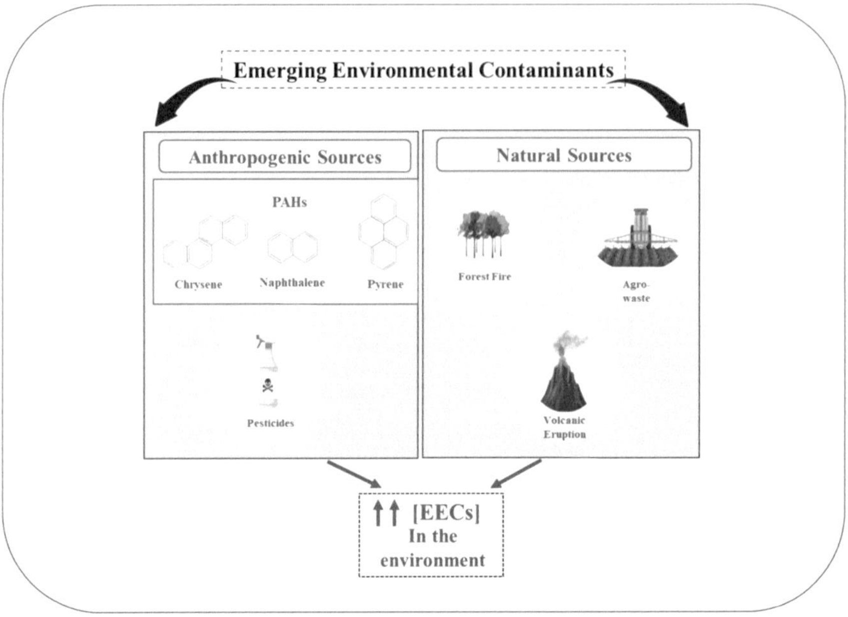

Fig. 1 Different types of emerging environmental contaminants

inflammation in animals. The growing frequency of nanoplastics in ecosystems could have cascade impacts on plant health, agriculture and ecological stability (Shah et al. 2025). PFAS can stay in surroundings for many years without degrading and have been discovered in 99% people in the United States when a blood test was conducted (Brambilla et al. 2015). With worldwide uses of three million tons annually, pesticides are rather common in agriculture. Studies reveal traces of these compounds in 75% of honey samples worldwide, therefore contaminating the environment (Mitchell et al. 2017). Public health issues have been raised by links between human contact to some pesticides and neurotoxicity (Rodríguez et al. 2025). Studies find phthalates and BPA in 90% of urine samples taken from Americans (Wang and Qian 2021). Artificial sweeteners, which are several times sweeter than sugar, are used widely to control diabetes and weight and are present in a variety of consumer goods, including food and beverages, but their detrimental effects have only recently noticed (Ghusn et al. 2023). PAHs, consisting of two or more fused rings and generated from both natural and anthropogenic sources, are ubiquitous organic molecules (Patel et al. 2020). The main contributors of PAHs are increased use of fossil fuels and incomplete combustion. Nanoscience is an emerging discipline; thus, many types of nanomaterials are identified in batteries, solar cells, and personal care products, which may progressively leach into the environment (Zhang et al. 2022). Toxic effects of some silver nanoparticles have been observed in aquatic organisms (Singha et al. 2024).

9.2.2 Toxicity of EEC in the Environment

Toxicity of different EEC in plants and animals are significant and concerning (Table 1). Toxicity of microplastics is a consequence of their ability to release hazardous compounds, including phthalates and BPA, and to induce immunological reactions including cytokine generation (Smith et al. 2018). The toxicity mechanism of BPA consists in disturbance of gut flora and interference with cytochrome P450 (Vandenberg et al. 2009; Zhang et al. 2019). Wheat crops showed significant decrease in the rate of photosynthesis when treated with microplastics (Qi et al. 2018). Long term effects of microplastics include soil infertility and bioaccumulation along food chains, hence disrupting ecosystem dynamics (Rochman et al. 2015). Airborne nanoplastics penetrate deep into lung alveoli, enter the bloodstream and trigger systemic inflammation (Lehner et al. 2019). Because of occupational contact with PFAS-laden products, firefighters and industrial personnel are at a higher risk of exposure (Weatherly and Gosse 2017). PFAS persistently contaminate drinking water due to environmental resilience and pose a risk of Arctic ecological collapse (Sonne et al. 2025). Toxic mechanism of PFAS is the disturbance of important soil enzymes, particularly dehydrogenase, which is absolutely vital for microbial metabolic activity of plants (Zhang et al. 2024a, b). Atrazine and BPA exposure to crop lands may alter the soil microbial activity, blocking photosystem II, delayed germination and variations in the auxin signalling (Jablonowski et al. 2011; Pan et al. 2013). Pesticides lead to reduced crop yields and the loss of important ecosystem services including pollination and nutrient cycling follow from the severe effects on bees, butterflies, soil fungus, and amphibians (Goulson et al. 2015; Relyea 2005; Das et al. 2024). Goulson et al. (2015) also reported that *Apis mellifera*-pollinated plants showed reduced pollen viability and impaired reproduction when exposed to imidacloprid. Infants, the elderly and immunocompromised people experience severe side effects due to pesticide exposure (Cimino et al. 2017). Silver nanoparticle was found to induce toxic responses in *Fejervarya limnocharis* and altered protein content in *Eruca sativa* (Singha et al. 2024; Vannini et al. 2013). Triclosan interferes with thyroid function in Danio rerio by interacting with thyroglobulin (Ding et al. 2025). Common harmful reactions to parabens and triclosan include asthma and hey-fever in children (Lee-Sarwar et al. 2018). PAHs exhibit toxicity to daphnia leading to increased mortality rates, while simultaneously impeding seed germination in ryegrass (Abdel-Shafy and Mansour 2016). Artificial sweeteners can have toxic effects on the gastrointestinal tract, the nervous system, gustatory changes, allergic reactions, insulin metabolism and the cardiovascular system (Ghusn et al. 2023). The implications of these findings highlight the extensive ecological hazards associated with these pollutants.

9.3 A Brief History of Biochar

The methods for making biochar and its broad range of applications are acknowledged worldwide. Still, its origin and the inspiration behind it are equally crucial since 'For the understanding of the future, the past must be looked into.' Biochar is a comparatively newer term, yet the matter is historical. It has a complex long history that is believed to have started in the Amazon Basin from 8000 BC as identified by various archaeologists. The soil of the Amazon Basin has a substance with high levels of carbon and higher nutrient content nitrogen- phosphorus-potassium (NPK), which probably gives it its characteristic dark color, this soil is typically known as *de Indio*/terra preta/anthropogenic dark earths (Groot et al. 2016). Terra preta soil is created by indigenous people of the places, the existence of such soil can go almost 2500 years back.Global natural calamities, such as grassland and forest fires, produce biochar, which occurs naturally; for instance, North America possesses some of the most fertile soils globally. (Hunt et al. 2010).

Low-density 'biochar' enriched soil or terra preta sites are also found in parts of Amazonian basins, Columbia, Venezuela, and parts of Peru along with some regions in Brazil (Bezerra et al. 2019). The initial mention of biochar's "discovery" occurred in the 1800s by a European geologist, who noted regions in South America characterized by unusually productive soils, in contrast to the generally shallow, acidic soils that support only brief periods of productivity (Glaser and Birk 2012). Radiocarbon dates for terra preta sites in the Upper Xingu region range from 60 to 1640 AD, while those in Central Amazonia range from 2500 to 500 BP (Groot et al. 2016). Currently, terra preta receives interest primarily because of its high level of carbon, unique soil fertility, ability to store carbon, its anthropogenic origins, and its historically dynamic relationship with biochar (Bezerra et al. 2019). Following the discovery of terra preta, the fertile anthropogenic soil in the Amazon, it was believed that humans contributed in the formation of the soil and landscape (news.mit.edu). During the early 2000s, terra preta gained acknowledgment for its role in soil management, climate change mitigation, and sustainable development (Allohverdi et al. 2021), but it was soon replaced by biochar because of its capacity to store carbon and soil fertility enhancement (Varkolu et al. 2025). In contrast to the terra preta Nova which was articulated as a way to reproduce or replicate the terra preta soil from the Amazon, biochar on the other hand was a new technology inspired by the soil and culture of the Amazon and pre-Colombian practices (Glaser and Birk 2012). It is important to recognize that terra preta and biochar differs from each other in terms of composition and carbon structure (Glaser and Birk 2012). Biochar can be artificially created by combining charcoal of low temperature with plant or animal biomass including plant or animal residues and compost (Kamarudin et al. 2022).

Since 2015, a significant rise in biochar research has been recorded, with a total of 12,162 articles published on the topic over the last two decades (Varkolu et al. 2025). The history of biochar reveals a remarkable journey from that of the Amazon basins to its rediscovery in the modern era for mitigating climate change. Despite its long tradition of raising soil carbon, biochar has just gained notification in the last two decades in modern agriculture. As research continues to uncover the various applications of biochar, it stands as a tribute to the persistent value of integrating conventional knowledge with contemporary science for the advantage of both people and the planet.

9.4 Biomaterials Used in Biochar Production

Biochar has attracted a significant interest in recent decades as a simple yet effective solution to the persistent climate problems. It offers a variety of advantages, in a wide range of domains apart from alleviating climate changes, like facilitating carbon sequestration, enhancing soil quality, and even purifying water (Yu et al. 2019). The breakdown and decomposition of organic matter can result in the release of greenhouse gases into the environment. However, by converting these organic substances to a stable carbon form by heat and incorporating it into the soil, carbon can be securely stored and the emissions will be thereby reduced (Fig. 4) (Liang et al. 2008). The manufacturing of biochar is done in such specific settings so that it has the maximum desirable properties for agricultural purposes, such as nominal resin residues and increased surface area (Lehmann and Rondon 2006). The specific thermal treatment of biomaterials with or without oxygen in biochar generation gives it a substantial surface area as well as prevents it from decay by microbes, which facilitates its preservation in the environment for extended duration (Lehmann 2007).

The distinctive features of biochar are primarily influenced by the nature of the biomaterial (feedstocks) as well as the specific manufacturing processes employed (Murtaza et al. 2025). Variations in the feedstock composition can yield biochar with distinct physiochemical properties even when produced under duplicate conditions. The generation of sustainable biochar for waste management and absorbing environmental contaminants can be done from plenty of readily available cost-effective resources in the environment (Yargicoglu et al. 2015). Such sustainable biochar manufacturing mainly involves the application of agricultural wastes and by-products, forestry wastes, municipal wastes, etc. (Kuncara et al. 2024). A huge range of carbon-rich materials—varying from wood, and coal to fruit shells, animal manure, food, and beverage processing wastes—all can serve as biomaterials to produce biochar (Fig. 4) (Suliman et al. 2016; Wang et al. 2018).

Recent techniques have developed our comprehension of thermochemical conversion methods—especially the slow pyrolysis of lignocellulosic biomass—for producing biochar with diverse industrial and engineering applications (Mukherjee et al. 2022). Aromatic polymers, including lignin, and carbohydrate polymers such as hemicellulose and cellulose constitute the lignocellulosic biomass utilized in the production of biochar (Yaashikaa et al. 2019; Li et al. 2020a, b; Yaashikaa et al. 2020). Feedstock components such cellulose, lignin, hemicellulose, lipids, and starches are being used in the pyrolysis process which influence the properties of biochar (Suliman et al. 2016). Notably, biochar from lignocellulosic biomass tends to be more suitable

Table 2 A comparative table on the properties of different lignocellulosic biomass

Characteristic	Cellulose	Hemicellulose	Lignin
Structure	Linear, unbranched polysaccharide of β-D-glucose units	Heterogeneous, branched polysaccharide of sugar monomers	Aromatic, branched, amorphous, heterogeneous, linked to cellulose and hemicellulose polymers
Abundance	Most abundant	Second most abundant polysaccharide in plants	Second most abundant organic polymer after cellulose
Main Function	Provides structural strength and rigidity to plant cell walls	Acts as a linkage between cellulose and lignin; adds flexibility	Provides rigidity, acts as a glue/binder, and prevents rotting
Location in Plant	Major component of primary cell wall	Surrounds cellulose fibers, links to lignin	Fills spaces between cellulose and hemicellulose, mainly in secondary cell wall
Monomer Units	Glucose (β-1,4 linkages)	Glucose, mannose, galactose, arabinose, xylose, etc.	Phenylpropanoid units (aromatic alcohols)

as a solid fuel, while that from non-lignocellulosic sources may be preferable for soil amendment (Nzediegwu et al. 2021).

Lignocellulosic biomass is widely available in plant materials like straw, wood, corn stalks, sugarcane bagasse and other agricultural residues (Table 2) (Mujtaba et al. 2023). These abundant, renewable resources are promising feedstocks for biofuels, bio-plastics, and other value-added items, and are capable of being processed through biochemical, thermochemical, or mechanical means (Patel et al. 2024). As biogenic materials with both organic and inorganic fractions, their complex structure—including lignocellulose, extractives, and ash—affects pyrolytic degradation pathways and biochar properties (Rangabhashiyam and Balasubramanian 2019). The usual biomass of feedstock [cellulose (32–45%), hemicellulose (19–25%), and lignin (14–26%)], carbon content, type of feedstock, the conversion process, etc., may determine the biochar properties (Anca-Couce 2016).

The characteristics of different lignocellulosic biomass have been sourced from Patel et al. (2024).

9.4.1 Agricultural Waste and Forestry Biomass as Biochar Feedstock

A variety of crop residues (energy crops, and starch energy crops), agricultural waste and forestry byproducts generate daily, necessitating appropriate disposal. Common disposal procedures, such as open combustion and composting, result in significant amount of pollutant emissions (Barati et al. 2022). However, utilizing these feedstocks for production of biochar leads to minimizing pollutant release and also has other different applications. A diverse array of resources, including rice straw, rice husk, coconut shell and fibre, wheat stalk, cotton stalk, and banana leaves, were examined for their biochar yield. The analysis revealed that crop residues, in comparison to woody biomass and organic waste, exhibit lower ash proportion and a higher percentage of nitrogen and carbon (Li et al. 2023). A higher ash content is deemed undesirable as it affects the adsorption-related property and limits the accessibility of adsorption sites (Liu 2019). The higher nitrogen and carbon content in crop residue biochar enhances its efficacy as fertilizer. Among various crop residues rice husk and rice straw tend to have the highest biochar yield when pyrolysis is done at 300 °C (Li et al. 2023). The analysis of the physiochemical properties of biochar developed from various feedstocks revealed that the highest carbon output was achieved from apple tree biomass, followed by oak tree and sawdust-derived biochar (Mishra and Mohanty 2023). The carbon content of sugarcane bagasse biochar is 83.91% when produced through pyrolysis at 450 °C (Kuncara et al. 2024). Corn is a ubiquitous crop that has huge production all over the world, but only a small percentage of corncob is used in industrial processes (Panwar and Pawar 2022). It was observed that the overall biochar percentage obtained from spruce wood was lower than corn cob and, which contradicts the typical trend of higher yields from woody biomass compared to herbaceous biomass (Wang et al. 2022). Forest residues are another crucial biomaterial for biochar production. One such forest waste is pine wastes including pine cone, pine wood and pine needles. The biochar production from pine waste products can go upto 60% when slow conventional pyrolysis is done (Sharma et al. 2023).

9.4.2 *Municipal Waste as Biochar Feedstock*

Municipal or urban solid waste has received much interest currently as a viable method for managing waste among the various feedstock options (Jayawardhana et al. 2018). Industrial and domestic garbage in developing nations is released into open landfills without prior treatment, becoming significant contributors of pollutants and gaseous emissions (Ladislao and Turrion-Gomez 2010). In economically advanced countries, trash from municipalities is regarded as a resource and can be utilised to produce fuel, heat, and energy, primarily through gasification and pyrolysis methods (Kumar et al. 2016). Physicochemical, biochemical and thermal conversion, are the three primary approaches for the transformation of urban solid wastes for sustainable applications (Gumisiriza et al. 2017). Under certain environmental conditions, biochar derived from municipal soil residues is effective in the removal of benzene from water. (Jayawardhana et al. 2018). In comparison to conventional activated sludge biochar, aerobic granular sludge biochar has a more compact framework, a larger surface area and higher carbon and microbial community contents, all of which facilitate to the improvement of contaminant adsorption (Ding et al. 2019).

9.4.3 Algae as Biochar Feedstock

Algal biomass has been recognized globally for biochar synthesis due to its elevated nutritional value, ion-exchange potential, and quick colonization of nutrient-rich habitats (Liu 2019). Biochar generated from microalgae is rich in nutrients, serves as a fertilizer for crops, exhibits adsorption potential for wastewater treatment, and facilitates long-term storage of carbon (Yu et al. 2017). Bird et al. (2011) utilized eight species of green tide algae for the production of biochar and observed that biochar formed from macroalgae possesses features ideal for application as an agricultural ameliorant and as a means for long-term retention of carbon (Bird et al. 2011).

The results for biomaterial application as feedstock for biochar production emphasize role that feedstock selection play in determining the physical and chemical characteristics of biochar. The lignocellulosic biomass i.e. lignin, hemicellulose and cellulose are the most popular and effective precursors for biochar due to their characteristics including being renewable, high carbon amount and, abundance in the environment (Table 2). Biochar manufactured from various feedstocks ranging from agricultural wastes to industrial wastes directly influences its yield and properties that of adsorption capabilities, porosity and, also nutrient content thus affecting its applications in soil amendment, carbon storage and remediation of contaminants.

9.5 Techniques of Biochar Production

The growing interest in biochar and its wide range of applications has prompted towards through increased biomass conversion into biochar. The process of thermochemical conversion has been acknowledge as one of the most widely utilized methods for production of biochar. This method encompasses several processes, including pyrolysis (fast and slow), torrefaction, hydrothermal carbonization and gasification (Lin et al. 2016; Pang 2019). To achieve maximum yield of biochar, it is essential to choose a production technique suited to the specific type of biomass and to optimize technique parameters such as temperature, heating rate and time of residence. These are pivotal parameters, as they significantly influence the physical and chemical features of the resulting biochar (Yaashikaa et al. 2020). Morphologically, biochar produced from biomass of plant vary depending on the procedure conditions, primarily due to biomass mass loss during thermal treatment. Initially, weight reduction occurs due to moisture evaporation around 100 °C. This is followed by thermal degradation of components like hemicellulose, cellulose and lignin at temperatures above 220 °C. Ultimately, additional mass loss occurs due to the combustion or volatilization of carbonaceous residues at higher temperatures. These transformations are essential in determining the structural and compositional features of the final biochar product (Yaashikaa et al. 2020).

9.5.1 Pyrolysis

Pyrolysis is a widely adopted thermochemical technique for production of biochar. It entails the breakdown of organic matter thermally in oxygen-deficient setting, usually at temperatures ranging from 300 to 700 °C. The process yields three primary products: syngas, bio-oil and primarily biochar (Rangabhashiyam and Balasubramanian 2019; Senthil and Lee 2021). The distribution of these products is influenced by different factors, such as the rate of heating, pyrolysis temperature, time of residence, physical and chemical characteristics of the feedstock. Lower operating temperatures paired with extended residence periods generally enhance biochar yield, as extended processing allows greater polymerization, promoting the formation of solid carbon residues (Fig. 2). On the contrary, the generation of liquid bio-oil is typically favoured by moderate temperatures and shorter residence durations (Fig. 2) (Yaashikaa et al. 2020). Thermochemical pyrolysis processes can vary significantly in their operational parameters, including residence periods that can fluctuate from < 1 s to several hours, heating rates spanning from under 11 °C/s to over 1000 °C/s, and reaction temperatures exceeding 700 °C (Santoso et al. 2020).

Given that different pyrolysis conditions yield different product distributions, selecting the appropriate pyrolysis method is essential to achieving the desired outcome. The pyrolysis of biomass can be described through three primary reaction pathways: char formation, depolymerization and then fragmentation (Daful and Chandraratne 2020; Rangabhashiyam and Balasubramanian 2019; Yaashikaa et al. 2020). Formation of char formation is typically driven by both intramolecular and intermolecular rearrangements, leading to a solid material with increased thermal stability. This pathway involves the creation of benzene rings, which subsequently merge to form stable polycyclic aromatic structures (Karimi et al. 2018). These structural transformations are usually accompanied by the emission of gases or liquids. Depolymerization refers to the breakdown of polymer chains into smaller

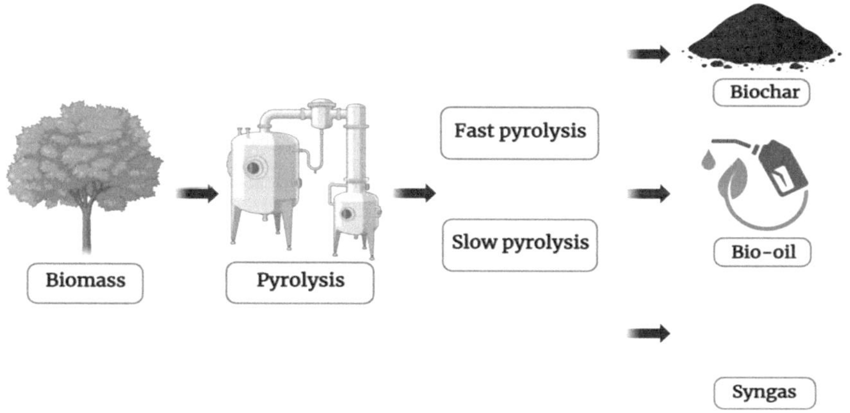

Fig. 2 Schematic overview of pyrolysis process in biochar production

units—monomers, dimers, and trimers—followed by stabilization. These volatile compounds condense at room temperature, forming part of the liquid product. In contrast, fragmentation involves the cleavage of both polymer and monomer bonds, resulting in the production of gases and a range of condensable organic liquids that are also recoverable at ambient temperatures (Lam et al. 2019). A variety of reactors are used for biochar manufacturing such as paddle kilns, bubbling fluidized beds, agitated sand rotating kilns and wagon reactors. The amount of biochar produced is largely dependent on the characteristics and type of the biomass used. Among the operational parameters, temperature is a critical factor in regulating the efficacy of the process. Generally, a rise in pyrolysis temperature results in a lower biochar yield while boosting syngas production (Yaashikaa et al. 2020).

Depending on the temperature, residence duration, pressure, and heating rate, pyrolysis is generally classified as either a fast or slow type. Because fast pyrolysis has shorter residence times and faster heating rates, it yields a larger percentage of bio-oils (Fig. 2). Slow pyrolysis, however, yields a larger percentage of biochar because of the longer residency times and slower heating rates (Fig. 2). The crucial difference between rapid and slow pyrolysis is shown by the primary and secondary processes that take place during the heat decomposition of biomass (Yaashikaa et al. 2020; Amalina et al. 2022).

9.5.2 Fast Pyrolysis

Fast pyrolysis is a type of thermochemical conversion process defined by extremely high temperatures, involving rapid heating rates between 10 and 1000 °C per second, along with short vapour residence times typically between 0.5 and 2 s (Tomczyk et al. 2020). A pivotal aspect of fast type of pyrolysis technology is minimizing the residence time of vapours in the section of hot zone to preserve good quality of bio-oil and is typically achieved by rapidly quenching or cooling the pyrolysis vapours (Wang et al. 2014a, b; Mutsengerere et al. 2019). The yields and composition of products—bio-oil, biochar, and syngas—depend on the feedstock's characteristics, temperature, and the rate of heating. To optimize bio-oil yield, fast pyrolysis is usually conducted at temperatures ranging from 425 to 600 °C, the higher limit not exceeding 650 °C. If gas production is preferred, operating temperatures can rise to around 1000 °C (Ge et al. 2021). Achieving such rapid heating requires the use of finely ground biomass, generally smaller than 1 mm in particle size, to enhance heat transfer and reduce resistance to mass and energy flow (Ravindran et al. 2018). Because of the limited opportunity for secondary reactions, fast pyrolysis tends to produce low amounts of biochar, mainly composed of primary carbon. The process is also largely endothermic due to the rapid nature and absence of exothermic secondary reactions. For optimal performance, the feedstock should contain less than 10% moisture by weight. Low moisture not only minimizes water content in the resulting bio-oil but also facilitates grinding the biomass to the fine particle sizes necessary for effective thermal decomposition (Lee et al. 2020).

9.5.3 *Slow Pyrolysis*

Slow pyrolysis is a thermochemical conversion process characterized by a gradual heating rate, and is recognised as the highly effective techniques for greater yield of biochar compared to different pyrolysis and carbonization approaches. The biochar produced is widely acknowledged for its potential application in soil amendment, enhancing soil quality and agricultural productivity (Arni 2018). The process typically operates at moderate temperatures between 300 and 550 °C, the rates of heating ranging between 0.1 and 0.8 °C/s. The contact time in the reactor can vary significantly, from as short as 5–30 min to as long as 25–35 h (El-Naggar et al. 2019), depending on the operational conditions. The prolonged vapour residence time promotes secondary reactions, enhancing the overall biochar yield. These reactions facilitate to the development of two types char structures (primary and secondary), increasing the carbon retention in the final product (Das et al. 2021). The efficiency of biochar production in slow pyrolysis is affected by various factors, which include the physicochemical attributes of the feedstock, reactor configuration, and process parameters such as temperature and heating rate. In particular, feedstocks with high mineral content tend to produce lower amounts of biochar due to the catalytic influence of inorganic matter on volatile evolution and decomposition pathways (Sonu et al. 2020). Moreover, the process may become exothermic due to the sequential development of secondary reactions, which further influences energy balance and system design. Slow pyrolysis is capable of processing feedstock particles within a size range of 5–50 mm, offering flexibility in raw material preparation (Sakhiya et al. 2020). This adaptability, combined with its high biochar yield, makes slow pyrolysis a favourable option for sustainable biomass valorization and carbon sequestration strategies.

9.5.4 *Torrefaction*

Torrefaction is a thermal pre-treatment process classified as a mild form of pyrolysis, performed at moderate temperatures which generally ranges from 200 to 300 °C, heating rate typically below 11 °C/s, under atmospheric pressure and in an inert or oxygen-deficient condition (Manyà et al. 2020). This process facilitates the removal of moisture and light volatiles, while partially decomposing biopolymers like cellulose, hemicellulose, and lignin through the generation of organic volatile chemicals. Unlike high-temperature pyrolysis, torrefaction predominantly yields solid outputs rather than liquid or gaseous ones. The resulting solid, referred to as torrefied biomass, generally exhibits a higher molar oxygen to carbon ratio than the threshold value of 0.4 defined by the European Commission in terms of biochar (Daful and Chandraratne 2020). Consequently, torrefied biomass does not meet the criteria to be classified as biochar. Torrefied biomass demonstrates physicochemical properties that are intermediate between raw biomass and char. It is primarily used as a pre-treatment step to reduce water content of biomass and increase bulk energy density, which in turn improves the efficiency of transportation and enhances fuel quality (Benis et al. 2020; Yadav et al. 2022). Furthermore, torrefaction improves grindability, hydrophobicity and resistance to microbial degradation of biomass, thereby extending its storage life. The process typically results in yields ranging from 70 to 80% for mass and from 80 to 90% for energy (Enaime et al. 2020).

9.5.5 *Hydrothermal Carbonization*

Most of the biomass feedstocks contain high water content, reaching often up to 95 weight percent. Biomass that has water content higher than 30% needs to be dried before pyrolysis (Ha and Lee 2020). The process of hydrothermal carbonization offers an energy- efficient alternative for processing wet biomass without the need for extensive drying. This technique utilizes heat and pressure in presence of water for the conversion of biomass into carbon-rich solid fuel, known as hydrochar (Zhang et al. 2018). In this process, water functions as a reactive as well as a solvent medium. When biomass with intrinsic moisture is heated at temperatures below 200 °C in a sealed reactor under self-generated (autogenous) pressure, hydrochar is produced as the primary solid product (Mbarki et al. 2019). Other hydrothermal processes include hydrothermal liquefaction and hydrothermal gasification. Hydrothermal liquefaction operates at moderate temperatures (200–350 °C) and converts biomass predominantly into liquid bio-oil. In contrast, hydrothermal gasification is carried out close to the critical point of water (374 °C, 22.1 MPa), facilitating transformation of biomass into gaseous products. Hydrochar generally possesses a higher hydrogen to carbon molar ratio as compared to biochar, reflecting its distinct physicochemical properties and potential applications in bioenergy and soil amendment (You et al. 2017; Ponnusamy et al. 2020; Liu et al. 2021).

9.5.6 *Gasification*

Gasification is a process of thermochemical conversion that transforms carbon-rich materials into gaseous products through the application of heat in presence of a gasifying agent like air, steam or oxygen. This process typically occurs at temperatures exceeding 750 °C and can be conducted under atmospheric or elevated pressures. Under oxygen-deficient conditions, gasification produces a mixture of gases—commonly referred to as syngas—along with limited amounts of biochar (Nidheesh et al. 2021). When oxidizing agent is the air, the resulting syngas typically comprises approximately 85% of the gas output. Steam gasification, by contrast, tends to yield a higher concentration of hydrogen and results in syngas with greater value of heating (Nidheesh et al. 2021). The process of gasification generally involves four sequential stages: drying, pyrolysis followed by partial oxidation and finally reduction

(Umenweke et al. 2022). Based on the gas–solid contact mode, gasification systems are commonly classified into entrained flow reactors, fixed bed and fluidized bed. Although syngas is the primary product of gasification, it generates relatively low biochar yields (Yaashikaa et al. 2020), making it less suitable for dedicated production of biochar. Similarly, combustion is not considered a viable method for biochar synthesis, as optimal combustion conditions aim for complete oxidation, leaving negligible solid residue. One of the major drawbacks of gasification (and combustion) in this context is the minimal yield of biochar and generation of greenhouse gas (GHG) emissions (Nidheesh et al. 2021). Moreover, gasification parameters—such as temperature, pressure, and choice of gasifying agent must be tailored to the characteristics specific to the feedstock to optimize energy recovery (Zuber et al. 2020). While gasification holds promise for energy production, pyrolysis remains the most preferred, sustainable and efficient technique for the targeted biochar production.

9.6 Biochar Characterization Techniques

Characterizing biochar utilizes a multi-technique approach to assess its structural, chemical and physical attributes. From simple proximal examination to sophisticated NMR and XPS, every technique offers special information on biochar‘s environmental safety, stability and usefulness (Fig. 3) (Yaashikaa et al. 2020). Correct description guarantees most effective biochar selection for carbon sequestration, pollution control and agriculture thereby promoting sustainable uses (Chaudhry et al. 2025). Understanding the possible uses of biochar depends on knowing its physical, chemical and structural features. These features are analyzed by several methods in biochar characterizing.

9.6.1 Proximate and Ultimate Examination

Proximate and ultimate analyses are fundamental methods for assessing the composition and thermal characteristics of biochar, providing essential insights into its stability and suitability for various applications (Wood et al. 2024). Proximate analysis determines humidity, fixed carbon, volatile matter, and ash content, offering an understanding of thermal stability and organic matter breakdown. High volatile matter indicates incomplete carbonization, whereas high fixed carbon suggests greater recalcitrance (Racero-Galaraga et al. 2024). Ultimate analysis, conducted using elemental analyzers (e.g., CHNS/O analyzers), quantifies the elemental composition enabling evaluation of biochar‘s carbon sequestration capacity and purity. Derived molar ratios reflect biochar‘s aromaticity and degree of carbonization, with lower ratios signifies increased stability (Rúa-Díaz et al. 2023). Additionally, thermogravimetric analysis complements these techniques by tracking weight loss under

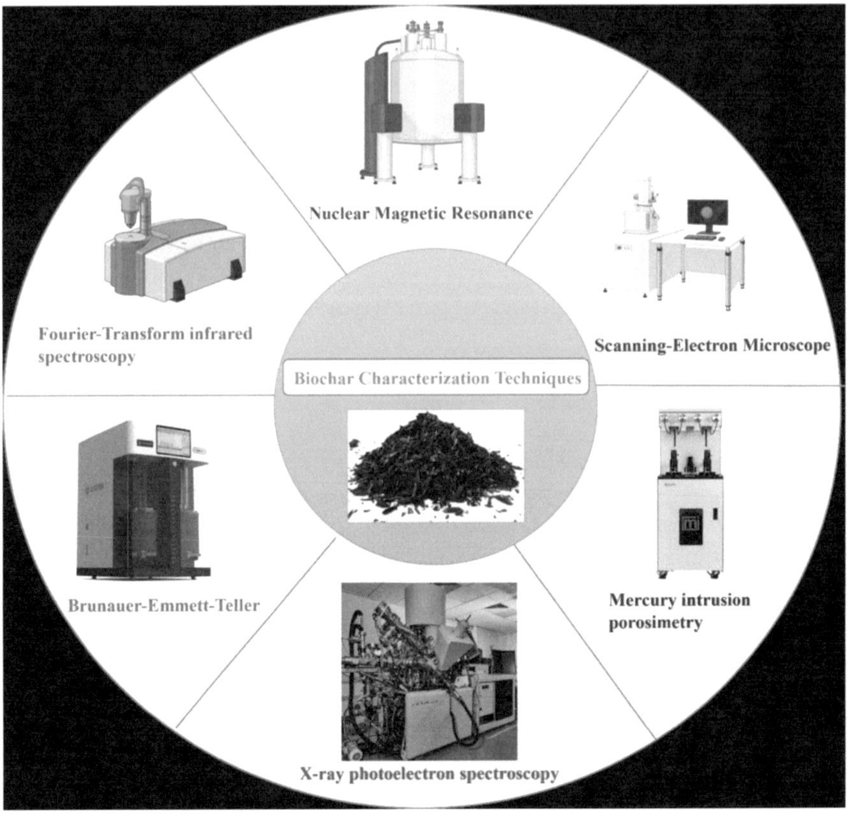

Fig. 3 Different techniques of biochar characterization

regulated temperatures, giving more information about how thermal breakdown works (Manyà 2012).

9.6.2 *Energy-Dispersive X-ray Spectroscopy (EDS) and Scanning Electron Microscopy (SEM)*

SEM is an essential instrument for analyzing the surface morphology of biochar at high resolution, elucidating pore structure, particle size distribution and texture—attributes that directly influence water retention, microbial habitat, and adsorption capability (Khan et al. 2024). When integrated with EDS, SEM enhances its functionality by facilitating elemental mapping which delineates the distribution of vital nutrients (e.g., potassium, calcium, and phosphorus) crucial for soil fertility, while concurrently identifying toxic heavy metals, thus safeguarding environmental integrity (Gabetto et al. 2025; Agrafioti et al. 2014). The integration of SEM-EDS facilitates the examination of mineral and inorganic element distribution in biochar, rendering it an essential technology for investigating biochar's roles in nutrient cycling and heavy metal immobilization (Chia et al. 2012). SEM-EDS offers comprehensive structural and compositional analysis facilitating the evaluation of biochar's agricultural and environmental uses. Hence, validating its efficacy in improving soil characteristics and reducing contamination hazards.

9.6.3 *Fourier-Transform Infrared Spectroscopy (FTIR) and X-ray Diffraction (XRD)*

FTIR is commonly employed to identify functional groups on biochar surfaces by examining the vibrational modes of chemical bonds, thereby detecting hydroxyl (-OH), carboxyl (-COOH), phenolic and aromatic (C=C) structures which characterize biochar's reactivity (Yaashikaa et al. 2020). Cation exchange capacity and interactions with organic contaminants are crucial elements in nutrient retention and pollutant adsorption (Yao et al. 2012). X-ray diffraction (XRD) identifies crystalline mineral components in biochar including silicates, phosphates and carbonates hence augmenting its liming capacity and phosphorus availability. XRD is especially useful for examining mineralogical changes during pyrolysis such as the transition of calcium carbonate from calcium oxalate (Ali et al. 2022).

9.6.4 *Analysis of Porosity and Surface Area (BET and MIP)*

MIP and BET analysis are complimentary methodologies employed to assess the surface area and porosity of biochar. BET study utilizes gas adsorption—commonly nitrogen (N_2) or carbon dioxide (CO_2)—to quantify specific surface area (SSA) and pore volume, essential for evaluating biochar's ability to adsorb gases, nutrients, and contaminants (Yaashikaa et al. 2024). A high SSA above 300 m^2/g is particularly beneficial for applications such as water filtration and gas capture, since it improves pollutant adsorption effectiveness (Pugazhendhi 2024). Conversely, MIP quantifies bigger macropores (> 50 nm), which affect water absorption and root penetration in soil applications. Collectively, these techniques enhance pyrolysis parameters to customize biochar for certain applications, including environmental remediation, where elevated porosity is crucial (Haider et al. 2022).

9.6.5 X-ray Photoelectron Spectroscopy (XPS) with Nuclear Magnetic Resonance (NMR)

Modern analytical techniques such as NMR spectroscopy and XPS provide detailed molecular-level insights into biochar's carbon structure and surface chemistry (Hu and Wei 2023). Solid-state ^{13}C NMR spectroscopy distinguishes between aromatic, aliphatic, and carboxylic carbon fractions, revealing the degree of carbonization and biochar's resistance to microbial degradation, with a high aromaticity index (low H/C ratio) indicating greater durability (Wang et al. 2013). Similarly, XPS analyses surface chemical states by measuring electron binding energies, detecting oxygen-containing functional groups (e.g., carbonyl, quinone) and metal oxidation states which influence redox processes and heavy metal binding (Li et al. 2020a, b).

9.6.6 Tests of Thermal and Chemical Stability

Tests of thermal and chemical stability evaluate biochar's resistance to breakdown in environmental settings, with accelerated aging studies—such as oxidation using hydrogen peroxide or thermal treatment in controlled atmospheres (Tomczyk et al. 2020) simulating long-term behaviour in soil. These studies, along with leaching assays, assess the release of nutrients (e.g., K^+, PO_4^{3-}) and potential pollutants, ensuring biochar's safety and effectiveness in environmental and agricultural applications (Zhang et al. 2024a, b). Combined with microbial degradation tests, these characterization methods yield a detailed insight of biochar‘s properties, predicting its longevity in soil and its impact on ecosystem health, thereby guiding its optimization for sustainable uses (Khan et al. 2024).

9.7 Sustainable Application of Biochar in Environmental Remediation

Biochar is regarded as a great resource for a range of sustainable applications, including environmental remediation (soil, air, water), wastewater treatment, enhancing soil fertility, and biodiesel production etc. (Varkolu et al. 2025) (Fig. 4). This material is utilized in energy storage and as a catalyst, due to its distinct properties, cost-effectiveness, and broad surface area (Li et al. 2024). Biochar significantly influences soil physicochemical and microbiological characteristics, enhancing resilience and crop productivity. Biochar enhances soil properties and mitigates the emission of hazardous gasses into the environment. The microporous geometry of biochar accelerates the adsorption process by reacting with different functional groups of pollutants (Yaashikaa et al. 2024).

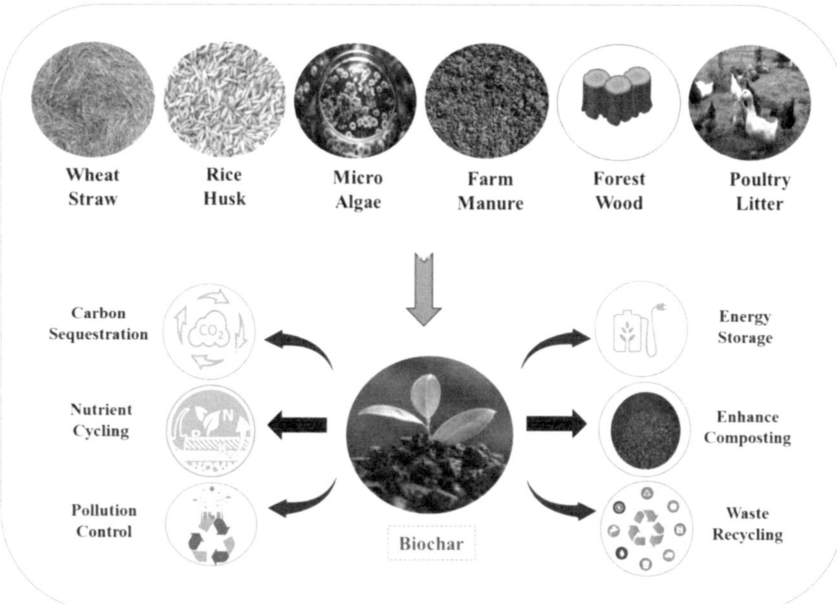

Fig. 4 Different Sources and Applications of Biochar

9.7.1 Biochar in EEC Remediation from Environment

Biochar has been found to remediate Various forms of EEC, including triclosan, PFAS, microplastic/nanoplastic, PAHs, pesticides, herbicides, phthalates, and BPA etc. (Table 3). A worldwide increase in triclosan, an ingredient in toothpaste and soap, has been observed following the COVID-19 epidemic (Dai et al. 2022). Modified biochar has been documented to remediate triclosan from groundwater through several types of interactions with the pollutant (Tong et al. 2024; Czech et al. 2021). The significant industrial use and persistent nature of per- and polyfluoroalkyl compounds have resulted in their proliferation in the environment, potentially causing serious health complications, including hepatic and renal cancer (Priyadarshini and Neelancherry 2024; Dasgupta et al. 2020). Biochar has demonstrated a remarkable capacity to adsorb per- and polyfluoroalkyl compounds, achieving levels of up to 634 mg/g from aqueous solutions (Zhang et al. 2023; Deng et al. 2023). The widespread plastic products utilization, since 1950, contributes significantly to the extensive distribution of microplastics (Li et al. 2024). The existence of this contaminant may influence the physicochemical and microbiological characteristics of both aquatic and terrestrial environments (Li et al. 2024). The current research established that biochar not only adsorbs microplastic/nanoplastic from soil and water, but also enhances the soil's microbial population (Wang et al. 2021; Ganie et al. 2021; Palansooriya et al. 2024). The excessive utilization of fossil fuels has resulted in a rising trend of PAHs in the environment, potentially effecting the physiology of various animals, plants and microbes (Paul et al. 2024; Paul et al. 2025; Patel et al. 2020). Studies demonstrate that biochar effectively removes PAHs from soil and water through the initiation of various chemical reactions (Ni et al. 2017; Rao et al. 2017). The increasing need for food has resulted in a surge in agricultural practices, along with an elevated dependence for the pesticides and herbicides applications. Pesticides and herbicides in terrestrial and aquatic systems have been documented to potentially harm the organisms of these ecosystems (Das et al. 2024; Alengebawy et al. 2021). Biochar has been reported to adsorb these chemicals from such contaminated areas (Rao et al. 2017; Ma et al. 2024). Biochar has the capacity to remediate and restore soil which was previously contaminated with organic pollutants. Esters of phthalic acid represent a class of organic compounds that are released from plastic products into the environment (Cao et al. 2024). Their lipophilic property results in bioaccumulation, subsequently functioning as endocrine-disrupting chemicals, reproductive toxicants, and carcinogens (Xu et al. 2019). Li et al. (2025) found that biochar has significantly adsorbed dimethyl phthalate and diethyl phthalate that are co-contaminated with cadmium. The adsorption of phthalates was diminished because of the coexistence pollutants, while the adsorption of cadmium was not significantly affected (Li et al. 2025). Dimethyl phthalate can be adsorbed and desorbed from wheat straw biochar which indicates that biochar influences the transportation and natural attenuation of xenobiotics, in addition to its enduring effects on soil ecosystems (Yan and Quan 2020). Reports of the presence of BPA, an endocrine-disrupting chemical, in soil have been noticed (Stewart et al. 2014). BPA acts as a basic component in everyday products for consumers, including sports equipment, food packaging, plastic bottles and dental sealants (Rochester 2013). Liu et al. (2020) reported that BPA was effectively decomposed after treatment with biochar activated by persulfate in both soil and aqueous solutions. A rise in living standards has led to a notable surge in the consumption of artificial sweeteners, which are linked to various detrimental health outcomes, including obesity, heart disease, and increased mortality rates (Subedi et al. 2014). Combination of laccase and biochar on synthetic sweeteners such as sucralose and acesulfame resulted in roughly 70% degradation of these compounds (Jiang et al. 2024a, b). When applied in environmental pollution mitigation, biochar is the solution for more sustainable utilization of biomass, thereby reigning in a new era in the field of organic waste management. Though nanoparticles of metals have been identified as an EEC with harmful effects on organisms (Singha et al. 2024), the present literature fails to show any remediation of these components using biochar.

9.7.2 Effects of Biochar Physical, Chemical and Microbial Parameters of Soil

Biochar promotes soil structure, assists in carbon storage and hence improves soil stability (Varkolu et al. 2025). Biochar pores serve as microbial habitats, facilitating their growth, metabolism, and the degradation of pollutants. Biochar also enhances

Table 3 Types of biochar and their remediation mechanisms of emerging environmental contaminants

S. No.	Type of EEC	Biochar type	Remediation mechanism	References
1	Triclosan from groundwater	Biochar produced from saccharification residues and sewage sludge	Pi-pi interactions, hydrogen bonding, oxidation reaction	Tong et al. (2024), Czech et al. (2021)
2	PFAS from water	Biochar derived from sludge; biochar loaded with sulfobetaine polymer	Electrostatic interaction, hydrophobic interaction, hydrogen bonding, pore diffusion and complexation	Zhang et al. (2023), Deng et al. (2023)
3	Microplastic/nanoplastic in soil and water	Pinewood sawdust; sugarcane bagasse; foodwaste biochar	Electrostatic interaction, chemical bonding and pore-retention	Wang et al. (2021), Ganie et al. (2021), Palansooriya et al. (2024)
4	Phenanthrene in soil and water; PAHs in soil	Poplar and conifer; corn straw, bamboo	Partitioning, adsorption, aromatic-pi-pi interaction, and electrostatic interactions	Rao et al. (2017), Ni et al. (2017)
5	Pentachlorophenol pesticide; 2,4-dichlorophenoxyacetic acid	Poplar and conifer (BC) biochars; Ginkgo leaves	Partitioning, adsorption, aromatic-pi-pi interaction, and electrostatic interactions	Rao et al. (2017), Ma et al. (2024)
6	Phthalates plasticizers co contaminated with cadmium; dimethyl phthalate from soil	Corn straw; wheat straw	Co-sorption, chemisorption, physisorption, pi- pi interaction, pore-filling and hydrophobic interactions	Li et al. (2025), Yan and Quan (2020)
7	BPA both soil and aqueous solution	Lychee branch biochar activated with persulfate	Adsorption and oxidation	Liu et al. (2020)
8	Artificial Sweeteners (sucralose, acesulfame)	Biochar (purchased) combined with laccase	Degradation	Jiang et al. (2024a, b)

soil microbial populations by supplying organic matter, hence improving the soil ecology and facilitating better nutrient cycling (Jing et al. 2020). Addition of biochar in coastal saline-alkali soil improved soil physicochemical properties by modifying the soil bacterial community and improving phosphorous and nitrogen cycling, hence facilitating the growth of a bioenergy crop, *Miscanthus lutarioriparius* (He et al. 2020). Utilization of biochar not only augmented the removal of PAHs from soil but additionally enhanced bacterial population, diversity, and metabolic activity, hence improving soil characteristics (Amoah-Antwi et al. 2020). These findings assist the implementation of in situ remediation strategies for PAHs contaminated lands (Li et al. 2020a, b). Application of biochar alleviated heat stress (33 °C) in rice plants and decreased bulk density of soil by augmenting organic matter and improving the ratio of Proteobacteria to Acidobacteria (Huang et al. 2021). An experiment in southwest China fields assessed microbial populations in biochar-amended karst soil, confirming that biochar enhances microbial ecology and soil nutrients (Yan et al. 2021). Soil enriched in biochar has a healthier microbial population that facilitates the breakdown of organochlorines and hydrocarbons (high molecular weight) (Gregory et al. 2015; Saeed et al. 2021). These findings suggest that biochar possesses the ability to enhance microbial populations, hence facilitating pollutant breakdown in soil and modifying soil properties.

In a study on soil supplemented with biochar for six years shown a notable enhancement in pore structure and rice yield in the karst soil of southwest China (Zhang et al. 2021). These findings suggest that pyrolyzed biomass is a great option for the restoration of karst soil. Replacing a small quantity of inorganic fertilizers with biochar resulted in enhanced quality of and increased rice yield (An et al. 2022), additionally improvements in organic elements and enzyme functioning in paddy soil (Jing et al. 2020). Biochar plays a significant role in improving the long-term stability of soil, depending on the feedstock and pyrolysis temperature (Singh et al. 2012). Biochar has been documented more resilient in comparison to other soil additives, and it significantly improves soil health when employed for a longer period (Varkolu et al. 2025). Use of biochar facilitates carbon retention in soil and enhances the quantity of naturally occurring binding elements in brown earth, hence stabilizing soil aggregates (Wu et al. 2016; Sun et al. 2021). Because of its slower degradation, biochar absorbs carbon from the atmosphere and serves a significant role in diminishing greenhouse gasses like methane and nitrous oxide, hence aiding in climate change mitigation (Azad et al. 2023). A wide range of benefits of biochar includes enhancing soil structure and promoting microbial diversity, rendering it a viable solution for emerging environmental challenges. Biochar alleviates climate change by sequestering CO2 from atmosphere. Soil enriched with biochar can hold greater amounts of water owing to the porous characteristics of biochar, which is advantageous in arid or drought-prone regions. Biochar enriches the soil environment by facilitating nutrient cycling and fostering microbial growth. Biochar can neutralize both acidic and alkaline soils, hence maintaining an optimal soil pH favourable to plant development and agricultural productivity.

9.7.3 Mechanisms of Remediation by Biochar

Porous structure, surface organic molecules and active functional groups of biochar significantly contribute to contaminant removal. Biochar interacts with pollutants by several mechanisms, including chemical and physical adsorption, electrostatic interactions, hydrophobic interactions, hydrogen bonding, oxidation reactions, complexation, pore diffusion, and pi - pi interactions (Table 3). In physisorption, a reversible weak van der Waals bond is established between biochar and the contaminants, whereas in the chemisorption process, an irreversible ionic or covalent interaction occurs between the contaminants and biochar (Yan and Quan 2020). Biochar that possesses a large surface area enhances adsorption capabilities, while biochar abundant in organic matter on its surface tends to promote chemisorption (Murtaza et al. 2023). Interactions of both physisorption and chemisorption are observable between dimethyl phthalate and biochar (Yan and Quan 2020). Electrostatic interactions represent a well-recognized mechanism for the immobilization of contaminants, in which the negatively charged surface of biochar interacts with positively charged contaminants and soil particles (Murtaza et al. 2023). PFAS, microplastics, PAHs, pesticides, and herbicides exhibit electrostatic interactions with biochar (Table 3). Hydrogen bonding represents a form of electrostatic interaction occurring between hydrogen atoms and electronegative atoms. In this context, organic pollutants associate with oxygen-containing functional groups on biochar via hydrogen bonding (Guo et al. 2020). Hydrogen bonding has been observed between biochar and per- and polyfluoroalkyl substances, as well as triclosan (Table 3). The existence of a non- polar group within a molecule makes it water-repellent or hydrophobic in nature. Rice husk biochar demonstrates a notable decrease of crude oil through the process of hydrophobic interaction (Kandanelli et al. 2018). Biochar, when activated with various organic chemicals, facilitates degradation of pollutants via diverse oxidation processes. The addition of persulfate and $H2O2$ with biochar can initiate the oxidation of adsorbed bisphenol A and sulfamethazine, respectively (Liu et al. 2020; Deng et al. 2020). Complexation refers to ligand exchange accompanied by electron transfer, resulting in the generation of a bridge with ligands that facilitates the adsorption of contaminants such as PFAS (Verma and Devi 2025) (Table 3). Pore diffusion is a physical adsorption mechanism in which micropores and mesopores of the biochar surface encapsulate organic contaminants (PFAS) within these pores (Ambaye et al. 2021). Pore diffusion, also referred to as pore filling, is conditional upon the dispersion potential within molecular dimensional spaces (Verma and Devi 2025). The pi-pi electron donor-acceptor interaction between biochar and contaminants occurs by raising the negative charge of the pi orbital in the aromatic structure of biochar, thereby increasing the adsorption of ionic species through ion exchange (Islam et al. 2021). The observation of pi-pi electron transfer is evident in an array of emerging contaminants and biochar (Table 3). A detailed comprehension of the adsorption mechanism will provide valuable insights for optimizing remediation processes for a range of pollutants in the future.

9.7.4 Other Sustainable Applications of Biochar

In addition to its roles in waste management and bioremediation, biochar serves multiple other purposes, including biodiesel production, energy storage, and functioning as a biocatalyst. The environmental challenges related to fossil fuel usage have triggered the production of carbon-neutral biodiesel. Various catalysts are employed in the generation of biodiesel through transesterification, categorized as either homogeneous or heterogeneous (Kedia et al. 2023). Homogeneous catalysts present certain challenges, including their lack of reusability and the significant effort needed for their purification from reaction products (Wang et al. 2017). In recent years, biochar has gained particular attention as an economical catalyst for the production of biodiesel. Significant considerations in biochar for its role as a catalyst in biodiesel production include a substantial surface area and specifically designed surface functional groups (Lee et al. 2017). Biochar derived from microalgae biomass, when supported with nickel, serves as a catalyst for the hydrodeoxygenation of bio-oil to facilitate diesel production (Nguyen et al. 2016). Biochar-supported metal has also been utilized in various reactions including hydrogenation, reduction, methanation, isomerization, and oxidation. The isomerization of glucose to fructose was facilitated by catalysts biochar-supported MgO (Tsang et al. 2020). The methanation of bio-syngas was observed via Ru/activated biochar catalyst (Wang et al. 2014a, b). An array of chemical hydrocarbons were generated from lignin through hydrogenation reaction facilitated by a biochar-supported Ni-Mo2 catalyst (Wang et al. 2016). The oxidation of methane to methanol has been documented through the use of various lignin- derived biochar supported metal catalysts (Wang et al. 2024a, b). 4-nitrophenol was reduced to 4-amino phenol through the application of Biochar-supported metal oxides (Ramirez et al. 2023). Due to its elevated porosity and surface area, biochar may effectively absorb and retain heat (Liu et al. 2022). Bio-composite materials can be synthesized with biochar, including phase- change substances for solar thermal power storage applications (Das et al. 2020). The inclusion of biochar has been discovered to improve the capacity of lithium-sulphur batteries (Gu et al. 2015). Cladophora glomerata and metal-free tannery waste were utilized in the development of lithium-ion batteries, which demonstrated longevity and cost-effective energy storage (Salimi et al. 2017). The various beneficial applications of biochar exhibit its significant influence on future environmental sustainability.

9.8 Nano-biochar

Nanobiochar is an innovative material situated at the convergence of nanotechnology, environmental science, and sustainable agriculture. It is synthesized from biomass via pyrolysis, yielding biochar—a porous, carbon-rich material traditionally applied to enhance soil fertility and sequester carbon. When engineered at the nanoscale (i.e., particles < 100 nm), biochar exhibits enhanced physicochemical properties such as increased surface area, greater reactivity, and improved sorption capabilities (Bhandari et al. 2023; Xiao et al. 2023). These attributes render nanobiochar a promising alternative in various applications, including environmental remediation, precision agriculture, and potentially biomedical fields (Bhandari et al. 2023). In agricultural systems, nanobiochar contributes to improved nutrient retention and delivery, thereby fostering enhanced plant growth and reducing dependence on synthetic fertilizers. Its high capacity for cation exchange facilitates the adsorption and gradual release of essential nutrients, thus mitigating nutrient leaching and runoff. Additionally, nanobiochar is capable of adsorbing toxic substances such as pesticides and heavy metals, thereby contributing to the detoxification of contaminated soils and water bodies (Chausali et al. 2021). The material also demonstrates significant potential in water purification. Due to its nanoscale structure and surface functionality, nanobiochar effectively removes a range of pollutants— including heavy metals, organic compounds, and pathogenic microorganisms—from aqueous environments. As a result, it presents a sustainable, cost-effective alternative to conventional nanomaterials in water treatment technologies (Chausali et al. 2021; Bhandari et al. 2023). Despite its numerous advantages, the use of nanobiochar warrants careful evaluation regarding its ecological safety. Concerns remain about its long-term interactions with aquatic organisms and soil microbial populations. Comprehensive studies are essential to elucidate its environmental fate, transport mechanisms, and potential toxicological effects (Chausali et al. 2021). In summary, nanobiochar represents a transformative development in sustainable materials science. By integrating the natural benefits of biochar with nanoscale engineering, it offers substantial potential to address critical challenges in agriculture, environmental management, and pollution mitigation. Continued research and regulatory oversight are essential to fully realize its benefits while ensuring environmental safety.

9.9 Conclusion and Future Perspectives

Biochar technology has progressed from a basic soil supplement to an advanced environmental cleanup technology, exhibiting efficacy against several new pollutants. Current scientific understanding affirms that biochar possesses multifunctional features, including adjustable surface chemistry, high surface area, and catalytic activity (Lee et al. 2017). This renders it uniquely equipped to tackle intricate pollution issues (Varkolu et al. 2025). However, other significant research deficiencies require focus to completely harness biochar's potential. Initially, standardized techniques for biochar production and modification must be established to guarantee uniform performance across various applications (Barbhuiya et al. 2024). Secondly, in-depth field applications are necessary to validate laboratory results, especially concerning the impacts of biochar aging and the hazards of contaminant re-mobilization (Wang et al. 2021). The environmental trade-offs of extensive biochar implementation, encompassing life cycle effects and possible adverse effects on soil biota, necessitate thorough evaluation. Emerging potential encompasses the creation of smart biochar composites with stimuli- responsive characteristics and their integration with additional remediation technologies (e.g., bioaugmentation, phytoremediation) (Kordrostami and Ghasemi-Soloklui 2025). Consequently, this has extensive applicability in emerging fields such as air pollution mitigation and stormwater management. Recent advancements in nano-biochar and catalytic biochar systems enable the active elimination of contaminants rather than just sequestration (Jadhav et al. 2025).

Current research focuses on surface modification strategies to enhance stability and reduce ecotoxicity, such as coatings made of organic substances or embedding in polymer matrices (Verma and Devi 2025). Regulatory frameworks for nano-biochar applications are still developing, with particular attention needed regarding dosage thresholds, mobility control, and long-term monitoring requirements (Bhandari et al. 2023). Future studies should emphasize circular economy methodologies. This encompasses waste-derived biochar specifically tailored for targeted remedial tasks prior to their eventual use as soil supplements. Consequently, establishing sustainable value chains and promoting waste management solutions. As legal frameworks align with technical advancements, biochar-based solutions are set to assume a more significant position in worldwide environmental protection plans. Thus, providing a waste management strategy for pollution remediation that corresponds with climate change mitigation objectives.

Acknowledgements Authors sincerely acknowledge the laboratory facilities provided by the Department of Life science and Bioinformatics, Assam University, Silchar.

References

Abdel-Shafy HI, Mansour MS (2016) A review on polycyclic aromatic hydrocarbons: source, environmental impact, effect on human health and remediation. Egypt J Pet 25(1):107–123

Agrafioti E, Kalderis D, Diamadopoulos E (2014) Ca and Fe modified biochars as adsorbents of arsenic and chromium in aqueous solutions. J Environ Manag 146:444–450

Alengebawy A, Abdelkhalek ST, Qureshi SR, Wang MQ (2021) Heavy metals and pesticides toxicity in agricultural soil and plants: Ecological risks and human health implications. Toxics 9(3):42

Ali A, Chiang YW, Santos RM (2022) X-ray diffraction techniques for mineral characterization: a review for engineers of the fundamentals, applications, and research directions. Minerals 12(2):205. https://doi.org/10.3390/min12020205

Allohverdi T, Mohanty AK, Roy P, Misra M (2021) A review on current status of biochar uses in agriculture. Molecules 26(18):5584

Amalina F, Abd Razak AS, Krishnan S, Sulaiman H, Zularisam AW et al (2022) Biochar production techniques utilizing biomass waste-derived materials and environmental applications—a review. J Hazard Mater Adv 7:100134

Ambaye TG, Vaccari M, van Hullebusch ED, Amrane A, Rtimi S (2021) Mechanisms and adsorption capacities of biochar for the removal of organic and inorganic pollutants from industrial wastewater. Int J Environ Sci Technol 18:1–22

Amoah-Antwi C, Kwiatkowska-Malina J, Thornton SF, Fenton O, Malina G, Szara E (2020) Restoration of soil quality using biochar and brown coal waste: a review. Sci Total Environ 722:137852

Amusat SO, Kebede TG, Dube S, Nindi MM (2021) Ball-milling synthesis of biochar and biochar–based nanocomposites and prospects for removal of emerging contaminants: a review. J Water Process Eng 41:101993
An N, Zhang L, Liu Y, Shen S, Li N, Wu Z, Yang J, Han W, Han X (2022) Biochar application with reduced chemical fertilizers improves soil pore structure and rice productivity. Chemosphere 298:134304
Anca-Couce A (2016) Reaction mechanisms and multi-scale modelling of lignocellulosic biomass pyrolysis. Prog Energy Combust Sci 53:41–79
Ancient Amazonians 2025 Ancient Amazonians intentionally created fertile "dark earth". MIT News. Massachusetts Institute of Technology. https://news.mit.edu/2023/ancient-amazonians-intentionally- created-fertile-dark-earth-0920. Accessed 16 June 2025
Arni SA (2018) Comparison of slow and fast pyrolysis for converting biomass into fuel. Renew Energy 124:197–201
Azad M, Javeed IA, Bhat SAS (2023) Potential of biochar to sequester carbon and mitigate greenhouse gas emissions. Curr J Appl Sci Technol 42:24–31
Barati B, Zafar FF, Wang S (2022) Different waste management methods, applications, and limitations. In: Abomohra AEF, Wang Q, Huang J (eds) Waste-to-energy. Springer, Cham. https://doi.org/10.1007/978-3-030-91570-4_2
Barbhuiya S, Das BB, Kanavaris F (2024) Biochar-concrete: a comprehensive review of properties, production and sustainability. Case Stud Constr Mater 20:02859
Benis KZ, Damuchali AM, Soltan J et al (2020) Treatment of aqueous arsenic—a review of biochar modification methods. Sci Total Environ 739:139750
Bezerra J, Turnhout E, Vasquez IM, Rittl TF, Arts B, Kuyper TW (2019) The promises of the Amazonian soil: shifts in discourses of Terra Preta and biochar. J Environ Policy Plan 21(5):623–635
Bhandari G, Gangola S, Dhasmana A, Rajput V, Gupta S, Malik S et al (2023) Nano-biochar: recent progress, challenges, and opportunities for sustainable environmental remediation. Front Microbiol 14:1214870
Bird MI, Wurster CM, de Paula Silva PH, Bass AM, De Nys R (2011) Algal biochar–production and properties. Bioresour Technol 102(2):1886–1891
Brambilla G, D'Hollander W, Oliaei F, Stahl T, Weber R (2015) Pathways and factors for food safety and food security at PFOS contaminated sites within a problem-based learning approach. Chemosphere 129:192–202
Buck RC, Franklin J, Berger U, Conder JM, Cousins IT, De Voogt P et al (2011) Perfluoroalkyl and polyfluoroalkyl substances in the environment: terminology, classification, and origins. Integr Environ Assess Manag 7(4):513 541
Cao J, Gao X, Zhang S, Wei Z, Chen X, Ma N, Li C, Zhao X (2024) Migration patterns of phthalic acid esters from mulch plastic film in the soil-plant-atmosphere continuum system. J Hazard Mater 480
Chaudhry UK, Asim A, Iqbal J, Ashraf A, Yaqoob S (2025) Biochar for climate change adaptation and mitigation. In: Biochar revolution, pp 83–97
Chausali N, Saxena J et al (2021) Nanobiochar and biochar based nanocomposites: advances and applications. J Agric Food Res 5:100191
Chen Q, Song Y, An Y, Lu Y, Zhong G (2024a) Soil microorganisms: their role in enhancing crop nutrition and health. Diversity 16(12):734
Chen Y, Li Y, Liang X, Lu S, Ren J, Zhang Y et al (2024b) Effects of microplastics on soil carbon pool and terrestrial plant performance. Carbon Res 3(1):37
Chia CH, Gong B, Joseph SD, Marjo CE, Munroe P, Rich AM (2012) Imaging of mineral-enriched biochar by FTIR, Raman and SEM–EDX. Vib Spectrosc 62:248–257
Cimino AM, Boyles AL, Thayer KA, Perry MJ (2017) Effects of neonicotinoid pesticide exposure on human health: a systematic review. Environ Health Perspect 125(2):155–162
Czech B, Kończak M, Rakowska M, Oleszczuk P (2021) Engineered biochars from organic wastes for the adsorption of diclofenac, naproxen and triclosan from water systems. J Clean Prod 288:125686

Daful AG, Chandraratne MR (2020) Biochar production from biomass waste-derived material. In: Hashmi S, Choudhury IA (eds) Encyclopedia of renewable and sustainable materials. Elsevier, pp 370–378

Dai H, Gao J, Li D, Wang Z, Cui Y, Zhao Y (2022) Family Sphingomonadaceae as the key executor of triclosan degradation in both nitrification and denitrification systems. Chem Eng J 442(1):136202

Dann AB, Hontela A (2011) Triclosan: environmental exposure, toxicity and mechanisms of action. J Appl Toxicol 31(4):285–311

Dar OI, Aslam R, Pan D, Sharma S, Andotra M, Kaur A, Jia AQ, Faggio C (2022) Source, bioaccumulation, degradability and toxicity of triclosan in aquatic environments: a review. Environ Technol Innov 25:102122

Das D, Bordoloi U, Muigai HH, Kalita P (2020) A novel form stable PCM based bio composite material for solar thermal energy storage applications. J Energy Storage 30:101403

Das SK, Ghosh GK, Avasthe R (2021) Applications of biomass derived biochar in modern science and technology. Environ Technol Innov 21:101306

Das I, Patar A, Singha ER et al (2024) Toxic effects of fenvalerate in tadpoles of Fejervarya limnocharis: a multiparametric study. Environ Sci Pollut Res 31:49656–49669

Dasgupta S, Reddam A, Liu Z, Liu J, Volz DC (2020) High-content screening in zebrafish identifies perfluorooctanesulfonamide as a potent developmental toxicant. Environ Pollut 256:113550

Deng R, Luo H, Huang D, Zhang C (2020) Biochar-mediated Fenton-like reaction for the degradation of sulfamethazine: role of environmentally persistent free radicals. Chemosphere 255:126975

Deng J, Han J, Hou C, Zhang Y, Fang Y, Du W et al (2023) Efficient removal of per- and polyfluoroalkyl substances from biochar composites: cyclic adsorption and spent regenerant degradation. Chemosphere 341:140051

Ding Y, Gao X, Qu Z, Jia Y, Hu M, Li C (2019) Effects of biochar application and irrigation methods on soil temperature in farmland. Water 11(3):499

Ding P, Zhang J, Li X, Ma P, Hu G, Zhang L, Yu Y (2025) Transgenerational thyroid hormone disruption in zebrafish induced by environmentally relevant concentrations of triclosan. Environ Pollut 374:126242. https://doi.org/10.1016/j.envpol.2025.126242

Ehalt Macedo H, Lehner B, Nicell JA, Khan U, Klein EY (2025) Antibiotics in the global river system arising from human consumption. PNAS Nexus 4(4):096

El-Naggar A, Lee SS, Rinklebe J, Farooq M, Song H, Sarmah AK et al (2019) Biochar application to low fertility soils: a review of current status, and future prospects. Geoderma 337:536–554

Enaime G, Baçaoui A, Yaacoubi A, Lübken M (2020) Biochar for wastewater treatment—conversion technologies and applications. Appl Sci 10(10):3492

Eskenazi B, Chevrier J, Rauch SA, Kogut K, Harley KG, Johnson C et al (2013) In utero and childhood polybrominated diphenyl ether (PBDE) exposures and neurodevelopment in the CHAMACOS study. Environ Health Perspect 121(2):257–262

Gabetto FP, Tenelli S, Netto-Ferreira JB, Martins Junior J, Almeida OAC, Cosenza ML et al (2025) Biochar from crop residues mitigates N_2O emissions and increases carbon content in tropical soils. Biofuels Bioprod Biorefin 19(3):759–775

Ganie ZA, Khandelwal N, Tiwari E, Singh N, Darbha GK (2021) Biochar-facilitated remediation of nanoplastic contaminated water: effect of pyrolysis temperature induced surface modifications. J Hazard Mater 417:126096

Ge S, Yek PNY, Cheng YW, Xia C, Mahari WAW, Liew RK et al (2021) Progress in microwave pyrolysis conversion of agricultural waste to value-added biofuels: a batch to continuous approach. Renew Sust Energ Rev 135:110148

Ghusn W, Naik R, Yibirin M (2023) The impact of artificial sweeteners on human health and cancer association: a comprehensive clinical review. Cureus 15(12):51299

Glaser B, Birk JJ (2012) State of the scientific knowledge on properties and genesis of Anthropogenic Dark Earths in Central Amazonia (terra preta de Índio). Geochim Cosmochim Acta 82:39–51

Goulson D, Nicholls E, Botías C, Rotheray EL (2015) Bee declines driven by combined stress from parasites, pesticides, and lack of flowers. Science 347(6229):1255957

Gregory SJ, Anderson CWN, Camps-Arbestain M, Biggs PJ, Ganley ARD, O'Sullivan JM, McManus MT (2015) Biochar in co-contaminated soil manipulates arsenic solubility and microbiological community structure, and promotes organochlorine degradation. PLoS One 10:0125393

Groot H, Fernholz K, Frank M, Howe J, Bowyer J, Bratkovich S (2016) Biochar 101: an introduction to an ancient product offering modern opportunities. Dovetail Partners INC 9(1):1–6

Gu X, Wang Y, Lai C, Qiu J, Li S, Hou Y et al (2015) Microporous bamboo biochar for lithium-sulfur batteries. Nano Res 8:129–139

Guerrero LA, Maas G, Hogland W (2013) Solid waste management challenges for cities in developing countries. Waste Manag 33(1):220–232

Gumisiriza R, Hawumba JF, Okure M, Hensel O (2017) Biomass waste-to-energy valorisation technologies: a review case for banana processing in Uganda. Biotechnol Biofuels 10:1–29

Guo M, Song W, Tian J (2020) Biochar-facilitated soil remediation: mechanisms and efficacy variations. Front Environ Sci 8:521512. https://doi.org/10.3389/fenvs.2020.521512

Ha JH, Lee IG (2020) Study of a method to effectively remove char byproduct generated from fast pyrolysis of lignocellulosic biomass in a bubbling fluidized bed reactor. PRO 8(11):1407

Haider FU, Wang X, Zulfiqar U, Farooq M, Hussain S, Mehmood T et al (2022) Biochar application for remediation of organic toxic pollutants in contaminated soils. An update. Ecotoxicol Environ Saf 248:114322

Hassan M, Wang B, Wu P, Wang S (2024) Engineered biochar for in-situ and ex-situ remediation of contaminants from soil and water. Sci Total Environ 957:177384

Hayes TB, Khoury V, Narayan A, Nazir M, Park A, Brown T et al (2010) Atrazine induces complete feminization and chemical castration in male African clawed frogs (Xenopus laevis). Proc Natl Acad Sci 107(10):4612–4617

He K, He G, Wang C, Zhang H, Xu Y, Wang S et al (2020) Biochar amendment ameliorates soil properties and promotes Miscanthus growth in a coastal saline-alkali soil. Appl Soil Ecol 155:103674

Hu Z, Wei L (2023) Review on characterization of biochar derived from biomass pyrolysis via reactive molecular dynamics simulations. J Compos Sci 7(9):354. https://doi.org/10.3390/jcs7090354

Huang J, Liu C, Tong H, Li W, Wu D (2012) Theoretical studies on pyrolysis mechanism of xylopyranose. Comput Theor Chem 1001:44–50

Huang M, Yin X, Chen J, Cao F (2021) Biochar application mitigates the effect of heat stress on rice (Oryza sativa L.) by regulating the root-zone environment. Front Plant Sci 12:711725

Hunt J, DuPonte M, Sato D, Kawabata A (2010) The basics of biochar: a natural soil amendment. Soil Crop Manag 30(7):1–6

International Biochar Initiative (IBI) 2025 International Biochar Initiative (IBI). https://biochar-international.org. Accessed 25th April 2025

Islam T, Li Y, Cheng H (2021) Biochars and engineered biochars for water and soil remediation: a review. Sustainability 13(17):9932

Jablonowski ND, Schäffer A, Burauel P (2011) Still present after all these years: persistence plus potential toxicity raise questions about the use of atrazine. Environ Sci Pollut Res 18:328–331

Jadhav V, Ahire B, Pawar A, Roy A, Kumar A, Sharma K et al (2025) Nanobiochar: a sustainable solution for environmental remediation. Environ Nanotechnol Monit Manag 23:101061

Jayawardhana Y, Kumarathilaka P, Mayakaduwa S, Weerasundara L, Bandara T, Vithanage M (2018) Characteristics of municipal solid waste biochar: its potential to be used in environmental remediation. In: Utilization and management of bioresources: proceedings of 6th Icon SWM 2016, pp 209–220

Jiang J, Cui H, Tang Z, Wang S, Cheng K, Zhang C et al (2024a) Synergistic effects of biochar and laccase on nitrogen conversation and degradations of two artificial sweeteners during the sewage sludge composting. Chem Eng J 498:155732

Jiang X, Liu J, Han X, Wang F, Zhang D, Fan P, Yi W (2024b) Amino-grafted biochar as a novel photocatalyst for degradation of high concentration dye. J Environ Manag 352:119989

Jin H, Capareda S, Chang Z, Gao J, Xu Y, Zhang J (2014) Biochar pyrolytically produced from municipal solid wastes for aqueous As (V) removal: adsorption property and its improvement with KOH activation. Bioresour Technol 169:622–629

Jing Y, Zhang Y, Han I, Wang P, Mei Q, Huang Y (2020) Effects of different straw biochars on soil organic carbon, nitrogen, available phosphorus, and enzyme activity in paddy soil. Sci Rep 10:8837

Kamarudin NS, Dahalan FA, Hasan M, An OS, Parmin NA, Ibrahim N (2022) Biochar: a review of its history, characteristics, factors that influence its yield, methods of production, application in wastewater treatment and recent development. Biointerface Res Appl Chem 12(6):7914–7926

Kandanelli R, Meesala L, Kumar J, Raju CSK, Peddy VR, Gandham S, Kumar P (2018) Cost effective and practically viable oil spillage mitigation: comprehensive study with biochar. Mar Pollut Bull 128:32–40

Karimi S, Mahboobi Soofiani N, Mahboubi A, Taherzadeh MJ (2018) Use of organic wastes and industrial by-products to produce filamentous fungi with potential as aqua-feed ingredients. Sustainability 10(9):3296

Kedia AG, Dutta A, Kumar P (2023) Dimethyl carbonate as a cost-effective substitute of methanol for biodiesel production via transesterification of nonedible oil. Bioenergy Res 16:1134–1142

Khan S, Irshad S, Mehmood K, Hasnain Z, Nawaz M, Rais A et al (2024) Biochar production and characteristics, its impacts on soil health, crop production, and yield enhancement: a review. Plants 13(2):166

Kordrostami M, Ghasemi-Soloklui AA (2025) Innovative applications of biochar in nuclear remediation and catalysis. Biochar 7(1):1–29

Kumar A, Schreiter IJ, Wefer-Roehl A, Tsechansky L, Schüth C, Graber ER (2016) Production and utilization of biochar from organic wastes for pollutant control on contaminated sites. In: Prasad MNV, Shih K (eds) Environmental materials and waste. Academic Press, pp 91–116

Kumari M, Mohapatra D, Bhushan M (2022) Emerging contaminants: sources, effects, and treatment by new adsorption methods. In: Kumar M, Mohapatra S (eds) Impact of COVID-19 on emerging contaminants. Springer transactions in civil and environmental engineering. Springer, Singapore, pp 97–121

Kuncara J, Setyawan M, Hakika DC (2024) Manufacture of industrial scale bagasse biochar: effect of temperature and residence time and biochar characterization. Equilibrium J Chem Eng 8(2):33–44

Ladislao BA, Turrion-Gomez JL (2010) Challenges, options and future research needs. In: Microbes at work: from wastes to resources, pp 313–321

Lam SS, Mahari WAW, Ok YS, Peng W, Chong CT, Ma NL et al (2019) Microwave vacuum pyrolysis of waste plastic and used cooking oil for simultaneous waste reduction and sustainable energy conversion: recovery of cleaner liquid fuel and techno-economic analysis. Renew Sust Energ Rev 115:109359

Lee J, Kim KH, Kwon EE (2017) Biochar as a catalyst. Renew Sust Energ Rev 77:70–79

Lee DJ, Lu JS, Chang JS (2020) Pyrolysis synergy of municipal solid waste (MSW): a review. Bioresour Technol 318:123912

Lee-Sarwar K, Hauser R, Calafat AM, Ye X, O'Connor GT, Sandel M, Bacharier LB, Zeiger RS, Laranjo N, Gold DR, Weiss ST, Litonjua AA, Savage JH (2018) Prenatal and early- life triclosan and paraben exposure and allergic outcomes. J Allergy Clin Immunol 142(1):269–278

Lehmann J (2007) Bio-energy in the black. Front Ecol Environ 5(7):381–387

Lehmann J, Rondon M (2006) Bio-char soil management on highly weathered soils in the humid tropics. Biol Approaches Sustain Soil Syst 113(517):530

Lehner R, Weder C, Petri-Fink A, Rothen-Rutishauser B (2019) Emergence of nanoplastic in the environment and possible impact on human health. Environ Sci Technol 53(4):1748–1765

Leng L, Xiong Q, Yang L, Li H, Zhou Y, Zhang W et al (2021) An overview on engineering the surface area and porosity of biochar. Sci Total Environ 763:144204

Li Y, Tasnady D (2023) Biochar for soil carbon sequestration: current knowledge, mechanisms, and future perspectives. J Carbon Res 9(3):67

Li S, Shao L, Zhang H, He P, Lü F (2020a) Quantifying the contributions of surface area and redox-active moieties to electron exchange capacities of biochar. J Hazard Mater 394:122541
Li X, Yao S, Bian Y, Jiang X, Song Y (2020b) The combination of biochar and plant roots improves soil bacterial adaptation to PAH stress: insights from soil enzymes, microbiome, and metabolome. J Hazard Mater 400:123227
Li H, Yang L, Mao Q, Zhou H, Guo P, Agathokleous E, Wang S (2023) Modified biochar enhances soil fertility and nutrient uptake and yield of rice in mercury-contaminated soil. Environ Technol Innov 32:103435
Li X, Shen X, Jiang W, Xi Y, Li S (2024) Comprehensive review of emerging contaminants: detection technologies, environmental impact, and management strategies. Ecotoxicol Environ Saf 278:116420
Li H, Li L, Sun H, Zhang H, Zhan W, Zuo W et al (2025) Machine learning enhancing biochar abatement predictions: advancing China climate goals for food production and promoting application. Resour Conserv Recycl 218:108248
Liang B, Lehmann J, Solomon D, Sohi S, Thies JE, Skjemstad JO et al (2008) Stability of biomass-derived black carbon in soils. Geochim Cosmochim Acta 72(24):6069–6078
Lin Y, Ma X, Peng X, Yu Z, Fang S, Lin Y, Fan Y (2016) Combustion, pyrolysis and char CO2-gasification characteristics of hydrothermal carbonization solid fuel from municipal solid wastes. Fuel 181:905–915
Liu K (2019) Effects of sample size, dry ashing temperature and duration on determination of ash content in algae and other biomass. Algal Res 40:101486. https://doi.org/10.1016/j.algal.2019.101486
Liu J, Jiang S, Chen D, Dai G, Wei D, Shu Y (2020) Activation of persulfate with biochar for degradation of bisphenol A in soil. Chem Eng J 381:122637
Liu Z, Wang Z, Chen H, Cai T, Liu Z (2021) Hydrochar and pyrochar for sorption of pollutants in wastewater and exhaust gas: a critical review. Environ Pollut 268:115910
Liu S, Peng S, Zhang B, Xue B, Yang Z, Wang S et al (2022) Effects of biochar pyrolysis temperature on thermal properties of polyethylene glycol/biochar composites as shape-stable biocomposite phase change materials. RSC Adv 12:9587–9598
Ma W, Song R, Wang Y, Cui X, Yan Y, Liu Z et al (2024) Optimized Ginkgo leaf biochar: an efficient adsorbent for 2,4-D herbicide removal from wastewater. Water Environ Res 96(9):11124
Manyà JJ (2012) Pyrolysis for biochar purposes: a review to establish current knowledge gaps and research needs. Environ Sci Technol 46(15):7939–7954
Manyà JJ, García-Morcate D, González B (2020) Adsorption performance of physically activated biochars for postcombustion CO_2 capture from dry and humid flue gas. Appl Sci 10(1):376
Mason SA, Welch VG, Neratko J (2018) Synthetic polymer contamination in bottled water. Front Chem 6:389699
Mbarki F, Selmi T, Kesraoui A, Seffen M, Gadonneix P, Celzard A, Fierro V (2019) Hydrothermal pre-treatment, an efficient tool to improve activated carbon performances. Ind Crop Prod 140:111717
Mishra RK, Mohanty K (2023) A review of the next-generation biochar production from waste biomass for material applications. Sci Total Environ 904:167171
Mitchell EA, Mulhauser B, Mulot M, Mutabazi A, Glauser G, Aebi A (2017) A worldwide survey of neonicotinoids in honey. Science 358(6359):109–111
Mitra T, Gulati R, Ramachandran K, Rajiv R, Enninga EAL, Pierret CK et al (2024) Endocrine disrupting chemicals: gestational diabetes and beyond. Diabetol Metab Syndr 16(1):95
Mu W, Ben H, Ragauskas A, Deng Y (2013) Lignin pyrolysis components and upgrading—technology review. Bioenergy Res 6:1183–1204
Mujtaba M, Fraceto LF, Fazeli M, Mukherjee S, Savassa SM, de Medeiros GA, Pereira AES, Mancini SD, Lipponen J, Vilaplana F (2023) Lignocellulosic biomass from agricultural waste to the circular economy: a review with focus on biofuels, biocomposites and bioplastics. J Clean Prod 402:136815. https://doi.org/10.1016/j.jclepro.2023.136815

Mukherjee A, Patra BR, Podder J, Dalai AK (2022) Synthesis of biochar from lignocellulosic biomass for diverse industrial applications and energy harvesting: effects of pyrolysis conditions on the physicochemical properties of biochar. Front Mater 9:870184
Murtaza G, Ahmed Z, Eldin SM, Ali I, Usman M, Iqbal R et al (2023) Biochar as a green sorbent for remediation of polluted soils and associated toxicity risks: a critical review. Separations 10(3):197. https://doi.org/10.3390/separations10030197
Murtaza G, Ahmed Z, Iqbal R, Deng G (2025) Biochar from agricultural waste as a strategic resource for promotion of crop growth and nutrient cycling of soil under drought and salinity stress conditions: a comprehensive review with context of climate change. J Plant Nutr:1–52
Mutsengerere S, Chihobo CH, Musademba D, Nhapi I (2019) A review of operating parameters affecting bio-oil yield in microwave pyrolysis of lignocellulosic biomass. Renew Sust Energ Rev 104:328–336
Neves EG, Petersen JB, Bartone RN, Augusto Da Silva C (2003) Historical and socio-cultural origins of Amazonian dark earth. In: Amazonian dark earths: origin properties management, pp 29–50
Nguyen HKD, Pham VV, Do HT (2016) Preparation of Ni/biochar catalyst for hydrotreating of bio-oil from microalgae biomass. Catal Lett 146:2381–2391
Ni N, Song Y, Shi R, Liu Z, Bian Y, Wang F et al (2017) Biochar reduces the bioaccumulation of PAHs from soil to carrot (Daucus carota L.) in the rhizosphere: a mechanism study. Sci Total Environ 601:1015–1023
Nidheesh PV, Gopinath A, Ranjith N, Akre AP, Sreedharan V, Kumar MS (2021) Potential role of biochar in advanced oxidation processes: a sustainable approach. Chem Eng J 405:126582
Nishmitha PS, Akhilghosh KA, Aiswriya VP, Ramesh A, Muthuchamy M, Muthukumar A (2025) Understanding emerging contaminants in water and wastewater: a comprehensive review on detection, impacts, and solutions. J Hazard Mater Adv 100755
Nzediegwu C, Arshad M, Ulah A, Naeth MA, Chang SX (2021) Fuel, thermal and surface properties of microwave-pyrolyzed biochars depend on feedstock type and pyrolysis temperature. Bioresour Technol 320:124282
Palansooriya KN, Withana PA, Jeong Y, Sang MK, Cho Y, Hwang G et al (2024) Contrasting effects of food waste and its biochar on soil properties and lettuce growth in a microplastic-contaminated soil. Appl Biol Chem 67:3
Pan WJ, Xiong C, Wu QP, Liu JX, Liao HM, Chen W et al (2013) Effect of BPA on the germination, root development, seedling growth and leaf differentiation under different light conditions in Arabidopsis thaliana. Chemosphere 93(10):2585–2592
Pang S (2019) Advances in thermochemical conversion of woody biomass to energy, fuels and chemicals. Biotechnol Adv 37(4):589–597
Panwar NL, Pawar A (2022) Influence of activation conditions on the physicochemical properties of activated biochar: a review. Biomass Convers Biorefinery 12(3):925–947
Patel AB, Shaikh S, Jain KR, Desai C, Madamwar D (2020) Polycyclic aromatic hydrocarbons: sources, toxicity, and remediation approaches. Front Microbiol 11:562813
Patel R, Babaei-Ghazvini A, Dunlop MJ, Acharya B (2024) Biomaterials-based concrete composites: a review on biochar, cellulose and lignin. Carbon Capture Sci Technol 12:100232
Paul S, Singh BK, Singha ER et al (2024) A comparative account of phototoxicity of anthracene and pyrene in the tadpoles of the anuran amphibian Fejervarya limnocharis using multiple toxicological end points. Ecotoxicology 33:737–749
Paul S, Dey A, Singh BK, Giri A (2025) Anthracene: multiparametric toxicity assessment in Pisum sativum, Vigna radiata and Epipremnum aureum, and remediation efficacy by Epipremnum aureum. Environ Sci Pollut Res 32(20):12513–12523
Ponnusamy VK, Nagappan S, Bhosale RR, Lay CH, Nguyen DD, Pugazhendhi A et al (2020) Review on sustainable production of biochar through hydrothermal liquefaction: physico-chemical properties and applications. Bioresour Technol 310:123414
Priyadarshini U, Neelancherry R (2024) Application of biochar in removal of per- and polyfluoroalkyl substances from aqueous medium. In: Neelancherry R, Gao B, Wisniewski A Jr (eds) Agricultural waste to value-added products. Springer, Singapore

Pugazhendhi A (2024) Biochar for wastewater treatment: Addressing contaminants and enhancing sustainability: challenges and solutions. J Hazard Mater Adv 100504

Qi Y, Yang X, Pelaez AM, Lwanga EH, Beriot N, Gertsen H et al (2018) Macro-and micro- plastics in soil-plant system: effects of plastic mulch film residues on wheat (Triticum aestivum) growth. Sci Total Environ 645:1048–1056

Racero-Galaraga D, Rhenals-Julio JD, German SS, Mendoza JM, Silvera AB (2024) Proximate analysis in biomass: standards, applications and key characteristics. Results Chem 101886

Ramirez LA, Dennehy M, Alvarez M (2023) Metal oxide-biochar supported recyclable catalysts: A feasible solution for the reduction of 4-nitrophenol in water. Catal Commun 181:106723

Rangabhashiyam S, Balasubramanian PJIC (2019) The potential of lignocellulosic biomass precursors for biochar production: performance, mechanism and wastewater application—a review. Ind Crop Prod 128:405–423

Rao MA, Simeone GDR, Scelza R, Conte P (2017) Biochar based remediation of water and soil contaminated by phenanthrene and pentachlorophenol. Chemosphere 186:193–201

Ravindran R, Hassan SS, Williams GA, Jaiswal AK (2018) A review on bioconversion of agro-industrial wastes to industrially important enzymes. Bioengineering 5(4):93

Relyea RA (2005) The lethal impact of roundup on aquatic and terrestrial amphibians. Ecol Appl 15(4):1118–1124

Richardson SD, Kimura SY (2017) Emerging environmental contaminants: challenges facing our next generation and potential engineering solutions. Environ Technol Innov 8:40–56

Rochester JR (2013) Bisphenol A and human health: a review of the literature. Reprod Toxicol 42:132–155

Rochman CM, Browne MA, Halpern BS, Hentschel BT, Hoh E, Karapanagioti HK et al (2013) Classify plastic waste as hazardous. Nature 494(7436):169–171

Rochman CM, Tahir A, Williams SL, Baxa DV, Lam R, Miller JT et al (2015) Anthropogenic debris in seafood: Plastic debris and fibers from textiles in fish and bivalves sold for human consumption. Sci Rep 5(1):14340

Rodríguez A, Castrejón-Godínez ML, Monterrosas-Brisson N (2025) Pesticides: environmental stressors implicated in the development of central nervous system disorders and neurodegeneration. Stress 5(2):31

Rúa-Díaz S, Forjan R, Lago-Vila M, Cerqueira B, Arco-Lázaro E, Marcet P et al (2023) Pyrolysis temperature influences the capacity of biochar to immobilize copper and arsenic in mining soil remediation. Environ Sci Pollut Res 30(12):32882–32893

Saeed M, Ilyas N, Jayachandran K, Gaffar S, Arshad M, Sheeraz Ahmad M et al (2021) Biostimulation potential of biochar for remediating the crude oil contaminated soil and plant growth. Saudi J Biol Sci 28:2667–2676

Sakhiya AK, Anand A, Kaushal P (2020) Production, activation, and applications of biochar in recent times. Biochar 2:253–285

Salimi P, Javadian S, Norouzi O, Gharibi H (2017) Turning an environmental problem into an opportunity: Potential use of biochar derived from a harmful marine biomass named Cladophora glomerata as anode electrode for Li-ion batteries. Environ Sci Pollut Res 24:27974–27984

Santoso E, Ediati R, Kusumawati Y, Bahruji H, Sulistiono DO, Prasetyoko D (2020) Review on recent advances of carbon based adsorbent for methylene blue removal from waste water. Mater Today Chem 16:100233

Sarkar D, Panicker TF, Mishra RK, Kini MS (2024) A comprehensive review of production and characterization of biochar for removal of organic pollutants from water and wastewater. Water-Energy Nexus 7:243–265

Schiavon M, Torretta V, Rada EC, Ragazzi M (2016) State of the art and advances in the impact assessment of dioxins and dioxin-like compounds. Environ Monit Assess 188:1–20

Senthil C, Lee CW (2021) Biomass-derived biochar materials as sustainable energy sources for electrochemical energy storage devices. Renew Sust Energ Rev 137:110464

Shah S, Ilyas M, Liu ZY, Zhang DJ, Wu G, Huang QY et al (2025) Interaction of micro and nanoplastics (MNPs) with agricultural stored products and their pests. Sci Total Environ 963:178534

Sharma AK, Ghodke PK, Goyal N, Bobde P, Kwon EE, Lin KYA, Chen WH (2023) A critical review on biochar production from pine wastes, upgradation techniques, environmental sustainability, and challenges. Bioresour Technol 387:129632

Shoudho KN, Khan TH, Ara UR, Khan MR, Shawon ZBZ, Hoque ME (2024) Biochar in global carbon cycle: towards sustainable development goals. Curr Res Green Sustain Chem 100409

Singh BP, Cowie AL, Smernik RJ (2012) Biochar carbon stability in a clayey soil as a function of feedstock and pyrolysis temperature. Environ Sci Technol 46:11770–11778

Singha ER, Das I, Patar A, Paul S, Giri S, Giri A (2024) Effects of changed water regime on the toxicity of silver nanoparticles (AgNPs) in tadpoles of *Fejervarya limnocharis*. Environ Sci Pollut Res 31:54873–54886. https://doi.org/10.1007/s11356-024-34832-3

Smith M, Love DC, Rochman CM, Neff RA (2018) Microplastics in seafood and the implications for human health. Curr Environ Health Rep 5:375–386

Solanki P, Jain S, Mehrotra R, Mago P, Dagar S (2024) Microplastics in agricultural soil and their impact: a review. Nat Environ Pollut Technol 23(4):2143–2155

Sonne C, Gustavson K, Bossi R, Søndergaard J, Desforges JP, Bonefeld-Jørgensen EC, Dietz R (2025) Ubiquitous global use of persistent PFAS threatens Arctic Indigenous peoples for decades to come. Cell Rep Sustain 2(3)

Sonu K, Sogani M, Syed Z, Dongre A, Sharma G (2020) Enhanced decolorization and treatment of textile dye wastewater through adsorption on acid modified corncob derived biochar. Chemistry Select 5(39):12287–12297

Stewart M, Olsen G, Hickey CW, Ferreira B, Jelić A, Petrović M, Barcelo D (2014) A survey of emerging contaminants in the estuarine receiving environment around Auckland, New Zealand. Sci Total Environ 468-469:202–210

Subedi B, Lee S, Moon HB, Kannan K (2014) Emission of artificial sweeteners, select pharmaceuticals, and personal care products through sewage sludge from wastewater treatment plants in Korea. Environ Int 68:33–40

Suliman W, Harsh JB, Abu-Lail NI, Fortuna AM, Dallmeyer I, Garcia-Perez M (2016) Modification of biochar surface by air oxidation: role of pyrolysis temperature. Biomass Bioenergy 85:1–11

Sun Q, Meng J, Lan Y, Shi G, Yang X, Cao D et al (2021) Long-term effects of biochar amendment on soil aggregate stability and biological binding agents in brown earth. Catena 205:105460. https://doi.org/10.1016/j.catena.2021.105460

Sunderland EM, Hu XC, Dassuncao C, Tokranov AK, Wagner CC, Allen JG (2019) A review of the pathways of human exposure to poly-and perfluoroalkyl substances (PFASs) and present understanding of health effects. J Expo Sci Environ Epidemiol 29(2):131–147

Tomczyk A, Sokołowska Z, Boguta P (2020) Biochar physicochemical properties: pyrolysis temperature and feedstock kind effects. Rev Environ Sci Biotechnol 19(1):191–215

Tong WK, Dai C, Jia C, Hu J, Gao M, Li J et al (2024) Eco-friendly and stable triclosan removal from groundwater using peroxyacetic acid activated with biochar produced from saccharification residues. Chem Eng J 481:148422

Tsang DCW, Wu KCW, Chen SS, Cao Y, Tessonnier JP, Shang J et al (2020) Effective dispersion of MgO nanostructure on biochar support as a basic catalyst for glucose isomerization. ACS Sustain Chem Eng 8:6990–7001

Umenweke GC, Afolabi IC, Epelle EI, Okolie JA (2022) Machine learning methods for modeling conventional and hydrothermal gasification of waste biomass: a review. Bioresour Technol Rep 17:100976

Van Bruggen AH, He MM, Shin K, Mai V, Jeong KC, Finckh MR, Morris JG Jr (2018) Environmental and health effects of the herbicide glyphosate. Sci Total Environ 616:255–268

Vandenberg LN, Maffini MV, Sonnenschein C, Rubin BS, Soto AM (2009) Bisphenol-A and the great divide: a review of controversies in the field of endocrine disruption. Endocr Rev 30(1):75–95

Vannini C, Domingo G, Onelli E, Prinsi B, Marsoni M, Espen L, Bracale M (2013) Morphological and proteomic responses of Eruca sativa exposed to silver nanoparticles or silver nitrate. PLoS One 8(7):68752

Varkolu M, Gundekari S, Palla VCS, Kumar P, Bhattacharjee S, Vinodkumar T (2025) Recent advances in biochar production, characterization, and environmental applications. Catalysts 15(3):243

Verma N, Devi NL (2025) Removal of atmospheric pollutants using biochar: preparation, application, regeneration and its future research. Air Qual Atmos Health 18:1205–1224

Wang Y, Qian H (2021) Phthalates and their impacts on human health. Healthcare (Basel) 9(5):603

Warner GR, Flaws JA (2018) Bisphenol A and phthalates: how environmental chemicals are reshaping toxicology. Toxicol Sci 166(2):246–249

Wang T, Camps-Arbestain M, Hedley M (2013) Predicting C aromaticity of biochars based on their elemental composition. Org Geochem 62:1–6

Wang S, Wang H, Yin Q, Zhu L, Yin S (2014a) Methanation of bio-syngas over a biochar supported catalyst. New J Chem 38:4471–4477

Wang Y, Yin R, Liu R (2014b) Characterization of biochar from fast pyrolysis and its effect on chemical properties of the tea garden soil. J Anal Appl Pyrolysis 110:375–381

Wang Y, Ling LL, Jiang H (2016) Selective hydrogenation of lignin to produce chemical commodities by using a biochar supported Ni-Mo2C catalyst obtained from biomass. Green Chem 18:4032–4041

Wang S, Zhao C, Shan R, Wang Y, Yuan H (2017) A novel peat biochar supported catalyst for the transesterification reaction. Energy Convers Manag 139:89–96

Wang S, Xu Y, Norbu N, Wang Z (2018) Remediation of biochar on heavy metal polluted soils. IOP Conf Ser Earth Environ Sci 108:042113

Wang J, Sun C, Huang QX, Chi Y, Yan JH (2021) Adsorption and thermal degradation of microplastics from aqueous solutions by Mg/Zn modified magnetic biochars. J Hazard Mater 419:126486

Wang L, Olsen MN, Moni C, Dieguez-Alonso A, de la Rosa JM, Stenrød M et al (2022) Comparison of properties of biochar produced from different types of lignocellulosic biomass by slow pyrolysis at 600 °C. Appl Energy Combust Sci 12:100090

Wang F, Xiang L, Leung KSY, Elsner M, Zhang Y, Guo Y et al (2024a) Emerging contaminants: a one health perspective. Int J Hydrogen Energ 5(4):2666–6758

Wang Y, Wang H, Chen M, Liang D, Lu Z, Qin B et al (2024b) Selective oxidation of methane to methanol over lignin-derived biochar supported transition metals catalysts. J Anal Appl Pyrolysis 177:106359

Weatherly LM, Gosse JA (2017) Triclosan exposure, transformation, and human health effects. J Toxicol Environ Health B 20(8):447–469

Wood R, Mašek O, Erastova V (2024) Developing a molecular-level understanding of biochar materials using public characterization data. Cell Rep Phys Sci 5(7)

Wu M, Han X, Zhong T, Yuan M, Wu W (2016) Soil organic carbon content affects the stability of biochar in paddy soil. Agric Ecosyst Environ 22:59-66. https://doi.org/10.1016/j.agee.2016.02.033

Xiao J, Li X, Cao Y, Chen G (2023) Does micro/nano biochar always good to phytoremediation? A case study from multiple metals contaminated acidic soil using Salix jiangsuensis' 172′. Carbon Res 2(1):21

Xu T, Chen J, Wang Z, Tang W, Xia D, Fu Z et al (2019) Development of prediction models on base-catalyzed hydrolysis kinetics of phthalate esters with density functional theory calculation. Environ Sci Technol 53(10):5828–5837

Yaashikaa PR, Kumar PS, Varjani SJ, Saravanan A (2019) Advances in production and application of biochar from lignocellulosic feedstocks for remediation of environmental pollutants. Bioresour Technol 292:122030

Yaashikaa PR, Kumar PS, Varjani S, Saravanan A (2020) A critical review on the biochar production techniques, characterization, stability and applications for circular bioeconomy. Biotechnol Rep 28:00570
Yaashikaa PR, Karishma S, Kamalesh R, Saravanan A, Vickram AS, Anbarasu K (2024) A systematic review on enhancement strategies in biochar-based remediation of polycyclic aromatic hydrocarbons. Chemosphere 355:141796
Yadav M, Joshi C, Paritosh K, Thakur J, Pareek N, Masakapalli SK, Vivekanand V (2022) Reprint of organic waste conversion through anaerobic digestion: a critical insight into the metabolic pathways and microbial interactions. Metab Eng 71:62–76
Yan J, Quan G (2020) Sorption behavior of dimethyl phthalate in biochar-soil composites: implications for the transport of phthalate esters in long-term biochar amended soils. Ecotoxicol Environ Saf 205:111169
Yan T, Xue J, Zhou Z, Wu Y (2021) Biochar-based fertilizer amendments improve the soil microbial community structure in a karst mountainous area. Sci Total Environ 794:148757
Yao Y, Gao B, Zhang M, Inyang M, Zimmerman AR (2012) Effect of biochar amendment on sorption and leaching of nitrate, ammonium, and phosphate in a sandy soil. Chemosphere 89(11):1467–1471
Yargicoglu EN, Sadasivam BY, Reddy KR, Spokas K (2015) Physical and chemical characterization of waste wood derived biochars. Waste Manag 36:256–268
You S, Ok YS, Chen SS, Tsang DC, Kwon EE, Lee J, Wang CH (2017) A critical review on sustainable biochar system through gasification: energy and environmental applications. Bioresour Technol 246:242–253
Yu KL, Lau BF, Show PL, Ong HC, Ling TC, Chen WH et al (2017) Recent developments on algal biochar production and characterization. Bioresour Technol 246:2–11
Yu H, Zou W, Chen J, Chen H, Yu Z, Huang J, Gao B (2019) Biochar amendment improves crop production in problem soils: a review. J Environ Manag 232:8–21
Yu X, Catanescu CO, Bird RE, Satagopan S, Baum ZJ, Lotti Diaz LM, Zhou QA (2023) Trends in research and development for CO_2 capture and sequestration. ACS Omega 8(13):11643–11664
Zhang H, Xue G, Chen H, Li X (2018) Magnetic biochar catalyst derived from biological sludge and ferric sludge using hydrothermal carbonization: preparation, characterization and its circulation in Fenton process for dyeing wastewater treatment. Chemosphere 191:64–71
Zhang L, Rana I, Shaffer RM, Taioli E, Sheppard L (2019) Exposure to glyphosate-based herbicides and risk for non-Hodgkin lymphoma: a meta-analysis and supporting evidence. Mutat Res Rev Mutat Res 781:186–206
Zhang M, Liu Y, Wei Q, Gou J (2021) Biochar enhances the retention capacity of nitrogen fertilizer and affects the diversity of nitrifying functional microbial communities in karst soil of southwest China. Ecotoxicol Environ Saf 226:112819
Zhang N, Xiong G, Liu Z (2022) Toxicity of metal-based nanoparticles: challenges in the nano era. Front Bioeng Biotechnol 10:1001572
Zhang Y, Tan X, Lu R, Tang Y, Qie H, Huang Z et al (2023) Enhanced removal of polyfluoroalkyl substances by simple modified biochar: adsorption performance and theoretical calculation. ACS ES&T Water 3(3):817–826
Zhang S, Han W, Liu T, Feng C, Jiang Q, Zhang B et al (2024a) Tetracycline inhibits the nitrogen fixation ability of soybean (Glycine max (L.) Merr.) nodules in black soil by altering the root and rhizosphere bacterial communities. Sci Total Environ 908:168047
Zhang X, Sun Y, Zhang Q, Tian W, Khan E, Tsang DC (2024b) Leaching characteristics of nutrients in food waste digestate-derived biochar. Bioresour Technol 399:130634
Zhang D, Bo W, Wang T, Yılmaz M, Sharma G, Kumar A, Shi H (2025) Multi-mechanism synergistic adsorption of lead and cadmium in water by structure-functionally adapted modified biochar: a review. Desalination Water Treat 101156
Zuber MA, Yahya WJ, Ithnin AM, Sugeng DA, Kadir HA, Ahmad MA (2020) A brief review of palm oil liquid waste conversion into biofuel. Environ Rev 28(1):67–76

Chapter 10
Biochar: A Promising Tool for Regenerative Agriculture

Rajashree Khatua, A. Nanda, N. Pradhan, S. Mahapatra, A. Mohanty, A. Mahapatra, P. Praveenkumar, and V. Guhan

Abstract The major challenges of the twenty-first century include managing soil health, remediating polluted wastewater, combating global warming, and finding sustainable alternative energy sources. Biochar, also known as biomass, is a material that resembles charcoal and is created by burning organic material from forestry and agricultural wastes. Pyrolysis is the controlled heat breakdown of biomass in an oxygen-limited environment. The dual benefits of waste management and its use in many industries are provided by biochar made from waste biomass, such as crop leftovers, algal biomass, municipal trash, etc. The addition of biochar to soil enhances its aeration, porosity, and health, all of which reduce greenhouse gas emissions. Biochar can serve as a catalyst to achieve the Sustainable Development Goals (SDGs) and has great potential to address global concerns. In addition to lowering waste and generating energy as a byproduct, biochar technology has the potential to improve soil quality and mitigate climate change. Although biochar may appear to be a straightforward substance, it may simultaneously aid in the solution of several global issues. For example, the manufacturing method might help store a billion tons of carbon every year, where it will be most useful for thousands of years in the soil. Burning fossil fuels, which has contributed to global warming by releasing greenhouse gases into the sky, may be replaced with the clean, renewable energy generated as a byproduct of the manufacture of biochar. Reduced groundwater contamination, cheaper water filtration, less waste, and increased farmer profitability are a few other environmental advantages of biochar. By boosting agricultural yields and holding onto water in drought-prone areas, this technology also helps ensure

R. Khatua (✉) · A. Nanda · N. Pradhan · S. Mahapatra · A. Mohanty
Department of Agricultural Meteorology, Odisha University of Agriculture and Technology, Bhubaneswar, India
e-mail: rajashreekhatua.agmet@ouat.ac.in

A. Mahapatra
Department of Agronomy, Odisha University of Agriculture and Technology, Bhubaneswar, India

P. Praveenkumar
Centre for Climate Change and Disaster Management, Anna University, Chennai, India

V. Guhan
India Meteorological Department, Hyderabad, India

D. Moulick et al. (eds.), *Biochar: Achieving Agro-Environmental Sustainability*, Sustainability Solutions, https://doi.org/10.1007/978-3-032-19855-6_10

food security. Since the process of making biochar is carbon-negative, it actually lowers atmospheric CO_2. The unstable carbon found in decomposing plant matter is transformed into a stable form during the biochar-making process, which is then stored in the biochar. Applying biochar to the soil allows the carbon to be safely stored for hundreds or even thousands of years. In other words, if the feedstocks used to make biochar were allowed to break down naturally, more carbon dioxide would be released into the sky. In the end, biochar technology lowers atmospheric carbon dioxide by heating the feedstocks and converting their carbon content into a stable structure that doesn't react to oxygen. By improving soil quality and lowering the demand for chemical fertilizers, biochar also helps to mitigate climate change by lowering greenhouse gas emissions. Additionally, the increased fertility of the soil promotes the development of plants, which absorb carbon dioxide. Biochar is a promising tool for regenerative agriculture because of its several advantages for agricultural systems and the climate.

Keywords Regenerative agriculture · Cabon sequestration · Global warming · Green house gases · Renewable energy · Carbon & climate change

10.1 Introduction

From the present and 2050, when the UN predicts there will be 10 billion people on the planet, the need for food is expected to quadruple. The world's arable land supply is limited. Man has been successfully turning forests into deserts for 6000 years, but we now have technology that can bring depleted areas back to productive levels. To produce more food more efficiently, agribusiness businesses will need to leverage these important factors. Global warming has been exacerbated by the rise in atmospheric carbon dioxide (CO_2) concentration, which has gone from 290 parts per million (ppm) in pre-industrial times to 425 ppm today (Gui et al. 2025). With 25% of the world's CO_2 exchange occurring in soil, the earth's largest terrestrial surface carbon reservoir is extremely important to the global CO_2 cycle (Fa et al. 2016). In addition to receiving carbon from the canopy through plant roots and litter, soil also sequesters CO_2 through inorganic abiotic processes. The concentration of soil gas changes as CO_2 dissolves in soil solution, which alters the soil's internal atmospheric pressure gradient and ultimately, the amount of atmospheric CO_2 that is pumped into the soil.

One of the major issues of the today is climate change, which is caused by CO_2 emissions. For instance, the air's CO_2 content has risen by more than 54.4% since the preindustrial era. Reducing carbon emissions and sequestering atmospheric CO_2 are the primary means of preventing the rise in global temperatures following the 2015 United Nations Climate Change Conference. In order to attain carbon neutrality, it is crucial to develop technologies that limit CO_2 emissions and lower atmospheric CO_2 concentrations. Along with rich carbon, biochar is characterized by high alkalinity, high porosity, and an abundance of minerals. The addition of biochar to soil for carbon sequestration may alter the soil's pH, CEC, and mineral composition, which might impact the soil's ability to absorb CO_2. One prominent carbon sink avenue for lowering atmospheric CO_2 is biochar technology, which is based on the pyrolysis of biomass and then stores the resulting biochar in soil. According to reports, biochar technology has the potential to reduce emissions by 3.4–6.3 Pg CO_2 per year and last for hundreds to thousands of years in soil (Lehmann et al. 2021, Joseph et al. 2021). Other non-climate benefits of this biochar carbon sequestration method include improved agricultural yields, improved soil, and reduced pollution.

Aqueous alkali and liquid amine are two examples of liquid adsorbents that are often used in industries. However, the main issues with liquid adsorbents are corrosion to the instrument and volatile loss (Zhang et al. 2023). Therefore, solid materials have been utilized in the carbon capture process to investigate the further viability of CO_2 adsorbents. Because of their high CO_2 capacity, ease of modification, affordability, and stability, solid adsorbents including metal-organic framework (MOF), zeolite, and carbonous materials show promise (Zhang 2023). Of the known solid adsorbents, carbonous compounds are the most ecologically benign. Specifically, biomass pyrolysis under inert circumstances produces biochar, one of the carbonous compounds. The production of biomass from urban and agricultural waste, which is the precursor of biochar, is significant worldwide, encouraging the manufacture of biochar in huge amounts. Furthermore, biochar has benefits for sustainable development, application stability, and environmental preservation. As a result of its exceptional production and versatility, biochar has been used in a variety of applications, including CO_2 capture, water purification, soil amendment, and catalytic support/catalysis. In contrast to other common solid CO_2 adsorbents, biochar primarily uses its physical characteristics to accomplish CO_2 adsorption, transit, and storage (Zhang et al. 2023). If biomass and biochar are used for carbon neutrality, it is anticipated that biochar may collect almost one gigaton of greenhouse gas annually, reducing emissions by almost 10% worldwide.

Large-scale biomass production worldwide, which provides sufficient sources for biochar ingredients, is primarily responsible for this exceptional environmental advantage. Biochar has virtually no negative environmental effects, as it is derived from biomass, a carbon-neutral substance that can accomplish carbon fixation. Biochar's stability and recyclability enhance its effectiveness in CO_2 capture. Additionally, one of the prerequisites for CO_2 collection is the porosity of biochar, and it is well known that micropores affect how well biochar captures CO_2. Activating the biochar is therefore a crucial step in improving the porous structure. Chemical and physical activation are the two primary techniques for activating biochar. By adding gases (such as CO_2 and H2O) to react with biochar at high temperatures and then removing the volatile to produce copious porosity, physical activation is achieved. Physical biochar activation has a lower activation strength than chemical activation, although being more mild and less polluting. On the other hand, in order to accomplish the goal of porosity creation, chemical activation of biochar depends on the reaction between active chemical substances (such as alkali, acid, or molten salts) and carbon. However, because of the high activity of activation agents, chemical activation conditions need to be carefully regulated to prevent biochar from being destroyed and having its structure collapsed (Zhang et al. 2023). Additionally, while biochar's physical adsorption of CO_2 is the primary mechanism, surface modification of biochar is often used to improve CO_2 chemical adsorption. The nitrogen deficiency in charcoal materials is compensated for by either nitrogen dope or amine grafting.

However, there are still issues with carbon capture using biochar. Interestingly, the procedure mostly relies on the biochar's porous nature. The impact of biochar's textural characteristics on CO_2 capture is unclear. However, due to the diverse feedstock and precursor components, biochar exhibits less homogeneity than hydrothermal char. Furthermore, a large body of research on the selection of biochar activation methods claims that chemical activation is superior than physical activation, but it does not compare the two. To support technological advancement, correlation study of the characteristics of biochar and its relationship to CO_2 collection performance may be crucial.

10.2 Biochar: The Black Gold

A "porous carbonaceous solid" created by thermochemically converting organic materials in an oxygen- depleted atmosphere, biochar possesses physiochemical characteristics that make it suited for long-term, safe carbon storage in the environment as well as possible, soil enhancement. It is a material that resembles charcoal and is created by burning organic waste from forestry and agriculture. The process of producing biochar converts carbon from certain biomasses into stable carbon that may be absorbed by the soil (Bombail 2022). Apart from its long-term carbon sequestration function, biochar also enhances soil performance by promoting the retention and diffusion of nutrients and water. Biochar is becoming more and more well-known as a soil supplement for agricultural use. It has been shown to improve soil fertility by altering the dynamics of soil nutrients and enzymes. Biochar can serve as a source of nutrients directly or indirectly (Wang et al. 2021).

Biochar, also known as "**Biological Charcoal**", acts as a carbon sink on agricultural fields by absorbing carbon from the atmosphere. Because it is biologically inaccessible, biochar may be used as a technique to capture net carbon from the atmosphere by storing fixed carbon in the soil for generations to millennia. Additionally, biochar slows down water runoff and reduces the requirement for fertilizer.

Agriculture fixes 30 gigatons of carbon annually, but there is little net effect 30 gigatons of carbon are released back into the atmosphere when the plants die (Urs 2024). Biochar recovers and stores a considerable portion of carbon in the ground when combined with compost, soil, and plants, resulting in a continuous and notable decrease in atmospheric GHG levels. Additionally, biochar increases agricultural output in marginal soils and lessens the requirement for fertilizer. Additionally, because biochar functions as a sponge, it greatly lowers runoff, benefiting rivers streams, and the seas while lowering the demand for petrochemical fertilizers. Burning or gasifying biomass creates biochar, a very resilient soil additive.

Waste (manure, agricultural or forestry wastes) is transformed into a solid, porous, carbon-rich, and stable material after being treated in high-temperature furnaces in an oxygen-limited environment. From a few decades to several millennia, this stability may occur. Because of its porous physical structure, biochar can store carbon and hold onto water and nutrients when applied to soil. It lessens water runoff and aids in restoring soil fertility.

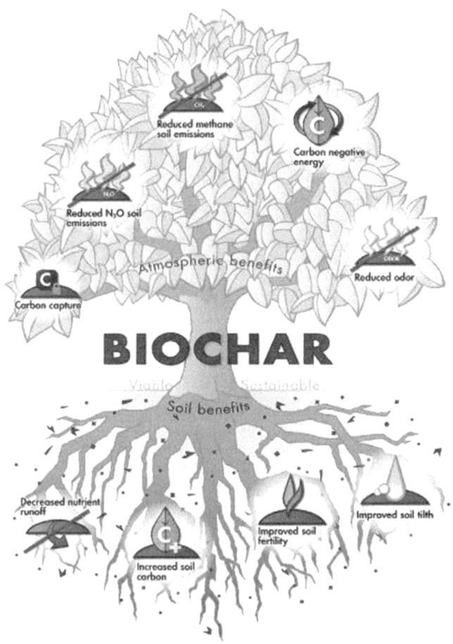

Source Biochar International

Although biochar is a fantastic product that may be very helpful in the early stages of the shift to regenerative agriculture, we must avoid the same error that industrial agriculture made, which was to make farmers reliant on outside inputs. A system that allows farmers to become as self-sufficient as feasible in terms of agricultural inputs, such as external amendments, nutrients, or chemical inputs, is a crucial component of regenerative agriculture. The characteristics of biochar produced at different temperatures (between 300° and 700 °C) vary, and it can change the soil in a number of ways. By buffering the soils, biochar can increase their resistance to crop diseases.

Pyrolysis is the process of heating organic material, such as biomass, without the presence of oxygen. The typical temperature range is between 300 °C and 800 °C (572 °F and 1472 °F) (Chatterjee et al. 2020). The organic material decomposes during pyrolysis, generating syngas, biochar, and bio-oil, depending on factors like temperature and heating rate. A variety of organic resources, including wood, crop leftovers, agricultural waste, and even municipal solid waste, can be used as feedstock for pyrolysis. The decision is influenced by local circumstances, intended final goods, and availability. The first step in making biochar is gathering biomasses, which might include things like wood chips, sawdust, agricultural leftovers, and even animal

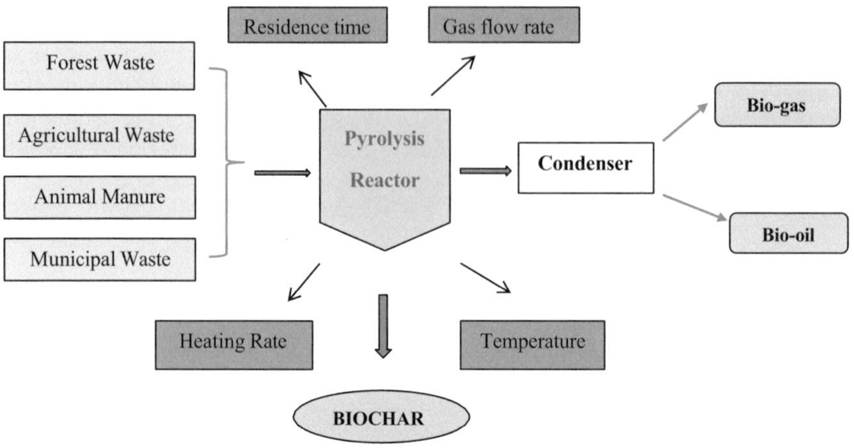

Fig. 10.1 Production of biochar

dung. Based on their availability and appropriateness for the pyrolysis process, these biomass sources were chosen. After being collected, the biomass is dried to lower its moisture level before being broken up into smaller pieces.

In a specialized reactor or kiln, the pyrolysis process takes place. The biomass undergoes a complicated chemical breakdown of organic materials when heated to a high temperature, usually between 400 and 700 °C, in an oxygen-free atmosphere. Volatile chemicals are as gases, which can be used to generate renewable energy. The remaining solid material further breaks down and carbonizes as the temperature rises, eventually producing biochar (Fig. 10.1).

10.3 Biochar in Agriculture

1. Improving soil fertility increases the soil's capacity to sustain agricultural production and encourage robust plant growth. It involves several steps, including as adding organic matter (such compost or biochar), adjusting nutrient levels, optimizing pH, and controlling soil erosion.
2. Nutrient availability and retention in soil refers to the soil's capacity to retain vital nutrients such as potassium, phosphorus, and nitrogen while making them accessible to plants (Akhil et al. 2021). Compost and biochar are examples of organic amendments that enhance nutrient retention by increasing cation exchange capacity and decreasing nutrient leaching.
3. Improving a soil's capacity to retain and exchange vital nutrients (cations) with plant roots is known as CEC improvement. Adding organic amendments like compost and biochar raises CEC, which increases the amount of nutrients accessible to plants for improved growth and nutrient use.

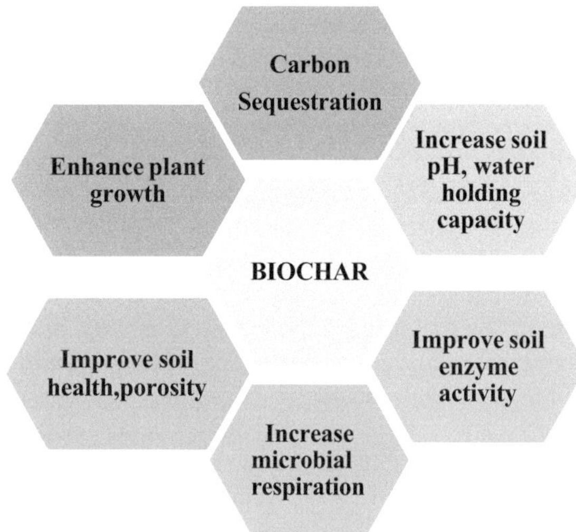

Fig. 10.2 Benefits of biochar application in agriculture

4. By lowering the concentration of CO_2, a greenhouse gas that causes global warming, it plays a critical role in limiting climate change (Lehmann et al. 2012). Reforestation., afforestation, and the use of carbon-storing materials like biochar are examples of practices that help store carbon and mitigate climate change.
5. Biochar improves soil structure by improving soil aggregation and porosity, which encourages water penetration and retention. Managing soil moisture improves the overall health of the soil and reduces the likelihood that plants may experience drought stress.
6. By changing the microbial composition and structure of the soil, biochar can lessen soil-borne illnesses by making the soil less conducive to pathogens (Parmar et al. 2014). This improves the general health of the soil and shields plants from diseases.
7. Biochar improves crop yield and quality by strengthening soil structure, enhancing nutrient availability, and enhancing water retention. Healthy, more productive plants result from higher yields and better quality, including enhanced flavor, nutrition, and disease resistance (Fig. 10.2).

10.4 Application Techniques of Biochar in Agriculture

An ancient technique for enhancing soil health is biochar. The Amazon Basin's indigenous pre-Columbian ancestors used biochar in agriculture for the first time more than 2000 years ago. The soil in the region has been rich for more than 800 years, according to tests conducted by specialists. This is due to the fact that the native inhabitants of the Amazon Basin created extremely rich and deep soils known as terra preta (or "dark earth") by using modified residual charcoal from cooking fires in the soil. Incorporation involves adding biochar or other organic amendments to the top layer of the soil using mechanical techniques like tilling or ploughs. This technique optimizes the benefits of improved soil structure, nutrient retention, and moisture control by guaranteeing uniform distribution. Biochar can be incorporated into the soil using a variety of equipment and methods, such as Plows, rototillers, or specialist incorporation machines. By using these methods to mix it in evenly, biochar's advantages for soil health and plant development are successfully incorporated into the soil. The kind of soil, the crop, and the goals all affect how much biochar should be added to the soil. Generally speaking, heavier soils may benefit from greater dosages of biochar (5–10%), whereas lighter soils may only require a little quantity of biochar (1–5% by weight). This improves the compost's overall quality, microbial activity, and ability to retain nutrients. By increasing the compost's ability to hold onto nutrients and reducing nutrient loss, biochar may be used as a useful soil addition rich in organic matter once the composting process is complete.

Biochar-amended compost is produced by adding biochar to the composting process. Charcoal and decaying organic materials, such as manure, yard waste, and kitchen scraps, are frequently combined to achieve this. Biochar composting has several benefits for plant and soil health. The inclusion of biochar improves the overall quality of compost by increasing its ability to retain nutrients and store water. When applied to soil, this combination enhances microbial activity, moisture management, and nutrient availability, leading to healthier plants and improved soil structure. When biochar or other ingredients are applied topically, they are not mixed into the soil first. To control soil temperature, reduce weeds, and retain moisture, a layer of biochar can be spread over the soil as a mulch (Fig. 10.3).

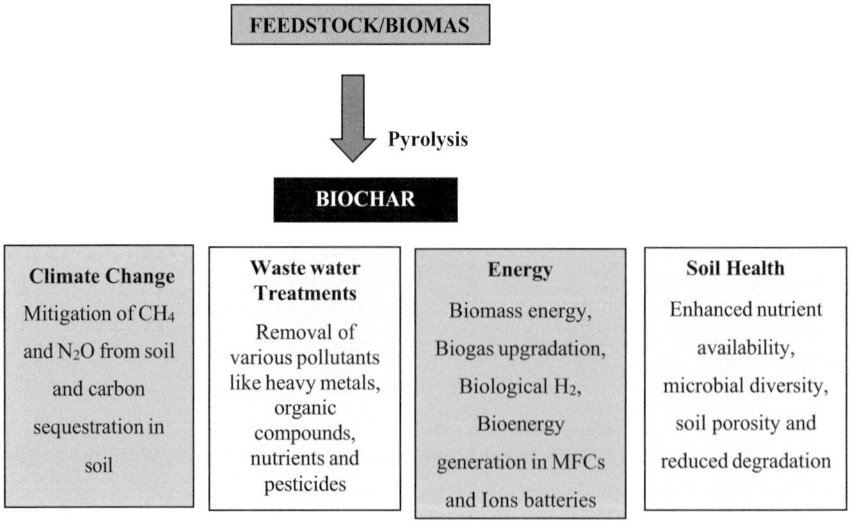

Fig. 10.3 Different purposes of biochar

10.5 Biochar's Impact on Soil Microorganisms

Because of its porous structure and improved habitat conditions, biochar can influence the microbial populations in soil by fostering the growth of beneficial bacteria and mycorrhizal fungi, which in turn improves nutrient uptake and increases agricultural yields. Biochar can also alter the pH and chemical composition of the soil, which can affect the activity of microbial communities. The porous design of biochar can also act as a haven for beneficial bacteria and fungi. The kind of biochar and the starting pH of the solution determine how much of this activity takes place, although biochar can act as a buffer, decreasing the acidity of the soil to help.

Biochar as a Carbon Credit Solution: Unlocking Carbon Sequestration Potential

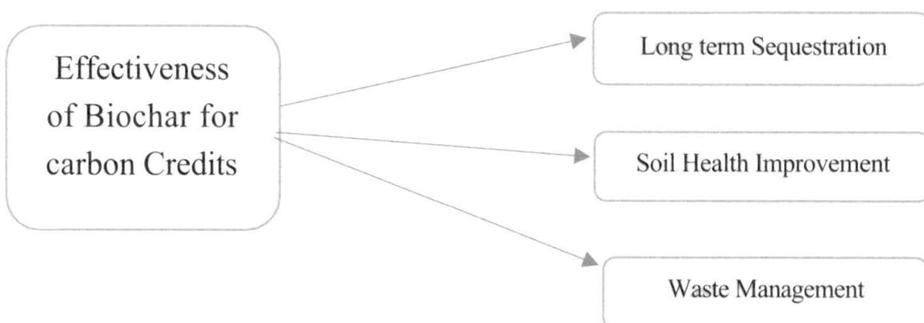

In the battle against climate change, biochar is becoming more and more acknowledged as a potent carbon credit option. Biochar, which is created by pyrolyzing organic waste at high temperatures in an oxygen-poor atmosphere, is a stable method of sequestering carbon that keeps it in the soil for hundreds to thousands of years. Because of this feature, biochar may be used to offset carbon emissions and create carbon credits.

10.6 Biochar Is Effective for Carbon Credits

(a) **Long-Term Carbon Sequestration:** Biochar stays stable and sequesters carbon in the
(b) Soil for an extended period of time, in contrast to other organic matter that breaks down and releases carbon back into the atmosphere (Farooq et al. 2025).
(c) **Improving Soil Health:** In addition to storing carbon, biochar increases soil fertility, retains water better, and stimulates microbial activity, all of which improve crop yields.
(d) It is a desirable option for the forestry and agriculture industries because of these two advantages.

(e) **Waste Management:** The manufacture of biochar recycles organic waste, forestry waste, and agricultural residues, lowering methane emissions from decomposing trash and offering an ecologically acceptable waste management option.

10.7 Biochar Generates Carbon Credits

(a) **Quantifiable Carbon Sequestration:** Carbon credit standards such as the Verified Carbon Standard (VCS) or Gold Standard, which evaluate the quantity of carbon dioxide sequestered throughout the pyrolysis process and over the biochar's lifetime, can be used to verify and certify biochar initiatives.
(b) **Certification Process**: Producers of biochar must monitor their production procedures, measure the amount of carbon sequestered, and submit to third-party certification in order to get carbon credits. Businesses looking to offset their emissions can then purchase these credits.
(c) **Market Demand**: The need for premium carbon credits is rising as governments and businesses work to achieve net-zero goals and adhere to laws like the Corporate Sustainability Reporting Directive (CSRD). Biochar provides a concrete, quantifiable solution to this problem. Biochar provides a concrete, quantifiable solution to this problem. Biochar improves regenerative agricultural methods and enhances climate resilience by turning organic waste into a carbon-negative resource. Because of this, it is an essential tool for companies trying to satisfy sustainability targets, lessen their carbon impact, and get carbon credits in a cutthroat market.

10.8 Biochar Sequester Carbon

(a) **Capturing Carbon from Biomass:** The carbon held in biomass, such as wood or plant waste, gets "locked" into a stable form when it is converted into biochar. This stops it from breaking down again into CO_2, which is what would normally happen if the biomass were allowed to break down organically.
(b) **Soil Carbon Storage:** Biochar may efficiently trap carbon that would otherwise contribute to greenhouse gas emissions by remaining stable in the soil for hundreds to thousands of years after it is introduced (Fawzy et al. 2021). Its porous structure also benefits the environment by enhancing soil health.
(c) **Reducing Emissions:** Potential pollutants can be converted into energy sources by designing the biochar-making process to catch and use the gases emitted during pyrolysis. This feature considerably lowers the carbon footprint associated with the manufacture of biochar (Fig. 10.4).

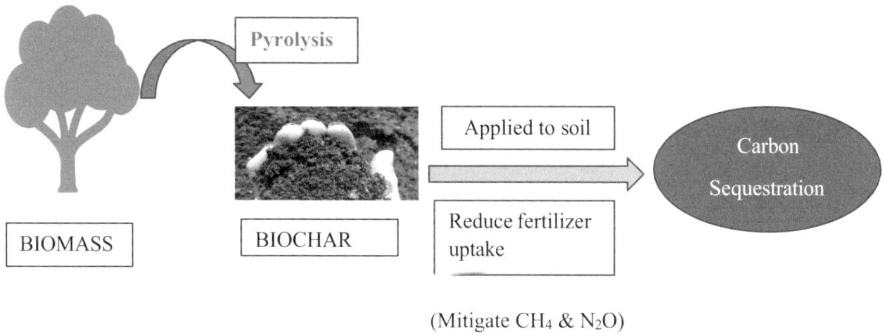

Fig. 10.4 Biochar for carbon sequestration

10.9 Biochar Fuelling Agriculture

(a) **pH:** By acting as a buffer in acidic soils, biochar can raise pH and enhance plant nutrient absorption.
(b) **Water retention capacity:** Compared to regular soils, soils modified with biochar have a greater capacity to retain water, which helps plants survive droughts for longer.
(c) **Carbon content:** In order for fertilizer usage efficiency to be high enough to recover its original investment, soil typically has to contain at least 3% carbon. Carbon in biochar stays in the soils for decades, whereas Kenyan soils have very little carbon.
(d) **Nutritional absorption:** Using biochar as a nutritional substrate can improve the effectiveness of fertilizer use (Nair et al. 2017). We are investigating the possibility of adsorbing nutrients, such as those obtained via urine collection, onto the surface of biochar.

10.10 Carbon Sequestration

By effectively absorbing and storing CO_2 from the atmosphere, carbon sequestration is an essential technique that contributes to the reduction of dangerous carbon emissions. Long-term storage of carbon in soils, plants, geologic formations, and the ocean is known as carbon sequestration. Carbon sequestration, which usually refers to the storage of carbon, can happen naturally or as a result of human activity.

10.11 Types of Carbon Sequestration

(a) **Terrestrial Carbon Sequestration**: The process by which CO_2 from the atmosphere is taken up by plants and trees through photosynthesis and stored as carbon in soils and biomass (tree trunks, branches, leaves, and roots) is known as terrestrial carbon sequestration.
(b) **Geologic Carbon Sequestration**: Oil and gas reserves, unmakeable coal seams, saline formations, and shale formations with a high organic content are all potential storage sites for CO_2.
(c) **Ocean Carbon Sequestration**: Large volumes of CO_2 from the atmosphere are absorbed, released, and stored by oceans. There are two approaches to do this: adding CO_2 to the deep ocean and using iron fertilization to increase the productivity of ocean biological systems. Iron deposition promotes the growth of phytoplankton, which in turn causes these microbes to produce more photosynthesis and aid in the absorption of CO_2.

10.12 Methods

(a) **Natural Carbon Sequestration**: It is how nature has balanced the amount of carbon dioxide in our atmosphere so that life can exist. Carbon dioxide is released at night by both plants and animals. Trees, the seas, the soil, and animals themselves are examples of natural carbon sinks, or sponges. Carbon is the foundation of all biological. Life on Earth, and when plants and animals die, a large portion of their carbon returns to the earth, where it has no effect on global warming.
(b) **Artificial Carbon Sequestration**: The term "artificial carbon sequestration"describes a variety of methods that absorb carbon emissions at the site of production (such as factories or chimneys) and bury them afterwards. Ocean sequestration is one suggested technique that involves injecting carbon dioxide deep into the ocean to create CO_2 lakes. Theoretically, because of the surrounding water's warmth and pressure, the CO_2 will remain deep below and eventually dissolve into it. Another example is geological sequestration, which involves pumping carbon dioxide into subterranean spaces like aquifers, coal seams that cannot be exploited, and former oil reserves.

10.13 A Brief History of Carbon Sequestration

The idea of carbon sequestration first surfaced in the early 1920s, when CO_2 scrubbers were developed to effectively remove contaminants from methane. Although the first carbon-capture facility was conceived in 1938, there was still no actual demonstration of the sequestration process (Marsh 2022). The marketed version of improved oil recovery introduced carbon sequestration and storage in the 1970s. The recovered carbon from oil production was injected and trapped in gas and oil reservoirs (Warwick and Demas 2023). Marchetti did, however, solidify the first contemporary idea of carbon sequestration. The Sleipner Carbon Sequestration and Storage Project was the first time it was appropriately applied to lower CO_2 generation from human activity. Another significant project that exemplifies large-scale CO_2 sequestration, use, and storage is the Weyburn–Midale CO_2 Monitoring and Storage Project (also known as the Weyburn Project), which was initiated in 2000 by the IEA Greenhouse Gas (IEAGHG) Research and Development Programme (Riding and Rochelle 2005)

10.14 Applications of Carbon Sequestration

Carbon sequestration technology is highly valued by researchers and experts, who estimate that it might cut carbon emissions by around 7 gigatons by 2030 (Sundquist et al. 2008). The market is expected to achieve a valuation of over 800 billion US dollars.

Recently, CO_2 has been captured and transformed into fuels for energy production through a process called carbon sequestration. This is revolutionary for the aviation and heavy transportation sectors in terms of reaching net-zero emissions.

10.15 The Future of Carbon Sequestration

For the direct collection of air and subsequent use of carbon as fuel, carbon sequestration is a crucial technique. It is anticipated that a number of carbon sequestration-focused firms would succeed. Globally, governments are making significant investments in carbon capture and storage technology, especially in Europe and Asia-Pacific (Surampalli et al. 2015).

To make carbon sequestration technology feasible for all countries, it is necessary to make sure that environmentally benign and reasonably priced carbon capture and storage solutions are required, as well as to investigate ways to stop carbon leaks.

10.16 Soil Carbon Sequestration

The process by which plants and trees absorb carbon dioxide from the atmosphere is known as soil carbon sequestration. Numerous biological activities, such as the assimilation of leaf and other plant litter by soil-dwelling organisms, cause carbon to get locked up in the soil. Plant residues are occasionally absorbed into ploughing and tilling as part of this process, which occurs as part of the agricultural cycle. The process of locking carbon into soils is known as soil carbon sequestration.

Sequestration is the result of several processes. Harvest leftovers can be used in agriculture as part of the cultivation process, such as during ploughing and discing. Other methods used in farming to provide nutrients and carbon-rich materials into the surface layers include the application of manures and composts as top dressings or via injection. Natural processes, such as those of fungus and wildlife, decompose leaves, plant litter, and fire debris in forests and uncultivated areas and integrate them into the top layers.

Through the preceding plant development that produced the plant, crop litter, or animal feed, these activities ultimately remove carbon dioxide from the atmosphere by incorporating carbon into soils. Carbon dioxide is not completely and permanently eliminated by these methods. Instead, they are a component of a dynamic balance between the soil processes that return carbon dioxide and methane (as well as nitrogen oxides) to the atmosphere and the aforementioned examples of carbon incorporations. Carbonaceous rich material does accumulate in soil where agricultural disturbance is minimal. Rising global temperatures raise the possibility that these equilibria may be upset and that more carbon dioxide will be released than is being stored. This might then result in a feedback loop.

The carbon in biochar is of recalcitrant carbon, which is resistant to and can in the environment for hundreds or even thousands of years**.** Biomass is stabilized and from being released back into the atmosphere as CO_2, which would the biomass feedstock is of growing biomass for of biochar is that atmospheric CO_2 is sequestered in the soil in a stable form.

According to research, converting 1% of annual plant biomass into biochar could help offset approximately 10% of carbon emissions caused by human activities (Dattappa et al. 2024). So, biochar can potentially indirectly reduce anthropogenic greenhouse gas emissions by improving soil quality through its enhancement of nutrient and water retention. This results in a reduction in the acidity of the soil and provides a more favourable habitat for beneficial soil microorganisms. These improvements can increase agricultural productivity and potentially reduce the need for synthetic fertilizers, a source of greenhouse gas emissions (Celignis Analytical).

Additionally, biochar production can utilize various types of waste biomass, including agricultural residues, forestry waste, and urban green waste, so diverting these materials from landfills or open burning, both of which can produce greenhouse gases.

10.17 Effects of Pyrolysis Conditions on Biochar Carbon Sequestration Potential

A. **Temperature:** One important element affecting biochar's capacity to sequester carbon is the pyrolysis temperature. Because more volatile chemicals are released and the residual carbon structure becomes more condensed and stable, higher pyrolysis temperatures usually result in the creation of biochar with higher carbon contents and greater recalcitrance. However, extremely high temperatures have the potential to reduce the carbon sequestration capability of biochar by causing significant carbon loss through the creation of gaseous products.

B. **Heating Rate**: The capacity of biochar to sequester carbon can also be influenced by the heating rate, or the rate at which the biomass is heated during pyrolysis. Biochar's with increased carbon contents may be produced by heating materials at rapid rates. However, uneven pyrolysis and the production of biochar's with a decreased capacity to sequester carbon might result from overly rapid heating rates.

C. **Residence Time:** increased thorough pyrolysis brought about by longer residence durations may produce biochars with higher carbon levels and increased stability. However, because more biomass is volatilized during extended residence durations, biochar with a lesser capacity to sequester carbon may be produced.

D. **Choice of Feedstock:** The quantity of carbon trapped in the final biochar might vary depending on the carbon content of the feedstock. More carbon feedstocks, such woody biomass, will often provide biochar with more potential for sequestering carbon than lower carbon feedstocks, like herbaceous or aquatic plants.

10.18 Assess Carbon Sequestration Potential of Biochar

(a) **Proximate and Elemental Analysis**: The amount of volatile matter and fixed carbon in biochar may be analyzed. Higher fixed carbon content biochars are generally more stable over longer time spans and are less likely to decompose and release CO_2 back into the environment. Using Duma's research, we can also identify the carbon content of biochar and distinguish between organic and inorganic carbon. There is more potential for carbon sequestration when the amount of organic carbon is larger.

(b) **Stability and Decomposition Tests:** These experiments evaluate biochar's potential lifetime in the environment as well as its resistance to microbial breakdown. In stability studies, CO_2 emissions from soil supplemented with biochar can be tracked over time or measured using methods like the incubation method, which involves mixing biochar with soil and allowing it to incubate under controlled circumstances.

(c) **Carbon Sequestration Efficiency:** This measure contrasts the overall carbon content of the initial biomass feedstock with the quantity of carbon content sequestered in the biochar. It may be used to assess how well various feedstocks and pyrolysis conditions work to produce biochar with a high potential for sequestering carbon.

10.19 The Benefits of Carbon Farming and Regenerative Agriculture

The process by which plants and trees absorb carbon dioxide from the atmosphere is known as soil carbon sequestration. Numerous biological activities, such as the assimilation of leaf and other plant litter by soil-dwelling organisms, cause carbon to get locked up in the soil. Plant residues are occasionally absorbed into ploughing and tilling as part of this process, which occurs as part of the agricultural cycle.

(a) **No tillage**: Minimize top soil disturbance, use tiny drills, and plant straight into the earth. More water storage, microbial development, and healthier, more robust plants are the results of this.
(b) **Cover crops:** In order to support the soil biology and create a diversified, healthy soil, plant cover crops in between seasons.
(c) **Agroforestry—planting perennials and trees**: This is the process of fusing agriculture with trees. This might involve growing trees close to agricultural areas, planting trees directly into crop fields, or farming in an existing forest setting. In addition to bringing in pollinators and biodiversity above ground, this maximizes the positive interactions between trees and crops below ground.
(d) **Brief intervals of grazing (movement of livestock):** Soils can be greatly enhanced by animals. They consume the topmost layer of organic matter, excrete nitrogen-rich dung and urine, and use their hooves to break up the topmost layer of material, all of which promote development. It's critical that they leave as soon as possible so that the area can recover. The additional preparation and effort required is worthwhile.
(e) **Mulching with organic matter**: Give the earth something to break down so that it may absorb nutrients and microbes.

All of these methods have the potential to increase localized rainfall, remove carbon from the atmosphere through photosynthesis, and lessen the likelihood of desertification, which turns the soil into dust and leaves it susceptible to erosion by wind and water.

10.20 Future Carbon Capture and Storage (CCS) Technologies

(1) **Eliminating carbon dioxide from industrial operations**: There is also increasing interest in removing carbon dioxide from industrial processes such the production of steel, cement, and electricity. In order to create a concentrated stream of pure carbon dioxide that can be dried and compressed for piping or shipping to a well and pumped as a super critical fluid into a deep sub-surface formation (rock) where it would be permanently stored, this usually uses a chemical system to capture the carbon dioxide in one step and release it in a second.

Another method is to use a membrane that selectively transmits carbon dioxide while preventing the transmission of other gases. After drying and compressing, the carbon dioxide would be piped, transported, and injected into the reservoir. These seemingly straightforward procedures conceal a number of obstacles that must be removed in order for these procedures to function. These procedures are expected to be quite expensive. Although the ground's carbon dioxide is worthless, it poses a long-term risk due to the requirement to continue monitoring for leaks and well and reservoir stability.

(2) **Direct air capture:** Some people, including Microsoft, believe that removing CO_2 directly from the atmosphere can help make up for previous CO_2 emissions. Others believe that employing the recovered CO_2 in greenhouse horticulture has a business benefit. Although they aim for $100 per tonne, Climeworks recently announced that the cost of stripping off the gas was $800 per tonne. This would include the expense of sequestering the collected gas in a subsurface aquifer, for example. The DAC option seems to be highly costly with the present methods. About $2900/tonne of carbon is equal to $800/tonne of CO_2 capture. Biochar would be a great deal on this premise.

(3) **The mineralization of atmospheric carbon dioxide:** All of these methods come with hefty prices, a lot of technological challenges, and massive scale requirements. Biochar has the potential to be utilized as a carbon dioxide emission offset. Biochar is a good offsetting medium since it doesn't break down in the environment. Furthermore, it is a valuable commodity that has the potential to be used on a very big scale in horticulture, forestry, urban areas, and agriculture. By employing electrical heating with renewable power, the carbon dioxide emissions from the conventional techniques of producing biochar may be drastically decreased. For this possibility to be successful, the biochar market must be developed quickly. With all these approaches, the costs are high, there are many technical obstacles to overcome and the scales required are enormous. There is an opportunity for biochar to be used to offset carbon dioxide emissions.

Biochar is resistant to degradation in the environment making it a suitable vehicle for offsetting. What's more it is a product with value and prospectively value at very large scale through its use in agriculture, horticulture, forestry and in the urban environment. The traditional methods for making biochar are themselves carbon dioxide intensive but using electrical heating with renewable electricity this can be radically reduced. The market in biochar needs to be expanded with some urgency to enable this opportunity to flourish (Biochar.co.uk).

10.21 Biochar and Regenerative Agriculture

Regenerative agriculture and biochar work together as natural means of enhancing soil health while using less water and fertilizer. Utilizing waste is another advantage of integrating biochar with regenerative agriculture as biochar can be made from biomass waste, including wood waste and agricultural waste. Producing biochar makes sense when farms and ranches need to heat things like greenhouses and shop buildings in the winter. This is because using trash to create heat, biochar, and carbon credits may all have a big financial impact while raising soil carbon levels.

The carbon credits from biochar are worth well over $100 per ton, which is far more than the majority of conventional soil-based carbon credits (Jennifer 2025). This is because biochar is an extremely stable form of carbon that lasts for thousands of years and is straightforward to weigh.

There is growing pressure on agriculture to minimize its environmental effect while producing more food with less resources. It provides a potent remedy that enables farmers to improve soil fertility, increase carbon capture, and adopt genuinely sustainable farming practices. Farmers can solve urgent issues like soil deterioration and expensive fertilizer prices by using these biochar solutions, all the while making the earth healthier.

10.22 Modern Farming Challenges in Agriculture

Farmers now deal with a variety of challenges:

(a) **Soil Degradation:** Excessive chemical usage and over farming can deplete vital nutrients, lowering yields and endangering the health of the soil.
(b) **Climate Risks:** Crops are at risk from droughts, floods, and other extreme weather occurrences, which increases unpredictability and expenses.

10.23 Biochar and Regenerative Agriculture

(c) **Managing Sustainability and Productivity:** While traditional techniques may increase yields in the short term, they frequently have a negative impact on the ecosystem and soil fertility over the long run.

Thus, without sacrificing sustainable agricultural methods, growers are looking for innovative ways to protect their property and increase income.

10.24 Biochar Use as a Soil Amendment

As a soil supplement, biochar can improve productivity, decrease fertilizer runoff, and increase the soil's ability to store water. This can be done in a number of ways, such as the widely used approach of mixing biochar with manure before application. Biochar can be added to the grow mixtures used by greenhouse farmers. In soil, biochar promotes the development of live microorganisms. It turns decomposing soil into alive soil by establishing a microbial super hotel.

(a) **Biochar Use in Seed Mixes Including Cover Crops**: The benefits of using biochar adjacent to the seed, which helps retain moisture and enhances germination without the need for additional tools or procedures, may be maximized by mixing it in with the seed during planting. Benefits will be increased in the future by supplementing the charcoal with nutrients and biology from sources like compost tea.
(b) **Biochar Use in Manure Management**: Biochar can be added to manure to decrease odour and preserve more nutrients. This may be done by either adding biochar to the manure before spreading it, or by putting it to the pens and allowing foot activity to incorporate it into the manure.
(c) **Biochar and Bale Grazing:** Bale grazing and biochar may be used to improve the soil by incorporating dung and biochar with hoof activity. This works well on unproductive fields where a bale may be placed immediately on top of biochar that has been put to the surface. With hoof motion, the hay's seed, dung, and biochar are all mixed into the soil.
(d) **(d)Biochar Use in Animal Feed**: Animals given tiny doses of biochar can benefit from a wide range of advantages. Numerous research studies have shown that removing accumulated toxins, strengthening immune systems, increasing milk production, lowering methane, and more.
(e) Importance of Biochar for Regenerative Farming:
(f) Adding pyrolyzed organic matter (biochar) to the soil is the main idea behind biochar for regenerative farming. This method combines carbon capture, soil fertility, and prudent resource management into a single, cohesive system.
(g) **Better Soil Structure:** Because biochar is porous, it improves aeration and water retention, promoting strong root development and making it easier for crops to withstand droughts and heavy rains
(h) **Decreased Chemical Dependency**: Biochar reduces the demand for synthetic fertilizers and pesticides by better retaining nutrients, which saves money and lessens its environmental impact.
(i) **Long-term Carbon Storage**: Biochar helps with carbon capture, which is a crucial step in achieving net-zero objectives, by locking carbon into the soil rather than releasing it into the atmosphere.
(j) **Enhanced Microbial Life:** Soils rich in biochar support robust microbial communities that thrive, increasing soil fertility and crop resilience.
(k) **Impact on the Real World:** When using biochar for regenerative farming, farmers and growers may encounter the following issues: Increased Crop Yields: Stronger crops produce more plentiful harvests when nutrient availability and soil structure are optimized.
(l) **Pest and Disease Resistance:** Plants with better natural defences thrive in healthier soils that have been farmed using biochar solutions.

(m) **Reduced Input Costs:** Farms become more sustainable and lucrative when less fertilizer and insecticide are used. Whether you oversee large-scale operations or are a smallholder, using biochar can yield long-term benefits for soil health as well as immediate profits.

10.25 Biochar Use as a Climate Mitigation and Adaptation Strategy

There are two primary types of mitigation: balance and reduction. Reducing, or cutting present emissions of different greenhouse gases (GHGs), appears to be the main focus of attention and investment. There are several methods to do this, but among the most often accepted suggestions are to use cleaner energy sources (such as solar, wind, geothermal, hydro, etc.) and/or less energy (Surampalli et al. 2015). Reducing is far from enough, but it is definitely necessary to mitigate the worst scenarios. Perhaps even more crucial is rebalancing. Rebalancing entails returning surplus atmospheric CO_2 to the earth's surface (I'll let others talk about ocean sequestration and release).

There are several ways that biochar can aid in the reduction process. The process of making biochar generates renewable energy in the form of heat that may be transformed into electrical power. The great majority of this heat may be utilized to replace energy obtained from fossil fuels, while occasionally some of it is used to dry feedstock before carbonization.

On the reduction side, we may also look at the usage of fertilizers, notably nitrogen (N), which has a particularly high carbon footprint due to its energy-intensive manufacture and even higher carbon footprint during application. In an attempt to increase yearly production in soils that have long ago run out of natural nutrients, farmers worldwide are using excessive amounts of nitrogen. Nitrous oxide (N_2O), which is 300 times more harmful to the planet than carbon dioxide (CO_2) (Kessel et al. 2013), is produced when microbes metabolize nitrogen. Because it gives the nutrients a bonding structure that releases them gradually and stops unwanted leaching into groundwater or as greenhouse gases into the atmosphere, biochar can be blended with either synthetic fertilizer or organic nutrients (such as manure, urine, or compost) to reduce the amount of fertilizer required.

Additionally, biochar can lower methane (CH4) emissions from at least two important sources: trash and cows. The 1.5 billion cows and other cattle throughout the world emit a startling quantity of CH4, a greenhouse gas that is 25–80 times more powerful than CO_2 (Draper 2017). After natural gas, livestock is the second-largest source of CH4 emissions in the United States. According to research, including biochar into livestock feed can significantly lower enteric methane emissions while also improving the health of the animals. Additionally, it may be used as an inexpensive method of introducing biochar into soils (via manure). Likewise, studies have demonstrated that the use of biochar in landfills, either as a daily cover or as a capping material once landfills are retired, can similarly lower CH4 emissions.

By keeping organic waste out of landfills, which are the third-largest source of CH4 emissions in the US, biochar may also drastically cut emissions. Carbonization reduces the amount of biomass by 75–90%, depending on the conditions of production (Draper 2017). Under the correct conditions, this can drastically cut down on transportation and stop CH4 from ever being produced at landfills.

By increasing photosynthesis and disrupting or significantly slowing down the regular carbon cycle, biochar can help mitigate climate change in two important ways. It has been demonstrated that applying biochar to crops or trees improves plant growth and survival, which can increase total photosynthetic activity. Plants keep the CO_2 they absorb until they die, at which point it typically breaks down again into CO_2 and rejoins the carbon cycle. The natural carbon cycle is disrupted, though, if the trees or plants are turned into durable goods (like furniture) or carbonized into biochar. Up to 50% of the carbon absorbed by the tree or plant over its lifespan is transformed into stable carbon when biomass is carbonized. If the resultant biochar is stored in soil or incorporated into other long-lasting products (like construction materials), it won't return to the atmosphere for many centuries. This is an illustration of what is often called carbon sequestration.

10.26 Climate Change Adaptation

The focus of adaptation is on methods that may be used to address the effects of climate change that are already being felt. Relocating people affected by rising sea levels, altering planting dates and/or crop kinds, improving drainage, or implementing more effective integrated pest management are just a few of the increasing number of adaptation techniques being used. Similar to mitigation, there are several ways that biochar might aid in climate change adaptation, some of which are included here.

Global food security is being severely impacted by climate change. There are more frequent and severe storms, floods, and droughts. In addition to climate change, additional concerns to food security include increased heat, illness, and land degradation. Building more resilient food production systems may be achieved by using biochar to increase crop yields, increase plant resilience to pests and diseases, and prolong water retention in drought-prone locations. Additionally, biochar aids in lowering irrigation water requirements, which is crucial in a world where water resources are becoming scarcer. According to certain studies, it can aid in replenishing and purifying groundwater resources, which might result in more robust crops and better human health. Certain biochar situations fall in between adaptation and mitigation. These include reforestation, often known as afforestation, urban agricultural techniques including green roofs and tree installation, and regenerative agriculture, which aims to improve soil fertility and biodiversity.

10.27 Biochar as a Future for Sustainable Agriculture

The push for sustainable agriculture is essential, not just a fad. Farmers must optimize their operations in ways that safeguard the environment as climatic pressures increase and the world's population rises. Biochar is a scalable and useful method for regenerative farming.

(a) **Climate-Resilient Farming**: Soils enhanced with biochar are better equipped to withstand temperature fluctuations and unpredictable rainfall.
(b) **Economies that are circular:** Resource loops are closed and waste is decreased when forestry or agricultural waste is turned into biochar.
(c) **Corporate Responsibility**: More companies understand the importance of biochar solutions for ethical supply chains and ESG plans.

Despite its apparent simplicity, biochar has the ability to address a number of global issues, including. A billion tons of carbon could be sequestered each year by the pyrolysis process, which would then remain in the soil for thousands of years, where it would be most useful. An alternative to burning fossil fuels is the clean, renewable energy that is created as a byproduct of the manufacture of biochar. Increases agricultural yields and retains water in drought-prone areas, so promoting food security (Rai and Saraswat 2023). Among the many environmental advantages of biochar are its ability to reduce groundwater contamination and the expense of water purification.

Four main points may be used to summarize the advantages of biochar:

(a) **Simple to use:** Applying biochar is simple.
(b) **Long-term consequences:** Because of its high carbon content, biochar doesn't decompose over time.
(c) **Eco-friendly:** Three tons of carbon dioxide are permanently removed from the atmosphere for every tonne of biochar buried in the ground.
(d) **Enhances the health of plants:** It has been demonstrated that biochar improves the health and vigour of almost all plant species.

10.28 Conclusion

Because it combines soil health improvement with carbon sequestration, biochar has a revolutionary potential in agriculture. As a sustainable approach, it enhances soil fertility and agricultural productivity in addition to mitigating climate change by long-term carbon capture. Farmers may build resilient ecosystems, support international sustainability initiatives, and make major progress toward a carbon-neutral future by implementing biochar processes. Adopting biochar is not just a positive move for individual farms; it is also an essential component of the larger initiative to encourage responsible land management and fight climate change. Regenerative farming using biochar provides a dynamic way to strike a balance between ecological health and productivity. Biochar is a powerful tool in sustainable agriculture, offering significant carbon capture advantages and enhancing soil fertility. Consider collaborating with industry leaders like Carbon Gold if you're prepared to investigate how biochar solutions might improve the resilience and profitability of your farm. By working together, we can improve soil health, increase yields, and ensure that agriculture throughout the world has a more sustainable future.

References

Akhil D, Lakshmi D, Kartik A et al (2021) Production, characterization, activation and environmental applications of engineered biochar: a review. Environ Chem Lett 19:2261–2297

Biochar Analysis, Production and evaluation. Celignis Analytical

Bombail S, Yogo G, Rafael RD (2022) Biochar: a carbon sequestration tool for rural communities. Ecoact

Carbon Sequestration Biochar.co.uk, https://biochar.co.uk/soil-carbon-sequestration/

Chatterjee R, Sajjadi B, Chen WY, Mattern DL, Hammer N, Raman V, Dorris A (2020) Effect of pyrolysis temperature on Physico chemical properties and acoustic-based amination of biochar for efficient CO_2 adsorption. Front Energy Res 8:85

Dattappa R, Chethan BL, Rohan PL, Ali S, Arunkumar CM (2024) Biochar: a green revolution in forestry. Vigyan Varta 5(11):148–150

Draper K (2017) Biochar's unmatched versatility in mitigating & adapting to climate change. Biochar J

Fa, K. Y., Zhang, Y.Q., Wu, B., Qin, S.G, Liu, Z., She, W.W. (2016). Patterns and possible mechanisms of soil CO_2 uptake in sandy soil. Sci Total Environ 544, 587–594.

Farooq S, Fatima A, H., Riaz, U., Rizwan, N. (2025) Soil health improvement and pollution remediation using biochar. Springer Science and Business, Media LLC

Fawzy S, Osman AI, Yang H, Doran J, Rooney DW (2021) Industrial biochar systems for atmospheric carbon removal: a review. Environ Chem Lett

Gui X, Xu X, Zhang Z (2025) Biochar-amended soil can further sorb atmospheric CO_2 for more carbon sequestration. Commun Earth Environ 6(5)

Initiate IB (2012) Standardized product definition and product testing guidelines for biochar that is used in soi. Int Biochar Initiate

Jennifer L (2025) The biochar gold rush: why companies are scrambling to lock in carbon credits, Carbon Credits

Joseph S, Annette L, Zwieten LV, Bolan N, Budai A, Buss W, Cayuela ML, Graber ER, Ippolito JA, Kuzyakov Y, Lehmann J (2021) How biochar works, and when it doesn't: a review of mechanisms controlling soil and plant responses to biochar. GCB Bioenergy 13(11):1731–1764

Lehmann J, Cowie A, Masiello CA, Kammann C, Woolf D, Amonette JE (2021) Biochar in climate change mitigation. Nat Geosci 14:883–892

Lehmann J, Gaunt J, Rondon M (2006) Bio-char sequestration in terrestrial ecosystems–a review. Mitigation Adapt Strat Glob Change 11:403–427

Lehmann J, Joseph S (2012) Biochar for environmental management. Earthscan

Kessel CV, Venterea R, Six J, Adviento-Borbe MA, Linquist B, van Groenigen KJ (2013) Climate, duration, and N placement determine N_2O emissions in reduced tillage systems: a meta-analysis. Glob Change Biol 19:33–44

Marsh J (2022) History and pitfalls of carbon sequestration. Renew Energy Mag

Nair VD, Nair PKR, Dari B, Freitas AM, Chatterjee N, Pinheiro FM (2017) Biochar in the agroecosystem-climate-change-sustainability nexus. Front Plant Sci 8:2051

Parmar A, Nema PK, Agarwal T (2014) Biochar production from agro-food industry residues: a sustainable approach for soil and environmental management. Curr Sci 107(10):1673–1682

Rai JPN, Saraswat S (2023) Green technologies for waste management—a wealth from waste approach. CRC Press

Riding JB, Rochelle CA (2005) The IEA Weyburn CO_2 monitoring and storage project. Final report of the European research team. British geological survey research report RR/05/03, pp 54

Sundquist E, Burruss R, Faulkner S, Gleason R, Harden J, Kharaka Y, Tieszen L, Waldrop M (2008) Carbon sequestration to mitigate climate change. USGS Publication

Surampalli RY, Zhang TC, Tyagi RD, Naidu R et al (2015) Carbon capture and storage. American Society of Civil Engineers (ASCE)

Urs K (2024) Assessing the mitigation potential of biochar as a permanent carbon storage solution: a comparative analysis of natural and artificial storage methods. India BioChar and BioResources Network

Wang F, Harindintwali JD, Yuan Z, Wang M et al (2021) Technologies and perspectives for achieving carbon neutrality. The Innovation

Warwick PD, Demas A (2023) Using petroleum reservoirs to store carbon. U.S, Geological Survey

Zhang C (2023) Porous biochar material for carbon capture and wastewater treatment. School of Chemistry and Chemical Engineering

Zhang C, Ji Y, Li C, Zhang Y, Sun S, Xu Y, Jiang L, Wu C (2023) The application of biochar for CO_2 capture: influence of biochar preparation and CO_2 capture reactors. Ind Eng Chem Res 62(42):17168–17,181

Chapter 11
Advances on Preparation and Applications of Biochar and Its Nanocomposite Materials

Puja Saikia, Jafrin Hasnu, and Chira R. Bhattacharjee

Abstract Biochar, a porous substance and part of the carbon family, is quite new to research area for variety of use. This carbon rich charred compound can be produced from carbonization of waste biomass in oxygen-limited condition. The attention for biochar has continuously been increasing due to its high specific surface area, excellent adsorption capacity, high ion exchange capacity, and most importantly low cost production in high scale from waste biomass and also directs towards elimination of further carbon footprint. However, studies have found out the fact that certain limitations associated with biochar can be addressed and some properties can be enhanced substantially when combined with other nanostructured materials. Thus produced biochar nanocomposites (BNC) by surface modification of biochar with nanomaterials have more extensive application area over pure biochar due to the increased porosity, modified surface functional groups, more active sites. This chapter includes compilation of previously published literature mainly focused on synthesis protocol of biochar, BNC, characterization using different techniques, investigation of the properties of biochar and its composites including antimicrobial, waste water treatment (adsorption of heavy metals, dye, antibiotics, other organic contaminants removal), soil treatment, sensing etc. Along with detailed discussion, suitable schemes and tables have also been incorporated to shed light on the potentialities of biochar and BNC over tradition carbon based materials.

Keywords Biochar · Nanocomposite · Pyrolysis · Degradation · Adsorption

List of Abbreviations

NPs Nanoparticles

P. Saikia · J. Hasnu · C. R. Bhattacharjee (✉)
Department of Chemistry, Assam University, Silchar 788011, India
e-mail: crbhattacharjee2@gmail.com

J. Hasnu
Department of Chemistry, Dakshin Kamrup College, Mirza Guwahati 781124, Assam, India

D. Moulick et al. (eds.), *Biochar: Achieving Agro-Environmental Sustainability*, Sustainability Solutions, https://doi.org/10.1007/978-3-032-19855-6_11

BC	Biochar
BNC	Biochar nanocomposite
RH	Rice husk
TEM	Transmission electron microscopy
WC	Wood chip
PW-BC	Plant waste biochar
AW-BC	Animal waste biochar
ATB	Acid treated biochar
MBC	Magnetic biochar composite
CV	Crystal violet
SEM	Scanning electron microscopy
XRD	X-ray diffraction
EDS	Energy dispersive X-ray analysis
TGA	Thermogravimetric analysis
CNT	Carbon nanotube
SDBS	Sodium dodecylbenzene sulfonate
ZVI	Zero valent ion
TOC	Total organic carbon
TSS	Total suspended solid
TOrCs	Trace organic contaminants
TN	Total nitrogen
TP	Total phosphate
PLB	Poultry litter biiochar
CEC	Cation exchange capacity
COD	Chemical oxygen demand
TKN	Total Kjeldahl Nitrogen
ZVF	Zero valent iron
PCP	Pentachlorophenol
HMs	Heavy metals
CPE	Carbon paste electrode
MFC	Microbial fuel cell
BPA	Bisphenol A
PCBs	Polychlorinated biphenyls
PAHs	Polycyclic aromatic hydrocarbons
MIPs	Molecularly imprinted polymers
DPV	Differential pulse voltammetry
DBP	Dibutyl phthalate
LOD	Limit of detection
MCLR	Cyanotoxin microcystin-LR
N-CDs	Nitrogen-doped carbon dots
ZIF	Zeolitic imidazolate framework
DOC	Doxycycline
TC	Tetracycline
CTC	Chlortetracycline
MOFs	Metal organic framework

NSC	Nitrogen doped Soulangeana sepal derived biochar
GCE	Glassy carbon electrode
FMBC	Ferromanganese oxide-impregnated biochar composites
EDTA	Ethylenediaminetetraacetic acid
GHG	Greenhouse gas
β-CD	β-cyclodextrin

11.1 Introduction

Carbon based materials have been under limelight due to its varied importance including academic curiosity since 1985, when the fullerene was discovered and subsequently its application in various fields were explored. Since then, there is a great impetus in research on various carbon based materials like nanocarbon, matrix materials. Biochar (BC) and its diverse application is a relatively newer field of scientific pursuits, developed only in past few decades (Khodakovskaya et al. 2012; Mohamed et al. 2018). The earliest evidence of biochar usage for agricultural soil amendment has been documented in Amazon basin some > 2500 years ago, where 'terra preta' (dark soil or black earth) has been shown to possess enhanced fertility owing to the occurrence of high carbon content (Marris 2006). It, however, is also produced naturally in soil due to vegetation fires. The term 'biochar' is of more recent origin. Biochar is basically a composite material comprising predominantly a form of porous solid carbonaceous material combined with varying amounts elements such as H, O and inorganic ashes. It is produced from renewable and abundantly available biomass in a sustainable, inexpensive way leading to reutilizing the waste into resources (Bolan et al. 2022). The biological feedstocks used for biochar production constitute sources such as agricultural waste material, animal litters, and other solid waste biomass (Panahi et al. 2020; Qin et al. 2020) (Fig. 11.1). In simple terms, the biomass is pyrolyzed under a controlled condition in limited oxygen environment to produce the black coloured solid BC. High stability of BC is attributed to their unique nature of carbon structure. The presence of high proportion of aromatic carbon and condensed aromatic structure makes it more stable and different from other organic matter (Lehmann et al. 2011; Nguyen et al. 2010). Depending on the pyrolyzing temperature, the presence of condensed aromatic structure vary. Predominantly, amorphous carbon is present when pyrolyzed at low temperature, while turbostratic carbon with irregularly stacked layered structure and graphitic carbon are produced as primary organic matter at high pyrolyzing temperature (Schmidt and Noack 2000; Keiluweit et al. 2010). Apart from organic carbon, it also contains other micro- or macro-elements such as potassium, magnesium, silica, iron, copper etc. in minute quantities. The morphology of biochar is characterized to have high surface area tucked with several functional groups with numerous pores of sizes (Lonappan et al. 2016).

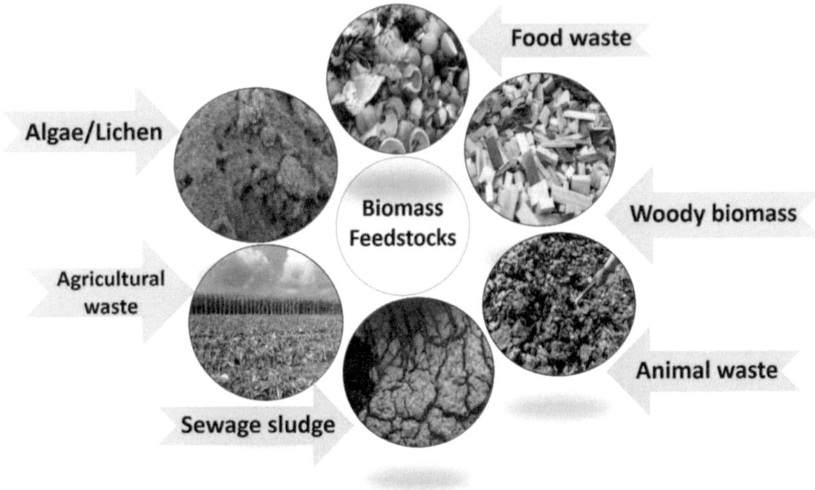

Fig. 11.1 Different types of biomass feedstocks for biochar production

These pores in biochar aid in improving the soil quality by providing space for symbiotic microorganisms.

For several years in the beginning, application of biochar has been confined primarily to improve soil quality and enhance crop yield. The internal structure of biochar enables it to preserve the nutrients, retain ground water for longer time which ultimately improve the soil quality (Nagaraju et al. 2023; Sohi et al. 2010). Alongside, biochar facilitates increased carbon sequestration in soil, lowering the emission of ammonia and carbon dioxide gases, maintaining a balanced pH level in the soil thus aiding the development of bacteria, fungi necessary for soil amendment and plant growth (Haefele et al. 2011; Hernandez-Mena et al. 2014). It has been observed that the effectiveness of biochar for soil amendment depends on factors like soil type, crop type and nutrients balance. An earlier report revealed the increased production of peanut and corn by simply using bark biochar (Yamato et al. 2006). Great influence on soil composition and microbial properties were noted by using biochar (Liang et al. 2010). Effectiveness of biochar for restoring fertility in desert area and toxic waste site is also on record (Van Zwieten et al. 2010; Ding et al. 2013). Recently, considerable scientific interest has been directed to explore the potential application of biochar in different fields like as adsorbents for waste water treatment, photocatalyst, sensing application (Wurzer and Mašek 2021). The physicochemical properties of BC based materials such as high specific surface area, extended porosity, high abundance, great adsorption capacity makes it suitable candidates for its outgrowth application. Biochar has been extensively applied as adsorbent for polluted aquatic system which gives double benefit as way of "C" sequestration and waste water remediation (Chen et al. 2017; Saxena et al. 2013). Though the advantages associated with biochar such as abundant feedstocks, economically viable synthesis, suitable physicochemical structures enabled its various applications, but there are some limitations linked with it. The constraints related with raw biochar comes from the pyrolyzing condition, pyrolysis temperature and feedstocks which greatly influence its physical and chemical characteristics. Separation of powdered biochar after treating the polluted system has turned out to be another challenge that limits its application (Chen et al. 2011; Yao et al. 2013a, b; Reddy and Lee 2014). Modifications of pristine biochar biologically, chemically or physically can be an alternative to overcome these issues. Chemical modifications involve treating biochar with organic functional groups, acids, alkali, nanomaterials (Wang and Wang 2019; Ngah et al. 2008; Chin et al. 2022). Engineered nanomaterial and biochar-composites not only improve the capacity of biochar via surface modification, but also evolve as superior composite material having the synergistic advantages of both nanomaterial and biochar. Hence, efforts have been made towards synthesis of tailored biochar-based nanocomposite materials to overcome the aforementioned limitations in order to broaden its application field (Shakoor et al. 2021; Chausali et al. 2021). The combination of nanoparticles and biochar integrates their individual advantages to create superior nanocomposite material. Introducing nanomaterials into biochar matrix helps in enhancing the functional groups, active sites, catalytic degradation ability, pore size of biochar by providing higher surface to volume ratio (Lehmann and Joseph 2015; Sohi 2012). Similarly, biochar as substrate material possess unique benefits like economical, large scale production, great matrix material that simultaneously increases the property of nanocomposite (Meyer et al. 2011).

Yan et al. (2015a, b) observed that by combining ZnS nanoparticle to BC matrix increases the sorption capacity of BC by 10 times (Yan et al. 2015a, b). In yet another approach, CeO_2-based BC nanocomposite accessed by hydrothermal method, exhibited higher degradation of Reactive Red 84 dye than the pristine biochar showing higher degradation rate with increasing amount of CeO_2-H@BC nanocomposite (Khataee et al. 2018). In another study, MnO_2 loaded biochar nanocomposite material has been used for successful removal of heavy metals from waste water (Zhang et al. 2020a, b). Nanocomposite synthesized under different reaction condition, using different biomass lead to variation in physicochemical properties which have significant influence on the property of nancomposite material. For instance, changing MnO_2 with NiO metal oxide to synthesize BC composite abruptly increases the adsorption capacity of Pb(II) (Tan et al. 2018; Saravanakumar et al. 2019). Similarly, Physical modification of BC has also been carried out using nitrogen, carbon dioxide doping to enlarge its applicability. The number of published work in scopus indexed journal on biochar based nanocomposite in the last one and half decade reflect an growing trend sharply rising since 2018 (Fig. 11.2).

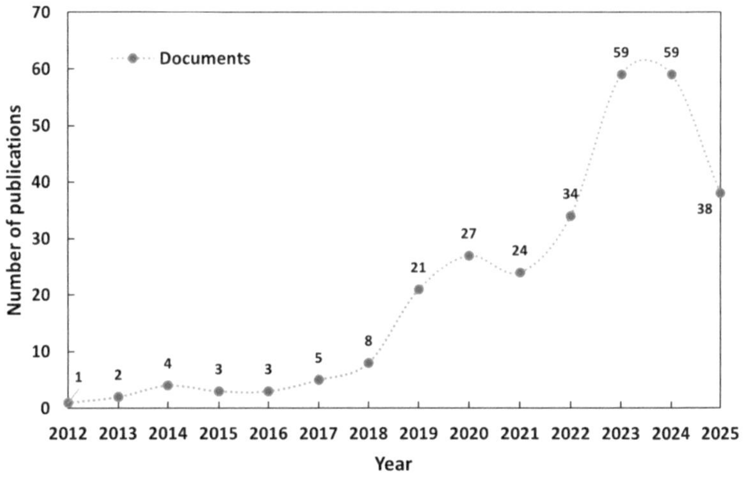

Fig. 11.2 Publication number of selenium-based nanocomposites in last 14 years (Scopus database, searched keywords: "biochar and nanocomposite and materials", June 2025)

11.2 Preparation and Characterization of Biochar and Biochar-Based Nanocomposite Material

11.2.1 Production of Biochar/Nanobiochar

The production of biochar normally involves thermochemical decomposition of biomass at the temperature range of 300–700 °C in limited oxygen environment which is called pyrolysis process of biochar formation. Agricultural biomass are more preferably used in this process which undergoes thermo-chemical process like gasification, slow/fast pyrolysis, torrefaction, and carbonization (Ghodake et al. 2021; Zabot et al. 2020). The different synthesis process produces different proportion of biochar, bio-oil and syngas. Rapid pyrolysis and gasification of biomass predominantly yield biofuel and syngas, whereas, slow pyrolysis (O_2-free environment) yields almost 35% of biochar with minimum bio-oil (Mohan et al. 2014a, b, c; Pourkarimi et al. 2019). Hence, slow pyrolysis method is most commonly preferred for biochar production. Biomass with high lignocellulosic material produces high amount of biochar than the solid municipal waste (Ashiq et al. 2019). In order to improve the efficacy of biochar, chemical modification such as acid/base treatment, CO_2 activation, spin coating, and physical modification such as ball milling have been used (Ramanayaka et al. 2020). Conversion of raw biochar to nano-size effectively increases its property which led to development of various techniques for the preparation of nanobiochar. Various methods are now available for nanobiochar preparation viz., mechanical grinding, flash ignition, ultrasonic vibration, among which mechanical grinding/mashing is commonly used (Yang et al. 2018; Anupama and Khare 2021). In ball milling, nanobiochar is formed by grinding, collision of nanoparticle that results in nanosized biochar with agglomeration and large size. Due to costly operational procedure, double-disc milling for nanobiochar production is not that popular (Ramanayaka et al. 2020). In another hydrothermal method suggested by Guo et al., agricultural biomass were used to produce bulk biochar which was then treated with acid in hydrothermal reaction in high pressure to get BC nanodots of size 4–5 nm (Guo et al. 2020). Some previously reported methods of biochar production using different biomass feedstocks, pyrolysis temperature and application is listed in Table 11.1.

Table 11.1 Sources of biomass, method of preparation and application of different biochar

Type of BC	Biomass used	Method of preparation	Application	References
Pruning waste based biochar	Peat	Partial decomposition	Antimicrobial and phytotoxic properties	Yao et al. (2020)
Sugarcane biochar	Sugar cane bagasse	Pyrolysis	Antimicrobial activity of the PHBV-AgNP	Gonçalves et al. (2018)
Sugarcane biochar	Sugar cane bagasse	Ball milling	Removal of methylene blue dye	Lyu et al. (2018)
AgNPs incorporated rosin-derived biochar	Rosin	Rapid pyrolysis	Antibacterial application against *E. coli* and *S. aureus*	Gale et al. (2021)
Dairy-manure derived biochar	Dairy-manure	Pyrolysis	Removal of Pb and atrazine	Cao et al. (2009)
Biosolid biochar	Biosolid	Microwave assisted pyrolysis	Removal of Ag from aqueous solution	Antunes et al. (2017)
Rice-husk and wood-chip biochar	Rice husk (RH) and wood chip (WC) of pine tree	Thermal decomposition	Removal of levofloxacin	Yi et al. (2016)
Plant and animal waste biochars	Plant waste (PW-BC) and animal waste (AW-BC)	Pyrolysis	Improvement of soil quality and the removal of organic contaminants	Mohan et al. (2014a, b, c)
Activated and magnetized biochar	Exhausted husk and dry tannin of black wattle	Pyrolysis	Removal of pesticides	Matos et al. (2017)
Tea waste biochar	Tea waste	Pyrolysis	Removal of sulfamethazine	Uzun et al. (2010)

11.2.2 Synthesis of Biochar Based Nanocomposite Material

The biochar-based nanocomposite materials can be categorized into three types based on the types of nanoparticles used in the synthesis process which are: oxide/hydroxide biochar nanocomposite, functional material such as graphitic carbon nitride, graphene, chitosan etc. coated biochar nanocomposite and magnetic biochar nanocomposite (Tukimin et al. 2018). According to the types of nanocomposite, the fabrication method also varies as shown in Fig. 11.3.

11.2.2.1 Nano Metal Oxide/Hydroxide Biochar Nanocomposite Material

Target Element Fortification via Bio-accumulation

In this method, selected target elements are bio-accumulated upon biomass which was then processed to produce engineered biochar nanocomposite with enriched properties. The selective elements present in high amount in the biomass are converted to hydroxide or nano-metal oxide after performing thermal treatment (Thakre et al. 2021). Following an innovative method, tomato plants were grown in a special medium treated with 25 mM Mg solution to increase Mg content (Yao et al. 2013a, b). The MgO and $Mg(OH)_2$ present in nanoscale amount in the Mg-enriched biochar composite acted as potential adsorbent for phosphate. The recorded XRD spectra confirmed the presence of oxides and hydroxide of Mg along with Mg-P in the form of $MgHPO_4$ and $Mg(H_2PO_4)_2$ (Fig. 11.4a). Till date, scant research have been carried

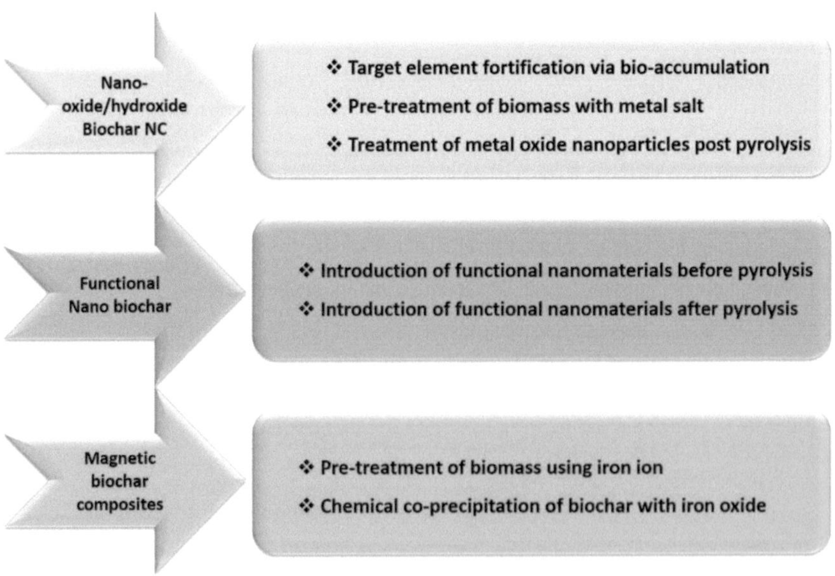

Fig. 11.3 Different synthesis methods used for the preparation biochar nanocomposite

out on the foreign elements fortification by bio-accumulation to produce nanocomposite. Therefore, one of the future scope of research lies in this domain where hyper accumulator plants can be used to enhance the bio-accumulation of targeted elements.

Pre-treatment of Biomass with Metal Salt

In this method, biomass is pretreated with metal salt, as chemical reagents, before pyrolysis. Once the biomass is dipped into the salt medium, the metal ions might get accumulated on the surface or might get penetrated to the interior. After pyrolysis, the metal ions impregnated biomass will convert to hydroxide/metal nano-oxide BC based nanocomposite (Yao et al. 2013a, b). This method was applied to synthesize MgO-BC nanocomposite where biomass pretreated with $MgCl_2$ salt pyrolyzed to get highly porous nanocomposite which can effectively remove phosphates and nitrates (Zhang et al. 2012a). Similarly, performing thermal pyrolysis to $AlCl_3$ pretreated biomass resulting AlO(OH) blended biochar nanocomposite formation where grown AlO(OH) flakes on biochar increased the reactive area and site for successful adsorption of heavy metals, organic contaminants (Zhang and Gao 2013). It was reported that when $ZnCl_2$-pretreated biomass was pyrolyzed to obtain ZnO-BC nanocomposite, the BET specific surface area, pore size and Cr(VI) removal efficiency got increased many folds (BET = 21.28 $m^2\ g^{-1}$ for composite; pristine biochar BET = 1.98 $m^2\ g^{-1}$) (Gan et al. 2015).

Treatment of Metal Oxide Nanoparticles Post Pyrolysis

In this technique, biomass is pyrolyzed in the first place where metal oxide nanoparticles are attached. Several methods are used in this technique viz.: heat treatment, evaporative, classical wet impregnation technique and directly hydrolyzes (Abdelhameed et al. 2022). Applying heat treatment to the mixture of $KMnO_4$ and BC results in the formation of nano/micro-MnO_X-BC composite with enhanced removal capacity of Cu^{2+} (Song et al. 2014). In this study, three different $KMnO_4$ percentages were applied to form three different biochar composite. The SEM images revealed that increased percentage of $KMnO_4$, led to the carbon surface getting gradually covered with agglomerated MnO_X, where composite with 2.5% $KMnO_4$ have more porous and smoother surface than the other composites (Fig. 11.4b, c, d). The EDS study also confirmed the presence of Mn and O on the surface; the atomic ratio increases with increased amount of $KMnO_4$ (Fig. 11.4e, f, g). Evaporative method was employed for the synthesis of iron oxide embedded BC composite using $Fe(NO_3)_3.9H_2O$ (Cope et al. 2014). Manganese oxide loaded biochar nanocomposite produced via classical wet impregnation method showed higher sorption ability for Pb than pristine material (Wang et al. 2015b). Directly hydrolyzing iron with hickory biochar produced Fe-drenched biochar where iron hydroxide nanoparticles lie on the biochar surface acting as primary site for adsorption (Hu et al. 2015) (Table 11.2).

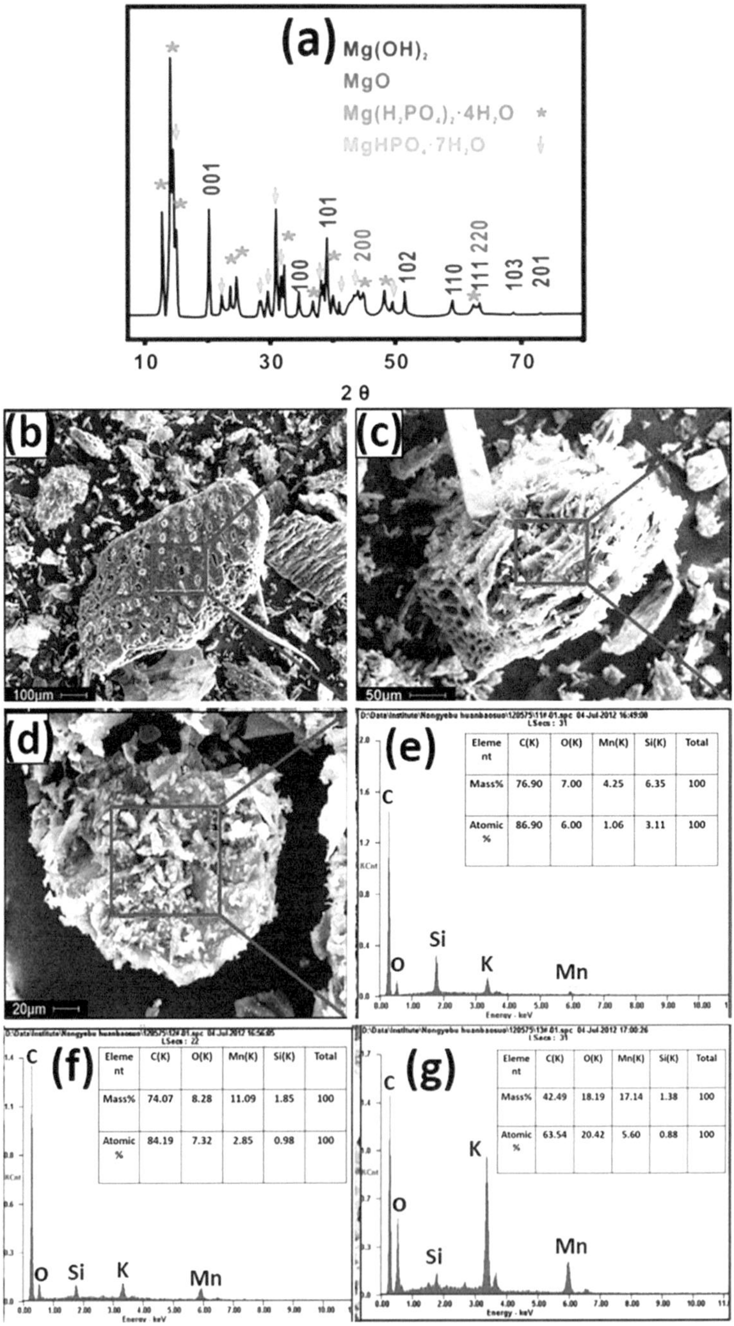

Fig. 11.4 **a** XRD pattern of Mg enriched BC composite (adapted from: Yao et al. 2013a). **b**, **c** and **d** SEM images of the MnO_X-loaded BCs of 2.5% MBC, 10% MBC and 60% MBC, respectively, and corresponding EDS spectra (**e**, **f**, **g**) (adapted from: Song et al. 2014)

Table 11.2 Type of biomass, nanocomposite, technique used and application of nano-metal oxide/hydroxide biochar composite

Biomass used	Nanocomposite	Technique used	Application	References
Tomato plant tissue	$Mg(OH)_2$, MgO-BC	Pyrolysis of Mg-enriched biomass at 600 °C for 1 h	Removal of phosphate	Yao et al. (2013a, 2013b)
Carya tree waste	MnOx-Hickory biochar	Pretreated biomass treated with $KMnO_4$ at 300–600 °C for 1 h	Removal of Pb(II), Cu(II), Cd(II)	Wang et al. (2015a)
Corn waste	Mg nanoparticle-BC	Pyrolysis of $MgCl_2$-pre treated biomass at 300–600 °C for 1 h	Removal of phosphate	Fang et al. (2014)
Cotton wood	AlOOH/BC	$AlCl_3$ pre-treated biomass pyrolyzed at 600 °C for 1 h	Arsenic, methylene blue, phosphate	Zhang and Gao (2013)
Peeled pine wood	Amorphous hydrous manganese oxide-biochar	Classical wet impregnation method	Removal of Pb(II)	Wang et al. (2015b)
Sugarcane bagasse	ZnN-BC	Pyrolysis of zinc nitrate-pretreated biomass at 450 °C for 1 h	Removal of Cr(VI)	Gan et al. (2015)
Rice husks	FeO-BC	Pyrolysis of ferric nitrite-BC at 450–1050 °C for 30 min	Removal of arsenic	Cope et al. (2014)
Marine macroalgae (*Laminaria japonica*)	Mg–Al oxide/hydroxide-BC	Pyrolysis of biomass treated by with $MgCl_2$ as an electrolyte	Removal of phosphate	Jung et al. (2015)

11.2.2.2 Synthesis of Magnetic Biochar Nanocomposite

Basically two synthesis methods have been used for the conversion of biochar to magnetic material: (i) prior treatment of feedstock with iron ion (Fe^{3+}, Fe^{2+}), (ii) co-precipitation of biomass with iron oxide. By producing magnetic nanoparticle impregnated biochar composite, magnetism and more active sites are introduced which enhanced its contaminant removal efficiency as well the separation of magnetic biochar, after treatment, becomes more feasible (Yang et al. 2020; Wan et al. 2020).

Pre-treatment of Biomass Using Iron Ion

This method involves impregnating biomass with Fe^{3+}, Fe^{2+} or chemical co-precipitation of these ions followed by pyrolysis or microwave heating under specific conditions. Zhang et al. (2013) performed thermal pyrolysis to $FeCl_3$ pretreated biomass which gives rise to Fe_2O_3 incorporated biochar composite with good ferro-magnetic properties and high adsorption capacity for arsenic. Magnetic biochar nanocomposite produced via chemical co-precipitation technique using Fe^{3+}, Fe^{2+} also reported on orange peel biochar before pyrolysis (Chen et al. (2011). Another approach of magnetic biochar synthesis has been reported where $FeCl_3$ infused palm oil fruit waste was treated in microwave which can effectively remove methylene blue with high adsorption capacity (Mubarak et al. 2014). Biomass can also be pretreated with Fe as well as other ions to form magnetic BC composite. For example, when $Fe(NO_3)_3.9H_2O$ and $Co(NO_3)_2.6H_2O$ were slow pyrolysed with pretreated biomass it resulted cobalt ferrite particles embedded on biochar surface with improved adsorption capacity for heavy metals (Pb(II),Cd(II)) (Reddy and Lee 2014). Moreover, magnetic biochar composite can also be produced by treating biomass by naturally occurring hematite mineral. Hematite modified biochar so produced possess enriched property than pristine biochar (Bakshi et al. 2018).

Chemical Co-precipitation of Biochar with Iron Oxide

Chemical co-precipitation of iron oxides on biochar is another commonly used strategy for magnetic biochar formation. Herein, the biomass is pyrolyzed in first place to produce biochar, then aqueous suspension of it is treated with ferrous/ferric solution and aqueous NaOH (Ren et al. 2015; Sun et al. 2015). Sun et al. 2015 reported the synthesis of a magnetic biochar using Fe_3O_4 nanoparticle coating. The acid treated-biochar (ATB) was previously prepared and the resultant magnetic biochar composite (MBC) was used for crystal violet (CV) dye removal. The FT-IR spectra (Fig. 11.5a) revealed that functional groups in the ATB are comparatively less compared to previously reported biochar. The presence of Fe_3O_4 in the sample was indicated by the presence of peak at 572 cm^{-1} caused for Fe–O of iron oxide, which was not present in other spectra. Upon adsorption of CV dye onto the composite, slight changes in the FT-IR spectra were observed. Recording the morphological features of ATB and MBC by TEM, coating of biochar with Fe_3O_4 nanoparticle was not limited on the external surface, but also resides in its interior. This suggested the ability of the synthesis method to produce Fe_3O_4 coated magnetic biochar composite (Fig. 11.5b, c). Some examples of magnetic biochar synthesis using this procedure have been reported from oak bark, oak wood and eucalyptus leaf biochar (Mohan et al. 2014a; Wang et al. 2014, 2015c). Of these two methods, magnetic biochar produced via chemical co-precipitation of iron oxide on biomass prior to pyrolysis possessed superior physicochemical properties than co-precipitation after pyrolysis (Baig et al. 2014).

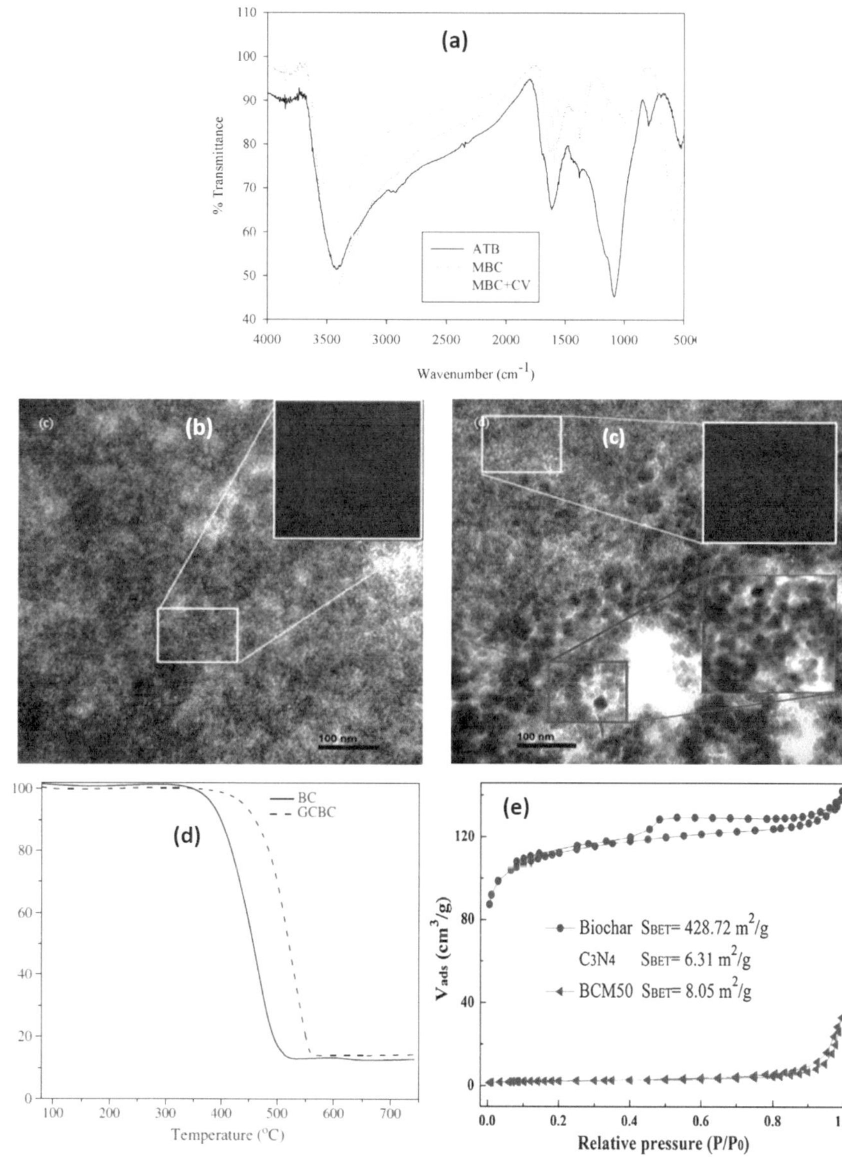

Fig. 11.5 **a** FT-IR spectra of acid-treated biochar (ATB), BC coated with Fe_3O_4 (MBC), and crystal violet-adsorbed Fe_3O_4 coated biochar (MBC + CV). **b** TEM image of acid-treated biochar, **c** TEM image of Fe_3O_4 coated biochar (adapted from: Sun et al. 2015), **d** TGA of BC and graphene coated BC (Zhang et al. 2012b), **e** BET isotherm of BC, C_3N_4 and its composite (adapted from: Pi et al. 2015)

11.2.2.3 Biochar Nanocomposite Coated with Functional Material

Formation of nanocomposite material by coating of biochar with functional material such as graphitic carbon nitride, graphene, graphene oxide, chitosan, ZnS nanocrystals, layered double hydroxide lead to generation of functional material coated biochar composite with some combined advantages and properties. Coating with functional nanoparticles lead to improvement of surface rea, surface functional groups, porosity, thermal stability of composite material (Kumar et al. 2020a, b; Qin et al. 2020).

Introduction of Functional Nanomaterials Before Pyrolysis
In this method, deep coating technique is applied for conversion of biomass to functional nanoparticle coated biochar formation. In general, functional nanoparticles of choice are added to deionized water and a homogeneous suspension is made using ultrasonication. In this suspension, biomass is dipped and dried in oven which is later pyrolyzed at specific temperature (He et al. 2018). A graphene coated biochar was prepared using this method via annealing graphene/pyrene-derivative coated biomass (Zhang et al. 2012b). Attaching graphene layer to biochar remarkably increases its thermal stability as observed from TGA. The graphene layer imparted thermal stability to biochar composite which showed about 64 °C higher degradation temperature than pristine biochar (Fig. 11.5d). Another graphene/biochar composite accessed by slow pyrolysis of graphene coated wheat straw biomass has been reported in which the graphene layer was wrapped in biochar via π–π interaction. This contributed to larger surface area, high thermal stability, more number of functional groups, enhanced removal efficiency of phenanthrene and mercury compared to pristine biochar (Tang et al. 2015). Similar method has also been employed to synthesize multi-walled carbon nanotube (CNT) coated biochar by dipping the biomass into the solution of carboxyl functionalized CNT before pyrolysis (Inyang et al. 2014, 2015).

Introduction of Functional Nanomaterials after Pyrolysis
In this method, nanomaterials are impregnated to raw biochar after pyrolysis to fabricate functional nanomaterial coated biochar nanocomposite with synergistic advantages of both the constituents. Some of the common functional nanomaterials used in this process are: chitosan, ZnS nanocrystals, Mg/Al LDH, and hydrogel (Wang et al. 2020a, b; Zhou et al. 2013). The enhancement of active sites upon incorporating functional materials to biochar occurs after deposition of carbon surface within biochar matrix which increases the contaminant removal efficiency. An example of functional nanomaterial coated biochar composite synthesis include coating of chitosan onto biochar surface. The as-synthesized chitosan-biochar composite has superior property owing to porous network structure of biochar and high surface area and superior chemical property of chitosan. One of the drawbacks noted in this system was clogging of pores of biochar by the functional nanomaterials. This ultimately reduces the specific surface area of biochar (Zhou et al. 2013). Similar result was observed in case of C_3N_4-BC composite prepared from chestnut leaf biomass and melamine as C_3N_4 precursor. The BET isotherm of C_3N_4, BC and C_3N_4-BC

composite (Fig. 11.5e) demonstrated that combining C_3N_4 and BC subsequently decreased the specific surface area and thereby, adsorption capacity (Pi et al. (2015). Therefore, more research is warranted to address these negative attributes associated with this synthesis process (Table 11.3).

Table 11.3 Sources of biomass, method of preparation, properties and application of some functional nanomaterial based biochar composite

Biomass used	Method of preparation	Properties	Application	Ref.
Cotton wood	Graphene/pyrene derivative-pretreated biomass (pyrolysis at 600 °C)	Thermal stability of graphene coated BC composite was increased	Removal of methylene blue	Zhang et al. (2012b)
Wheat straw	Graphene-pretreated biomass (slow pyrolysis at 600 °C for 1 h)	Graphene coated BC with extended surface area, more functional groups and thermal stability	Removal of phenanthrene	Tang et al. (2015)
Hickory chips and sugarcane bagasse	Biomass dip-coated in CNT suspensions with sodium dodecylbenzenesulfonate (SDBS)-aided dispersion before pyrolysis	CNT coated BC, stabilization provided by SDBS surfactant	Removal of Pb(II) and sulfapyridine	Inyang et al. (2015)
Eichornia crassipes	Chitosan covered on biochar/γ-Fe_2O_3	Enhanced magnetism and more functional groups	Removal of Cr(VI)	Zhang et al. (2015)
Chicken droppings, wood, and old car tire	Natural hydrogel loaded with biochar matrix	Hydrogel-BC composite	Removal of phenol	Karakoyun et al. (2011)
Bamboo	Used chitosan attached the ZVI particles onto bamboo biochar surfaces	Increased number of chitosan and ZVI particle	Removal of Pb(II), Cr(VI), As(V), phosphate and methylene blue	Zhou et al. (2014)
Chestnut leaves	Melamine condensed with biomass at 520–600 °C	C3N4 coated-BC composite	Removal of methylene blue	Pi et al. (2015)
Paper mill sludge	Insertion of the zero-valent iron (ZVI) on biochar surface	ZVI-BC composite with improved pore structure	Removal of pentachlorophenol	Devi and Saroha (2014)
Rice hull	Insertion of ZnS nanocrystals on magnetic biochar	ZnS encapsulated on magnetic BC	Removal of Pb(II)	Yan et al. (2015b)

11.3 Application of Biochar Based Nanocomposite Material

Now-a-days, biochar based nanocomposite are being extensively explored in applications covering multiple fields owing to their diverse and superior property (Fig. 11.6). Some of the potential applications are discussed in the following section.

11.3.1 Application in Waste Water Treatment

Biochar has shown significantly potential as a sustainable material for remediating pollutants across various sectors, including industrial and agricultural for enhancing environmental quality. Wastewater originating from domestic, industrial, commercial, and agricultural sources continues to be a global environmental issue (Wang et al. 2017). This section primarily explores the use of biochar in treating different types of wastewater, including industrial, municipal, agricultural, and stormwater (Fig. 11.7). Industrial wastewater is generated from large volumes of industrial wastewater are discharged into natural water bodies such as rivers, lakes, and ponds without undergoing proper treatment, contributing significantly to environmental degradation. Moreover, the high cost of adsorbents can hinder both commercialization and the advancement of circular economy practices. Major sources of industrial wastewater include smelting operations, chemical and pharmaceutical industries, dye and leather production, battery manufacturing, mining, and more. The predominant pollutants in industrial effluents are heavy metals and organic compounds. Moreover, the insoluble chromate compound Cr(VI) has been successfully recovered from

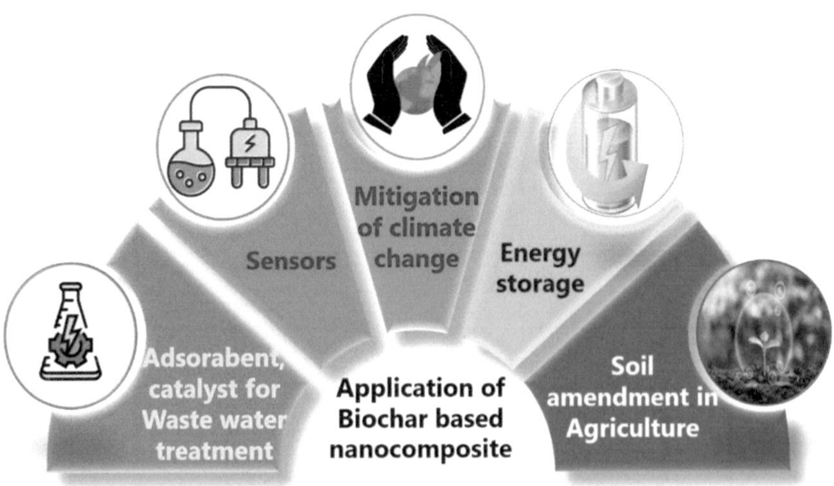

Fig. 11.6 Different application of biochar based nanocomposites

wastewater in certain cases. Biochar has been extensively studied for its potential in treating industrial wastewater. When combined with chitosan and cross-linked, it can be transformed into membranes, beads, or solutions, making it an effective adsorbent for removing heavy metals such as copper, lead, arsenic, and cadmium (Hossain et al. 2011). The adsorption efficiency depends significantly on the biochar and chitosan ratio. Additionally, biochar has shown promise in removing toxic dye from aqueous solutions, with the adsorption process influenced by factors like surface area, pH and pore volume (Wathukarage et al. 2019). Bagasse-derived biochar has been used to adsorb lead from effluents in the battery manufacturing industry, achieving a maximum capacity of 12.7 mg/g, with pH, contact time, and dosage playing key roles (Poonam and Kumar 2018). Moreover, biochar has been applied to recover nutrients like ammonium and phosphate from dairy wastewater, with adsorption rates ranging from 20–43% for ammonium and 19–65% for phosphate within 24 h (Ghezzehei et al. 2014). While these results are promising, most studies to date have been conducted in controlled laboratory environments only, highlighting the need for further investigation and validation under real-world industrial conditions. Utilizing biochar in combination with biofilters and other treatment technologies for municipal wastewater management, renders nitrogen and phosphorus readily available (Cole et al. 2017). In other approach, engineered biochar enhanced with aluminum oxyhydroxides ($AlHO_2$) has been applied to reclaim and reuse phosphorus from secondary treated wastewater (Zhang and Gao 2013). The primary mechanism for phosphorus adsorption on engineered biochar is electrostatic attraction. Once adsorbed, the phosphorus can be repurposed as a slow-release fertilizer to support crop growth. Biochar derived from digested sludge has been effectively used as an adsorbent for removing ammonium from municipal wastewater. Biochar produced at 450 °C demonstrated the highest ammonium removal efficiency, attributed to its increased surface area and abundance of functional groups, with the adsorption process primarily governed by chemisorption (Tang et al. 2019). Moreover, biochar made from waste sludge has been shown to serve as a catalyst in the ozonation of refinery wastewater, achieving significant total organic carbon removal. This enhanced performance is due to the presence of functional metal oxides, Si/O structures and carbon groups which facilitate oxidation by generating hydroxyl radicals and breaking down petroleum based contaminants (Chen et al. 2019).

In municipal wastewater treatment, biochar produced from municipal bio-waste has been applied during the biofiltration phase. Due to its high porous surface area, biochar functions efficiently as a biofilter, reducing COD by 90%, TSS by 89%, TKN by 64%, and TP by 78%. Furthermore, residential wastewater from areas lacking connection to municipal sewage systems has been treated using biochar treatment within on-site sewage facility (Blum et al. 2019). Furthermore, biochar significantly improves the removal of various polar and hydrophilic compounds, making it a cost-effective upgrade for on-site sewage facility. Stormwater runoff, containing degradation levels of metals, organic compounds, and biological contaminants, can severely impact the quality of natural water bodies. The quality required properly treated before being discharged into the environment (Gray 2016; Ashoori et al. 2019). Biochar and its modified nanocomposite have been successfully utilized

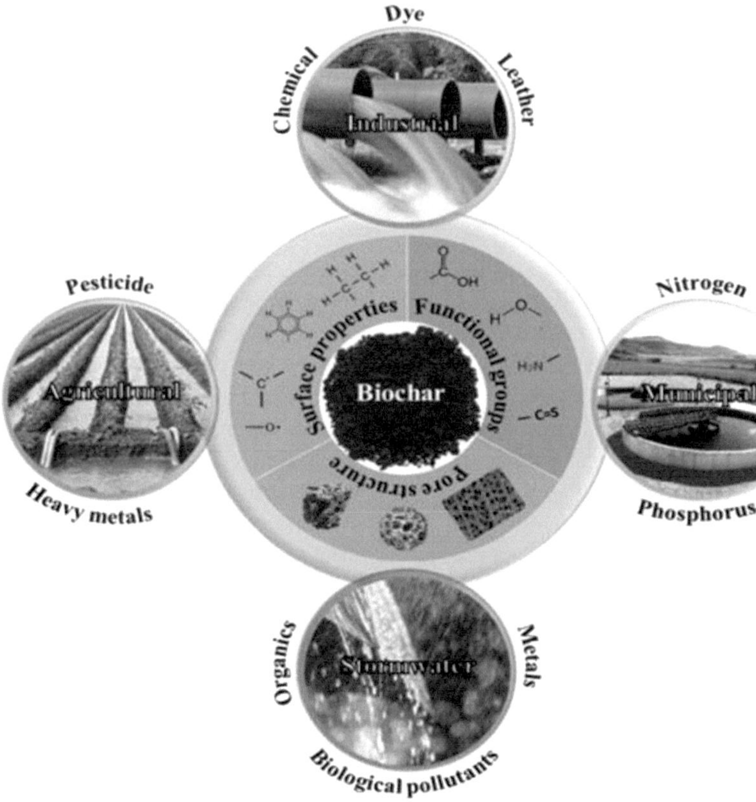

Fig. 11.7 Biochar application in wastewater treatment (adapted from: Xiang et al. 2020)

as efficient media in stormwater treatment systems. Recent research indicates that aluminum modified biochar can effectively eliminate pollutants such as As^{5+}, Pb^{2+}, Zn^{2+}, Cu^{2+}, and PO_4^{3-} from contaminated urban runoff (Liu et al. 2019). Biochar based filtration system achieved removal efficiencies exceeding 85% for copper and 95% for zinc in stormwater. However, it is essential to thoroughly test and precisely design biochar filtration media to ensure they meet the specific standards required for stormwater treatment (Gray 2016).

Biochar has been incorporated into biofilter systems to enhance the removal of bisphenol A from stormwater. Wood dust-derived biochar, with its high specific surface area and substantial pore volume, demonstrates excellent bisphenol adsorption capacity. This also supports in growth and enhances the removal efficiency of total organic carbon (TOC), total suspended solids (TSS), nitrogen, *E. coli* and phosphorus (Ashoori et al. 2019). Furthermore, biochar has significantly improved the performance of stormwater biofilters, especially in removing toxic trace organic contaminants (TOrCs), which are often inadequately treated in conventional filtration systems (Ulrich et al. 2017).

Biochar-amended biofilter columns have demonstrated the ability to maintain over 99% removal efficiency for trace organic contaminants that have been removed from conventional systems. Additionally, biochar-enhanced biofilters have achieved removal rates exceeding 60% for total organic carbon (TOC), total nitrogen (TN), and total phosphorus (TP). Poultry litter biochar (PLB), produced at a pyrolysis temperature of 500 °C, has been utilized for ammonium adsorption in stormwater treatment applications (Ulrich et al. 2017). A strong positive correlation has been observed between ammonium (NH_4^+) adsorption and the cation exchange capacity (CEC) of the biochar, with cation exchange identified as the dominant mechanism in adsorption experiments (Tian et al. 2016). Further, Zn coated sewage sludge-based activated carbon has proven effective in removing phosphate (PO_4-P) and nitrate (NO_3-N) from stormwater leachate, although the efficiency of removal declined at increased pH (Yue et al. 2018).

Incorporating, biochar and zero-valent iron (ZVI) into bioretention cells has been shown to enhance nitrate (NO_2^-) removal in stormwater systems, new approach for improving nitrate treatment efficiency in bioretention applications (Tian et al. 2019). Biochar-amended biofilters or bioretention systems are effective in reducing microbial contamination in stormwater (Lau et al. 2017). Biofilters containing 5% biochar have demonstrated the ability to retain *E. coli* at levels up to three orders of magnitude higher than unamended systems, while also minimizing their remobilization during intermittent flow events. Biochar amendments significantly improve bacterial removal from stormwater (Mohanty et al. 2014). Moreover, sulfuric acid-modified wood biochar serves as an efficient bioretention filter medium for *E. coli* removal from water. The transport and retention of *E. coli* are strongly influenced by the structural morphology and hydrophobic properties of the biochar used (Lau et al. 2017). The rapid growth of the agricultural industry, contamination from this sector has become increasingly severe, as large amounts of pesticides and toxic heavy metals are released into farmland (Wei et al. 2018). Various studies have explored the use of biochar and its modified variants for treating agricultural wastewater (Mandal and Singh 2017). Among the commonly applied agricultural pesticides are pentachlorophenol and atrazine. Rice straw biochar, along with its phosphoric acid-modified form, has demonstrated show high adsorption capacities for removing imidacloprid and atrazine from agricultural wastewater (Liu et al. 2015a, b). Biochar derived from soybean and corn straw have found high efficiency in removing atrazine, with their adsorption capacities primarily influenced by their pore volume and pH levels. Steam activated biochar has also proven effective in eliminating sulfamethazine, with its removal performance dependent on the solution's pH (Devi and Saroha 2014). Afterword, zero-valent iron (ZVF) modified magnetic biochar made from paper mill sludge has been utilized to treat pentachlorophenol (PCP) in wastewater. This biochar not only adsorbs but also dechlorinates PCP, leading to its complete removal. Studies have also explored the removal of pesticides such as diuron, glyphosate, and carbaryl from agricultural wastewater using biochar. The pesticide adsorption by biochar is influenced by factors including the type of biomass feedstock, the functional modifications applied, and the specific characteristics of the contaminants (Wei et al. 2018).

11.3.2 Biochar Based Composite in Sensing Application

Heavy metals (HMs) are commonly found in contaminated food, drinking water, and environments with occupational exposure, where their accumulation can severely disrupt metabolic processes in the human body. The redox reactions of heavy metal ions generate oxidative stress, which may lead to various diseases. These reactions produce free radicals that can cause oxidative damage to DNA and proteins (Fu and Xi 2020). Therefore, accurate detection of heavy metals in environmental, water, beverage samples and food is crucial. Carbon-based sensors are widely used for heavy metal detection, primarily through techniques such as photoluminescence sensing and electrochemical sensing (Bilge et al. 2021). Heavy metals can be detected by analyzing quenching effects and corresponding peaks. Biochar, with its diverse chemical, electronic and vibrational characteristics, exhibits different quenching behaviors depending on the metal ion (Shtepliuk et al. 2017). The interaction between heavy metals and biochar mainly occurs through functional groups such as C=O, C–O, and C–OH bonds via carboxylic and phenolic groups (Plácido et al. 2019a, b, c). Literature reports strong binding affinities between carboxylic acids and metal ions like Cu^{2+}, Pb^{2+}, Hg^{2+} and Ni^{2+} (Plácido et al. 2019a, b, c). A high sensitivity toward detecting heavy metals such as Hg^{2+}, Cd^{2+}, Cu^{2+} and Pb^{2+} has been achieved using electrodes doped with walnut shell biochar modified with poly-tyrosine. The walnut shells were pyrolyzed to obtained biochar, which was incorporated into carbon paste electrodes (CPE), followed by electrodeposition of poly-tyrosine onto the biochar surface. This modification resulted in electrodes with excellent sensitivity, selectivity and found reproducibility. In another study, sodium alginate was used to immobilize biochar on the anode of a microbial fuel cell (MFC) biosensor. The modified electrode produced significantly higher power output compared to unmodified normal electrodes (Zhang et al. 2023a, b). This biochar-MFC sensor was successfully employed to detect Cu, Zn, Ni and Cr in constructed wetland environments. Enhanced detection limits were associated with the proliferation of microbial communities such as Proteobacteria and Shewanella (Zhang et al. 2024). Additionally, artificial neural networks have been utilized to optimize biochar-based sensors for the detection of Hg^{2+} and Pb^{2+} (Zou et al. 2021). Nitrogen coated biochar derived from water mushroom bran demonstrated effective detection of Cd^{2+} and Pb^{2+} both individually and simultaneously. Modified biochars have been frequently employed in the fabrication of electrodes across these investigations.

Organic pollutants such as antibiotics, hypochlorite and nitrophenol are notorious and it is quite challenging to eliminate due to their chemical stability and resistance to bioremediation. Compounds like heptachlor, bisphenol A (BPA), polychlorinated biphenyls (PCBs), polycyclic aromatic hydrocarbons (PAHs), and are particularly harmful to human health, even at very low concentrations. To detect such contaminants, advanced techniques like photoluminescent nanoprobes and electrochemical sensing devices have been employed (Thakur and Kumar 2022). Among these, electrochemical sensors based on molecularly imprinted polymers (MIPs) have emerged as highly effective tools (Zhang et al. 2022). In one study, ZnO–MoO_3 mushroom biochar nanosheets and $Zn_{1.5}PMo_{12}O_{40}$ polyoxometalate were synthesized and used as the foundation for MIP development. Using differential pulse voltammetry (DPV), an MIP electrochemical sensor was created to detect acetaminophen with high sensitivity (Liu et al. 2021). Additionally, for the detection of dibutyl phthalate (DBP), a chitosan-MIP electrochemical sensor was fabricated using a corncob biochar layer. This sensor demonstrated excellent sensitivity within a linear detection range of 0–1.8 μM and achieved a limit of detection (LOD) of 2.6 Nm (Zhou et al. 2021). A paper-based electrochemical immunosensor was developed for the detection of the cyanotoxin microcystin-LR (MCLR) in water samples (Yao et al. 2021). The sensor surface was modified with nano-biochar particles derived from sugarcane waste, pyrolyzed at 800 °C. Anti-MCLR monoclonal antibodies were immobilized onto the paper substrate using a dip-drying technique, followed by the application of bovine serum albumin to block non-specific binding sites. Upon binding of MCLR to its specific antibody, a decrease in current was observed due to hindered electron transfer on the sensor surface. Amperometric measurements revealed a low detection limit of 17 pM (0.017 μg/L) and a rapid response time of under 5 min. The incorporation of nano-biochar significantly enhanced the electrochemical conductivity of the paper sensor, demonstrating its potential for high-performance analyte detection (Yao et al. 2021; Dong et al. 2018).

A novel fluorescent probe was developed by encapsulating nitrogen-doped carbon dots (N-CDs) into a zeolitic imidazolate framework (ZIF), enabling sensitive detection of antibiotic residues such as doxycycline (DOC), tetracycline (TC), and chlortetracycline (CTC) in complex matrices like milk and honey. This probe demonstrated effective fluorescence quenching upon interaction with these antibiotics, exhibiting good linearity across concentration ranges of 0–50 μM for DOC and TC, and 0–45 μM for CTC. The corresponding detection limits were 0.033 μM for DOC, 0.025 μM for TC, and 0.027 μM for CTC. Typically, such organic drug residues are identified using absorption methods followed by electrochemical redox reactions, with Differential Pulse Voltammetry (DPV) being a commonly applied technique. Biochar and N-doped carbon dot composites were utilized in fabricating the sensing platform (Brito et al. 2021; Zhang et al. 2023a, b). Ractopamine, commonly used to enhance pork quality, has raised health concerns due to its potential harmful effects on humans. To detect ractopamine residues in food, an eco-friendly electrochemical biosensor was developed using a biochar-supported Cu^{2+}/Cu^{+} composite. This sensor demonstrated excellent sensitivity and a very low detection limit within the range of 0.1–1.75 μM. Its applicability was confirmed through successful detection of ractopamine in pork sausage samples (Cao et al. 2021). Acetaminophen, a widely used anti-inflammatory and antipyretic drug, has been extensively released into the environment, raising concerns about water contamination. In 2023, Yu and co-workers designed an electrochemical sensor for both detection and removal of acetaminophen from environmental samples (Yu et al. 2023). The sensor was fabricated by combining β-cyclodextrin (β-CD) for selective drug recognition with metal-organic frameworks (MOFs) to enhance porosity. This β-CD/MOFs composite was further integrated with nitrogen-doped Soulangeana sepal-derived biochar (NSC), which improved conductivity and catalytic activity. The resulting NSC/β-CD-MOFs sensor exhibited a wide linear detection range (1.0–30.0 μM) and a low detection limit of 0.3 nM (Yu et al. 2023). Tetracycline, a broad-spectrum antibiotic derived from Streptomyces, is widely found in aquatic systems, where it poses a risk to ecological balance. To detect tetracycline residues in water, an electrochemical sensor was constructed using a glassy carbon electrode (GCE) modified with peanut shell-derived biochar. This biochar enhanced the electrode's conductivity, resulting in a highly sensitive sensor with an impressive detection limit of 3.6×10^{-9} nM (Chou et al. 2023). Chloramphenicol detection in dairy products was achieved using a simple, rapid, and sensitive voltammetry sensor. The electrode was fabricated by doping a glassy carbon electrode with a biochar-polydopamine composite derived from corn straw biogas residue, followed by electrodeposition of β-cyclodextrin. The strong hydrogen bonding between hydroxyl groups on the electrode surface and chloramphenicol contributed to a robust electrochemical signal. This sensor demonstrated excellent sensitivity, reproducibility, and reliability, highlighting the potential of digestate biochar materials for practical applications in electrochemical sensing (Wang et al. 2024).

Pesticides include herbicides, fungicides, insecticides, repellents and rodenticides for birds and animals. According to the U.S. Environmental Protection Agency (USEPA), the primary classifications of pesticides are organophosphates, carbamates, and organochlorines (Bashir et al. 2022). In efforts to develop high-performance biochar-based nanozyme sensors, agricultural waste has been employed as a raw material. Various algae types have been utilized to produce autogenous nitrogen-doped biochar, which possesses abundant surface functional groups and exhibits strong peroxidase-like activity. These sensors demonstrated the capability to detect pesticides in matrices such as tomatoes, peaches, apples, seawater, cabbages, soil and lake water within a concentration range of 1–50 μM (Yue et al. 2024). Specifically, a voltammetric sensor based on *Moringa oleifera*-derived biochar, prepared through pyrolysis at 400 °C, was applied to determine carbendazim and carbonyl pesticides. This sensor displayed excellent selectivity, effectively detecting carbendazim in the range of 0.29–6.00 μM and carbonyl compounds from 29.9–502 μM (Silva et al. 2024a, b). Electrode surface modification significantly enhanced sensor performance by altering the improving reactivity, interfacial properties, and increasing voltammetric signals. This enhancement was also reported who observed over a 70% increase in paraquat detection efficiency when using a sensor composed of Eichhornia crassipes biochar combined with reduced graphene oxide, due to a synergistic effect (Sant'Anna et al. 2022). For the dual detection of hydroquinone and catechol, a sensor was developed using biochar derived from myoga ginger, modified with gold nanoparticles on a glassy carbon electrode (GCE). This electrode achieved high sensitivity and an exceptionally low detection limit (Wang et al. 2020a, b). In another study targeting paraquat detection, activated carbon electrodes were used. Rice husk biochar, chemically activated with phosphoric acid and pyrolyzed at 450 °C, was incorporated into a carbon paste electrode (CPE). The best paraquat sensing performance was observed with a 10% activated carbon content in the CPE. Using an electrolysis potential of 0.9 V and an accumulation time of 8 min, the modified biochar-based CPE demonstrated enhanced sensitivity and stability. These results indicate its strong potential for the electroanalytical detection of xenobiotics (Tagne et al. 2021).

11.3.3 Energy Storage

In view of the current escalating global demand of energy coupled with rapidly declining fossil fuels, the need for development of storage and transformation of renewable energy using cost efficient superior material is considered very crucial. Biochar and biochar-based nanomaterials have drawn growing interest to meet this requirement. As already mentioned that pristine biochar is somewhat less efficient for energy storage performance vis-à-vis engineered biochar or their composite materials (Suh et al. 2012, Plácido et al. 2019c). This applications include supercapacitors, energy storage devices, batteries etc. The advantages associated with biochar based materials lies in using of waste material as the sources, the renewable nature, high natural abundance, the possibilities of producing various kinds of carbon based materials with different properties (Plácido et al. 2019a, b, c).

Strategies developed for storage and usage of hydrogen meets are transformed into commercialization as fuel cells. Apart from different H_2 storage methods, physical adsorption is most and porous carbon based materials are widely adopted (Sethia and Sayari 2016). Inspite of highly porous nature and fast adsorption-desorption kinetics, the capacity of biochar to adsorb H_2 is confined at the limit of 1 wt% at room temperature under 100 bar of H_2 pressure only. The metal decorated porous carbon eliminates this drawbacks and enhance the adsorption capacities (Xia et al. 2013). For this purpose, sepiolite-derived biochar was treated with Pd and the so engineered material showed four times higher H_2 adsorption capacity than the raw biochar (Bhat et al. 2010). In another study, an agricultural biomass corncob was used for production of biochar which was activated by two steps using KOH, K_2CO_3 and NaOH. These activated biochar materials with specific surface area exhibited H2 storage capacity of 2 wt% at 77 K and 0.6 wt% at 298 K (Sun and Webley 2010). A biochar composite derived from pine needle also reported to be used in electrocatalytic H_2 evolution reaction. The as-synthesized pine needle derived nitrogen doped carbon material was activated by KOH at 700–1000 °C. The material accessed at 800 °C showed more porous character achieving onset potential of ~ 4 mV and a Tafel slope of ~ 45.9 mV/dec (Zhu et al. 2017).

Super capacitors are another type of energy storage devices where biochar based materials are also employed due to their high energy density, reversibility, long life cycles. Oxygen rich functional groups in biochar tend to hinder performance of supercapacitors at high current densities due to the high resistance and decreased capacitance of functional groups. Therefore, reformations of porosity, surface functional groups, surface area, conductivity of biochar results in increasing capacitance in supercapacitor applications (Gupta et al. 2022). When the biochar was activated using H_3PO_4, $ZnCl_2$ or KOH, it increases its porosity and surface area. This results in increased capacitance due to ease of ion transfer and diffusion (Rawat et al. 2023). Doping of biochar and nitrogen and phosphorus boosts its capacitance by promoting Faradaic pseudo-capacitive reaction, improving electron transfer within the carbon matrix. Thus, heteroatom doped biochar electrodes offers high conductivity, cycling stability and increased efficiency in supercapacitor applications (Kumar et al. 2020a, b). Transition metal oxides nanoparticles can also be anchored on biochar to increase its capacitance. Wan et al. 2016 developed wood derived biochar and MnO2 composite by in situ reaction with $KMnO_4$ and observed that the composite has higher specific capacitance (101 Fg^{-1}) than the pristine biochar (19 Fg^{-1}) with superior cyclic stability.

A microbial fuel cells (MFC) is a bio-electrochemical device that converts chemical energy of biomass to electrical energy by harnessing the catalytic activity of microorganism. The electrodes in MFCs are separated by proton or ion exchange membrane that only allows the transport of proton, not the oxygen from oxygenated cathode (Bhatia et al. 2021). Biochar act as an outstanding electrode material due to its high specific surface area to grow microorganisms, biocompatibility, capacity to donate electrons. Biochar synthesized from Datura metel peel has been used in bacteria-based MFCs (Vempaty et al. 2023). It demonstrated superior electrolyte diffusion exhibiting good performance as good anode material for MFCs.

11.3.4 Soil Amendment for Agriculture

To reduce the mobility and bioavailability of heavy metal contaminants in soil, several remediation techniques including flotation, phytoremediation, chemical stabilization, soil flushing, adsorption and electrokinetic remediation (Zhai et al. 2018). Among them, magnetic nanoparticles supported metal, metal oxide or metal sulfide supported on biochar nanocomposite (MNPs-BC) have emerged as a promising and environmentally friendly soil amendment due to their high efficiency in immobilizing heavy metals and mitigating their toxicity, mobility and bioavailability (Yu et al. 2017; Wu et al. 2017; Yoo et al. 2018). Furthermore, the application of MNPs-BC for in-situ soil remediation can significantly alter soil's physical and chemical characteristics, including changes in organic matter content, enzyme activities, redox potential, pH levels and microbial communities. The primary mechanisms through which MNPs-BC immobilize heavy metals involve precipitation, chemical reduction, surface complexation, adsorption and ion exchange (Beiyuan et al. 2017) (Fig. 11.8). Literature study found that addition of 8 g/kg of zero-valent iron (Fe^0)-biochar composite to chromium (Cr) contaminated soil for 15 days, resulted in immobilization efficiencies of 92.9% for total chromium (Cr) and 100% for Cr(VI). Furthermore, the pot experiment demonstrated that chromium phytotoxicity was significantly reduced through remediation, which was supported by the growth of cabbage mustard (Su et al. 2016). In another study, a sulfur-iron modified biochar was developed to remediate cadmium (Cd)-contaminated soil. This amendment effectively reduced cadmium (Cd) accumulation in rice and facilitated the transformation of exchangeable cadmium (Cd) into forms bound to organic matter and ferromanganese oxides, thereby decreasing its mobility in the soil (Rajendran et al. 2019). Similarly, ferromanganese oxide-impregnated biochar composites (FMBC) were synthesized for the remediation of arsenic (As) contaminated soil (Lin et al. 2017). Reduce arsenic accumulation in rice, limit its bioavailability and transport from soil to plant, and enhance the formation of Fe–Mn plaques on roots, largely due to the oxidation of As(III) to As(V) by ferromanganese oxides (Wang et al. 2019). The application of Fe–Mn-modified biochar nanocomposite has been reported to effectively immobilize cadmium (Cd) and antimony (Sb) in contaminated soils. This amendment reduced the bioavailability of Cd and Sb in plants and enhanced several soil properties, including pH, organic matter content, nutrient level and enzyme activity. Furthermore, the treatment facilitated the transformation of exchangeable Cd and Sb into more stable, less bioavailable forms. The use of metal-biochar nanocomposite for remediating heavy metal-contaminated soils can significantly influence microbial community composition, soil pH, enzyme activities, redox potential, heavy metal fractionation, and organic matter content all of which are critical for soil health, plant development and effective remediation. The soil pH and redox potential, in particular, play key roles in regulating metal solubility and plant growth. It was observed that incorporation of Fe–Mn-modified biochar slightly decreased soil pH, indicating its capacity to alter soil acidity. This pH shift subsequently influenced arsenic (As) speciation and mobility. A notable increase in soil redox potential was recorded following biochar application, which substantially impacted microbial behavior. The amendment promoted the conversion of exchangeable As into Fe–Mn oxides, residual fractions and organic matter-bound. Soil enzyme activity also increased post-treatment, reflecting improved soil biological function. Moreover, microbial community analysis revealed that Fe–Mn-modified biochar enhanced the abundance of Firmicutes and Proteobacteria, while reducing the presence of Bacteroidetes, contributing to a more favorable microbial environment for soil recovery (Lin et al. 2019). Addition of FMBC to contaminated soil can enhance growth by increasing amino acid concentrations and overall biomass. FMBC amendment also led to a rise in soil pH by approximately 0.2 to 1.0 units, which played a significant role in influencing the migration and transformation of cadmium (Cd). Consequently, the accumulation and bioavailability of Cd in rice were notably reduced. The treatment also facilitated the transformation of exchangeable Cd into more stable residual fractions. Additionally, FMBC promoted the development of Fe–Mn plaques on rice root surfaces, further contributing to Cd immobilization. In another study, remediation using biochar-supported nanoscale zero-valent iron (n-ZVI-BC) was found to significantly alter the speciation of chromium (Cr) in soil. The amendment converted more mobile forms of Cr such as carbonate bound and exchangeable fractions into less bioavailable forms, including those bound to organic matter and Fe–Mn oxides. The application of n-ZVI-BC also increased soil pH, primarily due to the reduction of Cr(VI) and the corrosion process of zero-valent iron (Zhou et al. 2018; Qian et al. 2017). Furthermore, this treatment not only improved soil fertility but also enhanced the biomass of cabbage plants, indicating its potential benefits for both soil health and crop productivity. A recent report demonstrated that the application of biochar supported iron phosphate nanocomposite for the remediation of cadmium (Cd) contaminated soil effectively reduced Cd toxicity, enhanced soil nutrient content, and improved vegetable yields. Additionally, the treatment significantly decreased Cd bioaccumulation in cabbage mustard, indicating its potential for safe crop production. In the remediating arsenic (As) contaminated paddy soils, another investigation assessed the influence of pine sawdust biochar on the mobility and phyto-availability of As under varying redox conditions (Beiyuan et al. 2017). The study revealed that changes in redox potential (EH) played a critical role in regulating As behavior in the soil. Specifically, lower EH conditions were found to increase As mobility, thereby affecting its availability to plants. Furthermore, Fe/Ni bimetallic with biochar supported (BC-Fe/Ni) nanocomposite were developed for the remediation of soil contaminated with decabromodiphenyl. The composite achieved a removal efficiency of 69%, which was conspicuously higher than that achieved by either Fe/Ni nanoparticles or biochar alone. The enhanced removal performance of BC-Fe/Ni was attributed to a combination of adsorption and reductive degradation. Biochar primarily facilitated the adsorption of decabromodiphenyl from the soil, while the Fe/Ni nanoparticles played a crucial role in degrading the adsorbed compound into less toxic byproducts. These degradation products were subsequently captured on the surface of BC-Fe/Ni (Wu et al. 2018). The product was identified by GC-MS analysis which conformed that debromination was the primary pathway for reductive degradation. Functioning as an adsorbent or reductant, magnetic nanoparticles supported on biochar nanocomposite are frequently used as catalysts to activate oxidants for breaking down organic pollutants in soil (Wu et al. 2016). For instance, in the degradation of 2,4-dichlorophenol, biochar-supported nanoscale zero-valent iron (B-nZVI) was used to activate persulfate, resulting in a degradation efficiency of 91%. The process was particularly effective under aerobic and acidic conditions, and the presence of ethylenediaminetetraacetic acid (EDTA) enhanced the degradation, whereas humic acid inhibited it. A portion of the 2,4-dichlorophenol was also removed via dechlorination by the nZVI component of B-Nzvi (Diao et al. 2020).

11.3.5 Climate Change Mitigation

In alignment with the Paris Agreement goal, to meet the true essence of climate change mitigation, reduction of greenhouse gas (GHG) emission and withdrawal of atmospheric CO_2 has been emphasised. Use of biochar in this context is a relatively new concept and its potential utilization is widely studied for past decade (Wu et al. 2019).

Carbon sequestration, in other words, carbon capture and prevention of emission to environment by storing it in inert passive carbon pool strategic goal for climate change issue. Biochar, being biologically and chemically stable form of carbon, act as route for transferring carbon from active to passive pool, making it difficult to be released as CO_2 (Duku et al. 2011; Lehmann 2009). Redirecting only 1% of net annual carbon uptake by plants into biochar could be helpful to mitigate around 10% of present anthropogenic carbon emission. It is anticipated that if all the 60.6 billion tons of biomass is completely converted to biochar via pyrolysis, then 3 billion tons

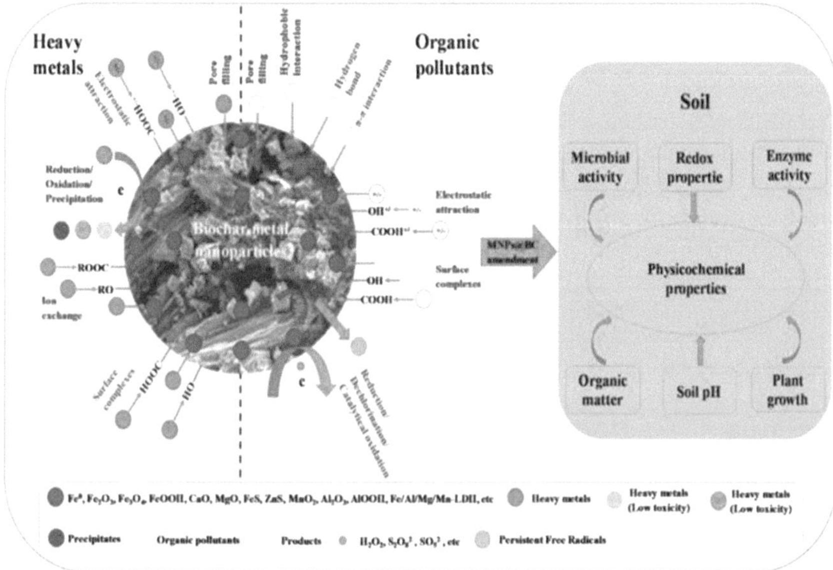

Fig. 11.8 Remediation mechanism of the contaminants in soil by MNPs@BC (adapted from: Liu et al. 2020)

of annual biochar production could effectively eliminate 3 billion tons of atmospheric carbon emission. Biochar serves a long time carbon sink in soil that last for extended period due its persistence towards chemical or microbial breakdown thus increasing the terrestrial carbon stocks. It is anticipated that converting biomass into biochar can lead to capture of 20% of its total carbon content (Lehmann and Joseph 2009; Kleiner 2009).

After the mineralization and microbial respiration to the decayed biomass in soil, CO_2 is emitted to the atmosphere. But upon addition of biochar, due to its recalcitrant nature, it stays for prolonged period reducing GHG emission. A comparative study was made which used agricultural waste, poultry litter, cattle manure, sewage sludge, food waste and checked global warming potential (Alhashimi and Aktas 2017). Almost all the biochar cycles exhibited negative values average of − 0.9 kg CO_2eq kg^{-1} indicating higher amount of GHG consumption than emission. As per estimate, if the total biochar output from Hubei Pyrolysis Plant in China over a 20-years is applied to soil, with an assumption of 30% loss in carbon emission, then the net reduction of GHG emission would be − 32 Mt. CO_2eq (Yang et al. 2016a, b). Apart from CO_2, soil amendment with biochar application can successfully reduce the other non-CO_2 GHG emission such as CH_4, and N_2O.

11.4 Conclusion and Future Perspectives

This chapter has mostly focused on the preparation of biochar, nanobiochar, variety of fabrication technique of biochar based nanocomposite, based on their category, and their potential application. It includes the different sources of biomass available for biochar preparation. Transforming biomass into biochar welcomes superior properties in the new synthesized material as well as eliminates possible environmental depletion caused from waste biomass. The adsorption capacity, pore size, surface area and other properties have been greatly enhanced by converting BC to its nanosize (nanobiochar). Blending nanoparticle to pristine biochar give rise to new class of material called biochar based nanocomposite material. This new composite emerged as superior material, than their precursors, in terms of property, by combining the advantages associated with both the material. Basically two fabrication method-pre-treatment of biomass with nanoparticle and post treatment of nanoparticle with biochar has been adopted based on the category of nanocomposite formed. The nanomaterial based biochar composite has widened its application area in many field including wastewater remediation, agricultural application, sensing, fertilizer application etc. A new term has also came up called "the biochar effect" which demonstrates the alteration of microbial activity caused due to the interaction of BC with soil-plant system. However, there are few gaps in this area where lies the potential perspective of future research interest. Targeting the appropriate biomass with suitable composition and an optimized fabrication method in order to generate low cost nanocomposite material in larger quantity having superior property is still is matter of concern. While applying the BC composite for waste water remediation, more focused to be given on the recycling possibilities. Studies on the sewage treatment using these also been lacking for years. In real case scenario, polluted water contains a variety of pollutants, selective remediation and recycle of the material is demanding. Therefore, biochar composite should be designed in such a way that it has specific adsorption as well as desorption capacity for contaminants. This helps in reuse of the adsorbent for the next cycle of remediation and the desorbed contaminants can also be reused further. Hence, addressing this recyclability issue can be helpful and profitable for large scale industrial usage. Most of reported studies focused on the synthesis and application part of composite, whereas, virtually no study addressed the toxicity aspects of biochar-based composite materials. Until the toxicity of these materials on plants and mammalian species is thoroughly understood and properly disclosed, their application in biological and environmental fields will remain highly risky. Hence, it is desirable that future research focuses on optimizing quality and quantity production of biochar-based nanocomposite materials, practical experiments of prototype in industrial scale, toxicity assessment, health care application.

References

Abdelhameed RM, Taha M, Abdel-Gawad H, Emam HE (2022) Purification of soybean oil from diazinon insecticide by iron-based metal organic framework: effect of geometrical shape and simulation study. J Mol Struct 1250:131914–131929

Alhashimi HA, Aktas CB (2017) Life cycle environmental and economic performance of biochar compared with activated carbon: a meta-analysis. Resour Conserv Recyl 118:13–26

Antunes E, Jacob MV, Brodie G, Schneider PA (2017) Silver removal from aqueous solution by biochar produced from biosolids via microwave pyrolysis. J Environ Manag 203:264–272

Anupama, Khare P (2021) A comprehensive evaluation of inherent properties and applications of nano-biochar prepared from different methods and feedstocks. J Clean Prod 320:128759–128781

Ashiq A, Adassooriya NM, Sarkar B, Rajapaksha AU, Ok YS, Vithanage M (2019) Municipal solid waste biochar-bentonite composite for the removal of antibiotic ciprofloxacin from aqueous media. J Environ Manag 236:428–435

Ashoori N, Teixido M, Spahr S, LeFevre GH, Sedlak DL, Luthy RG (2019) Evaluation of pilot-scale biochar-amended woodchip bioreactors to remove nitrate, metals, and trace organic contaminants from urban stormwater runoff. Water Res 154:1–11

Baig SA, Zhu J, Niaz M, Sheng T, Xu X (2014) Effect of synthesis methods on magnetic Kans grass biochar for enhanced As(III, V) adsorption from aqueous solutions. Biomass Bioenergy 71:299–310

Bashir O, Bhat SA, Basharat A, Qamar M, Qamar SA, Bilal M, Iqbal HMN (2022) Nano-engineered materials for sensing food pollutants: technological advancements and safety issues. Chemosphere 292:133320–133343

Beiyuan J, Awad YM, Beckers F, Tsang DCW, Ok YS, Rinklebe J (2017) Mobility and phytoavailability of As and Pb in a contaminated soil using pine sawdust biochar under systematic change of redox conditions. Chemosphere 178:110–118

Bhat VV, Contescu CI, Gallego NC (2010) Kinetic effect of Pd additions on the hydrogen uptake of chemically-activated ultramicroporous carbon. Carbon 48:2361–2364

Bhatia SK, Palai AK, Kumar A, Bhatia RK, Patel AK, Thakur VK, Yang YH (2021) Trends in renewable energy production employing biomass-based biochar. Bioresour Technol 340:125644–125662

Bilge S, Karadurmus L, Sınağ A, Ozkan SA (2021) Green synthesis and characterization of carbon-based materials for sensitive detection of heavy metal ions. TrAC Trends Anal Chem 145:116473–116496

Blum KM, Gallampois C, Andersson PL, Renman G, Renman A, Haglund P (2019) Comprehensive assessment of organic contaminant removal from on-site sewage treatment facility effluent by char-fortified filter beds. J Hazard Mater 361:111–122

Bolan N, Hoang SA, Beiyuan J, Gupta S, Hou D, Karakoti A, Joseph S, Jung S, Kim K-H, Kirkham MB, Kua HW, Kumar M, Kwon EE, Ok YS, Perera V, Rinklebe J, Shaheen SM, Sarkar B, Sarmah AK, Singh BP, Singh G, Tsang DCW, Vikrant K, Vithanage M, Vinu A, Wang H, Wijesekara H, Yan Y, Younis SA, Van Zwieten L (2022) Multifunctional applications of biochar beyond carbon storage. Int Mater Rev 67(2):150–200

Bakshi S, Banik C, Rathke SJ, Laird DA (2018) Arsenic sorption on zero-valent iron-biochar complexes. Water Res 137:153–163

Brito LRD, Ganiyu SO, Dos Santos EV, Oturan MA, Martínez-Huitle CA (2021) Removal of antibiotic rifampicin from aqueous media by advanced electrochemical oxidation: role of electrode materials, electrolytes and real water matrices. Electrochim Acta 396:139254–139277

Cao X, Ma L, Gao B, Harris W (2009) Dairy-manure derived biochar effectively sorbs lead and atrazine. Environ Sci Technol 43(9):3285–3291

Cao L, Ding Q, Liu M, Lin H, Yang D-P (2021) Biochar-supported Cu^{2+} /Cu^{+} composite as an electrochemical ultrasensitive interface for ractopamine detection. ACS Appl Bio Mater 4(2):1424–1431

Chausali N, Saxena J, Prasad R (2021) Nanobiochar and biochar based nanocomposites: advances and applications. J Agr Food Res 5:100191–100114
Chen B, Chen Z, Lv S (2011) A novel magnetic biochar efficiently sorbs organic pollutants and phosphate. Bioresour Technol 102(2):716–723
Chen L, Chen XL, Zhou CH, Yang HM, Ji SF, Tong DS, Zhong ZK, Yu WH, Chu MQ (2017) Environmental-friendly montmorillonite-biochar composites: facile production and tunable adsorption-release of ammonium and phosphate. J Clean Prod 156:648–659
Chen C, Yan X, Xu Y, Yoza BA, Wang X, Kou Y, Ye H, Wang Q, Li QX (2019) Activated petroleum waste sludge biochar for efficient catalytic ozonation of refinery wastewater. Sci Total Environ 651:2631–2640
Chin JF, Heng ZW, Teoh HC, Chong WC, Pang YL (2022) Recent development of magnetic biochar crosslinked chitosan on heavy metal removal from wastewater—modification, application and mechanism. Chemosphere 291:133035–133058
Chou C-M, Dai Y-D, Yuan C, Shen Y-H (2023) Preparation of an electrochemical sensor utilizing graphene-like biochar for the detection of tetracycline. Environ Res 236:116785–116797
Cole AJ, Paul NA, De Nys R, Roberts DA (2017) Good for sewage treatment and good for agriculture: algal based compost and biochar. J Environ Manag 200:105–113
Cope CO, Webster DS, Sabatini DA (2014) Arsenate adsorption onto iron oxide amended rice husk char. Sci Total Environ 488–489:554–561
Devi P, Saroha AK (2014) Synthesis of the magnetic biochar composites for use as an adsorbent for the removal of pentachlorophenol from the effluent. BioresourTechnol 169:525–531
Diao Z-H, Yan L, Dong F-X, Qian W, Deng Q-H, Kong L-J, Yang J-W, Lei Z-X, Du J-J, Chu W (2020) Degradation of 2,4-dichlorophenol by a novel iron based system and its synergism with Cd(II) immobilization in a contaminated soil. Chem Eng J 379:122313–122325
Ding G-C, Pronk GJ, Babin D, Heuer H, Heister K, Kögel-Knabner I, Smalla K (2013) Mineral composition and charcoal determine the bacterial community structure in artificial soils. FEMS Microbiol Ecol 86(1):15–25
Dong X, He L, Liu Y, Piao Y (2018) Preparation of highly conductive biochar nanoparticles for rapid and sensitive detection of 17β-estradiol in water. Electrochim Acta 292:55–62
Duku MH, Gu S, Hagan EB (2011) Biochar production potential in Ghana—a review. Renew Sust Energy Rev 15(8):3539–3551
Fang C, Zhang T, Li P, Jiang R, Wang Y (2014) Application of magnesium modified corn biochar for phosphorus removal and recovery from swine wastewater. IJERPH 11(9):9217–9237
Fu Z, Xi S (2020) The effects of heavy metals on human metabolism. Toxicol Mech Methods 30(3):167–176
Gale M, Nguyen T, Moreno M, Gilliard-AbdulAziz KL (2021) Physiochemical properties of biochar and activated carbon from biomass residue: influence of process conditions to adsorbent properties. ACS Omega 6(15):10224–10233
Gan C, Liu Y, Tan X, Wang S, Zeng G, Zheng B, Li T, Jiang Z, Liu W (2015) Effect of porous zinc–biochar nanocomposites on Cr(VI) adsorption from aqueous solution. RSC Adv 5(44):35107–35115
Ghezzehei TA, Sarkhot DV, Berhe AA (2014) Biochar can be used to capture essential nutrients from dairy wastewater and improve soil physico-chemical properties. Solid Earth 5(2):953–962
Ghodake GS, Shinde SK, Kadam AA, Saratale RG, Saratale GD, Kumar M, Palem RR, AL-Shwaiman HA, Elgorban AM, Syed A, Kim D-Y (2021) Review on biomass feedstocks, pyrolysis mechanism and physicochemical properties of biochar: state-of-the-art framework to speed up vision of circular bioeconomy. J. Clean. Produc. 297:126645–126669
Gonçalves SPC, Strauss M, Martinez DST (2018) The positive fate of biochar addition to soil in the degradation of PHBV-silver nanoparticle composites. Environ Sci Technol 52:13845–13853
Gray M (2016) Black is green: biochar for Stormwater management. Proc Water Environ Fed 2016(6):2108–2123

Guo F, Bao L, Wang H, Larson SL, Ballard JH, Knotek-Smith HM, Zhang Q, Su Y, Wang X, Han F (2020) A simple method for the synthesis of biochar nanodots using hydrothermal reactor. MethodsX 7:101022

Gupta M, Savla N, Pandit C, Pandit S, Gupta PK, Pant M, Khilari S, Kumar Y, Agarwal D, Nair RR, Thomas D, Thakur VK (2022) Use of biomass-derived biochar in wastewater treatment and power production: a promising solution for a sustainable environment. Sci Total Environ 153892

Haefele SM, Konboon Y, Wongboon W, Amarante S, Maarifat AA, Pfeiffer EM, Knoblauch C (2011) Effects and fate of biochar from rice residues in rice-based systems. Field Crop Res 121(3):430–440

Reddy DHK, Lee S-M (2014) Magnetic biochar composite: Facile synthesis, characterization, and application for heavy metal removal. Colloids Surf. A: Physicochem. Eng. Asp. 454:96–103.

He Z, Hu Y, Nie T, Tang H, Zhu J, Chen K, Liu L, Leong KW, Chen Y, Mao H-Q (2018) Size-controlled lipid nanoparticle production using turbulent mixing to enhance oral DNA delivery. Acta Biomater 81:195–207

Hernandez-Mena L, Pecora A, Beraldo A (2014) Slow pyrolysis of bamboo biomass: analysis of biochar properties. Chem Eng Trans 37:115–120

Hossain MK, Strezov V, Chan KY, Ziolkowski A, Nelson PF (2011) Influence of pyrolysis temperature on production and nutrient properties of wastewater sludge biochar. J Environ Manag 92(1):223–228

Hu X, Ding Z, Zimmerman AR, Wang S, Gao B (2015) Batch and column sorption of arsenic onto iron-impregnated biochar synthesized through hydrolysis. Water Res 68:206–216

Inyang M, Gao B, Zimmerman A, Zhang M, Chen H (2014) Synthesis, characterization, and dye sorption ability of carbon nanotube–biochar nanocomposites. Chem Eng J 236:39–46

Inyang M, Gao B, Zimmerman A, Zhou Y, Cao X (2015) Sorption and cosorption of lead and sulfapyridine on carbon nanotube-modified biochars. Environ Sci Pollut Res 22(3):1868–1876

Joseph S, Lehmann J (eds) (2009) Biochar for environmental management: science and technology. Earthscan, London; Sterling, VA

Jung K-W, Jeong T-U, Hwang M-J, Kim K, Ahn K-H (2015) Phosphate adsorption ability of biochar/Mg–Al assembled nanocomposites prepared by aluminum-electrode based electro-assisted modification method with $MgCl_2$ as electrolyte. Bioresour Technol 198:603–610

Karakoyun N, Kubilay S, Aktas N, Turhan O, Kasimoglu M, Yilmaz S, Sahiner N (2011) Hydrogel–biochar composites for effective organic contaminant removal from aqueous media. Desalination 280(1–3):319–325

Panahi HKS, Dehhaghi M, Ok YS, Nizami A-S, Khoshnevisan B, Mussatto SI, Aghbashlo M, Tabatabaei M, Lam SS (2020) A comprehensive review of engineered biochar: production, characteristics, and environmental applications. J Clean Produc 270:122462–122485

Keiluweit M, Nico PS, Johnson MG, Kleber M (2010) Dynamic molecular structure of plant biomass-derived black carbon (biochar). Environ Sci Technol 44(4):1247–1253

Khataee A, Gholami P, Kalderis D, Pachatouridou E, Konsolakis M (2018) Preparation of novel CeO_2-biochar nanocomposite for sonocatalytic degradation of a textile dye. Ultrason Sonochem 41:503–513

Khodakovskaya MV, De Silva K, Biris AS, Dervishi E, Villagarcia H (2012) Carbon nanotubes induce growth enhancement of tobacco cells. ACS Nano 6(3):2128–2135

Kleiner K (2009) The bright prospect of biochar. Nat Rep Clim Change 1:72–74

Kumar M, Xiong X, Sun Y, Yu IKM, Tsang DCW, Hou D, Gupta J, Bhaskar T, Pandey A (2020a) Critical review on biochar-supported catalysts for pollutant degradation and sustainable biorefinery. Adv Sustain Syst 4(10):1900149–1900172

Kumar R, Strezov V, Weldekidan H, He J, Singh S, Kan T, Dastjerdi B (2020b) Lignocellulose biomass pyrolysis for bio-oil production: a review of biomass pretreatment methods for production of drop-in fuels. Renew Sust Energ Rev 123:109763–110988

Lau AYT, Tsang DCW, Graham NJD, Ok YS, Yang X, Li X (2017) Surface-modified biochar in a bioretention system for Escherichia coli removal from stormwater. Chemosphere 169:89–98

Lehmann J (2009) Biological carbon sequestration must and can be a win-win approach. Clim Chang 97(3):459–463
Lehmann J, Joseph S (2009) Biochar for environmental management: an introduction. In: Lehmann J, Joseph S (eds) Biochar for environmental management: science and technology. London, UK, Earthscan
Lehmann J, Joseph S (eds) (2015) Biochar for environmental management: science, technology and implementation, 2nd edn. Routledge, Taylor & Francis Group, London; New York
Lehmann J, Rillig MC, Thies J, Masiello CA, Hockaday WC, Crowley D (2011) Biochar effects on soil biota—a review. Soil Biol Biochem 43(9):1812–1836
Liang B, Lehmann J, Sohi SP, Thies JE, O'Neill B, Trujillo L, Gaunt J, Solomon D, Grossman J, Neves EG, Luizão FJ (2010) Black carbon affects the cycling of non-black carbon in soil. Org Geochem 41(2):206–213
Lin L, Gao M, Qiu W, Wang D, Huang Q, Song Z (2017) Reduced arsenic accumulation in indica rice (*Oryza sativa* L.) cultivar with ferromanganese oxide impregnated biochar composites amendments. Environ Pollut 231:479–486
Lin L, Li Z, Liu X, Qiu W, Song Z (2019) Effects of Fe-Mn modified biochar composite treatment on the properties of As-polluted paddy soil. Environ Pollut 244:600–607
Liu N, Charrua AB, Weng C-H, Yuan X, Ding F (2015a) Characterization of biochars derived from agriculture wastes and their adsorptive removal of atrazine from aqueous solution: a comparative study. Bioresour Technol 198:55–62
Liu W-J, Jiang H, Yu H-Q (2015b) Development of biochar-based functional materials: toward a sustainable platform carbon material. Chem Rev 115(22):12251–12285
Liu Q, Wu L, Gorring M, Deng Y (2019) Aluminum-impregnated biochar for adsorption of arsenic(V) in urban stormwater runoff. J Environ Eng 145(4):04019008
Liu J, Jiang J, Meng Y, Aihemaiti A, Xu Y, Xiang H, Gao Y, Chen X (2020) Preparation, environmental application and prospect of biochar-supported metal nanoparticles: a review. J Hazard Mater 388:122026–122049
Liu X, Huang L, Qian K (2021) Nanomaterial-based electrochemical sensors: mechanism, preparation, and application in biomedicine. Adv NanoBiomed Res 1(6):2000104–2000129
Lonappan L, Rouissi T, Das RK, Brar SK, Ramirez AA, Verma M, Surampalli RY, Valero JR (2016) Adsorption of methylene blue on biochar microparticles derived from different waste materials. Waste Manag 49:537–544
Lyu H, Gao B, He F, Zimmerman AR, Ding C, Tang J, Crittenden JC (2018) Experimental and modeling investigations of ball-milled biochar for the removal of aqueous methylene blue. Chem Eng J 335:110–119
Mandal A, Singh N (2017) Optimization of atrazine and imidacloprid removal from water using biochars: designing single or multi-staged batch adsorption systems. Int J Hyg Environ Health 220(3):637–645
Marris E (2006) Putting the carbon back: black is the new green. Nature 442:624–626
Matos T, Schultz J, Khan M, Zanoelo E, Mangrich A, Araújo B, Navickiene S, Romão L (2017) Using magnetized (Fe_3O_4/biochar nanocomposites) and activated biochar as adsorbents to remove two neuro-active pesticides from waters. J Braz Chem Soc 28(10):1–13
Meyer S, Glaser B, Quicker P (2011) Technical, economical, and climate-related aspects of biochar production technologies: A literature review. Environ Sci Technol 45(22):9473–9483
Mohamed MA, Hashim AF, Alghuthaymi MA, Abd-Elsalam KA (2018) Nano-carbon: plant growth promotion and protection. In: Abd-Elsalam KA, Prasad R (eds) Nanobiotechnology applications in plant protection. Springer International Publishing, Cham, pp 155–188
Mohan D, Kumar H, Sarswat A, Alexandre-Franco M, Pittman CU (2014a) Cadmium and lead remediation using magnetic oak wood and oak bark fast pyrolysis bio-chars. C Eng J 236:513–528
Mohan D, Kumar S, Srivastava A (2014b) Fluoride removal from ground water using magnetic and nonmagnetic corn Stover biochars. Ecol Eng 73:798–808

Mohan D, Sarswat A, Ok YS, Pittman CU (2014c) Organic and inorganic contaminants removal from water with biochar, a renewable, low cost and sustainable adsorbent—a critical review. Bioresour Technol 160:191–202

Mohanty SK, Cantrell KB, Nelson KL, Boehm AB (2014) Efficacy of biochar to remove *Escherichia coli* from stormwater under steady and intermittent flow. Water Res 61:288–296

Mubarak NM, Kundu A, Sahu JN, Abdullah EC, Jayakumar NS (2014) Synthesis of palm oil empty fruit bunch magnetic pyrolytic char impregnating with $FeCl_3$ by microwave heating technique. Biomass Bioenergy 61:265–275

Nagaraju KK, Prasad TNVKV, Naidu MVS, Chari MS, Ramu YR, Murthy BR (2023) Exploring the benefits of rice husk waste: synthesis and characterization of biochar and nanobiochar for agricultural and environmental sustainability. Int J Environ Clim Chang 13:715–725

Nguyen BT, Lehmann J, Hockaday WC, Joseph S, Masiello CA (2010) Temperature sensitivity of black carbon decomposition and oxidation. Environ Sci Technol 44(9):3324–3331

Pi L, Jiang R, Zhou W, Zhu H, Xiao W, Wang D, Mao X (2015) G-C3N4 modified biochar as an adsorptive and photocatalytic material for decontamination of aqueous organic pollutants. Appl Surf Sci 358:231–239

Plácido J, B-Lopez S, Meissner KE, Kelly DE, Kelly SL (2019a) Multivariate analysis of biochar-derived carbonaceous nanomaterials for detection of heavy metal ions in aqueous systems. Sci Total Environ 688:751–761

Plácido J, B-Lopez S, Meissner KE, Kelly DE, Kelly SL (2019b) Microalgae biochar-derived carbon dots and their application in heavy metal sensing in aqueous systems. Sci Total Environ 656:531–539

Plácido J, B-López S, Meissner KE, Kelly DE, Kelly SL (2019c) Comparative study of the characteristics and fluorescent properties of three different biochar derived-carbonaceous nanomaterials for bioimaging and heavy metal ions sensing. Fuel Process Technol 196:106163–106183

Poonam BSK, Kumar N (2018) Kinetic study of lead (Pb^{2+}) removal from battery manufacturing wastewater using bagasse biochar as biosorbent. Appl Water Sci 8(4):119

Pourkarimi S, Hallajisani A, Alizadehdakhel A, Nouralishahi A (2019) Biofuel production through micro- and macroalgae pyrolysis—a review of pyrolysis methods and process parameters. J Anal Appl Pyrol 142:104599–1045123

Qian L, Zhang W, Yan J, Han L, Chen Y, Ouyang D, Chen M (2017) Nanoscale zero-valent iron supported by biochars produced at different temperatures: synthesis mechanism and effect on Cr(VI) removal. Environ Pollut 223:153–160

Qin C, Wang H, Yuan X, Xiong T, Zhang J, Zhang J (2020) Understanding structure-performance correlation of biochar materials in environmental remediation and electrochemical devices. Chem Eng J 382:122977–1229100

Rajendran M, Shi L, Wu C, Li W, An W, Liu Z, Xue S (2019) Effect of sulfur and sulfur-iron modified biochar on cadmium availability and transfer in the soil–rice system. Chemosphere 222:314–322

Ramanayaka S, Tsang DCW, Hou D, Ok YS, Vithanage M (2020) Green synthesis of graphitic nanobiochar for the removal of emerging contaminants in aqueous media. Sci Total Environ 706:135725–135747

Rawat S, Boobalan T, Sathish M, Hotha S, Thallada B (2023) Utilization of CO_2 activated litchi seed biochar for the fabrication of supercapacitor electrodes. Biomass Bioenergy 171:106747–106758

Ren J, Li N, Li L, An J-K, Zhao L, Ren N-Q (2015) Granulation and ferric oxides loading enable biochar derived from cotton stalk to remove phosphate from water. Bioresour Technol 178:119–125

Sant'Anna MVS, Silva JDOS, Gevaerd A, Lima LS, MDS M, Carregosa ISC, Wisniewski A, Marcolino-Junior LH, Bergamini MF, Sussuchi EM (2022) Selective carbonaceous-based (nano)composite sensors for electrochemical determination of paraquat in food samples. Food Chem 373:131521–131546

Saravanakumar R, Muthukumaran K, Selvaraju N (2019) Enhanced Pb (II) ions removal by using magnetic NiO/biochar composite. Mater Res Express 6(10):105504–105527

Saxena J, Rana G, Pandey M (2013) Impact of addition of biochar along with *Bacillus* sp. on growth and yield of French beans. Sci Hortic 162:351–356

Schmidt MWI, Noack AG (2000) Black carbon in soils and sediments: analysis, distribution, implications, and current challenges. Glob Biogeochem Cycles 14(3):777–793

Sethia G, Sayari A (2016) Activated carbon with optimum pore size distribution for hydrogen storage. Carbon 99:289–294

Shakoor MB, Ye Z-L, Chen S (2021) Engineered biochars for recovering phosphate and ammonium from wastewater: a review. Sci Total Environ 779:146240–146262

Shtepliuk I, Caffrey NM, Iakimov T, Khranovskyy V, Abrikosov IA, Yakimova R (2017) On the interaction of toxic heavy metals (Cd, Hg, Pb) with graphene quantum dots and infinite graphene. Sci Rep 7(1):3934–3945

Silva JOS, Dos Santos JF, Granja HS, Almeida WS, TFL L, Freitas LS, Bergamini MF, Marcolino-Junior LH, Sussuchi EM (2024a) Simultaneous determination of carbendazim and carbaryl pesticides in water bodies samples using a new voltammetric sensor based on *Moringa oleifera* biochar. Chemosphere 347:140707–140727

Silva JDOS, Dos Santos JF, Granja HS, Almeida WS, Loeser TFL, Freitas LS, Bergamini MF, Marcolino-Junior LH, Sussuchi EM (2024b) Simultaneous determination of carbendazim and carbaryl pesticides in water bodies samples using a new voltammetric sensor based on *Moringa oleifera* biochar. Chemosphere 347:140707–140719

Sohi SP (2012) Carbon storage with benefits. Science 338(6110):1034–1035

Sohi SP, Krull E, Lopez-Capel E, Bol R (2010) A review of biochar and its use and function in soil. In: Advances in agronomy. Elsevier, pp 47–82

Song Z, Lian F, Yu Z, Zhu L, Xing B, Qiu W (2014) Synthesis and characterization of a novel MnO_x-loaded biochar and its adsorption properties for Cu^{2+} in aqueous solution. Chem Eng J 242:36–42

Su H, Fang Z, Tsang PE, Fang J, Zhao D (2016) Stabilisation of nanoscale zero-valent iron with biochar for enhanced transport and in-situ remediation of hexavalent chromium in soil. Environ Pollut 214:94–100

Suh MP, Park HJ, Prasad TK, Lim D-W (2012) Hydrogen storage in metal–organic frameworks. Chem Rev 112(2):782–835

Sun Y, Webley PA (2010) Preparation of activated carbons from corncob with large specific surface area by a variety of chemical activators and their application in gas storage Chem. Eng J 162(3):883–892

Sun P, Hui C, Azim Khan R, Du J, Zhang Q, Zhao Y-H (2015) Efficient removal of crystal violet using Fe_3O_4-coated biochar: the role of the Fe_3O_4 nanoparticles and modeling study their adsorption behavior. Sci Rep 5(1):12638–12649

Tagne RFT, Ndifor-Angwagor NG, Temgoua RCT, Tchuifon DRT, Vintila T, Ngueabouo AS, Anagho SG (2021) Development of an electroanalytical method using activated rice husk-derived carbon for the detection of a paraquat herbicide. Carbon Trends 4:100060–100072

Tan G, Wu Y, Liu Y, Xiao D (2018) Removal of Pb(II) ions from aqueous solution by manganese oxide coated rice straw biochar A low-cost and highly effective sorbent. J Taiwan Inst Chem Eng 84:85–92

Tang J, Lv H, Gong Y, Huang Y (2015) Preparation and characterization of a novel graphene/biochar composite for aqueous phenanthrene and mercury removal. Bioresour Technol 196:355–363

Tang Y, Alam MS, Konhauser KO, Alessi DS, Xu S, Tian W, Liu Y (2019) Influence of pyrolysis temperature on production of digested sludge biochar and its application for ammonium removal from municipal wastewater. J. Clean. Produc. 209:927–936

Thakre KG, Barai DP, Bhanvase BA (2021) A review of graphene-TiO2 and graphene-ZnO nanocomposite photocatalysts for wastewater treatment. Water Environ Res 93(11):2414–2460

Thakur A, Kumar A (2022) Recent advances on rapid detection and remediation of environmental pollutants utilizing nanomaterials-based (bio)sensors. Sci Total Environ 834:155244

Tian J, Miller V, Chiu PC, Maresca JA, Guo M, Imhoff PT (2016) Nutrient release and ammonium sorption by poultry litter and wood biochars in stormwater treatment. Sci Total Environ 553:596–606

Tian J, Jin J, Chiu PC, Cha DK, Guo M, Imhoff PT (2019) A pilot-scale, bi-layer bioretention system with biochar and zero-valent iron for enhanced nitrate removal from stormwater. Water Res 148:378–387

Tukimin N, Abdullah J, Sulaiman Y (2018) Electrodeposition of poly(3,4-ethylenedioxythiophene)/reduced graphene oxide/manganese dioxide for simultaneous detection of uric acid, dopamine and ascorbic acid. J Electroanal Chem 820:74–81

Ulrich BA, Loehnert M, Higgins CP (2017) Improved contaminant removal in vegetated stormwater biofilters amended with biochar. Environ Sci: Water Res Technol 3(4):726–734

Uzun BB, Apaydin-Varol E, Ateş F, Özbay N, Pütün AE (2010) Synthetic fuel production from tea waste: characterisation of bio-oil and bio-char. Fuel 89(1):176–184

Van Zwieten L, Kimber S, Morris S, Chan KY, Downie A, Rust J, Joseph S, Cowie A (2010) Effects of biochar from slow pyrolysis of papermill waste on agronomic performance and soil fertility. Plant Soil 327(1–2):235–246

Vempaty A, Kumar A, Pandit S, Gupta M, Mathuriya AS, Lahiri D, Nag M, Kumar Y, Joshi S, Kumar N (2023) Evaluation of the Datura peels derived biocharbased anode for enhancing power output in microbial fuel cell application. Biocatal Agric Biotechnol 47:102560–102573

Ngah WSW, Hanafiah MAKM (2008) Removal of heavy metal ions from wastewater by chemically modified plant wastes as adsorbents: a review. Bioresour Technol 99(10):3935–3948

Wan C, Jiao Y, Li J (2016) Core–shell composite of wood-derived biochar supported MnO_2 nanosheets for supercapacitor applications. RSC Adv 6:64811–64817

Wan X, Li C, Parikh SJ (2020) Simultaneous removal of arsenic, cadmium, and lead from soil by iron-modified magnetic biochar. Environ Pollut 261:114157

Wang J, Wang S (2019) Preparation, modification and environmental application of biochar: a review. J Clean Prod 227:1002–1022

Wang S, Tang Y, Li K, Mo Y, Li H, Gu Z (2014) Combined performance of biochar sorption and magnetic separation processes for treatment of chromium-contained electroplating wastewater. Bioresour Technol 174:67–73

Wang H, Gao B, Wang S, Fang J, Xue Y, Yang K (2015a) Removal of Pb(II), Cu(II), and Cd(II) from aqueous solutions by biochar derived from $KMnO_4$ treated hickory wood. Bioresour Technol 197:356–362

Wang MC, Sheng GD, Qiu YP (2015b) A novel manganese-oxide/biochar composite for efficient removal of lead(II) from aqueous solutions. Int J Environ Sci Technol 12(5):1719–1726

Wang S, Tang Y, Chen C, Wu J, Huang Z, Mo Y, Zhang K, Chen J (2015c) Regeneration of magnetic biochar derived from eucalyptus leaf residue for lead(II) removal. Bioresour Technol 186:360–364

Wang S, Zhou Y, Gao B, Wang X, Yin X, Feng K, Wang J (2017) The sorptive and reductive capacities of biochar supported nanoscaled zero-valent iron (nZVI) in relation to its crystallite size. Chemosphere 186:495–500

Wang Y-Y, Ji H-Y, Lyu H-H, Liu Y-X, He L-L, You L-C, Zhou C-H, Yang S-M (2019) Simultaneous alleviation of Sb and cd availability in contaminated soil and accumulation in *Lolium multiflorum* Lam. after amendment with Fe–Mn-modified biochar. J Cleaner Prod 231:556–564

Wang J, Yang J, Xu P, Liu H, Zhang L, Zhang S, Tian L (2020a) Gold nanoparticles decorated biochar modified electrode for the high-performance simultaneous determination of hydroquinone and catechol. Sens Actuators B: Chem 306:127590–1275100

Wang L, Hou D, Cao Y, Ok YS, Tack FMG, Rinklebe J, O'Connor D (2020b) Remediation of mercury contaminated soil, water, and air: a review of emerging materials and innovative technologies. Environ Int 134:105281–1052104

Wang R, Lan X, Zhou T, Qian X, Qu B, Lv P, Wang Y (2024) Detection of chloramphenicol in dairy products based on biogas residue biochar based electrochemical sensor. J Food Compos Anal 125:105824–105847

Wathukarage A, Herath I, Iqbal MCM, Vithanage M (2019) Mechanistic understanding of crystal violet dye sorption by woody biochar: implications for wastewater treatment. Environ Geochem Health 41(4):1647–1661

Wei D, Li B, Huang H, Luo L, Zhang J, Yang Y, Guo J, Tang L, Zeng G, Zhou Y (2018) Biochar-based functional materials in the purification of agricultural wastewater: fabrication, application and future research needs. Chemosphere 197:165–180

Wu J, Yi Y, Li Y, Fang Z, Tsang EP (2016) Excellently reactive Ni/Fe bimetallic catalyst supported by biochar for the remediation of decabromodiphenyl contaminated soil: reactivity, mechanism, pathways and reducing secondary risks. J Hazard Mater 320:341–349

Wu H, Feng Q, Yang H, Alam E, Gao B, Gu D (2017) Modified biochar supported Ag/Fe nanoparticles used for removal of cephalexin in solution: characterization, kinetics and mechanisms. Colloids Surf A Physicochem Eng Asp 517:63–71

Wu J, Yi Y, Fang Z, Tsang EP (2018) Effects of biochar on phytotoxicity and translocation of polybrominated diphenyl ethers in Ni/Fe bimetallic nanoparticle-treated soil. Environ Sci Pollut Res 25(3):2570–2579

Wu P, A-U-Karim ST, Singh BP, Wang H, Wu T, Liu C, Fang G, Zhou D, Wang Y, Chen W (2019) A scientometric review of biochar research in the past 20 years (1998–2018). Biochar 1:23–43

Wurzer C, Mašek O (2021) Feedstock doping using iron rich waste increases the pyrolysis gas yield and adsorption performance of magnetic biochar for emerging contaminants. Bioresour Technol 321:124473–124489

Xia Y, Yanga Z, Zhu Y (2013) Porous carbon-based materials for hydrogen storage: advancement and challenges. J Mater Chem A 1:9365–9381

Xiang W, Zhang X, Chen J, Zou W, He F, Hu X, Tsang DCW, Ok YS, Gao B (2020) Biochar technology in wastewater treatment: a critical review. Chemosphere 252:126539–126561

Yamato M, Okimori Y, Wibowo IF, Anshori S, Ogawa M (2006) Effects of the application of charred bark of *Acacia mangium* on the yield of maize, cowpea and peanut, and soil chemical properties in South Sumatra. Indonesia Soil Sci Plant Nutrit 52(4):489–495

Yan J, Han L, Gao W, Xue S, Chen M (2015a) Biochar supported nanoscale zerovalent iron composite used as persulfate activator for removing trichloroethylene. Bioresour Technol 175:269–274

Yan L, Kong L, Qu Z, Li L, Shen G (2015b) Magnetic biochar decorated with ZnS Nanocrytals for Pb (II) removal. ACS Sustain Chem Eng 3(1):125–132

Yang G, Wu L, Xian Q, Shen F, Wu J, Zhang Y (2016a) Removal of Congo red and methylene blue from aqueous solutions by Vermicompost-derived biochars. PLoS One 11(5):e0154562

Yang Q, Han F, Yingquan C, Haiping Y, Chen H (2016b) Greenhouse gas emissions of a biomass-based pyrolysis plant in China. Renew Sust Energ Rev 53:1580–1590

Yang M, Li C, Zhang Y, Jia D, Zhang X, Hou Y, Shen B, Li R (2018) Microscale bone grinding temperature by dynamic heat flux in nanoparticle jet mist cooling with different particle sizes. Mater Manuf Process 33(1):58–68

Yang H, Ye S, Zeng Z, Zeng G, Tan X, Xiao R, Wang J, Song B, Du L, Qin M, Yang Y, Xu F (2020) Utilization of biochar for resource recovery from water: a review. Chem Eng J 397:125502–125526

Yao Y, Gao B, Chen J, Yang L (2013a) Engineered biochar reclaiming phosphate from aqueous solutions: mechanisms and potential application as a slow-release fertilizer. Environ Sci Technol 47(15):8700–8708

Yao Y, Gao B, Chen J, Zhang M, Inyang M, Li Y, Alva A, Yang L (2013b) Engineered carbon (biochar) prepared by direct pyrolysis of Mg-accumulated tomato tissues: characterization and phosphate removal potential. Bioresour Technol 138:8–13

Yao X, Ji L, Guo J, Ge S, Lu W, Cai L, Wang Y, Song W, Zhang H (2020) Magnetic activated biochar nanocomposites derived from wakame and its application in methylene blue adsorption. Bioresour Technol 302:122842–122865

Yao L, He L, Yang Y, Zhang Y, Liu Z, Liang L, Piao Y (2021) Nanobiochar paper based electrochemical immunosensor for fast and ultrasensitive detection of microcystin-LR. Sci Total Environ 750:141692–1416115

Yi S, Gao B, Sun Y, Wu J, Shi X, Wu B, Hu X (2016) Removal of levofloxacin from aqueous solution using rice-husk and wood-chip biochars. Chemosphere 150:694–701

Yoo J-C, Beiyuan J, Wang L, Tsang DCW, Baek K, Bolan NS, Ok YS, Li X-D (2018) A combination of ferric nitrate/EDDS-enhanced washing and sludge-derived biochar stabilization of metal-contaminated soils. Sci Total Environ 616–617:572–582

Yu Z, Qiu W, Wang F, Lei M, Wang D, Song Z (2017) Effects of manganese oxide-modified biochar composites on arsenic speciation and accumulation in an indica rice (*Oryza sativa* L.) cultivar. Chemosphere 168:341–349

Yu Q, Zou J, Yu C, Peng G, Fan G, Wang L, Chen S, Lu L, Wang Z (2023) Nitrogen doped porous biochar/β-CD-MOFs heterostructures: bi-functional material for highly sensitive electrochemical detection and removal of acetaminophen. Molecules 28(6):2437–2451

Yue C, Li LY, Johnston C (2018) Exploratory study on modification of sludge-based activated carbon for nutrient removal from stormwater runoff. J Environ Manag 226:37–45

Yue N, Wu J, Qi W, Su R (2024) Algae-derived biochar nanozyme array for discrimination and detection of multiple pesticides in soil, water and food. Food Chem 438:137946–137968

Zabot GL, Tres MV, Ferreira PAA, Dal Molin A, Mazutti MA (2020) Power the future with bioenergy from organic wastes. In: Recent developments in bioenergy research. Elsevier, pp 85–114

Zhai X, Li Z, Huang B, Luo N, Huang M, Zhang Q, Zeng G (2018) Remediation of multiple heavy metal-contaminated soil through the combination of soil washing and in situ immobilization. Sci Total Environ 635:92–99

Zhang M, Gao B (2013) Removal of arsenic, methylene blue, and phosphate by biochar/AlOOH nanocomposite. Chem Eng J 226:286–292

Zhang M, Gao B, Yao Y, Xue Y, Inyang M (2012a) Synthesis of porous MgO-biochar nanocomposites for removal of phosphate and nitrate from aqueous solutions. Chem Eng J 210:26–32

Zhang M, Gao B, Yao Y, Xue Y, Inyang M (2012b) Synthesis, characterization, and environmental implications of graphene-coated biochar. Sci Total Environ 435–436:567–572

Zhang M, Gao B, Varnoosfaderani S, Hebard A, Yao Y, Inyang M (2013) Preparation and characterization of a novel magnetic biochar for arsenic removal. Bioresour Technol 130:457–462

Zhang M, Liu Y, Li T, Xu W, Zheng B, Tan X, Wang H, Guo Y, Guo F, Wang S (2015) Chitosan modification of magnetic biochar produced from *Eichhornia crassipes* for enhanced sorption of Cr(VI) from aqueous solution. RSC Adv 5(58):46955–46964

Zhang H, Xu F, Xue J, Chen S, Wang J, Yang Y (2020a) Enhanced removal of heavy metal ions from aqueous solution using manganese dioxide-loaded biochar: behavior and mechanism. Sci Rep 10(1):6067–6080

Zhang X, Xiang W, Wang B, Fang J, Zou W, He F, Li Y, Tsang DCW, Ok YS, Gao B (2020b) Adsorption of acetone and cyclohexane onto CO2 activated hydrochars. Chemosphere 245:125664–125680

Zhang Y, Maierdan Y, Guo T, Chen B, Fang S, Zhao L (2022) Biochar as carbon sequestration material combines with sewage sludge incineration ash to prepare lightweight concrete. Construc Build Mater 343:128116–128140

Zhang K, Cao H, Luo H, Chen W, Chen J (2023a) Enhanced MFC sensor performances and extracellular electron transport efficiency mediated by biochar and underlying biochemical mechanisms. J Environ Manag 332:117282–1172103

Zhang S, Mao Y, Wei L, Song Z, Zhao X, Wang W (2023b) Full-value preparation of biochar and 2D N-doped CDs@ZIF-L from fermentation residues for sensitive sensing tetracyclines in food samples. Colloids Surf A Physicochem Eng Asp 676:132073–132084

Zhang K, Cao H, Li Y, Shan S, Chen J, Luo H, Chen W, Huang X (2024) Enhanced performance of biochar-biosensor applied to heavy metals detection in constructed wetlands and biological mechanisms. J Clean Product 436:140339–140362

Zhou Y, Gao B, Zimmerman AR, Fang J, Sun Y, Cao X (2013) Sorption of heavy metals on chitosan-modified biochars and its biological effects. Chem Eng J 231:512–518
Zhou Y, Gao B, Zimmerman AR, Chen H, Zhang M, Cao X (2014) Biochar-supported zerovalent iron for removal of various contaminants from aqueous solutions. Bioresour Technol 152:538–542
Zhou Q, Lin L, Qiu W, Song Z, Liao B (2018) Supplementation with ferromanganese oxide-impregnated biochar composite reduces cadmium uptake by indica rice (*Oryza sativa* L.). J Clean Prod 184:1052–1059
Zhou Q, Guo M, Wu S, Fornara D, Sarkar B, Sun L, Wang H (2021) Electrochemical sensor based on corncob biochar layer supported chitosan-MIPs for determination of dibutyl phthalate (DBP). J Electroanal Chem 897:115549–115573
Zhu G, Ma L, Lv H, Hu Y, Chen T, Chen R, Liang J, Wang X, Wang Y, Yan C, Tie Z, Jin Z, Liu J (2017) Pine needle-derived microporous nitrogen-doped carbon frameworks exhibit high performances in electrocatalytic hydrogen evolution reaction and supercapacitors. Nanoscale 9:1237–1243
Zou J, Qian W, Li Y, Yu Q, Yu Y, Chen S, Qu F, Gao Y, Lu L (2021) Multilayer activated biochar/UiO-66-NH2 film as intelligent sensing platform for ultra-sensitive electrochemical detection of Pb^{2+} and Hg^{2+}. Appl Surf Sci 569:151006–151017

Chapter 12
Biochar for Sustainable Farming: Advancing Soil Health, Livestock Productivity and Climate Resilience

Umalaxmi Thingujam, Subrata Gorain, Suman Dutta, Animesh Ghosh Bag, and Victor Thingujam

Abstract Biochar has gained prominence as a multifunctional material supporting sustainable agricultural practices, livestock management, and climate adaptation strategies. This chapter presents an integrated overview of biochar's production techniques, including slow pyrolysis, fast pyrolysis, and hydrothermal carbonization, and examines the influence of feedstock types on its physicochemical properties. We describe how biochar enhances soil fertility by improving nutrient retention, reducing nutrient leaching, increasing microbial biomass and diversity, and augmenting soil structure and water-holding capacity. Its role in carbon sequestration is outlined, emphasizing mechanisms such as chemical stability, negative priming effects, and aggregate formation that contribute to long-term soil organic carbon enhancement. In livestock management, biochar's use as a feed additive to improve digestion efficiency, reduce enteric methane emissions, and detoxify harmful substances is discussed, along with its application in manure management to stabilize nutrients, suppress odors, and reduce pathogen proliferation. The chapter further highlights biochar's contributions to climate change mitigation by lowering emissions of CO_2, CH_4, and N_2O, and enhancing agricultural resilience against droughts

U. Thingujam
Department of Soil Science and Agricultural Chemistry, Palli Siksha Bhavana (Institute of Agriculture), Visva-Bharati (Central University), Sriniketan, West Bengal, India

S. Gorain (✉)
Department of Agricultural Economics, Palli Siksha Bhavana (Institute of Agriculture), Visva-Bharati (Central University), Sriniketan, West Bengal, India
e-mail: subrata.gorain@visva-bharati.ac.in

S. Dutta
Department of Genetics and Plant Breeding, Ramakrishna Mission Vivekananda Educational and Research Institute, Kolkata, West Bengal, India

A. G. Bag
Department of Agriculture, Swami Vivekananda University, Barrackpore, Kolkata, India

V. Thingujam
ICAR Research Complex for Eastern Region-Farming System Research Centre for Hill and Plateau Region, Ranchi, Jharkhand, India

D. Moulick et al. (eds.), *Biochar: Achieving Agro-Environmental Sustainability*, Sustainability Solutions, https://doi.org/10.1007/978-3-032-19855-6_12

and soil degradation. Linkages between biochar applications and the achievement of multiple United Nations Sustainable Development Goals (SDGs), including food security, clean energy, responsible consumption, and climate action, are elaborated. Challenges such as production scalability, cost-effectiveness, and policy integration are also identified. By compiling recent scientific advancements and future directions, the chapter underscores biochar's pivotal role in fostering sustainable farming systems, promoting circular economy principles, and advancing global environmental stewardship.

Keywords Biochar · Sustainable agriculture · Soil fertility · Livestock management · Climate resilience · Sustainable development goals

12.1 Introduction

The terrestrial environment's largest carbon storage is soil (Friedlingstein et al. 2022). The stability of the soil environment and the amount of organic carbon in the soil affect how well our efforts to become carbon neutral work. The need to preserve soil organic carbon is further highlighted by the fact that soil is not renewable (Antoneli et al. 2022). However, irreversible carbon loss is being caused by degradation of billions of hectares of land worldwide (Gibbs and Salmon 2015; Rickson et al. 2015; UN 2022). Appropriate soil use and management techniques are essential. A promising substance for soil and non-soil uses in the pursuit of carbon neutrality is biochar. Biochar is a light, black material similar to charcoal which commonly made from anaerobic oxidation or burning organic biomasses. Pyrolysis, a process that includes heating organic materials, produces biochar, a type of charcoal. Biochar, bio-oil, and syn-gas are produced via the process of pyrolysis, which involves heating biomass without the presence of oxygen. Since its synthesis transforms labile biomass into refractory carbon that endures in the environment for centenary timeframes, it is, on the one hand, an inherently carbon-negative soil amendment. However, a number of recent studies indicate that biochar has enormous potential to fight climate change in non-soil applications such as energy storage, chemical engineering aquaculture (Man et al. 2021), construction, and carbon capture, utilisation, and storage (CCUS) (Shafawi et al. 2021). Biochar can be used as an important soil amendment but novel non soil for non-soil applications are also getting popular recently. An ancient substance that has gained new life in the perspective of carbon neutrality is biochar. Pre-columbian Indians used biological wastes to create highly fertile amazonian dark earths as early as 500–2500 years B.P. (Denevan 1992). Pyrolysis, or the thermal conversion of biomass into charcoal under oxygen-limited circumstances, was the fundamental concept behind converting trash into extremely fertile soils (Maschio et al. 1992). It is described as a carbon-rich substance produced by thermally converting biomass feedstock under oxygen-limited conditions, including pyrolysis and hydrothermal carbonisation (IBI 2015). Pyrolysis is the process of thermally converting biomass feedstock at temperatures typically ranging from 250 to 900 °C in an oxygen-limited atmosphere. On the other hand, hydrothermal carbonisation involves heating biomass in a closed reactor below 250 °C after combining it with water (Masoumi et al. 2021). For a long time, biochar has been added to soil to improve its fertility. As a green immobilisation agent, it also demonstrated the capacity to immobilise organic pollutants and heavy metals (Hou et al. 2022). In addition to its conventional use as a soil amendment to increase soil fertility and immobilise pollutants, biochar has recently been used in non-soil applications that have shown significant promise in reducing the effects of climate change and has been used as a green catalyst in biorefineries.

Reusing CO, a greenhouse gas (GHG), would cause its concentration to gradually rise, endangering human health and the environment. The United Nations (UN) established the 17 Sustainable Development Goals in order to address these issues and promote sustainable development globally. Many issues can be successfully resolved by reducing carbon emissions into the atmosphere. A key element in accomplishing the global carbon cycle, which is necessary to meet the Sustainable Development Goals (SDGs), is biochar (Bhandari et al. 2023). There is a high degree of stability and long-term persistence of the carbon to biochar in the soil. Regarding the sustainable development goals, biochar has the ability to efficiently alleviate food shortages, preserve health, clean wastewater, and aid in the production of medications. The production of biochar may be impacted by the pyrolysis settings and feedstock type. It affects the economy and soil remediation, among other things. The latest use of biochar is as a component of polymeric building materials, primarily as a sustainable substitute for cement, sand, and other energy-intensive mixtures used to make concrete. Because it can store carbon for extended periods of time, it can also help slow down climate change. As a result, overall GHG emissions have decreased. In keeping with Sustainable Development Goals 2 (Zero Hunger), 11and 15 (Life on Land) (Sparrevik et al. 2013), the use of this approach in agricultural soils not only improves soil fertility, water retention, and nutrient cycling, but it also encourages increased crop output. Furthermore, the renewable energy by products produced by the synthesis of biochar contribute significantly to SDG 7 (Affordable and Clean Energy), offering a sustainable alternative to traditional fossil fuels. The problems of disposing of organic waste are successfully addressed by integrating biochar into waste management practices, which also helps to mitigate climate change. This aligns with both SDG 13 (addressing climate change) and SDG 12 (supporting responsible consumption and production).Additionally, biochar's microbial activity and porous structure can benefit water filtration and purification for agricultural usage, which aligns with SDG 6 (Clean Water and Sanitation). The manufacture of biochar exhibits promise, but cost-effectiveness and scalability remain challenges (Singhal 2023). Therefore, more investigation and improvement are needed to completely reap its benefits. This chapter discusses an effective way to combat climate change, enhance agricultural sustainability, and support the framework of sustainable development goals by incorporating biochar into sustainable land management techniques and also discussing the properties, diverse applications, socio-economic and technical impacts of biochar in the global carbon cycle towards achieving the sustainable development goals.

12.2 Biochar Production and Composition

Biochar is defined as a carbon enrich, light, porous, solid compound produced by the thermal decomposition of organic biomass like wood, animal manure, straw or crop residues etc. In recent years it has grained significant attention globally due to its potential in different sectors like carbon sequestration for environmental sustainability and agricultural sector. Apart from that, biochar is a very rich source of essential plant nutrients like Calcium (Ca), Phosphorus (P), Potassium (K) and other micronutrients. Here the production process of biochar is elaborated below.

12.2.1 Methods of Production for Biochar

Biochar mainly produced or synthesized by heating biomass (plant or animal origin) in absence of oxygen (anaerobic), popularly known as pyrolysis. Production process of biochar may be classified in different categories based on the operating condition which are elaborated below.

(i) **Slow pyrolysis process**: These types of pyrolysis required low temperature (300–700 °C) but longer time which may be several hours to prepare biochar from organic biomasses but recovery or yield is higher than other processes (Do and Nguyen 2024).
(ii) **Fast pyrolysis process**: It is the opposite process of slow pyrolysis, where biomass is pyrolysised for higher temperature (600–1200 °C) for shorter period of time to produce biochar (Do and Nguyen 2024). First pyrolysis can produce more biochar within shorter period of time.
(iii) **Hydrothermal carbonization (HTC)**: Hot water is reacted with biomass in this process to produce biochar. Here hot pressured water is used in spite of anaerobic heating and the produce is commonly known as hydrochar (Hoekman et al. 2018).

12.2.2 Composition of Biochar

Composition of biochar mostly vary due to use of different types of feedstock i.e. plant or animal based biomasses. It can also differ if different pyrolysis methods are selected. The main compositions of biochar are stated below.

(i) **Carbon content**-Carbon constructs the main structure for biochar and enhance long term sequestration of carbon. Carbon content of biochar varies from 20 to 85% depending upon the type of feedstock used for the process. As example, biochar prepared from sewage and sledge contains only 25% C, while biochar prepared from wheat or maize straw contains near about 60% C (Ippolito et al. 2015).

(ii) **Nutrient content of biochar**- Apart from carbon, biochar is rich in essential plant nutrients like potassium, phosphorus and calcium. Biochar also improve the physical and chemical properties of soil hence improve the nutrient availability (Zhang et al. 2018).

12.2.3 Properties of Biochar

Biochar is a globally recognized soil amendment which can improve the physical and chemical properties of soil and improve environment. These properties may vary depending on the feed stock, retention time and temperature given at the time of production. Major physical and chemical properties of biochar have been discussed below.

(i) **Particle size of biochar**

Production process of biochar i.e. pyrolysis significantly reduces the particle size and converts the biomass into black tiny particles. According to He et al. (2018) particle size of biochar can vary between 5 and 5000 μm depending on several factors.

(ii) **Surface area**

Due to small particle size, specific surface area of biochar is very high which promotes better nutrient and herbicide adsorption. Surface area of biochar can be as low as 3.6 m^2g^{-1} to as high as 242 m^2g^{-1} (Graber et al. 2012).

(iii) **Density and Porosity**

Particle density of biochar varies from 1.34 to 1.96 gm^{-3} while bulk density mostly lies between 0.25 and 0.6 gm^{-3} which indicating lower density of the particles (Brewer et al. 2014). Due to low density, porosity of the biochar molecule become higher which enables it to be used as a physical soil amendment.

(iv) **Moisture content and water holding capacity**

Biochar has very low moisture content (1–5% weight basis) but it can increase water holding capacity if applied in sandy soil (Verheijen et al. 2019).

12.2.3.1 Chemical Properties

(i) **Cation Exchange Capacity (CEC)**

Mostly biochar has higher CEC which increase the ability to retain cations in higher quantity. CEC of biochar mostly varies from 5 to 50 cmol(+) Kg^{-1} but can be as high as 204 cmol(+) kg^{-1} depending upon the biomass (Munera-Echeverri et al. 2018).

(ii) **pH of Biochar**

pH of biochar can vary from acidic to alkaline range depending on the nature of biomass availability. Mostly biochar has pH in alkaline range which can remediates the acidic soil (Fidel et al. 2017).

(iii) **Nutrient retention**

Biochar have high CEC hence are able to retain essential plant nutrients in sufficient amount.

12.2.4 Biochar and Sustainable Farming Practices

Field experiments conducted across different regions have shown that biochar applications substantially enhance biomass production and crop yields by improving soil quality and nutrient retention (Iswaran et al. 1980; Lehmann et al. 2003; Kimetu et al. 2008). Table 12.1 compiles evidence from global studies demonstrating the positive yield responses of multiple crops to biochar amendments across diverse soil types and climatic conditions.

A. **Enhancing Soil Fertility and Crop Productivity**

Biochar is commonly used as a soil amendment in the crop fields which improves the fertility of soil and ultimately enhance the crop productivity. Biochar plays a vital role to improve soil health through these processes described below.

(i) **Nutrient Retention and Reduced Leaching**

High surface area of biochar leads to sufficient nutrient retention of the essential plant nutrients mainly due to these following reasons.

(a) Direct adsorption: Biochar have high CEC as well as AEC and when applied to the soil it can binds the essential elements (Hailegnaw et al. 2019) like N (ammonium and nitrate), phosphate, potassium and other secondary (Ca, Mg, S) and micronutrients (Zn, Cu, Fe).
(b) Enhanced water retention: Biochar applied in light textured soil slower the drainage capacity and enhance water retention. Higher water holding capacity helps in absorbing more water for a prolonged period of time which gradually enhance absorption of water-soluble nutrients. Researches showed that biochar application enhance available water capacity (AWC) higher in course textured (25.6%) soil than fine textured soil (11.5%) (Wei et al. 2023).
(c) Microbial interactions: Biochar enhances microbial activity by proving foods to the microorganisms (bacteria and fungi). Microbes convert insoluble nutrient fractions to soluble forms which improves plant growth.

Apart from that biochar application reduces leaching of phosphorus and nitrogen by absorbing significant amount of nutrients. Research studies showed that application of biochar reduces 37% nitrogen leaching in sandy soil and 20% phosphorus in clay

Table 12.1 Effect of biochar on various crop growth and yield in different parts of the world

Study area and crop	Yield response	References
Pea, India	Biochar @ 0.5 t/ha increased biomass by160%	Iswaran et al. (1980)
Mungbean, India	Biochar @0.5 t/ha increased biomass by122%	
Soybean on volcanic ash loam, Japan	Biochar @ 0.5 t/ha increased yield by 151%, and biochar @ 15 t/ha decreased yield by 29%	Kishimoto and Sugiura (1985)
Sugitrees on clay loam, Japan	Wood charcoal, bark charcoal and activated charcoal at 0.5 t/ha increased biomass by 249, 324 and 244%, respectively	
Bauhinia trees on Ultisol	Biochar application increased biomass yield by 13% and height by24%	Chidumayo (1994)
Cowpea on *Xanthic Ferralsol*	Char at 67 t/ha increased biomass by 150% char at 135 t/ha increased biomass by 200%	Glaser et al. (2002)
Cowpea and rice at Embrapa Amazonia Ocidental, Manaus, Brazil	Biochar additions significantly increased biomass production by 38 to 45% (no yield reported)	Lehmann et al. (2003)
Maize, Kotokosu watershed, Ghana	Grain and biomass yield was 91 and 44% higher on charcoal site than control	Oguntunde et al. (2004)
Maize, cowpea and peanut	Acacia bark charcoal plus fertilizer increased maize and peanut yields (but not cowpea)	Yamamoto et al. (2006)
Enhanced biological nitrogen fixation (BNF) by bean biochar in Colombia	Bean yield increased by 46% and biomass production by 39%comparedto control	Rondon et al. (2007)
Radish yield in heavy soil (with and without N)	Biocharat100t/ha increased yield 3 times; but no effect without added N	Chan et al. (2007)
Rice and sorghum cropping system	Biochar from chicken manure gave highest system yield (12.4 t/ha)	Steiner et al. (2007)
Maize yields in degraded soils in Kenya	Doubling of maize grain yield in the highly degraded soils from 3 to 6 t/ha	Kimetu et al. (2008)

loam soil. (Kabir et al. 2023). Reduction in nutrient leaching enhances the nutrient use efficiency in arable lands.

12.2.4.1 Enhanced N-Leaching and Utilization

Overuse of nitrogenous fertilizers in agriculture, primarily urea, is extremely concerning because it causes issues with water quality and releases nitrous oxide (N_2O) into the atmosphere from the soil-crop continuum (Gonzalez et al. 2015). Global environmental sustainability has been hampered by low N use efficiency (about 30–40%), risk from N leaching, N_2O emissions, and ammonia volatilisation (Li et al. 2018). Therefore, it is crucial for sustainable agriculture that fertiliser application reduces nitrogen release while increasing plant utilisation of nitrogen. The increased surface area and porosity of biochar aid in the retention of nutrients (such as N) in soil and soil solution. The pyrolysis temperature and feedstock types control how well biochar retains nitrogen. When combined with commercial N fertiliser, biochar amendments provide a comparatively gradual release of nitrogen from agricultural soil (Wen et al. 2017). The minerals bentonite and sepiolite were shown to boost adsorption and encourage biochar-urea agglomeration, which further improves N retention. Additionally, biochar made from agricultural waste could be helpful in combining with urea to create organo-mineral mixed urea, which would reduce the need for nitrogen by replacing the exclusive form of mineral urea. To understand the possible effects of biochar combined with urea on the N process in agricultural systems, more thorough research must be conducted. By increasing and improving soil CEC, biochar has been shown to have an impact on soil N cycling, nitrification rates, ammonia adsorption, and ammonium storage (Gupta et al. 2019). Its use aids in lowering gaseous N losses and nitrate leaching.

(ii) **Enhanced Microbial Activity and Diversity**

Biochar serves as food and habitat both for beneficial microbes, enhancing the biomass and diversity of soil microbes. Researches showed application of biochar in soil increases microbial biomass (10–40%) specifically when derived from nutrient rich soil (Li et al. 2020). Research showed that in arid soils of UAE use of biochar enhance salt-tolerant and plant growth-promoting microbes like *Bacillus*, *Pseudomonas* etc. which can improve nutrient cycling under saline conditions (Khan et al. 2023). Apart from that biochar application enhances the fungal and bacterial biodiversity in all types of soil.

(iii) **Improved Soil Aeration and Water Retention**

Application of biochar in improving soil aeration and water retention have already been discussed in the physical property section. Experiment showed that application of biochar in sandy enhance water holding capacity by 10–30%, and water retention by 23% (Brtnicky et al. 2021). Application of biochar to improve water retention is mostly effective in arid region. Biochar also reduce bulk density in heavy textured or clay soil. Studies showed that biochar can reduce bulk density by 9% and improve drainage in clay soil whereas in sandy soil biochar reduce rapid drying (Brtnicky et al. 2021). Application of biochar in poor fertility and arid soil improved crop yield by 15–30%.

12.2.5 *Carbon Sequestration and Soil Carbon Dynamics*

Biochar is chemically stable carbon molecule having aromatic structure which resist chemical decomposition and oxidation for a long time (hundreds of years). Conserving biomass into biochar is a great tool for stabilizing carbon for a long time. This stability enables biochar to act as a long-term carbon sink which can conserve carbon potentially with a global sequestration potential of 0.7–1.8 Giga ton CO_2 annually (Vetter et al. 2022).

Mechanisms of carbon sequestration are described below.

(i) **Inherent Chemical Resistance**:

Biochar is a polycyclic aromatic hydrocarbon which minimizes enzymatic breakdown, with half-life more than 500 years. Research in Germany showed that in loamy soils can retain 38 Mg ha^{-1} of biochar with continuous application for 11 while sandy soils showed low retention of biochar due to limited physical protection (Gross et al. 2024).

(ii) **Negative Priming Effect**:

Biochar reduces mineralization of soil organic matter by 3.8% and stabilize rhizospheric deposits and microbial byproducts through stable aggregate formation. Continuous application of biochar for prolonged time (more than 5 years) in maize and wheat field increased organic carbon by 39% (Babu et al. 2023)

(iii) **Aggregate Dynamics**:

Application of biochar increases the number of micro aggregates (< 0.25 mm) by significantly reducing macroaggregate (> 0.25 mm) which improve to stabilize carbon for a longer period of time. Biochar improves aggregate stability of soil which controls the carbon dynamics of our environment.

12.2.6 *Biochar and Soil Health*

Biochar significantly modifies soil physical, chemical, and biological properties, leading to improvements in water holding capacity, nutrient availability, microbial activity, and overall soil fertility (Hearth et al. 2013; Abrol et al. 2016; Singh et al. 2022). Table 12.2 presents a synthesis of research findings on how different biochar types influence critical soil parameters and contribute to enhanced crop productivity under various environmental conditions.

Table 12.2 Effect of Biochar on different soil parameters and crop productivity

Biochar	Impact on soil	References
Physical properties		
Corn Stover, 350 °C	Increase macro aggregates	Hearth et al. (2013)
Corn Stover, 350 and 550 °C	Increase water holding capacity and hydraulic conductivity	Hearth et al. (2013)
Mixedwood, 620 °C	Reduced bulk density, increase infiltration	Abrol et al. (2016)
Miscanthus, 450 °C	Increase water content of soil	Duarte et al. (2019)
Rice husk	Reduce bulk density of soil upto1.5%	Sharma et al. (2021)
Bio char (overall)	Increase CEC upto 45%	Singh et al. (2022)
Rice husk	Reduce water use efficiency (WUE)	Abrol et al. (2024)
Chemical properties		
Poultry litter (450°–550 °C)	Increase soil carbon content	Chan et al. (2008)
Rice straw (250–450 °C)	Increased soil pH by 8.48–79.25%	Peng et al. (2011)
Cacao shell, oil palm shell, rice husk; (250–350 °C)	Soil pH increased by 0.5 (cacao shell biochar); 0.05 (oil palm shell biochar); 0.04 (Rice husk biochar)	Martinsen et al. (2015)
Tobacco stalk biochar	Increase soil pH	Bindu et al. (2016)
Oil palm biochar	Increased the soil available K (37% over RDF)	Bindu et al. (2020)
Eucalyptus wood, bamboo and rice husk biochar	Decrease soluble and exchangeable Al by 34.38–95.66%	Geng et al. (2022)
Soil biological properties		
Hardwood (500 °C for 2 h)	Increase in population of bacteria	Prommer et al. (2014)
Lignin rich wood (350–500 °C)	Enhance phosphorous solubilizing microbes'(PSB)	Deb et al. (2016)
Eucalyptus waste (650–750 °C)	Improve microbial population	Abujabhah et al. (2016)

A. Improving Soil Structure and Water Retention

Application of biochar improves the soil structure as well as porosity of soil. It also enhances the infiltration of water and water retention mostly in course textured soil. Here the mechanisms to improve physical health of soil by application of biochar described elaborately with scientific evidences.

(i) **Soil Aggregation and Compaction Reduction**:

Biochar, rich in carbon, acts as a binding material for soil particles and forms water stable soil aggregates. Formation of proper aggregates in soil improves the aeration and gradually reducing compaction. Overall improvement in soil structure helps in better root penetration resulting higher yield. Researches showed that application of biochar in soil can reduce bulk density 3–31% and increase porosity by 14–64% depending on the rate of application (Blanco-Canqui 2017).

(ii) **Water infiltration and retention**:

Biochar application primarily in heavy soil enhances infiltration of water due to reduction in compaction. Hydraulic conductivity also be enhanced by application of biochar primarily in compacted clay soil (Wong et al. 2018). Porous structure of biochar (contains micropores < 10 μm) traps water, which gradually helps in retaining water particularly in arid regions.

Moreover, combining effect of biochar and water retaining agents like polyacrylamide can further boosts water holding capacity, slowing water loss under drought conditions.

12.2.6.1 Enhancing Microbial Activity and Nutrient Cycling

Biochar is an organic substance mainly contains carbon in its structure and other essential elements which also helps in growth and proliferation of soil microbes i.e. bacteria, fungi and actinomycetes. These microbes are also responsible for nutrient cycling which ultimately provide available forms to the plant roots. Here the beneficial effects of biochar on microbes and nutrient availability has elaborated below.

(i) **Habitat for Beneficial Microorganisms**:

Low bulk density and high porosity of biochar provides convenient environment to grow and replicate beneficial soil microbes. Evidences showed that biochar gives a favourable habit for nitrogen-fixing bacteria and mycorrhizal fungi which are beneficial for plants (Gou et al. 2023). Besides higher population of the microbes, biochar also improves microbial diversity of soil which ultimately leading towards better soil biological health. Moreover, it provides support to the organisms which produces phytohormones, different soil enzymes and antifungal compounds which improve the quality of soil.

(ii) **Nutrient Availability**:

Nutrient cycles are mostly governed by soil microbes i.e. bacteria and fungi which converts immobile nutrients to mobile form and promotes availability of nutrients. Application of biochar directly improves soil microbial activity hence increase the bioavailability of nitrogen (N), phosphorus (P), and other essential nutrients (Song et al. 2018). Application of biochar in soil also benefits arbuscular mycorrhizal fungi to improve nutrient uptake for phosphorus while reducing harmful emissions like N_2O.

Evidences showed that addition of biochar in soil exhibit higher microbial enzyme activities leading to carbon, nitrogen, and phosphorus cycling. Improved nutrient availability results in better crop growth and yield in stressed environment.

12.2.7 *Biochar in Livestock Management*

Biochar is also used for serving several purposes for raring animals and managing livestock also in India.

12.2.7.1 Biochar as a Feed Supplement

In recent years use of biochar as a feed additive for ruminants (cattle and goats), pigs, poultry (chickens and ducks) and fish has been increased (Man et al. 2021). Here role of biochar is discussed elaborately as feed supplement with scientific evidences.

(i) **Improved Nutrient Absorption and Digestion**

Biochar having porous structure and high nutrient adsorption capacity through which it able to bind nutrients effectively within the digestive system of the ruminants. Effective adsorption of nutrients in cattle leads to improve the feed efficiency (Schmidt et al. 2019). This improves nutrient availability and absorption, leading to enhanced feed efficiency. Researches have proven that addition of biochar in feed supplements impacts positively by improving digestion, blood values, egg yield for birds and overall productivity of the animal (Man et al. 2021). Variation in biochar due to different types of feedstock or pyrolysis can also cause variation in animal productivity.

(ii) **Reduction of Methane Emissions**

The major cause for global warming and climate change is emission of potent greenhouse gases like methane which mostly produced from ruminant animals like bovine, sheep and goats. Addition of biochar in the feeds of ruminant animals can significantly alter the microbial fermentation into the gut of the animals. As a result, activity of methanogenic bacteria is suppressed leading to significant reduction in methane production. Research studies indicate that biochar consumed animals exhibit lower methane emissions compared to controls i.e. no biochar in feed (Dittmann et al. 2024). So, it is obvious from the text that reduction in methane emission not only improve the environment but also enhance energy utilization within the animal.

(iii) **Detoxification and Toxin Binding**

Biochar is chemically activated carbon which has been used for hundreds of years to adsorbs toxins in traditional medical treatments. When biochar is applied with feed additive, it binds harmful substances like mycotoxins, pesticides residues and heavy metals (As, Cd, Hg etc) (Nair et al. 2023). Detoxifying effect of biochar significantly reduces the chance of digestive disorders hence improves animal health. Moreover, biochar can reduce butyric acid formation in silage which improves feed quality.

12.2.7.2 Benefits of Co-Composting Manure with Biochar

Manures are prepared from the organic substances (animal or plant origin) through composting process. Application of biochar in this process reduce toxins and enhance the quality of the manure. Benefits of composting with biochar has been discussed below.

(i) **Improved Compost Quality**

Application of biochar at the time of composting significantly increases the quality of the produce. Due to smaller size and higher surface area, biochar can adsorb and retain huge number of essential ions (both anion and cation) like NH_4^+, $H_2PO_4^-$, K^+ etc. (Helal et al. 2019). High CEC and AEC of biochar retains the ions so tightly that they cannot be lost during the process of composting. This ultimately results in preparation of more stable and nutrient enrich compost than the ordinary ones.

(ii) **Odour Reduction**

Composting and manure formation is associated with strong and foul odor due to anaerobic decomposition. Biochar has an excellent adsorptive property which can mitigate the also mitigate the strong odours associated with manure decomposition (Kaikiti et al. 2021). Huge surface area and porous structure of biochar binds volatile organic compounds (VOCs) through physisorption which gradually reduces the foul smell that often produced at the time of manure handling and storage.

(iii) **Ammonia Volatilization Reduction**

Volatilization of ammonia (NH_3) is a vital problem in manure preparation as it has two bad impacts, firstly it reduces the nitrogen content of the manure and secondly it releases NH_3 gas which ultimately leads to air pollution. Biochar, a super absorbent, can significantly absorb ammonia leading to reduction in volatilization and enriching nitrogen in the compost (Agyarko-Mintah et al. 2017). This not only enhances the fertilizer value of the compost but also decreases environmental impacts.

(iv) **Pathogen Proliferation Control**

Biochar has atypical property to inhibit the growth and proliferation of pathogenic microbes at the time of composting (Chung et al. 2021). Its higher pH (alkaline) and redox properties can create an unfavourable environment which does not allow the pathogens to survive and ultimately making the compost safer for use in the crop fields.

12.2.8 Biochar for Climate Mitigation and Adaptation

Biochar in recent days emerged as a critical tool to fight against climate change. Its dual role in mitigating greenhouse gas (GHG) emissions and enhancing agricultural resilience makes it a valuable natural solution. This text contains the potential of biochar in sequestering carbon, reducing GHG emissions and improving agricultural systems' adaptability to climate change.

12.2.8.1 Mitigating Greenhouse Gas Emissions

Application of biochar in agricultural soils significantly reduces emission of greenhouse gasses like CO_2, N_2O and CH_4. Mechanisms by which biochar reduces the harmful gases are elaborated below.

(i) **Carbon-dioxide reduction in atmosphere**

Biochar, a highly stable carbon compound, can store carbon in soils for long period of time and as a result CO_2 is removed from the atmosphere and stored in it. Locking of atmospheric carbon for prolonged time by storing it in soil in the form of biochar known as soil carbon sequestration. Scientific studies showed that biochar has a potential to conserve 3.7 GT (giga ton) of CO_2 equivalent emission annually which can significantly reduce global warming (Burns 2025).

(ii) **Reduction of Nitrous Oxide (N_2O) Emissions**

Nitrous oxide, a potential greenhouse gas, mostly released in the process of denitrification from agricultural soils when excess nitrogen fertilizer is applied in waterlogged condition. Biochar can reduce N_2O emission by altering soil nitrogen dynamics and adsorbing ammonium or nitrate ions which are the main raw material for the process. Research studies showed application of biochar can lower N_2O emissions by 65–68% in certain conditions, particularly in the cereal crops like barley which enhance denitrification efficiency (Agegnehu et al. 2016).

(iii) **Lowering Methane (CH_4) Emissions**

Reduction of CH_4 emission by application of biochar rich feed from ruminants has already been discussed previously. Biochar, having high adsorption power reduces the fermentation process in the gut of the animals which ultimately leads to low CH_4 emission.

12.2.9 Enhancing Climate Resilience in Agriculture

Climate change and global warming exerts a huge pressure to modern agriculture. Biochar can reduce the ill effects of climate change and give a sustainable approach to conserve the environment by following mechanisms.

(i) **Improved Soil Moisture Retention**

Biochar's porous structure can increase water holding capacity (WHC) of soil by improving porosity and reducing evaporation as already discussed earlier. This improves the resilience of the soil in terms of moisture when crops are grown in arid and semi-arid regions, where the main limiting crop factor is water scarcity (Verheijen et al. 2019).

(ii) **Enhanced Nutrient Availability**

Huge adsorbing capacity of biochar enables it to adsorb huge amount of chemical synthetic fertilizer when they are applied in the soil. Later stages of the plant growth, biochar acts as a slow releasing source of nutrients like N and P when fertilizer is not applied. This increases the nutrient resilience of the crop which not only enhances yields but also reduces the negative environmental impacts associated with fertilizer overuse (Alkharabsheh et al. 2021).

(iii) **Adaptation to Climatic Changes**

Biochar acts as an amendment in different soils which enhances the resilience to extreme weather events like severe droughts and heavy rainfall. Improved soil structure due to biochar application reduces erosion risk during heavy rains, while enhanced moisture retention in it supports crops during dry spells. These typical properties make biochar a crucial tool for ensuring food security under changing climatic conditions (Kumar et al. 2025).

12.3 Application of Biochar in Agriculture for Achieving Sustainable Development Goals

Present issues and future directions while introducing the latest biochar technology for agricultural applications to achieve sustainable development goals has to be addressed. Recent information abundantly based on the production and characterisation of biochar, the use of biochar for soil improvement and agricultural application, the functions of biochar for farmland environmental improvement to alleviate water and waste management issues and climate change, the use of biochar to promote the bio-economy and clean energy, and future prospects. Agricultural engineers, researchers, and anyone looking to use biochar to enhance soil and environmental conditions in focusses on the use of biochar in agricultural operations, with the goal of further developing biochar as an affordable and sustainable resource for field-scale agricultural applications. Emphasises how biochar can help the bio economy grow, which has enormous potential to advance a circular economy and optimise social, economic and environmental advantages. Linking the use of biochar to sustainable development objectives is an important aspect of agricultural waste management.

12.3.1 SOC Sequestration and C-Use Efficiency in Agriculture

The slow degradation of biochar in soil facilitates the progressive accumulation of SOC, or an increase in soil organic C status over the years. Because organic carbon is firmly bonded to soil particles, comparatively less CO_2 is released into the atmosphere from the soil. Accordingly, using biochar in agriculture lowers the greenhouse gas emissions and the effects of climate change (Liu et al. 2018). When biochar is incorporated into soil, it may act as a carbon sink for long-term carbon storage. The mean residence period of biochar is also influenced by the type and quality of the soil. Biochar has beneficial effects on SOC dynamics and helps to create a refractory SOC pool, depending on management techniques, agriculture can either be a net source or a sink for greenhouse gases. Climate change can be somewhat mitigated by agricultural management techniques that promote soil carbon sequestration. According to research, SOC is protected from use by soil amendment. The high surface area and porosity of the biochar help to control the stabilisation of native SOC and reduce microbial activity. By altering the properties and microbial diversity of soil, the addition of biochar reduces CO_2 emissions. With improved C-use efficiency, it can aid in raising soil microbial biomass and SOC build-up because biochar converts carbon sinks into carbon stores that may be sustained for a long time, its use is often regarded as "C negative" (Brassard et al. 2017).

12.3.2 Biochar in Controlling Pests and Stress

Controlling biotic stress and maximising agricultural yields must be sustainable and environmentally beneficial processes. This situation is appropriate for the use of biochar as an organic soil amendment (OSA). The application of biochar stimulates the suppression of pathogens and enriches beneficial soil microbes. Diseases brought on by nematodes, fungus, bacteria, and viruses are prevented from developing by suppressing the soil. This can be attributed to a decrease in the possibility for pathogen inoculum, an increase in the population of preferred biocontrol agents, and stimulation of soil biota activity (Chung et al. 2021). For example, biochar reduces infections from plant parasitic nematodes by suppressing the pathogenic activities and infection-causing capabilities of soil. Additionally, it is thought that a greater number of free-living nematodes are good for the health of the soil (Wu et al. 2020). Through pesticide adsorption and absorption, biochar reduces the harmful effects of pesticides on the environment. Considering its larger surface area and higher porosity, it aids in lowering the bioavailability of pesticides. Additionally, by decomposing plant residue and adding soil amendments in the form of agro-waste products, it shields the roots of plants in the soil from phytotoxic chemicals emitted by other plant roots (Xia et al. 2022).

12.3.3 Role of Biochar in Bioremediation of Environmental Pollutants

Many dangerous pollutants, including heavy metals, poisonous azo-dyes, and pharmaceuticals and personal care products (PPCPs), have been removed using biochar. Biochar has demonstrated remarkable potential in remediating contaminated soils by interacting with pollutants through mechanisms such as inner complexation, hydrophobic interactions, electrostatic binding, and precipitation processes (Cao and Harris 2010; Gholami and Rahimi 2020; Azeem et al. 2021). Table 12.3 summarizes experimental findings on biochars produced from various feedstocks under different pyrolysis conditions, showcasing their pollutant removal efficiencies and dominant sorption mechanisms.

Regarding the use of biochar in the removal of various pollutants can be discussed below:

12.3.3.1 Removal of Heavy Metals

Physical adsorption, ion-exchange interactions, electrostatic attraction, functional group combination, surface complexation and precipitation are the primary adsorption processes of heavy metals on biochar. Every metal has a unique adsorption mechanism, and the pertinent biochar qualities also have an impact on the adsorption properties. The high sorption efficacy of biochar for heavy metals has also been demonstrated by a number of papers based on characterisations such as SEM, TEM, FTIR, and XRD analysis. To study the adsorption of heavy metals, biochar colloids-mycelial pellets (BC-MP) has also been developed and batch tests using the model pollutant Cd (II) related work has also been conducted. Since biochar has a great deal of potential for heavy metal removal, it can help cure water pollution by breaking down harmful heavy metals into their most basic form, which promotes environmental sustainability.

12.4 Biochar for Climate Change Mitigation

12.4.1 Reducing the GHGs Emissions and Neutral Carbon Footprint

The soil C/N ratio is impacted by the concentration of C in biochar, as a larger fraction is resistant and a lesser percentage is labile. Incorporating biochar reduces carbon emissions from burning fossil fuels and stores biogenic carbon in soil. Applying biochar can reduce CO_2 emissions from soil to the atmosphere, which in turn reduces CH_4 and N_2O emissions, making progress in reducing the greenhouse effect.

Table 12.3 Biochar and its interaction with different soil pollutants

Feedstock	Pyrolysis		Pollutants	Initial conc. (mg/kg)	Removal efficiency (%)	Interaction mechanism	References
	Temperature (°C)	Applied dose (%)					
Pig-carcass	650 °C	2.0	Zn	48.21	76.4	Inner complexation	Cao and Harris (2010)
Rice husk	700 °C	4.0	PCB'S	0.08	91.0	Hydrophobic interaction	Silvani et al. (2019)
Carrot pulp	550 °C	8.0	Cu	29.32	90.1	Electrostatic forces, covalent bonding	Gholami and Rahimi (2020)
Wood	450 °C	1.5	Thiamethoxam	6.0	22.8	Oxygen-containing groups, reactive oxygen species, persistent free radicals	You et al. (2020)
Pig manure	700 °C	2.0	Clothianidin	4.95	90.5	Hydrophobic interaction, H–bonding, p/-interactions	Zhang et al. (2020)
			Imidacloprid		81.4		
Sheep bone	800 °C	10.0	Zn	265	57.0	Precipitation, ion-exchange, surface complexation	Azeem et al. (2021)
			Cd	5.83	60.0		
Rice husk	500 °C	5.0	Cd	6.10	25.0	Surface complexation	Karmaker et al. (2021)
Sewage manure	700 °C	2.0	PAHS	0.04	74.0	Pore filling, hydrophobic interaction, -interactions	Godlewska et al. (2022)
Rice straw	500 °C	5.0	Zn	37.98	36.9	Hydroxide precipitation, cation-interaction	Liu et al. (2022)

However, some other studies has reported no discernible impact on soil CO_2 efflux, but it was noted that the soil changed with biochar exhibited a decrease in CO_2 emissions. The addition of biochar to paddy soil has in-fact reduced CO_2 and CH_4 emissions (Yin et al. 2021) observing that rice grown on soil amended with biochar lowers CO_2 emissions. Numerous studies have established that biochar amendments reduce greenhouse gas emissions, particularly CH_4, N_2O, and CO_2, by altering soil microbial dynamics, enhancing soil aeration, and stabilizing soil organic carbon (Feng et al. 2012; Wang et al. 2013; Trinh et al. 2017). Table 12.4 compiles experimental data demonstrating the impact of biochar feedstocks on soil GHG emissions and their associated mitigation strategies across diverse cropping systems.

For estimating GHG emissions from agricultural productions, the carbon footprinting technique has been frequently used (Cooper and McCullough 2021) which refers to the quantity of GHG generated by various products and activities which can be used to assess and compare the greenhouse gas emissions of different agricultural activities and to pinpoint areas where safe environmental efficiency could be increased. Due to the impact of biochar, comparatively very little CO_2 and CH_4 emission was noted in a study based on Mediterranean wheat (Liu et al. 2011).

The addition of biochar and inorganic N, which supply the labile C and ammonium, has aided in nitrate synthesis mediated by the nitrifying bacteria. Under sporadic aerobic and anaerobic circumstances, denitrifying bacteria convert nitrate

Table 12.4 Soil Biochar application and its effects on GHGs emission and mitigation strategies

Biochar feedstock	Effects	References
Corn stalk	Increases SOC, grain yield and enhancement of soil productivity, CH_4 emission reduction ($\sim$ 63%)	Feng et al. (2012)
Rice husk	Biochar incorporation significantly decreased N_2O emissions and promoted crop production	Wang et al. (2013)
Canola straw	Biochar plus calcium superphosphate decreased loss of NH_3 by $\sim$ 39%, biochar alone reduced N_2O emission by $\sim$ 24%	Sun et al. (2016)
Rice straw and husk	Increases SOC and decreased CO_2 and CH_4 emissions	Sui et al. (2016)
Rice straw	Significantly reduces CH_4 and N_2O emissions	Trinh et al. (2017)
	Increase in rice yield, CO_2 emissions reduced by $\sim$ 2.22% under water-saving (irrigation)management in comparison to flooded (irrigation) system	Yang et al. (2018)
	Significantly decreased N_2O emissions from irrigated fields	Sun et al. (2019)
	Reduction in CH_4 emission by $\sim$ 28–680% as compared to manure amended soils	Nguyen et al. (2020)
Wheat straw	$\sim$ 43.1% decrease in CH_4 emissions attributable to improved soil aeration	Chen et al. (2018)

to N_2O/N_2. Thus, the inhibitory effect of biochar on N_2O emissions has been documented in a number of investigations. By reducing the amount of labile C and N fertiliser added, the addition of biochar to soil also lowers the intensity of GHGs per unit of agricultural product while preserving production (Qian et al. 2014). According to the International Biochar Initiative (IBI), a significant factor in the stability of biochar in soil was its oxygen and carbon ratio (Budai et al. 2013). Additionally, it was shown that approximately 79.6% of biochar C may last for over a century (Dharmakeerthi et al. 2015). Given the vast array of potential uses for biochar, it is equally important to talk about the environmental concerns involved. The surface characteristics of the biochar, such as the presence of different functional groups, pH variations, and the oxidation of aromatic C rings, occasionally led to the production of dangerous substances (free radicals, dioxins, heavy metals, and polycyclic aromatic hydrocarbons, or PAHs).

12.4.2 Biochar for Boosting Bio-Economy

Crop dregs are actually leftovers after harvest with the world's rice crop alone producing almost 800 million tonnes straw a year. The availability of logical, scientifically developed methods for the efficient and profitable utilisation of massive biomass that is rich in nutrients and its appropriate breakdown is crucial. The production of biochar, biogas, and concurrent renewable energy generation enable the economic viability and environmentally sustainable use of agricultural waste. Utilising biomass in a way that ensures zero waste and zero environmental impact is essential from the perspective of environmental sustainability (Bhattacharyya et al. 2020).

Through the preparation of biochar, biomass can be converted into a useful carbonised product. Effective biomass identification and optimisation, appropriate biochar production technology, doses, and applications in crop fields may guarantee a notable increase in soil carbon contents and its stocks along various soil profiles, reducing greenhouse gas emissions from the soil-crop system to the atmosphere (Atkinson 2018). The technique for producing biochar from biomass waste and its numerous uses in agriculture, remediation, and energy creation appears to be promising as a useful strategy in the context of climate amelioration, preserving soil health, producing sustainable crops, producing renewable energy, and improving environmental management all contribute to the development of a circular bio-economy.

Even in tiny amounts, functionally improved biochar may prove to be a cost-effective and highly efficient option in this situation. The new nano-biochar production that can assist in addressing soil metal contaminations is worth mentioning discussing. Small amounts of biochar have also been shown to significantly improve agronomic nutrient usage efficiency and plant growth-promoting properties when carefully combined with chemical nutrients. The use of this biochar-based compound fertiliser (BCF) technology could be beneficial as we move away from the use of mineral fertilisers and towards promoting more and more green agriculture. Furthermore, through improved crop yields, less chemical fertiliser use, and decreased greenhouse gas emissions from the crop and soil system, BCF technology may help farmers enhance their net income (Hu et al. 2019). Since biochar is often produced from bio-waste, the waste products are continuously being used and recycled. Therefore, standardised and optimised biochar manufacturing method and application are seen as an economically and environmentally sound way to mitigate climate change.

12.5 Biochar Aids Towards Attaining SDGs

In one way or another, the economical synthesis of biochar from waste and its application for soil enrichment, wastewater treatment, and bio-energy production has helped to accomplish the sustainable development goals. Reducing costs and dependency on outside resources (such as using biochar made from biomass feedstock) and increasing crop production will enable farmers become self-sufficient and make money in order to achieve SDG 1 (no poverty). In order to achieve SDG 2 (zero hunger), biochar in the form of mineral-rich composite fertilisers may be effective for a variety of applications in boosting crop yields in damaged soils (Gao et al. 2020). Increased yields and increased agro-ecosystem resilience will increase food security. Application of biochar to the soil can improve crop yields, remediate the soil, purify the water, and improve people's nutritional health, all of which contribute to SDG 3 (excellent health and well-being).

To holistically capture the interconnected benefits and systemic functions of biochar, a conceptual framework has been developed (Fig. 12.1). This figure illustrates the entire pathway from feedstock utilization through pyrolysis-based biochar production to its diverse agricultural and environmental applications. These include soil amendment, enhanced crop productivity, improved livestock performance, manure composting, and bioremediation of pollutants. Each of these applications contributes to critical outcomes such as greenhouse gas (GHG) reduction, carbon sequestration, and the development of a sustainable circular economy. Additionally, the framework highlights how these outcomes are aligned with global policy goals, particularly several United Nations Sustainable Development Goals (SDGs), thereby reinforcing biochar's strategic importance in climate-smart and resilient agriculture.

Applying iron-biochar to soil and using biochar with microbial inoculation can help decompose organic pollutants and immobilise metals (Meili et al. 2019). In order to reduce leaching and water pollution, biochar has been crucial in adsorbing soil contaminants for SDG 6 (clean water and sanitation). There have been attempts to enhance the sorption of pollutants (Atinafu et al. 2020). Pyrolysis of feedstock biomass serves as a source of comparatively less expensive, affordable renewable energy for SDG 7 (affordable and clean energy). SDG 7 can be achieved by using biochar composites as an affordable alternative to costly chemicals for the production of bioenergy.

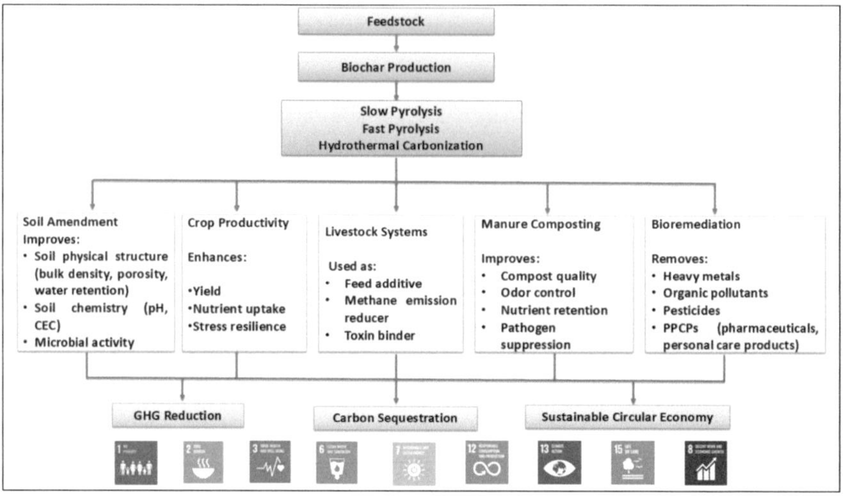

Fig. 12.1 Biochar: a sustainable circular economy tool for future agriculture

Achieving SDGs 1 and 2 is linked to improving the living conditions of farmers in rural regions, which is related to SDG 8 (decent job and economic growth). The development of low-cost paralyser technologies, such as earth-dug, for use by small-holder farmers in rural areas contributes to the achievement of SDG 9 (industry, innovation, and infrastructure) (Hamid et al. 2020). Due to clean cooking technology, biochar can help smallholder farmers in rural areas become self-sufficient and improve female health and well being, which will help achieve SDG 10 (reduced inequality). It will also increase female income, which will reduce poverty and hunger and, ultimately, reduce inequality between rural and urban areas. Urban abandoned polluted ground rehabilitation using iron-biochar and clay-biochar composites may be possible for SDG 11 (sustainable cities and communities) (Wang et al. 2020). This SDG can be attained in part by using biochar in urban planting. Urban plants aid in preparing cities for upcoming heat waves and other adverse weather conditions. According to Kammann et al. (2017), the creation of biochar by biomass pyrolysis can promote ground-based carbon storage (SDG 13, climate action) and alter the usage of non-renewable resources that are rich in carbon (SDG 12 which strives to assure responsible consumption and production). Applying biochar to the soil can help meet SDG 12 by enhancing crop yields. The utilization of biochar in animal husbandry may help cut down on the use of antibiotics, which would benefit both human and livestock health. The comparatively high recalcitrant nature of biochar organic C and the reduction of N_2O, CH_4, and CO_2 emissions from modified soils contribute to climate amelioration for SDG 13 (climate action) (Ahmad et al. 2019).

Increasing the resilience of agro-ecosystems aids in adaptation and mitigation of the expected future consequences of climate change. C stability and adverse priming effects can be improved by addition of minerals during pyrolysis or the application of iron-biochar composites. By increasing soil fertility and reducing soil-water pollutants, adding biochar can support beneficial soil microbes and improve soil health, which is important for SDG 15 (life on land). When viewed and balanced from the perspective of poverty alleviation and environmental preservation, biochar technologies, their many uses, and their potentials prove to be crucial in achieving the Sustainable Development Goals (SDGs) (Yu et al. 2020).

12.6 Socio-Economic Impact and Future Prospects

As a promising alternative force to negative emission technologies (NETs) and a viable material to reduce hazardous elements in the surroundings for sustainable environmental reclamation and remediation, biochar has attracted scientific attention from all around the world. Applying biochar to the soil has been shown to increase crop yield while reducing the need for chemical fertilisers. By storing soil carbon, its application to soil reduces greenhouse gas emissions to the environment on a comparatively long-term basis. Fossil fuel energy usage was significantly reduced by even paired biochar-bio-energy production systems. Biochar is a promising economic resource that can be applied to environmental sustainability, according to several researchers. Runoff, leaching, and vaporisation squander around 60% of nitrogen fertilisers. In addition to raising fertiliser costs, this nitrogen loss results in eutrophication, oxygen loss, and greenhouse gas emissions (Cen 2021).

Praveen et al. (2021) recorded that the cost of biochar made from feed stocks such rice husk (RH), groundnut shell (GS), and coconut shell (CS) is approximately 0.61 USD, 0.57 USD, and 0.57 USD per kilogramme, respectively. Biochar production, use, and storage are predicted to be remarkably carbon negative, with an estimated 0.3–2 Gt CO_2 sequestration annually by 2050. The costs of removing carbon from the atmosphere using techniques such as direct air capture (CO_2), forestation, soil-based C sequestration, wood-based building materials, and biochar ranged from 13.78 to 1233.5 USD per tonne CO_2 whereas the price of biochar ranges from 71.67 to 180.54 USD per tonne CO_2. This information indicates the necessity of producing biochar on a big scale (Fawzy et al. 2021). The minimum selling price of biochar and bio-based fuels is said to be lowered with the aid of renewal energy and C-credits. The main area that has to be improved in order to increase the financial benefits of biochar is farmers' ignorance of the regulations and the usage of biochar for soil enrichment. As a result, biochar has become a cost-effective substitute for a number of traditional techniques for reducing and remediating toxicity and attaining environmental sustainability.

12.7 Conclusion

There is always a need for a cost-effective, environmentally responsible, and highly effective way to handle the massive amount of solid waste has led to the development of biochar-based strategies and technologies which have several advantages, such as improving soil productivity and health overall and enhancing the sustainability of agricultural systems for the production of food and biofuel.

Biochar is known to aid in mitigating climate change by enhancing the sequestering of carbon in the soil; as it serves as the major sink of carbon for longer periods and hereby reducing GHG emissions. Biochar improves soil fertility, water retention capacity, and nutrient circulation and also promoting higher soil productivity Biochar and its by-products can be harvested and utilized purposely as renewable energy source. Besides, biochar integration in agricultural systems can be effective in waste management strategies that will mitigate and reduce the challenges of organic waste disposal effectively in a sustainable manner. Biochar acting as an efficient water purification element with a diverse application helps in the global carbon cycle towards sustainable development goals. However more thorough research is needed to determine the long-term effects of biochar-based technologies on human existence. In order to develop wise intervention strategies in line with sustained crop productivity, enhancing soil health and environmental management, reducing carbon footprints, and proper waste disposals arising from agricultural activities, the impact of various biochar production technologies in relation to a variety of feedstocks should be measured and evaluated.

High specific surface area, large pore size, facile separation of constructed biochar, and an abundance of functional groups are characteristics of biochar and engineered biochar that can also be tailored to achieve high adsorption capabilities. Sustainable development is a process of constant change, and biochar has a role to play in reaching the UNSDGs and creating a circular bio-economy. As a result, certain potential possibilities for achieving the SDGs within the period can be taken into consideration. To optimise the design of biochar production systems for their widespread adoption among agricultural communities, studying the carbon cycle assessment and cost-benefit analysis techniques must be coupled and encouraged.

References

Abrol V, Ben-Hur M, Verheijen FG, Keizer JJ, Martins MA, Tenaw H, Tchehansky L, Graber ER (2016) Biochar effects on soil water infiltration and erosion under seal formation conditions: rainfall simulation experiment. J Soils Sediments 16:2709–2719. https://doi.org/10.1007/s11368-016-1448-8

Abrol V, Sharma P, Haziq S, Kumar A, Brar A, Srinivasarao C, Lado M (2024) Synergistic benefits of biochar and polymer integration in rice–wheat system: enhancing productivity, soil health, water use efficiency, and profitability. J Soil Sci Plant Nutr 24:4984. https://doi.org/10.1007/s42729-024-01886-8

Abujabhah IS, Pranamuda H, Agustian E (2016) Effects of biochar application on nutrient availability and microbial activity in a tropical Ultisol. J Trop Soils 21(2):123–130

Agegnehu G, Nelson PN, Bird MI (2016) The effects of biochar, compost and their mixture and nitrogen fertilizer on yield and nitrogen use efficiency of barley grown on a nitisol in the highlands of Ethiopia. Sci Total Environ 569–570:869–879. https://doi.org/10.1016/j.scitotenv.2016.05.033

Agyarko-Mintah E, Cowie A, Van Zwieten L, Singh BP, Smillie R, Harden S, Fornasier F (2017) Biochar lowers ammonia emission and improves nitrogen retention in poultry litter composting. Waste Manag 61:129–137. https://doi.org/10.1016/j.wasman.2016.12.009

Ahmad M, Ahmad M, Usman ARA, Al-Faraj AS, Abduljabbar A, Ok YS, Al-Wabel MI (2019) Date palm waste-derived biochar composites with silica and zeolite: synthesis, characterization and implication for carbon stability and recalcitrant potential. Environ Geochem Health 41(4):1687–1704. https://doi.org/10.1007/s10653-017-9980-1

Alkharabsheh HM, Seleiman MF, Battaglia ML, Shami A, Jalal RS, Alhammad BA, Almutairi KF, Al-Saif AM (2021) Biochar and its broad impacts in soil quality and fertility, nutrient leaching and crop productivity: a review. Agronomy 11(5):993. https://doi.org/10.3390/agronomy11050993

Antoneli V, Bednarz JA, Thomaz EL (2022) Tobacco harvest phase is critical to runoff and soil loss in conventional tillage system. Soil Use Manag 38(1):1–11. https://doi.org/10.1111/sum.12785

Atinafu DG, Wi S, Yun BY, Kim S (2020) Engineering biochar with multiwalled carbon nanotube for efficient phase change material encapsulation and thermal energy storage. Energy 216:119294. https://doi.org/10.1016/j.energy.2020.119294

Atkinson CJ (2018) How good is the evidence that soil-applied biochar improves water-holding capacity? Soil Use Manag 34(2):177–186. https://doi.org/10.1111/sum.12413

Azeem M, Ali A, Arockiam Jeyasundar PGS, Li Y, Abdelrahman H, Latif A, Li R, Basta N, Li G, Shaheen SM, Rinklebe J, Zhang Z (2021) Bone-derived biochar improved soil quality and reduced Cd and Zn phytoavailability in a multi-metal contaminated mining soil. Environ Pollut 277:116800. https://doi.org/10.1016/j.envpol.2021.116800

Babu S, Singh R, Kumar S, Rathore SS, Yadav D, Yadav SK, Ansari MA, Das A, Rajanna GA, Wani OA (2023) Biochar implications in cleaner agricultural production and environmental sustainability. Environ Sci Adv 2(8):1042–1059. https://doi.org/10.1039/D2VA00324D

Bhandari G, Gangola S, Dhasmana A, Rajput V, Gupta S (2023) Nano-biochar: recent progress, challenges and opportunities for sustainable environmental remediation. Front Microbiol 14:1214870. https://doi.org/10.3389/fmicb.2023.1214870

Bhattacharyya P, Bhaduri D, Adak T, Munda S, Satapathy BS, Dash PK, Padhy SR, Pattanayak A, Routray S, Chakraborti M, Baig MJ (2020) Characterization of rice straw from major cultivars for best alternative industrial uses to cut off the menace of straw burning. Ind Crop Prod 143:111919. https://doi.org/10.1016/j.indcrop.2019.111919

Bindu MS, Prakash M, Ranganna G (2016) Influence of biochar on soil properties and growth of tomato. Arch Agron Soil Sci 62(4):486–500

Bindu MS, Prakash M, Ranganna G (2020) Biochar application influences soil properties and nutrient dynamics in an oil palm plantation. J Plant Nutr 43(4):465–478

Blanco-Canqui H (2017) Biochar and soil physical properties. Soil Sci Soc Am J 81(4):687–711. https://doi.org/10.2136/sssaj2017.01.0017

Brassard P, Godbout S, Raghavan V (2017) Pyrolysis in auger reactors for biochar and bio-oil production: a review. Biosyst Eng 161:80–92. https://doi.org/10.1016/j.biosystemseng.2017.06.020

Brewer CE, Chuang VJ, Masiello CA, Gonnermann H, Gao X, Dugan B, Driver LE, Panzacchi P, Zygourakis K, Davies CA (2014) New approaches to measuring biochar density and porosity. Biomass Bioenergy 66:176–185. https://doi.org/10.1016/j.biombioe.2014.03.059

Brtnicky M, Hammerschmiedt T, Elbl J, Kintl A, Skulcova L, Radziemska M, Latal O, Baltazar T, Kobzova E, Holatko J (2021) The potential of biochar made from agricultural residues to increase soil fertility and microbial activity: impacts on soils with varying sand content. Agronomy 11(6):1174. https://doi.org/10.3390/agronomy11061174

Budai A, Zimmerman AR, Cowie AL, Webber JBW, Singh BP, Glaser B, Masiello CA, Andersson D, Shields F, Lehmann J, Camps Arbestain M, Williams M, Sohi S, Joseph S (2013) Biochar carbon stability test method: an assessment of methods to determine biochar carbon stability. Int Biochar Initiat:1–10. https://biochar-international.org/wp-content/uploads/2018/04/IBI_Biochar_Stability_Test_Method.pdf

Burns W (2025) Biochar and atmospheric carbon dioxide removal. In: The Palgrave Encyclopedia of sustainable resources and ecosystem resilience. Springer Nature, Switzerland, pp 1–11. Advance online publication. https://doi.org/10.1007/978-3-030-67776-3_64-1

Cao X, Harris W (2010) Properties of dairy-manure-derived biochar pertinent to its potential use in remediation. Bioresour Technol 101(14):5222–5228. https://doi.org/10.1016/j.biortech.2010.02.052

Cen Z (2021) Develop biochar-based controlled release nitrogen fertilizers. Doctoral dissertation, South Dakota State University. ProQuest Dissertations Publishing

Chan KY, Zwieten LV, Meszaros I, Downie A, Joseph S (2007) Agronomic values of green waste biochar as a soil amendment. Aust J Soil Res 45:629–634

Chen D, Wang C, Shen J, Li Y, Wu J (2018) Response of CH_4 emissions to straw and biochar applications in double-rice cropping systems: insights from observations and modeling. Environ Pollut 235:95–103

Chidumayo EN (1994) Effects of wood carbonization on soil and initial development of seedlings in miombo woodland, Zambia. For Ecol Manag 70:353–357

Chung WJ, Chang SW, Chaudhary DK, Shin J, Kim H, Karmegam N, Govarthanan M, Chandrasekaran M, Ravindran B (2021) Effect of biochar amendment on compost quality, gaseous emissions and pathogen reduction during in-vessel composting of chicken manure. Chemosphere 283:131129. https://doi.org/10.1016/j.chemosphere.2021.131129

Cooper JA, McCullough BP (2021) Bracketing sustainability: carbon footprinting march madness to rethink sustainable tourism approaches and measurements. J Clean Prod 318:128475. https://doi.org/10.1016/j.jclepro.2021.128475

Deb S, Shukla MK, Adhya TK (2016) Influence of biochar on soil microbial biomass and microbial activities in a saline Alfisol. Biol Fertil Soils 52(4):465–476

Denevan WM (1992) The pristine myth: the landscape of the Americas in 1492. Ann Assoc Am Geogr 82(3):369–385. https://doi.org/10.1111/j.1467-8306.1992.tb01965.x

Dharmakeerthi RS, Hanley K, Whitman T, Woolf D, Lehmann J (2015) Organic carbon dynamics in soils with pyrogenic organic matter that received plant residue additions over seven years. Soil Biol Biochem 88:268–274. https://doi.org/10.1016/j.soilbio.2015.06.005

Dittmann MT, Baki C, Terranova M, Amelchanka SL, Dubois S, Wiget A, Leiber F, Krause HM, Baumann S (2024) The effect of biochar supplementation on feed utilization, milk production and methane emission in lactating dairy cows. Anim Feed Sci Technol 318:116127. https://doi.org/10.1016/j.anifeedsci.2024.116127

Do PTM, Nguyen LX (2024) A review of thermochemical decomposition techniques for biochar production. Environ Dev Sustain. Advance online publication. https://doi.org/10.1007/s10668-024-05841-6

Duarte S, Glaser B, Lima RP, Cerri CEP (2019) Chemical, physical and hydraulic properties as affected by one year of *Miscanthus* biochar interaction with sandy and loamy tropical soils. Soil Syst 3(2):24. https://doi.org/10.3390/soilsystems3020024

Fawzy S, Osman AI, Yang H, Doran J, Rooney DW (2021) Industrial biochar systems for atmospheric carbon removal: a review. Environ Chem Lett 19(4):3023–3055. https://doi.org/10.1007/s10311-021-01210-1

Feng Y, Xu Y, Yu Y, Xie Z, Lin X (2012) Mechanisms of biochar decreasing methane emission from Chinese paddy soils. Soil Biol Biochem 46:80–88

Fidel RB, Laird DA, Thompson ML, Lawrinenko M (2017) Characterization and quantification of biochar alkalinity. Chemosphere 167:367–373. https://doi.org/10.1016/j.chemosphere.2016.09.151

Friedlingstein P, Jones MW, O'Sullivan M, Andrew RM, Bakker DC, Hauck J, Le Quéré C, Peters GP, Peters W, Pongratz J, Sitch S (2022) Global carbon budget 2021. Earth Syst Sci Data 14(4):1917–2005. https://doi.org/10.5194/essd-14-1917-2022

Gao X, Peng Y, Guo L, Wang Q, Guan CY, Yang F, Chen Q (2020) Arsenic adsorption on layered double hydroxides biochars and their amended red and calcareous soils. J Environ Manag 271:111045. https://doi.org/10.1016/j.jenvman.2020.111045

Gholami L, Rahimi G (2020) Chemical fractionation of copper and zinc after addition of carrot pulp biochar and thiourea–modified biochar to a contaminated soil. Environ Technol 42(22):3523–3532. https://doi.org/10.1080/09593330.2020.1733101

Gibbs H, Salmon JM (2015) Mapping the world's degraded lands. Appl Geogr 57:12–21. https://doi.org/10.1016/j.apgeog.2014.11.024

Glaser B, Lehmann J, Zech W (2002) Ameliorating physical and chemical properties of highly weathered soils in the tropics with charcoal-a review. Biol Fertil Soils 35:219–230

Godlewska P, Jośko I, Oleszczuk P (2022) Ecotoxicity of sewage sludge-or sewage sludge/willow-derived biochar-amended soil. Environ Pollut 305:119235. https://doi.org/10.1016/j.envpol.2022.119235

González ME, Cea M, Medina J, González A, Diez MC, Cartes P, Monreal C, Navia R (2015) Evaluation of biodegradable polymers as encapsulating agents for the development of a urea controlled-release fertilizer using biochar as support material. Sci Total Environ 505:446–453. https://doi.org/10.1016/j.scitotenv.2014.10.014

Gou Z, Zheng H, He Z, Su Y, Chen S, Chen H, Ma NL, Sun Y (2023) The combined action of biochar and nitrogen-fixing bacteria on microbial and enzymatic activities of soil N cycling. Environ Pollut 317:120790. https://doi.org/10.1016/j.envpol.2022.120790

Graber ER, Tsechansky L, Gerstl Z, Lew B (2012) High surface area biochar negatively impacts herbicide efficacy. Plant Soil 353(1–2):95–106. https://doi.org/10.1007/s11104-011-1012-7

Gross A, Bromm T, Polifka S, Fischer D, Glaser B (2024) Long-term biochar and soil organic carbon stability—evidence from field experiments in Germany. Sci Total Environ 954:176340. https://doi.org/10.1016/j.scitotenv.2024.176340

Gupta N, Yadav KK, Kumar V, Kumar S, Chadd RP, Kumar A (2019) Trace elements in soil-vegetables interface: translocation, bioaccumulation, toxicity and amelioration—a review. Sci Total Environ 651(pt 2):2927–2942. https://doi.org/10.1016/j.scitotenv.2018.10.047

Hailegnaw NS, Mercl F, Pračke K, Száková J, Tlustoš P (2019) Mutual relationships of biochar and soil pH, CEC, and exchangeable base cations in a model laboratory experiment. J Soils Sediments 19(5):2405–2416. https://doi.org/10.1007/s11368-019-02264-z

Hamid Y, Tang L, Hussain B, Usman M, Gurajala HK, Rashid MS, He Z, Yang X (2020) Efficiency of lime, biochar, Fe containing biochar and composite amendments for Cd and Pb immobilization in a Co-contaminated alluvial soil. Environ Pollut 257:113609. https://doi.org/10.1016/j.envpol.2019.113609

He P, Liu Y, Shao L, Zhang H, Lü F (2018) Particle size dependence of the physicochemical properties of biochar. Chemosphere 212:385–392. https://doi.org/10.1016/j.chemosphere.2018.08.106

Hearth SK, Ippolito JA, Laird DA (2013) Biochar's effect on soil water content and hydraulic conductivity. Soil Sci 178(10):471–479

Helal MI, Husein ME, Walaa G, Mostafa ED (2019) Characterization of agricultural residues-based nano biochar and its efficiency in adsorption/desorption of nutrients. Int J Environ 8(2):130–141

Hoekman SK, Leland A, Felix L (2018) Hydrothermal carbonization (HTC) of biomass for energy applications. In: Biomass preprocessing and pretreatments for production of biofuels. CRC Press, pp 196–254. https://doi.org/10.1201/9781315153735

Hou R, Wang L, O'Connor D, Rinklebe J, Hou D (2022) Natural field freeze-thaw process leads to different performances of soil amendments towards cd immobilization and enrichment. Sci Total Environ 831:154880. https://doi.org/10.1016/j.scitotenv.2022.154880
Hu P, Zhang Y, Liu L, Wang X, Luan X, Ma X, Chu PK, Zhou J, Zhao P (2019) Biochar/struvite composite as a novel potential material for slow release of N and P. Environ Sci Pollut Res 26(17):17152–17162. https://doi.org/10.1007/s11356-019-05076-9
International Biochar Initiative (2015) Standardized product definition and product testing guidelines for biochar that is used in soil (version 2). International Biochar Initiative. https://biochar-international.org/wp-content/uploads/2018/04/IBI_Biochar_Standards_V2.1_Final.pdf
Ippolito JA, Spokas KA, Novak JM, Lentz RD, Cantrell KB (2015) Biochar elemental composition and factors influencing nutrient retention. In: Biochar for environmental management. Routledge, pp 139–163. https://doi.org/10.4324/9780203762264
Iswaran V, Jauhri KS, Sen A (1980) Effect of charcoal, coal and peat on the yield of moong, soybean and pea. Soil Biol Biochem 12:191–192
Kabir E, Kim KH, Kwon EE (2023) Biochar as a tool for the improvement of soil and environment. Front Environ Sci 11:1324533. https://doi.org/10.3389/fenvs.2023.1324533
Kaikiti K, Stylianou M, Agapiou A (2021) Use of biochar for the sorption of volatile organic compounds (VOCs) emitted from cattle manure. Environ Sci Pollut Res 28(42):59141–59149. https://doi.org/10.1007/s11356-020-09545-y
Kammann C, Ippolito J, Hagemann N, Borchard N, Cayuela ML, Estavillo JM, Fuertes-Mendizabal T, Jeffery S, Kern J, Novak J, Rasse D, Wrage-Mönnig N (2017) Biochar as a tool to reduce the agricultural greenhouse-gas burden—knowns, unknowns and future research needs. J Environ Eng Landsc Manag 25(2):114–139. https://doi.org/10.3846/16486897.2017.1319375
Karmaker SC, Eljamal O, Saha BB (2021) Response surface methodology for strontium removal process optimization from contaminated water using zeolite nanocomposites. Environ Sci Pollut Res 28(40):56535–56551. https://doi.org/10.1007/s11356-021-14503-3
Khan MA, Salman AZ, Khan ST (2023) Indigenously produced biochar retains fertility in sandy soil through unique microbial diversity sustenance: a step toward the circular economy. Front Microbiol 14:1158784. https://doi.org/10.3389/fmicb.2023.1158784
Kimetu JM, Lehmann J, Ngoze SO, Mugendi DN, Kinyangi JM, Riha S, Verchot L, Recha JW, Pell AN (2008) Reversibility of soil productivity decline with organic matter of differing quality along a degradation gradient. Ecosystems 11:726–739
Kishimoto S, Sugiura G (1985) Charcoal as a soil conditioner. Int Achieve Futur 5:12–23
Kumar SS, Wani OA, Malik AR, Kumar S, Patel R (2025) Biochar for food security and environmental sustainability under current climate change scenario. In: Agricultural diversification for sustainable food production. Springer Nature Singapore, pp 59–88. Advance online publication. https://doi.org/10.1007/978-981-97-7517-0_3
Lehmann J, Pereira da Silva J, Steiner C, Nehls T, Zech W, Glaser B (2003) Nutrient availability and leaching in an archaeological Anthrosol and a Ferralsol of the Central Amazon basin: fertilizer, manure and charcoal amendments. Plant Soil 249:343–357
Li R, Wang JJ, Zhang Z, Awasthi MK, Du D, Dang P, Huang Q, Zhang Y, Wang L (2018) Recovery of phosphate and dissolved organic matter from aqueous solution using a novel CaO–MgO hybrid carbon composite and its feasibility in phosphorus recycling. Sci Total Environ 642:526–536. https://doi.org/10.1016/j.scitotenv.2018.06.capturing074
Li X, Wang T, Chang SX, Jiang X, Song Y (2020) Biochar increases soil microbial biomass but has variable effects on microbial diversity: a meta-analysis. Sci Total Environ 749:141593. https://doi.org/10.1016/j.scitotenv.2020.141593
Liu Y, Yang M, Wu Y, Wang H, Chen Y, Wu W (2011) Reducing CH_4 and CO_2 emissions from waterlogged paddy soil with biochar. J Soils Sediments 11(6):930–939. https://doi.org/10.1007/s11368-011-0376-x
Liu Y, Chen Y, Wang Y, Lu H, He L, Yang S (2018) Negative priming effect of three kinds of biochar on the mineralization of native soil organic carbon. Land Degrad Dev 29(11):3985–3994. https://doi.org/10.1002/ldr.3157

Liu M, Ke X, Liu X, Fan X, Xu Y, Li L, Solaiman ZM, Pan G (2022) The effects of biochar soil amendment on rice growth may vary greatly with rice genotypes. Sci Total Environ 810:152223. https://doi.org/10.1016/j.scitotenv.2021.152223

Man KY, Chow KL, Man YB, Mo WY, Wong MH (2021) Use of biochar as feed supplements for animal farming. Crit Rev Environ Sci Technol 51(2):187–217. https://doi.org/10.1080/10643389.2020.1721980

Martinsen V, Alling V, Nurida NL, Mulder J, Hale SE, Ritz C, Cornelissen G (2015) pH effects of biochar in acidic soil: studies in a laboratory and in a field. Soil Sci Soc Am J 79(1):99–108

Maschio G, Koufopanos C, Lucchesi A (1992) Pyrolysis, a promising route for biomass utilization. Bioresour Technol 42(3):219–231. https://doi.org/10.1016/0960-8524(92)90025-S

Masoumi S, Borugadda VB, Nanda S, Dalai AK (2021) Hydrochar: a review on its production technologies and applications. Catalysts 11(8):939. https://doi.org/10.3390/catal11080939

Meili L, Lins PV, Zanta CLPS, Soletti JI, Ribeiro LMO, Dornelas CB, Vieira MGA (2019) MgAl-LDH/biochar composites for methylene blue removal by adsorption. Appl Clay Sci 168:11–20. https://doi.org/10.1016/j.clay.2018.10.012

Munera-Echeverri JL, Martinsen V, Strand LT, Zivanovic V, Cornelissen G, Mulder J (2018) Cation exchange capacity of biochar: an urgent method modification. Sci Total Environ 642:190–197. https://doi.org/10.1016/j.scitotenv.2018.06.017

Nair PS, Sivani MP, Suresh S, Sreekanth AJ, Sivasabari K, Adithya KS, Anuranj PR, Nayana K, Parvathy S, Chakraborty S, Chopra H (2023) Beneficial impacts of biochar as a potential feed additive in animal husbandry. Indian J Anim Res 11(3):479–499. https://doi.org/10.18006/2023.11(3).479.499

Oguntunde PG, Fosu M, Ajayi AE, Van de Giesen N (2004) Effects of charcoal production on maize yield, chemical properties and texture of soil. Biol Fertil Soils 39:295–299

Peng X, Ye LL, Wang CH, Zhou H, Sun B, Yuan T (2011) Changes in properties of an acidic soil with rice straw incorporation and humification. Geoderma 160(1):91–98

Praveen S, Gokulan R, Pushpa TB, Jegan J (2021) Techno-economic feasibility of biochar as biosorbent for basic dye sequestration. J Indian Chem Soc 98(8):100107. https://doi.org/10.1016/j.jics.2021.100107

Prommer J, Wanek W, Hofhansl F, Trojan D, Offre P, Urich T, Schleper C (2014) Biochar decelerates soil organic nitrogen cycling but stimulates soil nitrification in a temperate arable field trial. PLoS One 9(1):e86388. https://doi.org/10.1371/journal.pone.0086388

Qian L, Chen L, Joseph S, Pan G, Li L, Zheng J, Wang J (2014) Biochar compound fertilizer as an option to reach high productivity but low carbon intensity in rice agriculture of China. Carbon Manag 5(2):145–154. https://doi.org/10.1080/17583004.2014.912865

Rickson RJ, Deeks LK, Graves A, Harris JAH, Kibblewhite MG, Sakrabani R (2015) Input constraints to food production: the impact of soil degradation. Food Secur 7(2):351–364. https://doi.org/10.1007/s12571-015-0437-x

Rondon MA, Lehmann J, Ramirez J, Hurtado M (2007) Biological nitrogen fixation by common beans (Phaseolus vulgaris L.) increases with biochar additions. Biol Fertil Soils 43:699–708

Schmidt HP, Hagemann N, Draper K, Kammann C (2019) The use of biochar in animal feeding. PeerJ 7:e7373. https://doi.org/10.7717/peerj.7373

Shafawi AN, Mohamed AR, Lahijani P, Mohammadi M (2021) Recent advances in developing engineered biochar for CO_2 capture: an insight into the biochar modification approaches. J Environ Chem Eng 9(6):106869. https://doi.org/10.1016/j.jece.2021.106869

Sharma P, Abrol V, Sharma V, Chaddha S, Rao CS, Ganie AQ, Mansoor S (2021) Effectiveness of biochar and compost on improving soil hydro-physical properties, crop yield and monetary returns in inceptisol subtropics. Saudi J Biol Sci 28(12):7539–7549. https://doi.org/10.1016/j.sjbs.2021.09.043

Silvani L, Hjartardottir S, Bielská L, Škulcová L, Cornelissen G, Nizzetto L, Hale SE (2019) Can polyethylene passive samplers predict polychlorinated biphenyls (PCBs) uptake by earthworms and turnips in a biochar amended soil? Sci Total Environ 662:873–880. https://doi.org/10.1016/j.scitotenv.2019.01.202

Singh BP, Hatton BJ, Singh B, Cowie AL, Kathuria A, Mobbs C, Cowie A (2022) Biochar and soil physical properties: a meta-analysis. Soil Tillage Res 225:105028
Singhal S (2023) Biochar as a cost-effective and eco-friendly substitute for binder in concrete: a review. Eur J Environ Civ Eng 27(2):984–1009. https://doi.org/10.1080/19648189.2022.2068655
Song D, Tang J, Xi X, Zhang S, Liang G, Zhou W, Wang X (2018) Responses of soil nutrients and microbial activities to additions of maize straw biochar and chemical fertilization in a calcareous soil. Eur J Soil Biol 84:1–10. https://doi.org/10.1016/j.ejsobi.2017.11.003
Sparrevik M, Field JL, Martinsen V, Breedveld GD, Cornelissen G (2013) Life cycle assessment to evaluate the environmental impact of biochar implementation in conservation agriculture in Zambia. Environ Sci Technol 47(3):1206–1215. https://doi.org/10.1021/es302720k
Steiner C, Teixeira WG, Lehmann J, Nehls T, de Macedo JLV, Blum WEH, Zech W (2007) Long term effects of manure, charcoal and mineral fertilization on crop production and fertility on a highly weathered central Amazonian upland soil. Plant Soil 291:275–290
Sui Y, Gao J, Liu C, Zhang W, Lan Y, Li S, Meng J, Xu Z, Tang L (2016) Interactive effects of straw-derived biochar and N fertilization on soil C storage and rice productivity in rice paddies of Northeast China. Sci Total Environ 544:203–210
Sun H, Zhang H, Min J, Feng Y, Shi W (2016) Controlled-release fertilizer, floating duckweed, and biochar affect ammonia volatilization and nitrous oxide emission from rice paddy fields irrigated with nitrogen-rich wastewater. Paddy Water Environ 14:105–111
Sun X, Atiyeh HK, Adesanya YA, Zhang H, Okonkwo C, Ezeji T (2019) Enhanced acetone-butanol-ethanol production by Clostridium beijerinckii using biochar. In: ASABE annual international meeting, American Society of Agricultural and Biological Engineers
Trinh MV, Tesfai M, Borrell A, Nagothu US, Bui TPL, Quynh VD, Thanh LQ (2017) Effect of organic, inorganic and slow-release urea fertilisers on CH_4 and N_2O emissions from rice paddy fields. Paddy Water Environ 15:317–330
United Nations (2022) Global land outlook, 2nd edn. United Nations Convention to Combat Desertification. https://www.unccd.int/resources/publications/global-land-outlook-second-edition
Verheijen FG, Zhuravel A, Silva FC, Amaro A, Ben-Hur M, Keizer JJ (2019) The influence of biochar particle size and concentration on bulk density and maximum water holding capacity of sandy vs sandy loam soil in a column experiment. Geoderma 347:194–202. https://doi.org/10.1016/j.geoderma.2019.03.044
Vetter SH, Abdalla M, Kuhnert M, Smith P (2022) Soil carbon sequestration and biochar. In: Biochar in agriculture for achieving sustainable development goals. Royal Society of Chemistry, pp 194–210. https://doi.org/10.1039/9781839165245-00194
Wang C, Lu H, Dong D, Deng H, Strong PJ, Wang H, Wu W (2013) Insight into the effects of biochar on manure composting: evidence supporting the relationship between N_2O emission and denitrifying community. Environ Sci Technol 47:7341–7349
Wang L, O'Connor D, Rinklebe J, Ok YS, Tsang DCW, Shen Z, Hou D (2020) Biochar aging: mechanisms, physicochemical changes, assessment, and implications for field applications. Environ Sci Technol 54(23):14797–14814. https://doi.org/10.1021/acs.est.0c04033
Wei B, Peng Y, Lin L, Zhang D, Ma L, Jiang L, Li Y, He T, Wang Z (2023) Drivers of biochar-mediated improvement of soil water retention capacity based on soil texture: a meta-analysis. Geoderma 437:116591. https://doi.org/10.1016/j.geoderma.2023.116591
Wen P, Wu ZS, Han YJ, Cravotto G, Wang J, Ye BC (2017) Microwave-assisted synthesis of a novel biochar-based slow-release nitrogen fertilizer with enhanced water-retention capacity. ACS Sustain Chem Eng 5(8):7374–7382. https://doi.org/10.1021/acssuschemeng.7b01721
Wong JTF, Chen Z, Wong AYY, Ng CWW, Wong MH (2018) Effects of biochar on hydraulic conductivity of compacted kaolin clay. Environ Pollut 234:468–472. https://doi.org/10.1016/j.envpol.2017.11.079
Wu H, Qin X, Wu H, Li F, Wu J, Zheng L, Wang J, Chen J, Zhao Y, Lin S, Lin W (2020) Biochar mediates microbial communities and their metabolic characteristics under continuous monoculture. Chemosphere 246:125835. https://doi.org/10.1016/j.chemosphere.2020.125835

Xia H, Riaz M, Zhang M, Liu B, Li Y, El-Desouki Z, Jiang C (2022) Biochar-N fertilizer interaction increases N utilization efficiency by modifying soil C/N component under N fertilizer deep placement modes. Chemosphere 286(pt 2):131594. https://doi.org/10.1016/j.chemosphere.2021.131594

Yamamoto M, Okimori Y, Wibowo IF, Anshori S, Ogawa M (2006) Effects of the application of charred bark of Acacia mangium on the yield of maize, cowpea and peanut, and soil chemical properties in South Sumatra, Indonesia. Soil Sci Plant Nutr 52:489–495

Yang S, Jiang Z, Sun X, Ding J, Xu J (2018) Effects of biochar amendment on CO_2 emissions from paddy fields under water-saving irrigation. Int J Environ Res Publ Health 15:2580

Yin X, Peñuelas J, Sardans J, Xu X, Chen Y, Fang Y, Wu L, Singh BP, Tavakkoli E, Wang W (2021) Effects of nitrogen-enriched biochar on rice growth and yield, iron dynamics, and soil carbon storage and emissions: a tool to improve sustainable rice cultivation. Environ Pollut 287:117565. https://doi.org/10.1016/j.envpol.2021.117565

You X, Jiang H, Zhao M, Suo F, Zhang C, Zheng H, Sun K, Zhang G, Li F, Li Y (2020) Biochar reduced Chinese chive (*Allium tuberosum*) uptake and dissipation of thiamethoxam in an agricultural soil. J Hazard Mater 390:121749. https://doi.org/10.1016/j.jhazmat.2019.121749

Yu X, Liu Y, Wang Y, Feng X, Tu M, Chen J (2020) Role of bioengineering and laborers in integration of farmland resources toward to improve dimension of sustainable agriculture in China. Bioengineered 11(1):559–571. https://doi.org/10.1080/21655979.2020.1765898

Zhang C, Liu L, Zhao M, Rong H, Xu Y (2018) The environmental characteristics and applications of biochar. Environ Sci Pollut Res 25(22):21525–21534. https://doi.org/10.1007/s11356-018-2521-1

Zhang Y, Jiao X, Liu N, Lv J, Yang Y (2020) Enhanced removal of aqueous Cr (VI) by a green synthesized nanoscale zero-valent iron supported on oak wood biochar. Chemosphere 245:125542. https://doi.org/10.1016/j.chemosphere.2019.125542

Chapter 13
Biochar in Water and Wastewater Treatment

Kabyashree Buragohain, Dhruba Jyoti Bordoloi, and Ratul Nath

Abstract Water pollution and wastewater management remain critical global challenges, necessitating the development of cost-effective and sustainable treatment solutions. Biochar, a carbon-rich material derived from the pyrolysis of biomass, has gained significant attention as a promising adsorbent for water and wastewater treatment. Its unique physicochemical properties, including high surface area, porosity, and functionalized surface groups, enable the efficient removal of various contaminants such as heavy metals, organic pollutants, nutrients, and microbial pathogens. This chapter aims to elucidate the mechanisms by which biochar interacts with pollutants, including adsorption, ion exchange, catalytic degradation, and microbial interactions. Additionally, it will discuss the influence of feedstock selection, production conditions, and surface modifications on biochar's performance in water treatment applications. Comparisons with conventional treatment technologies highlight its advantages in terms of cost-effectiveness, environmental sustainability, and potential for large-scale implementation. While biochar presents numerous benefits, challenges such as regeneration, long-term stability, and optimization for specific contaminants need further research. The chapter concludes by outlining future research directions, including advancements in biochar functionalization, integration with hybrid treatment systems, and its role in circular economy frameworks for sustainable water management.

13.1 Introduction

Biochar, an environmentally friendly and cost-effective material typically derived from organic waste like agricultural by-products, forestry residues, and municipal waste, has gained growing interest due to its expanding use in various environmental applications. These organic wastes can be transformed into char through several methods, including pyrolysis, hydrothermal carbonization (HTC), gasification, and

K. Buragohain · D. J. Bordoloi · R. Nath (✉)
Microbiology Laboratory, Department of Life Sciences, Dibrugarh University, Dibrugarh, Assam, India
e-mail: ratulnath@dibru.ac.in

D. Moulick et al. (eds.), *Biochar: Achieving Agro-Environmental Sustainability*, Sustainability Solutions, https://doi.org/10.1007/978-3-032-19855-6_13

torrefaction (Enaime et al. 2020). It is a porous, carbon-rich solid material produced by pyrolyzing biomass at temperatures above 250 °C in an environment with little or no oxygen. It is currently being explored as a method for carbon capture and storage (CCS) to help reduce the negative environmental impacts of atmospheric CO_2 buildup, which contributes to climate change (Kamali et al. 2021). The biomass used can include a wide range of organic waste materials, such as agricultural and forestry residues, wood chips, algae, sewage sludge, animal manures, and organic components of municipal solid waste (Xiang et al. 2020).

Various thermochemical processes, such as pyrolysis, hydrothermal carbonization, gasification, torrefaction, and microwave heating, differ in terms of the temperatures and time durations have proven to be effective methods for producing biochar (Tripathi et al. 2016). Biochar is a promising material for environmental cleanup due to its high surface area, significant porosity, and rich functional groups, which make it effective as both an adsorbent and catalyst for removing various organic and inorganic pollutants from wastewater (Gopinath et al. 2021). Interest in biochar primarily stems from its two key advantages: Firstly, the process of producing biochar helps reduce greenhouse gas emissions by converting biomass into a stable carbon form, thereby preventing the release of these gases that would otherwise occur during natural biomass decomposition (Creamer and Gao 2016; Yang et al. 2018). Secondly, biochar serves as an efficient, affordable, and environmentally sustainable adsorbent (Cha et al. 2016; Inyang et al. 2016), that is due to its comparatively high surface area and the presence of numerous surface functional groups (Wang et al. 2017; Zhang et al. 2017). Biochar can serve as an effective adsorbent for metals and metalloids, making it useful for water purification, used in soils to enhance fertility and boost crop yields (Yoo et al. 2018), and also utilized in clean energy generation as a partial alternative to fossil fuels (Cao et al. 2019; Fang et al. 2018), and applied as adsorbents and catalysts for removing various pollutants and mitigating greenhouse gas emissions (Xiong et al. 2017). Consequently, biochar is gaining significance as a potential solution to major global challenges, including climate change, environmental contamination, and the deterioration of soil quality (Creamer and Gao 2016).

It is well established that the type of feedstock, the thermochemical decomposition process, along with its temperature and duration, have a substantial influence on the physical and chemical characteristics of biochar (Yu et al. 2019). Numerous earlier review articles have explored the decomposition techniques, characterization methods, and various applications of biochar for eliminating diverse contaminants from water-based solutions (Cha et al. 2016; Mohan et al. 2014; Tan et al. 2016). The properties of biochar can also be influenced by the pre-treatment of the feedstock and the post-treatment processes applied to the biochar (Xiang et al. 2020). Pre-treatment methods differ based on the type of feedstock and the intended application of the biochar. These may include physical approaches (such as drying, crushing, sieving, and washing), chemical processes (involving treatment with chemicals or functional materials, or loading with precursors and active agents), and biological techniques (like bacterial treatment). Post-treatment primarily involves physical methods (e.g., ball milling, magnetization) and chemical processes (such as treatment with corrosive substances) (Xiang et al. 2020). To date, only a limited number of review articles have focused on the influence of pre-treatment strategies on feedstock decomposition and the resulting biochar characteristics, as well as the impact of post-treatment technologies on biochar properties and their effectiveness in wastewater treatment (Xiang et al. 2020).

Water pollution is a widespread global concern. According to the United Nations Water Resources Development Agency, approximately 80% of industrial wastewater is released into the environment without undergoing any treatment. This untreated wastewater often contains a variety of harmful organic and inorganic contaminants, including pesticides, herbicides, heavy metals, dyes, and other micropollutants, all of which pose significant risks to ecosystems and human health (Kamarudin 2022). A range of biological and physicochemical techniques have been created to treat contaminated water and wastewater, offering industries effective and cost-efficient solutions for managing their effluents and complying with strict environmental standards (Kamali et al. 2021).

Several studies published in recent years have explored the use of biochar as a method for wastewater treatment (Kamarudin 2022). Biochar holds significant promise for use in wastewater treatment; its potential negative effects must also be considered. The type of feedstock used and the production method can result in biochar containing heavy metals and other pollutants, which may leach into water during application (Enaime et al. 2020).

13.2 Biochar: Properties and Production

Biochar is a material with high carbon content, porosity, specific surface area, and adsorption properties, having the capability of remediation and soil improvement (Feijie et al. 2022). Due to having some unique physical, chemical, and biological properties of biochar, there is a wide range of applications that mainly include soil amendments, treatment of wastewater, and remediation of environmental pollutants. Furthermore, biochar is used for the manufacturing of metals, generation of heat, flue gas purification, and useful materials (Xie et al. 2022).

13.2.1 Properties of Biochar

In the soil, biochar, is a stable form of carbon that persists for thousands of years. It is mainly prepared with the intention of enhancing the quality of soil and sequestering carbon by adding to the soil. The various characteristics of biochar depend on the materials used and the conditions of pyrolysis (Rawat et al. 2019).

13.2.1.1 Physical Properties of Biochar

(a) **Surface area and porosity**: Biochar has a high surface area with a porous structure; this property increases its ability to retain water and nutrients in the soil. The surface area of biochar can be further increased by the conversion of biochar into biochar-based nanocomposites (Samoylova et al. 2025). Biochar that is made from other sources (e.g., manure, agricultural wastes, and seaweed) has higher nutrient content, a higher pH, and less stable carbon as compared to biochar made from woody biomass (Kabir et al. 2023).
(b) **Water retention**: The porous structure of biochar contributes water retention capacity of soils; this property is beneficial in sandy soils (Ntuka et al. 2024).

13.2.1.2 Chemical Properties

The specific chemical properties of biochar can vary depending on the production process and the source of the material used.

(a) **Nutrient content**: Biochar generally has lower nutrient content, with potassium (K) typically having a higher content compared to nitrogen (N) and phosphorus (P). In biochar, the content of N, P, and K can vary greatly depending on the initial biomass and production conditions, and the content of sodium (Na) can be comparatively higher than the content of calcium (Ca) and magnesium (Mg) (Elangovan et al. n.d.).
(b) **Alkalinity**: Many biochars are alkaline in nature, which helps in balancing out acidic soils. The kind and quality of functional groups, along with the content of minerals of the feedstock, affect the pH and the electrical conductivity (Li et al. 2013a, b).
(c) **pH Modification**: surface oxidation can lower biochar pH by introducing acidic functional groups, such as carboxylic acids (Kharel et al. 2019).
(d) **Cation exchange capacity**: The Majority of biochars have high cation exchange capacity (CEC), partly due to their negative surface charge and consequent affinity for soil cations, including the heavy metals (e.g. Cr^{3+}, Pb^{2+}) (Xie et al. 2015).

For the production of biochar, the availability and composition of the biomass are among the most crucial elements for cost-effective and efficient manufacturing. A wide range of feedstocks, including agricultural leftovers, urban trash, woody biomass, paper waste, aquatic biomass, animal and human excreta, industrial waste, food and kitchen waste, dairy and paper mill waste, poultry waste, etc., are mainly used in the production of biochar (Gabhane et al. 2020a, b). Different types of residues from the agricultural crop or the agricultural byproducts (wastes) have been utilized for the production of biochar, which includes various types of waste wood, corn stalk and corn cob, residues of wheat, sugar beet tailings, and rice straw. The properties and the composition of different elements, including cellulose, hemicelluloses, lignin, and small amounts of extractives, vary with the type of biomass (Xie et al. 2015).

13.2.2 Production Methods

The production of biochar occurs under anaerobic conditions means there must be a limited supply of oxygen (Gabhane et al. 2020a). A variety of waste products have been tested as raw material for the synthesis of biochar, including straw, faces, and sludge. There are several processes of biochar production (Fig. 1), including pyrolysis, gasification, and hydrothermal carbonization, etc. (Ambaye et al. 2021a, b). Although numerous biochar production methods are documented in the literature, a clear classification system is lacking. Therefore, this section focuses on outlining the various techniques used for biochar production. These methods are broadly categorized into traditional and modern approaches, based on the level of technological advancement and innovation involved (Gabhane et al. 2020a, b).

13.2.2.1 Traditional Approaches

Archaeological findings indicate that humans began producing and using biochar more than 2500 years ago. The earliest evidence, discovered in the Amazon Basin of South America, is known as *Terra Preta*, which exhibited soil organic matter and nutrient levels up to three times higher than surrounding soils. Ancient people would stack wood in soil-covered pits and burn it slowly under limited or no air supply. In another method, people would burn biomass in open areas and then quickly cover the partially burned material with soil. In ancient times, biochar production served

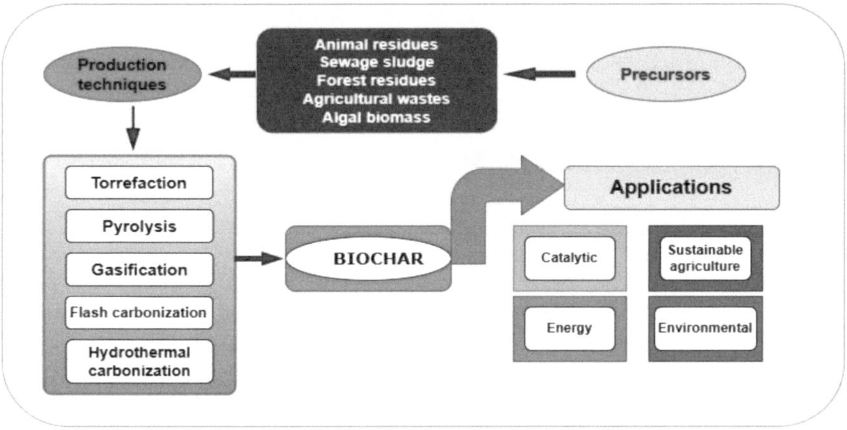

Fig. 1 Processes of biochar production

more than just as a soil enhancer. Alongside the solid char, a liquid byproduct was generated from burning wood, which had diverse uses ranging from preserving meat and dead bodies to painting houses, sealing wooden barrels (caulking), building ships, and fastening arrowheads to spear shafts. As human civilization has progressed and scientific knowledge has advanced, many conventional methods of producing biochar have been replaced by more modern techniques (Gabhane et al. 2020a, b).

Pyrolysis

Based on different parameters like temperature, heating rate, and residence time, the pyrolysis process has developed over the decades and is divided into traditional and modern methods of pyrolysis. Slow and fast pyrolysis come under traditional methods. Modern pyrolysis technique includes vacuum pyrolysis, microwave pyrolysis, and flash pyrolysis (Gabhane et al. 2020a, b).

Pyrolysis is a process of thermochemical decomposition in which the low-density biomass (~ 1.5 GJ/m^3) is converted into high-density solids called biochar and also produces liquids having high density called bio-oil(~ 22 GJ/m^3 or 17 MJ/kg) (Sakhiya et al. 2020a, b). In this thermal degradation process, there is an anaerobic condition, or the supply of oxygen should be limited. Although the process of pyrolysis is anaerobic, it requires additional heating methods, including oxidation, partial oxidation processes, hot gases, hot solids, and liquid heat transfer media (Gabhane et al. 2020a). To produce char, the pyrolysis of biomass occurs in an oxygen-free atmosphere at temperatures between 300 to 500 °C. The complex physical and chemical reactions of pyrolysis rely on the properties of biomass and process variables (Sakhiya et al. 2020a, b). For the production of significant amounts of biochar, slow pyrolysis is recommended since it produces more biochar than bio-oil and syngas due to gradual temperature increase and decomposition of biomass slowly with a longer residence time (Chi et al. 2021). Biomass pyrolysis, the complex reaction mechanism, can be simplified into three reactions. Dehydration of biomass is the first step, followed by the formation of primary biochar with the leftover biomass, where there is a loss of some volatile components and gas. After that, by gradual decomposition of the primary char, secondary char is formed, which is a carbon-rich material (Chi et al. 2021).

13.2.2.2 Modern Approaches

Torrefaction

Heating biomass at relatively low temperature (typically between 230 °C and 300 °C) enhances its properties in a process commonly known as torrefaction (Bourgeois and Doat 1984; Pentananunt et al. 1990). Alternatively, torrefaction serves as a pretreatment method that enhances the physical, chemical, and biochemical properties of raw biomass, thereby improving its performance in combustion, gasification, and co-firing applications (Almeida et al. 2010; Tumuluru et al. 2011). During this process, the hemicellulose component of the wood breaks down, resulting in the formation of torrefied wood and volatile compounds, producing more stable, carbon-rich solid materials. Torrefaction is a thermochemical process applied to biomass in an oxygen-free environment at atmospheric pressure. Typically, this process is conducted using a fixed-bed reactor. Numerous studies have shown that combining torrefaction with densification produces torrefied wood pellets of higher quality than untreated, standard pellets (Gabhane et al. 2020a, b).

Gasification

Gasification is a thermodynamic process in which there is transformation thermal energy found in carbon-based materials is transformed into a combination of biochar, tar, and syngas (made up of hydrogen and carbon monoxide). This process of conversion generally takes place in an environment where there is a restricted supply of oxygen and at very high temperatures, which is above 500 °C. Drying, pyrolysis, partial oxidation, and reduction are the four consecutive steps that often make up gasification. There are three primary categories of gasification technologies-fixed bed, fluidized bed, and entrained flow. These are differentiated based on the way that gas and solid materials interact. Depending on different variables like the type of biomass being utilized and the desired composition of the ultimate product, the type of gasifier is selected (Samuel Olugbenga et al. 2024).

Flash Pyrolysis

Flash pyrolysis is an enhanced and refined version of fast pyrolysis. In this process, biomass breaks down rapidly at very high temperatures, typically exceeding 1000 °C, within a very short time usually under one minute. The heating rate can reach over 1000 °C per second. Flash pyrolysis is conducted at temperatures between 900 and 1200 °C, which can be achieved in just a fraction of a second. A fast heating rate combined with high temperature and short vapor residence time results in a high bio-oil yield but decreases the biochar yield during the process (Demirbas and Arin 2002; Li et al. 2013a, b).

Vacuum Pyrolysis

Vacuum pyrolysis involves the thermal breakdown of biomass under reduced pressure or vacuum conditions, without the presence of oxygen. During this process, the pressure is maintained between 0.05 and 0.20 MPa, while the temperature is controlled within the range of 450 to 600 °C. Like pyrolysis, vacuum pyrolysis involves a low heating rate. Although the heating conditions and rates are comparable to those in slow pyrolysis, the resulting products differ notably. This difference arises from the efficient removal of vapor during vacuum pyrolysis. In this process, vacuum or low pressure is applied to extract the vapor produced, which positively influences the quality and yield of the products by preventing the loss of devolatilized inorganic components (Gabhane et al. 2020a, b).

Hydrothermal Carbonization (HTC)

Feedstocks with high moisture content, like sewage sludge, animal waste, and compost, are transformed into biochar using the hydrothermal carbonization (HTC) technique (Titirici et al. 2007). In the HTC method, drying the biomass beforehand is unnecessary. Instead, the wet biomass mixture is heated to temperatures between 220 and 240 °C under high pressure (2–10 MPa) within a reactor for several hours. The HTC process typically takes place in rotary drums, kilns, or stoves. Most of the organic material during HTC either stays dissolved or is converted into a form similar to brown coal (Libra et al. 2011). The HTC process is effective for producing a variety of carbon-based materials that differ in size, shape, surface characteristics, and functional groups. Nevertheless, a large portion of the original carbon stays dissolved in a soluble form. Key features of biochar produced through hydrothermal carbonization (HTC) include the preservation of nutrients like nitrogen (N) and phosphorus (P), which benefit soil fertility. Compared to torrefaction or pyrolysis methods, the HTC process offers several advantages such as a notable decrease in the oxygen-to-carbon (O/C) ratio, higher calorific value, enhanced grindability, and increased hydrophobic properties (Gabhane et al. 2020a, b).

Microwave Pyrolysis

Microwave heating is a leading advanced method for biochar production. Unlike conventional heating, microwave heating produces thermal energy via dielectric heating, delivering energy directly into the reactor without any physical contact between the energy source and the reaction mixture. It is a faster heating method that selectively targets materials more efficiently than the traditional technique. Microwave technology has gained interest in both academic and industrial sectors because of its excellent thermal properties, providing fast, selective, and even heating. It also lowers the sintering temperature, which improves steam gasification. Compared to traditional pyrolysis, biochar produced using microwave heating offers significant benefits, reducing the necessary temperature by 200 °C while delivering comparable outcomes. Many researchers have concentrated on using microwave processing for the production of biochar (Gabhane et al. 2020a, b).

Magnetic Biochar

The improved adsorption efficiency combined with the full recoverability of the adsorbent has attracted considerable global research interest, encouraging the development of new adsorbent materials. Magnetic biochar demonstrates excellent adsorption capabilities along with easy separation and recovery from contaminated water or pollutant sources. Produced from different biomass types, magnetic biochar possesses strong magnetic properties, a large surface area, and distinctive morphology, achieved through various fabrication techniques. One method involves preparing magnetic biochar by incorporating iron ions onto the biochar surface using a binding agent derived from chestnut shells, processed at 450 °C via microwave heating (Gabhane et al. 2020a, b) (Table 1).

13.3 Biochar as an Efficient Adsorbent for Organic and Inorganic Pollutants

13.3.1 Organic Pollutants

Biochar produced from various types of biomasses has been widely and effectively utilized as a sorbent for removing organic pollutants from water and wastewater. Among these contaminants, phenols and dyes have attracted the most attention due to their prevalence and impact in aqueous environments. For example, biochar produced through the pyrolysis of macroalgae has been successfully employed to eliminate textile dyes such as malachite green, crystal violet, and Congo red, which are difficult to break down because of their resistance to light, oxidizing agents, and aerobic degradation (Chen et al. 2018). Switchgrass-based biochars produced at elevated pyrolysis temperatures were effectively utilized to remove methylene blue, orange G, and Congo red (Park et al. 2019). The findings showed that methylene blue, a cationic dye, was adsorbed more effectively than the anionic dyes Orange G and Congo red. This greater adsorption was attributed to methylene blue's lower molecular weight, favorable electrostatic interactions, and strong π–π bonding with the biochar surface. Additionally, biochars produced at 900 °C demonstrated higher adsorption capacities compared to those produced at 600 °C, which is likely due to the significantly increased surface area at the higher pyrolysis temperature. Biochar has also been recognized as an efficient adsorbent for removing phenols from wastewater (Mohammed et al. 2018). Thang et al. (2019), conducted a study to explore the effectiveness of biochar made from chicken manure in removing toxic phenol and 2,4-dinitrophenol from water (Enaime et al. 2020).

Table 1 Different biochar production approaches

Production techniques	Conditions	Applications	References
Traditional approaches			
1. Pyrolysis	Biomass is often burned in a pit covered with soil to limit the oxygen supply, which helps control the combustion process. In some methods, biomass is burned in an open area, and the partially burned material is immediately covered with soil to restrict further burning. Additionally, handmade reactors such as firebrick pits, clay burners, brick kilns, and iron retorts are commonly used to facilitate controlled biomass Combustion.	Soil amendment is one common use, while the liquid product serves various purposes such as preserving dead bodies and meat, house painting, sealing wood barrels through caulking, shipbuilding, and even attaching arrowheads to spear shafts.	Barrow 2012; Emrich 1985; Reilly, 1925; Thines et al. 2017
	Temperature: 300–600 °C; heating rate: 5 to 7 °C min^{-1}	Biomass conversion produces valuable products such as soil amendments, bio-oil, and syngas, which can be utilized as energy fuels.	Lai et al. 2013
Modern approaches			
2. Gasificat ion	Temperature: > 700 °C with gasifying agents (steam)	Biochar finds diverse applications including dye removal from wastewater, functioning as an efficient adsorbent for various pollutants, aiding in carbon sequestration, enhancing soil fertility as a soil amendment, and serving as a byproduct source of syngasduringitsproduction process.	Al-Rahbi and Williams 2017; Gonzálezet al. 2008

(continued)

Table 1 (continued)

Production techniques	Conditions	Applications	References
3. Torrefact ion	Temperature: 230–300 °C	It is considered a pre-treatment step aimed at enhancing the physical, chemical,and biochemical properties of raw biomass, thereby improving its suitability for further processing or utilization.	Bourgeois and Doat 1984; Pentananunt et al. 1990
4. Flash pyrolysis	Temperature: 900–1200 °C; heating rate: 800–1000 °C sec^{-1}	Biochar production yields multiple valuable byproducts and applications, including its use as a soil amendment to improve soil health and fertility, the generation of bio-oil which can serve as a renewable fuel or chemical feedstock, and the production of syngas, a combustible gas mixture that can be utilized for energy generation or as a chemical precursor.	Demirbas and Arin 2002; Li et al., 2013a, b
5. Vacuum pyrolysis	Pressure: 0.05–0.20 MPa; temperature: 450–600 °C; heating rate: 300 min^{-1}	High-porosity biochar is particularly effective due to its large surface area, making it ideal for the adsorption of minerals and other contaminants from water or soil. in addition to its adsorptive capabilities, it is widely used as a soil amendment, where it enhances soil structure, nutrient retention, andmicrobial activity, contributing to improved agricultural productivity and environmental sustainability.	Gabhane et al. 2020a, b

(continued)

Table 1 (continued)

Production techniques	Conditions	Applications	References
6. Hydroth ermal carboniz ation	Pressure: 2 to 10 MPa and temperature: 220 to 240 °C	Biochar offers several advantageos properties, including enhanced nutrient retention which supports soil fertility, a high calorific value that makes it suitable as a renewable energy source, improved grindability for easier handling and application, and increased hydrophobicity, which contributes to its stability and resistance to moisture absorption during storage and use.	Gabhane et al. 2020a, b
7. Microwa ve pyrolysis	Temperature: 450–800 °C; microwave power: 400–600 W and microwave frequency: 2450 MHz	Biochar production results in several valuable outputs, including its primary use as a soil amendment to enhance soil health and productivity. Additionally, the pyrolysis process generates bio-oil, which can be refined into renewable fuels or industrial chemicals, and syngas, a combustible gas mixture useful for energy production or as a chemical feedstock.	Gabhane et al. 2020a, b
8. Magnetic biochar	Temperature: 450–1000 °C; material soaked with a solution of Fe_2O_3˙ $FeSO_4.7H_2O$, or $FeCl_3.6H_2O$ before pyrolysis	Biochar acts as an effective adsorbent for pollutants and also helps retain nutrients like nitrogen and phosphorus, reducing leaching and improving soil fertility.	Gabhane et al. 2020a, b

13.3.2 Heavy Metals

Recent research has demonstrated that biochar is highly effective at removing inorganic contaminants, including heavy metals, from wastewater. It is well established that the ability of biochar to adsorb heavy metals is greatly influenced by its own properties and the specific characteristics of the metals involved (Enaime et al. 2020). Biochar is highly effective for heavy metal adsorption because of its rich surface chemistry, including functional groups like phenolic, hydroxyl, and carboxyl groups, as well as its porous structure and extensive surface area (Chen et al. 2014; Zhao et al. 2020). Pan et al. (2013) conducted a comparison of biochars derived from the straws of canola, rice, soybean, and peanut for their ability to adsorb trivalent chromium. The adsorption capacities correlated with the levels of acidic functional groups present in the biochars, following the order: peanut > soybean > canola > rice. In a separate study conducted by Zhao et al. (2020), the characteristics of biochars derived from rice straw, chicken manure, and sewage sludge were examined, along with their effectiveness in removing Pb^{2+} and Zn^{2+} from solutions (Enaime et al. 2020).

Compared to traditional adsorbents like activated carbon, biochar shows potential as an effective alternative for removing heavy metals through adsorption. Kołodyńska et al. (2017), found that although biochar produced by gasification had a lower surface area (115.5 m^2/g) than commercial activated carbon (759.8 m^2/g), it was more effective at removing heavy metal ions from aqueous solutions. This enhanced performance was attributed to the presence and abundance of oxygen-containing functional groups on the biochar surface. Furthermore, activating biochar significantly enhanced its ability to adsorb heavy metals. In particular, biochar activated with potassium hydroxide demonstrated more than a threefold increase in Cu^{2+} adsorption capacity compared to non-activated biochar (Enaime et al. 2020).

13.4 Mechanisms of Biochar in Water Treatment

13.4.1 Adsorption

The surface of biochar is highly heterogeneous, enabling various sorption mechanisms to take place. The adsorption process is influenced by both the type of contaminants and the chemical characteristics of the adsorbent surface. The surface of biochar is diverse, allowing for different sorption mechanisms to occur. The adsorption process is influenced by the properties of the contaminants as well as the chemical nature of the adsorbent surface (Enaime et al. 2020).

Biochar has the ability to absorb pollutants (such as ammonia, nitrate, phosphorus, and antibiotics) because of its high surface area and negative charge (Li et al. 2019; Xu et al. 2018; Zhou et al. 2019). The high adsorption of pollutants to biochar was primarily attributed to the strong intermolecular π interactions, hydrogen bonding, and electrostatic attraction (Peiris et al. 2020; Zhou et al. 2018). The ammonia adsorption capacity of biochar varied from a few to several tens of mg/g, that may be improved through biochar modification and rising ammonia levels. Biochar derived from sorghum distillers grain can adsorb ammonia at a rate of 2.06 mg/g. When activated with NaOH, the biochar exhibits a greater surface area and more functional groups, enhancing its ammonia adsorption capacity by 2.93 times (Zhuang et al. 2022).

The mechanism by which biochar adsorbs organic and inorganic pollutants involves processes such as electrostatic interactions, ion exchange, pore filling, and precipitation. These processes are influenced by the physicochemical properties of the biochar, including its dosage, pyrolysis temperature, and the pH of the surrounding medium or effluent (Ambaye et al. 2021a, b). However, the swift pace of urbanization and industrialization in developing nations has led to the production of substantial amounts of wastewater containing hazardous chemicals that pose significant environmental threats, potentially impacting human health and socio-economic activities. As a result, there is growing interest in the use of biochar for wastewater treatment and water purification, despite limited data on its effectiveness as an adsorbent for industrial wastewater (Ambaye et al. 2021a, b).

In adsorption, the adsorbate remains attached to the surface of the adsorbent until a state of equilibrium is reached. The adsorption process involves several stages: (a) physical adsorption, where the adsorbate adheres to the surface of the adsorbent; (b) precipitation and complexation, where the adsorbate forms deposits or complexes on the surface; and (c) pore filling, where the adsorbate accumulates within the pores of the adsorbent (Fagbohungbe et al. 2017). This process unfolds in three distinct stages, each corresponding to a specific zone: The first stage, known as the clean zone, is where no adsorption occurs. The second stage, called the mass transfer zone, is where adsorption actively takes place. The final stage, referred to as the exhausted zone, is when equilibrium is reached (De Ridder et al. 2012). During the process, the saturated or spent zone expands, while the clean zone shrinks. The mass transfer zone is influenced by an increase in the adsorbate concentration; if the concentration remains unchanged, the zone is unaffected. This pattern persists until the breakthrough point is reached, where the adsorbent becomes fully saturated (Ambaye et al. 2021a, b).

It can be inferred that the adsorption efficiency of biochar depends on the feedstock used, the type of pollutant, and its concentration. It can also be enhanced by treatment with acids, bases, or metal ions. Unmodified biochar showed a limited ability to adsorb pollutants, particularly when the concentration of pollutants in the incoming flow was low. Additionally, it became saturated over time with the continuous introduction of pollutants into the system (Zhuang et al. 2022).

13.4.1.1 Biochar Adsorption of Wastewater Organic Pollutants

In recent years, the contamination of water by organic substances has become increasingly serious. Typical organic pollutants include, but are not limited to, dyes and antibiotics (Utami et al. 2023). Biochar has also demonstrated strong capabilities in adsorbing or breaking down organic pollutants present in water. Among these pollutants, dyes commonly released into wastewater from industries such as leather, pharmaceuticals, textiles, paints, and paper are particularly significant (Gao et al. 2021). They possess intricate organic compositions that may lead to health issues affecting the skin, digestive system, and respiratory tract (Rajarathinam et al. 2020). Furthermore, the discharge of dyes into water bodies can elevate the chemical oxygen demand and reduce light penetration, thereby significantly harming the aquatic ecosystem and disrupting the habitat of plants and animals. Hence, it is essential to eliminate dye compounds or transform them into non-toxic substances. Studies have shown that biochar is effective in removing dyes of various chemical structures. Hydrogen bonding commonly occurs between the oxygen-containing functional groups on biochar and those of the dye, contributing to its adsorption capacity. Additionally, biochar removes dyes through mechanisms such as electrostatic attraction and π–π interactions. These removal pathways can vary depending on the specific type of dye and the characteristics of the biochar involved (Wang et al. 2024).

13.4.1.2 Adsorption of Heavy Metals in Wastewater

Heavy metals present in wastewater pose significant risks to human health, animal life, and plant growth. Research indicates that biochar is highly effective in adsorbing heavy metal ions from contaminated water (Qiu et al. 2021). The primary mechanisms by which biochar adsorbs heavy metals from water include ion exchange, precipitation, and complexation. Additionally, surface adsorption plays a significant role in the removal process (Zulfiqar et al. 2022). The surface of biochar typically carries a negative charge, making it highly effective at attracting and adsorbing positively charged heavy metal ions through electrostatic interactions (Zhang et al. 2019a, b). The effective elimination of heavy metals by biochar results from the combined action of several mechanisms, primarily including functional group surface complexation, surface precipitation, ion exchange, and electrostatic interactions (Rouibah et al. 2023). Ion exchange involves the replacement of ions in solution with organic oxygen-based functional groups such as hydroxyl and carboxyl groups present on the surface of biochar, enabling the removal of various anions and cations. Chemical precipitation, on the other hand, is a widely recognized and frequently employed method for extracting heavy metals from water bodies. Complexation is the process in which metal ions bind with organic functional groups present on the surface of biochar adsorbents, particularly oxygen-containing groups like hydroxyl, carbonyl, and carboxyl groups, forming stable complexes (Wang et al. 2024).

13.4.1.3 Biochar Adsorption of Wastewater Inorganic Salt

Elevated levels of nutrients like phosphates, nitrates, and ammonium can lead to significant environmental damage to ecosystems (Goala et al. 2021). For instance, they can stimulate the proliferation of photosynthetic organisms, trigger eutrophication in aquatic environments, and result in the deterioration of water quality (Gizaw et al. 2021). Among the range of affordable adsorbent materials, biochar stands out as a promising and eco-friendly option. Its ability to remove inorganic salts from wastewater is largely influenced by its characteristics, including surface functional groups, porous architecture, mineral content, and any incorporated active compounds. The underlying adsorption mechanisms involve interactions with surface functional groups, electrostatic attraction, co-precipitation on the surface, complex formation, and hydrogen bonding (Wang et al. 2024).

13.4.2 *Ion Exchange Capacity*

The primary mechanism involves the exchange of protons and ionized cations with dissolved salts on the surface of the biochar. The ability of biochar to adsorb and remove heavy metals is influenced by the extent of contamination and the nature of its surface functional groups. Ali et al. (2019) observed that an increase in the cation exchange capacity (CEC) of biochar correlates with enhanced metal adsorption. However, they also noted that the CEC tends to decline when the pyrolysis temperature exceeds 350 °C. El-Shafey (2010) studied the removal of Hg^{2+} and Zn^{2+} from contaminated water using biochar derived from rice husk, produced at a pyrolysis temperature of 180 °C. The results indicated that Hg^{2+} ions were adsorbed more effectively than Zn^{2+} ions. Trakal et al. (2016) explored the removal of cadmium (Cd) and lead (Pb) using biochar derived from various feedstocks, including wheat straw, grape stalk, grape husk, plum stone, and nutshell. Their findings indicated that feedstocks containing iron oxides exhibited greater efficiency in removing Pb and Cd. This improvement was attributed to the iron content, which enhanced the biochar's cation exchange capacity (Ambaye et al. 2021a, b).

13.5 Environmental Application of Biochar for Wastewater Treatment

13.5.1 *Use in Different Treatment Systems*

Earlier biochar has been used as soil amendments shows the potential to improve the quality of soil, boost crop productivity, sequester of carbon, and lower the greenhouse gas emission. In recent years the advancement in the manufacturing of several types of biochar have enhanced their functionality and broadened its application in a diverse multidisciplinary fields (Wang et al. 2020). Due to the inexpensive cost, relative availability, and comparable sorptive capabilities, biochar is considered as one of most suitable substitute for activated carbon in the process of water treatment and remediation of the environment (Inyang and Dickenson 2015). The treatment of water and wastewater is one of the new emerging sides of biochar application. Biochar exhibits a strong sorption capacity for both organic and inorganic pollutants in the aqueous solutions due to its large surface area and pore volume, high organic carbon content and mineral components, and numerous and abundant kinds of functional groups (Wang et al. 2020). Adsorbents that are based on biochar have demonstrated significant potential in the efficient treatment of industrial wastewater, which usually contains a high amount of organic contaminants and different heavy metals (Ambaye et al. 2021a, b).

13.5.2 Removal of Contaminants

13.5.2.1 Heavy Metals

In contrast to biodegradable organic pollutants, inorganic pollutants -mainly the heavy metals are not easily biodegradable and they incorporated into the food chain through bioaccumulation and biomagnification. To protect humans and ecosystems from such harmful biological effects, it is critical to remove these long-persistent contaminants from water and wastewater bodies. Biochar has shown a superior capacity to remove both the organic and inorganic particles from the contaminated water sources. Biochar has been widely promoted as a beneficial substitute technique for the removal of metals from water and wastewater (Dashti et al. 2023). Heavy metals can be removed by a variety of processes, which include complexation, physical sorption, precipitation, and electrostatic interactions, depending on the kind of biochar (Inyang et al. 2016). Direct reduction, complexation, and electrostatic attraction are a few of the removal processes for chromium (Cr(VI)), which is the most dominant heavy metal contaminant. Direct reduction is the main method which is primarily used for the removal of Cr(VI)(Liang et al. 2021). The adsorption capacities for heavy metals can be increased by the engineering and functionalized of biochar through the surface modification as well as earlier treatment of the feed stocks which are similar or may even better than the activated carbon (Dashti et al. 2023). In studies conducted by Venkatraman et al. (2023) evaluate the effectiveness of rice straw biosorbent to remove the lead (Pb) and chromium (Cr) heavy metal ions from the wastewater and the biochar can be reused up to three cycles for the removal of these heavy metal without any significant loss of its removal capacity.

13.5.2.2 Organic Pollutants

Wastewater usually contains organic pollutants. Dyes, phenols, antibiotics and insecticides have earned a lot of interest because of their complicated aromatic structure, high levels of toxicity, and resistance environmental biodegradation (Liang et al. 2021). Diverse studies indicate that biochar has the capability to absorb organic contaminants comparably to commercial activated carbon. Large surface area, environmental abundance, high micropores, superior adsorption capacity, and structural features are some of the physical and chemical characteristics of the surfaces of products of biochar. The adsorption of different contaminants like drugs, minerals and vitamins is occurring due the characteristics that are mentioned above (Qiu et al. 2022).

13.6 Factors Influencing Biochar Performance

Biochar production is shaped by several key factors that collectively impact its yield, quality, and properties. Among these, the most influential are the highest treatment temperature (HTT), residence time, type of biomass feedstock, heating rate, and moisture content. These parameters interact in intricate ways, with HTT playing a particularly dominant role in defining the characteristics of the resulting biochar. Effective management of these variables is crucial for optimizing the pyrolysis process and tailoring the biochar to meet the requirements of targeted applications (Zhao et al. 2022).

13.6.1 Pyrolysis Conditions and Feedstock Type

The type of raw biomass feedstock plays a crucial role in determining the chemical and physical characteristics of biochar, particularly its functional groups (Mohammadian 2023). Carbon (C), hydrogen (H), and oxygen (O) are the main elemental components of biomass feedstocks. However, their proportions can vary based on the type of biomass used, the resulting biochar yield, and the development of functional groups (Liu et al. 2023). The characteristics of biochar such as functional groups, surface area, zeta potential, stability, hydrophobicity, and pH are largely influenced by the type of feedstock used in its production (Haider et al. 2024).

The type of feedstock used can influence the pollutant removal efficiency and underlying mechanisms by altering the physicochemical characteristics of biochar. Commonly used feedstocks include plant-based materials, animal-derived wastes, and sewage sludge (Song et al. 2022). Earlier studies have indicated that plant-based biochar typically exhibits a higher specific surface area (SSA) and more developed pore structure compared to biochar from other sources. This is primarily attributed to the volatilization of lignin and cellulose present in plant materials during pyrolysis (Dissanayake et al. 2020). Biochar derived from sludge typically exhibits the lowest porosity, primarily because its high ash content tends to clog the pores (Zielińska et al. 2015). Moreover, the pore size distribution of biochar varies depending on the type of feedstock used. For instance, biochar produced from plant materials and sludge typically contains a higher proportion of micropores and mesopores, whereas biochar derived from animal sources predominantly features mesopores (Song et al. 2022). Variations in lignin and cellulose content among plant feedstocks influence the pore structure of the resulting biochar. Biochars derived from lignin-rich materials tend to exhibit larger pore sizes, whereas those produced from feedstocks with higher cellulose content typically display smaller pores (Li et al. 2017). Variations in lignin and cellulose content among plant feedstocks influence the pore structure of the resulting biochar. Biochars derived from lignin-rich materials tend to exhibit larger pore sizes, whereas those produced from feedstocks with higher cellulose content typically display smaller pores (Zhao et al. 2022).

13.6.2 Conditions of Activation

The effectiveness of biochar in contaminant removal depends on the type and amount of activating agent used, along with the activation temperature and duration. Thus, selecting suitable activation conditions, both temperature and time, is crucial for maximizing its removal efficiency (Ding and Liu 2020; Zhang et al. 2019a, b). Low temperatures and brief activation periods can result in incomplete activation, whereas excessively high temperatures or prolonged activation durations may alter the biochar's structure, degrade surface functional groups (SFGs), or reduce the overall biochar yield (Song et al. 2017; Zhang et al. 2019a, b). Using CO_2-activated biochar derived from palm oil sludge, Iberahim et al. (2019) found that its sulfur dioxide (SO_2) removal efficiency improved steadily as the activation temperature increased from 300 °C to 600 °C, but began to decrease beyond this point. A similar trend was observed with activation time: SO_2 removal capacity increased from 30 to 60 minutes, then declined after 90 minutes. Therefore, optimizing activation parameters is essential to achieve optimal performance (Zhao et al., 2022).

13.6.3 Influence of Pre-Post Treatments

The physicochemical properties of biochar are greatly affected by the pre-treatment of the feedstock and the post-treatment of the biochar itself, highlighting a strong interdependence between these processes (Adeniyi et al. 2024). The pre- and post-treatment methods can be tailored according to the specific feedstock and the desired application area (Jagadeesh and Sundaram 2023). Pre-treatment techniques are generally classified into physical, chemical, and biological approaches. Procedures like washing, drying, and crushing significantly influence the physical characteristics of biochar (Haider et al. 2024).

Post-treatment processes primarily focus on inducing physical and chemical modifications (Priyanka et al. 2023) to enhance pore volume, surface area, surface chemistry, and the presence of functional groups (Dai et al. 2017). An example of post-treatment modification is the magnetization and elimination of corrosive agents, as demonstrated by Xiang et al. (2020). This involves converting biochar into a magnetic material by embedding magnetic iron oxide particles such as γ-Fe_2O_3, Fe_3O_4, or $CoFe_2O_4$ into its structure, so biochar can be easily separated from the liquid solution. Nitrogen doping methods can be used to introduce nitrogen atoms into the structure of biochar (Haider et al. 2024).

13.6.4 Other Influencing Factors

While nitrogen (N_2) is commonly used as the carrier gas for pyrolysis, carbon dioxide (CO_2) has also been employed for this role. In fact, biochar produced using CO_2 typically exhibits greater specific surface area (SSA), total pore volume (TPV), and microporosity compared to biochar generated with N_2. Kim et al. (2019) observed that oak biochar produced under CO_2 pyrolysis exhibited twice the specific surface area (SSA) and total pore volume (TPV) compared to that produced under N_2. However, CO_2 also enhanced the release of volatile organic compounds (VOCs) during the pyrolysis process. Therefore, additional studies are needed to fully evaluate the implications of using CO_2 as a carrier gas (Zhao et al. 2022).

The ability of biochar to adsorb pollutants is influenced by several variables, including solution pH, adsorbent concentration, and temperature. Studies have shown that pH plays a critical role in altering the physical and chemical characteristics of biochar, as well as its distribution behavior and overall adsorption efficiency. Additionally, temperature is another key factor that can significantly impact the adsorption performance of biochar (Yao et al. 2022). Temperature variations significantly influence the morphological distribution of pollutants in wastewater and their adsorption behavior on the adsorbent surface, thereby affecting the overall removal efficiency. In the case of biochar adsorbents, differences in adsorbent types and adsorption mechanisms lead to varying levels of heat generation during the process. This can hinder effective ion exchange between functional groups on the adsorbent and the pollutants, ultimately reducing adsorption performance (Yan et al. 2022). The ability of biochar to adsorb pollutants is typically influenced by the amount used; higher dosages usually enhance pollutant removal. Additionally, during the adsorption–desorption cycle, factors such as the type of desorbing agent, its concentration, temperature, and duration significantly affect the efficiency of desorption (Xiong et al. 2022). In related studies on desorption and regeneration, researchers observed that the direct use of solvents often results in limited desorption efficiency. To address this issue, several investigations have explored combining solvent-based techniques with alternative methods such as microwave treatment, high-temperature calcination, ultrasound application, and microbial processes which have shown improved regeneration performance of the adsorbents (Wang et al. 2024).

13.7 Environmental and Economic Concerns

While biochar offers significant potential for water and wastewater treatment due to its high adsorption capacity and sustainable nature, several environmental and economic concerns must be addressed. Environmentally, the production of biochar involves pyrolysis, which can emit greenhouse gases and potentially harmful volatile organic compounds if not properly managed. Additionally, the sourcing of feedstock raises sustainability issues, especially if it leads to deforestation or competes with agricultural land use. Post-treatment disposal of spent biochar is another concern, as it may concentrate adsorbed pollutants that could leach back into the environment if not safely handled or regenerated.

Economically, large-scale implementation faces challenges related to production costs, scalability, and consistency in biochar quality due to variability in raw materials and processing conditions. The cost of setting up pyrolysis units, transporting feedstock, and integrating biochar into existing treatment systems can be significant. Furthermore, there is limited regulatory clarity and market incentive for using biochar in water treatment, which hinders its commercial viability. Addressing these concerns through lifecycle assessments, standardized production methods, and supportive policies is essential to realizing biochar's full potential as an eco-friendly and cost-effective water treatment solution.

13.8 Future Perspectives

New production methods like controlled pyrolysis, hydrothermal carbonization, and microwave-assisted synthesis are allowing for more precise customization of biochar's characteristics. Advances in surface modification, nanoparticle doping, and activation techniques are further enhancing its capacity to selectively remove specific contaminants. Going forward, research will likely emphasize sustainable, energy-efficient, and scalable approaches to improve the commercial potential of biochar.

Emerging biochar materials are evolving beyond basic adsorption capabilities. By integrating features such as catalytic, magnetic, or photocatalytic properties, researchers are creating multifunctional biochar composites suited for advanced treatment applications like contaminant breakdown, microbial disinfection, and targeted recovery of valuable substances. Developing biochars with these combined functionalities will be essential for enhancing treatment performance and reducing overall operational expenses.

Expanding the use of biochar from laboratory research to full-scale industrial and municipal water treatment systems poses significant challenges. Conducting pilot-scale experiments and demonstration projects is vital to assess its effectiveness in real-world scenarios. Key concerns such as regeneration capacity, durability, regulatory compliance, and economic viability must be addressed to facilitate broader implementation. Collaboration among scientists, industry stakeholders, and policymakers will be critical in moving biochar technologies from experimental phases to practical, large-scale applications.

Biochar presents valuable potential for advancing circular economy goals by converting biomass waste into high-value treatment resources. Its incorporation into sustainable water management systems can promote resource recycling, minimize dependence on synthetic chemicals, and decrease the environmental impact of treatment processes. Looking ahead, biochar could serve multiple functions not just in removing pollutants, but also in recovering nutrients, generating energy, and sequestering carbon enhancing its contribution to comprehensive and resilient water management strategies.

13.9 Conclusion

Numerous biological and physicochemical methods have been developed in recent years for treating contaminated wastewater. While biochar (BC)-based technologies are advancing swiftly, their progress has largely been confined to laboratory and small-scale settings, with only a few having been successfully scaled up for broader, real-world implementation.

Biochar has gained recognition as a highly adaptable and promising material for water and wastewater treatment, providing eco-friendly, cost-effective, and efficient methods for removing various contaminants. Its distinctive physicochemical characteristics, including large surface area, high porosity, and numerous functional groups, allow it to effectively adsorb a broad spectrum of pollutants such as heavy metals, organic substances, nutrients, and microbes. Recent progress in modifying biochar and creating biochar-based composites has significantly improved its effectiveness, expanding its potential use in a wide range of treatment applications. It is gaining significant attention in wastewater adsorption research due to its large surface area and diverse functional groups. From a sustainability standpoint, biochar holds promising long-term potential. Its physicochemical and adsorption properties can be further improved through surface modification techniques. However, despite its vast application prospects, research on biochar modification is still at an early stage. More extensive studies are needed to fully explore and optimize the various modification methods. Additionally, recent advancements in nanotechnology have created numerous opportunities for the development of innovative water treatment technologies.

Although laboratory findings are promising, implementing these results in real-world applications faces notable challenges, such as regeneration of materials, risks of secondary pollution, and ensuring cost-effective scalability. To address these hurdles, future studies should prioritize refining production techniques, customizing biochar characteristics for targeted contaminants, and incorporating biochar systems into current water treatment frameworks. With sustained innovation and cross-disciplinary cooperation, biochar could play a pivotal role in advancing global water sustainability initiatives.

References

Adeniyi AG, Iwuozor KO, Emenike EC, Ajala OJ, Ogunniyi S, Muritala KB (2024) Thermochemical co-conversion of biomass-plastic waste to biochar: a review. Green Chemical Engineering 5(1):31–49. https://doi.org/10.1016/j.gce.2023.03.002

Ali A, Guo D, Arockiam Jeyasundar PGS, Li Y, Xiao R, Du J, Li R, Zhang Z (2019) Application of wood biochar in polluted soils stabilized the toxic metals and enhanced wheat (*Triticum aestivum*) growth and soil enzymatic activity. Ecotoxicol Environ Saf 184:109635. https://doi.org/10.1016/j.ecoenv.2019.109635
Almeida G, Brito JO, Perré P (2010) Alterations in energy properties of eucalyptus wood and bark subjected to torrefaction: the potential of mass loss as a synthetic indicator. Bioresour Technol 101(24):9778–9784. https://doi.org/10.1016/j.biortech.2010.07.026
Al-Rahbi AS, Williams PT (2017) Hydrogen-rich syngas production and tar removal from biomass gasification using sacrificial Tyre pyrolysis char. Appl Energy 190:501–509. https://doi.org/10.1016/j.apenergy.2016.12.099
Ambaye TG, Vaccari M, van Hullebusch ED, Amrane A, Rtimi S (2021) Mechanisms and adsorption capacities of biochar for the removal of organic and inorganic pollutants from industrial wastewater. Int J Environ Sci Technol 18(10):3273–3294. https://doi.org/10.1007/s13762-020-03060-w
Barrow CJ (2012) Biochar: potential for countering land degradation and for improving agriculture. Appl Geogr 34:21–28. https://doi.org/10.1016/j.apgeog.2011.09.008
Bourgeois JP, Doat J (1984) Torrefied wood from temperate and tropical species. Advantages and prospects. https://www.cabidigitallibrary.org/doi/full/10.5555/19850605434
Cao L, Yu IKM, Cho D-W, Wang D, Tsang DCW, Zhang S, Ding S, Wang L, Ok YS (2019) Microwave-assisted low-temperature hydrothermal treatment of red seaweed (*Gracilaria lemaneiformis*) for production of levulinic acid and algae hydrochar. Bioresour Technol 273:251–258. https://doi.org/10.1016/j.biortech.2018.11.013
Cha JS, Park SH, Jung S-C, Ryu C, Jeon J-K, Shin M-C, Park Y-K (2016) Production and utilization of biochar: a review. J Ind Eng Chem 40:1–15. https://doi.org/10.1016/j.jiec.2016.06.002
Chen T, Zhang Y, Wang H, Lu W, Zhou Z, Zhang Y, Ren L (2014) Influence of pyrolysis temperature on characteristics and heavy metal adsorptive performance of biochar derived from municipal sewage sludge. Bioresour Technol 164:47–54. https://doi.org/10.1016/j.biortech.2014.04.048
Chen Y, Lin Y-C, Ho S-H, Zhou Y, Ren N (2018) Highly efficient adsorption of dyes by biochar derived from pigments-extracted macroalgae pyrolyzed at different temperature. Bioresour Technol 259:104–110. https://doi.org/10.1016/j.biortech.2018.02.094
Chi NTL, Anto S, Ahamed TS, Kumar SS, Shanmugam S, Samuel MS, Mathimani T, Brindhadevi K, Pugazhendhi A (2021) A review on biochar production techniques and biochar based catalyst for biofuel production from algae. Fuel 287:119411. https://doi.org/10.1016/j.fuel.2020.119411
Creamer AE, Gao B (2016) Carbon-based adsorbents for Postcombustion CO_2 capture: a critical review. Environ Sci Technol 50(14):7276–7289. https://doi.org/10.1021/acs.est.6b00627
Dai L, Fan L, Liu Y, Ruan R, Wang Y, Zhou Y, Zhao Y, Yu Z (2017) Production of bio-oil and biochar from soapstock via microwave-assisted co-catalytic fast pyrolysis. Bioresour Technol 225:1–8. https://doi.org/10.1016/j.biortech.2016.11.017
Dashti A, Raji M, Riasat Harami H, Zhou JL, Asghari M (2023) Biochar performance evaluation for heavy metals removal from industrial wastewater based on machine learning: application for environmental protection. Sep Purif Technol 312:123399. https://doi.org/10.1016/j.seppur.2023.123399
De Ridder M, De Jong S, Polchar J, Lingemann S (2012) Risks and opportunities in the global phosphate rock market: robust strategies in times of uncertainty. Hague Centre for Strategic Studies Hague, The Netherlands. https://library.wur.nl/WebQuery/titel/2016340
Demirbas A, Arin G (2002) An overview of biomass pyrolysis. Energy Sources 24(5):471–482. https://doi.org/10.1080/00908310252889979
Ding S, Liu Y (2020) Adsorption of CO2 from flue gas by novel seaweed-based KOH—activated porous biochars. Fuel 260:116382. https://doi.org/10.1016/j.fuel.2019.116382
Dissanayake PD, You S, Igalavithana AD, Xia Y, Bhatnagar A, Gupta S, Kua HW, Kim S, Kwon J-H, Tsang DCW, Ok YS (2020) Biochar-based adsorbents for carbon dioxide capture: a critical review. Renew Sust Energ Rev 119:109582. https://doi.org/10.1016/j.rser.2019.109582

Elangovan R, Rangasami SS, Murugaragavan R Sekaran NC (n.d.). Characteristics of biochar: a review
El-Shafey EI (2010) Removal of Zn(II) and hg(II) from aqueous solution on a carbonaceous sorbent chemically prepared from rice husk. J Hazard Mater 175(1):319–327. https://doi.org/10.1016/j.jhazmat.2009.10.006
Emrich W (1985) Handbook of charcoal making. Springer Netherlands. https://doi.org/10.1007/978-94-017-0450-2
Enaime G, Baçaoui A, Yaacoubi A, Lübken M (2020) Biochar for wastewater treatment—conversion technologies and applications. Appl Sci 10(10):10. https://doi.org/10.3390/app10103492
Fagbohungbe MO, Herbert BMJ, Hurst L, Ibeto CN, Li H, Usmani SQ, Semple KT (2017) The challenges of anaerobic digestion and the role of biochar in optimizing anaerobic digestion. Waste Manag 61:236–249. https://doi.org/10.1016/j.wasman.2016.11.028
Fang J, Zhan L, Ok YS, Gao B (2018) Minireview of potential applications of hydrochar derived from hydrothermal carbonization of biomass. J Ind Eng Chem 57:15–21. https://doi.org/10.1016/j.jiec.2017.08.026
Feijie C, Huijuan XIA, Fude LIU, Weijing K, Shaoyong LU (2022) Characteristics of biochar and its effects and mechanism on soil properties. *Journal of environmental*. Engineering Technology 12(1):161–172. https://doi.org/10.12153/j.issn.1674-991X.20210067
Gabhane JW, Bhange VP, Patil PD, Bankar ST, Kumar S (2020) Recent trends in biochar production methods and its application as a soil health conditioner: a review. SN Appl Sci 2(7):1307. https://doi.org/10.1007/s42452-020-3121-5
Gao Y, Zhang J, Chen C, Du Y, Teng G, Wu Z (2021) Functional biochar fabricated from waste red mud and corn straw in China for acidic dye wastewater treatment. J Clean Prod 320:128887. https://doi.org/10.1016/j.jclepro.2021.128887
Gizaw A, Zewge F, Kumar A, Mekonnen A, Tesfaye M (2021) A comprehensive review on nitrate and phosphate removal and recovery from aqueous solutions by adsorption. AQUA—Water Infrastructure, Ecosystems and Society 70(7):921–947. https://doi.org/10.2166/aqua.2021.146
Goala M, Yadav KK, Alam J, Adelodun B, Choi KS, Cabral-Pinto MMS, Hamid AA, Alhoshan M, Ali FAA, Shukla AK (2021) Phytoremediation of dairy wastewater using *Azolla pinnata*: application of image processing technique for leaflet growth simulation. J Water Process Eng 42:102152. https://doi.org/10.1016/j.jwpe.2021.102152
González JF, Román S, Bragado D, Calderón M (2008) Investigation on the reactions influencing biomass air and air/steam gasification for hydrogen production. Fuel Process Technol 89(8):764–772. https://doi.org/10.1016/j.fuproc.2008.01.011
Gopinath A, Divyapriya G, Srivastava V, Laiju AR, Nidheesh PV, Kumar MS (2021) Conversion of sewage sludge into biochar: a potential resource in water and wastewater treatment. Environ Res 194:110656. https://doi.org/10.1016/j.envres.2020.110656
Haider MIS, Liu G, Yousaf B, Arif M, Aziz K, Ashraf A, Safeer R, Ijaz S, Pikon K (2024) Synergistic interactions and reaction mechanisms of biochar surface functionalities in antibiotics removal from industrial wastewater. Environ Pollut 356:124365. https://doi.org/10.1016/j.envpol.2024.124365
Iberahim N, Sethupathi S, Goh CL, Bashir MJK, Ahmad W (2019) Optimization of activated palm oil sludge biochar preparation for Sulphur dioxide adsorption. J Environ Manag 248:109302. https://doi.org/10.1016/j.jenvman.2019.109302
Inyang M, Dickenson E (2015) The potential role of biochar in the removal of organic and microbial contaminants from potable and reuse water: a review. Chemosphere 134:232–240. https://doi.org/10.1016/j.chemosphere.2015.03.072
Inyang MI, Gao B, Yao Y, Xue Y, Zimmerman A, Mosa A, Pullammanappallil P, Ok SY, Cao X (2016) A review of biochar as a low-cost adsorbent for aqueous heavy metal removal. Crit Rev Environ Sci Technol 46(4):406–433. https://doi.org/10.1080/10643389.2015.1096880
Jagadeesh N, Sundaram B (2023) Adsorption of pollutants from wastewater by biochar: a review. Journal of Hazardous Materials Advances 9:100226. https://doi.org/10.1016/j.hazadv.2022.100226

Kabir E, Kim K-H, Kwon EE (2023) Biochar as a tool for the improvement of soil and environment. Front Environ Sci 11. https://doi.org/10.3389/fenvs.2023.1324533

Kamali M, Appels L, Kwon EE, Aminabhavi TM, Dewil R (2021) Biochar in water and wastewater treatment—a sustainability assessment. Chem Eng J 420:129946. https://doi.org/10.1016/j.cej.2021.129946

Kamarudin NS (2022) Biochar: a review of its history, characteristics, factors that influence its yield, methods of production, application in wastewater treatment and recent development. *Biointerface research in applied*. Chemistry 12:10.33263/BRIAC126.79147926

Kharel G, Sacko O, Feng X, Morris JR, Phillips CL, Trippe K, Kumar S, Lee JW (2019) Biochar surface oxygenation by ozonization for super high cation exchange capacity. ACS Sustain Chem Eng 7(19):16410–16418. https://doi.org/10.1021/acssuschemeng.9b03536

Kim Y, Oh J-I, Vithanage M, Park Y-K, Lee J, Kwon EE (2019) Modification of biochar properties using CO_2. Chem Eng J 372:383–389. https://doi.org/10.1016/j.cej.2019.04.170

Kołodyńska D, Krukowska J, Thomas P (2017) Comparison of sorption and desorption studies of heavy metal ions from biochar and commercial active carbon. Chem Eng J 307:353–363. https://doi.org/10.1016/j.cej.2016.08.088

Lai W-Y, Lai C-M, Ke G-R, Chung R-S, Chen C-T, Cheng C-H, Pai C-W, Chen S-Y, Chen C-C (2013) The effects of woodchip biochar application on crop yield, carbon sequestration and greenhouse gas emissions from soils planted with rice or leaf beet. J Taiwan Inst Chem Eng 44(6):1039–1044. https://doi.org/10.1016/j.jtice.2013.06.028

Li C, Zhu X, He H, Fang Y, Dong H, Lü J, Li J, Li Y (2019) Adsorption of two antibiotics on biochar prepared in air-containing atmosphere: influence of biochar porosity and molecular size of antibiotics. J Mol Liq 274:353–361. https://doi.org/10.1016/j.molliq.2018.10.142

Li H, Dong X, da Silva EB, de Oliveira LM, Chen Y, Ma LQ (2017) Mechanisms of metal sorption by biochars: biochar characteristics and modifications. Chemosphere 178:466–478. https://doi.org/10.1016/j.chemosphere.2017.03.072

Li L, Rowbotham JS, Greenwell HC, Dyer PW (2013a) An Introduction to Pyrolysis and Catalytic Pyrolysis: Versatile Techniques for Biomass Conversion. https://doi.org/10.1016/b978-0-444-53878-9.00009-6

Li X, Shen Q, Zhang D, Mei X, Ran W, Xu Y, Yu G (2013b) Functional groups determine biochar properties (pH and EC) as studied by two-dimensional 13C NMR correlation spectroscopy. PLoS One 8(6):e65949. https://doi.org/10.1371/journal.pone.0065949

Liang L, Xi F, Tan W, Meng X, Hu B, Wang X (2021) Review of organic and inorganic pollutants removal by biochar and biochar-based composites. Biochar 3(3):255–281. https://doi.org/10.1007/s42773-021-00101-6

Libra JA, Ro KS, Kammann C, Funke A, Berge ND, Neubauer Y, Titirici MM, Fühner C, Bens O, Kern J, Emmerich K-H (2011) Hydrothermal carbonization of biomass residuals: a comparative review of the chemistry, processes and applications of wet and dry pyrolysis. Biofuels 2(1):71–106. https://doi.org/10.4155/bfs.10.81

Liu J, Zhang T, Den J, Lu Q, Zhao Y, Yuan S (2023) Effect of sulfur and carbon in the pyrolysis volatiles on the activity of Fe-Ni/biochar. Fuel Process Technol 244:107720. https://doi.org/10.1016/j.fuproc.2023.107720

Mohammadian N (2023) A Review of Biochar as a Stable Adsorbent for Removing Antibiotics from Wastewater 33(6)

Mohammed NAS, Abu-Zurayk RA, Hamadneh I, Al-Dujaili AH (2018) Phenol adsorption on biochar prepared from the pine fruit shells: equilibrium, kinetic and thermodynamics studies. J Environ Manag 226:377–385. https://doi.org/10.1016/j.jenvman.2018.08.033

Mohan D, Sarswat A, Ok YS, Pittman CU (2014) Organic and inorganic contaminants removal from water with biochar, a renewable, low cost and sustainable adsorbent—a critical review. Bioresour Technol 160:191–202. https://doi.org/10.1016/j.biortech.2014.01.120

Ntuka JL, Clinquart P, Kifu C, Isulu G, Kibal I, Ekuya A, Kabambi F, Botula Y-D, Mafuka P (2024) Contribution of biochar in maintaining the productivity of sandy soils after acacia sp plantations under maize (Zea mays) cultivation at Ntsio in DRC. Journal of Applied Biosciences 200:21113–21124. https://doi.org/10.35759/JABs.200.1

Pan J, Jiang J, Xu R (2013) Adsorption of Cr(III) from acidic solutions by crop straw derived biochars. J Environ Sci 25(10):1957–1965. https://doi.org/10.1016/S1001-0742(12)60305-2

Park J-H, Wang JJ, Meng Y, Wei Z, DeLaune RD, Seo D-C (2019) Adsorption/desorption behavior of cationic and anionic dyes by biochars prepared at normal and high pyrolysis temperatures. Colloids Surf A Physicochem Eng Asp 572:274–282. https://doi.org/10.1016/j.colsurfa.2019.04.029

Peiris C, Nawalage S, Wewalwela JJ, Gunatilake SR, Vithanage M (2020) Biochar based sorptive remediation of steroidal estrogen contaminated aqueous systems: a critical review. Environ Res 191:110183. https://doi.org/10.1016/j.envres.2020.110183

Pentananunt R, Rahman ANMM, Bhattacharya SC (1990) Upgrading of biomass by means of torrefaction. Energy 15(12):1175–1179. https://doi.org/10.1016/0360-5442(90)90109-F

Priyanka, Vashisht D, Taylor MJ, Mehta SK (2023) Evaluating the pre-treatment protocol required to produce an effective carbonized waste adsorbent for organic pollution control. Front Environ Sci 11. https://doi.org/10.3389/fenvs.2023.1224388

Qiu B, Shao Q, Shi J, Yang C, Chu H (2022) Application of biochar for the adsorption of organic pollutants from wastewater: modification strategies, mechanisms and challenges. Sep Purif Technol 300:121925. https://doi.org/10.1016/j.seppur.2022.121925

Qiu B, Tao X, Wang H, Li W, Ding X, Chu H (2021) Biochar as a low-cost adsorbent for aqueous heavy metal removal: a review. J Anal Appl Pyrolysis 155:105081. https://doi.org/10.1016/j.jaap.2021.105081

Rajarathinam N, Arunachalam T, Raja S, Selvasembian R (2020) Fenalan yellow G adsorption using surface-functionalized green nanoceria: an insight into mechanism and statistical modelling. Environ Res 181:108920. https://doi.org/10.1016/j.envres.2019.108920

Rawat J, Saxena J, Sanwal P, Rawat J, Saxena J, Sanwal P (2019) Biochar: a sustainable approach for improving plant growth and soil properties. In: Biochar—An Imperative Amendment for Soil and the Environment Intech Open. https://doi.org/10.5772/intechopen.82151

Reilly J (1925) The Technology of Wood Distillation: with special reference to the methods of obtaining the intermediate and finished products from the primary distillate. Nature 116(2926):779–780. https://doi.org/10.1038/116779a0

Rouibah K, Ferkous H, Delimi A, Himeur T, Benamira M, Zighed M, Darwish AS, Lemaoui T, Yadav KK, Bhutto JK, Ahmad A, Chaiprapat S, Benguerba Y (2023) Biosorption of zinc (II) from synthetic wastewater by using Inula Viscosa leaves as a low-cost biosorbent: experimental and molecular modeling studies. J Environ Manag 326:116742. https://doi.org/10.1016/j.jenvman.2022.116742

Sakhiya AK, Anand A, Kaushal P (2020) Production, activation, and applications of biochar in recent times. Biochar 2(3):253–285. https://doi.org/10.1007/s42773-020-00047-1

Samoylova YV, Sorokina KN, Parmon VN (2025) Biochar from microalgae: production approaches and properties. Catal Ind 17(1):84–90. https://doi.org/10.1134/S2070050424700430

Samuel Olugbenga O, Goodness Adeleye P, Blessing Oladipupo S, Timothy Adeleye A, Igenepo John K (2024) Biomass-derived biochar in wastewater treatment—a circular economy approach. Waste Management Bulletin 1(4):1–14. https://doi.org/10.1016/j.wmb.2023.07.007

Song G, Qin F, Yu J, Tang L, Pang Y, Zhang C, Wang J, Deng L (2022) Tailoring biochar for persulfate-based environmental catalysis: impact of biomass feedstocks. J Hazard Mater 424:127663. https://doi.org/10.1016/j.jhazmat.2021.127663

Song X, Ning P, Wang C, Li K, Tang L, Sun X, Ruan H (2017) Research on the low temperature catalytic hydrolysis of COS and CS2 over walnut shell biochar modified by Fe–cu mixed metal oxides and basic functional groups. Chem Eng J 314:418–433. https://doi.org/10.1016/j.cej.2016.11.162

Tan X, Liu Y, Gu Y, Xu Y, Zeng G, Hu X, Liu S, Wang X, Liu S, Li J (2016) Biochar-based nano-composites for the decontamination of wastewater: a review. Bioresour Technol 212:318–333. https://doi.org/10.1016/j.biortech.2016.04.093

Thang PQ, Jitae K, Giang BL, Viet NM, Huong PT (2019) Potential application of chicken manure biochar towards toxic phenol and 2,4-dinitrophenol in wastewaters. J Environ Manag 251:109556. https://doi.org/10.1016/j.jenvman.2019.109556

Thines KR, Abdullah EC, Mubarak NM, Ruthiraan M (2017) Synthesis of magnetic biochar from agricultural waste biomass to enhancing route for waste water and polymer application: a review. Renew Sust Energ Rev 67:257–276. https://doi.org/10.1016/j.rser.2016.09.057

Titirici M-M, Thomas A, Antonietti M (2007) Back in the black: hydrothermal carbonization of plant material as an efficient chemical process to treat the CO 2 problem? New J Chem 31(6):787–789

Trakal L, Veselská V, Šafařík I, Vítková M, Číhalová S, Komárek M (2016) Lead and cadmium sorption mechanisms on magnetically modified biochars. Bioresour Technol 203:318–324. https://doi.org/10.1016/j.biortech.2015.12.056

Tripathi M, Sahu JN, Ganesan P (2016) Effect of process parameters on production of biochar from biomass waste through pyrolysis: a review. Renew Sust Energ Rev 55:467–481. https://doi.org/10.1016/j.rser.2015.10.122

Tumuluru JS, Wright CT, Hess JR, Kenney KL (2011) A review of biomass densification systems to develop uniform feedstock commodities for bioenergy application. Biofuels Bioprod Biorefin 5(6):683–707. https://doi.org/10.1002/bbb.324

Utami M, Wang S, Musawwa MM, Purbaningtias TE, Fitri M, Yuspita I, Abd-Elkader OH, Yadav KK, Munusamy-Ramanujam G, Bang D, Chang SW, Ravindran B (2023) Simultaneous photocatalytic removal of organic dye and heavy metal from textile wastewater over N-doped TiO2 on reduced graphene oxide. Chemosphere 332:138882. https://doi.org/10.1016/j.chemosphere.2023.138882

Venkatraman Y, Arunkumar P, Kumar NS, Osman AI, Muthiah M, Al-Fatesh AS, Koduru JR (2023) Exploring modified Rice straw biochar as a sustainable solution for simultaneous Cr(VI) and Pb (II) removal from wastewater: characterization, mechanism insights, and application feasibility. ACS Omega 8(41):38130–38147. https://doi.org/10.1021/acsomega.3c04271

Wang B, Gao B, Fang J (2017) Recent advances in engineered biochar productions and applications. Crit Rev Environ Sci Technol 47(22):2158–2207. https://doi.org/10.1080/10643389.2017.1418580

Wang X, Guo Z, Hu Z, Zhang J (2020) Recent advances in biochar application for water and wastewater treatment: a review. PeerJ 8:e9164. https://doi.org/10.7717/peerj.9164

Wang Y, Chen L, Zhu Y, Fang W, Tan Y, He Z, Liao H (2024) Research status, trends, and mechanisms of biochar adsorption for wastewater treatment: a scientometric review. Environ Sci Eur 36(1):25. https://doi.org/10.1186/s12302-024-00859-z

Xiang W, Zhang X, Chen J, Zou W, He F, Hu X, Tsang DCW, Ok YS, Gao B (2020) Biochar technology in wastewater treatment: a critical review. Chemosphere 252:126539. https://doi.org/10.1016/j.chemosphere.2020.126539

Xie T, Reddy KR, Wang C, Yargicoglu E, Spokas K (2015) Characteristics and applications of biochar for environmental remediation: a review. Crit Rev Environ Sci Technol 45(9):939–969. https://doi.org/10.1080/10643389.2014.924180

Xie Y, Wang L, Li H, Westholm LJ, Carvalho L, Thorin E, Yu Z, Yu X, Skreiberg Ø (2022) A critical review on production, modification and utilization of biochar. J Anal Appl Pyrolysis 161:105405. https://doi.org/10.1016/j.jaap.2021.105405

Xiong J, Liang L, Shi W, Li Z, Zhang Z, Li X, Liu Y (2022) Application of biochar in modification of fillers in bioretention cells: a review. Ecol Eng 181:106689. https://doi.org/10.1016/j.ecoleng.2022.106689

Xiong X, Yu IKM, Cao L, Tsang DCW, Zhang S, Ok YS (2017) A review of biochar-based catalysts for chemical synthesis, biofuel production, and pollution control. Bioresour Technol 246:254–270. https://doi.org/10.1016/j.biortech.2017.06.163

Xu K, Lin F, Dou X, Zheng M, Tan W, Wang C (2018) Recovery of ammonium and phosphate from urine as value-added fertilizer using wood waste biochar loaded with magnesium oxides. J Clean Prod 187:205–214. https://doi.org/10.1016/j.jclepro.2018.03.206

Yan S, Yu W, Yang T, Li Q, Guo J (2022) The adsorption of corn stalk biochar for Pb and cd: preparation, characterization, and batch adsorption study. Separations 9(2):2. https://doi.org/10.3390/separations9020022

Yang F, Xu Z, Yu L, Gao B, Xu X, Zhao L, Cao X (2018) Kaolinite enhances the stability of the dissolvable and Undissolvable fractions of biochar via different mechanisms. Environ Sci Technol 52(15):8321–8329. https://doi.org/10.1021/acs.est.8b00306

Yao Z, Shao P, Fang D, Shao J, Li D, Liu L, Huang Y, Yu Z, Yang L, Yu K, Luo X (2022) Thiol-rich, porous carbon for the efficient capture of silver: understanding the relationship between the surface groups and transformation pathways of silver. Chem Eng J 427:131470. https://doi.org/10.1016/j.cej.2021.131470

Yoo J-C, Beiyuan J, Wang L, Tsang DCW, Baek K, Bolan NS, Ok YS, Li X-D (2018) A combination of ferric nitrate/EDDS-enhanced washing and sludge-derived biochar stabilization of metal-contaminated soils. Sci Total Environ 616–617:572–582. https://doi.org/10.1016/j.scitotenv.2017.10.310

Yu H, Zou W, Chen J, Chen H, Yu Z, Huang J, Tang H, Wei X, Gao B (2019) Biochar amendment improves crop production in problem soils: a review. J Environ Manag 232:8–21. https://doi.org/10.1016/j.jenvman.2018.10.117

Zhang J, Shao J, Jin Q, Li Z, Zhang X, Chen Y, Zhang S, Chen H (2019a) Sludge—based biochar activation to enhance Pb(II) adsorption. Fuel 252:101–108. https://doi.org/10.1016/j.fuel.2019.04.096

Zhang X, Gao B, Creamer AE, Cao C, Li Y (2017) Adsorption of VOCs onto engineered carbon materials: a review. J Hazard Mater 338:102–123

Zhang Z, Cano ZP, Luo D, Dou H, Yu A, Chen Z (2019b) Rational design of tailored porous carbon-based materials for CO2 capture. J Mater Chem A 7(37):20985–21003. https://doi.org/10.1039/C9TA07297G

Zhao M, Dai Y, Zhang M, Feng C, Qin B, Zhang W, Zhao N, Li Y, Ni Z, Xu Z, Tsang DCW, Qiu R (2020) Mechanisms of Pb and/or Zn adsorption by different biochars: biochar characteristics, stability, and binding energies. Sci Total Environ 717:136894. https://doi.org/10.1016/j.scitotenv.2020.136894

Zhao Z, Wang B, Theng BKG, Lee X, Zhang X, Chen M, Xu P (2022) Removal performance, mechanisms, and influencing factors of biochar for air pollutants: a critical review. Biochar 4(1):30. https://doi.org/10.1007/s42773-022-00156-z

Zhou L, Xu D, Li Y, Pan Q, Wang J, Xue L, Howard A (2019) Phosphorus and nitrogen adsorption capacities of biochars derived from feedstocks at different pyrolysis temperatures. Water 11(8):8. https://doi.org/10.3390/w11081559

Zhou X, Liang C, Jia L, Feng L, Wang R, Wu H (2018) An innovative biochar—amended substrate vertical flow constructed wetland for low C/N wastewater treatment: impact of influent strengths. Bioresour Technol 247:844–850. https://doi.org/10.1016/j.biortech.2017.09.044

Zhuang L-L, Li M, Li Y, Zhang L, Xu X, Wu H, Liang S, Su C, Zhang J (2022) The performance and mechanism of biochar-enhanced constructed wetland for wastewater treatment. J Water Process Eng 45:102522. https://doi.org/10.1016/j.jwpe.2021.102522

Zielińska A, Oleszczuk P, Charmas B, Skubiszewska-Zięba J, Pasieczna-Patkowska S (2015) Effect of sewage sludge properties on the biochar characteristic. J Anal Appl Pyrolysis 112:201–213. https://doi.org/10.1016/j.jaap.2015.01.025

Zulfiqar M, Rabat NE, Bahadar A, Lashari N, Mahnashi MH, Alqarni AO (2022) Development of Elaeis guineensis/polyvinyl alcohol/carbon nanotube composites for efficient adsorption of dye: experimental and theoretical approach. Int J Environ Sci Technol 19(7):6499–6520. https://doi.org/10.1007/s13762-021-03520-x

Chapter 14
Role of Biochar in NPK Cycle in Agro-Ecosystem

Md. Parvez Kabir, Mohammed Zia Uddin Kamal, Shams-Al-Mahmud, Falguni Akter Shraboni, Muhammad Waliur Rahman, and Akbar Hossain

Abstract Biochar application has received much attention in view of its promise to enhance nutrient bioavailability in soils. This chapter focuses on the role of biochar in the nitrogen (N), phosphorus (P), and potassium (K) cycles in agro-ecosystems. Biochar produces from biomass through the pyrolysis process, as black, porous material with high carbon content, opens up possibilities for improving soil properties, soil fertility and minimizes heavy metal contamination and other environmental impacts. The chapter addresses the effect of the interactions of biochar with soil properties on the N, P, K availability and mineralization, and subsequent N, P, K use efficiency in crops. The paper describes different mechanisms such as adsorption, microbial stimulation, changes in soil structure and pH alterations. Also, the chapter reviews how biochar is effective at reducing fertilizers and soil fertility as well as how we can be involved in carbon sequestration through biochar as a long-term sustainable agricultural practice. Finally, this chapter highlights the existing knowledge gaps and also provides future perspectives towards maximizing biochar's role in the quest for sustainable food production.

Keywords Biochar · Carbon sequestration · Nutrient retention · NPK cycle · Soil fertility · Sustainable agriculture

---

Md. Parvez Kabir · A. Hossain (✉)
Soil Science Division, Bangladesh Wheat and Maize Research Institute, Dinajpur, Bangladesh
e-mail: akbarhossainwrc@gmail.com

M. Z. U. Kamal
Department of Soil Science, Gazipur Agricultural University, Gazipur, Bangladesh
e-mail: zia@bsmrau.edu.bd

Shams-Al-Mahmud
Plant Breeding Division, Bangladesh Institute of Nuclear Agriculture, Mymensingh, Bangladesh

F. A. Shraboni
Wheat Breeding Division, Bangladesh Wheat and Maize Research Institute, Dinajpur, Bangladesh

M. W. Rahman
Breeder Seed Production Centre, Bangladesh Agricultural Research Institute, Debiganj, Panchagarh, Bangladesh

D. Moulick et al. (eds.), *Biochar: Achieving Agro-Environmental Sustainability*, Sustainability Solutions, https://doi.org/10.1007/978-3-032-19855-6_14

14.1 Introduction

Agro-ecosystems are dynamic landscapes where agriculture interacts with natural processes, resulting in complex interactions among soils, plants, animals, and humans. The functioning of these ecosystems is vital for global food production, and their productivity depends on efficient nutrient cycling, notably for nitrogen (N), phosphorus (P), and potassium (K), which are important for plant growth and soil fertility (Geremew 2021). However, modern agriculture intensification has disrupted the cycles with nutrient imbalances and threats of crop yield and environmental stability (Kopittke et al. 2019).

The right NPK balance keeps plants healthy, nitrogen promotes superior growth of foliage, phosphorus is particularly designed for energy transfer and roots development, and potassium regulates water movement and stomatal opening and closing and also support to fight against the drought stress (Wang et al. 2024). Soil imbalances in any of these elements can make soil less productive because of the lack of or overabundance of these nutrients in the soil (McKenzie and Williams 2015). As the world's food demand increases, effectively managing these nutrients becomes imperative for food security and environmental sustainability.

Modern agriculture is confronted with contradictions; chemical fertilizers have been overused, soil degradation and nutrient imbalance have become quite obvious, and environmental pollution is happening and becoming more serious every day (Bisht and Chauhan 2020; Saha et al. 2024; Kumar et al. 2019). Chemical fertilizers deplete the soil of organic matter; they kill off the beneficial micro-organisms in the soil and get into the water systems (Reda and Hailu 2017; Weldeslassie et al. 2017). Sustainable practices, including crop rotation, and additions of organic matter, maintain nutrient cycling and soil health (Lynch et al. 2015; Jiang et al. 2022). In this context, biochar appears to be an alternative way forward for better nutrient retention and soil sustainability.

Biochar is a stable, carbon-enriched solid material that is developed by heating organic matter, for example, agricultural waste (ranging from wood chips) under oxygen-limiting conditions (Alam et al. 2020). That process turns carbon into a type that can be stored in the soil for hundreds of years (Shackley et al. 2016). Biochar has been regarded as a potential measure to improve soil fertility and water retention in soil and reduce GHGs emissions (Li et al. 2018; Zhang et al. 2016a, 2016b). It improves microbiological activity and nutrients absorption and also related to the sustainability in cropping system for its porous structure (Alkharabsheh et al. 2021; Khan et al. 2024). As a means of cycling nutrients and especially NPK efficiently, biochar works wonders to revive the resiliency and sustainability of agro-ecosystems.

The chapter focuses on the biochar contribution in the improvement of NPK cycle in the agroecosystems in terms of the soil properties, nutrient immobilization and availability of the fertilizer. Moreover, it highlights the ecological services of biochar including C sequestration and greenhouse gas (GHG) reduction. Actual application techniques, best examples of practice, case studies will be offered looking at biochar use in varying soil types with practical hands-on demonstrations of the best approaches to utilize biochar. Lastly, this chapter describes the critical issues, knowledge gaps, and the way forward and underlines the biochar's prospective in sustainable agriculture.

14.2 Mechanisms Underlying the Effects of Biochar on Soil

14.2.1 Physical Properties of Biochar

The most important benefit of biochar as a sustainable agricultural product is soil improvement. When biochar is incorporated into the soil, it changes the soil's structure, and porosity (Blanco-Canqui 2017; Baiamonte et al. 2019). Its porosity and spongy soil structure means that the soil can hold more water, which makes it more drought tolerant in dry seasons as well as less prone to waterlogging during heavy rains. These spongy skeletons are, indeed, micro-reservoirs, holding water and then slowly letting it go at the time of water scarcity (Li et al. 2021). Furthermore, biochar improves soil aeration, which facilitates greater root penetration and higher accessibility of oxygen to plant roots (Obia et al. 2018). This in fact, helps better roots development and results in robust, healthier plant growth. When soil porosity increases, further compaction is mitigated (hence mainly influencing soil porosity), which is believed to contribute positively to root growth and nutritional uptake, which are critical for crop productivity (Shaheb et al. 2021). The following figure (Fig. 14.1) illustrates how biochar enhances soil structure by improving physical properties and promoting aggregate stability, contributing to healthier, more resilient soil.

14.2.2 Chemical Properties of Biochar

In addition to these physical improvements, biochar becomes an important modifier of chemical properties of the soil. One of its most noticeable impacts is the alteration in the pH of the soil. Due to its alkaline nature, biochar can contribute to balancing soil acidity and improving the availability of nutrients for plant growth depending on the feedstock used and the pyrolysis conditions (Huang et al. 2023). This pH change has particular advantages in regions where soil acidity is a limitation to crop production. In addition, the presence of high surface area and negatively charged functional groups such as hydroxyl (-OH), carboxyl (-COOH) contribute to higher cation-exchange capacity (CEC) of soil and increase the capacity of retaining essential nutrients (e.g., N, P, and K) by soil (Hossain et al. 2020). This enables the increased Level of availability of these nutrients to plants and therefore lower amounts of nutrients leached to ground water, minimizing the possibilities of water-bodies contamination (Major et al. 2012). Besides, biochar has also a high capacity to absorb nutrients, especially N, P and K, which could be a promising approach in

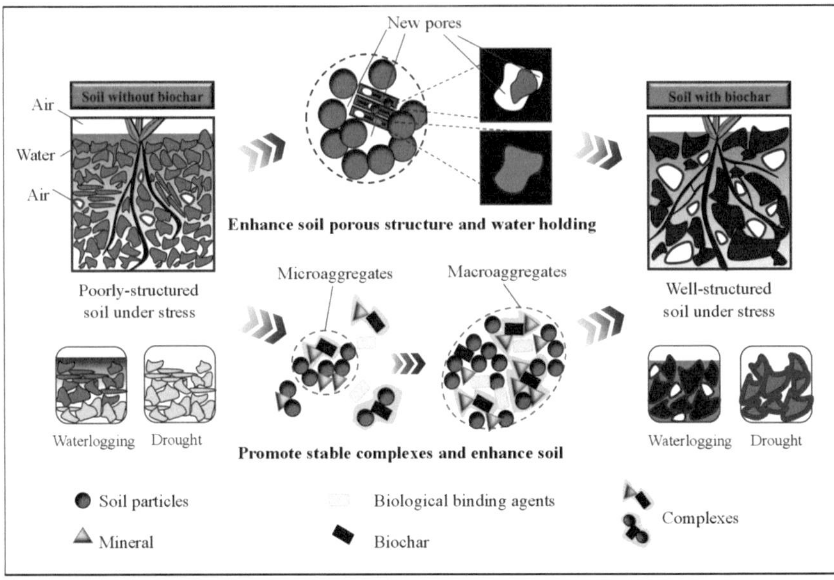

Fig. 14.1 The effect of biochar on promoting soil structure by improving soil physical properties and aggregate stability. (*Source* Chi et al. 2024, adapted from Hardie et al. 2014; Yang and Lu 2021; Zong and Lu 2021; Ali and Elshaikh 2022)

increasing nutrient use efficiency and decreasing the application of chemical fertilizer (Yin et al. 2017).

14.2.3 Biochar and Soil Microbial Community

The potential for biochar is not in its chemical and physical properties but also with its biological properties in relation to soil. Biochar is porous and supports colonization by beneficial soil organisms such as bacteria, fungi and others, that play a role in cycling nutrients (Dai et al. 2021). These micro-organisms have a symbiotic association with the root of plants where organic material decompose and nutrients mineralize and are released to plants (Condron et al. 2010). Additionally, biochar applications have also been demonstrated to enhance microbial diversity, and the stability and productivity of soil ecosystem (Singh et al. 2022). The soil microbial community in biochar-amended soil was reported to have raised the nitrogen fixation, phosphorus and potassium cycling, and the total nutrient mobilization, in turn leading to soil fertility improvement (Xu et al. 2019; Gao and DeLuca 2020). However, biochar can also act as a microbial inoculant carrier by predominantly adhering to the surface substrates where microorganisms can colonize in addition to protecting microorganisms from adverse environmental conditions such as high temperature and drought (Egamberdieva et al. 2018). These microorganisms therefore have an important role in the mineralization of organic materials and in the release of locked-up nutrients necessary for the sustainable cycling of nutrients in agro-ecosystems (Purakayastha et al. 2023).

By fostering these beneficial microbial communities, biochar contributes to soil health and long-term sustainability of agricultural systems. Such stimulation of microbial activities accelerates the turnover of nutrients from soil (mineralization), making organic and inorganic fertilizers used more efficiently and reducing the need of external inputs, contributing to greater sustainability of agricultural practices.

14.3 Biochar's Role in the NPK Cycle

14.3.1 Impact on Nitrogen Cycle

One of the most important contributions that biochar plays in soil health is its influence on the nitrogen cycle (Clough and Condron 2010). Nitrogen, an important plant nutrient, is susceptible to losses from soils through leaching, volatilization, and denitrification (Mahmud et al. 2021). Due to its high specific surface area and porous structure, biochar is a potential candidate to adsorb NH_4^+ ions, thereby reduce NO_3^- ions leaching into the groundwater and surrounding ecosystems (Wang et al. 2015; Lv et al. 2021). This retention, which conserves nitrogen as well as improving fertilizer use efficiency, decreasing the synthetic input requirements and mitigating environmental contamination (Ghorbani et al. 2022).

In addition, biochar contributes N cycling by affecting soil microorganisms, especially N-fixing bacteria (Gou et al. 2023). It has been reported that biochar can promote activity of these bacteria to improve biological N fixation in soil (Lin et al. 2023; Yin et al. 2021; Farhangi-Abriz et al. 2024). This process turns atmospheric nitrogen into forms that are readily available to plants. Moreover, biochar can impact the nitrogen transformation in soil, which are relatively less leachable (Davys et al. 2023). In other words, biochar contributes to increased stability of the N cycle, enhances nitrogen efficiency and minimizes the potential and occurrence of N induced environmental problems. The following figure (Fig. 14.2) demonstrates how the application of biochar in the soil can directly or indirectly impact various processes within the nitrogen (N) cycle, enhancing nutrient availability and soil health.

14.3.2 Impact on Phosphorus Cycle

Phosphorus is another essential nutrient for plant development, but it is usually deficient in the regions, particularly in acidic or low-phosphorus soils (Malhotra et al. 2018). Biochar is known to have a strong affinity for phosphate ions, therefore

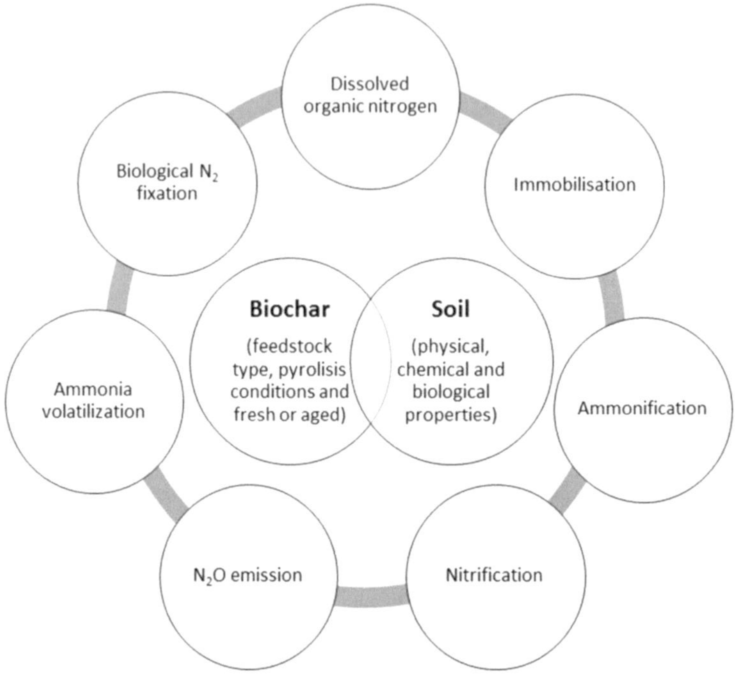

Fig. 14.2 Application of biochar in the soil may directly or indirectly influence many processes involved in the N cycle. (*Source* Jindo et al. 2020)

adsorbing phosphorus and help to reduce its loss by runoff. This adsorption thus minimizes the risk of eutrophication of adjacent waters bodies; an important environmental issue related to agriculture (Luo et al. 2023). By retaining phosphorus into the soil, biochar also increases its availability to plants, so that plant can make use of this vital nutrient at critical times (Zhang et al. 2016a, 2016b).

Apart from the mitigation of phosphorus leaching, the role of biochar also is significant in preserving the phosphorus retention in soils having the low content of intrinsic soil phosphorus content (Zhang et al. 2016a, 2016b). Due to its high surface area and porous structure, biochar is likely to be effective for the adsorption and slow release of P (especially in soils with low organic matter) (Schneider and Haderlein 2016). This increased phosphorus availability helps to maintain sustainable crop production and constantly assures phosphorus to plant uptake over extended period of time, thus rationalizing the use of external phosphorus fertilizers and promoting a more efficient phosphorus cycling. The following figure (Fig. 14.3) highlights the role of biochar in the phosphorus cycle, showing how it influences phosphorus availability in the soil for plant uptake.

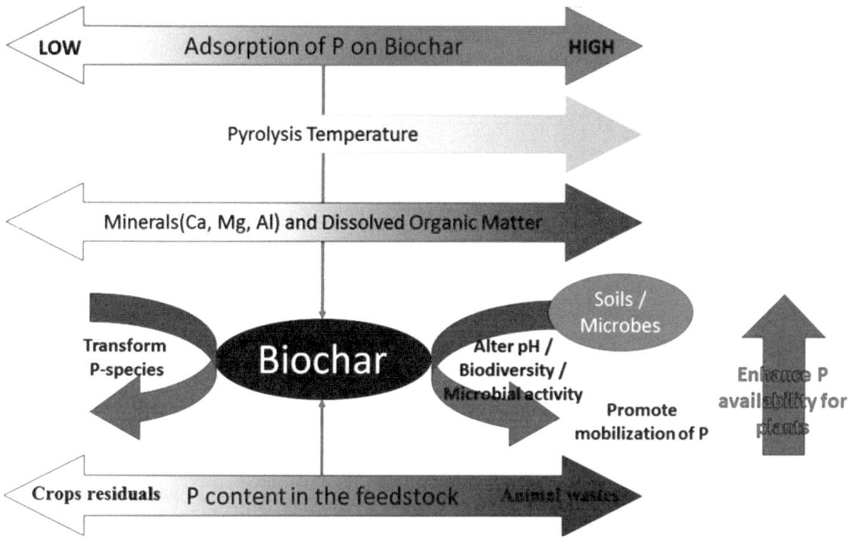

Fig. 14.3 The role of biochar in phosphorus cycle. (*Source* Pan et al. 2021)

14.3.3 Impact on Potassium Cycle

Potassium is indispensable to many physiological processes in plants, such as water regulation, enzyme activation and stress tolerance (Hasanuzzaman et al. 2018). In potassium deficient soils, biochar serves as a reservoir of potassium. Biochar, made from feedstocks (e.g. wood or crop residues), typically have high potassium that is gradually available to the soil (Akbar et al. 2023). This product offers a continuous supply of potassium to the crop, especially in low-nutrient soils where other sources of potassium availability may not be available (Xiu et al. 2023).

Biochar also helps to retain potassium, keeping this vital element from leaching away from the soil. Biochar is an effective soil K sink via cation-exchange capacity (CEC), while the potential for K leaching losses due to heavy precipitation events or irrigation is reduced (Cao and Harris 2010). Furthermore, biochar exhibits a more synergistic K release from biochar-soil mixture compared to K release can be obtained in different soil systems. Biochar application can even stabilize the leached K in the low CEC soils to make it available to crops in the soils and to increase K availability in the soils during the crop growing season. This synergy improves potassium use efficiency and eco-friendly environment of soil potassium management in the agro-ecosystem (Liu et al. 2013). The following figure (Fig. 14.4) illustrates how the incorporation of biochar into soil can influence various stages of the potassium (K) cycle, either directly or indirectly, by affecting nutrient availability and uptake.

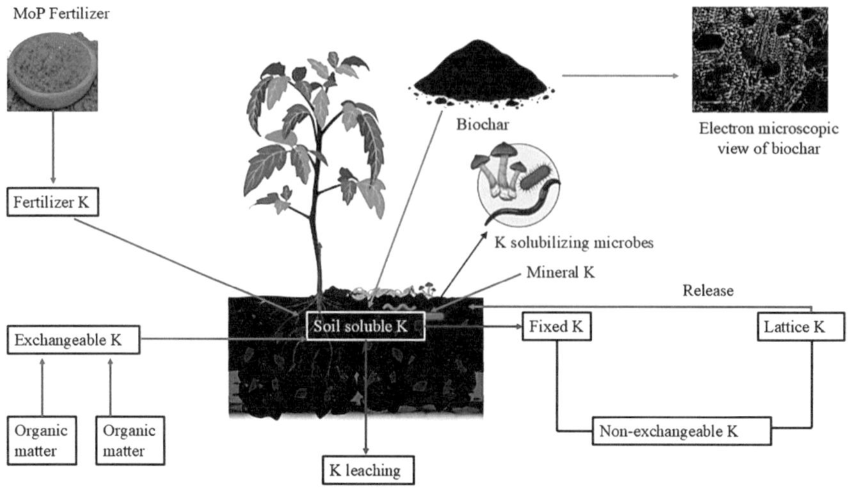

Fig. 14.4 Effects of biochar on several processes within the potassium (K) cycle

14.4 Enhancing Nutrient Use Efficiency in Crops

Biochar improves nutrient use efficiency by increasing nutrient retention, reducing nutrient leaching, and stimulating nutrient uptake (Alkharabsheh et al. 2021). It naturally breaks down and feeds depleted soil with nitrogen, phosphorus and potassium over time (Hussain et al. 2023). As biochar works more effectively with fertilizer, it enhances the efficiency of any fertilizer, leading to decreased application rates (Jia et al. 2021; Zhao et al. 2022; Pogorzelski et al. 2020; Wang et al. 2018). Through the promotion of synthetic and organic fertilizer effective use, the technology will contribute directly to cost reduction and strengthening the environmental sustainability and indirect contribution will be eutrophication control and reduction of greenhouse gas emission (Zheng et al. 2017; Gu et al. 2025; Das and Ghosh 2023). The durable fertilization effect of biochar on soil fertility results in sustainable agriculture, nutrient recycling and crop production under climate stress and complicate a more productive and climatically resilient agriculture. The following table (Table 14.1) presents the impact of biochar on nutrient use efficiency and key attributes across different crops, highlighting its potential to improve nutrient availability and uptake in various soil types.

Table 14.1 Effects of biochar on NPK use efficiency in various crops

Feedstock	Pyrolysis temperature (°C)	Crop	Impacts	References
Rice husk	350 °C	Upland rice (*Oryza sativa*)	The combination of rice husk biochar and nitrogen fertilizer enhanced rice yield, agronomic efficiency, and soil properties, with 3–6 t/ha^{-1} biochar and 30 kg/ha^{-1} nitrogen recommended for optimal growth	Oladele et al. (2019)
Rice husk, oil palm empty fruit bunch	300 °C, 450 °C	Rice	Rice husk biochar and oil palm empty fruit bunch biochar improved rice yield, fertilizer N uptake, and water productivity, particularly with alternate wetting and drying irrigation, enhancing soil N retention and efficiency	Haque et al. (2022)
Poultry litter biochar, macadamia nut shell biochar	550 °C, 460 °C	Wheat	Biochar reduced ammonia volatilization by about 70%, particularly from soils with higher pH, and increased wheat dry weight and nitrogen uptake by up to 24.24% and 76.11%, respectively, enhancing nitrogen use efficiency	Mandal et al. (2016)
Wheat straw	400 °C	Rice, wheat	Biochar application in a rice-wheat rotation increased crop biomass, nitrogen and phosphorus use efficiencies, and soil carbon stocks, while improving soil structure and aggregate stability, making it an effective long-term strategy for enhanced crop yield and nutrient use efficiency	Zhang et al. (2020a, 2020b)
Corncobs and green waste	500 °C	Maize	Acidified green waste biochar (AGWB) significantly improved soil pH, organic carbon, phosphorus bioavailability, and maize growth, outperforming sole phosphorus fertilization in enhancing soil properties, nutrient uptake, and crop yield	Ahmed et al. (2021)

(continued)

Table 14.1 (continued)

Feedstock	Pyrolysis temperature (°C)	Crop	Impacts	References
Peanut shells	300 °C	Cotton	The application of 1% biochar increased cotton yield, growth parameters, and potassium use efficiency, reducing the need for chemical K fertilizer, with the optimal K application at 75 mg/kg K_2O for improved yield and environmental sustainability	Wu et al. (2019)
Upper branches of eucalyptus trees	350 °C	Sugarcane	Eucalyptus biochar application at 6.2 mg ha^{-1} enhanced nutrient availability, increased sugarcane yield by 41%, improved nitrogen use efficiency by 118%, and boosted phosphorus and potassium uptake, making it a promising practice for improving crop productivity and soil fertility in sandy soils	Butphu et al. (2020)
Corn stalk	400–450 °C	Soybean	Biochar significantly increased available potassium in soil, with the highest increase (60.90%) in sandy soils at 30 g kg^{-1}, enhancing soybean potassium accumulation through both direct enrichment and improved soil properties, though its effect on potassium supply capacity varied by soil type	Xiu et al. (2025)

(continued)

Table 14.1 (continued)

Feedstock	Pyrolysis temperature (°C)	Crop	Impacts	References
Prunings of the acacia tree	450 °C	Wheat, maize	Biochar applied at 10 t/ha, combined with organic and inorganic phosphorus sources, significantly improved maize and wheat yields, soil organic carbon, and phosphorus use efficiency (PUE), enhancing soil quality and crop productivity in nutrient-poor soils	Arif et al. (2017)
Rice straw	600 °C	Rapeseed	The co-application of biochar (30 g pot^{-1}) and nitrogen (800 mg pot^{-1}) significantly improved soil organic carbon, total nitrogen accumulation, leaf and stem biomass, and photosynthesis in rapeseed, highlighting its potential to enhance nitrogen utilization efficiency and promote sustainable crop production	Khan et al. (2020)

14.5 Environmental Benefits of Biochar

Biochar offers a range of environmental benefits, from sequestering carbon in soil for hundreds of years to fighting climate change (Lorenz and Lal 2014). Moreover, this prevents the emission of CO_2 into the air and the cause of a possible greenhouse effect there too. The use of biochar also has the co-benefit of mitigating other greenhouse-gas (GHG) emissions such as nitrous oxide (N_2O) and methane (CH_4), through enhancement of nitrogen stabilization in the soil, with concomitant improvement in soil aeration by diminishing anaerobic conditions which encourage methane production (Tisserant and Cherubini 2019; Anaya de la Rosa 2013). On top of all that, biochar also reduces nutrient leaching, a huge problem in modern agriculture. It also adsorbs excess nutrients, such as nitrogen and phosphorous, so they don't end up in the water bodies to potentially cause eutrophication (Ngatia et al. 2019; Yin et al. 2018). This improves water quality and ecological defenses, thereby biochar is an effective way to sustainable farming for benefiting farming productivity and environmental health.

14.6 Application of Biochar in Agro-Ecosystems

Biochar utilization as soil amendment can be applied in many ways, such as direct soil application, through compost and with fertigation (Zhang et al. 2020a, 2020b; Naeem et al. 2018). Direct soil application enhanced physical (soil structure and water-holding capacity), and chemical (availability of nutrients) properties and was particularly efficient for nutrient limited soils (Kang et al. 2022; Buchkina et al. 2017). However, biochar combined with organic materials like compost are bound by nutrients which controls the release of nutrients (slow release) to some extent (Wang et al. 2022). When used in fertigation, biochar is combined with liquid fertilizers and dispensed through irrigation systems to deliver a consistent distribution of nutrients at the roots for plant uptake (Sarma et al. 2023).

In order to harness the benefits of biochar, it is crucial to pay attention to its best practices including the best application rates (5–10 tons per hectare), the timing (ensure it is before planting) and understanding the soil types. The benefits of biochar are particularly important in soils that are low in nutrients, acidic or waterlogged (Dong et al. 2015; Das et al. 2025). In Philippines, Brazil and USA, there are dozens of case studies showing that biochar can boost soil fertility, enhance crop yield, and cut chemical fertilizer use.

14.7 Challenges and Research Gaps

However, there are still a number of constraints on the use of biochar on a larger scale, such as the heterogeneity of biochar quality (Zhao et al. 2013), which is determined by the feedstock and production process, lack of finance and health risks (Gwenzi et al. 2015). More studies are required in order to fine-tune biochar for agricultural purposes. The effectiveness of biochar also depends on site-specific conditions such as soil type, climate and agricultural systems. Understanding these interactions will contribute to effectively using biochar potential.

In addition, existing knowledge is limited on long-term effects of biochar on soil health, microbial communities, and crop productivity. There's a need for large field trials and further testing to determine the validity of these lab results and biochar's potential for various scales of farming (Vijay et al. 2021). These challenges will be addressed through additional research and interdisciplinary working together in order to develop biochar applications and to exploit the full range of their potential.

14.8 Conclusion

Biochar is a revolutionary means to enhance soil fertility, ensuring the optimization of the NPK cycle in agro-ecosystems. Because of its high efficiency in increasing nutrient retention and reducing leaching and promoting microbial activity, it is a potential source for the sustainability of nutrient cycling in agriculture. However, its efficacy varies with the feedstock, methods of production, and environmental conditions. Biochar still offers good opportunities for sustainable farming systems for improved nutrient transformations, mitigation of soil degradation, and better agricultural management. It will not be long before biochar becomes an essential tool for maintaining a new generation of agricultural management on a world scale to sustain global food security without detriment to soil health.

References

Ahmed N, Basit A, Bashir S, Bashir S, Bibi I, Haider Z, Ali MA, Aslam Z, Aon M, Alotaibi SS, El-Shehawi AM, Li Y (2021) Effect of acidified biochar on soil phosphorus availability and fertilizer use efficiency of maize (*Zea mays* L.). J King Saud Univ-Sci 33(8):101635

Akbar WA, Ilyas M, Arif M, Rahim HU, Munsif F, Mudassir M, Fahad S, Jalal F, Zaheer S (2023) Biochar from on-farm feedstocks for sustainable potassium management in soils. In: Sustainable agriculture reviews 61: Biochar to improve crop production and decrease plant stress under a changing climate, pp 219–229

Alam SN, Khalid Z, Sweta, Singh B, Guldhe A, Shahi DK, Bauddh K (2020) Application of biochar in agriculture: a sustainable approach for enhanced plant growth, productivity and soil health. In: Ecological and practical applications for sustainable agriculture. Springer Singapore, Singapore, pp 107–130

Ali AB, Elshaikh NA (2022) Performance of biochar under diminish water stress in plants. Commun Soil Sci Plant Anal 53(1):1–16

Alkharabsheh HM, Seleiman MF, Battaglia ML, Shami A, Jalal RS, Alhammad BA, Almutairi KF, Al-Saif AM (2021) Biochar and its broad impacts in soil quality and fertility, nutrient leaching and crop productivity: a review. Agronomy 11(5):993

Anaya de la Rosa, R. K. (2013). Biochar systems for carbon finance—an evaluation based on life cycle assessment studies in New Zealand: a thesis presented in partial fulfilment of the requirements of doctor of philosophy in science at Massey University, Wellington, New Zealand. Doctoral dissertation, Massey University

Arif M, Ilyas M, Riaz M, Ali K, Shah K, Haq IU, Fahad S (2017) Biochar improves phosphorus use efficiency of organic-inorganic fertilizers, maize-wheat productivity and soil quality in a low fertility alkaline soil. Field Crop Res 214:25–37

Baiamonte G, Crescimanno G, Parrino F, De Pasquale C (2019) Effect of biochar on the physical and structural properties of a sandy soil. Catena 175:294–303

Bisht N, Chauhan PS (2020) Excessive and disproportionate use of chemicals cause soil contamination and nutritional stress. In: Soil contamination. IntechOpen

Blanco-Canqui H (2017) Biochar and soil physical properties. Soil Sci Soc Am J 81(4):687–711

Buchkina NP, Balashov EV, Šimanský V, Igaz D, Horák J (2017) Changes in biological and physical parameters of soils with different texture after biochar application. Selskokhozyaistvennaya Biologiya (Agric Biol) 52(3):471–477

Butphu S, Rasche F, Cadisch G, Kaewpradit W (2020) Eucalyptus biochar application enhances ca uptake of upland rice, soil available P, exchangeable K, yield, and N use efficiency of sugarcane in a crop rotation system. J Plant Nutr Soil Sci 183(1):58–68
Cao X, Harris W (2010) Properties of dairy-manure-derived biochar pertinent to its potential use in remediation. Bioresour Technol 101(14):5222–5228
Chi W, Nan Q, Liu Y, Dong D, Qin Y, Li S, Wu W (2024) Stress resistance enhancing with biochar application and promotion on crop growth. Biochar 6(1):43
Clough TJ, Condron LM (2010) Biochar and the nitrogen cycle: introduction. J Environ Qual 39(4):1218–1223
Condron L, Stark C, O'Callaghan M, Clinton P, Huang Z (2010) The role of microbial communities in the formation and decomposition of soil organic matter. In: Soil microbiology and sustainable crop production. Springer Netherlands, Dordrecht, pp 81–118
Dai Z, Xiong X, Zhu H, Xu H, Leng P, Li J, Tang C, Xu J (2021) Association of biochar properties with changes in soil bacterial, fungal and fauna communities and nutrient cycling processes. Biochar 3(3):239–254
Das R, Ghosh S, Rani K, Phonglosa A, Ghosh BN (2025) Mitigating soil acidity and waterlogging impacts on soil health in India. In: Soil Health and Sustainability in India, vol 385
Das SK, Ghosh GK (2023) Developing biochar-based slow-release NPK fertilizer for controlled nutrient release and its impact on soil health and yield. Biomass Convers Biorefinery 13(14):13051–13063
Davys D, Rayns F, Charlesworth S, Lillywhite R (2023) The effect of different biochar characteristics on soil nitrogen transformation processes: a review. Sustainability 15(23):16446
Dong D, Feng Q, Mcgrouther K, Yang M, Wang H, Wu W (2015) Effects of biochar amendment on rice growth and nitrogen retention in a waterlogged paddy field. J Soils Sediments 15(1):153–162
Egamberdieva D, Hua M, Reckling M, Wirth S, Bellingrath-Kimura SD (2018) Potential effects of biochar-based microbial inoculants in agriculture. Environ Sustain 1(1):19–24
Farhangi-Abriz S, Ghassemi-Golezani K, Torabian S, Qin R (2024) A meta-analysis to estimate the potential of biochar in improving nitrogen fixation and plant biomass of legumes. Biomass Convers Biorefinery 14(3):3293–3303
Gao S, DeLuca TH (2020) Biochar alters nitrogen and phosphorus dynamics in a western rangeland ecosystem. Soil Biol Biochem 148:107868
Geremew D (2021) Review on nutrient management, cycles, flows and balances in different farming systems. J Biol, Agric Healthc 11
Ghorbani M, Konvalina P, Neugschwandtner RW, Kopecký M, Amirahmadi E, Bucur D, Walkiewicz A (2022) Interaction of biochar with chemical, green and biological nitrogen fertilizers on nitrogen use efficiency indices. Agronomy 12(9):2106
Gou Z, Zheng H, He Z, Su Y, Chen S, Chen H, Chen G, Ma NL, Sun Y (2023) The combined action of biochar and nitrogen-fixing bacteria on microbial and enzymatic activities of soil N cycling. Environ Pollut 317:120790
Gu K, Gao K, Guan S, Zhao J, Yang L, Liu M, Su J (2025) The impact of the combined application of biochar and organic fertilizer on the growth and nutrient distribution in wheat under reduced chemical fertilizer conditions. Sci Rep 15(1):5285
Gwenzi W, Chaukura N, Mukome FN, Machado S, Nyamasoka B (2015) Biochar production and applications in sub-Saharan Africa: opportunities, constraints, risks and uncertainties. J Environ Manag 150:250–261
Hardie M, Clothier B, Bound S, Oliver G, Close D (2014) Does biochar influence soil physical properties and soil water availability? Plant Soil 376(1):347–361
Haque ANA, Uddin MK, Sulaiman MF, Amin AM, Hossain M, Solaiman ZM, Aziz AA, Mosharrof M (2022) Combined use of biochar with ^{15}Nitrogen labelled urea increases rice yield, N use efficiency and fertilizer N recovery under water-saving irrigation. Sustainability 14(13):7622
Hasanuzzaman M, Bhuyan MB, Nahar K, Hossain MS, Mahmud JA, Hossen MS, Masud AA, Moumita, Fujita M (2018) Potassium: a vital regulator of plant responses and tolerance to abiotic stresses. Agronomy 8(3):31

Hossain MZ, Bahar MM, Sarkar B, Donne SW, Ok YS, Palansooriya KN, Kirkham MB, Chowdhury S, Bolan N (2020) Biochar and its importance on nutrient dynamics in soil and plant. Biochar 2(4):379–420

Huang K, Li M, Li R, Rasul F, Shahzad S, Wu C, Shao J, Huang G, Li R, Almari S, Hashem M, Aamer M (2023) Soil acidification and salinity: the importance of biochar application to agricultural soils. Front Plant Sci 14:1206820

Hussain, A. J., Al-Taey, D. K., & Kadhum, H. J. (2023) Biochar application increases the amount of nitrogen, phosphorus and potassium in the soil: a review. In IOP conference series: earth and environmental science (Vol. 1213 No. 1, p 012023 IOP Publishing

Jia Y, Hu Z, Ba Y, Qi W (2021) Application of biochar-coated urea-controlled loss of fertilizer nitrogen and increased nitrogen use efficiency. Chem Biol Technol Agric 8(1):3

Jiang Y, Zhang J, Manuel DB, De Beeck MO, Shahbaz M, Chen Y, Deng X, Xu Z, Li J, Liu Z (2022) Rotation cropping and organic fertilizer jointly promote soil health and crop production. J Environ Manag 315:115190

Jindo K, Audette Y, Higashikawa FS, Silva CA, Akashi K, Mastrolonardo G, Sánchez-Monedero MA, Mondini C (2020) Role of biochar in promoting circular economy in the agriculture sector. Part 1: a review of the biochar roles in soil N, P and K cycles. Chem Biol Technol Agric 7(1):15

Kang MW, Yibeltal M, Kim YH, Oh SJ, Lee JC, Kwon EE, Lee SS (2022) Enhancement of soil physical properties and soil water retention with biochar-based soil amendments. Sci Total Environ 836:155746

Khan S, Irshad S, Mehmood K, Hasnain Z, Nawaz M, Rais A, Gul S, Wahid MA, Hashem A, Abd Allah EF, Ibrar D (2024) Biochar production and characteristics, its impacts on soil health, crop production, and yield enhancement: a review. Plants 13(2):166

Khan Z, Zhang K, Khan MN, Fahad S, Xu Z, Hu L (2020) Coupling of biochar with nitrogen supplements improve soil fertility, nitrogen utilization efficiency and rapeseed growth. Agronomy 10(11):1661

Kopittke PM, Menzies NW, Wang P, McKenna BA, Lombi E (2019) Soil and the intensification of agriculture for global food security. Environ Int 132:105078

Kumar, R., Kumar, R., & Prakash, O. (2019). Chapter-5 the impact of chemical fertilizers on our environment and ecosystem, Chief Ed 35(69), 1173–1189

Li L, Zhang YJ, Novak A, Yang Y, Wang J (2021) Role of biochar in improving sandy soil water retention and resilience to drought. Water 13(4):407

Li Y, Hu S, Chen J, Muller K, Li Y, Fu W, Lin Z, Wang H (2018) Effects of biochar application in forest ecosystems on soil properties and greenhouse gas emissions: a review. J Soil Sediments 18(2):546–563

Lin X, Xie Z, Hu T, Wang H, Chen Z, Zhou R, Jin P (2023) Biochar application increases biological nitrogen fixation in soybean with improved soil properties in an Ultisol. J Soil Sci Plant Nutr 23(3):3095–3105

Liu X, Zhang A, Ji C, Joseph S, Bian R, Li L, Pan G, Paz-Ferreiro J (2013) Biochar's effect on crop productivity and the dependence on experimental conditions—a meta-analysis of literature data. Plant Soil 373(1):583–594

Lorenz K, Lal R (2014) Biochar application to soil for climate change mitigation by soil organic carbon sequestration. J Plant Nutr Soil Sci 177(5):651–670

Luo D, Wang L, Nan H, Cao Y, Wang H, Kumar TV, Wang C (2023) Phosphorus adsorption by functionalized biochar: a review. Environ Chem Lett 21(1):497–524

Lv R, Wang Y, Yang X, Wen Y, Tan X, Zeng Y, Shang Q (2021) Adsorption and leaching characteristics of ammonium and nitrate from paddy soil as affected by biochar amendment. Plant, Soil Environ 67(1)

Lynch DH (2015) Nutrient cycling and soil health in organic cropping systems-importance of management strategies and soil resilience. Sustain Agric Res 4(3)

Mahmud K, Panday D, Mergoum A, Missaoui A (2021) Nitrogen losses and potential mitigation strategies for a sustainable agroecosystem. Sustainability 13(4):2400

Major J, Steiner C, Downie A, Lehmann J (2012) Biochar effects on nutrient leaching. In: Biochar for environmental management. Routledge, pp 303–320
Malhotra H, Vandana S, S., & Pandey, R. (2018) Phosphorus nutrition: plant growth in response to deficiency and excess. In: Plant nutrients and abiotic stress tolerance. Springer Singapore, Singapore, pp 171–190
Mandal S, Thangarajan R, Bolan NS, Sarkar B, Khan N, Ok YS, Naidu R (2016) Biochar-induced concomitant decrease in ammonia volatilization and increase in nitrogen use efficiency by wheat. Chemosphere 142:120–127
McKenzie FC, Williams J (2015) Sustainable food production: constraints, challenges and choices by 2050. Food Secur 7(2):221–233
Naeem MA, Khalid M, Aon M, Abbas G, Amjad M, Murtaza B, Khan WUD, Ahmad N (2018) Combined application of biochar with compost and fertilizer improves soil properties and grain yield of maize. J Plant Nutr 41(1):112–122
Ngatia LW, Grace JM III, Moriasi D, Bolques A, Osei GK, Taylor RW (2019) Biochar phosphorus sorption-desorption: potential phosphorus eutrophication mitigation strategy. In: Biochar-an imperative amendment for soil and the environment. InTech Open
Obia A, Mulder J, Hale SE, Nurida NL, Cornelissen G (2018) The potential of biochar in improving drainage, aeration and maize yields in heavy clay soils. PLoS One 13(5):e0196794
Oladele S, Adeyemo A, Awodun M, Ajayi A, Fasina A (2019) Effects of biochar and nitrogen fertilizer on soil physicochemical properties, nitrogen use efficiency and upland rice (*Oryza sativa*) yield grown on an Alfisol in southwestern Nigeria. Int J Recycl Org Waste Agric 8(3):295–308
Pan SY, Dong CD, Su JF, Wang PY, Chen CW, Chang JS, Kim H, Huang CP, Hung CM (2021) The role of biochar in regulating the carbon, phosphorus, and nitrogen cycles exemplified by soil systems. Sustainability 13(10):5612
Pogorzelski D, Lustosa Filho JF, Matias PC, Santos WO, Vergütz L, Melo LCA (2020) Biochar as composite of phosphate fertilizer: characterization and agronomic effectiveness. Sci Total Environ 743:140604
Purakayastha TJ, Bhaduri D, Kumar D, Yadav R, Trivedi A (2023) Soil and plant nutrition. In: Trajectory of 75 years of Indian agriculture after Independence. Springer Nature Singapore, Singapore, pp 365–411
Reda A, Hailu AH (2017) Extensive utilization of inorganic fertilizers in Ethiopian agriculture and its possible consequences on soil quality. J Integr Agric Sci 13(4):155–171
Saha B, Fatima A, Saha S, Sahoo SK, Poddar P (2024) Environmental pollution due to improper use of chemical fertilizers and their remediation. In: Environmental Contaminants. Apple Academic Press, pp 203–219
Sarma HH, Paul A, Kakoti M, Talukdar N (2023) Fertigation: a modern approach for enhancing nutrient use efficiency. Front Agric Sustain Integr Publ TM 1:111–126
Schneider F, Haderlein SB (2016) Potential effects of biochar on the availability of phosphorus—mechanistic insights. Geoderma 277:83–90
Shackley S, Ruysschaert G, Zwart K, Glaser B (2016) Biochar in European soils and agriculture. Sci Pract
Shaheb MR, Venkatesh R, Shearer SA (2021) A review on the effect of soil compaction and its management for sustainable crop production. J Biosyst Eng 46(4):417–439
Singh H, Northup BK, Rice CW, Prasad PV (2022) Biochar applications influence soil physical and chemical properties, microbial diversity, and crop productivity: a meta-analysis. Biochar 4(1):8
Tisserant A, Cherubini F (2019) Potentials, limitations, co-benefits, and trade-offs of biochar applications to soils for climate change mitigation. Land 8(12):179
Vijay V, Shreedhar S, Adlak K, Payyanad S, Sreedharan V, Gopi G, Sophia van der Voort T, Malarvizhi P, Yi S, Gebert J, Aravind PV (2021) Review of large-scale biochar field-trials for soil amendment and the observed influences on crop yield variations. Front Energy Res 9:710766
Wang C, Luo D, Zhang X, Huang R, Cao Y, Liu G, Zhang Y, Wang H (2022) Biochar-based slow-release of fertilizers for sustainable agriculture: a mini review. Environ Sci Ecotechnol 10:100167

Wang H, Yang Y, Yao C, Feng Y, Wang H, Kong Y, Riaz U, Zaman QU, Sultan K, Fahad S, Deng G (2024) The correct combination and balance of macronutrients nitrogen, phosphorus and potassium promote plant yield and quality through enzymatic and antioxidant activities in potato. J Plant Growth Regul 43(12):4716–4734

Wang L, Xue C, Nie X, Liu Y, Chen F (2018) Effects of biochar application on soil potassium dynamics and crop uptake. J Plant Nutr Soil Sci 181(5):635–643

Wang Z, Guo H, Shen F, Yang G, Zhang Y, Zeng Y, Wang L, Xiao H, Deng S (2015) Biochar produced from oak sawdust by lanthanum (La)-involved pyrolysis for adsorption of ammonium (NH_4^+), nitrate (NO_3^-), and phosphate (PO_4^{3-}). Chemosphere 119:646–653

Weldeslassie T, Naz H, Singh B, Oves M (2017) Chemical contaminants for soil, air and aquatic ecosystem. In: Modern age environmental problems and their remediation. Springer International Publishing, Cham, pp 1–22

Wu X, Wang D, Riaz M, Zhang L, Jiang C (2019) Investigating the effect of biochar on the potential of increasing cotton yield, potassium efficiency and soil environment. Ecotoxicol Environ Saf 182:109451

Xiu L, Gu W, Sun Y, Wu D, Wang Y, Zhang H, Zhang W, Chen W (2023) The fate and supply capacity of potassium in biochar used in agriculture. Sci Total Environ 902:165969

Xiu L, Sun Y, Han X (2025) Effects of biochar amendment on potassium supply capacity and potassium accumulation in soybean across diverse soils. Plants 14(13):1959

Xu M, Gao P, Yang Z, Su L, Wu J, Yang G, Zhang X, Ma J, Peng H, Xiao Y (2019) Biochar impacts on phosphorus cycling in rice ecosystem. Chemosphere 225:311–319

Yang CD, Lu SG (2021) Effects of five different biochars on aggregation, water retention and mechanical properties of paddy soil: a field experiment of three-season crops. Soil Tillage Res 205:104798

Yin Q, Wang R, Zhao Z (2018) Application of Mg–Al–modified biochar for simultaneous removal of ammonium, nitrate, and phosphate from eutrophic water. J Clean Prod 176:230–240

Yin Q, Zhang B, Wang R, Zhao Z (2017) Biochar as an adsorbent for inorganic nitrogen and phosphorus removal from water: a review. Environ Sci Pollut Res 24(34):26297–26309

Yin S, Suo F, Kong Q, You X, Zhang X, Yuan Y, Yu X, Cheng Y, Sun R, Zheng H, Li Y (2021) Biochar enhanced growth and biological nitrogen fixation of wild soybean (*Glycine max subsp. soja* Siebold & Zucc.) in a coastal soil of China. Agriculture 11(12):1246

Zhang D, Pan G, Wu G, Kibue GW, Li L, Zhang X, Zheng J, Zheng J, Cheng K, Joseph S, Liu X (2016a) Biochar helps enhance maize productivity and reduce greenhouse gas emissions under balanced fertilization in a rainfed low fertility inceptisol. Chemosphere 142:106–113

Zhang H, Chen C, Gray EM, Boyd SE, Yang H, Zhang D (2016b) Roles of biochar in improving phosphorus availability in soils: a phosphate adsorbent and a source of available phosphorus. Geoderma 276:1–6

Zhang Q, Song Y, Wu Z, Yan X, Gunina A, Kuzyakov Y, Xiong Z (2020a) Effects of six-year biochar amendment on soil aggregation, crop growth, and nitrogen and phosphorus use efficiencies in a rice-wheat rotation. J Clean Prod 242:118435

Zhang X, Qu J, Li H, La S, Tian Y, Gao L (2020b) Biochar addition combined with daily fertigation improves overall soil quality and enhances water-fertilizer productivity of cucumber in alkaline soils of a semi-arid region. Geoderma 363:114170

Zhao H, Xie T, Xiao H, Gao M (2022) Biochar-based fertilizer improved crop yields and N utilization efficiency in a maize–Chinese cabbage rotation system. Agriculture 12(7):1030

Zhao L, Cao X, Mašek O, Zimmerman A (2013) Heterogeneity of biochar properties as a function of feedstock sources and production temperatures. J Hazard Mater 256:1–9

Zheng J, Han J, Liu Z, Xia W, Zhang X, Li L, Liu X, Bian R, Cheng K, Zheng J, Pan G (2017) Biochar compound fertilizer increases nitrogen productivity and economic benefits but decreases carbon emission of maize production. Agric Ecosyst Environ 241:70–78

Zong Y, Xiao Q, Lu S (2021) Biochar derived from cadmium-contaminated rice straw at various pyrolysis temperatures: Cadmium immobilization mechanisms and environmental implication. Bioresour Technol 321:124459

Chapter 15
Algal Biochar: Production Strategies and Sustainable Application Perspectives

Pushpita Nath, Yumnam Nganthoibi Chanu, Tanushree Paul, and Jayashree Rout

Abstract Algal biochar is a charcoal or coke like dark carbonaceous material produced as an end product after prolonged slow heating of algal biomass at high temperatures under oxygen-free condition. It is derived from algae through thermochemical processes such as hydrothermal carbonisation, pyrolysis and microwave assisted conversion. In the last decade biochar, in general, has garnered significant worldwide attention as a sustainable and multifunctional material characterised by unique physicochemical properties like high porosity, abundant surface functional groups, high surface area and a nutrient-rich composition that enables its application across diverse domains. This chapter provides a comprehensive review of algal biochar production strategies, emphasising on the influence of feedstock selection and production parameters on biochar's final structural characteristics and functional performance. An analysis of global research output over the past decade or so elucidates evolving trends, major contributing countries and research themes thereby identifying emerging frontiers and knowledge gaps. Beyond its established role in wastewater treatment particularly in the adsorption of heavy metals, nutrients, and pharmaceutical contaminants, algal biochar has demonstrated potential in soil remediation, carbon sequestration, and sustainable agriculture as a biofertiliser. Its application extends to the pharmaceutical and dermal cosmetics industries, as well as to advanced material science serving as a support for nanomaterials in catalytic systems and in energy storage technologies such as fuel cells. The chapter concludes by addressing the current challenges related to standardisation of algal biochar source and production process, large-scale production and applications, regulatory considerations and necessity for interdisciplinary research to advance algal biochar technologies within circular bioeconomy and global sustainability frameworks.

Keywords Algal biochar, Bioremediation, Energy storage, Nanocatalysts, Renewable resource, Waste management

P. Nath · Y. N. Chanu · T. Paul · J. Rout (✉)
Department of Ecology and Environmental Science, Assam University, Silchar, Assam, India
e-mail: jayashreerout8@gmail.com

D. Moulick et al. (eds.), *Biochar: Achieving Agro-Environmental Sustainability*, Sustainability Solutions, https://doi.org/10.1007/978-3-032-19855-6_15

List of Abbreviations

VOCs	Volatile organic compounds
CO_2	Carbon dioxide
SO_2	Sulfur dioxide
NO_x	Nitrogen oxides
MFC	Microbial fuel cells
FESEM-SEM	Scanning electron microscopy
EDS/EDX	Field emission scanning electron microscopy with energy-dispersive spectrometer/energy-dispersive X-ray spectroscopy
XRD	X-ray diffraction
XPS	X-ray photoelectron spectroscopy
TGA	Thermogravimetric analysis
BET	Brunauer-Emmett-Teller analysis
TEM	Transmission electron microscopy
FT-IR	Fourier transform infrared spectroscopy
C	Carbon
HTC	Hydrothermal carbonisation
HTL	Hydrothermal liquefaction
N	Nitrogen
O	Oxygen
H	Hydrogen
S	Sulphur

15.1 Introduction

Biochar is a coke or charcoal-like dark carbonaceous porous composite material produced upon prolonged slow heating of algal biomass at high temperatures (< 900 °C) in the absence of oxygen (Sohi 2012). Its biomass feedstock encompasses a variety of organic materials including algae, plant biomass, agriculture residues, sewage sludge, manure and organic municipal solid wastes (Xiong et al. 2019). Due to its large specific surface area, high porosity, high cation exchange capacity, stability and recyclability, biochar is widely used across various fields (Osman et al. 2022). Its role in carbon sequestration, soil improvement, and renewable fuels is well recognised (Kambo and Dutta 2015). Petroleum-derived materials are not carbon neutral (Adeleye et al. 2019), however, biochar derived from biomass is a carbon neutral and renewable resource, as it sequesters carbon during photosynthesis (Yang et al. 2018) thus, playing a role in mitigating greenhouse gas emissions. The transformation of biomass into biochar is typically accomplished through various processes like pyrolysis, gasification, torrefaction, hydrothermal carbonisation and flash carbonisation (Hirst et al. 2018; Li et al. 2020).

Algae are mostly aquatic photosynthetic organisms, including both prokaryotic cyanobacteria and some thalloid eukaryotes, which are crucial primary producers in most global ecosystems, contributing to global CO_2 fixation and climate regulation· They are adapted to diverse environments such as seas, lakes, rivers, and coastal waters (Karthik et al. 2021). Algal biochar is any biochar that is prepared using micro or macroalgal biomass as source. The use of algal biomass is ecologically sustainable and cost-effective as it can be cultured and is highly abundant. Algal biomass is advantageous over other biomass sources (including plants) owing to its fast proliferation and exceptional photosynthetic efficiency leading to high biomass production, capable of growing in nutrient rich wastewater environments (Ramesh et al. 2023; Paul et al. 2025). Algal biochars have smaller surface areas but higher nitrogen and oxygen content than biochars derived from plant-based biomass (Bird et al. 2011). This composition results in low C/N and high O/C ratios, aiding heavy metal and inorganic pollutant adsorption (Amin and Chetpattananondh 2019). Although microalgal biochars generally exhibit lower carbon content and surface area than lignocellulosic types, they are rich in nutrients and possess higher pH, making them effective for environmental and agricultural use (Yu et al. 2017a).

Furthermore, algae derived biochar is recognized as a negative carbon (C) emission technology that not only helps mitigate greenhouse gas emissions but also enables stable retentions of biogenic C in soils (Neogi et al. 2022). For degraded soil restoration, biochar can be used as a soil enrichment agent to restore soil fertility (Liu et al. 2019), as an adsorbent in air filters to purify emitted gases (Ma et al. 2022) as an effective material for removing pollutants from water (e.g., pesticides, heavy metals, and organic waste) (Ahmaruzzaman 2021). Currently biochar is being extensively researched for energy storage and conversion (Dutta et al. 2023), such as in biochar-based electrodes (Ehsani nd Parsimehr 2020) and biofuels (Amesho et al. 2023; Mohamed and Li 2023). It can also serve as a catalyst carrier in green chemistry formulations (Anto et al. 2021) and as a fertiliser for crops (Buss et al. 2020). In the bio-medical field, biochar can be used as a filler medium for wound regeneration and as a drug-delivery agent (Bielicka et al. 2013; de Gunzburg et al. 2015), thereby enhancing therapeutic efficiency. These versatile applications position algal biochar as a promising material platform within the domains of agriculture, environmental remediation, pharmacology, nanotechnology and material science.

Scientometric analysis of past 15 years (Scopus, March 2025) using 'algal biochar' as keyword show exponentially rising global interest in this field since 2019, with India, China, and the USA leading (Fig. 15.1). Previously biochar research focused more on waste remediation and agriculture but recent developments clearly reveal its importance in areas related to material science including energy related technologies like fuel cells, supercapacitators and energy storage devices. In this context, the present chapter critically examines the production strategies, physicochemical characteristics, and multifunctional applications of algal biochar, offering insights into emerging research trends, key challenges and future perspectives for its integration into sustainable development.

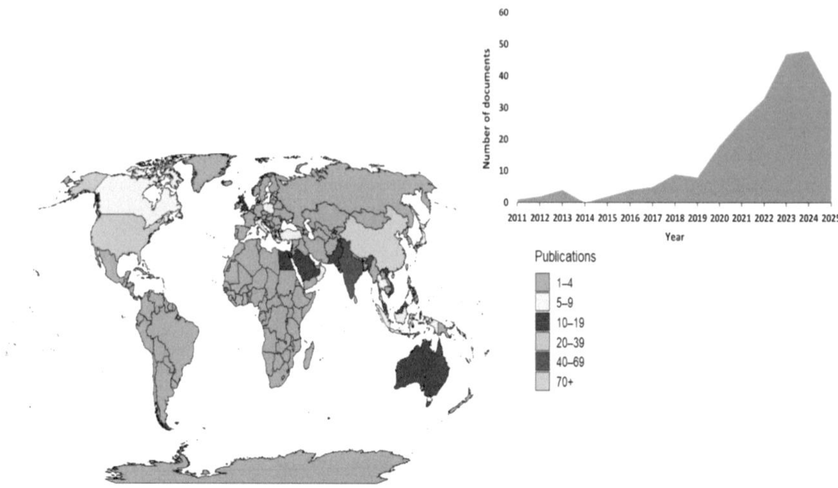

Fig. 15.1 Global analysis of algal biochar research (Scopus, up to March-2025) showing country-wise publication (different colours indicate different publication numbers) and year-wise growth trends in the number of publications from 2010 to 2025

15.2 Suitability of Algae as Feedstock for Biochar Production

Algae offer several advantages over conventional terrestrial plants due to their rapid reproduction, doubling time, high growth rates, suitable biochemical composition and higher photosynthetic efficiency (Binda et al. 2020). These attributes make algae an attractive biomass source for biochar production.

15.2.1 Algae as Biochar Feedstock

Algae are increasingly being recognised for their ecological and economic significance, as they deliver vital ecosystem services and serve as valuable sources of biomass for applications in nutraceuticals, food products, soil enhancers, animal feed, bioplastics and biofuels (Nath et al. 2025). After extracting high-value compounds, the remaining biomass presents an opportunity for conversion into biochar (Mondal et al. 2024).

Owing to their simple unicellular structure, algae exhibit significantly higher photosynthetic efficiency, up to 20 times greater than that of terrestrial plants, resulting in enhanced carbon fixation and elevated biomass yields (Leong And Chang 2023). Thus, promising a stable supply of biochar feedstock with a shorter growth period. Structurally, microalgae-based biochar is characterised by sizable particle clusters ranging from 10 to 100 μm in size, with uneven pore structures around

1 μm, which contrasts with the more fibrous and porous morphology observed in lignocellulosic biochar after pyrolysis (Bach and Chen 2017). Biochar obtained from macroalgae have emerged as a potentially significant material in environmental remediation, particularly for heavy metal adsorption, carbon sequestration, and improvement of degraded soils (Inyang et al. 2016). In the past decade, macroalgal biochar have gained enormous interests due to high porosity, surface functional groups, low production cost, strong ion exchange capabilities and favorable aromatic structure (Gurav et al. 2021).

15.2.2 Cultivation of Algal Biomass

For efficient biochar production, it is essential to select fast-growing algal species such as *Spirulina* sp., *Chlorella* sp., *Saccharina* spp., *Ulva* spp., *Gracilaria* spp., etc. due to their high reproductive capacity (García-Poza et al. 2020), high photosynthetic efficiency capable of accumulating substantial quantity of bioactive compounds within their cells, making them ideal candidates as raw materials (Randrianarison and Ashraf 2017). The cultivation of macroalgae, such as seaweeds like *Ulva*, typically involves long-line or mesh-based farming systems in nutrient-rich coastal waters. These methods support rapid growth by facilitating the uptake of dissolved carbon, nitrogen and phosphorus ultimately resulting in high biomass yields. While microalgae, can be grown in nutrient-rich eutrophic environments, including sewage and effluents. For large-scale microalgal cultivation, open pond systems, especially raceway ponds are widely used due to their cost-effectiveness and reliance on readily available resources such as sunlight, water, and nutrients (Sarwer et al. 2022).

15.2.3 Effect of Algal Biomass Composition on Biochar Characteristics

As a fast-growing organism, microalgae accumulate high levels of lipids, proteins, carbohydrates and other metabolites (Tan et al. 2023). The major biochemical components of algae (Table 15.1) proteins, carbohydrates and lipids interact during pyrolysis to produce nutrients rich biochar, chemically versatile and suitable for various applications. However, the specific composition and structural characteristics of proteins, carbohydrates and lipids can differ markedly between algal species (Wang et al. 2024). Recognising the difference in biochemical composition and mineral content of algae will allow for the refinement and optimisation of biochar production strategies resulting in production of specific types of algal biochar matching specific end-use goals.

In addition to their compositional variability across species, algal lipids and polysaccharides play key roles in shaping the functional and structural properties

Table 15.1 Biochemical composition of different algal genera which are used for biochar production

Algae	Biochemical components			References
Microalgae	Protein (%)	Carbohydrate (%)	Lipid (%)	
Isochrysis	32.9	13.9	39.3	Wang et al. 2015
Spirulina sp.	57.8	23.4	2.9	Hong et al. 2017
Schizochytrium limacinum	14	24	51	Anand et al. 2017
Chlorella ellipsoidea	56	25	35	Sarmah et al. 2019
Nannochloropsis sp.	40.8	18.67	30.0	Tang et al. 2020
Chlorella vulgaris	45.45	22.18	24.57	Nejati et al. 2020
Scenedesmus sp.	35.7	9.0	5.4	Van Wychen et al. 2021
Chlamydomonas reinhardtii	61.7	3.3	13	Wang et al. 2022
Kirchneriella obesa	26.28	38.74	14.07	Sinha et al. 2024
Macroalgae				
Ulva lactuca	12.17	32.61	1.45	Verma et al. 2017
Porphyra sp.	12.12	33.61	2.39	Verma et al. 2017
Gracilaria gracilis	13.7	28.6	1.7	Parsa et al. 2018
Cystoseira trinodis	12.59	33.18	1.3	Dixit et al. 2018
Laminaria digitata	14.1	41.4	0.9	Kostas et al. 2019
Enteromorpha prolifera	42.06	20.21	4.31	Xu et al. 2020
Sargassum vulgare	10.32	39.07	4.02	de Melo et al. 2021
Cladophora sp.	24.6	24.8	5.8	Wang et al. 2022
Fucus vesiculosus	13.71	41.9	3.74	Bayomy and Alamri 2024

of biochar. Macroalgae contain polysaccharides like alginates, carrageenans, cellulose and are minerals rich (K, Na, Ca, P, Mg) (Hentati et al. 2020). Biochars from macroalgae are found to be disordered yet structurally porous, have high carbon and nitrogen content, elevated ash content and alkaline pH, ideal for soil liming and nutrient rich amendments (Mondal et al. 2024). Lipids contribute to pore development during thermal conversion, while polysaccharides such as alginates and carrageenans enhance surface functionalities that improve adsorption and catalytic performance (Li et al. 2025). Carbohydrates are primarily transformed into phenolic compounds and ketones during pyrolysis (Azizi et al. 2018), while the presence of proteins can result in nitrogen-enriched bio-oils formed as byproduct during biochar production (Debiagi et al. 2017). The higher protein content in microalgae compared to macroalgae contributes to formation of N-doped functional groups in the resulting biochar, enhancing its performance in soil amendment, catalysis and adsorption (Sun et al. 2022).

15.3 Algal Biochar Production Strategies

Algal biochar is produced when algal biomass is slowly heated under oxygen-limited condition at high temperatures (< 900 °C) (Godlewska et al. 2017). Of the several strategies (Fig. 15.2), pyrolysis process is the most widely followed because of its rapid processing time, cost effectiveness, productivity and scalability (Bhatnagar et al. 2021). The various algal biochar production strategies operate under different specific thermal conditions and are suitable for diverse biomass feedstock types with different characteristics, which greatly affect the properties and yield of biochar product (Yu et al. 2017a). In the following sub-sections, we discuss three most important techniques for biochar production.

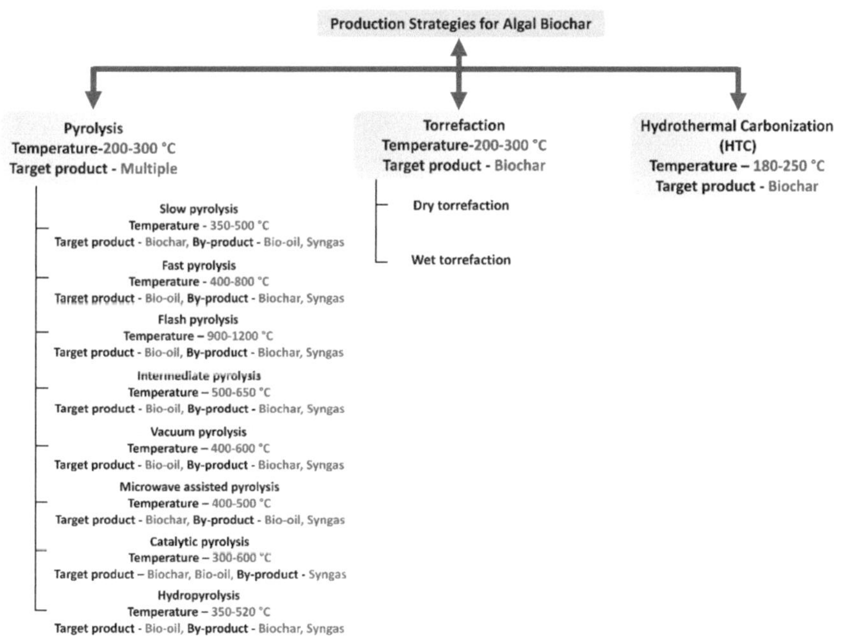

Fig. 15.2 Different methods for producing algal biochar

15.3.1 *Pyrolysis*

Pyrolysis is a conversion process which thermally decomposes algal biomass in an oxygen limited environment, leading to the formation of solid (algal biochar), gaseous products (pyrogas or syngas) and liquid (bio-oil) (Aravind et al. 2020). It is considered one of the most effective thermochemical conversion technologies for biochar production, involving thermally decomposition of algal biomass at temperatures typically ranging from 300 to 1000 °C (Wang et al. 2024). The process encompasses key steps including moisture evaporation, breakdown of organic compounds and deoxidation.

There are several types of pyrolysis each type producing products having varied composition depending on the operating parameters, like heating rate, temperature and residence time (Tan et al. 2023). Based on the heating rate and residence time, pyrolysis can be categorised mainly into four processes, viz., slow, fast, flash and intermediate pyrolysis. Besides these conventional methods, through the integration of advanced techniques other types of pyrolysis, like catalytic pyrolysis and microwave assisted pyrolysis, are also adopted to improve energy efficiency and product quality. The variation in operational parameters across these methods leads to distinct differences in the characteristics and yields of the biochar produced (Nguyen et al. 2023). A brief account of different pyrolysis methods are furnished in the following sections.

15.3.1.1 Slow Pyrolysis

Slow pyrolysis or carbonisation occur at a temperature range of 350–500 °C, slow increase in heating rate of 0.1–1 °C/s with a long residence period of about 5–30 min (Sekar et al. 2021). Under these conditions, the process primarily yields charcoal or biochar as the dominant product with good carbon content and porosity. The slow heating and extended residence time promote secondary reactions facilitating the condensation and removal of volatile compounds, allowing for more complete thermal decomposition and enhanced solid product formation. Bio-oil and syngas are also generated, but they are produced in lesser quantities (Kavan Kumar and Panwar 2024; Yu et al. 2017a). Slow pyrolysis is the most appropriate method for biochar production, as it prioritises algal biochar formation over other coproducts.

15.3.1.2 Fast Pyrolysis

Fast pyrolysis typically operates at elevated temperatures (400–800 °C) with a brief residence time (1–5 s) and rapid heating rate (> 100 °C/s) (Mustapha and Isa 2025; Tan et al. 2023; Aravind et al. 2020). These conditions tend to favor the production of liquid fuels during the biomass pyrolysis process. The byproducts of the fast pyrolysis process include 10–30% biochar, 10–20% syngas and 60–80% of bio-oil (Afshar and Mofatteh 2024; Wang et al. 2024).

15.3.1.3 Flash Pyrolysis

Flash pyrolysis is an advanced and intensified form of rapid pyrolysis utilising hot gases or superheated liquids to achieve the swift thermal decomposition of algal biomass. It operates at high temperatures (900–1200 °C), characterised by very high heating rate (1000 to 10,000 °C/s) and very brief residence time, (0.1 to 1 s) (Kavan Kumar and Panwar 2024; Mustapha and Isa 2025). Rapid heating in flash pyrolysis minimises biochar formation as it converts biomass into condensable vapours and prioritises bio-oil production (65–75%) over biochar (10–25%) (Afshar and Mofatteh 2024).

15.3.1.4 Intermediate Pyrolysis

The intermediate pyrolysis technique is employed to achieve a balance among the outputs of solid, liquid and gaseous products, thereby ensuring uniformity in the product output ratios. The operational parameters are maintained between those of slow and fast pyrolysis, typically ranging from 500–650 °C, with a heating rate of 0.1–10 °C/s and a residence time of 300–1000 s (Kavan Kumar and Panwar 2024).

15.3.1.5 Vacuum Pyrolysis

The vacuum pyrolysis procedure is a slow pyrolysis technique, operating at a low-pressure range of 0.05–0.20 MPa and temperatures of 400–600 °C for the conversion of feedstocks while maintaining all other essential process parameters. The primary distinction between this and other pyrolysis methods lies in the approach to vapour removal from the reaction zone. Unlike conventional biomass pyrolysis processes that typically employ an inert gas, vacuum pyrolysis utilises a vacuum to evacuate the vapours. Operating under low pressure allows biomass to decompose at lower temperatures, which is particularly advantageous for heat-sensitive feedstocks like algae. The biochar produced through vacuum pyrolysis typically possesses a highly porous structure (Dong et al. 2024).

15.3.1.6 Microwave Assisted Pyrolysis

Microwave assisted pyrolysis represents an efficient thermochemical process that employs electromagnetic waves to achieve uniform and rapid internal heating of wet biomass. This method is particularly advantageous for processing larger particles and moist feedstocks, such as algae. In contrast to conventional pyrolysis, microwave assisted pyrolysis benefits from the strong microwave absorption properties of water and allows for the incorporation of catalysts to enhance yield. Microwave assisted pyrolysis operates typically within a frequency range of 3×10^8 to 3×10^{11} Hz and wavelengths of 1 to 10^3 m, enabling consistent heating at 400–500 °C (Wang et al. 2024). The process yields 25% biochar with high porosity and 47% bio-oil (Yu et al. 2017a).

15.3.1.7 Catalytic Pyrolysis

Catalytic pyrolysis is a highly effective method for improving the quality and yield of bio-oil by reducing its oxygen and nitrogen content, thereby improving its stability and energy value. Conducted at temperatures ranging from 300 to 600 °C in fixed-bed reactors, this method involves either the staged addition of a catalyst bed or the premixing of biomass and catalysts such as zeolites, metal oxides etc. Catalysts facilitate the cracking of biomass into lighter compounds, resulting in more stable and less acidic bio-oil with a higher content of aromatic hydrocarbons. Acidic catalysts tend to produce more biochar, whereas basic catalysts are more favorable for bio-oil production (Sekar et al. 2021).

15.3.1.8 Hydropyrolysis

Hydropyrolysis represents an innovative thermochemical conversion technology that converts biomass feedstock into high quality bio-oil under high pressures, typically ranging from 5 to 20 MPa, in the presence of hydrogen or hydrogen rich gases. Operating at a temperature range of 350–520 °C, this process enhances the production of liquid hydrocarbons while significantly reducing oxygen content and minimising biochar formation. By facilitating hydrogenation reactions during biomass decomposition, hydropyrolysis not only improves the quality and energy content of the bio-oil but also increases hydrogen utilisation efficiency (Oh et al. 2021).

15.3.2 Hydrothermal Carbonisation (HTC)

Hydrothermal liquefaction (HTL) represents an effective thermochemical conversion technique for valorising algal biomass into biochar. Conducted in aqueous media under subcritical conditions (180–250 °C, ~2 MPa), HTL facilitates rapid depolymerisation and restructuring of organic constituents, resulting in the formation of biochar as the primary solid product, along with co-products such as bio-oil and biogas. A principal advantage of HTL is its utilisation of water, an environmentally benign solvent whose properties such as density, surface tension, and dielectric constant alter under high temperature and pressure conditions, thereby promoting the breakdown of biomass at the molecular level (Afshar and Mofatteh 2024; Mustapha and Isa 2025; Tan et al. 2023).

15.3.3 Torrefaction

Torrefaction represents a crucial thermochemical pathway for transforming biomass into biochar, encompassing both wet and dry processes. Dry torrefaction, also known as mild pyrolysis, typically occurs at temperatures between 200 to 300 °C under an oxygen free nitrogen environment. This process significantly improves the critical properties of the resultant biochar, such as energy density, calorific value, and carbon content, primarily through moisture removal. While essential biomolecules, including carbohydrates, proteins, and lipids, are retained within algal cells, the torrefaction process disrupts internal structures and modifies the surface morphology of biochar. It also influences the diversity of surface functional groups such as carboxyl and hydroxyl, as well as the material's porosity, hydrophilicity/hydrophobicity, and the formation of volatile by-products like benzene and acetic acid. Torrefaction both dry and wet presents a promising and energy efficient thermochemical approach for producing high quality algal biochar with enhanced hydrophobicity, calorific value and reduced ash content (Afshar And Mofatteh 2024; Mustapha and Isa 2025; Yu et al. 2017a).

15.4 Characteristics of Algal Biochar

Each algal species possesses distinct cell sizes and biochemical components (such as carbohydrates, lipids, protein and cellulose) which contribute to the distinct and unique properties of biochar derived from algal biomass (Kumar et al. 2017). Algal biochar is generally characterised using several physical and chemical properties (Fig. 15.3), all of which significantly impact its applicability (Nguyen et al. 2023). These properties are reliant upon the biomass sources and preparation techniques employed (Sohi et al. 2010; Tan et al. 2023). Physical properties encompass porosity, total pore volume/bulk density, pore size, pore structure, specific surface area, particle size, surface charge, pH, electrical conductivity, hydrophilicity or hydrophobicity, polarity and water retention capacity. Chemical properties include ash content, carbon content, volatile chemicals, moisture content, aromaticity, organic and inorganic elemental composition, functional groups and cation exchange capacity (Tan et al. 2023).

Research has demonstrated that biochar produced from algal feedstocks exhibit distinct structural and compositional characteristics compared to the original biomass prior to pyrolysis. The content of nitrogen (N), oxygen (O), hydrogen (H), carbon (C) and sulphur (S) in algal derived biochar is noteworthy. During the pyrolysis process, it has been observed that the C content in algal biochar increases, while the contents

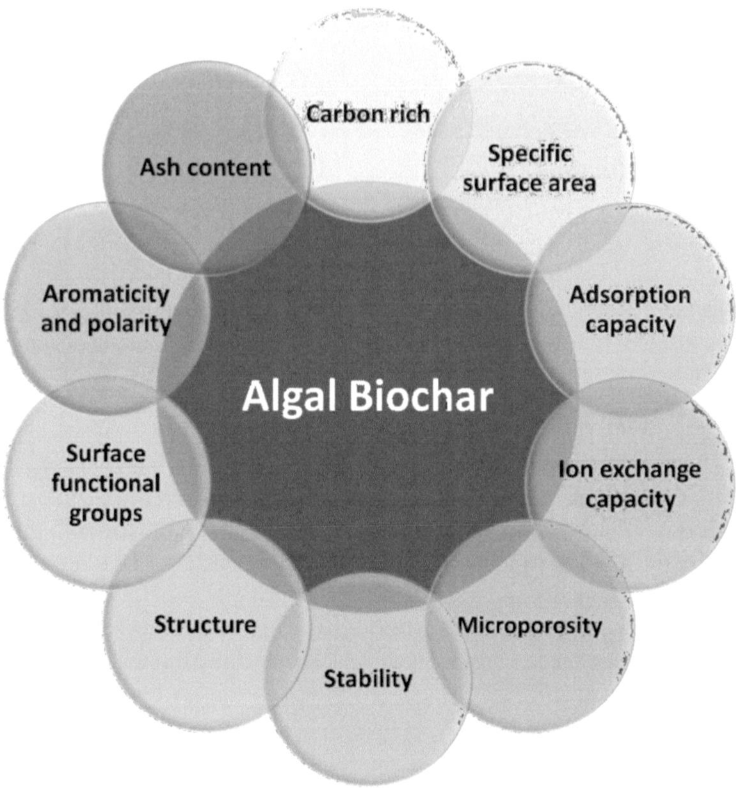

Fig. 15.3 Chemical and physical properties of algal biochar

of S, O and H decrease significantly. This is mainly due to the thermal degradation of volatile components, where compounds containing H, O and S are released as gases (e.g., H_2O, CO_2, H_2S), resulting in a carbon enriched residue. However, N content remains relatively stable, as many nitrogenous compounds in algae decompose into thermally stable structures such as pyrroles and pyridines (Tan et al. 2023). The O/C and H/C ratios of algal biochar serve as indicators of its stability and structural composition. Biochars with an O/C ratio below 0.2 are considered very stable, with an estimated half-life exceeding 1000 years. While, biochars with O/C ratios ranged from 0.2 and 0.6 display moderate stability (half-life of 100–1000 years). In contrast, those above 0.6 are relatively unstable, degrading within 100 years. Biochar with H/C ratios below 0.7 exhibits a high degree of aromatisation characterised by fused aromatic ring structures, indicating a greater extent of thermochemical transformation and chemical stability (Ippolito et al. 2020). These properties underscore the importance of elemental composition and atomic ratios in predicting the environmental persistence and functional performance of algal biochar.

Regarding physical characteristics, algal biochar possesses a smaller surface area and higher cation exchange capacity in comparison to biochar produced from lignocellulosic biomass (Nguyen et al. 2023). Macroalgae possess a fixed carbon content and a higher ash content compared to microalgae and the ash content of macroalgae being approximately twice that of microalgae (Lee et al. 2020). In addition to the feedstock, the methods of biochar production and the pyrolysis temperature are critical factors influencing the properties of algal biochar (Yu et al. 2017a). The pyrolysis temperature affects algal biochar properties: pH, porosity, surface area, degree of graphitisation, electrical conductivity and zeta potential. Increasing the pyrolysis temperature results in an increase in the surface area of algal biochar. Algal biomass both micro and macroalgae can produce comparatively higher biochar yields per unit mass (Table 15.2) than lignocellulosic sources, with reported ranges between 25–63% for microalgae and 20–83% for macroalgae.

15.5 Characterisation of Algal Biochar

Depending on the nature of algal feedstock and the preparation process, the morphological features including porosity, surface area features and surface chemical entities etc., may differ. The stable carbon structure of biochar enables it as a long-term carbon sink mitigating climate change. Therefore, it is extremely important to characterise the as-obtained biochar before exploring or realising their potential applications. A brief discussion on the key characterisation techniques for biochar and their insights is presented (Fig. 15.4).

15.5.1 UV-Vis Spectroscopy

It helps to evaluate the optical properties of algal biochar and can be used as a tool to monitor the adsorption experiments (Mondal et al. 2024).

15.5.2 Scanning Electron Microscopy (SEM) and Transmission Electron Microscopy (TEM)

SEM can be used to determine surface morphology, particle size and porosity of the algal biochar (Mondal et al. 2024). Whereas, TEM provides a closer view of the internal structure and crystallinity of the algal biochar at the nanometer scale (Zhu and Zou 2021). A SEM representative image of biochar obtained from a marine alga (*Sargassum crassifolium)* is provided in Fig. 15.5.

Table 15.2 Characteristics of algal biochar derived from different algal feedstock using different production processes

Algae	Yield (%)	C (%)	O (%)	N (%)	H (%)	H/C	O/C	N/C	Fixed carbon (%)	Ash content (%)	HHV (MJ/kg)	Volatile matter (%)	pH	Moisture (%)	Surface area (m^2/g)	Pore volume (cm^3/g)	Production process	References
Scenedesmus dimorphus	27	50.2	35.8	6	8	–	–	–	52.4	43.3	–	4.3	7.9	0.06	123	–	Pyrolysis, 300–600 °C, 40 min	Bordoloi et al. (2016)
Arthrospira platensis	25	51	18	7.7	2.5	–	–	–	42	29.5	–	28	–	–	–	–	Pyrolysis, 500 °C, 20 min	Conti et al. (2016)
Undaria pinnatifida	83.95 67.70 47.75 39.29	30.6 31.9 36.6 40.5	27.1 13.6 11.8 10.0	3.4 2.3 2.3 2.0	3.8 1.7 10.7	0.12 0.05 0.03 0.02	0.89 0.42 0.32 0.25	0.11 0.07 0.06 0.05	–	32.83 41.85 48.71 50.41	–	–	7.4 8.4 10.4 11.1	–	1.91 70.29 61.81 44.49	0.04 0.11 0.08 0.06	Pyrolysis, N2, 5 °C/min-2 h, 200 °C 400 °C 600 °C 800 °C	Jung et al. (2016)
Nannochlor opsis sp.	28.6 22.7	54.9 27.3	32.2 66.4	5.9 2.7	6.4 3.3	0.12 0.12	0.59 2.43	0.11 0.19	19.3 11.5	11.2 53.7	–	66.0 33.0	–	–	1.38 12.56	0.01 0.01	Hydrothermal carbonisation, Ar, 1 h, 210 °C 250 °C	Saber et al. (2018)
Cladophora glomerata	63 47	–	–	–	–	–	–	–	28.8 51.1	30.3 40.1	–	68.9 50.0	5	1.5 1.5	–	0.56 0.77	Pyrolysis, 300 °C 450 °C	Michalak et al. (2019)
Enteromorpha	20.5	70.2	23.1	2.2	4.5	–	–	–	–	–	–	–	–	–	29.7	–	Hydrothermal liquefication, 250 °C, 40 min	Yang et al. (2019)
Spirulina sp.	–	61.6 65.4 66.6	18.7 14.2 12.3	11.7 11.1 9.4	6.6 3.7 1.3	1.29 0.68 0.14	0.23 0.16 0.14	–	42.26 64.92 70.0	1.27 5.53 10.4	–	56.47 29.55 19.6	6.5 7.0 8.3	–	0.31 1.55 2.63	–	Pyrolysis-hydrothermal process, 2 h, 350 °C 550 °C 750 °C	Choi et al. (2020)

(continued)

Table 15.2 (continued)

Algae	Yield (%)	C (%)	O (%)	N (%)	H (%)	H/C	O/C	N/C	Fixed carbon (%)	Ash content (%)	HHV (MJ/kg)	Volatile matter (%)	pH	Moisture (%)	Surface area (m^2/g)	Pore volume (cm^3/g)	Production process	References
Chlorella vulgaris	54.5	57.4	30.0	3.8	8.6	–	–	–	15.5	4.7	23.56–32.35	77.9	–	1.8	–	–	Microwave-assisted acid hydrolysis, wet torrefaction	Yu et al. (2020)
Enteromorpha prolifera	39.0 9.5 8.4 7.5	28.6 32.3 29.3 20.7	26.0 19.5 18.6 19.5	2.0 1.9 1.6 1.5	1.9 1.2 1.2 1.2	0.07 0.04 0.04 0.06	0.91 0.61 0.63 0.94	0.07 0.06 0.05 0.07	–	33.64 37.78 43.96 52.71	–	–	–	–	12.02 38.49 99.34 134.19	0.01 0.01 0.02 0.03	Pyrolysis, N2, 5 °C/min 4 h, 500 °C 600 °C 700 °C 800 °C	Lin et al. (2023)
Laminaria digitata	56.71	27.7	–	2.28	4.58	–	–	–	33.16	10.72	–	52.33	11–13	3.78	37	–	Pyrolysis 300–800 °C	Pak et al. (2023)
Saccharina latissimi	62.09	22.7	–	2.02	3.08	–	–	–	16.04	18.96	–	60.8	11–13	4.20	3	–	Pyrolysis 300–800 °C	Pak et al. (2023)
Saccorhiza polyschides	66.56	21.9	–	1.98	3.98	–	–	–	15.56	22.49	–	57.35	11–13	4.61	12	–	Pyrolysis 300–800 °C	Pak et al. (2023)
Chlorella pyrenoidosa	–	64.7	23	9.8	2.2	–	–	–	25.04	3.42	19.7	69.35	–	2.19	–	–	Fast pyrolysis 750 °C	Rohman et al. (2024)
Oedogonium sp.	28.2	70.2	22.5	7.91	–	–	–	8.87	68.86	12.60	–	14.57	10.9	3.97	–	–	Pyrolysis 600 °C for 40 min, N_2, 30 °C/min	Mondal et al. (2024)
Ulva sp.	30.46	55.7	38.5	3.78	–	–	–	14.7	42.41	31.10	–	22.87	10.5	3.62	–	–	Pyrolysis 600 °C for 40 min, N_2, 30 °C/min	Mondal et al. (2024)
Eucheuma denticulatum	46.96	40.1	28.3	1.50	–	–	–	26.7	2.63	52.12	–	43.92	9.86	1.33	–	–	Pyrolysis 600 °C for 40 min, N_2, 30 °C/min	Mondal et al. (2024)
Asparagopsis	40	51.8	30.3	4.97	–	–	–	10.4	32.50	30.86	–	34.21	9.74	2.43	–	–	Pyrolysis 600 °C for 40 min, N_2, 30 °C/min	Mondal et al. (2024)

(–) Denotes data not available

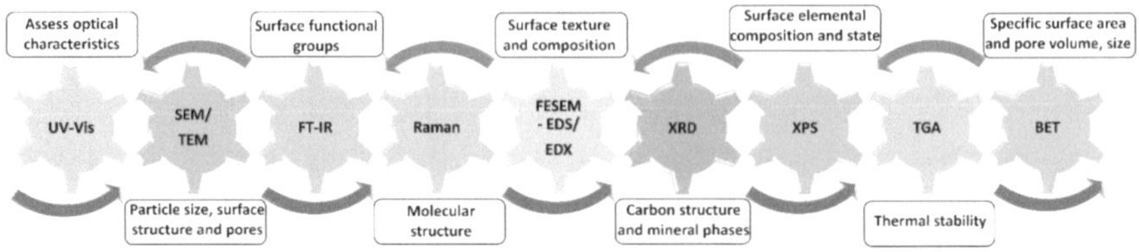

Fig. 15.4 Different characterisation techniques for algal biochar properties

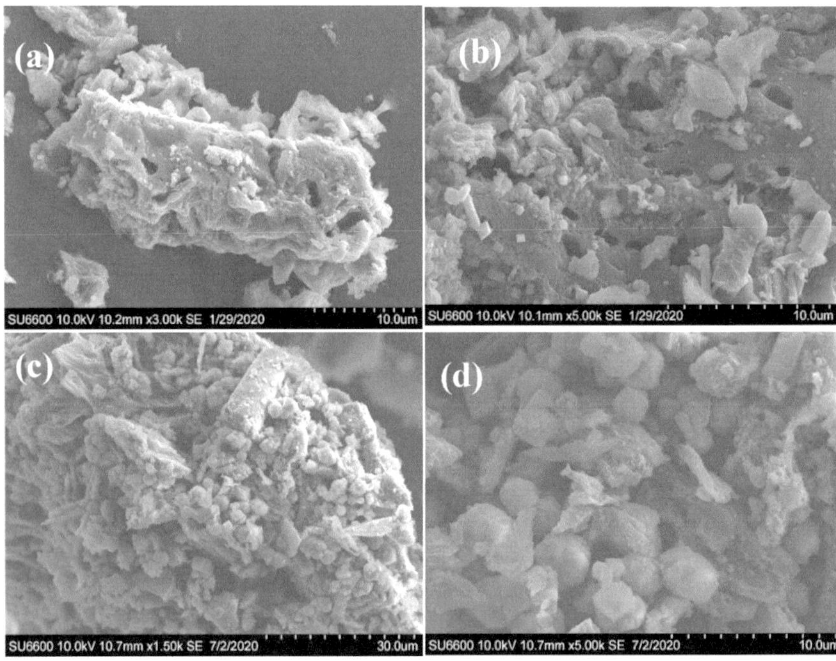

Fig. 15.5 SEM images of a marine alga biochar *Sargassum crassifolium* (a and b) and modified algal biochar with zeolite (c and d) (reproduced with permission from Atugoda et al. 2021)

15.5.3 Fourier Transform Infrared spectroscopy (FT-IR)

It is used to detect functional groups on the algal biochar surface (e.g., O-H, C-H, C-C, C=O, etc.) within the wavenumber range of 4000 to 500 cm^{-1} (Maleki and Bozorg 2023), which influence the reactivity and adsorption properties. As the time of pyrolysis increases, functional groups undergo some changes (Zhu and Zou 2021). This variation provides valuable information when compared to algal biochar produced under different environments (Yu et al. 2017a).

15.5.4 Raman Spectroscopy

This analysis provides insights into the molecular structure of algal biochar, particularly in the wavenumber range of 500–2000 cm^{-1} (Zhu and Zou 2021), where key peaks are observed. It helps in the quantification of carbon crystallinity and structural defects by distinguishing between graphitic (G-band) and disordered (D-band) carbon components (Mondal et al. 2024).

15.5.5 Field Emission Scanning Electron Microscopy with Energy-Dispersive Spectrometer/ Energy-Dispersive X-Ray Spectroscopy (FESEM-EDS/EDX)

FESEM provides high-resolution images of the algal biochar surface, revealing detailed morphology, pore structure, and particle distribution. EDS/EDX coupled with FESEM, detects the elemental composition and distribution across the surface. This combined analysis helps to understand surface texture and the presence of key elements such as C, O, N, K, Ca, Si, Al, Mg, S, etc. (Maleki and Bozorg, 2023; Mondal et al. 2024).

15.5.6 X-Ray Diffraction (XRD)

It determines crystalline and amorphous phases in algal biochar, indicating the degree of carbon ordering and presence of mineral phases like calcite, silica or graphite (Mondal et al. 2024).

15.5.7 X-Ray Photoelectron Spectroscopy (XPS)

The technique analyse surface elemental composition (C, O, N and trace elements), oxidation states and chemical forms of carbon atoms of algal biochar. It also helps in the determination of surface functional groups (Li et al. 2023).

15.5.8 Thermogravimetric Analysis (TGA)

It measures thermal stability and decomposition patterns, helping determine fixed carbon content, ash content, moisture content and volatile matter of algal biochar

(Mondal et al. 2024). The TGA curve for biochar usually shows an initial weight loss starting around 600–800 °C due to the evaporation of moisture and the breakdown of volatile organic compounds. As the temperature continues to rise, the thermal decomposition of carbon-rich solids starts and eventually leaves behind the residual ash content (Li et al. 2023; Mondal et al. 2024).

15.5.9 Brunauer-Emmett-Teller Analysis (BET)

BET analysis help ascertain the textural properties of biochar, including its pore volume, pore size, specific surface area and micro and mesoporosity (Yu et al. 2017a; Zhu and Zou 2021) by measuring nitrogen gas adsorption (Maleki and Bozorg 2023).

15.6 Applications of Algal Biochar

In the last decade algal biochar has gained recognition as a potential tool and versatile material for diverse applications (Fig. 15.6). It has found key applications in pollution control, environmental remediation, biofertiliser, green chemistry, nanotechnology, material science. Some of the applications have been elaborated in the following sections.

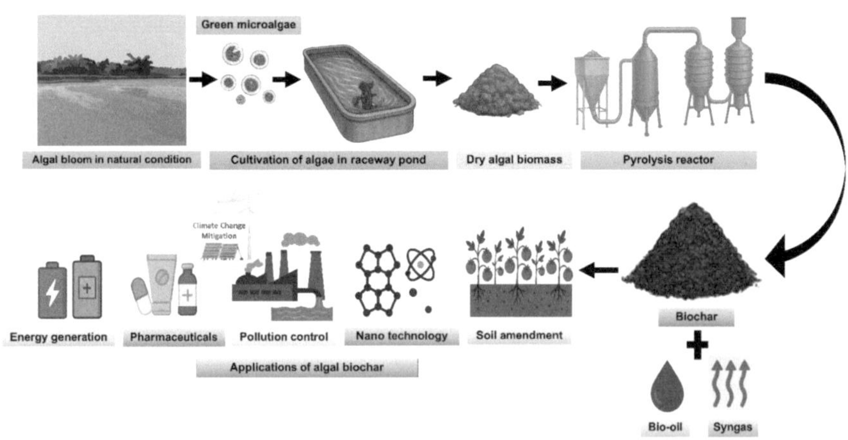

Fig. 15.6 Production and different applications of algal biochar

15.6.1 Pollution Control

Due to growing human population, there has been a corresponding surge in waste generation, leading to serious environmental problems regarding its disposal. Prompting the need to explore efficient and cost-effective treatment methods (Lu et al. 2013). Biochar has emerged as a promising solution, especially in wastewater treatment, due to its versatility and effectiveness (Enaime et al. 2020). It absorbs and adsorbs harmful substances like heavy metals, dyes, pharmaceutical residues and persistent organic pollutants (González Fernández et al. 2025).

Among various pollutants, heavy metals such as mercury, cadmium and lead pose some of the greatest threats. They are toxic, tend to accumulate in living organisms, and can severely harm aquatic life by disrupting food chains and causing mass fish deaths (Briffa et al. 2020). Algal biochar especially when activated with chemicals like KOH or $ZnCl_2$ offers a high surface area and porous structure, making it highly efficient at binding and removing these metals from contaminated water (González Fernández et al. 2025). Moreover, N-doped biochars derived from algae, when used along with peroxymonosulfate, have shown excellent performance in removing complex pollutants such as antibiotics from water (Yu et al. 2017b). Biochar has also been tested successfully for reducing excess nitrogen and phosphorus, the nutrients that lead to eutrophication in aquatic ecosystems (Ambaye et al. 2021). Biochar systems can also reduce global C emissions by approximately 3.4–6.3 Pg CO_2e, with about 50% CO_2 removal, showing emission reduction and supporting climate change mitigation through sustained carbon storage (Lehmann et al. 2021).

15.6.2 Soil Amendment and Biofertiliser

Algal biochar plays a crucial role in enhancing soil quality by improving nutrient retention, boosting microbial activity, and increasing water-holding capacity. It contains a comprehensive range of micro and macronutrients essential for plant development, with the high concentrations of P, N and K in algal biochar serving as fertiliser (Bird et al. 2012). The combination of biochar into soil can improve its physical structure, adjust pH levels, enrich carbon content and fertility, immobilise heavy metals and other contaminants. These effects positively influence the soil microbial communities and microenvironment, ultimately refining soil health and boosting crop yield (Yang et al. 2022; Kapoor and Zdarta 2024). Biochar also increases the soil's cation exchange capacity and has shown to improve mineral uptake in plants by approximately 13% (James et al. 2022).

15.6.3 Pharmaceutical and Dermal Cosmetics

The applications of biochar in human healthcare remains underexplored. Algae-based biochar is rich in N-containing functional groups such as graphitic-N, oxidised-N and pyrrolic-N and byproducts of protein pyrolysis which significantly enhance its adsorption capacity (Tan et al. 2023). These chemical groups, along with the material's porous structure, make algal biochar highly effective in eliminating microbial toxins, pollutants and oxidative agents from the skin or wounds. This makes it well suited for use in facial masks, detoxifying skin cleansers and wound-care products.

When integrated with bioactive compounds derived from algae such as alginates, carrageenans, phlorotannins and carotenoids algal biochar can be transformed into multifunctional composites. These natural biopolymers/biocomposites contribute towards moisturising, film-forming, viscosity-enhancing and UV-protective qualities (Parikh et al. 2024). Furthermore, the N and O rich groups from the pyrolysed proteins and polysaccharides are endowed biochar with antioxidant properties and antimicrobial potential (Tan et al. 2023). Together, these features highlight it as a sustainable, bio-based material for use in advanced cosmetic products aimed at improving skin health and pharmaceuticals that deliver drugs in a targeted and efficient manner.

15.6.4 Nanomaterials and Catalysis

Transforming algal biochar into nanoparticles significantly improves its porosity, surface area and ample amount of reactive surface functional groups, enhancing its effectiveness as a catalytic material. Nano-scale algal biochar serves as an ideal material for enzyme immobilisation and acts as a nanocatalyst in bioremediation. Methods such as ball milling and chemical treatments are commonly used to produce these nanoparticles (Bhandari et al. 2023).

Catalysts such as Fe-impregnated and TiO_2-supported composites derived from algal biochar have been successfully employed in various systems for contaminant degradation in wastewater treatment, this includes advanced oxidation processes such as Fenton-type reactions and photocatalysis (Sutar et al. 2022). N-doped graphitic biochar derived from *Spirulina* residue (Ho et al. 2019) and kelp biomass (Huang et al. 2021) can be utilised to activate peroxymonosulfate for degrading ofloxacin. Algal biochar based photocatalyst composites have demonstrated strong photocatalytic performance because biochar acts as an effective support matrix, offering more active sites for catalyst attachment, improving interactions between pollutants and photocatalysts, facilitating rapid electron transfer, minimising electron-hole recombination, and narrowing the band gap of the photocatalyst material (Sutar et al. 2022). Algae derived biochar catalysts are significantly less expensive to synthesise than conventional Pt-C cathode catalysts (Chakraborty et al. 2020). Another study highlighted the development of a biochar based photocatalyst (Ag@BC-rGO), synthesised using *Trentepohlia* sp. as a green reducing agent which enabled effective degradation of antibiotic (Saikia et al. 2024).

15.6.5 Energy Storage and Fuel Cells

In the field of material science algal biochar is also being used in energy storage and fuel cell applications. Microbial fuel cells (MFC) present a promising, low-cost, eco-friendly, and sustainable approach for bioelectricity generation (Lefebvre et al. 2011). Biochar can be integrated into MFC systems in three main ways: as electrode materials, as electrocatalysts and in the fabrication of polymer electrolyte membranes (Chakraborty et al. 2020). The development of nickel-adsorbed algal biochar has proven to be an efficient cathode catalyst for promoting oxygen reduction reactions within microbial carbon-capture cells (Neethu et al. 2024). Microalgae harvested from HABs have been converted into biochar and used to create anodes that improved electron transfer by *Shewanella oneidensis* MR-1 in MFCs (Wang et al. 2018).

15.7 Challenges and Future Prospectives

Algal biochar presents significant potential for enhancing environmental sustainability. However, numerous technical and practical challenges must be addressed, to fully exploit its multifunctional benefits. One major limitation is its underdeveloped porosity, often due to pore blockage and narrow pore size distribution, which limits its adsorption efficiency and catalytic performance compared to lignocellulosic biochar (Li et al. 2025). Furthermore, the high ash and mineral content inherent in algal biomass, due to the presence of inorganic elements such as sodium, calcium, potassium and iron, can dilute carbon concentration and compromise structural stability.

Algal biochar prepared from algal blooms can also introduce risks such as phytotoxicity or heavy metal accumulation, particularly when applied to soils (Chu et al. 2024). The heterogeneity in algal species, cultivation methods, and biochemical composition presents additional challenges for standardisation and reproducibility in large-scale production (Mustapha and Isa 2025). Moreover, based on species the pyrolysis process itself requires careful optimisation of key parameters such as heating rate, residence time and temperature which are often resource intensive and complex to control. In this context, machine learning and artificial intelligence tools are emerging as promising approaches for predictive optimisation of process variables, aiming to improve biochar yield and desired functional properties.

Environmental concerns also persist, including the potential release of persistent organic pollutants or secondary contaminants during production and application, necessitating comprehensive risk assessments and regulatory frameworks (Chu et al. 2024; Mustapha and Isa 2025). Despite these barriers, the future of algal biochar remains highly promising. Advances in thermochemical conversion methods, including microwave assisted pyrolysis and hydrothermal carbonisation, are being explored to enhance material quality, reduce energy input, and enable cleaner production pathways. Strategic surface modifications, whether through physical activation, chemical treatment, or biological functionalisation, are critical for enhancing algal biochar's application-specific performance. These tailored biochars demonstrate improved efficiency in environmental remediation, encompassing the sequestration of toxic metals, recovery of nutrients, and elimination of pharmaceutical or organic pollutants (Li et al. 2025).

Beyond water treatment, algal biochar contributes significantly to soil fertility augmentation, carbon sequestration, and mitigation of greenhouse gas emissions (Sun and Masek 2021). Innovatively, the integration of algal biochar with nanomaterials, microbial consortia, or agro-inputs can lead to the development of multifunctional platforms with enhanced nutrient bioavailability (e.g., fertilisers), pollutant capture, and biocompatibility. Additionally, biochar-based composites and nanomaterial-infused systems are gaining attention in air purification technologies, leveraging their high surface area and reactive functional groups for the effective capture of air pollutants such as nitrogen oxides (NO_x), sulfur dioxide (SO_2) and volatile organic compounds (VOCs) (Wang et al. 2024).

From a sustainability perspective, the valorisation of algal biomass, especially from wastewater-grown algae or algal blooms, into biochar aligns with the principles of the circular bioeconomy and waste-to-resource strategies. Moving forward, interdisciplinary collaboration in research and robust government policy support will be essential to scale up algal biochar technologies for wide-ranging environmental and industrial applications, effectively connecting experimental research with practical, real-world use (Inyang et al. 2016).

15.8 Conclusion

Algal biochar constitutes a promising, multifunctional carbon-neutral material platform with immense application potential in environmental remediation, pollution mitigation, sustainable agriculture and advanced material science. Its production from algal biomass is consistent with circular economy principles, facilitating zero waste production, carbon neutrality and pollution control. It offers high productivity, rich nutrient profile and rapid growth on non-arable land. Both freshwater and marine algae offer tremendous scope in terms of easy and rather quick harvesting for biochar production in large-scale. Despite some existing challenges related to standardisation, scalability and environmental risks, ongoing advancements in processing and usage of biochar are broadening its scope of applicability. Future research should prioritise and put emphasis on interdisciplinary studies and real-world implementation to fully realise the potential of algal biochar in advancing global sustainability goals and cutting-edge technologies like microbial fuel cells, nano-catalysts in a cost-effective manner.

References

Adeleye AT, Louis H, Akakuru OU, Joseph I, Enudi OC, Michael DP (2019) A review on the conversion of levulinic acid and its esters to various useful chemicals. Aims Energy 7(2):165–185. https://doi.org/10.3934/energy.2019.2.165

Afshar M, Mofatteh S (2024) Biochar for a sustainable future: environmentally friendly production and diverse applications. Results Eng 102433. https://doi.org/10.1016/j.rineng.2024.102433

Ahmaruzzaman M (2021) Biochar based nanocomposites for photocatalytic degradation of emerging organic pollutants from water and wastewater. Mater Res Bull 140:111262. https://doi.org/10.1016/j.materresbull.2021.111262

Ambaye TG, Vaccari M, van Hullebusch ED, Amrane A, Rtimi SJIJOES (2021) Mechanisms and adsorption capacities of biochar for the removal of organic and inorganic pollutants from industrial wastewater. Int J Environ Sci Technol 18(10):3273–3294. https://doi.org/10.1007/s13762-020-03060-w

Amesho KT, Lin YC, Mohan SV, Halder S, Ponnusamy VK, Jhang SR (2023) Deep eutectic solvents in the transformation of biomass into biofuels and fine chemicals: a review. Environ Chem Lett 21(1):183–230. https://doi.org/10.1007/s10311-022-01521-x

Amin M, Chetpattananondh P (2019) Biochar from extracted marine *chlorella* sp. residue for high efficiency adsorption with ultrasonication to remove Cr (VI), Zn (II) and Ni (II). Bioresour Technol 289:121578. https://doi.org/10.1016/j.biortech.2019.121578

Anand V, Gautam R, Vinu R (2017) Non-catalytic and catalytic fast pyrolysis of *Schizochytrium limacinum* microalga. Fuel 205:1–10. https://doi.org/10.1016/j.fuel.2017.05.049

Anto S, Sudhakar MP, Ahamed TS, Samuel MS, Mathimani T, Brindhadevi K, Pugazhendhi A (2021) Activation strategies for biochar to use as an efficient catalyst in various applications. Fuel 285:119205. https://doi.org/10.1016/j.fuel.2020.119205

Aravind S, Kumar PS, Kumar NS, Siddarth N (2020) Conversion of green algal biomass into bioenergy by pyrolysis. A review. Environ Chem Lett 18:829–849. https://doi.org/10.1007/s10311-020-00990-2

Atugoda T, Gunawardane C, Ahmad M, Vithanage M (2021) Mechanistic interaction of ciprofloxacin on zeolite modified seaweed (*Sargassum crassifolium*) derived biochar: kinetics, isotherm and thermodynamics. Chemosphere 281:130676. https://doi.org/10.1016/j.chemosphere.2021.130676

Azizi K, Moraveji MK, Najafabadi HA (2018) A review on bio-fuel production from microalgal biomass by using pyrolysis method. Renew Sust Energ Rev 82:3046–3059. https://doi.org/10.1016/j.rser.2017.10.033

Bach QV, Chen WH (2017) A comprehensive study on pyrolysis kinetics of microalgal biomass. Energy Convers Manag 131:109–116. https://doi.org/10.1016/j.enconman.2016.10.077

Bayomy HM, Alamri ES (2024) Biochemical assessments of six species of edible coastal algae collected from Tabuk region in Saudi Arabia. Molecules 29(3):639. https://doi.org/10.3390/molecules29030639

Bhandari G, Gangola S, Dhasmana A, Rajput V, Gupta S, Malik S, Slama P (2023) Nano-biochar: recent progress, challenges, and opportunities for sustainable environmental remediation. Front Microbiol 14:1214870. https://doi.org/10.3389/fmicb.2023.1214870

Bhatnagar P, Gururani P, Bisht B, Kumar V (2021) Algal biochar: an advance and sustainable method for wastewater treatment. Octa J Biosci 9(2)

Bielicka A, Wiśniewski M, Terzyk AP, Gauden PA, Furmaniak S, Roszek K, Kowalczyk P, Bieniek A (2013) Carbon materials as new nanovehicles in hot-melt drug deposition. J Phys Condens Matter 25(35):355002. https://doi.org/10.1088/0953-8984/25/35/355002

Binda G, Spanu D, Bettinetti R, Magagnin L, Pozzi A, Dossi C (2020) Comprehensive comparison of microalgae-derived biochar from different feedstocks: a prospective study for future environmental applications. Algal Res 52:102103. https://doi.org/10.1016/j.algal.2020.102103

Bird MI, Wurster CM, de Paula Silva PH, Bass AM, De Nys R (2011) Algal biochar-production and properties. Bioresour Technol 102(2):1886–1891. https://doi.org/10.1016/j.biortech.2010.07.106

Bird MI, Wurster CM, de Paula Silva PH, Paul NA, De Nys R (2012) Algal biochar: effects and applications. GCB Bioenergy 4(1):61–69. https://doi.org/10.1111/j.1757-1707.2011.01109.x

Bordoloi N, Narzari R, Sut D, Saikia R, Chutia RS, Kataki R (2016) Characterization of bio-oil and its sub-fractions from pyrolysis of *Scenedesmus dimorphus*. Renew Energy 98:245–253. https://doi.org/10.1016/j.renene.2016.03.081

Briffa J, Sinagra E, Blundell R (2020) Heavy metal pollution in the environment and their toxicological effects on humans. Heliyon 6(9). https://doi.org/10.1016/j.heliyon.2020.e04691

Buss W, Bogush A, Ignatyev K, Masek O (2020) Unlocking the fertilizer potential of waste-derived biochar. ACS Sustain Chem Eng 8(32):12295–12303. https://doi.org/10.1021/acssuschemeng.0c04336

Chakraborty I, Bhowmick GD, Ghosh D, Dubey BK, Pradhan D, Ghangrekar MM (2020) Novel low-cost activated algal biochar as a cathode catalyst for improving performance of microbial fuel cell. Sustain Energy Technol Assess 42:100808. https://doi.org/10.1016/j.seta.2020.100808

Choi YK, Choi TR, Gurav R, Bhatia SK, Park YL, Kim HJ, Kan E, Yang YH (2020) Adsorption behavior of tetracycline onto *spirulina* sp. (microalgae)-derived biochars produced at different temperatures. Sci Total Environ 710:136282. https://doi.org/10.1016/j.scitotenv.2019.136282

Chu Y, Wang Z, Wang W, Zeng Y, He S, Yan C, Qin F, Wu M, Zeng G, Zhou C (2024) A review on the algae-derived biochar catalysts: advanced oxidation processes and machine learning tools. Sep Purif Technol 129336. https://doi.org/10.1016/j.seppur.2024.129336

Conti R, Fabbri D, Vassura I, Ferroni L (2016) Comparison of chemical and physical indices of thermal stability of biochars from different biomass by analytical pyrolysis and thermogravimetry. J Anal Appl Pyrolysis 122:160–168. https://doi.org/10.1016/j.jaap.2016.10.003

de Gunzburg J, Ducher A, Modess C, Wegner D, Oswald S, Dressman J, Augustin V, Feger C, Andremont A, Weitschies W, Siegmund W (2015) Targeted adsorption of molecules in the colon with the novel adsorbent-based medicinal product, DAV132: a proof of concept study in healthy subjects. J Clin Pharmacol 55(1):10–16. https://doi.org/10.1002/jcph.359

de Melo NSM, Cardoso LG, de Castro Nunes JM, Brito GB, Caires TA, de Souza CO, Portz L, Druzian JI (2021) Effects of dry and rainy seasons on the chemical composition of *Ulva fasciata*, *Crassiphycus corneus* and *Sargassum vulgare* seaweeds in tropical environment. Braz J Bot 44:331–344. https://doi.org/10.1007/s40415-021-00700-4

Debiagi PEA, Trinchera M, Frassoldati A, Faravelli T, Vinu R, Ranzi E (2017) Algae characterization and multistep pyrolysis mechanism. J Anal Appl Pyrolysis 128:423–436. https://doi.org/10.1016/j.jaap.2017.08.007

Dixit DC, Reddy CRK, Balar N, Suthar P, Gajaria T, Gadhavi DK (2018) Assessment of the nutritive, biochemical, antioxidant and antibacterial potential of eight tropical macro algae along Kachchh coast, India as human food supplements. J Aquat Food Prod Technol 27(1):61–79. https://doi.org/10.1080/10498850.2017.1396274

Dong L, Zhi W, Li J (2024) In-situ reaction mechanism of vacuum pyrolysis for extracting valuable materials from waste excess biological sludge. J Environ Chem Eng 12(2):112096. https://doi.org/10.1016/j.jece.2024.112096

Dutta N, Usman M, Ashraf MA, Luo G, Gamal El-Din M, Zhang S (2023) Methods to convert lignocellulosic waste into biohydrogen, biogas, bioethanol, biodiesel and value-added chemicals: a review. Environ Chem Lett 21(2):803–820. https://doi.org/10.1007/s10311-022-01511-z

Ehsani A, Parsimehr H (2020) Electrochemical energy storage electrodes from fruit biochar. Adv Colloid Interf Sci 284:102263. https://doi.org/10.1016/j.cis.2020.102263

Enaime G, Baçaoui A, Yaacoubi A, Lübken M (2020) Biochar for wastewater treatment-conversion technologies and applications. Appl Sci 10(10):3492. https://doi.org/10.3390/app10103492

García-Poza S, Leandro A, Cotas C, Cotas J, Marques JC, Pereira L, Gonçalves AM (2020) The evolution road of seaweed aquaculture: cultivation technologies and the industry 4.0. Int J Environ Res Public Health 17(18):6528. https://doi.org/10.3390/ijerph17186528

Godlewska P, Schmidt HP, Ok YS, Oleszczuk P (2017) Biochar for composting improvement and contaminants reduction a review. Bioresour Technol 246:193–202. https://doi.org/10.1016/j.biortech.2017.07.095

González Fernández LA, Medellín Castillo NA, Sánchez Polo M, Navarro Frómeta AE, Vilasó Cadre JE (2025) Algal-based carbonaceous materials for environmental remediation: advances in wastewater treatment, carbon sequestration, and biofuel applications. https://doi.org/10.3390/pr13020556

Gurav R, Bhatia SK, Choi TR, Kim HJ, Lee HJ, Cho JY, Ham S, Shu MJ, Kim SH, Kim SK, Yoo DW, Yang YH (2021) Seafood processing chitin waste for electricity generation in a microbial fuel cell using halotolerant catalyst *Oceanisphaera arctica* YHY1. Sustainability 13(15):8508. https://doi.org/10.3390/su13158508

Hentati F, Tounsi L, Djomdi D, Pierre G, Delattre C, Ursu AV, Fendri I, Abdelkafi S, Michaud P (2020) Bioactive polysaccharides from seaweeds. Molecules 25(14):3152. https://doi.org/10.3390/molecules25143152

Hirst EA, Taylor A, Mokaya R (2018) A simple flash carbonization route for conversion of biomass to porous carbons with high CO_2 storage capacity. J Mater Chem A 6(26):12393–12403. https://doi.org/10.1039/C8TA04409K

Ho SH, Chen YD, Li R, Zhang C, Ge Y, Cao G, Ma M, Duan X, Wang S, Ren NQ (2019) N-doped graphitic biochars from C-phycocyanin extracted *spirulina* residue for catalytic persulfate activation toward nonradical disinfection and organic oxidation. Water Res 159:77–86. https://doi.org/10.1016/j.watres.2019.05.008

Hong Y, Chen W, Luo X, Pang C, Lester E, Wu T (2017) Microwave-enhanced pyrolysis of macroalgae and microalgae for syngas production. Bioresour Technol 237:47–56. https://doi.org/10.1016/j.biortech.2017.02.006

Huang YM, Li G, Li M, Yin J, Meng N, Zhang D, Cao XQ, Zhu FP, Chen M, Li L, Lyu XJ (2021) Kelp-derived N-doped biochar activated peroxymonosulfate for ofloxacin degradation. Sci Total Environ 754:141999. https://doi.org/10.1016/j.scitotenv.2020.141999

Inyang MI, Gao B, Yao Y, Xue Y, Zimmerman A, Mosa A, Pullammanappallil P, Ok YS, Cao X (2016) A review of biochar as a low-cost adsorbent for aqueous heavy metal removal. Crit Rev Environ Sci Technol 46(4):406–433. https://doi.org/10.1080/10643389.2015.1096880

Ippolito JA, Cui L, Kammann C, Wrage-Mönnig N, Estavillo JM, Fuertes-Mendizabal T, Cayuela ML, Sigua G, Novak J, Spokas K, Borchard N (2020) Feedstock choice, pyrolysis temperature and type influence biochar characteristics: a comprehensive meta-data analysis review. Biochar 2:421–438. https://doi.org/10.1007/s42773-020-00067-x

James A, Sánchez A, Prens J, Yuan W (2022) Biochar from agricultural residues for soil conditioning: technological status and life cycle assessment. Curr Opin Environ Sci Health 25:100314. https://doi.org/10.1016/j.coesh.2021.100314

Jung KW, Kim K, Jeong TU, Ahn KH (2016) Influence of pyrolysis temperature on characteristics and phosphate adsorption capability of biochar derived from waste-marine macroalgae (*Undaria pinnatifida* roots). Bioresour Technol 200:1024–1028. https://doi.org/10.1016/j.biortech.2015.10.016

Kambo HS, Dutta A (2015) A comparative review of biochar and hydrochar in terms of production, physico-chemical properties and applications. Renew Sust Energ Rev 45:359–378. https://doi.org/10.1016/j.rser.2015.01.050

Kapoor RT, Zdarta J (2024) Fabrication of engineered biochar for remediation of toxic contaminants in soil matrices and soil valorization. Chemosphere 142101. https://doi.org/10.1016/j.chemosphere.2024.142101

Karthik V, Kumar PS, Vo DVN, Sindhu J, Sneka D, Subhashini B, Saravanan K, Jeyanthi J (2021) Hydrothermal production of algal biochar for environmental and fertilizer applications: a review. Environ Chem Lett 19:1025–1042. https://doi.org/10.1007/s10311-020-01139-x

Kavan Kumar V, Panwar NL (2024) Pyrolysis technologies for biochar production in waste management: a review. Clean Energy 8:61–78. https://doi.org/10.1093/ce/zkae036

Kostas ET, Williams OS, Duran-Jimenez G, Tapper AJ, Cooper M, Meehan R, Robinson JP (2019) Microwave pyrolysis of *Laminaria digitata* to produce unique seaweed-derived bio-oils. Biomass Bioenergy 125:41–49. https://doi.org/10.1016/j.biombioe.2019.04.006
Kumar G, Shobana S, Chen WH, Bach QV, Kim SH, Atabani AE, Chang JS (2017) A review of thermochemical conversion of microalgal biomass for biofuels: chemistry and processes. Green Chem 19(1):44–67. https://doi.org/10.1039/C6GC01937D
Lee XJ, Ong HC, Gan YY, Chen WH, Mahlia TMI (2020) State of art review on conventional and advanced pyrolysis of macroalgae and microalgae for biochar, bio-oil and bio-syngas production. Energy Convers Manag 210:112707. https://doi.org/10.1016/j.enconman.2020.112707
Lefebvre O, Uzabiaga A, Chang IS, Kim BH, Ng HY (2011) Microbial fuel cells for energy self-sufficient domestic wastewater treatment-a review and discussion from energetic consideration. Appl Microbiol Biotechnol 89:259–270. https://doi.org/10.1007/s00253-010-2881-z
Lehmann J, Cowie A, Masiello CA, Kammann C, Woolf D, Amonette JE, Cayuela ML, Camps-Arbestain M, Whitman T (2021) Biochar in climate change mitigation. Nat Geosci 14(12):883–892. https://doi.org/10.1038/s41561-021-00852-8
Leong YK, Chang JS (2023) Microalgae-based biochar production and applications: a comprehensive review. Bioresour Technol 389:129782. https://doi.org/10.1016/j.biortech.2023.129782
Li F, Zimmerman AR, Hu X, Yu Z, Huang J, Gao B (2020) One-pot synthesis and characterization of engineered hydrochar by hydrothermal carbonization of biomass with $ZnCl_2$. Chemosphere 254:126866. https://doi.org/10.1016/j.chemosphere.2020.126866
Li P, Zhao T, Zhao Z, Tang H, Feng W, Zhang Z (2023) Biochar derived from Chinese herb medicine residues for rhodamine B dye adsorption. ACS Omega 8(5):4813–4825. https://doi.org/10.1021/acsomega.2c06968
Li L, Wang J, Sun W, Peng X, Qi X (2025) Production of algae-derived biochar and its application in pollutants adsorption-a mini review. Separations 12(2):32. https://doi.org/10.3390/separations12020032
Lin S, Shen Z, Pan D, Ji R, Bian Y, Han J, Jiang X, Song Y, Cheng H, Xue J (2023) Marine macroalgae-derived multielement-doped porous biochars for efficient removal of sulfamethoxazole from aqueous solution: sorption performance and governing mechanisms. J Anal Appl Pyrolysis 171:105963. https://doi.org/10.1016/j.jaap.2023.105963
Liu Z, Zhu M, Wang J, Liu X, Guo W, Zheng J, Bian R, Wang G, Zhang X, Cheng K, Liu X, Li L, Pan G (2019) The responses of soil organic carbon mineralization and microbial communities to fresh and aged biochar soil amendments. GCB Bioenergy 11(12):1408–1420. https://doi.org/10.1111/gcbb.12644
Lu H, Zhang W, Wang S, Zhuang L, Yang Y, Qiu R (2013) Characterization of sewage sludge-derived biochars from different feedstocks and pyrolysis temperatures. J Anal Appl Pyrolysis 102:137–143. https://doi.org/10.1016/j.jaap.2013.03.004
Ma C, Zhao Y, Chen H, Liu Y, Huang R, Pan J (2022) Biochars derived from by-products of microalgae pyrolysis for sorption of gaseous H_2S. J Environ Chem Eng 10(3):107370. https://doi.org/10.1016/j.jece.2022.107370
Maleki A, Bozorg A (2023) Algal biochar of unique structure as a robust alternative to manipulate mixed-matrix membranes performance and fouling resistance. Results Engi 18:101043. https://doi.org/10.1016/j.rineng.2023.101043
Michalak I, Baśladyńska S, Mokrzycki J, Rutkowski P (2019) Biochar from a freshwater macroalga as a potential biosorbent for wastewater treatment. Water 11(7):1390. https://doi.org/10.3390/w11071390
Mohamed BA, Li LY (2023) Biofuel production by co-pyrolysis of sewage sludge and other materials: a review. Environ Chem Lett 21(1):153–182. https://doi.org/10.1007/s10311-022-01496-9
Mondal AK, Hinkley C, Krishnan L, Ravi N, Akter F, Ralph P, Kuzhiumparambil U (2024) Macroalgae-based biochar: preparation and characterization of physicochemical properties for potential applications. RSC Sustainability 2(6):1828–1836. https://doi.org/10.1039/D4SU00008K

Mustapha SI, Isa YM (2025) Thermal conversion of microalgae into biochar: a review on processes, properties, and applications. ChemBioEng Rev 12(1):e202400035. https://doi.org/10.1002/cben.202400035

Nath P, Borah D, Paul P, Rout J (2025) Integrated biorefinery approach for sustainable production of biodiesel, bioplastics and high value bioproducts from a bloom forming alga *Botryococcus braunii*. Sci Total Environ 964:178599. https://doi.org/10.1016/j.scitotenv.2025.178599

Neethu B, Ihjas K, Chakraborty I, Ghangrekar MM (2024) Nickel adsorbed algae biochar based oxygen reduction reaction catalyst. Bioelectrochem 159:108747. https://doi.org/10.1016/j.bioelechem.2024.108747

Nejati B, Adami P, Bozorg A, Tavasoli A, Mirzahosseini AH (2020) Catalytic pyrolysis and bio-products upgrading derived from *Chlorella vulgaris* over its biochar and activated biochar-supported Fe catalysts. J Anal Appl Pyrolysis 152:104799. https://doi.org/10.1016/j.jaap.2020.104799

Neogi S, Sharma V, Khan N, Chaurasia D, Ahmad A, Chauhan S, Singh A, You S, Pandey A, Bhargava PC (2022) Sustainable biochar: a facile strategy for soil and environmental restoration, energy generation, mitigation of global climate change and circular bioeconomy. Chemosphere 293:133474. https://doi.org/10.1016/j.chemosphere.2021.133474

Nguyen TB, Nguyen VT, Hoang HG, Cao NDT, Nguyen TT, Vo TDH, Nguyen NKQ, Pham MDT, Nghiem DL, Vo TKQ, Bui XT (2023) Recent development of algal biochar for contaminant remediation and energy application: a state-of-the art review. Curr Pollut Rep 9(1):73–89. https://doi.org/10.1007/s40726-022-00243-6

Oh S, Lee J, Lam SS, Kwon EE, Ha JM, Tsang DC, Ok YS, Chen WH, Park YK (2021) Fast hydropyrolysis of biomass conversion: a comparative review. Bioresour Technol 342:126067. https://doi.org/10.1016/j.biortech.2021.126067

Osman AI, Fawzy S, Farghali M, El-Azazy M, Elgarahy AM, Fahim RA, Maksoud A, Ajlan AA, Yousry M, Saleem Y, Rooney DW (2022) Biochar for agronomy, animal farming, anaerobic digestion, composting, water treatment, soil remediation, construction, energy storage, and carbon sequestration: a review. Environ Chem Lett 20(4):2385–2485. https://doi.org/10.1007/s10311-022-01424-x

Pak T, Gomari KE, Bose S, Tonon T, Hughes D, Gronnow M, Macquarrie D (2023) Biochar from brown algae: production, activation, and characterisation. Bioresour Technol Rep 24:101688. https://doi.org/10.1016/j.biteb.2023.101688

Parikh HS, Singh PK, Tiwari A (2024) Algal biorefinery: focus on cosmeceuticals. Syst Microbiol Biomanuf 4(4):1239–1261. https://doi.org/10.1007/s43393-024-00287-3

Parsa M, Jalilzadeh H, Pazoki M, Ghasemzadeh R, Abduli M (2018) Hydrothermal liquefaction of *Gracilaria gracilis* and *Cladophora glomerata* macro-algae for biocrude production. Bioresour Technol 250:26–34. https://doi.org/10.1016/j.biortech.2017.10.059

Paul T, Nath P, Tapadar S, Sultana S, Deb Purkayastha S, Sharma H, Rout J (2025) Growth potential, biochemical properties and nutrient removal efficiency of some freshwater microalgae and their consortia from wastewater. Int J Phytoremediation 1-11. https://doi.org/10.1080/15226514.2024.2405001

Ramesh B, Saravanan A, Kumar PS, Yaashikaa PR, Thamarai P, Shaji A, Rangasamy G (2023) A review on algae biosorption for the removal of hazardous pollutants from wastewater: limiting factors, prospects and recommendations. Environ Pollut 327:121572. https://doi.org/10.1016/j.envpol.2023.121572

Randrianarison G, Ashraf MA (2017) Microalgae: a potential plant for energy production. Geol Ecol Landscapes 1(2):104–120. https://doi.org/10.1080/24749508.2017.1332853

Rohman GAN, Aziz MA, Nawaz A, Elgzoly MA, Hossain MM, Razzak SA (2024) High-performance biochar from *Chlorella pyrenoidosa* algal biomass for heavy metals removal in wastewater. Sep Purif Technol 341:126870. https://doi.org/10.1016/j.seppur.2024.126870

Saber M, Takahashi F, Yoshikawa K (2018) Characterisation and application of microalgae hydrochar as a low-cost adsorbent for cu (II) ion removal from aqueous solutions. Environ Sci Pollut Res 25(32):32721–32734. https://doi.org/10.1007/s11356-018-3106-8

Saikia P, Borah D, Gogoi D, Rout J, Ghosh NN, Choudhury S, Bhattacharjee CR (2024) Sustainable synthesis of biochar-rGO supported AgNPs nanohybrid as high performance photocatalyst for Cr (VI) ion reduction and antibiotic degradation. Mater Today Sustain 28:100970. https://doi.org/10.1016/j.mtsust.2024.100970

Sarmah P, Das S, Sharma H, Rout J (2019) Microalgal biomass generation by phycoremediation of sewage water: an integrated approach for production of antioxidant and value added products. Vegetos 32:556–563. https://doi.org/10.1007/s42535-019-00056-x

Sarwer A, Hamed SM, Osman AI, Jamil F, Al-Muhtaseb AAH, Alhajeri NS, Rooney DW (2022) Algal biomass valorization for biofuel production and carbon sequestration: a review. Environ Chem Lett 20(5):2797–2851. https://doi.org/10.1007/s10311-022-01458-1

Sekar M, Mathimani T, Alagumalai A, Chi NTL, Duc PA, Bhatia SK, Brindhadevi K, Pugazhendhi A (2021) A review on the pyrolysis of algal biomass for biochar and bio-oil-bottlenecks and scope. Fuel 283:119190. https://doi.org/10.1016/j.fuel.2020.119190

Sinha T, Borah D, Rout J (2024) Algal communities in urban freshwater ponds: prospects for biodiesel production. Vegetos 1-11. https://doi.org/10.1007/s42535-023-00804-0

Sohi SP (2012) Carbon storage with benefits. Sci 338(6110):1034–1035

Sohi SP, Krull E, Lopez-Capel E, Bol R (2010) A review of biochar and its use and function in soil. Adv Agron 105:47–82. https://doi.org/10.1016/S0065-2113(10)05002-9

Sun J, Masek O (2021) A comprehensive review of algal biochar for soil improvement: bottlenecks and opportunities. In: EGU general assembly conference abstracts, p EGU21-3558. https://doi.org/10.5194/egusphere-egu21-3558

Sun J, Norouzi O, Mašek O (2022) A state-of-the-art review on algae pyrolysis for bioenergy and biochar production. Bioresour Technol 346:126258. https://doi.org/10.1016/j.biortech.2021.126258

Sutar S, Otari S, Jadhav J (2022) Biochar based photocatalyst for degradation of organic aqueous waste: a review. Chemosphere 287:132200. https://doi.org/10.1016/j.chemosphere.2021.132200

Tan X, Zhang F, Wang H, Ho SH (2023) The magic of algae-based biochar: advantages, preparation, and applications. Bioeng 14(1):2252157. https://doi.org/10.1080/21655979.2023.2252157

Tang Z, Chen W, Hu J, Li S, Chen Y, Yang H, Chen H (2020) Co-pyrolysis of microalgae with low-density polyethylene (LDPE) for deoxygenation and denitrification. Bioresour Technol 311:123502. https://doi.org/10.1016/j.biortech.2020.123502

Van Wychen S, Rowland SM, Lesco KC, Shanta PV, Dong T, Laurens LM (2021) Advanced mass balance characterization and fractionation of algal biomass composition. J Appl Phycol 33(5):2695–2708. https://doi.org/10.1007/s10811-021-02508-x

Verma P, Kumar M, Mishra G, Sahoo D (2017) Multivariate analysis of fatty acid and biochemical constitutes of seaweeds to characterize their potential as bioresource for biofuel and fine chemicals. Bioresour Technol 226:132–144. https://doi.org/10.1016/j.biortech.2016.11.044

Wang X, Zhao B, Tang X, Yang X (2015) Comparison of direct and indirect pyrolysis of micro-algae *Isochrysis*. Bioresour Technol 179:58–62. https://doi.org/10.1016/j.biortech.2014.11.015

Wang YS, Li DB, Zhang F, Tong ZH, Yu HQ (2018) Algal biomass derived biochar anode for efficient extracellular electron uptake from *Shewanella oneidensis* MR-1. Front Environ Sci Eng 12:1–9. https://doi.org/10.1007/s11783-018-1072-5

Wang K, Wang Y, Zhang S, Chen YD, Wang R, Ho SH (2022) Tailoring a novel hierarchical cheese-like porous biochar from algae residue to boost sulfathiazole removal. Environ Sci Ecotechnol 10:100168. https://doi.org/10.1016/j.ese.2022.100168

Wang C, Lin X, Zhang X, Show PL (2024) Research advances on production and application of algal biochar in environmental remediation. Environ Pollut 123860. https://doi.org/10.1016/j.envpol.2024.123860

Xiong X, Iris KM, Tsang DC, Bolan NS, Ok YS, Igalavithana AD, Kirkham MB, Kim KH, Vikrant K (2019) Value-added chemicals from food supply chain wastes: state-of-the-art review and future prospects. Chem Eng J 375:121983. https://doi.org/10.1016/j.cej.2019.121983

Xu S, Cao B, Uzoejinwa BB, Odey EA, Wang S, Shang H, Li C, Hu Y, Wang Q, Nwakaire JN (2020) Synergistic effects of catalytic co-pyrolysis of macroalgae with waste plastics. Process Safety Environ Protection 137:34–48. https://doi.org/10.1016/j.psep.2020.02.001

Yang F, Xu Z, Yu L, Gao B, Xu X, Zhao L, Cao X (2018) Kaolinite enhances the stability of the dissolvable and undissolvable fractions of biochar via different mechanisms. Environ Sci Technol 52(15):8321–8329. https://doi.org/10.1021/acs.est.8b00306

Yang W, Wang Z, Song S, Han J, Chen H, Wang X, Sun R, Cheng J (2019) Adsorption of copper (II) and lead (II) from seawater using hydrothermal biochar derived from *Enteromorpha*. Mar Pollut Bull 149:110586. https://doi.org/10.1016/j.marpolbul.2019.110586

Yang F, Zuo X, Zhou Y, Wu S, Wang M (2022) Stability of biochar in five soils: effects from soil property. Environ Prog Sustain Energ 41(3):e13775. https://doi.org/10.1002/ep.13775

Yu KL, Lau BF, Show PL, Ong HC, Ling TC, Chen WH, Ng EP, Chang JS (2017a) Recent developments on algal biochar production and characterization. Bioresour Technol 246:2–11. https://doi.org/10.1016/j.biortech.2017.08.009

Yu KL, Show PL, Ong HC, Ling TC, Lan JCW, Chen WH, Chang JS (2017b) Microalgae from wastewater treatment to biochar-feedstock preparation and conversion technologies. Energy Convers Manag 150:1–13. https://doi.org/10.1016/j.enconman.2017.07.060

Yu KL, Chen WH, Sheen HK, Chang JS, Lin CS, Ong HC, Ng EP, Show PL, Ling TC (2020) Production of microalgal biochar and reducing sugar using wet torrefaction with microwave-assisted heating and acid hydrolysis pretreatment. Renew Energy 156:349–360. https://doi.org/10.1016/j.renene.2020.04.064

Zhu H, Zou H (2021) Characterisation of algae residue biochar and its application in methyl orange wastewater treatment. Water Sci Technol 84(12):3716–3725. https://doi.org/10.2166/wst.2021.473

Chapter 16
Biochar-Assisted Remediation of Heavy Metals, Volatile Organic Compounds and Nutrients from Water and Wastewater

Apurba Koley, Nitu Gupta, Vyshnavi A. Nair, Arunkumar Patchaiyappan, Sristi Debnath, Aishiki Banerjee, Deep Chakraborty, and Srinivasan Balachandran

Abstract Biochar is a carbon-rich material produced through the pyrolysis of organic feedstocks, which has gained significant attention in recent days as an effective adsorbent for removing pollutants, including emerging contaminants, from water and wastewater due to its high surface area and porous structure. The accumulation of these contaminants in industrial, municipal, and agricultural wastewater presents a critical environmental challenge, necessitating targeted remediation strategies. Common biochar sources used in wastewater treatment include corn straw, rice husk, wheat straw, coconut husk, sugarcane bagasse, sawdust, and scallion sheathings. Heavy metals (HMs) are of particular concern due to their environmental persistence and tendency to bioaccumulate in living organisms, leading to adverse health effects. Since these metals resist degradation, adsorption has emerged as one of the most

A. Koley · S. Debnath · S. Balachandran (✉)
Bioenergy Laboratory, Department of Environmental Studies, Siksha-Bhavana (Institute of Science), Visva-Bharati, Santiniketan, West Bengal, India
e-mail: s.balachandran@visva-bharati.ac.in

N. Gupta
Department of Environmental Science, Tezpur University, Napaam, Tezpur, Assam, India

V. A. Nair · A. Patchaiyappan
Integrated Science Education and Research Centre (ISERC), Siksha-Bhavana (Institute of Science), Visva-Bharati, Santiniketan, West Bengal, India

A. Patchaiyappan
Svarnim, Sri Aurobindo Society, Puducherry, India

A. Banerjee
Biomarker Discovery Laboratory, The University of Kansas Medical Centre, Kansas City, KS, USA

D. Chakraborty
Department of Environmental Health Engineering, Sri Ramachandra Faculty of Public Health, Sri Ramachandra Institute of Higher Education and Research (Deemed to Be University), Porur, Chennai, India

D. Moulick et al. (eds.), *Biochar: Achieving Agro-Environmental Sustainability*, Sustainability Solutions, https://doi.org/10.1007/978-3-032-19855-6_16

effective removal methods. Modified or activated biochar demonstrates enhanced adsorption capacities for metals such as arsenic (As), mercury (Hg), cadmium (Cd), lead (Pb), zinc (Zn) and copper (Cu). Additionally, biochar effectively adsorbs nitrate, phosphate, and ammonia from water and wastewater, particularly when modified. Beyond heavy metals and nutrients, biochar is also used for the removal of volatile organic compounds (VOCs) such as benzene, toluene, xylene, and formaldehyde from wastewater. This chapter explores the removal efficiencies of HMs, VOCs, and nutrients under different operational conditions using various biochar types. Furthermore, a bibliometric analysis using the Web of Science database provides insights into the research landscape of biochar-assisted green remediation.

Keywords Pyrolysis · Biochar modification · Water pollution · Potential toxic elements · NPK · BTEX

16.1 Introduction

Water is a fundamental component of the environment, which is essential for sustenance of life and to support ecological balance. Water plays a crucial role in for humans and industrial development. Increasingly people around the world recognise water will be one of the most valuable natural resources in the future (Das 2025). The rapid expansion of cities and industries lead to increased consumption of water leading to enhanced pressure on water resources (Wang et al. 2025). Water pollution is still a big problem all over the world. Waste water impacts animals, plants, and human health, and also affects economy as well (Tian et al. 2025). The Lancet Commission on Pollution and Health identified pollution as a leading environmental cause for sickness and early death. In 2019, around nine million people died early because of diseases caused by pollution (Münzel et al. 2025). Water pollution adversely affects the health of both women and men and cause hormonal changes accelerate aging process and lead to health problems at a younger age. The contaminants in the polluted water can be chemical (organic or inorganic), physical, or biological (Bochynska et al. 2024). Water contamination with heavy metals and metalloids can result from various factors, including rapid urbanization, industrial development, coal mining, improper disposal of metal-rich wastes, excessive use of fertilizers, human activities, and irrigation with untreated or poorly treated wastewater (Roy et al. 2024).

Heavy metal pollution in water is a of great environmental concern that needs urgent attention. Heavy metals are non-degradable and can be persistent in the environment. These metals, along with other harmful chemicals, can adversely affect aquatic organisms like fish and other water animals. Metals like nickel, chromium, cadmium, mercury, and lead are very dangerous for both people and the environment (Bayar et al. 2024). Volatile organic compounds (VOCs) are major water contaminants linked to serious health issues, including cancer, genetic mutations, eye and mucosal irritation, dizziness, headaches, and memory loss. When exposed to nitrogen oxides and UV light, VOCs also contribute to the formation of photochemical pollutants and acidic aerosols, leading to photochemical smog (Li et al. 2018a, b; Song and Chun 2021; Jin et al. 2022). The global rise in agricultural activities—such as fuel use, burning of agricultural residues, and the use of VOCs in pesticides—has led to widespread VOC contamination in the environment (Liang and Liao 2007; Abdullahi et al. 2014; David and Niculescu 2021). Additionally, nutrient pollution in water bodies and waterways poses a significant environmental challenge in many countries. Water quality is essential for human health, ecosystems, and overall environmental balance, it can be negatively impacted by elevated levels of nitrogen (N) and phosphorus (P) resulting from both natural processes and human activities (Nie et al. 2018; Shin et al. 2013). Nutrient leaching from fertilized soils is widely recognized as a primary contributor to the nutrient pollution. The process is intensified by the application of both synthetic and organic fertilizers, atmospheric nitrogen deposition, and biological nitrogen fixation by certain crops (Strokal et al. 2016). Nutrient pollution is classified as point or nonpoint source. Point source pollution comes from identifiable sources like factory discharge pipes, while non-point source pollution originates from diffuse areas such as agricultural runoff, urban surface runoff, and rainfall (Aniebone et al. 2024). A significant portion of hazardous VOCs is released by industries through the direct discharge of excessive organic solvents in wastewater, particularly from sectors like chemical, pharmaceutical, mining, printing, paint, and petrochemical industries (Fatima et al. 2022).

Wastewater treatment encompasses technical, economic, environmental, and social considerations, all of which are essential when designing or selecting a treatment method. The removal of pollutants like heavy metals, organic compounds etc. poses a significant challenge, as these pollutants are difficult to eliminate through conventional biological, physical, or chemical processes. Therefore, effective treatment requires both the immobilization and separation of these contaminants (Naz et al. 2016; Kaur and Roy 2021). Adsorption is a widely studied technique due to its low cost, operational simplicity, sludge-free process, and high efficiency. Effective adsorbents possess high activity, adjustable pore structures, and active sites with strong affinity for target pollutants (Esrafili et al. 2021; Kumar et al. 2023). Biochar has emerged as a cost-effective and sustainable adsorbent for water and wastewater treatment applications (Madadi and Bester 2021a; b).

Biochar is a porous, carbon-rich material produced through the thermochemical decomposition of biomass in limited or no oxygen. The feedstock used for biochar production can include organic waste such as crop and forest residues, wood chips, algae, sewage sludge, manures, and organic municipal solid waste (Colantoni et al. 2016; Xiang et al. 2020; Xiong et al. 2019). Biochar is produced by heating biomass at 300–600 °C in a closed reactor with little to no oxygen, causing thermochemical conversion of the biomass into stable carbon rich material called biochar (Panwar et al. 2019). The first use of biochar as a soil amendment was observed in *Terra Preta*, or "Indian black earth," found in Western Amazonia. This dark, nutrient-rich soil is known for its high carbon content, strong aggregate stability, and enhanced microbial activity (Glaser and Birk 2012; Allohverdi et al. 2021). Since the late 2010s, biochar has also been widely used in wastewater treatment for sludge conditioning and odor control. Compared to activated carbon, biochar offers several advantages like lower production costs, long-term carbon sequestration, and added environmental benefits through waste valorisation and co-production of renewable energy (He et al. 2022). Its high surface area, substantial pore volume, rich carbon content, presence of minerals, and diverse functional groups contribute to its strong adsorption capacity for both inorganic and organic pollutants in water (Wang et al. 2020a; b). The properties of the biochar vary depending on the feedstock and production method and it may contain heavy metals or contaminants that can leach into water during application (Enaime et al. 2020). Despite its potential limitation biochar has been extensively used for removing environmental contaminants. It serves as an adsorbent for immobilizing toxic elements like heavy metals and also acts as a catalyst in advanced oxidation processes to break down complex organic pollutants (Kamali et al. 2021). Biochar based adsorbents have demonstrated strong sorption capacity for agrochemicals, pharmaceuticals, personal care products, and endocrine-disrupting compounds (Thompson et al. 2016).

The objective of this study is to explore the application of biochar for the removal of heavy metals, nutrients, and VOCs from various types of wastewater. It also aims to examine the global research trends and insights related to the use of biochar in wastewater treatment process.

16.2 Biochar: An Adsorbent for Inorganic and Organic Pollutants

Biochar is typically produced through thermochemical conversion methods, with pyrolysis and hydrothermal carbonization being the most widely used. Other less common approaches include techniques like partial gasification (also referred to as thermo-oxidative pyrolysis) and microwave-assisted pyrolysis (MAP). Among these, slow pyrolysis is the most conventional and widely practiced technique. Slop pyrolysis operates at heating rates below 1 °C per second, uses moderate temperatures (generally under 600 °C), involves extended residence times, and handles feedstocks with relatively large particle sizes. The design of slow pyrolysis reactors is generally simple, making them suitable for small-scale applications. Studies on the pyrolysis of agricultural residues such as wheat straw, corn straw, rape straw, and rice straw within the temperature range of 300–600 °C have demonstrated that higher temperatures typically lead to lower biochar yields (Chun et al. 2021; Seo et al. 2022). In biochar production, the primary aim interest is obtaining the carbonaceous solid product. During the process water evaporation and the release of volatile components cause an increase in the relative fixed carbon content of the solid. It is believed that polymerization of organic compounds in vapours and gases may lead to secondary char formation and increase the solid yield. The process is characterised by slow heating rate, and the high residence time leading to maximum biochar production. In a traditional charcoal production process (charcoal pit), the carbonization takes several weeks to complete. Typical operating temperatures for slow pyrolysis are around 500 °C, but ultimately depend on the desired properties of final product. A very high carbon content of more than 95% may require treatment temperatures close to 1000 °C, which can be achieved for woody feedstocks, but poses problems for agricultural residues and other materials with low ash melting temperatures which are generally not treated beyond 700 °C. Pyrolysis conducted at temperature between 200 and 300 °C is referred to as torrefaction. The main goal is to retain and concentrate most of the energy content in the solid, and significantly improve mechanical properties of the biomass (such as grindability), which may otherwise be limiting for some applications (Weber and Quicker 2018).

Various types of biomass feedstocks have been used to produce biochar, resulting in a wide range of physicochemical properties. To better evaluate these differences, biochar can be classified into six categories based on the dominant biomass source: lignin-rich biochar derived from wood and tree components (Grottola et al. 2022); cellulose-rich biochar produced from grasses (Reza et al. 2020), straws (Muhammad et al. 2017); nut/shell biochar (Moreira et al. 2017) sourced from husks and shells; manure/waste biochar generated from manures (Tan et al. 2023), biosolids, or green waste; algae biochar obtained from freshwater or marine algae; and black carbon, which includes biochar from miscellaneous biomass sources or naturally formed carbon (Aller 2016). Table 16.1 presents various types of biochar produced from different feedstocks under varying operating conditions.

16.2.1 Mechanisms Involved in Waste Water Remediation Using Biochar

Biochar removes contaminants from wastewater through multiple mechanisms, including physical adsorption, ion exchange, electrostatic interactions, surface complexation, and precipitation (Fig. 16.1). These mechanisms are discussed as follows.

Ion exchange is a reversible process in which ions present in wastewater are exchanged with those bound to the surface of biochar. This interaction is primarily driven by the functional groups on the biochar surface and the specific characteristics of the target contaminants (Rizwan et al. 2016). Biochar swaps surface- bound ions with dissolved ions in solution, similar to materials like zeolites and clays, replacing surface- bound ions dissolved ions in aqueous phase. The process involves an initial rapid surface adsorption followed by slower diffusion into the pores. Ion exchangers can be classified as cationic, anionic, or amphoteric. The efficiency of this ion exchange is influenced by several factors like ion charge, hydration radius, and bonding affinity with the biochar surface (Odega et al. 2023; Kumi et al. 2020).

The adsorption of pollutants onto biochar surfaces is largely governed by physical interactions such as hydrogen bonding, pore filling, and π–π stacking. The structural properties of biochar particularly surface area, porosity, and degree of aromaticity

Table 16.1 Preparation of different biochar used for water and wastewater treatment

Name/Type of biochar	Preparation method and operational conditions	Uses in pollutant removal from wastewater	Reference
Empty fruit bunch biochar (EFBB)	The Fe-coated material was used at a pH of 6 with a contact time of 24 h	As (III) and As (V) removal	Sadegh-Zadeh and Seh-Bardan (2013)
Rice husk biochar (RHB)			
Bismuth impregnated biochar (BiBC500)	Wheat straw was used as the raw material and treated with a bismuth oxide (Bi_2O_3) solution for impregnation. Pyrolysis was carried out at 500 °C under a nitrogen (inert) atmosphere with a heating rate of 10 °C per minute and a holding time of 60 min. After pyrolysis, the product was washed with $NaHCO_3$ and distilled water as a post-treatment step	Removal of arsenic (As III), phosphorus (P), and chromium (Cr VI)	Zhu et al. (2016)
Magnesium oxide modified biochar (MgO-AC)	*Camellia oleifera* shells were used as the raw material, washed thoroughly, and dried at 90 °C for 4 h. The dried shells were then soaked in 2.5 M nitric acid (HNO_3) for acid treatment. For magnesium loading, the material was soaked in a magnesium nitrate ($Mg(NO_3)_2$) solution with a concentration ranging from 0.1 to 0.4 M. Heating and activation were performed using microwave roasting at 800 °C for 40 min under a nitrogen gas atmosphere. Magnesium oxide (MgO) was doped at a ratio of 1/20th of the activated carbon mass	Cadmium ($Cd2^+$) removal	Xu et al. (2022)

(continued)

Table 16.1 (continued)

Name/Type of biochar	Preparation method and operational conditions	Uses in pollutant removal from wastewater	Reference
Hydrogel-biochar composite (HBC-RH)	Rice husk biochar (RHB) was used as the biochar source and incorporated into a poly(acrylamide) (AAm) hydrogel matrix. N, N′-methylenebisacrylamide (MBA) was used as the crosslinker in varying amounts (0.001, 0.01, and 0.1 g), and ammonium persulfate (APS) served as the initiator. The hydrogel was synthesized via *in-situ* polymerization at 40 °C. After polymerization, the product was dried, washed, and stored	Arsenic (As) removal	Sanyang et al. (2014)
Tobacco petiole biochar	Pyrolysis was carried out at a temperature of 700 °C for 0.5 h	Removal of chromium (Cr)	Zhang et al. (2018)
Pinewood biochar	Pyrolysis was conducted at a temperature of 400 °C for 0.5 h		Sforza et al. (2020)
Corn stalk biochar	Pyrolysis was performed at a temperature of 800 °C for 2 h	Removal of lead (Pb)	Yang et al. (2018)
Rice husk biochar	The feedstock was first dried at 103 °C for 24 h and milled to a particle size of less than 0.5 mm. Pyrolysis was carried out at a temperature of 300 °C under a nitrogen atmosphere with a flow rate of 200 mL/min. The heating rate was maintained at 17 °C per minute, and the residence time was 60 min. In a separate set of conditions, pyrolysis was conducted over a temperature range of 300–600 °C for durations ranging from 1.5 to 6 h	Removal of As (V)	
Municipal solid waste biochar			
Removal of As(V), Cr (III), and Cr (VI)	Agrafioti et al. (2014)		
Sewage sludge biochar		Removal of As(V), Cr(VI)	

(continued)

Table 16.1 (continued)

Name/Type of biochar	Preparation method and operational conditions	Uses in pollutant removal from wastewater	Reference
Acacia bark		BTEX (benzene, toluene, ethylbenzene, xylene), MTBE, ETBE	
Eucalyptus (BEu-2)	High-temp pyrolysis (> 400 °C)	BTEX	Saiz-Rubio et al. (2019)
Corncob (BCc-2)	High-temp pyrolysis (>400°C)	BTEX, FO	
Chicken manure (BCm)	Low-temp pyrolysis (≤400 °C)	MTBE, benzene	
Activated rice husk biochar	Rice husk was carbonized by heating to 100 °C at a rate of 7 °C/min and held for 6–8 min, followed by heating to 550 °C at 37 °C/min for 30 min. Activation was performed using steam (H_2O) at 750 °C with a biochar-to-water mass ratio of 3:1 under a nitrogen atmosphere, held for 60 min. The resulting activated biochar had a surface area of 351.23 m^2/g and a pH of 9.7	Adsorption of n-hexane	Li et al. (2019)
Cobalt-doped coconut husk biochar (Co@CHBc)	The material was first washed with water and then dried at 100 °C for 48 h. It was then soaked in a cobalt solution at a 10:1 ratio for 24 h for metal doping. After doping, the material underwent a second drying step at 120 °C for 48 h. Pyrolysis was carried out at 600 °C for 2 h. Following pyrolysis, the product was washed with water to remove ash and finally dried again for 48 h	Methyl Orange (MO) dye (anionic azo dye)	Khajuria et al. (2024)
Wheat straw biochar	Wheat straw was pyrolyzed at 300 °C, 500 °C, and 700 °C under oxygen-limited conditions for typically 1–4 h	Polycyclic aromatic hydrocarbons	Cao et al. (2016)
Wood biochar	Biochar biomass pyrolyzed at 600–700 °C (O_2-limited), 300–500 m^2/g surface area, pH ~ 9–10	Organic micropollutants	Thompson et al. (2016)

(continued)

Table 16.1 (continued)

Name/Type of biochar	Preparation method and operational conditions	Uses in pollutant removal from wastewater	Reference
Sulfonated bagasse biochar	Sulfonated sugarcane bagasse biochar (pyrolyzed at 500–700 °C) was functionalized with alginate and ZIF-8 (Zeolitic imidazolate framework-8), yielding a high-surface-area nanocomposite optimized for Hg (II) removal at pH 5–7	Mercury (Hg(II))	Elamin et al. (2025)
Olive pomace biochar	Olive pomace biochar (pyrolyzed at 400–600 °C, moderate-high surface area) and raw olive pits were compared in packed-bed column filtration systems for wastewater treatment	Total phenolic compounds olive mill wastewater effluent	Ezzahi et al. (2025)
Magnetic potato residue biochar	Magnetic potato residue biochar (pyrolyzed at 500–700 °C) was modified with Fe_3O_4 for separation and MoS_2 for enhanced adsorption/catalysis, yielding a high- surface-area composite with magnetic recovery capability	Rhodamine B	Guo et al. (2025)
Silver-doped neem stem biochar	Silver-doped neem stem biochar (pyrolyzed at 400–600 °C) was chemically impregnated with Ag nanoparticles, enhancing its surface area and functional groups (oxygen-containing + Ag)	Hexavalent chromium (Cr (VI))	Baskaran (2025)
Nano-iron-modified biochar fiber-textured composite	Biomass-derived fiber-textured biochar was synthesized via controlled pyrolysis with preserved natural fiber structure, then modified with green nano-iron (Fe^0/Fe oxides) to create a porous composite	Hexavalent chromium (Cr (VI))	Ma et al. (2025)
Nickel-loaded crayfish shells biochar	Crayfish shell biochar (pyrolyzed) was modified with nickel nanoparticles, leveraging its inherent calcium content and Ni catalytic sites for enhanced anaerobic digestion of saline wastewater	Anaerobic digestion of high-salinity organic wastewater	Lu et al. (2025)

(continued)

Table 16.1 (continued)

Name/Type of biochar	Preparation method and operational conditions	Uses in pollutant removal from wastewater	Reference
Sewage sludge biochar	Sewage sludge biochar produced via pyrolysis (300–800 °C, oxygen-limited) or gasification (600–1000 °C, partial oxidation), optionally activated (steam/CO_2/acids), with high ash content (20–50%) from inorganics	Heavy metal immobilization (e.g., Cd, Pb); Organic pollutant adsorption (e.g., dyes, BPA); phosphorus recovery	Hu and Chen (2025)
reed biochar	waste-derived biochar (pyrolyzed at 400–700 °C from agricultural/industrial organics) was combined with mineral wastes (slag/clay) to create optimized constructed wetland substrates	nutrients (N, P); organic matter; pathogens	Sun et al. (2025)
Coconut shell biochar	Coconut shell biochar (pyrolyzed at 500–800 °C, activated for conductivity) served as particle electrodes in a 3D electrocatalytic oxidation system	Antibiotics (e.g., sulfonamides, tetracyclines)	Xie et al. (2024)
Rice husk biochar	Rice husk biochar produced via rapid microwave-assisted pyrolysis (400–600 °C)	Phosphate and nitrate	Shukla et al. (2019)
Banana peel biochar	Banana peel biochar produced via conventional pyrolysis (400–700 °C) exhibits moderate-high surface area and oxygen-rich functional groups (carboxyl/hydroxyl)	Nitrofurantoin (antibiotic)	Bernardino et al. (2024)
Banana waste biochar	Banana waste biochar (500–700 °C pyrolysis, 120–350 m^2/g surface area). Electroplating wastewater via ion exchange and complexation. At pH 4.5–6.0	Cu^{2+}, Ni^{2+}, Zn^{2+}, Cr^{3+}	Arriola-Villaseñor et al. (2023)
Sewage sludge biochar	Municipal sewage sludge biochar produced via slow pyrolysis (700 °C, 10 °C/min, 1 h residence) yielded a 215 m^2/g surface area material with 48.3% ash content	Carbamazepine (pharmaceutical); bisphenol A (BPA); triclosan (antimicrobial)	Regkouzas and Diamadopoulos (2019)

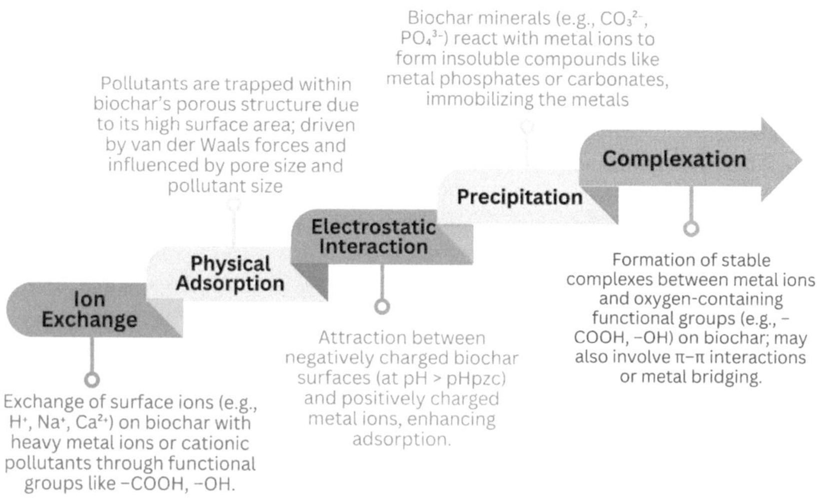

Fig. 16.1 Different mechanisms of wastewater remediation using biochar

are critical in facilitating these interactions. A higher surface area and pore volume improve the diffusion and retention of contaminants, while the aromatic domains promote π–π stacking interactions and hydrogen bond formation with various pollutants. Post-synthesis modifications of biochar can further enhance these physical interactions by improving pore accessibility and strengthening hydrogen bonding or multi-point adsorption mechanisms. In addition to these mechanisms, electrostatic interactions between oppositely charged molecules and biochar surfaces contribute to the overall adsorption capacity through non-covalent binding process (Pugazhendhi 2024).

Electrostatic interaction is a fundamental mechanism governing the adsorption of ionic organic contaminants onto biochar in aqueous systems (Zheng et al. 2013). These interactions occur when oppositely charged species attract—positively charged (cationic) pollutants bind to negatively charged functional groups (e.g., carboxyl, hydroxyl) on the biochar surface, while negatively charged (anionic) pollutants are drawn to positively charged sites, often associated with surface-modified or doped biochar (Qiu et al. 2009; Pandey et al. 2025).

Precipitation involves the formation of solid phases either within the solution or on the surface of the adsorbent, and it is widely recognized as a key mechanism for the immobilization of heavy metals by biochar. This process is strongly dependent on solution pH, which plays a critical role in modifying the surface charge, ionization state, and metal speciation during adsorption. Biochar is rich in surface functional groups, particularly oxygen-containing moieties such as hydroxyl (–OH) and carboxyl (–COOH) groups which undergo protonation or deprotonation in response to changes in pH variation. At lower pH levels, biochar surfaces tend to be protonated, leading to a net positive surface charge that can hinder metal cation interaction. As pH increases, these functional groups deprotonate, providing negatively charged sites that facilitate metal complexation and precipitation. Additionally, precipitation may occur through the formation of insoluble metal hydroxides or carbonates, especially at higher pH levels, there by contributing to the long-term stabilization of metals on the biochar surface. Furthermore, the pyrolysis temperature significantly influences the physicochemical properties of biochar, including its morphology, elemental composition, and the abundance of reactive surface sites, all of which affect its capacity to induce precipitation reactions (Kołodyńska et al. 2012; Zhang et al. 2013; Patra et al. 2017).

The removal and stabilization of pollutants by biochar are facilitated through both outer-sphere and inner-sphere complexation processes. Outer-sphere complexation involves physical adsorption, where ions are attracted to the biochar surface through electrostatic forces without forming direct chemical bonds. In contrast, inner-sphere complexation involves the formation of coordination complexes between pollutant ions and specific functional groups on the biochar surface, resulting in more stable, multi-atom interactions. Oxygen-containing functional groups—such as carboxyl, phenolic, and lactonic groups—are particularly important in forming strong binding interactions with various metal and inorganic ions. The presence and abundance of these functional groups can increase over time due to surface oxidation either during pyrolysis or environmental aging. This oxidation leads to the formation of additional reactive sites, especially carboxylic groups, which enhance the metal-binding capacity of biochar. Consequently, the pollutant removal efficiency of biochar may improve with time, increasing its effectiveness in long-term environmental applications (Xia et al. 2020; Li et al. 2020).

16.3 Application of Biochar in Wastewater Clean-Up

16.3.1 Removal of Heavy Metals

Heavy metal in the environment poses a severe threat to the ecosystem and public health owing to their toxicity, persistency and tendency to bioaccumulate. The major source of heavy metal contamination includes mining and industrial operations, untreated industrial discharge and agricultural runoff. Once released into the environment they enter the food chain and accumulate across the trophic and severe damage to the organisms (Ahmad et al. 2012). However, conventional treatment techniques like ion exchange, chemical precipitation, electrochemical method and reverse osmosis are persuaded but are often subjected to sludge generation, elevated operational costs and certain inefficiencies (Tan et al. 2015). In this context, carbon-rich biochar stemmed from pyrolysis of biomass has emerged as an affordable, sustainable and economical adsorbent for heavy metal remediation.

Biochar is produced through thermochemical conversion of organic biomass with no or limited oxygen. The commonly deployed methods include hydrothermal carbonization and slow pyrolysis. Hydrothermal carbonization is achieved in a closed aqueous environment by thermal treatment at 180–230 °C. The process generates increased surface functional groups like carbonyl and hydroxyl groups, enriching adsorption affinity for metal ions. In contrast, slow pyrolysis is achieved at 300–700 °C, delivering a porous solid surface with substantial surface area and stability. The physiochemical properties, such as surface functional groups, surface area, pore size, pH, ash content, aromaticity and cation exchange capacity, differ with parameters such as biomass used, pyrolysis temperature and pre/post treatment methods. In particular, biochar obtained from manure possesses high mineral content and basicity; however, wood wood-derived biochar demonstrates higher porosity and surface area, promoting physical adsorption (Xu et al. 2013b).

Removal of heavy metals by biochar happens through a combination of physicochemical adsorption mechanisms in a synergistic mode. Biochar encompasses negatively charged functional groups like COOH and OH, which functionally reciprocate their surface-bound cations (H^+, Na^+, K^+). The process is still effective for divalent metal ions like Pb^{2+} and Cd^{2+} (Tan et al. 2015). As pH influences the surface charge of biochar and speciation of metal ions, the process of adsorption is pH-dependent. When the pH levels rise above the biochar's point of recharge, functional groups like carbonyl, hydroxyl and phenolic moieties tend to deprotonate, resulting in a net negative charge. The subsequent negative surface charge amplifies electrostatic attraction and adsorption of positively charged metal ions. While rising pH increases the adsorption efficiency, beyond a certain pH threshold, metal ions undergo hydrolysis, resulting in metal precipitation and settling down as insoluble hydroxides (Lu et al. 2012).

Surface complexation is one of the dominant mechanisms in biochar's immobilization of heavy metals. The process of surface complexation is enabled through the formation of inner-sphere complexes. It is achieved through direct coordination between the donor atoms and the metal ions, leaving out water molecules in the coordination shell. The above mechanism facilitates long-term sequestration and stabilization of heavy metals, but is poorly reversible or irreversible due to the strong covalent-like interaction. The surface complexation efficiency is influenced by pH, pyrolysis temperature and feedstock type. By this process, the divalent metals like Pb^{2+}, Cu^{2+}, Zn^{2+}, and Cd^{2+} chemically bind to the oxygen-rich functional groups like carboxyl, hydroxyl and phenolic moieties (Kim et al. 2013a; b). These functional moieties emerge from the partial oxidation of lignocellulosic biomass during pyrolysis. Another significant pathway of heavy metal removal through biochar in aqueous systems is precipitation. This pathway is significant in alkaline biochar, which has higher ash content containing metal oxides, phosphates and carbonates. At pH levels above 7.5, heavy metal ions undergo hydrolysis and form insoluble metal hydroxides and carbonates by reacting with hydroxide and carbonates present in the biochar matrix. The precipitation process is generally governed by ash composition influenced by hydroxide and carbonate, solution pH, contact time and agitation (Xu et al. 2013b). Beyond metal removal, precipitation contributes to stabilization, as the precipitates are least bioavailable and less prone to leaching under stable environmental conditions.

In treatment systems where biochar is preferred for pre-treatment or short-term bioremediation, physical adsorption or physisorption plays a crucial role as a initial mechanism of for metal sequestration before long-term intervention. As opposed to chemisorption or precipitation, physisorption does not involve the formation of strong chemical bonds. Therefore, they are non-selective, reversible and highly influenced by surface area, concentration and pore size distribution of the biochar. Metal ions loosely bound onto the biochar surface facilitated through weak intermolecular forces like Van der Waals interaction which enables rapid assimilation. Albeit they are individual interactions are weak, while their cumulative effect can be substantial when under high metal concentrations or when the biochar porosity is increased through activation techniques. Most likely, physisorption commonly occurs alongside other mechanisms such as ion exchange or surface complexation (Ahmad et al. 2012).

Complementary to surface complexation and precipitation, another process known as π-metal interactions serve as an important complementary mechanism, particularly in biochar enriched with aromatic carbon structures. Biochar produced at higher pyrolysis temperature, tends to form delocalized π-electron systems which, facilitate non-covalent interactions with transition metals. The interplay happens through the formation of π-electron donation or π complex formation, in particular with transition metals having vacant d orbitals (eg. Fe^{3+}. Cu^{2+}, Ni^{2+}) can accept electron density. As a result of such π-metal which enhances stability and binding affinity of metals on the biochar matrix (Tan et al. 2015). The various types and applications of biochar for heavy removal from wastewater are summarized in Table 16.2.

16.3.2 Removal of Volatile Organic Compounds (VOCs)

VOCs are carbon-based chemicals that tend to evaporate at room temperature, thereby contributing to the indoor and outdoor environments. Given their harmful effects on the environment and human health, including carcinogenicity, mutagenicity and their role in the formation of ground-level ozone and secondary organic aerosols its effective remediation is crucial. Traditional VOC remediation methods like adsorption, photocatalysis and catalytic oxidation have their limitations with respect to efficiency, environmental impact, cost effectiveness, etc.

Recently biochar a carbon-rich porous material obtained from the pyrolysis has attracted attention as an promising alternative for VOCs removal. The efficacy of biochar in VOC abatement is dictated by multifaceted interactions between the physicochemical characteristics of the biochar surface and the VOC of the biochar.

Table 16.2 Removal of Heavy metals using different Biochars

Metal	Type of biochar used	Operational conditions	Initial concentration (mg/L)	Adsorbed amount (mg/g)	Reference
Arsenic	Empty fruit bunch	pH–8 Temperature-25 ± 1 °C Sorbent concentration-5 g/L^{-1}	3–300	18.9	Sadegh-Zadeh and Seh-Bardan (2013)
Arsenic	Rice husk	pH–8 Temperature-25 ± 1 °C	3–300	19.3	Sadegh-Zadeh and Seh-Bardan (2013)
Arsenic	Rice husk	pH–6	1–150	28.32	Sanyang et al. (2016)
Arsenic	Brown seaweed	pH–6 Contact time-90 min	0.25	1.64	Devrajani et al. (2024)
Arsenic	Scallion sheathing	pH–7 Temperature-303 K	0.5	104.55	Zhu et al. (2022)
Mercury	Iron-modified rice husk	pH–7.34–8.15 Contact time-60 min	1	36.50	Tichem et al. (2023)
Mercury	Chlorinated rice husk	pH–2–2.5 Contact time-48 h	0.01–20,000	620	Mochizuki et al. (2021)
Mercury	Sugarcane bagasse	pH–6	1	13.55	Giraldo et al. (2020)
Mercury	Wheat straw	pH–6 Biochar dosage-0.03 g/ 50 mL	2	5.85	Zhang et al. (2018)
Copper	Waste rapeseed cake	pH–5 Contact time-240 min	100*	52.2	Mazurek et al. (2021)
Copper	Sawdust	pH–5.0 Contact time-90 min	100	17.01	Yang and Jiang. (2014)
Copper	Oil palm frond	pH–3.0–4.5 Contact time-10–100 min	10–50	63.45	Shitu et al. (2024)
Copper	Pistachio green hulls	pH–3, 3.5, 4, and 5	70–270	19.84	Jalayeri and Pepe (2019)
Zinc	Rice husk	pH–4–10 Contact time-48 h	10 to 150	35.75	Sanyang et al. (2014)
Zinc	Paper mill sludge	pH–5 Contact time-24 h Adsorbent dosage-0.3 g/L	30	242.50	Xu et al. (2021)
Zinc	Rice straw	pH–5 Adsorbent dosage-10 mg/8 mL		39.7	Zhao et al. (2020)
Zinc	Sewage sludge	pH–5 Adsorbent dosage-10 mg/8 mL		4.25	Zhao et al. (2020)

In the case of non-polar VOCs, physical adsorption is dominant, which is facilited by high surface area and extensive network of micropore and mesopore which facilitate sites for entrapment.

In case aromatic VOCs like toluene, benzene and xylene, the presence of delocalized π-electrons in both biochar and VOCs enhance the adsorption affinity. The aromaticity in biochar is materialized only under low pyrolysis temperature. In these interactions, Van der Waals forces play a crucial role particularly in non-polar VOCs and their strength is modulated by the hydrophobic nature of the biochar surface.

In contrast polar VOCs containing functional groups like carboxyl, carbonyl and hydroxyl groups, through acid–base reaction or electron transfer which are essential for the capture of oxygen-containing VOCs. The modified biochar doped with mineral or metal oxide supports facilitates oxidative degradation through Lewis acid sites or redox cycling. In Fe-modified biochar, promote redox-mediated pathways for the effective removal of compounds such as formaldehyde and acetaldehyde.

The removal efficiency of VOC using biochar differs significantly depending on the VOC the instances property of biochar properties and their treatment conditions applied. Many studies have reported removal efficiency ranging from 60 to 99% proving the promising ability of biochar as a sustainable remediation material. For instance, Mg–Al modified biochar has reported over 95% removal efficiency for formaldehyde and acetone, linked to higher surface basicity and reactive metal centres. Similarly, chemically modified rice husk-derived biochar accounted for 91% removal efficiency for a VOC mixture in a continuous flow reactor, demonstrating scalability (Wu et al. 2020). Moreover, activated biochar, like sulphuric acid-treated walnut shell biochar, displayed promising adsorption capacities up to 166.72 mg/g for acetone and toluene. The acid treatment significantly improved pore formation, surface polarity and density of oxygenated functional groups (David, 2023).

The removal efficiency of VOCs by biochar is also strongly affected by experimental and environmental conditions particularly temperature. Generally elevated temperatures tend to reduce physisorption owing to the increased kinetic energy of VOC molecules which promotes desorption. However, in the case of catalytically active biochar, moderate heating of 100–150 °C enhances VOCs degradation by activating catalytic sites. So, the pyrolytic temperature during the biochar production plays a crucial role in determining its surface chemistry and adsorption behaviour. At lower temperature regimes (300–400 °C), biochar retains more oxygenated functional moieties, enhancing adsorption of polar VOCs, whereas higher temperature (500–700 °C) stimulates hydrophobicity and aromaticity and surface area, triggering higher adsorption for non-polar VOCs via Van der Waals and π–π interactions.

The adsorption efficacy of VOCs by biochar is influenced by a range of factors such as pH, flow dynamics, biomass source and regeneration strategies. The pH of the environment affects biochar surface chemistry and the speciation of VOCs. Under acidic condition the adsorption of basic VOC adsorption is favoured due to electrostatic attraction whereas in neutral conditions generally support efficient adsorption across a wide range of compounds.

In packed-bed reactor systems, increased airflow rates can reduce adsorption efficiency, impairing the removal efficiency. In dynamic packed column studies, higher feed concentration compromises the long- term adsorption efficiency due to the saturation of active sites. Biochar derived from higher lignocellulosic biomass tend to possess a well-developed pore structure and a stable aromatic framework, making them suitable for hydrophobic VOCs. In case of regeneration and reusability, biochar may experience saturation and surface fouling and impact the adsorption efficiency. Therefore, various techniques like thermal regeneration, mechanical modification and chemical regeneration can be tailored to restore and enhance functionalization. Thermal degeneration though mild heating (100–150 °C) can restore adsorption capacity by 80–90% by desorbing the physically bound VOCS. However, prolonged or excessive exposure to high temperatures may degrade surface properties and impact regeneration. Alternatively, chemical regeneration using solvents like NaOH or ethanol desorbs polar VOCs, while mechanical methods like ball milling may also integrate structural integrity and resist fouling. Zhang et al. (2021) proved that biochar can retain high adsorption efficiency even after five regeneration cycles indicating strong reusability potential. Despite these advantages, humidity interference, potential leaching, higher cost and scalability may pose challenges in VOC removal with biochar. Various applications of different types of biochar for VOCs removal from wastewater are summarized in Table 16.3.

16.3.3 Removal of Nutrients

Nutrient pollution poses a significant challenge to water quality and ecosystem healthprimarily driven by excessive presence of nitrogen and phosphorus. These nutrients are introduced into aquatic environment through wastewater discharge, storm water, industrial effluents and agricultural runoff (Nie et al. 2018). Carbon-rich biochar is often regarded as a potential alternative for nutrient adsorption due to its high porosity, abundance surface functional groups and considerable cation exchange capacity. Like other adsorbents widely used, biochar removes nutrients through mechanisms like ion exchange, adsorption and surface complexation (Vasseghian et al. 2024). For phosphorus removal, key mechanism involves precipitation reactions with calcium, iron, magnesium or ligand exchange process. In the case of Nitrogen species like ammonium are is retained through by cation exchange, while nitrate adsorption will be less effective unless biochar is modified with layered double hydroxides or metal oxides to enhance anion retention. The performance of biochar in nutrient removals depends on factors such as feedstock type, pyrolysis temperature, surface structure, pH, and ion exchange capacity. Additionally, water chemistry, particularly pH and competing ions, plays a crucial role in determining P adsorption efficiency (Nobaharan et al. 2021).

Biochar has emerged as a promising material for the removal of N and P from wastewater due to its porous structure, large surface area, and abundant functional groups that enhance adsorption. Wu et al. (2024) highlighted its potential to mitigate eutrophication and nutrient pollution, providing a comprehensive overview of its preparation, modification, adsorption mechanisms, and reusability. In the context of

Table 16.3 Removal of VOCs using different types of biochar

VOCs	Type of biochar used	Operational conditions	Initial concentration (mg/L)	Adsorbed amount	Reference
PAHs and petroleum-based compounds	UK sludge cake		10	46.8% to 65.4%	Sullivan et al. (2019)
Phenolic compounds			10	78.7% to 88.8%	Sullivan et al. (2019)
BTEX (Benzene, toluene, ethylbenzene, xylenes) and Fuel oxygenates (MTBE, ETBE)	Eucalyptus (BEu-2), corncob biochar (BCc-2), acacia bark biochar (BAcB), chicken manure biochar (BCm)	Contact time: 36 h	2–600	BEu-2 (toluene: 6.84 g/kg; ethylbenzene: 6.97 g/kg; mp-xylene: 15.18 g/kg; o-xylene: 5.62 g/kg); BCc-2 (toluene: 3.86 g/kg; ethylbenzene: 3.25 g/kg; mp-xylene: 7.63 g/kg; o-xylene: 3.32 g/kg)	Saiz-Rubio et al. (2019)
p-Cresol	Oak wood biochar (OWB)	Temp: 25 °C contact time: 24 h pH: neutral	~ 50	2.88 mg/g	Lee et al. (2023)
	Switchgrass biochar(SGB)		~ 50	3.71 mg/g	
	Pine chip biochar (PCB)		~ 50	2.64 mg/g	
Dibenzothiophene, 3-pentylthiophene, hexanethiol	Iron-rich biochar	Temperature: 150 °C Contact Time: 24 h		76%	Mousavi et al. (2023)
Toluene, acetone	Non-activated biochar from Rapeseed Cake (RSC)	Temp: 20 °C Adsorption time: 15 min carrier gas: N2 at 50 mL/min desorption temperature: 20– 150 °C	1000	16.76–26.65 mg/g	David (2023)

(continued)

Table 16.3 (continued)

VOCs	Type of biochar used	Operational conditions	Initial concentration (mg/L)	Adsorbed amount	Reference
Toluene, acetone	KOH-activated biochar from rapeseed cake (RSCK)		1000	51.28–73.46 mg/g	
Toluene, acetone	H2SO4-activated biochar from Rapeseed Cake (RSCS)		1000	153.34–166.72 mg/g	
Acetic acid	Vineyard pruning biochar	Temp: 25 °C Dose: 30 g/L Volume: 100 mL	1136	10.7 mg/g	Martínez et al. (2018)
Dimethyl sulfide	Walnut shell biochar (Ag + impregnated)	Temp: 25 °C Contact time: 24 h Agitation: 150 rpm Dose: 0.1 g Volume: 20 mL	1000	30.2 mg/g	Jalilvand et al. (2020)
Toluene	Municipal solid waste biochar (MSW-BC)	Pyrolysis Temp: 450 °C (30 min hold) pH 8.3 Contact time: up to 24 h Sorbent dose: 1 g/L Temperature: 25 °C, 35 °C, 45 °C	25–600 µg/L	1090 µg/g (25 °C), 876 µg/g (35 °C), 850 µg/g (45 °C)	Jayawardhana et al. (2019)
m-Xylene	Municipal solid waste biochar (MSW-BC)	Pyrolysis Temp: 450 °C (30 min hold) pH: 8.3 Contact time: up to 24 h Sorbent dose: 1 g/L Temperature: 25 °C, 35 °C, 45 °C	25–600 µg/L	555 µg/g (25 °C), 601 µg/g (35 °C), 612 µg/g (45 °C)	Jayawardhana et al. (2019)

(continued)

Table 16.3 (continued)

VOCs	Type of biochar used	Operational conditions	Initial concentration (mg/L)	Adsorbed amount	Reference
n-Hexane	Activated biochar (rice husk, H_2O activated)	Pyrolysis: 550 °C (BC); Activation: 750 °C (AC) N_2 atmosphere mass ratio-BC:H_2O = 3:1 BET surface area: 351.23 m^2/g; pore volume: 0.32 cm^3/g		29.33 mg/g (AC); 5.02 mg/g (BC)	Li et al. (2019)
BTEX, aromatic hydrocarbons, halohydrocarbons	Activated biochar (rice husk, H_2O activated)	Pyrolysis: 550 °C Activation: 750 °C N_2 atmosphere		110.62 μg/g (AC); 93.26 μg/g (BC)	Li et al. (2019)

constructed wetlands, Feng et al. (2020) examined biochar-modified aerated system treating anaerobically- digested swine wastewater. The study reported COD and NH_4-N removal efficiency of 63.06–77.18% and 87.19–96.54%, respectively. However, TN removal was lower (30.92–40.12%) compared to conventional systems (40.83–48.70%), possibly due to limited carbon sources and acidic conditions. Additionally, higher N_2O emissions (185.43–443.33 μg m^{-2} h^{-1}) were observed with biochar under higher influent strengths. Conversely in vertical flow constructed wetlands, Zhou et al. (2017) demonstrated that integrating biochar and intermittent aeration significantly improved removal efficiencies—COD (94.9%), NH_4^+-N (99.1%), and TN (52.7%)—while reducing N_2O emissions to 60.54 μg·m^{-2}·h^{-1}, supporting its suitability for low C/N wastewater. Sun et al. (2018) explored subsurface wastewater infiltration systems and found tthat biochar coupled with intermittent aeration yielded high removal efficiencies for COD (95.4 ± 0.2%), NH_4^+-N (96.2 ± 0.6%), and TN (86.4 ± 0.5%), with minimal N_2O emissions (18.4 mg/m^2·d), attributed to enhanced microbial gene expression related to nitrogen transformation. In a separate study Wang et al. (2024) investigated sludge-derived biochar in sequencing batch reactors (SBRs) treating low C/N municipal wastewater. Biochar produced at 400 °C exhibited optimal surface properties for electron transfer. At a dosage of 4 g/L, its improved COD, NH_4-N, and TN removal to 97.2 ± 1.2%, 99.2 ± 0.6%, and 83.8 ± 2.4%, respectively. Various applications of different types of biochar for nutrient removal from wastewater are summarized in Table 16.4.

16.4 Research Insights on Application of Biochar for Waste Water Treatment

16.4.1 Methodology

To understand the current research landscape, data were collected from the Web of Science Core Collection database (Koley et al. 2024; Gupta et al. 2024) using the advanced search feature interface available at https://www.webofscience.com/wos/woscc/advanced-search (Balagopal et al. 2025). The search was conducted to target article titles using a following structured keyword query: ((((((TI = (Biochar)) OR TI = (Charcoal)) AND TI = (Water)) OR TI = (Waste Water)) AND TI = (Removal)) OR TI = (Remediation)) OR TI = (Adsorption)). In this query, "TI" denotes the title field, thereby restricting the search document to such as journal articles, proceedings papers, books, or book chapters, where the specified term appear in the title. This keyword combination was designed to identify literature focused on the application of biochar or charcoal in the context of water and wastewater treatment, with particularly emphasis on removal, remediation, and adsorption processes. The search returned a total of 169,097 records, which served a foundation for subsequent bibliometric and content analysis.

Table 16.4 Removal of various nutrients using different types of biochar

Nutrients	Type of biochar used	Operational conditions	Initial concentration (mg/L)	Adsorbed amount	Reference
Phosphate	Rice husk biochar (RHC) from microwave pyrolysis	Pyrolysis temperature: 600 °C Microwave power: 900 W pyrolysis time: 15 min adsorption temperature: 25 ± 1 °C contact time: 24 h Biochar dose: 1 g/20 mL	0.5–6	71 mg/kg	Shukla et al. (2019)
nitrate			3–10	497 mg/kg	Shukla et al. (2019)
NH_3-N (Ammonium)	Eucalyptus wandoo (EW) biochar	Pyrolysis temp: 400 °C particle size: 2.36 mm Dosage: 2–10 g/100 mL pH 4–6.5 Contact time: up to 24 h	0.5–5	6.37–60.71 μg/g	Alam and Anwar (2020)
NO_2-N (Nitrite)			0.5–5	6.40–59.62 μg/g	
PO_4-P (Phosphate)	Alum sludge	Dosage: 10 g/100 mL pH ~ 6 Contact time: ≤ 10 min	0.5–5	25.40–249.85 μg/g	
NH_3-N, NO_2-N, PO_4-P (Simultaneous)	Mixture of EW biochar and alum sludge	Ratio: 8 g biochar: 2 g alum sludge/100 mL pH 4–8 Contact time: up to 6 h (NH_3-N, NO_2-N), ≤ 10 min (PO_4-P)	0.5–5	NH_3-N: 98.2% removal NO_2-N: 99.4% removal PO_4-P: 99.8% removal	
Phosphate (PO_4^{3-})	Pig manure biochar (PM 400 + PM 700 mixed as PM 4–7)	Pyrolysis at 400 °C and 700 °C Dosage-0.5 g pH 8 Contact time-12 h Temp-25–45 °C	5–100 mg/L	1.355 mg/g at 45 °C	Ji et al. (2023)
Ammonium (NH_4^+)			5–100 mg/L	2.478 mg/g at 45 °C	
Phosphorus	Mineral-enriched hemp biochar	Biochar porosity: 0.45 pH: 7.3 (biochar wetland) Contact period: 7 months	7.07–29.87	Removal efficiency: up to 97%	Bolton et al. (2019)

(continued)

Table 16.4 (continued)

Nutrients	Type of biochar used	Operational conditions	Initial concentration (mg/L)	Adsorbed amount	Reference
	$ZnCl_2$-activated pinewood biochar (BC–Zn)	Biochar dose: 0.5 g/50 mL pH: 9 Contact time: 24 h Temperature: 25 °C Agitation: 150 rpm	5–200	10.4 mg/g	Biswas et al. (2023)
	$ZnCl_2$-activated pinewood biochar		1.3	Removal efficiency:100%	
Phosphorus	(BC–Zn)–real wastewater				
	BC–KOH, BC–HNO_3, BC–H_2O_2, BC–Ca, BC–Ca/Zn (various modified biochars)		10	Removal efficiency: 36%	
Phosphorus	Mg-modified biochar (sugarcane straw + chicken manure	pH ~ 2	1282.14	129.79 mg/g	Fang et al. (2020)
	Fe-modified biochar (Thalia dealbata)	pH:10 Temp:25 °C Biochar dosage:12.5 g/L	40	15.86 mg/g	Xiao et al. (2019)
	Calcium-rich biochar (sedimentation sludge)	pH 4.5 Temp: 22 °C Biochar dosage: 10 g/L	1500	147.06 mg/g	Antunes et al. (2018)
	Pristine biochar (sedimentation sludge)	pH 4.0 Temp: 22 °C Biochar dosage: 10 g/L	1500	14.08 mg/g	
	Lanthanum-doped hydrochar (rice straw)	pH 4.5 Temp: 25 °C Biochar dosage: 1 g/L	40	61.57 mg/g	Dai et al. (2014)

(continued)

Table 16.4 (continued)

Nutrients	Type of biochar used	Operational conditions	Initial concentration (mg/L)	Adsorbed amount	Reference
	Core–shell γ-Al_2O_3/Fe_3O_4 biochar (C. korshinskii)	pH 5.0 Temp: 25 °C Biochar dosage: 1 g/L	50	205.70 mg/g	Cui et al. (2020)
	Iron-rich sludge biochar	pH 5.5 Temp: 25 °C Biochar dosage: 15 g/L	46	1.84 mg/g	Wang et al. (2020b)
	Dolomite-carbon composite (poplar tree sawdust)	pH 5.83 Temp:25 °C Biochar dosage: 1.25 g/L	25	207.00 mg/g	Li et al. (2018a; b)
	Coal gangue-biochar composite (oilseed rape straw)	pH 4.0 Temp:25 °C Biochar dosage: 12.5 g/L	50	7.90 mg/g	Wang et al. (2021)
	Pristine biochar (corn cob)	pH 2.8 Temp: 22 °C Biochar dosage: 16.7 g/L	100	6.50 mg/g	Eduah et al. (2020)
	Iron-modified biochar (waste activated sludge)	pH 2.0 Temp:22 °C Biochar dosage: 2 g/L	20	110.00 mg/g	Yang et al. (2018)
	Calcium-alginate beads, granular biochar (L. Japonica)	pH 6.0 Temp: 30 °C Biochar dosage: 1 g/L	1000	157.70 mg/g	Jung et al. (2016)

Microsoft Excel was employed to analyze the predicted values using the Exponential Triple Smoothing method, specifically Holt's Linear Trend model (Alpha = 0.5, Beta = 0.5, Gamma = 0, with 95% confidence interval). The time series data spanned from 1989 to 2035, with actual values recorded up to 2024 and forecasted values generated from 2025 to 2035. R Studio is used for plotting and visualising this dataset ensuring precision in graphical presentation. Additionally, a pie chart and country-wise map were prepared using OriginPro 2024 (Version: 10.1.0.178) and Microsoft Excel to illustrate bibliometric patterns and geographical trends.

16.4.2 Year-Wise Publication and Forecast

The year-wise publication data were retrieved from the WoS database, encompassing 169,097 documents, shows a strong upward trend in research output related to biochar, water treatment, adsorption, and remediation from 1989 to 2023. Beginning with 1624 publications in 1989, the number of annual publication gradually increased, surpassing 4000 by 2009 and accelerating significantly in the last decade, reaching 10,342 publications by 2023. This rise reflects growing global attention toward sustainable and low-cost environmental technologies. Forecasting from 2025 to 2035 based on Holt's Linear Trend Model suggests that this trend will continue, with the number of publications expected to increase each year. In 2025, the predicted publication count is 11,843 ± 652, rising to 12,289 ± 922 in 2026, 12,736 ± 1343 in 2027, 13,182 ± 1871 in 2028, and 13,629 ± 2489 in 2029. By 2030, the number the forecast is projected to reach 14,075 ± 3156, followed by 14,522 ± 3892 in 2031, 14,968 ± 4682 in 2032, 15,414 ± 5523 in 2033, 15,861 ± 6410 in 2034, and 16,307 ± 7171 in 2035. These projections indicate not only continued growth but also increasing variability and uncertainty in future research output, which may reflect the global research priorities, evolving environmental policies, and technological innovation in the field (Fig. 16.2).

16.4.3 Country-Wise Contribution of Biochar Research

China leads overwhelmingly with 52,545 publications, showing its dominant role and strong research focus in the field of biochar and wastewater treatment. The USA follows distantly with 26,669 publications, also demonstrating significant involvement. India (9922) and Japan (9284) also demonstrate robust research contributions from Asia, reflecting the region's growing environmental concerns and research investment. European countries such as Germany (7845), France (6505), England (5454), and Russia (4805) also show substantial research output, indicating active academic interest in sustainable water treatment methods. Countries like Iran (5279), South Korea (5442), Saudi Arabia (3587), and Brazil (3785) show emerging regional leadership, likely driven by environmental challenges and increasing wastewater

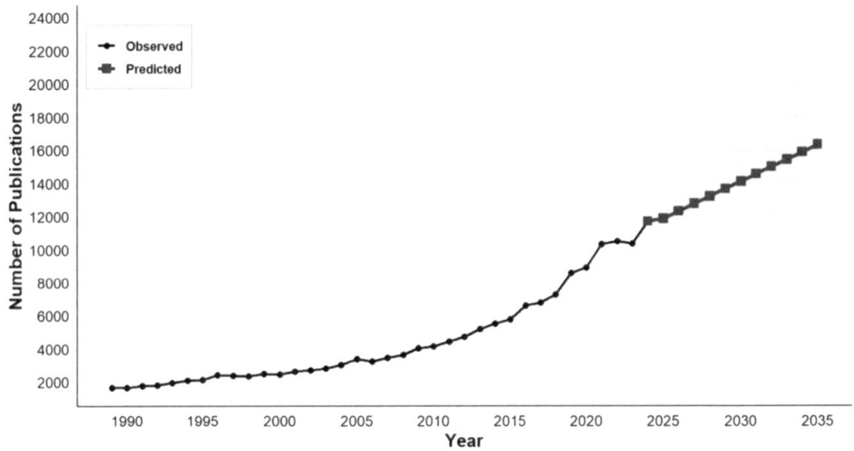

Fig. 16.2 Number of publications with forecast

loads. Other nations, including Poland, Turkey, Malaysia, and Egypt, show moderate levels of research activity, suggesting growing interest. Even small countries like Ukraine (1123), Portugal (1312), and Finland (893) are contributing to the global biochar research landscape (Fig. 16.3).

Countries are increasingly adopting biochar for wastewater treatment due to its low cost, environmental benefits, and alignment with global sustainability goals. Nations like China, India, and Brazil face high pollution and waste generation, making biochar

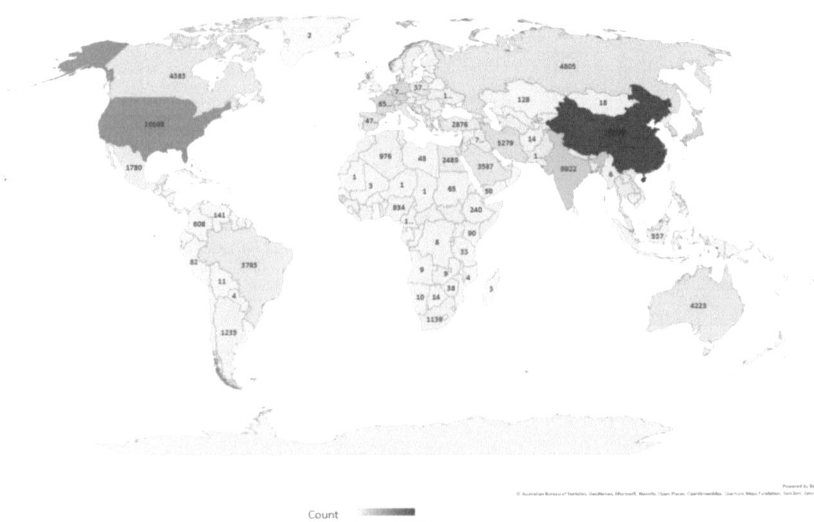

Fig. 16.3 Country-wise research activity on the use of biochar for wastewater treatment

an effective solution for nutrient removal and waste recovery. In contrast developed countries such as the USA, Germany, and Japan invest in biochar research through strong scientific infrastructure and environmental regulations. Biochar also supports circular economy practices, particularly in Europe, by recovering nutrients and reducing sludge. Its adaptability, affordability, and potential for carbon sequestration make it a valuable tool for both developed and developing regions in addressing water pollution and promoting sustainable development.

16.4.4 Linking Biochar Research to the SDGs

An analysis of 169,097 documents retrieved using the specified keyword combination focused on biochar, charcoal, water, wastewater, removal, remediation, and adsorption, reveal that that the researchers have aligned their work with various United Nations Sustainable Development Goals (SDGs). This trend highlights the broad applicability and impact of biochar research in addressing global sustainability challenges. SDG 6 (Clean Water and Sanitation) was the most represented, with 54,028 publications (48.8%), highlighting biochar's crucial role in water and wastewater treatment due to its high adsorption capacity for pollutants like heavy metals and pharmaceuticals. SDG 3 (Good Health and Well-being) followed with 24,845 records (22.4%), reflecting biochar's contribution to health by reducing environmental contaminants. Furthermore, the use of biochar in bioenergy production and decentralized energy solution aligns with SDG 7 (Affordable and Clean Energy), accounting for 8047 publications (7.3%). Similarly, SDG 13 (Climate Action) was also prominent with 8008 records underscored biochar's role in carbon sequestration and emission mitigation (Fig. 16.4).

Additionally, biochar's derived from agricultural waste and its alignment with sustainable consumption practices are linked to SDG 12 (Responsible Consumption and Production) accounted for 4491 publication. The soil-enhancing properties of biochar, improving fertility and yields, are connected to SDG 2 (Zero Hunger) with 3973 papers. Ecosystem protection goals—SDG 14 (Life Below Water) and SDG 15 (Life on Land)—were represented by 2464 and 1915 documents, respectively, emphasizing biochar's role in pollution mitigation and land restoration. Urban focussed applications like stormwater treatment and green infrastructure contributed to SDG 11 (Sustainable Cities and Communities) with 2092 publications. Meanwhile, goals such as SDG 4 (Quality Education), SDG 10 (Reduced Inequality), SDG 9 (Industry, Innovation, and Infrastructure), and SDG 5 (Gender Equality)receive compartitively low attention and were represented in 111 to 348 publications, indicating emerging but limited research engagement. Notably, fewer than 50 studies were directly linked to SDG 16 (Peace and Justice), SDG 8 (Decent Work), SDG 1 (No Poverty), and SDG 17 (Partnerships), revealing potential gaps and opportunities for future interdisciplinary exploration. This analysis illustrates biochar's multifaceted tool contribution to environmental sustainability, particularly in areas such as clean water access, climate mitigation, and resource efficiency. However, it

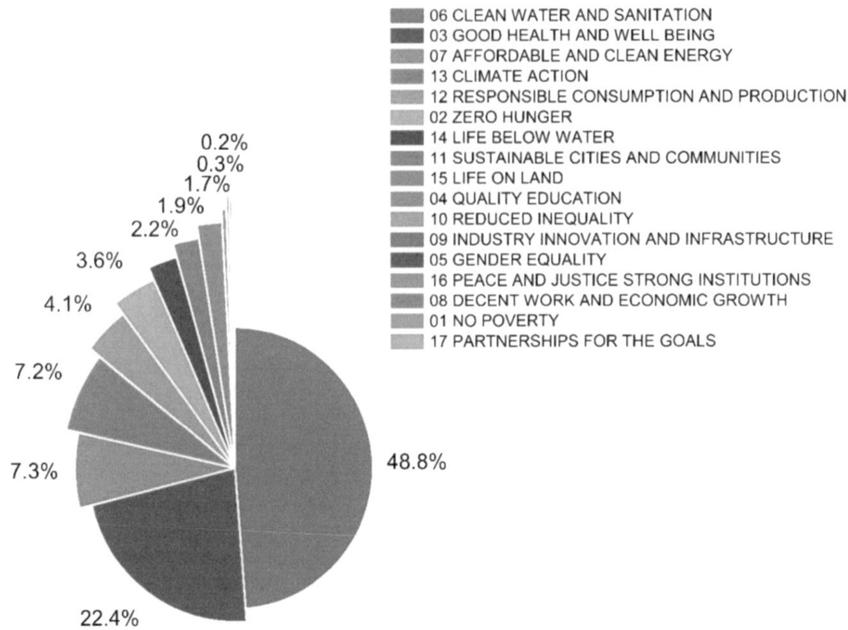

Fig. 16.4 SDGs linked with biochar

highlights the need for expanded enquiry into social and economic dimensions to more fully align with broader SDGs framework (Fig. 16.4).

16.5 Factors Affecting the Removal of Pollutants Using Biochar

Biochar, a widely recognized pyrolyzed carbon compound, has attracted multidisciplinary interest owing to its remarkable physicochemical properties, cost-effectiveness, and environmental sustainability. As a versatile carbonaceous material, biochar (BC) and its derivatives possess excellent properties and multifunctional capabilities. This leads to its variable applications in improving soil health (Majumder et al. 2019; Hossain et al. 2020; Zhang et al. 2021), mitigating greenhouse gas emissions from soil and carbon sequestration (Tan 2019), environmental remediation (Dong et al. 2023), and enhanced energy production (Liu et al. 2019). Biochar's ability to remove water contaminants stems from its abundant raw material sources, high surface area, well-developed pore structure, and low production cost. Several key factors, including the intrinsic properties of the biochar, deashing treatment, pH of the solution, dosage of the adsorbent, presence of competing anions, and temperature, influence the adsorption efficiency of biochar. The effectiveness of biochar in eliminating pollutants from air and water depends on various interrelated factors. Although biochar shows great potential as an adsorbent, its efficiency in removing contaminants from environmental media can be hindered by several challenges. If these limitations are not properly addressed, they may significantly reduce the overall adsorption capacity and limit the practical application of biochar in pollutant remediation. Some of the factors that affect the removal of pollutants using biochar are listed in the subsequent sections.

16.5.1 Biochar Preparation Parameters

Biochar's effective removal of pollutants is largely dependent on its physicochemical properties like pore structure, specific surface area (SSA), surface functional groups (SFG), and mineral content, which are governed by feedstock type, pyrolysis temperature, residence time, and preparation method. Higher pyrolysis temperatures generally increase carbonization, surface area, and microporosity, enhancing the adsorption capability of organic contaminants and some heavy metals. However, excessive temperature can reduce oxygen-containing functional groups, which lowers the adsorption capacity for pollutants like Cr (VI) and lead that rely on these groups. A key constraint in biochar application is the lack of functional surface groups such as hydroxyl, carboxyl, or carbonyl groups, which are essential for chemical interactions with pollutants. Without adequate functionalization, biochar exhibits limited affinity toward certain contaminants (Chen et al. 2012; Chen et al. 2008). Moreover, low surface area and poorly developed porosity—often resulting from suboptimal pyrolysis conditions—can hinder access to internal adsorption sites, reducing overall adsorption performance. For example, Kim et al. (2013a; b) reported that biochars produced at pyrolysis temperatures exceeding 500 °C exhibited increased pH and surface area, leading to enhanced cadmium (Cd) adsorption.

In contrast, Shen et al. (2012) reported a decrease in Cr(VI) adsorption with rising pyrolysis temperatures, as higher temperatures reduce the number of oxygen-containing functional groups critical for Cr(VI) reduction to Cr(III). Similar trends were noted in the adsorption of lead (Pb) by bagasse-derived biochars; Ding et al. (2014) found that Pb sorption capacity decreased from 21 to 6.1 mg/g as the pyrolysis temperature increased from 250 °C to 600 °C due to the loss of surface functional groups. Different preparation techniques impact biochar properties: hydrothermal carbonization (HTC) tends to produce biochar with more oxygen-containing groups and better adsorption for metals like copper (Cu) compared to conventional pyrolysis, which yields more carbonized but less functionalized surfaces. High ash content and inorganic impurities also present challenges. Ash can obstruct pores and occupy active sites, decreasing adsorption efficiency. The surrounding pH environment plays a critical role as well, since many pollutants adsorb effectively only within specific pH ranges. Optimizing preparation parameters such as temperature, feedstock choice, and processing method is crucial for maximizing biochar's adsorption efficiency and environmental remediation potential. Deviation from these optimal conditions can reduce the binding affinity of biochar. Most articles have drawn the same inferences that increasing pyrolytic temperatures play an active role in the adsorption of organic contaminants.

The effectiveness of biochar in eliminating pollutants from air and water depends on various interrelated factors. Although biochar shows great potential as an adsorbent, its efficiency in removing contaminants from environmental media can be hindered by several challenges. If these limitations are not properly addressed, they may significantly reduce the overall adsorption capacity and limit the practical application of biochar in pollutant remediation.

16.5.2 De-Mineralization and Deashing Treatment

While some minerals may enhance adsorption through ion exchange or surface precipitation, excessive ash content tends to block micropores, mask functional groups, and reduce surface area, ultimately limiting the adsorption capacity for both organic and inorganic pollutants. The efficacy of biochar in pollutant removal is significantly influenced by its surface chemistry, porosity, and active site availability—all of which can be altered through de-mineralization and deashing treatments (Zhang et al. 2013b). These post-treatment processes are specifically designed to remove inorganic impurities such as ash, metal oxides, and mineral salts from the biochar matrix. De-mineralization treatments aim to unblock pores and enhance specific surface area; expose and regenerate active functional groups (e.g., carboxyl, hydroxyl, carbonyl); improve overall porosity and accessibility to adsorption sites; and reduce pH in highly alkaline biochars caused by carbonate or oxide-rich ash. These treatments have been shown to significantly improve the removal efficiency of heavy metals like Pb^{2+}, Cd^{2+}, and Zn^{2+}. Removal of competing ions and increased functional group exposure enhance metal adsorption. While deashing improves biochar performance, it also introduces some trade-offs like loss of beneficial mineral components that aid in precipitation or co- adsorption mechanisms, structural changes or partial degradation of biochar matrix due to aggressive chemical treatments, and increased production costs and environmental footprint if not managed sustainably. Therefore, the choice and extent of de-mineralization should be tailored to the target pollutants and intended application, balancing performance enhancement with process feasibility.

De-mineralization and deashing are effective strategies to enhance biochar's adsorption performance by improving surface properties and accessibility to active sites. When properly optimized, these treatments can significantly increase the removal efficiency of both organic and inorganic pollutants, making biochar a more reliable and versatile material in environmental remediation.

16.5.3 Feedstock Composition

Biochars often contain substantial amounts of inorganic residues depending on the feedstock and pyrolysis conditions. Feedstock composition also significantly affects biochar performance under identical pyrolysis conditions. The natural mineral content of feedstocks—such as carbonates and phosphates—can influence adsorption properties (Cao et al. 2009; Xu et al. 2013b, c). Xu et al. (2011) found that biochars derived from different crop residues exhibited varying capacities for methyl violet removal, in the order: canola straw > peanut straw > soybean straw > rice hull, corresponding to differences in their surface charge. Xu et al. (2013b) also observed that dairy manure biochar was more effective than rice husk biochar in adsorbing heavy metals such as Pb, Cu, Zn, and Cd in both single- and multi-metal solutions.

16.5.4 Solution pH

The solution pH is a critical parameter in optimizing the adsorption process. Its influence varies depending on the type of biochar and the specific contaminants involved. pH affects not only the surface charge of the biochar but also the ionization state and speciation of the adsorbate (Kołodyńska et al. 2012; Regmi et al. 2012; Li et al. 2013; Zhang et al. 2013c). As a result, most studies on contaminant adsorption by biochars have included the effect of pH as a key factor. Biochar contains various surface functional groups, primarily oxygen-containing groups such as carboxyl (–COOH) and hydroxyl (–OH). The behavior of these groups changes with increasing solution pH. At low pH, most of these functional groups become protonated, resulting in a positively charged biochar surface (Oh et al. 2012; Abdel-Fattah et al. 2015).

16.6 Conclusion

Biochar has emerged as a highly promising material for the removal of various pollutants from wastewater, including heavy metals such as Pb, Cd, Cr, and As; VOCs like benzene, toluene, and phenol; and nutrients such as NO_3^-, NH_4^+, and PO_43^-. Bibliometric studies indicate a significant increase in global research interest focussed on the application of biochar in wastewater treatment reflecting growing recognition of its environmental and technological potential. The use of diverse waste biomass sources—such as agricultural residues, forestry waste, and sewage sludge—for biochar production not only offers a sustainable solution for waste management but also contributes to enhanced adsorption efficiency for contaminants. The efficacy of biochar is primarily attributed to its high surface area, porosity, and the presence of reactive surface functional groups. However, some limitations persist, including the variability in biochar properties due to inconsistent feedstock and pyrolysis conditions, potential leaching of contaminants from biochar, and limited adsorption capacity for certain emerging pollutants like pharmaceuticals and microplastics. To overcome these challenges, future research should focus on standardizing biochar production processes, developing engineered or functionalized biochars with tailored properties, and exploring regeneration and reuse strategies. Integrating biochar with other treatment technologies (e.g., constructed wetlands, membranes) and assessing its long-term environmental impact will also be essential for scaling up its application in real-world wastewater treatment systems. This multifaceted approach aligns with and supports the achievement of Sustainable Development Goals, especially SDG6 (Clean Water and Sanitation), SDG 3 (Good Health and Well-being), SDG 7 (Affordable and clean energy), and SDG 13 (Climate action).

References

https://doi.org/10.1016/j.chemosphere.2019.02.165

Abdel-Fattah TM, Mahmoud ME, Ahmed SB, Huff MD, Lee JW, Kumar S (2015) Biochar from woody biomass for removing metal contaminants and carbon sequestration. J Ind Eng Chem 22:103–109. https://doi.org/10.1016/j.jiec.2014.06.030

Abdullahi ME, Abu Hassan MA, Noor ZZ, Ibrahim RKR (2014) Application of a packed column air stripper in the removal of volatile organic compounds from wastewater. Rev Chem Eng 30(5):431–451. https://doi.org/10.1515/revce-2014-0003

Agrafioti E, Kalderis D, Diamadopoulos E (2014) Arsenic and chromium removal from water using biochars derived from rice husk, organic solid wastes and sewage sludge. J Environ Manag 133:309–314. https://doi.org/10.1016/j.jenvman.2013.12.007

Ahmad M, Rajapaksha AU, Lim JE, Zhang M, Bolan N, Mohan D, Vithanage M, Lee SS, Ok YS (2012) Biochar as a sorbent for contaminant management in soil and water: a review. Chemosphere 99:19–33. https://doi.org/10.1016/j.chemosphere.2013.10.071

Alam MZ, Anwar AHMF (2020) Nutrients adsorption onto biochar and alum sludge for treating stormwater. J Water Environ Technol 18(2):132–146. https://doi.org/10.2965/jwet.19-077

Aller MF (2016) Biochar properties: transport, fate, and impact. Crit Rev Environ Sci Technol 46(14–15):1183–1296. https://doi.org/10.1080/10643389.2016.1212368

Allohverdi T, Mohanty AK, Roy P, Misra M (2021) A review on current status of biochar uses in agriculture. Molecules 26(18):5584. https://doi.org/10.3390/molecules26185584

Aniebone VO, Imo DO, Shonde OO (2024) Nutrients pollution of the coastal marine ecosystems. IJIABR 12(2):82–95

Antunes E, Jacob MV, Brodie G, Schneider PA (2018) Isotherms, kinetics and mechanism analysis of phosphorus recovery from aqueous solution by calcium-rich biochar produced from biosolids via microwave pyrolysis. J Environ Chem Eng 6(1):395–403

Arriola-Villaseñor E, Barrera ZR, Hernández J (2023) Using banana waste biochar for simultaneous removal of heavy metals from raw real wastewater from the electroplating industry. Desalin Water Treat 314:88–102. https://doi.org/10.5004/dwt.2023.30079

Balagopal A, Alva A, Chandrasekhar HK, Thorat SA, Kaniyassery A, Koley A, Balachandran S, Govender N, Muthusamy A (2025) The fate of heavy metals and metalloids in plant systems–a comprehensive review. Toxicol Environ Chem 107(6):1114–1171

Baskaran T (2025) Reductive removal of hexavalent chromium from aqueous solutions by silver-doped biochar derived from Azadirachta indica stem. Water Air Soil Pollut 236(9):1–19. https://doi.org/10.1007/s11270-025-08241-2

Bayar J, Ali N, Dong Y, Ahmad U, Anjum MM, Khan GR, Zaib M, Jalal A, Ali R, Ali L (2024) Biochar-based adsorption for heavy metal removal in water: a sustainable and cost-effective approach. Environ Geochem Health 46(11):428. https://doi.org/10.1007/s10653-024-02214-w

Bernardino CAR, Mahler CF, Silva FWL et al (2024) Banana peel biochar from pyrolysis for the removal of nitrofurantoin in wastewater. Biomass Convers Biorefin https://doi.org/10.1007/s13399-024-06303-9

Biswas B, Rahman T, Sakhakarmy M, Jahromi H, Eisa M, Baltrusaitis J, Adhikari S (2023) Phosphorus adsorption using chemical and metal chloride activated biochars: isotherms, kinetics and mechanism study. Heliyon 9(9)

Bochynska S, Duszewska A, Maciejewska-Jeske M, Wrona M, Szeliga A, Budzik M, Szczesnowicz A, Bala G, Trzcinski M, Meczekalski B, Smolarczyk R (2024) The impact of water pollution on the health of older people. Maturitas 185:107981. https://doi.org/10.1016/j.maturitas.2024.107981

Bolton L, Joseph S, Greenway M, Donne S, Munroe P, Marjo CE (2019) Phosphorus adsorption onto an enriched biochar substrate in constructed wetlands treating wastewater. Ecol Eng 142:100005. https://doi.org/10.1016/j.ecoena.2019.100005

Cao X, Ma L, Gao B, Harris W (2009) Dairy-manure derived biochar effectively sorbs lead and atrazine. ES&T 43(9):3285–3291. https://doi.org/10.1021/es803092k

Cao Y, Yang B, Song Z, Wang H, He F, Han X (2016) Wheat straw biochar amendments on the removal of polycyclic aromatic hydrocarbons (PAHs) in contaminated soil. Ecotoxicol Environ Saf 130:248–255

Chen B, Zhou D, Zhu L (2008) Transitional adsorption and partition of nonpolar and polar aromatic contaminants by biochars of pine needles with different pyrolytic temperatures. Environ Sci Technol 42(14):5137–5143

Chen B, Chen Z, Lv S (2011) A novel magnetic biochar efficiently sorbs organic pollutants and phosphate. Bioresource Technol 102(2):716–723. https://doi.org/10.1016/j.biortech.2010.08.067

Chen Z, Chen B, Zhou D, Chen W (2012) Bisolute sorption and thermodynamic behavior of organic pollutants to biomass-derived biochars at two pyrolytic temperatures. Environ Sci Technol 46(22):12476–12483

Chun Y, Lee SK, Yoo HY, Kim SW (2021) Recent advancements in biochar production according to feedstock classification, pyrolysis conditions, and applications: a review. Bioresource technol 16(3):6512. https://doi.org/10.15376/biores.16.3.Chun

Colantoni A, Evic N, Lord R, Retschitzegger S, Proto AR, Gallucci F, Monarca D (2016) Characterization of biochars produced from pyrolysis of pelletized agricultural residues. Renew Sust Energ Rev 64:187–194. https://doi.org/10.1016/j.rser.2016.06.003

Cui Q, Xu J, Wang W, Tan L, Cui Y, Wang T, Zheng J (2020) Phosphorus recovery by core-shell γ-Al_2O_3/Fe_3O_4 biochar composite from aqueous phosphate solutions. Sci Total Environ 729:138892

Dai L, Wu B, Tan F, He M, Wang W, Qin H, Hu Q (2014) Engineered hydrochar composites for phosphorus removal/recovery: lanthanum doped hydrochar prepared by hydrothermal carbonization of lanthanum pretreated rice straw. Bioresour Technol 161:327–332

Das A (2025) Surface water potential zones delineation and spatiotemporal variation characteristics of water pollution and the cause of pollution formation in Brahmani River Basin, Odisha. HydroResearch 8:99–112. https://doi.org/10.1016/j.hydres.2024.10.002

David E (2022) Production of activated biochar derived from residual biomass for adsorption of volatile organic compounds. Mater 16(1):389. https://doi.org/10.3390/ma16010389

David M (2023) Activated biochar derived from rapeseed cake and walnut shell for toluene and acetone adsorption. Materials 16(389):1–17. https://doi.org/10.3390/ma16030389

David E, Niculescu VC (2021) Volatile organic compounds (VOCs) as environmental pollutants: occurrence and mitigation using nanomaterials. IJERPH 18(24):13147. https://doi.org/10.3390/ijerph182413147

Devrajani SK, Ahmed Z, Qambrani NA, Kanwal S, Sundaram UM, Mubarak NM (2024) Mechanism of arsenic removal using brown seaweed derived impregnated with iron oxide biochar for batch and column studies. Sci Rep 14(1):18102

Ding W, Dong X, Ime IM, Gao B, Ma LQ (2014) Pyrolytic temperatures impact lead sorption mechanisms by bagasse biochars. Chemosphere 105:68–74

Dong M, He L, Jiang M, Zhu Y, Wang J, Gustave W, Wang Z (2023) Biochar for the removal of emerging pollutants from aquatic systems: a review. Int J Environ Res Public Health 20(3):1679

Eduah JO, Nartey EK, Abekoe MK, Henriksen SW, Andersen MN (2020) Mechanism of orthophosphate (PO_4-P) adsorption onto different biochars. Environ Technol Innov 17:100572

Elamin NY, Elamin MR, Abdulkhair BY, Subaihi A, Shahat A, Hemdan M, Mubarak MF (2025) Eco-friendly alginate–sulfonated bagasse biochar/zeolitic imidazolate framework-8 nanocomposite for efficient mercury(II) removal: synthesis, mechanism, and regeneration performance. Colloids Surf A Physicochem Eng Asp 137557. https://doi.org/10.1016/j.colsurfa.2025.137557

Enaime G, Baçaoui A, Yaacoubi A, Lübken M (2020) Biochar for wastewater treatment—conversion technologies and applications. Chem Eng J 10(10):3492. https://doi.org/10.3390/app10103492

Esrafili L, Firuzabadi FD, Morsali A, Hu ML (2021a) Reuse of predesigned dual-functional metal organic frameworks (DF-MOFs) after heavy metal removal. J Hazard Mater 403:123696. https://doi.org/10.1016/j.jhazmat.2020.123696

Esrafili L, Firuzabadi FD, Morsali A, Hu ML (2021b) Reuse of predesigned dual-functional metal organic frameworks (DF-MOFs) after heavy metal removal. J Hazard Mater 403:123696

Ezzahi K, Rabichi I, Rochdi N et al (2025) Innovative dynamic filtration of olive mill wastewater: comparing the efficiency of biochar and olive stone. Water Air Soil Pollut 236:604. https://doi.org/10.1007/s11270-025-08231-4

Fang L, Li JS, Donatello S, Cheeseman CR, Poon CS, Tsang DC (2020) Use of Mg/Ca modified biochars to take up phosphorus from acid-extract of incinerated sewage sludge ash (ISSA) for fertilizer application. J Clean Prod 244:118853

Fatima S, Govardhan B, Kalyani S, Sridhar S (2022) Extraction of volatile organic compounds from water and wastewater by vacuum-driven membrane process: a comprehensive review. Chem Eng J 434:134664. https://doi.org/10.1016/j.cej.2022.134664

Feng L, Wang R, Jia L, Wu H (2020) Can biochar application improve nitrogen removal in constructed wetlands for treating anaerobically-digested swine wastewater? Chem Eng J 379:122273. https://doi.org/10.3390/molecules29051005

Giraldo S, Robles I, Ramirez A, Flórez E, Acelas N (2020) Mercury removal from wastewater using agroindustrial waste adsorbents. SN Appl Sci 2(1):1–17. https://doi.org/10.1007/s42452-020-2736-x

Glaser B, Birk JJ (2012) State of the scientific knowledge on properties and genesis of Anthropogenic dark earths in Central Amazonia (terra preta de Índio). Geochim Cosmochim Acta 82:39–51. https://doi.org/10.1016/j.gca.2010.11.029

Grottola CM, Giudicianni P, Stanzione F, Ragucci R (2022) Influence of pyrolysis temperature on biochar produced from lignin–rich biorefinery residue. ChemEngineering 6(5):76. https://doi.org/10.3390/chemengineering6050076

Guo J, Fu L, Gao R, Zhang Y, Yin C, Xue B (2025) Efficiency removal of rhodamine B by MoS_2/magnetic potato residues biochar: performance and mechanism. Ind Crops Prod 233:121390. https://doi.org/10.1016/j.indcrop.2025.121390

Gupta N, Banerjee S, Koley A, Bharali P, GhoshThakur R, Hoque RR, Balachandran S (2024) Strategies for remediation of polycyclic aromatic hydrocarbons in contaminated soil: a systematic review and bibliometric analysis. Appl Soil Ecol 204:105688

He M, Xu Z, Hou D, Gao B, Cao X, Ok YS, Rinklebe J, Bolan NS, Tsang DC (2022) Waste-derived biochar for water pollution control and sustainable development. Nat Rev Earth Environ 3(7):444–460. https://doi.org/10.1038/s43017-022-00306-8

Hossain MZ, Bahar MM, Sarkar B, Donne SW, Ok YS, Palansooriya KN, Kirkham MB, Chowdhury S, Bolan N (2020) Biochar and its importance on nutrient dynamics in soil and plant. Biochar 2:379–420

Hu Y, Chen Z (2025) Thermochemical conversion of sewage sludge: progress in pyrolysis and gasification. Water 17(12):1833. https://doi.org/10.3390/w17121833
Jalayeri H, Pepe F (2019) Novel and high-performance biochar derived from pistachio green hull biomass: production, characterization, and application to Cu (II) removal from aqueous solutions. Ecotoxicol Environ Saf 168:64–71. https://doi.org/10.1016/j.ecoenv.2018.10.058
Jalilvand H, Feyzi F, Dehghani MR (2020) Adsorption of dimethyl sulfide from model fuel on raw and modified activated carbon from walnut and pistachio shell origins: kinetic and thermodynamic study. Colloids Surf A Physicochem Eng Asp 593:124620
Jayawardhana Y, Gunatilake SR, Mahatantila K, Ginige MP, Vithanage M (2019) Sorptive removal of toluene and m-xylene by municipal solid waste biochar: simultaneous municipal solid waste management and remediation of volatile organic compounds. J Environ Manag 238:323–330. https://doi.org/10.1016/j.jenvman.2019.02.097
Ji S, Zhang F, Yao P, Li C, Faheem M, Feng Q, Wang B (2023) Optimization of pig manure-derived biochar for ammonium and phosphate simultaneous recovery from livestock wastewater. Environ Sci Pollut Res 30(34):82532–82546
Jin X, Wu Y, Santhamoorthy M, Le TTN, Le VT, Yuan Y, Xia C (2022) Volatile organic compounds in water matrices: recent progress, challenges, and perspective. Chemosphere 308:136182. https://doi.org/10.1016/j.chemosphere.2022.136182
Jung KW, Jeong TU, Kang HJ, Ahn KH (2016) Characteristics of biochar derived from marine macroalgae and fabrication of granular biochar by entrapment in calcium-alginate beads for phosphate removal from aqueous solution. Bioresour Technol 211:108–116
Kamali M, Appels L, Kwon EE, Aminabhavi TM, Dewil R (2021) Biochar in water and wastewater treatment-a sustainability assessment. Chem Eng J 420:129946. https://doi.org/10.1016/j.cej.2021.129946
Kaur RA (2021) Bioremediation of heavy metals from wastewater using nanomaterials. EDS 23(7):9617–9640. https://doi.org/10.1007/s10668-020-01078-1
Kaur S, Roy A (2021) Bioremediation of heavy metals from wastewater using nanomaterials. Environ Dev Sustain 23(7):9617–9640
Khajuria A, Kaushal J, Sudan S (2024) Encapsulated cobalt-doped coconut husk biochar (Co@CHBc) for the remediation of anionic dye from wastewater. Biomass Convers Biorefin 1–18
Kim P, Kim M, Yoo S, Chae KJ (2013a) Comparison of the adsorption characteristics of heavy metals by urban sewage sludge-derived biochars produced at different temperatures. J Ind Eng Chem 19(5):1730–1734. https://doi.org/10.1016/j.jiec.2013.01.001
Kim WK, Shim T, Kim YS, Hyun S, Ryu C, Park YK, Jung J (2013b) Characterization of cadmium removal from aqueous solution by biochar produced from a giant Miscanthus at different pyrolytic temperatures. Bioresour Technol 138:266–270
Kolcy A, Gupta N, Singh A, GhoshThakur R, Balachandran S (2024) Microalgae-based biofuel for sustainable bioenergy production. In: Application of microbial technology in wastewater treatment and bioenergy recovery. Springer Nature, Singapore, pp 359–384
Kołodyńska D, Wnętrzak R, Leahy JJ, Hayes MHB, Kwapiński W, Hubicki ZJCEJ (2012) Kinetic and adsorptive characterization of biochar in metal ions removal. Chem Eng J 197:295–305. https://doi.org/10.1016/j.cej.2012.05.025
Kumar R, Zhi S, Liu J (2011) Effect of hydrothermal carbonization on the physicochemical characteristics and copper adsorption potential of biochars derived from switchgrass. Est 45(17):7506–7513. https://doi.org/10.1021/es200730y
Kumar P, Patel AK, Singhania RR, Chen CW, Saratale RG, Dong CD (2023a) Enhanced copper (II) bioremediation from wastewater using nano magnetite (Fe_3O_4) modified biochar of Ascophyllum nodosum. Bioresource Technol 388:129654. https://doi.org/10.1016/j.biortech.2023.129654
Kumar P, Patel AK, Singhania RR, Chen CW, Saratale RG, Dong CD (2023b) Enhanced copper (II) bioremediation from wastewater using nano magnetite (Fe_3O_4) modified biochar of Ascophyllum nodosum. Biores Technol 388:129654

Kumi AG, Ibrahim MG, Nasr M, Fujii M (2020) September. Biochar synthesis for industrial wastewater treatment: a critical review. In: Materials science forum, vol 1008. Trans Tech Publications Ltd., pp 202–212
Lee J, Lee S, Park YK (2023) Reduction of odor-causing compounds in wastewater using biochar: a review. Bioresour Technol 385:129419
Li M, Liu Q, Guo L, Zhang Y, Lou Z, Wang Y, Qian G (2013) Cu (II) removal from aqueous solution by Spartina alterniflora derived biochar. Bioresource Technol 141:83–88. https://doi.org/10.1016/j.biortech.2012.12.096
Li R, Wang JJ, Zhang Z, Awasthi MK, Du D, Dang P, Wang L (2018a) Recovery of phosphate and dissolved organic matter from aqueous solution using a novel CaO–MgO hybrid carbon composite and its feasibility in phosphorus recycling. Sci Total Environ 642:526–536
Li Z, Chang F, Shi P, Chen X, Yang F, Zhou Q, Pan Y, Li A (2018b) Occurrence and potential human health risks of semi-volatile organic compounds in drinking water from cities along the Chinese coastland of the Yellow Sea. Chemosphere 206:655–662. https://doi.org/10.1016/j.chemosphere.2018.05.064
Li J, Li Q, Qian C, Wang X, Lan Y, Wang B, Yin W (2019) Volatile organic compounds analysis and characterization on activated biochar prepared from rice husk. Int J Environ Sci Technol 16:7653–7662
Li X, Wang C, Zhang J, Liu J, Liu B, Chen G (2020) Preparation and application of magnetic biochar in water treatment: a critical review. Sci Total Environ 711:134847
Liang HM, Liao CM (2007) Modeling VOC-odor exposure risk in livestock buildings. Chemosphere 68(4):781–789. https://doi.org/10.1016/j.chemosphere.2006.12.096
Liu WJ, Jiang H, Yu HQ (2019) Emerging applications of biochar-based materials for energy storage and conversion. Energy Environ Sci 12(6):1751–1779
Lu H, Zhang W, Wang S, Zhuang L, Yang Y, Qiu R, Chen J (2012) Characterization of sewage sludge-derived biochars from different feedstocks and pyrolysis temperatures. J Anal Appl Pyrolysis 94:201–207. https://doi.org/10.1016/j.jaap.2011.12.007
Lu CS, Qiu JM, Cao L, Jiang H, Kong Z, Hu Y, He S (2025) Role of nickel-loaded crayfish shells biochar in anaerobic digestion of high salinity organic wastewater. J Environ Chem Eng 117677. https://doi.org/10.1016/j.jece.2025.117677
Ma X, Chen H, Zhang C, Xu N, Sun S, Jiang W, Liu Y (2025) Construction of a green nano-iron-modified biochar fiber-textured composite using the biomass template method for enhanced hexavalent chromium reduction in water. J Environ Chem Eng 117662. https://doi.org/10.1016/j.jece.2025.117662
Madadi R, Bester K (2021a) Fungi and biochar applications in bioremediation of organic micropollutants from aquatic media. Mar Pollut Bull 166:112247. https://doi.org/10.1016/j.marpolbul.2021.112247
Madadi R, Bester K (2021b) Fungi and biochar applications in bioremediation of organic micropollutants from aquatic media. Mar Pollut Bull 166:112247
Majumder S, Neogi S, Dutta T, Powel MA, Banik P (2019) The impact of biochar on soil carbon sequestration: meta-analytical approach to evaluating environmental and economic advantages. J Environ Manage 250:109466. https://doi.org/10.1016/j.jenvman.2019.109466
Martínez EJ, Rosas JG, Sotres A, Moran A, Cara J, Sánchez ME, Gómez X (2018) Codigestion of sludge and citrus peel wastes: evaluating the effect of biochar addition on microbial communities. Biochem Eng J 137:314–325
Mazurek K, Drużyński S, Kiełkowska U, Szłyk E (2021) New separation material obtained from waste rapeseed cake for copper (II) and zinc (II) removal from the industrial wastewater. Mater 14(10):2566. https://doi.org/10.3390/ma14102566
Mochizuki Y, Bud J, Liu J, Takahashi M, Tsubouchi N (2021) Mercury (II) ion adsorption performance of Cl-loaded carbonaceous material prepared by chlorination of pyrolyzed rice husk char. J Clean Prod 305:127176. https://doi.org/10.1016/j.jclepro.2021.127176

Moreira R, dos Reis OR, Vaz JM, Penteado JC, Spinacé EV (2017) Production of biochar, bio-oil and synthesis gas from cashew nut shell by slow pyrolysis. WBV 8:217–224. https://doi.org/10.1007/s12649-016-9569-2

Mousavi M, Aldagari S, Crocker MS, Ackerman-Biegasiewicz LK, Fini EH (2023) Iron-rich biochar to adsorb volatile organic compounds emitted from asphalt-surfaced areas. ACS Sustain Chem Eng 11(7):2885–2896

Muhammad N, Aziz R, Brookes PC, Xu J (2017) Impact of wheat straw biochar on yield of rice and some properties of Psammaquent and Plinthudult. J Soil Sci Plant Nutr 17(3):808–823. https://doi.org/10.4067/S0718-95162017000300019

Münzel T, Hahad O, Lelieveld J, Aschner M, Nieuwenhuijsen MJ, Landrigan PJ, Daiber A (2025) Soil and water pollution and cardiovascular disease. Nat Rev Cardiol 22(2):71–89. https://doi.org/10.1038/s41569-024-01068-0

Naz T, Khan MD, Ahmed I, Rehman SU, Rha ES, Malook I, Jamil M (2016a) Biosorption of heavy metals by Pseudomonas species isolated from sugar industry. Toxicol Ind Health 32(9):1619–1627. https://doi.org/10.1177/0748233715569900

Naz T, Khan MD, Ahmed I, Rehman SU, Rha ES, Malook I, Jamil M (2016b) Biosorption of heavy metals by Pseudomonas species isolated from sugar industry. Toxicol Ind Health 32(9):1619–1627

Nie J, Feng H, Witherell BB, Alebus M, Mahajan MD, Zhang W, Yu L (2018) Causes, assessment, and treatment of nutrient (N and P) pollution in rivers, estuaries, and coastal waters. Curr Pollut Rep 4:154–161. https://doi.org/10.1007/s40726-018-0083-y

Nobaharan K, Bagheri Novair S, AsgariLajayer B, van Hullebusch ED (2021) Phosphorus removal from wastewater: the potential use of biochar and the key controlling factors. Water 13(4):517

Odega CA, Ayodele OO, Ogutuga SO, Anguruwa GT, Adekunle AE, Fakorede CO (2023) Potential application and regeneration of bamboo biochar for wastewater treatment: a review. Adv Bamboo Sci 2:100012. https://doi.org/10.1016/j.bamboo.2022.100012

Oh TK, Choi B, Shinogi Y, Chikushi J (2012) Effect of pH conditions on actual and apparent fluoride adsorption by biochar in aqueous phase. WASP 223:3729–3738. https://doi.org/10.1007/s11270-012-1144-2

Olugbenga OS, Adeleye PG, Oladipupo SB, Adeleye AT, John KI (2024) Biomass-derived biochar in wastewater treatment—a circular economy approach. Waste Manag Bull 1(4):1–14. https://doi.org/10.1016/j.wmb.2023.07.007

Pandey D, Singh SV, Savio N, Bhutto JK, Srivastava RK, Yadav KK, Sharma R, Nandipamu TMK, Sarkar B (2025) Biochar application in constructed wetlands for wastewater treatment: a critical review. JWPE 69:106713. https://doi.org/10.1016/j.jwpe.2024.106713

Panwar NL, Pawar A, Salvi BL (2019) Comprehensive review on production and utilization of biochar. SN Appl Sci 1:1–19. https://doi.org/10.1007/s42452-019-0172-6

Patra JM, Panda SS, Dhal NK (2017) Biochar as a low-cost adsorbent for heavy metal removal: a review. Int J Res Biosci 6(1):1–7

Pugazhendhi A (2024) Biochar for wastewater treatment: addressing contaminants and enhancing sustainability: challenges and solutions. J Hazard Mater Adv 100504. https://doi.org/10.1016/j.hazadv.2024.100504

Qiu Y, Zheng Z, Zhou Z, Sheng GD (2009) Effectiveness and mechanisms of dye adsorption on a straw-based biochar. Bioresource Technol 100(21):5348–5351. https://doi.org/10.1016/j.biortech.2009.05.054

Regkouzas P, Diamadopoulos E (2019) Adsorption of selected organic micro-pollutants on sewage sludge biochar. Chemosphere 224:840–851

Regmi P, Moscoso JLG, Kumar S, Cao X, Mao J, Schafran G (2012) Removal of copper and cadmium from aqueous solution using switchgrass biochar produced via hydrothermal carbonization process. J Environ Manage 109:61–69. https://doi.org/10.1016/j.jenvman.2012.04.047

Reza MS, Afroze S, Bakar MS, Saidur R, Aslfattahi N, Taweekun J, Azad AK (2020) Biochar characterization of invasive Pennisetum purpureum grass: effect of pyrolysis temperature. Biochar 2:239–251. https://doi.org/10.1007/s42773-020-00048-0
Rizwan M, Ali S, Qayyum MF, Ibrahim M, Zia-ur-Rehman M, Abbas T, Ok YS (2016) Mechanisms of biochar-mediated alleviation of toxicity of trace elements in plants: a critical review. ESPR 23:2230–2248. https://doi.org/10.1007/s11356-015-5697-7
Roy B, Bhunia B, Bandyopadhyay TK, Khan SA, Nandi NB, Nath PC (2024) Water pollution by heavy metals and their impact on human health. HWP 333–352. https://doi.org/10.1002/9781119904991.ch10
Sadegh-Zadeh F, Seh-Bardan BJ (2013) Adsorption of As(III) and As(V) by Fe coated biochars and biochars produced from empty fruit bunch and rice husk. J Environ Chem Eng 1(4):981–988. https://doi.org/10.1016/j.jece.2013.08.009
Saiz-Rubio R, Balseiro-Romero M, Antelo J, Díez E, Fiol S, Macías F (2019) Biochar as low-cost sorbent of volatile fuel organic compounds: potential application to water remediation. Environ Sci Pollut Res 26:11605–11617
Sanyang L, Ghani WAWAK, Idris A, Mansor A (2014) Zinc removal from wastewater using hydrogel modified biochar. Appl Mech Mater 625:842–846. https://doi.org/10.4028/www.scientific.net/AMM.625.842
Sanyang ML, Ghani WAWAK, Idris A, Ahmad MB (2016) Hydrogel biochar composite for arsenic removal from wastewater. Desalin Water Treat 57(8):3674–3688. https://doi.org/10.1080/19443994.2014.989412
Seo JY, Tokmurzin D, Lee D, Lee SH, Seo MW, Park YK (2022) Production of biochar from crop residues and its application for biofuel production processes–an overview. Bioresource Technol 361:127740. https://doi.org/10.1016/j.biortech.2022.127740
Sforza E, Kumkum P, Barbera E, Kumar S (2020) Bioremediation of industrial effluents: how a biochar pretreatment may increase the microalgal growth in tannery wastewater. J Water Process Eng 37:101431. https://doi.org/10.1016/j.jwpe.2020.101431
Shen YS, Wang SL, Tzou YM, Yan YY, Kuan WH (2012) Removal of hexavalent Cr by coconut coir and derived chars–the effect of surface functionality. Bioresour Technol 104:165–172
Shin JY, Artigas F, Hobble C, Lee YS (2013) Assessment of anthropogenic influences on surface water quality in urban estuary, northern New Jersey: multivariate approach. EMA 185:2777–2794. https://doi.org/10.1007/s10661-012-2748-0
Shitu IG, Katibi KK, Tafida RA, Iya SGD, Alotaibi KM, Babani S, Mallik S (2024) Synthesis and characterization of magnetic biochar nanocomposite from oil palm fronds for efficient copper (II) ion removal from leachate. Clean Technol Environ Policy 1–23. https://doi.org/10.1007/s10098-024-03041-4
Shukla N, Sahoo D, Remya N (2019) Biochar from microwave pyrolysis of rice husk for tertiary wastewater treatment and soil nourishment. J Clean Prod 235:1073–1079. https://doi.org/10.1016/j.jclepro.2019.07.042
Song MY, Chun H (2021) Species and characteristics of volatile organic compounds emitted from an auto-repair painting workshop. Sci Rep 11(1):16586. https://doi.org/10.1038/s41598-021-96163-4
Strokal M, Ma L, Bai Z, Luan S, Kroeze C, Oenema O, Velthof G, Zhang F (2016) Alarming nutrient pollution of Chinese rivers as a result of agricultural transitions. ERL 11(2):024014. https://doi.org/10.1088/1748-9326/11/2/024014
Sullivan GL, Prigmore RM, Knight P, Godfrey AR (2019) Activated carbon biochar from municipal waste as a sorptive agent for the removal of polyaromatic hydrocarbons (PAHs), phenols and petroleum based compounds in contaminated liquids. J Environ Manage 251:109551
Sun Y, Qi S, Zheng F, Huang L, Pan J, Jiang Y, Hou W, Xiao L (2018) Organics removal, nitrogen removal and N_2O emission in subsurface wastewater infiltration systems amended with/without biochar and sludge. Bioresource Technol 249:57–61. https://doi.org/10.1016/j.biortech.2017.10.004

Sun L, Di M, Jia S, Liang J, Zhang P, Zhang G (2025) Synergistic waste-based substrates: optimizing constructed wetland performance for rural wastewater remediation. J Environ Sci. https://doi.org/10.1016/j.jes.2025.06.027
Tan RR (2019) Data challenges in optimizing biochar-based carbon sequestration. Renew Sustain Energy Rev 104:174–177
Tan X, Liu Y, Zeng G, Wang X, Hu X, Gu Y, Yang Z (2015) Application of biochar for the removal of pollutants from aqueous solutions. Chemosphere 125:70–85. https://doi.org/10.1016/j.chemosphere.2014.12.058
Tan S, Zhou G, Yang Q, Ge S, Liu J, Cheng YW, Yek PNY, Mahari WAW, Kong SH, Chang JS, Sonne C (2023) Utilization of current pyrolysis technology to convert biomass and manure waste into biochar for soil remediation: a review. Sci Total Environ 864:160990. https://doi.org/10.1016/j.scitotenv.2022.160990
Thompson KA, Shimabuku KK, Kearns JP, Knappe DR, Summers RS, Cook SM (2016a) Environmental comparison of biochar and activated carbon for tertiary wastewater treatment. ES&T 50(20):11253–11262. https://doi.org/10.1021/acs.est.6b03239
Thompson KA, Shimabuku KK, Kearns JP, Knappe DR, Summers RS, Cook SM (2016b) Environmental comparison of biochar and activated carbon for tertiary wastewater treatment. Environ Sci Technol 50(20):11253–11262. https://doi.org/10.1021/acs.est.6b03239
Tian Y, Zhao Y, Yin Z, Deng N, Li S, Zhao H, Huang B (2025) Integrating spatial-temporal features into prediction tasks: a novel method for identifying the potential water pollution area in large river basins. J Environ Manage 373:123522. https://doi.org/10.1016/j.jenvman.2024.123522
Tichem TM, Wang Y, Gameli RB, Mbage B, Li B (2023) Multicomponent adsorption of pollutants from wastewater using low-cost eco-friendly iron-modified rice husk biochar in the era of green chemistry. Sustain 15(23):16348
Vasseghian Y, Nadagouda MM, Aminabhavi TM (2024) Biochar-enhanced bioremediation of eutrophic waters impacted by algal blooms. J Environ Manage 367:122044. https://doi.org/10.3390/w13040517
Wang H, Xiao K, Yang J, Yu Z, Yu W, Xu Q, Liu B (2020a) Phosphorus recovery from the liquid phase of anaerobic digestate using biochar derived from iron-rich sludge: a potential phosphorus fertilizer. Water Res 174:115629
Wang X, Guo Z, Hu Z, Zhang J (2020b) Recent advances in biochar application for water and wastewater treatment: a review. PeerJ 8:9164. https://doi.org/10.7717/peerj.9164
Wang B, Ma Y, Lee X, Wu P, Liu F, Zhang X, Chen M (2021) Environmental-friendly coal gangue-biochar composites reclaiming phosphate from water as a slow-release fertilizer. Sci Total Environ 758:143664
Wang Y, Tian L, Zheng J, Tan Y, Li Y, Wei L, Zhang F, Zhu L (2024) Enhancing nitrogen removal in low C/N wastewater with recycled sludge-derived biochar: a sustainable solution. Water Res 267:122551. https://doi.org/10.1016/j.watres.2024.122551
Wang C, Xiong W, Zhang G (2025) Application of deep learning models with spectral data augmentation and denoising for predicting total phosphorus concentration in water pollution. J Taiwan Inst Chem E 167:105852. https://doi.org/10.1016/j.jtice.2024.105852
Weber K, Quicker P (2018) Properties of biochar. Fuel 217:240–261. https://doi.org/10.1016/j.fuel.2017.12.054
Wu Z et al (2020) Continuous VOC capture with modified rice husk biochar. Environ Pollut 259:113897
Wu X, Quan W, Chen Q, Gong W, Wang A (2024) Efficient adsorption of nitrogen and phosphorus in wastewater by biochar. Molecules 29(5):1005. https://doi.org/10.3390/molecules29051005
Xia S, Song Z, Jeyakumar P, Bolan N, Wang H (2020) Characteristics and applications of biochar for remediating Cr(VI)-contaminated soils and wastewater. Environ Geochem Health 42:1543–1567
Xiang W, Zhang X, Chen J, Zou W, He F, Hu X, Tsang DC, Ok YS, Gao B (2020) Biochar technology in wastewater treatment: a critical review. Chemosphere 252:126539. https://doi.org/10.1016/j.chemosphere.2020.126539

Xiao L, Jiang H, Shen C, Li K (2019) Adsorption of phosphorus onto Fe-modified Thalia dealbata derived biochar. In: E3S web conference vol 118. EDP Sciences, p 01022
Xie Q, Li S, Lian Y, Yang M, Wang F, Ye C, Liu H, Lu Y, Fang C (2024) Mechanism of enhanced degradation of antibiotic wastewater by three-dimensional electrocatalytic oxidation system: coconut shell biochar as particle electrode. J Water Process Eng 64:105749. https://doi.org/10.1016/j.jwpe.2024.105749
Xiong X, Iris KM, Tsang DC, Bolan NS, Ok YS, Igalavithana AD, Kirkham MB, Kim KH, Vikrant K (2019) Value-added chemicals from food supply chain wastes: state-of-the-art review and future prospects. Chem Eng J 375:121983. https://doi.org/10.1016/j.cej.2019.121983
Xu RK, Xiao SC, Yuan JH, Zhao AZ (2011) Adsorption of methyl violet from aqueous solutions by the biochars derived from crop residues. Bioresource Technol 102(22):10293–10298. https://doi.org/10.1016/j.biortech.2011.08.089
Xu X, Cao X, Zhao L (2013a) Comparison of rice husk-and dairy manure-derived biochars for simultaneously removing heavy metals from aqueous solutions: role of mineral components in biochars. Chemosphere 92(8):955–961. https://doi.org/10.1016/j.chemosphere.2013.03.009
Xu X, Cao X, Zhao L, Wang H, Yu H (2013b) Removal of Cu, Zn, and Cd from aqueous solutions by the dairy manure-derived biochar. ESPR 20:358–368. https://doi.org/10.1007/s11356-012-1055-4
Xu X, Cao X, Zhao L, Wang H, Yu H, Gao B (2013c) Removal of Cu, Zn, and Cd from aqueous solutions by the dairy manure-derived biochar. ESPR 20:358–368. https://doi.org/10.1007/s11356-012-0873-5
Xu Z, Lin Y, Lin Y, Yang D, Zheng H (2021) Adsorption behaviors of paper mill sludge biochar to remove Cu, Zn and As in wastewater. Environ Technol Innov 23:101616. https://doi.org/10.1016/j.eti.2021.101616
Xu Y, Xia H, Zhang Q, Jiang G, Cai W, Hu W (2022) Adsorption of cadmium(II) in wastewater by magnesium oxide modified biochar. Arab J Chem 15(9):104059. https://doi.org/10.1016/j.arabjc.2022.104059
Yang GX, Jiang H (2014) Amino modification of biochar for enhanced adsorption of copper ions from synthetic wastewater. Water Res 48:396–405. https://doi.org/10.1016/j.watres.2013.09.050
Yang Q, Wang X, Luo W, Sun J, Xu Q, Chen F, Zeng G (2018) Effectiveness and mechanisms of phosphate adsorption on iron-modified biochars derived from waste activated sludge. Bioresour Technol 247:537–544
Zhang P, Sun H, Yu L, Sun T (2013) Adsorption and catalytic hydrolysis of carbaryl and atrazine on pig manure-derived biochars: impact of structural properties of biochars. J Hazard Mater 244:217–224. https://doi.org/10.1016/j.jhazmat.2012.11.046
Zhang P, Sun H, Yu L, Sun T (2013a) Adsorption and catalytic hydrolysis of carbaryl and atrazine on pig manure-derived biochars: impact of structural properties of biochars. J Hazard Mater 244:217–224. https://doi.org/10.1016/j.jhazmat.2012.11.046
Zhang ZB, Cao XH, Liang P, Liu YH (2013b) Adsorption of uranium from aqueous solution using biochar produced by hydrothermal carbonization. JRNC 295:1201–1208. https://doi.org/10.1007/s10967-012-2017-2
Zhang QZ, Wang XH, Du ZL, Liu XR, Wang YD (2013c) Impact of biochar on nitrate accumulation in an alkaline soil. Soil Res 51(6):521–528. https://doi.org/10.1071/SR13153
Zhang S, Abdalla MA, Luo Z, Xia S (2018) The wheat straw biochar research on the adsorption/desorption behaviour of mercury in wastewater. Desalin Water Treat 112:147–160. https://doi.org/10.5004/dwt.2018.21850
Zhang Y, Wang J, Feng Y (2021) The effects of biochar addition on soil physicochemical properties: a review. CATENA 202:105284
Zhao M, Dai Y, Zhang M, Feng C, Qin B, Zhang W, Qiu R (2020) Mechanisms of Pb and/or Zn adsorption by different biochars: biochar characteristics, stability, and binding energies. Sci Total Environ 717:136894. https://doi.org/10.1016/j.scitotenv.2020.136894

Zheng H, Wang Z, Zhao J, Herbert S, Xing B (2013) Sorption of antibiotic sulfamethoxazole varies with biochars produced at different temperatures. Environ Pollut 181:60–67. https://doi.org/10.1016/j.envpol.2013.05.056

Zhou X, Wang X, Zhang H, Wu H (2017) Enhanced nitrogen removal of low C/N domestic wastewater using a biochar-amended aerated vertical flow constructed wetland. Bioresource Technol 241:269–275. https://doi.org/10.1016/j.biortech.2017.05.072

Zhu N, Yan T, Qiao J, Cao H (2016) Adsorption of arsenic, phosphorus and chromium by bismuth impregnated biochar: adsorption mechanism and depleted adsorbent utilization. Chemosphere 164:32–40. https://doi.org/10.1016/j.chemosphere.2016.08.036

Zhu T, Zhang Y, Chen Y, Liu JL, Song XL (2022) Synthesis of novel hydrated ferric oxide biochar nanohybrids for efficient arsenic removal from wastewater. Rare Met 41(5):1677–1687

Zin MMT, Kim DJ (2021) Simultaneous recovery of phosphorus and nitrogen from sewage sludge ash and food wastewater as struvite by Mg-biochar. J Hazard Mater 403:123704. https://doi.org/10.1016/j.jhazmat.2020.123704

Chapter 17
Cattle Manure Biochar: A Paradigm Shift in Sustainable Agriculture

Samridhi Kapoor, Rishav Sharma, Rohit Bishist, Himani Sharma, Prashant Sharma, and Krishan Lal Gautam

Abstract Livestock farming remains a cornerstone of global agriculture, with cattle serving as a primary source of edible items, draught power and manure. The cattle manure production in India is about 344 million tonnes, with the daily average cow dung production of 5.5 kg. Majority of the manure produced is incorporated into the agricultural lands. However, the prodigious volumes of manure generated present a paradox. On one hand, manure is rich in essential nutrients required for plants, on the other hand, manure mismanagement exacerbates environmental degradation. Conventional manure applications can lead to nutrient imbalances, leaching, ammonia volatilization and the emission of potent greenhouse gases thereby, contributing significantly to the climate change. As a sustainable alternative, manure derived biochar, produced through pyrolysis of manure under oxygen-limited conditions offers a promising solution for mitigating these challenges. Manure-derived biochar is a recalcitrant carbon reservoir characterized by superior cation exchange capacity, enhanced water retention and long-term carbon sequestration potential thereby mitigating greenhouse gas emissions and improving soil resilience. Its incorporation into soil also fosters microbial proliferation, improves nutrient retention and stabilizes organic matter. Additionally, biochar immobilizes potentially toxic elements, suppresses pathogenic microorganisms and mitigates contamination risks from antibiotics and other micro-pollutants. However, nitrogen loss during pyrolysis remains a concern, necessitating pre-treatment strategies such as ammonia stripping and liquid–solid separation to enhance nitrogen retention. Despite its multi-faceted agronomic and environmental benefits, large-scale adoption of cattle manure biochar remains constrained by knowledge gaps in pyrolysis conditions and feedstock composition, nutrient mass balance assessments and lack of information on long term soil and crop response. Hence, an attempt has been made in this chapter

S. Kapoor (✉) · R. Sharma · R. Bishist · P. Sharma · K. L. Gautam
Department of Silviculture and Agroforestry, Dr. Y S Parmar University of Horticulture and Forestry, Nauni, Solan, Himachal Pradesh, India

H. Sharma
Department of Silviculture and Agroforestry, College of Horticulture and Forestry Thunag, Mandi, Himachal Pradesh, India

D. Moulick et al. (eds.), *Biochar: Achieving Agro-Environmental Sustainability*, Sustainability Solutions, https://doi.org/10.1007/978-3-032-19855-6_17

to highlight the importance of cattle manure biochar in sustainable soil management, emphasizing its potential to drive climate-resilient agricultural systems while mitigating the environmental footprint of livestock farming.

Keywords Biochar · Carbon sequestration · Manure · Pyrolysis and resilience

17.1 Introduction

The impediment to sustainable agriculture in India relies on efficient recycling of livestock waste and boosting soil health. The global livestock sector encompasses an extensive spectrum of domesticated species including bovines, caprines, ovines and poultry. Livestock sector has surfaced as a linchpin in the architecture of food systems and rural economies worldwide. Livestock forms an essential reservoir of high-quality nutrition and contributes to nutritional security. Livestock sector is also a fundamental impetus for employment generation, income diversification and socio-economic mobility particularly in agrarian societies. The current projections indicate a staggering worldwide footprint of approximately 1.7 billion cattle, 2.2 billion sheep and goats and over 25 billion poultry underscoring the sector's gigantic extent—equivalent to nearly six poultry birds and one ruminant per every two humans on the planet (FAO 2020). This gigantic extent of livestock sector is a result of the "livestock revolution", fueled by demographic expansion, rising affluence, urban migration and evolving consumer preferences across emerging and developing economies. The consumption of various products derived from livestock has also enhanced manifolds over the years. Forecasts predicts an 80% increase in global meat demand by 2030, escalating to more than 200% by 2050 if present consumption trends continue (FAO 2020). Historical data also supports this trend, with global goat populations tripling—from 349 million in the 1960s to 1.12 billion in the 2020s—and poultry stocks increasing more than six-fold, from 3.91 billion to 26 billion over the same period (FAO 2024).

Livestock systems today accounts for about 40% of the worldwide agricultural output by value. Further, this sector supports the subsistence and livelihoods of an estimated 1.3 billion people, including 600 million resource-constrained smallholders globally (FAO 2020; World Bank 2020). As of 2019, India harbors the world's largest livestock population, accounting for approximately 31% of the world's total livestock population. This includes approximately 192 million cattle and 110 million buffalo representing world's second largest and largest population of these species, respectively (DAHD 2019). The livestock sector is indispensable to India's rural economy. Nearly 30.5 million people in India (~ 2.3% of the population) is reliant on livestock for their livelihood with small farmers deducing around 26% of their income from livestock (vs. 24% for all rural households). Cumulatively, animal husbandry supports ~ 4.1% of India's GDP (and about 25.6% of agricultural GDP). These activities play a crucial role in providing food (dairy, meat, eggs) and revenue; India is the world's largest milk supplier with total milk production of 239.3 million tonnes in 2023–24 (BAHS 2024) and livestock products provide access to essential protein to rural diets. Additionally, animals also carry out key farm functions. Bullocks are the foundation of small-holder agriculture (ploughing fields and transporting goods). Cattle dung and urine are customarily used as fuel (dung cakes), fertilizer and even as building plaster. In many villages, the greater part of all rural households keeps at least a handful of livestock for these purposes (Mitra et al. 2024). In essence, cattle and their manure are pivotal to India's agrarian livelihoods and nutritional food security. Albeit beneficial, the magnitude of manure generation presents difficulties (bulk storage, odors and emissions), stimulating focus towards novel management practices such as biochar. This chapter critiques cattle manure and management issues and then explores how changing manure to biochar can revolutionize waste liability into a resource that improves soil fertility, sequesters carbon and mitigates greenhouse gases.

17.2 Cattle Manure and its Composition

Cattle manure, chiefly cow dung, serves as a major constituent of agrarian ecosystems in India. Historically regarded as waste, cow dung is now ratified as a potent bio resource due to its physio-chemical and microbiological properties (Onwosi et al. 2022). On an average, a single cow releases between 9 and 15 kg of dung on a daily basis which serves as a primary input for a plethora of agronomic, energy and environmental applications (Gupta et al. 2016). From chemical perspective, cow dung is comprised for the most part of undigested residues rich in lignocellulosic material-lignin, cellulose and hemicellulose and contains an array of macro and micronutrients. These consist of nitrogen, phosphorus, potassium, sulphur and trace elements like magnesium, copper and zinc. Prominently, indigenous Indian cattle breeds yield dung with paramount mineral content, particularly, higher concentrations of calcium, phosphorus and zinc compared to their cross-bred counterparts. Cattle manure is a compositionally diverse organic material whose composition varies with animal diet, bedding and handling (Font-Palma 2019) as follows:

(a) *Moisture content*: Moisture is a principal constituent of fresh cattle manure. The moisture content of cattle manure varies between 13 and 75%t by weight. This variation significantly influences its suitability for thermochemical conversion as higher moisture content necessitates additional energy for drying prior to processing.
(b) *Ash content*: The ash content on a dry basis range between 10.8 and 45.2% which is substantially higher than that of other biomass residues such as wheat straw. Higher ash levels can lead to operational challenges in thermochemical processes, including slagging, fouling and corrosion due to the presence of alkali metals and other low-melting point constituents.

(c) *Elemental composition*: On a dry ash-free basis, the cattle manure typically contains 45.5–59.1% carbon (C), 4.4–7.2% hydrogen (H), 28.9–60.6% oxygen (O), 2.1–4.3% nitrogen (N) and 0.45–1.36% sulphur (S). These values support its classification as an energy rich organic waste suitable for bioenergy production.
(d) *Volatile matter and fixed carbon*: The volatile matter content ranges from 50.2 to 64.6% on a dry basis while fixed carbon content lies between 2.2 and 11.2%. The high volatile content is particularly beneficial for gasification processes, enhancing syngas yield.
(e) *Heating value*: The cattle manure possesses a higher heating value ranging from 8.7 to 18.7 MJ kg^{-1} on dry basis with values dependent on ash content and organic matter composition. This calorific potential makes it a viable feedstock for energy generationvia pyrolysis or gasification.
(f) *Nutrient composition*: The nutrient analysis indicates the presence of essential macronutrients including nitrogen (2.8–8.1 g kg^{-1}), phosphorus (approximately 7.89 g kg^{-1}) and potassium (approximately 38.45 g kg^{-1}). These nutrients derived primarily from the animal's diet make cattle manure a valuable soil amendment when applied appropriately.
(g) *Heavy metal content*: Trace elements such copper (10.5–352 mg kg^{-1}), zinc (41–727 mg kg^{-1}) and cadmium (< 0.1–1.74 mg kg^{-1}) are also present. While these can enhance micronutrient availability in soils excessive accumulation poses environmental risks and must be monitored.
(h) *Chemical formula and C:N ratio*: The approximate empirical formula of cattle manure on a dry and ash-free basis is $CH_{1.686}O_{0.05}S_{0.006}$. The carbon-to-nitrogen ratio typically ranges from 17:1 in raw manure to approximately 11: following treatments such as composting or vermicomposting.
(i) *Microbial diversity*: Cattle manure harbors a rich microbial diversity, including different bacterial species such as *Bacillus*, *Corynebacterium* and *Lactobacillus*, as well as protozoa and yeast which play an important role in nutrient cycling and soil health.

17.3 Benefits and Disadvantages of Cattle Manure

17.3.1 Benefits

Cattle manure elevates the physical status of the soil by increasing soil porosity, decreasing bulk density and improving the aggregation of soil. These changes steer towards optimized water infiltration, retention and aeration, fundamental for root development and microbial activity. Furthermore, cow dung has been demonstrated to neutralize acidic soils due to its buffering capacity, thereby optimizing pH levels for nutrient uptake (Tagele et al. 2023). Cattle manure brings forth and supports an ample array of favorable microorganisms including nitrogen fixers, phosphate solubilizers and organic matter decomposers like *Bacillus*, *Pseudomonas* and *Lactobacillus* spp. These microbial communities improve nutrient cycling, suppress the plant pathogens and promote plant health (Dhiman et al. 2020). Cattle manure functions as a prerequisite for organic farming systems. Its use correlates with agro ecological principles by lessening dependence on synthetic inputs and augmenting thesoil biodiversity and long-term fertility. Additionally, composted cow-dung is frequentlyincorporated into bio-dynamic preparations, enhancing soil health and crop resilience (Sharma et al. 2022). By means of anaerobic digestion, cattle manure generates biogas-sustainable and clean energy source. The biogas plants not only supply household energy security but also create nutrient-rich slurry as a by-product which can be delivered back to the fields as high-quality organic fertilizers (Carotenuto et al. 2016). The microbial extracts isolated from cattle excreta are substantial in industrial biotechnology for enzyme production (cellulase, xylanase, and protease), biofuel generation and even pharmaceutical applications. These enzymes are used in textile processing, paper bleaching and bioremediation of pollutants. Cattle dung exhibits antifungal and antibacterial properties, having demonstrated efficacy against *Fusarium*, *Rhizoctonia* and *Xanthomonas* spp. This catalyzes advancements for its use as a biocontrol agent in integrated pest management strategies (Behera and Ray 2021).

17.3.2 Disadvantages

Improperly maintained and stored livestock waste can lead to the emission of significant amount of greenhouse gases like methane (CH_4) and nitrous oxide (N_2O). Additionally, volatile organic compounds like ammonia and hydrogen sulphide may engender odor nuisance and air quality degradation (Andraskar et al. 2021). Raw cattle dung may retain zoonotic pathogens like *Escherichia coli*, *Salmonella* and *Listeria monocytogenes* (Gupta et al. 2016). When inadequately managed or freshly applied, it entails risks of infecting crops, water bodies and public spaces, potentially leading to disease outbreaks. The nutritional profile of manure varies substantially with animal diet, age, health and housing conditions (Rayne and Aula 2020). This disparity can lead to erratic nutrient supply in the soil, impeding precision nutrient management in agriculture. As opposed to synthetic fertilizers, nutrients in manure are gradually liberated due to microbial mineralization. In cases of high crop requirement or unfavorable weather conditions, this may lead to nutrient inadequacies or disparities, impacting crop performance. Extensive aggregation, composting and application of cattle manure entails substantial land area and labor inputs. This can be a constraint for smallholder farmers or those devoid of mechanization, especially hilly or marginal regions. Research has revealed steroid hormones and trace heavy metals in cattle dung (Ray et al. 2013). Through repeated soil incorporation, these contaminants may stockpile, eliciting risks to food safety and soil health over time.

17.4 Greenhouse Gas Emission from Livestock

Since the pre-industrial epoch, the terrestrial mean surface air temperature has elevated profoundly above the global average surface temperature. In the interval traversing from 2011 to 2020, the land surface temperatures were approximately 1.59 °C higher than between 1850 and 1900, indicating a cumulative rise in global warming (IPCC 2021). This discernible shift in the Earth's mean surface temperature-largely attributable to anthropogenic greenhouse gas (GHG) emissions since the advent of industrialization-is referred to as 'global warming' (IPCC 2021). Greenhouse gas (GHG) emissions originate from four fundamental sectors: energy production and consumption, industrial activities and product utilization, agriculture, forestry and other land-use systems and waste management strategies (IPCC 2019). The Kyoto Protocol recognizes six greenhouse gases: CO_2, CH_4, N_2O, HFCs, PFCs and SF_6 (UNFCCC). Of these, CO_2 accounts for about 76% of the global GHG emissions in CO_2-equivalent terms, trailed by CH_4 at 16% and N_2O at 6%. The subsequent 2% is attributed to fluorinated gases. Cumulatively, CO_2, CH_4, N_2O account for nearly 98% of total anthropogenic emissions (C2ES 2023).

According to the UNFCCC's most recent inventory statistics for developed nations, the agriculture sector, excluding emissions from land-use change and forestry (LULUCF), contributes for around 9.2% of total GHG emissions. This implies, relative to other sectors such as energy and industry, agriculture accounts for a lesser proportion of total emissions in these countries (Nugrahaeningtyas et al. 2024). In this milieu, livestock production solely accounted for 4.91% of total anthropogenic GHG emissions when evaluated on the basis of CO_2-equivalent values. Livestock-related activities, such as feed production, enteric fermentation and manure management release around 3.7–3.8 gigatons of CO_2-equivalent annually. According to the recent estimates, the cattle industry might emit up to 3.75 GT CO_2-eq per year, accounting 7–8% of total anthropogenic GHG emissions globally (Manzano et al. 2023). The preeminent emission sources within the livestock cognizance are enteric fermentation-mainly emitting methane and manure management, which liberates both methane and nitrous oxide. Agriculture, especially livestock-related activities account for the majority of worldwide anthropogenic methane emissions. According to the most current global methane estimates, agricultural sources account for about 40% of human-induced methane, with fossil fuel activities accounting roughly 34% (GCP 2024). The ruminant cattle accounts for over one-third of the globalanthropogenic methane emissions.

17.4.1 Enteric Fermentation: Source of CH_4 Emission

The eructation process in ruminant animals is the predominant source of methane emissions from enteric fermentation. Enteric fermentation in ruminants contributes for roughly 33% of total human-induced methane emissions. Nearly 70% of these emissions come from developing regions, including Latin America, Asia and Africa. Notably, Asia solely accounts for around 30.3%, owing to the region's high livestock density (Nugrahaeningtyas et al. 2024). In the digestive systems of ruminants, methane is produced as metabolic byproduct during microbial fermentation (FAO 2019). This fermentation takes place chiefly in the rumen, the initial chamber within the ruminal stomach, which serves as a complex microbial ecosystem aiding the degradation of ingested carbohydrates. The microbial disintegration of plant matter liberates hydrogen (H_2), which is eventually used bymethanogenic archaea to produce methane (CH_4). This methanogenic pathway serves as theprincipal route of hydrogen discharge in the rumen with methanogens metabolizing H_2 by ablating carbon substrates. The production of CH_4 in the rumen is inextricably tied to the composition of volatile fatty acids (VFA) formed during fermentation (Weimer 2015). The synthesis of acetate and butyrate facilitates methanogenesis, whereas propionate formation acts as a hydrogen sink competing with methanogenesis and reducing CH_4 output. Dietary measures substantially influence ruminal pH, which subsequently impacts VFA patterns.

Several physiological and nutritional factors govern the volume of methane produced, comprising the structure and function of the animal's digestive system (FAO 2019), the amount and digestibility of organic matter, the interval of ruminal feed retention, feed intake levels and the chemical makeup of the carbon source. This highlights the pivotal role of feeding regimes and nutritional management in alleviating enteric CH_4 emissions. For instance, high-quality forages are inversely correlated with methane emissions (Jonker et al. 2016), whereas low-quality roughages augment CH_4 output. Furthermore, stringent feed intake limitations can escalate methane production by up to 10% (Goopy et al. 2020), likely due to the high fraction-over 50% of digestible organic matter undergoing fermentation in the rumen, which is directly related to CH_4 generation. The composition and properties of the feed also determine its retention time within the rumen, consequentlyinfluencing methane production. In conclusion, enteric methane emissions are controlled by an interwoven dynamic of microbial ecology, dietary composition, fermentation and host physiology. These aspects offer strategic opportunities for mitigating CH_4 emissions through targeted and nutritional and management interventions.

17.4.2 Manure Management: A Significant Source of Methane and Nitrous Oxide Emissions

Livestock manure is mainly composed of animal feces and urine. It also includes additional components like bedding material, surplus water and feed leftovers. As a repository of vital macro and micronutrients, such as nitrogen (N), phosphorus (P), potassium (K), copper (Cu), manganese (Mn) and zinc (Zn), it is rich in organic matter (Dhiman et al. 2021). In particular, urine contains nitrogen in the form of urea, which is hydrolyzed by an enzyme called urease-originating from feces or soil-resulting in the formation of ammonia (NH_3). This urinary nitrogen represents a crucial antecedent to nitrous oxide (N_2O) emissions associated with manure management. The nutrient output from manure is highly intertwinedwith the digestibility of the animal's diet as more digestible feed results in altered nutrient excretion sequences.

The organic content in manure may decompose through either aerobic or anaerobic microbial pathways after excretion, contingent on handling procedures and ambient conditions. CH_4 emissions from manure primarily stem from the anaerobic breakdown of organic substrates by methanogenic archaea (Dalby et al. 2021). The microbial alteration of volatile fatty acids (VFA) into CH_4 and CO_2 is contingent upon storage conditions. Systems such as anaerobic lagoons, tanks or deep pits maximize the generation of CH_4 by favoring oxygen-derived conditions, compared to solid-state systems promoting aerobic decomposition. In addition to management approaches, several abiotic factors-ambient temperatures, moisture content, storage duration and the physico-chemical properties of the manure have a major impact on methanogen activity and proliferation, which in turn affects CH_4 production.

Manure emits nitrous oxide (N_2O) via both direct and indirect mechanisms. Microbial processes such as nitrification and denitrification produce direct emissions, including gases like nitric oxide (NO) and nitrous oxide (N_2O) (Lazcano et al. 2021). These transformations are aided by specialized enzymes, notably nitrate reductase and nitrite reductase. In contrast, indirect emissions arise when nitrogen is lost from manure due to ammonia volatilization, nitrate leaching, or runoff. These nitrogen forms can transform into N_2O in the environment, leading to emissions beyond the original manure site. The factors affecting N_2O emissions include animal's diet, manure management procedures, storage time and environmental variables such as pH, temperature, moisture and aeration. A nutrient-rich diet and extended manure storage in oxygen-rich conditions can considerably boost N_2O production. Consequently, better feeding techniques and improving manure management are critical measures for lowering greenhouse gas emissions from livestock operations.

17.5 Strategies to Control Greenhouse Gas Emissions

The livestock sector plays a major role in human induced greenhouse gas emissions. Direct emissions from livestock mainly include enteric fermentation and losses from manure which contribute 11% of total anthropogenic GHG emissions (Llonch et al. 2017). To address the increasing demand for livestock products driven by population growth, it is essential to implement mitigation approaches that lower the greenhouse gas emission intensity (Table 17.1).

Table 17.1 Feasible mitigation options to control the sources of GHG emissions

Emission sources	Mitigation options	Mechanisms	Reference
Enteric fermentation	Plant bioactive compounds dietary lipids	Changing the VFA proportions in ruminal fluid results in a reduction in fermentation	Williams et al. (2020)
		Increasing fat concentration decreased mean ruminal pH and increased the duration of pH below 6	Alvarez-Hess et al. (2019)
	Concentrate inclusion	Increased starch intake reduces ruminal pH, affecting DM and NDF digestibility and favors propionate rather than acetate in the rumen	Herliatika et al. (2024)
	Improving forage quality and management	With lower quality of forage, the fibre contents increase, thus higher CH_4 production occurs	Eugene et al. (2021)
	Processing of low-quality feeds	Improving nutritive value of low-quality feeds could increase productivity, thus reducing herd size and concomitant reduction in herd GHG emissions	Hristov et al. (2013)
	Macro-supplementation (when deficient)	Improve animal performance by supplementing available N for microbial protein synthesis in the rumen and balancing rations for macro and micro nutrients	
Manure management	Dietary manipulation and nutrient balance	Lower urea-N in urine and TAN results in lower NH_3	Edouard et al. (2019)
	Housing	High porosity of bio filter media containing a mixture of organic and inorganic media allowed sufficient oxygen transfer for methaneoxidation	Tung et al. (2021)
	Anaerobic digestion	Manure composition changes. NPK are transformed from organic forms to inorganic forms, whereas C is transformed to biogas for use as fuel	Tratzi et al. (2023), Harrison and Ndegwa (2020)
	Decreased storage time	When storage time is decreased and manure is applied directly to land, less CH_4 occurs after land application of manure	Grossi et al. (2019), Hristov et al. (2013)
	Timing of application	Application of manure on land before rain can decrease emission spike	

(continued)

Table 17.1 (continued)

Emission sources	Mitigation options	Mechanisms	Reference
	Solid-liquid separation	Mechanical separation of solid manure is done from liquid manure resulting in predominantly aerobic conditions, thereby reducing CH_4 emission by approximately 35%	Holly et al. (2017)
Animal management	Breeding efficiency	Enhancing reproductive efficiency and increasein fertility rate may decrease the methane emission by approximately 10–24%	Grossi et al. (2019)
	Genetic improvement	The genetic and nutritional improvement significantly improve the nitrogen utilizationand thereby reduced GHG emissions	

17.6 Biochar as a Solution

Biochar is a form of black carbon which is produced by thermal decomposition of plant-derived biomass-such as wood residues, agricultural waste or animal manure-under oxygen limited conditions. This process sustains the carbon content, making it less vulnerable to rapid disintegration and thereby augmenting its potential for long-term carbon sequestration (Yin et al. 2022). Biochar is formed via a thermochemical conversion process known as pyrolysis, wherein carbon-rich organic biomass-the feedstock-is subjected to elevated temperatures varying between 350 and 700 °C in a severely oxygen-deprived atmosphere (Twagirayezu et al. 2024). This restrictive oxygen environment prevents combustion thereby ensuring the biomass transformation while minimizing gaseous emissions. During pyrolysis, the structural components of the biomass disintegrate into smaller solid fractions. Following cooling, a fraction of the evolved gases condenses to form bio-oil (a form of biofuel), while the residual non-condensable gases linger as synthesis gas (syngas). The resultant solid fraction, principally derived from the lignin component of the biomass is what is regarded as '**biochar**'. It is essential to remember that biochar encompasses a broad array of carbonaceous solids rather than a single, homogenous chemical entity (Wang et al. 2022). The kind of feedstock, pyrolysis technique, thermal parameters and subsequent cooling and storage protocols all have a substantial impact on its composition and structure (Yadav et al. 2023).

The ability of biochar to sequester atmospheric carbon and therefore mitigate climate change is the principal driving force behind its production. According to the historical evidence, a primitive form of this technology-known as "Terra Preta" or "dark earth" was employed by the ancient civilizations in the Amazon basin. Indigenous inhabitants are believed to have engaged in 'slash-and-char agriculture', whereby vegetation was burned under controlled conditions and then the resulting char was integrated into the soil to improve the soil fertility and uplift productivity of crop (Yadav et al. 2023). Since consuming biochar for energy would re-release the sequestered carbon back into the atmosphere, negating the climatic advantages of biochar, soil application remains the predominant use of biochar in modern contexts (Akhil et al. 2021).

17.6.1 Mechanism of Action in Soil Environments

Biochar's contribution to soil fertility has been ascribed to its role in enhancing soil organic carbon (SOC), nutrient concentrations (particularly N, P and K) and cation exchange capacity (CEC). These characteristics were initially reported in Terra Preta soils, which developed through the integration of charcoal, food scraps and organic waste into nutrient-depleted tropical soils (Pandian et al. 2024; Yadav et al. 2023). The soils produced, showcased enhanced fertility and productivity even millennia after their creation.

17.6.2 Physico-chemical Properties of Biochar

The inherent properties of biochar are profoundly impacted by both the type of feedstock and the thermal regime employed in pyrolysis. Biochar generally exhibits an alkaline pH, a highly porous structure and a composition rich in both fixed and labile forms of carbon (Leng et al. 2021). While most biochars have relatively low macronutrient content in terms of directly supporting plant nutrition, their high cation exchange capacity (CEC) significantly augments the fertilizer application efficiency by enhancing nutrient retention and availability within the soil matrix. Among the prominent physico-chemical attributes, the porosity and specific surface area of biochar are preeminent, as these directly impact key soil properties like adsorption potential, water-holding capacity and microbial habitat formation (Tomczyk et al. 2020; Blanco-Canqui 2017). However, the applicability of a particular biochar for soil amendment is not universal; its efficacy is context-dependent, necessitating modification based on soil type, crop requirement and desired agronomic outcome. Thus, strategic feedstock selection and precise control of pyrolysis conditions are required to essential to customize biochar properties to specific applications.

17.7 Biochar Production from Cattle Manure

Cattle manure, an inevitable by-product of livestock production, has long been considered as a beneficial organic amendment for enhancing the soil fertility and crop productivity. However, intensive livestock production, particularly in high-density agricultural zones has resulted in a localized surplus of manure, raising environmental concerns such as nutrient leaching, greenhouse gas emissions, pathogens spread and the accumulation of potentially toxic elements (PTEs) (Huygens et al. 2020). Transforming cattle manure into biochar via pyrolysis presents an integrative solution to these multi-faceted issues while valorizing waste into a high-value soil amendment and carbon sequestration agent (Rathnayake et al. 2023).

The feasibility of cattle manure for biochar production emerges from its biochemical makeup. It's mainly composed of organic matter including cellulose (14–35%), hemicellulose (11–32%), lignin (13–15%), proteins (8–30%) and lipids and fats (15–32%) (Saady et al. 2021). These chemicals breakdown under heat, producing solid biochar, liquid bio-oil and gaseous products. The relative abundance of nutrients (eg, N, P, K, Ca, Mg) and ash in comparison to lignocellulosic biomass enhances its value as a soil amendment but 357 complicates energy efficiency owing to high moisture and lower energy density.

17.7.1 Pyrolysis: Thermal Conversion of Cattle Manure into Biochar

Pyrolysis is a thermo-chemical process that converts organic, carbon-rich feedstocks into three primary products-solid (biochar), liquid (bio-oil) and gas (non-condensable syngas) by subjecting them to high temperatures (typically 400 °C) in an oxygen-deprived environment (Hagemann et al. 2018). Biochar is defined as a solid substance that meets specific thresholds for hydrogen-to-carbon ($H/C < 0.7$) and oxygen-to-carbon ($O/C < 0.4$) molar ratios, making it suitable for non-oxidative applications such as soil improvement or long-term carbon sequestration (EBC 2020). The bio-oil and gaseous proportions formed during pyrolysis are frequently combusted to fulfill the energy requirements of the pyrolysis process, which include biomass drying or heat recovery for auxiliary applications such as district heating. Notably, the production and application of biochar has emerged as a recognized negative emission strategy, called pyrogenic carbon capture (PyCCS) due to its ability to sequester atmospheric carbon while improving soil health and mitigating environmentaldamage caused by agricultural activities (Schmidt et al. 2019 2021). Early documentation and ethnobotanical analysis by Frausin and co-workers (2014) displays that wood-based biochar was not an intended agronomic input but rather a common by-product of open-fire cooking. Today, however, advanced pyrolysis methods enable the deliberate conversion of biochar from a variety of biomass sources. These comprise unprocessed agricultural by-products (such as straw, husks), virgin biomass (e.g., forest residues) and even marginal or processed wastes like demolished timber and digested food scraps (Buss et al. 2016; Rathnayake et al. 2021; Rodriguez Alberto et al. 2021). Interestingly, high-ash biomass like cattle manure is becoming acknowledged for generating biochar with comparable or better soil-enhancing qualities compared to traditional wood feedstocks (Jeffery et al. 2017; Schmidt et al. 2021).

17.7.2 *Feedstock Preparation*

The pyrolysis of raw, moisture-laden manure brings about a considerable energy challenge because of the high thermal load required to evaporate water content prior to thermochemical conversion (Ro et al. 2010). Consequently, numerous strategies have been devised to mitigate this energy load while simultaneously improving nutrient retention and biochar quality-particularly in the context of cattle manure.

17.7.2.1 Solid-Liquid Separation

Mechanical separation of manure into solid and liquid components is a frequently used method. The separated solids often possess adequate dry matter content to be subjected to pyrolysis directly or after minimal drying, significantly reducing energy demands. The pre-dewatering of swine manure to around 75% moisture can lower the energy input needed for pyrolysis by roughly 19 times (Ro et al. 2010). The nutrient-rich liquid fraction, rather of being discarded has significant agronomic potential. It may be reintroduced into the system by enriching ("or charging") the resultant biochar to create a value-added biochar-based fertilizer (Schmidt et al. 2017). Alternatively, the liquid may act as a stand-alone fertilizer, be used in fertigation or undergo further treatment procedures such struvite precipitation to recover phosphorus (Federolf et al. 2016; Katanda et al. 2016). However, it is crucial to highlight that the direct field application of this liquid fraction may negate some of the advantages of pyrolysis, including the degradation of antibiotic residues and other micropollutants (Fig. 17.1).

17.7.2.2 Drying

A more traditional strategy employed in many experimental researches is to pre-dry manure without prior phase separation. Air-drying, solar drying and oven drying are common approaches for reducing moisture levels in manure feedstocks (Gomez-Munoz et al. 2016; Meng et al. 2018). In operational settings, excess heat generated by the combustion of pyrolysis vapors (bio-oils and gases) can be recovered to fulfill drying requirements. However, depending on the manure's initial moisture content and energy density, supplementary energy input may be necessary (Sanford et al. 2022). Importantly, drying manure prior to pyrolysis concentrates nutrients in the solid phase, potentially producing biochar with a richer nutritional profile than solid-liquid separation (Fig. 17.2).

17.7.2.3 Co-pyrolysis with Drier Biomass

Co-pyrolysis is a novel approach which combines wet manure with dry, energy-dense biomass like wood chips or rice straw. This approach boosts the feedstock's total dry

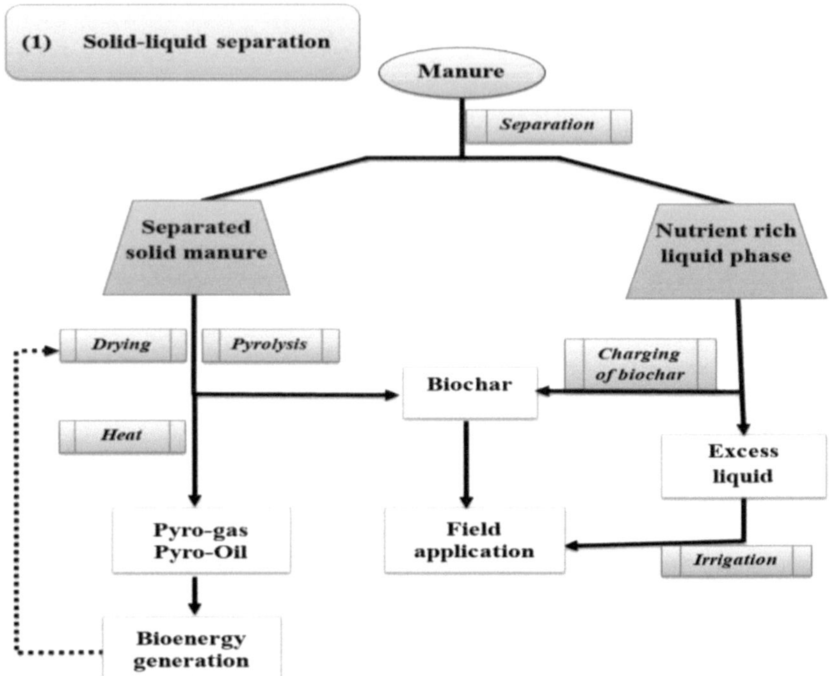

Fig. 17.1 Solid-liquid separation

Fig. 17.2 Drying

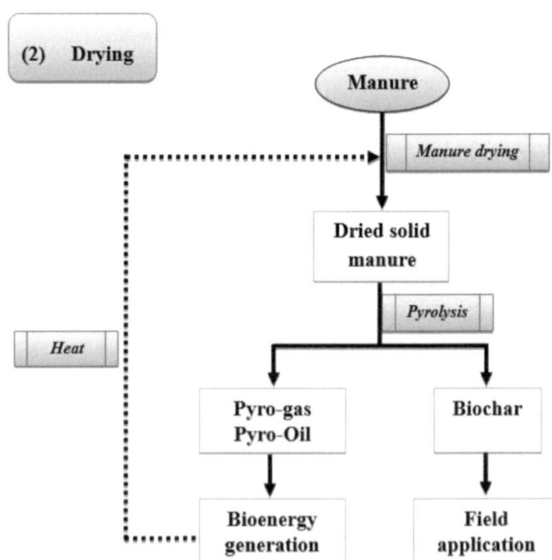

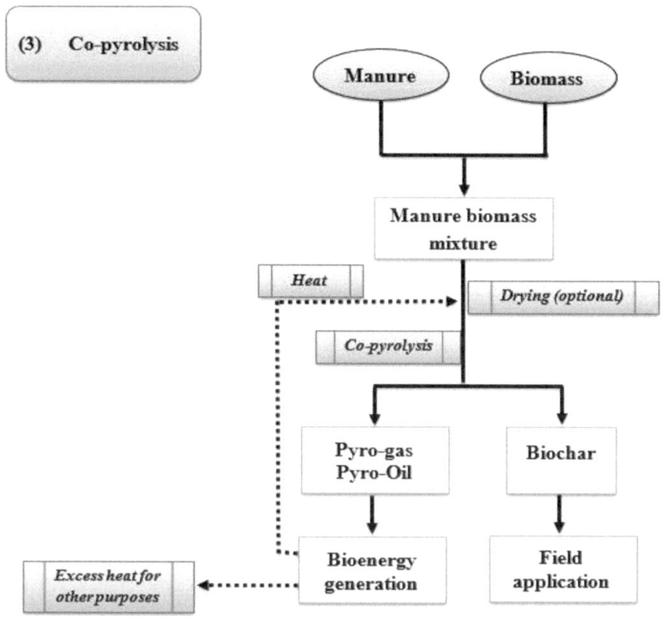

Fig. 17.3 Co-pyrolysis

matter content, thereby reducing or even eliminating the requirement for separate drying stages. According to studies, co-processing can increase biochar output, adjust the nutrient and potentially toxic element (PTE) content and customize the physico-chemical properties of the final-biochar product based on the co-feedstock used (Li et al. 2021; Meng et al. 2018; Qiu et al. 2023). Furthermore, co-pyrolysis promotes controlled nutrient release in soil applications and reduces the hazards of heavy metal accumulation in manure-derived biochar (Li et al. 2020; Qiu et al. 2023). Among all strategies, the combustion of non-condensable gases and vapors from pyrolysis offers a useful source of renewable energy, while bio-oils may also be employed in high-value applications (Leijenhorst 2020). Nonetheless, the primary goal remains the production of high-quality solid biochar, which is a potential tool for nutrient cycling, soil health improvement and long-term carbon storage (Fig. 17.3).

17.7.3 Effect of Feedstock and Pyrolysis Conditions on Biochar

17.7.3.1 Feedstock-Specific Characteristics

Manure derived biochar is generally rich in nutrients and displays a high CEC, making it ideal for nutrient-retentive applications. However, its high electrical conductivity may aggravate soil salinity and often contains lower levels of stable carbon, limiting its significance in long-term carbon sequestration (Rehman et al. 2020). Biochar produced from crop and grass residues has high potassium levels but possesses lower porosity and a noticeably higher concentration of organic carbon. Its application may be more suited for increasing soil fertility rather than improving structural qualities. Wood-based biochars, particularly those derived from hardwood or lignified biomass, stand out for their high fixed carbon content, which improves the structural stability and longevity of the soil. Their high porosity adds to better water retention.

17.7.3.2 Effect of pyrolysis Temperature

High temperature pyrolysis (500–650 °C) facilitates the development of ample microporosity, thereby escalating the adsorption capacity of biochar making it particularly suitable for amending contaminated soils or immobilizing heavy metals (Zou et al. 2024). Additionally, biochars formed at these temperatures tend to display increased pH values and greater mineral content though this is often coupled with a decline in CEC. Low temperature pyrolysis (350–500 °C) produces biochar rich in degradable organic carbon, which acts as substrate for soil microbial communities. These biochars have often higher CEC and are more conducive to fostering soil biological and nutrient cycling.

17.8 Effect of Cattle Manure Biochar on Agricultural Productivity

The escalating emission of livestock waste exhibits both a management barrier and a possibility for sustainable resource recycling. The conversion of cattle manure into biochar is a novel strategy which serves a dual-purpose: as soil amendment that delivers plant nutrients and enhances the long-term soil health while addressing environmental problems associated with raw manure applications. Recent analysis highlights the agronomic and ecological benefits of manure-based biochar. Its application boosts soil fertility, nitrogen cycling and plant productivity (Holatko et al. 2022). Notably, co-composting or fermenting manure with biochar and humic substances has been demonstrated to elevate its microbiological activity and nutrient efficiency, establishing it as a promising tool for regenerative agriculture (Brtnicky et al. 2021). Cattle manure biochar enhances the water dynamics as well as the structure of soil. Studies reveal that manure biochar discernibly augments soil porosity, hydraulic conductivity and water retention. For instance, Gavili et al. (2018) reported that incorporating 1.25% cattle manure biochar raised soil saturated hydraulic conductivity by 82–221% and total porosity by 2–12% under drought conditions. In essence, biochar's high porosity and low bulk density builds more pore space in the soil, boosting moisture retention by facilitating water infiltration and storage. Concisely put, integrating biochar made from cattle manure improves the aeration and soil bulk density by increasing pore connectivity and water-holding capacity.

Correspondingly, the manure derived biochar has pronounced effects on the chemistry of soil including pH, nutrient availability and soil organic carbon (SOC). Cattle manure biochar frequently increases the pH and cation exchange capacity (CEC) of soil. For instance, the increasing rate of cattle manure biochar amendment contributed to substantially higher soil pH and CEC (with 7 wt% application yielding highest values for pH and CEC). Furthermore, the cattle manure biochar incorporates stable carbon and nutrients to the soil. The higher biochar rates resulted in an increased soil organic carbon content, reflecting amplified soil carbon sequestration (Shin et al. 2024). The cattle manure biochar has displayed amplified soil exchangeable Mg and P and enhanced shoot uptake of Ca, Mg and P resulting in higher grain yields (Sapkota et al. 2025). Subsequently, cattle manure biochar fortifies the soil with organic carbon and plant nutrients, improving fertility and buffering capacity (pH and CEC). Cattle manure biochar also encourages soil biological activity. According to studies, biochar amendments improve microbial biomass and enzyme activities in the soil. Soils amended with manure biochar displayed improved soil enzymatic activities when compared to unamended soils, indicating a more active microbial population (Rathnayake et al. 2023). Biochar's porous structure facilitates biodiversity haven for microbes and can offer slow-release nutrients that support microbial growth. As reported in a study conducted by Sapkota and co-workers (2025), soils amended with cattle manure biochar exhibited changes in the makeup of microbial community, including enhanced fungal-to-bacterial ratios and microbial diversity. The manure-biochar can improve soil biological health by providing microorganisms with favorable circumstances (pH stability, moisture, substrate), even though it's less labile than raw manure. Overall, biochar derived from cattle manure tends to augment microbial biomass and activity, improving the biological fertility and resilience of soils.

Administering cattle manure biochar at optimal rates can enhance the crop yields. Plant growth is frequently enhanced by low to moderate dosages in water-limited environments. For instance, Gavili et al. (2018) reported that the integration of 1.25% cattle manure biochar improved the dry-matter production and water-use efficiency of spinach under drought, whereas higher amounts were ineffective or even harmful. The cattle manure biochar showed improved nutrient uptake and led to higher sorghum yield when compared to unamend soil (Sapkota et al. 2025). The combined physical and chemical changes in the soil (better moisture, more nutrients) that promote crop development are responsible for these yield gains. These impacts, however, might vary depending on the rate and context, excessively high biochar rates can cause nutritional imbalances or salt accumulation that negates the benefits. Overall, manure-biochar enhances agricultural productivity by augmenting crop biomass and yields in many situations when utilized sparingly (Table 17.2).

Table 17.2 Effect of cattle manure biochar on soil properties and crop growth

S.No.	Effect on soil properties	Effect on crop growth	References
1.	Enhanced phosphorus availability in the rhizosphere and non-rhizosphere soils; improved microbial diversity	Promoted root and shoot growth of *Populus euphratica* seedlings	Fan et al. (2023)
2.	Improved soil enzymatic activity and microbial biomass	Stimulated growthof *Salicornia herbacea* L.; reduced heavy metal uptake	Shin et al. (2024)
3.	Improved moisture content, nutrient retention and microbial population (bacteria, fungi, actinomycetes)	Increased growth and tuber yield of tiger nut (*Cyperus esculentus*)	Adekiya et al. (2020)
4.	Elevated soil pH and CEC; improved nitrogen, phosphorus and potassium levels	Enhanced lettuce growth parameters (fresh/dry weight, leaf number) and nitrogen use efficiency	Park et al. (2024)
5.	Improved water holding capacity by 1.5% per 1% biochar addition; nutrient retention; reduced leaching	Improved maize yield; enhanced water availability (upto 10–12 additional days at 12% biochar rate)	Rehman et al. (2020)

17.9 Manure Biochar as a Carbon Sink

The biochar which is derived from cattle manure plays an important role in pyrogenic carbon capture and storage (PyCCS), a well-established negative emission route that entails pyrolyzing biomass to create stable carbon forms and then integrating those forms into soils (Smith et al. 2019; Werner et al. 2018, 2022). The capture of atmospheric CO_2, its stability and quantifiable prolonged sequestration are necessary for the establishment of a functional carbon sink. Albeit transient sinks aid climatic relief, neither they nor the persistent sinks can completely neutralize fossil-based emissions (Leifield and Keel 2022). The carbon instilled in livestock manure emanates from animal feed and bedding materials-primarily atmospheric CO_2 fixed amid plant growth. This carbon is stabilized by pyrolysis into refractory, pyrogenic forms. Nonetheless, modern biochar production frequently emits 40–70% of the preliminary carbon back into the environment as CO_2, owing to the combustion of pyrolysis gases for energy (Schmidt et al. 2019). Consequently, only 30–60% of the initial biomass carbon is stably retained within the emergent biochar, defining its carbon potential.

The carbon footprint of biochar generation, comprising greenhouse gas emissions and the fraction of carbon predicted to mineralize must be taken into account for an appropriateestimate of net-negative emission (EBC 2020). According to Camps-Arbestian et al. (2015), biochars with hydrogen-to-organic carbon (H/Corg) ratios less than 0.4 are deemed adequately stable, with a speculated persistent carbon fraction of 74%-equating to a minimal annual degradation of 0.3%. The manure derived biochars nevertheless meet the stringent EBC requirements despite having reduced thermal stability which is linked to elevated ash and volatile carbon concentrations.

The carbon content of cattle manure biochar, which is produced at > 500 °C is typically between 47 and 69%, translating to a carbon offset potential of 1.07–1.67 tons of CO_2-equivalent (based entirely on the stable fraction) (Klasson 2017). Therefore, when thermochemically tailored, biochar evolves into a feasible and verified long-term carbon sink.

17.10 Conclusion

A major breakthrough in sustainable agriculture is cattle manure biochar which transforms a large amount of livestock waste into a high-efficiency input with multifaceted benefits. Its operationalization enhances soil health via improved structure, nutrient availability and microbial activity, while contemporaneously adding to long-term carbon sequestration. The integration of cattle manure biochar into farming systems can augment the crop productivity curtailing environmental externalities and advance climate resilience. Biochar made from manure is a quintessential exemplar of resource recovery and regenerative agriculture as it is a validated carbon sink and soil-quality enhancer. Its significance in closing nutrient loops and alleviating greenhouse gas emissions positions it as a central tenet in climate-smart agriculture frameworks. Henceforth, calibrating its generation and utilization across diverse agroecosystems will be paramount to harness its full potential as a tool for ecological and agronomic transformation.

References

Adekiya AO, Olaniran AF, Adenusi TT, Aremu C, Ejue WS, Iranloye YM, Gbadamosi A, Olayanju A (2020) Effects of cow dung and wood biochars and green manure on soil fertility and tiger nut (*Cyperus esculentus L.*) performance on a savanna Alfisol. Sci Rep 10(1). https://doi.org/10.1038/s41598-020-78194-5

Akhil D, Lakshmi D, Kartik A, Vo DVN, Arun J, Gopinath KP (2021) Production, characterization, activation and environmental applications of engineered biochar: a review. Environ Chem Lett 19:2261–2297. https://doi.org/10.1007/s10311-020-01167-7

Alvarez-Hess P, Williams S, Jacobs J, Hannah M, Beauchemin K, Eckard R, Wales W, Morris G, Moate P (2019) Effect of dietary fat supplementation on methane emissions from dairy cows fed wheat or corn. J Dairy Sci 102(3):2714–2723. https://doi.org/10.3168/jds.2018-14721

Andraskar J, Yadav S, Kapley A (2021) Challenges and control strategies ofodor emission from composting operation. Appl Biochem Biotechnol 193:2331–2356. https://doi.org/10.1007/s12010-021-03490-3568

BAHS (2024) Basic animal husbandry statistics, DADF, MoA, GoI. Available at: https://dahd.gov.in/schemes/programmes/animal-husbandry-statistics. Accessed on 20 June 2025

Behera SS, Ray RC (2021) Bioprospecting of cowdung microflora for sustainable agricultural, biotechnological and environmental applications. Curr Res Microb Sci 2:100018. https://doi.org/10.1016/j.crmicr.2020.100018

Blanco-Canqui H (2017) Biochar and soil physical properties. Soil Sci Soc Am J 81(4):687–711. https://doi.org/10.2136/sssaj2017.01.0017

Brtnicky M, Datta R, Holatko J, Bielska L, Gusiatin ZM, Kucerik J, Hammerschmiedt T, Danish S, Radziemska M, Mravcova L, Fahad S, Kintl A, Sudoma M, Ahmed N, Pecina V (2021) A critical review of the possible adverse effects of biochar in the soil environment. Sci Total Environ 796:148756. https://doi.org/10.1016/j.scitotenv.2021.148756

Buss W, Graham MC, Shepherd JG, Mašek O (2016) Risks and benefits of marginal biomass-derived biochars for plant growth. Sci Total Environ 569–570:496–506. https://doi.org/10.1016/j.scitotenv.2016.06.129

Camps-Arbestain M, Amonette JE, Singh B, Wang T, Schmidt H-P (2015) A biochar classification system and associated test methods. In: Lehmann J, Joseph S (eds) Biochar for environmental management. R outledge, pp 165–194

Carotenuto C, Guarino G, Morrone B, Minale M (2016) Temperature and pH effect on methane production from buffalo manure anaerobic digestion. Int J Heat Technol 34(S2):S425–S429. https://doi.org/10.18280/ijht.34s233

Center for Climate and Energy Solutions (C2ES) (2023) International emissions: global greenhouse gas emissions by gas. https://www.c2es.org/content/international-emissions/

Dalby FR, Hafner SD, Petersen SO, VanderZaag AC, Habtewold J, Dunfield K, Sommer SG (2021) Understanding methane emission from stored animal manure: a review to guide model development. J Environ Qual 50(4):817–835. https://doi.org/10.1002/jeq2.20252

DAHD (2019) Annual reports of Department of Animal Husbandry, Dairying and Fisheries. Available at: www.dahd.nic.in. Accessed on 20 June 2025

Dhiman S, Baliyan N, Maheshwari DK (2020) Buffalo dung-inhabiting bacteria enhance the nutrient enrichment of soil and proximate contents of *Foeniculum vulgare Mill*. Arch Microbiol 202(9):2461–2470. https://doi.org/10.1007/s00203-020-01969-x

Dhiman S, Kumar S, Baliyan N, Dheeman S, Maheshwari DK (2021) Cattle dung manure microbiota as a substitute for mineral nutrients and growth management practices in plants. Endophytes Min Nutr Manag 3:77–103. https://doi.org/10.1007/978-3-030-65447-4_4

EBC (2020) Certification of the carbon sink potential of biochar, version 2.1E, 1 Feb 2021. Ithaka Institute. http://european-biochar.org

Edouard N, Charpiot A, Robin P, Lorinquer E, Dollé J, Faverdin P (2019) Influence of diet and manure management on ammonia and greenhouse gas emissions from dairy barns. Animal 13(12):2903–2912. https://doi.org/10.1017/s1751731119001368

Eugène M, Klumpp K, Sauvant D (2021) Methane mitigating options with forages fed to ruminants. Grass Forage Sci 76(2):196–204. https://doi.org/10.1111/gfs.12540

Fan Y, Lv G, Chen Y, Chang Y, Li Z (2023) Differential effects of cow dung and its biochar on Populus euphratica soil phosphorus effectiveness, bacterial community diversity and functional genes for phosphorus conversion. Front Plant Sci 14. https://doi.org/10.3389/fpls.2023.1242469

FAO 2019 Food and Agriculture Organization of the United Nations. Climate change and the global dairy cattle sector: the role of the dairy sector in a low-carbon future. http://www.fao.org/3/ca2929en/ca2929en.pdf. Accessed on 20 June 2025

FAO: FAOSTAT Statistical Database (2024) https://www.fao.org/faostat/en/#home,

Federolf C, Westerschulte M, Olfs H, Broll G, Trautz D (2016) Enhanced nutrient use efficiencies from liquid manure by positioned injection in maize cropping in Northwest Germany. Eur J Agron 75:130–138. https://doi.org/10.1016/j.eja.2016.01.01622

Font-Palma C (2019) Methods for the treatment of cattle manure—a review. C—J Carbon Res 5(2):27. https://doi.org/10.3390/c5020027

Food and Agriculture Organization Report (2020) http://www.fao.org/faostat

Frausin V, Fraser JA, Narmah W, Lahai MK, Winnebah TR, Fairhead J, Leach M (2014) "God made the soil, but we made it fertile": gender, knowledge, and practice in the formation and use of African dark earths in Liberia and Sierra Leone. Hum Ecol 42:695–710. https://doi.org/10.1007/s10745-014-9686-0

Gavili E, Moosavi AA, Choghamarani FM (2018) Cattle manure biochar potential for ameliorating soil physical characteristics and spinach response under drought. Arch Agron Soil Sci 64(12):1714–1727. https://doi.org/10.1080/03650340.2018.1453925

Global Carbon Project (2024). Methane budget 2024—media highlights. https://www.globalcarbonproject.org/methanebudget/24/files/MethaneBudget2024_MediaHighlights.pdf
Gómez-Muñoz B, Case S, Jensen L (2016) Pig slurry acidification and separation techniques affect soil N and C turnover and N_2O emissions from solid, liquid and biochar fractions. J Environ Manag 168:236–244. https://doi.org/10.1016/j.jenvman.2015.12.018
Goopy JP, Korir D, Pelster D, Ali AIM, Wassie SE, Schlecht E, Dickhoefer U, Merbold L, Butterbach-Bahl K (2020) Severe below-maintenance feed intake increases methane yield from enteric fermentation in cattle. Br J Nutr 123(11):1239–1246. https://doi.org/10.1017/s0007114519003350
Grossi G, Goglio P, Vitali A, Williams AG (2019) Livestock and climate change: impact of livestock on climate and mitigation strategies. Anim Front 9(1):69–76. https://doi.org/10.1093/af/vfy034
Gupta KK, Aneja KR, Rana D (2016) Current status of cow dung as a bioresource for sustainable development. Bioresour Bioprocess 3(1). https://doi.org/10.1186/s40643-016-0105-9
Hagemann N, Spokas K, Schmidt H, Kägi R, Böhler M, Bucheli T (2018) Activated carbon, biochar and charcoal: linkages and synergies across pyrogenic carbon's ABCs. Water 10(2):182. https://doi.org/10.3390/w10020182
Harrison JH, Ndegwa PM (2020) Anaerobic digestion of dairy and swine waste. ASA special publication, pp 115–127. https://doi.org/10.2134/asaspecpub67.c13
Herliatika A, Widiawati Y, Jayanegara A, Harahap RP, Kusumaningrum DA, Shiddieqy MI, Adiati U (2024) Meta-analysis of the relationship between dietary starch intake and enteric methane emissions in cattle from in vivo experiments. J Adv VetY Anim Res 11(1):212. https://doi.org/10.5455/javar.2024.k767
Holatko J, Bielska L, Hammerschmiedt T, Kucerik J, Mustafa A, Radziemska M, Kintl A, Baltazar T, Latal O, Brtnicky M (2022) Cattle manure fermented with biochar and humic substances improve the crop biomass, microbiological properties and nutrient status of soil. Agronomy 12(2):368. https://doi.org/10.3390/agronomy12020368
Holly MA, Larson RA, Powell JM, Ruark MD, Aguirre-Villegas H (2017) Greenhouse gas and ammonia emissions from digested and separated dairy manure during storage and after land application. Agric Ecosyst Environ 239:410–419. https://doi.org/10.1016/j.agee.2017.02.007
Hristov AN, Oh J, Lee C, Meinen R, Montes F, Ott T, Firkins J, Rotz A, Dell C, Adesogan A, Yang W, Tricarico J, Kebreab E, Waghorn G, Dijkstra J, Oosting S (2013) Mitigation of greenhouse gas emissions in livestock production: a review of technical options for non-CO_2 emissions. In: Gerber PJ, Benjamin H, Makkar HPS (eds) FAO animal production and health (Book no. 177). FAO. https://www.uncclearn.org/wp-content/uploads/library/fao180.pdf
Huygens D, Orveillon G, Lugato E, Tavazzi S, Comero S, Jones A, Gawlik B, Saveyn H, Joint Research Centre (2020) Technical proposals for the safe use of processed manure above the threshold established for nitrate vulnerable zones by the nitrates directive (91/676/EEC). In: Publications office of the European Union [report]. Publications Office of the European Union. https://doi.org/10.2760/373351
Intergovernmental Panel on Climate Change (IPCC) (2019) 2019 refinement to the 2006 IPCC guidelines for national greenhouse gas inventories: Vol 4—Agriculture, forestry and other land use. http://www.ipcc-nggip.iges.or.jp/public/2019rf/vol4.html. Accessed on 20 June 2025
Intergovernmental Panel on Climate Change (IPCC) (2021) Climate change 2021: the physical science basis. Contribution of working group I to the sixth assessment report of the intergovernmental panel on climate change. Cambridge University Press. Accessed on 20 June 2025
Intergovernmental Panel on Climate Change (IPCC) (2021) Climate change 2021: the physical science basis. Contribution of working group I to the sixth assessment report. https://www.ipcc.ch/report/ar6/wg1/chapter/chapter-2
Jeffery S, Abalos D, Prodana M, Bastos AC, Van Groenigen JW, Hungate BA, Verheijen F (2017) Biochar boosts tropical but not temperate crop yields. Environ Res Lett 12(5):053001. https://doi.org/10.1088/1748-9326/aa67bd

Jonker A, Muetzel S, Molano G, Pacheco D (2016) Effect of fresh pasture forage quality, feeding level and supplementation on methane emissions from growing beef cattle. Anim Prod Sci 56(10):1714. https://doi.org/10.1071/an15022

Katanda Y, Zvomuya F, Flaten D, Cicek N (2016) Hog-manure-recovered struvite: effects on canola and wheat biomass yield and phosphorus use efficiencies. Soil Sci Soc Am J 80(1):135–146. https://doi.org/10.2136/sssaj2015.07.0280

Klasson KT (2017) Biochar characterization and a method for estimating biochar quality from proximate analysis results. Biomass Bioenergy 96:50–58. https://doi.org/10.1016/j.biombioe.2016.10.011

Lazcano C, Zhu-Barker X, Decock C (2021) Effects of organic fertilizers on the soil microorganisms responsible for N_2O emissions: a review. Microorganisms 9(5):983. https://doi.org/10.3390/microorganisms9050983

Leifeld J, Keel SG (2022) Quantifying negative radiative forcing of non-permanent and permanent soil carbon sinks. Geoderma 423:115971. https://doi.org/10.1016/j.geoderma.2022.115971

Leijenhorst E (2020) Pyrolysis of agro-digestate. In: Meers E, Velthof GL, Michels E, Rietra R (eds) Biorefinery of inorganics: recovering mineral nutrients from biomass and organic waste. Wiley, pp 133–146

Leng L, Xiong Q, Yang L, Li H, Zhou Y, Zhang W, Huang H (2021) An overview on engineering the surface area and porosity of biochar. Sci Total Environ 763:144204. https://doi.org/10.1016/j.scitotenv.2020.144204

Li J, Wang X, Cong C, Wan L, Xu Y, Li X, Hou F, Wu Y, Wang L (2020) Inoculation of cattle manure with microbial agents increases efficiency and promotes maturity in composting. 3 Biotech 10(3). https://doi.org/10.1007/s13205-020-2127-4

Li Q, Lin H, Fan H, Zhang S, Yuan X, Wang Y, Xiang J, Hu S, Kontchouo FMB, Hu X (2021) Co-pyrolysis of swine manure and pinewood sawdust: evidence of cross-interaction of the volatiles and profound impacts on product characteristics. Renew Energy 179:1370–1384. https://doi.org/10.1016/j.renene.2021.07.104

Llonch P, Haskell MJ, Dewhurst RJ, Turner SP (2017) Current available strategies to mitigate greenhouse gas emissions in livestock systems: an animal welfare perspective. Animal 11(2):274–284. https://doi.org/10.1017/S1751731116001440

Manzano P, Del Prado A, Pardo G (2023) Comparable GHG emissions from animals in wildlife and livestock-dominated savannas. Npj Clim Atmos Sci 6(1). https://doi.org/10.1038/s41612-023-00349-8

Mitra A, Tripathy AK, Singh RK, Bujarbaruah KM (2024) Transforming the Indian livestock sector. In: Transformation of agri-food systems. Springer Nature, Singapore, pp 173–191. https://doi.org/10.1007/978-981-99-8014-7_14

Meng J, Liang S, Tao M, Liu X, Brookes PC, Xu J (2018) Chemical speciation and risk assessment of Cu and Zn in biochars derived from co-pyrolysis of pig manure with rice straw. Chemosphere 200:344–350. https://doi.org/10.1016/j.chemosphere.2018.02.138

Nugrahaeningtyas E, Lee J, Park K (2024) Greenhouse gas emissions from livestock: sources, estimation, and mitigation. J Anim Sci Technol 66(6):1083–1098. https://doi.org/10.5187/jast.2024.e86

Onwosi CO, Ozoegwu CG, Nwagu TN, Nwobodo TN, Eke IE, Igbokwe VC, Ugwuodo CJ (2022) Cattle manure as a sustainable bioenergy source: prospects and environmental impacts of its utilization as a major feedstock in Nigeria. Bioresour Technol Rep 19:101151. https://doi.org/10.1016/j.biteb.2022.101151

Pandian K, Vijayakumar S, Mustaffa MRAF, Subramanian P, Chitraputhirapillai S (2024) Biochar–a sustainable soil conditioner for improving soil health, crop production and environment under changing climate: a review. Front Soil Sci 4:1376159. https://doi.org/10.3389/fsoil.2024.1376159

Park J, Cho H, Lee I, Kang S (2024) Effect of cow manure biochar on lettuce growth and nitrogen agronomy efficiency. Plants 13(23):3326. https://doi.org/10.3390/plants13233326

Qiu J, De Souza MF, Robles-Aguilar AA, Ghysels S, Ok YS, Ronsse F, Meers E (2023) Improving biochar properties by co-pyrolysis of pig manure with bio-invasive weed for use as the soil amendment. Chemosphere 312:137229. https://doi.org/10.1016/j.chemosphere.2022.137229

Rathnayake D, Ehidiamhen PO, Egene CE, Stevens CV, Meers E, Masek O, Ronsse F (2021) Investigation of biomass and agricultural plastic co-pyrolysis: effect on biochar yield and properties. J Anal Appl Pyrolysis 155:105029. https://doi.org/10.1016/j.jaap.2021.105029

Rathnayake D, Schmidt HP, Leifeld J, Mayer J, Epper CA, Bucheli TD, Hagemann N (2023) Biochar from animal manure: a critical assessment on technical feasibility, economic viability, and ecological impact. GCB Bioenergy 15(9):1078–1104. https://doi.org/10.1111/gcbb.13082

Ray P, Zhao Z, Knowlton KF (2013) Emerging contaminants in livestock manure: hormones, antibiotics and antibiotic resistance genes. In: Sustainable animal agriculture. CABI, Wallingford UK, pp 268–283. https://doi.org/10.1079/9781780640426.0268

Rayne N, Aula L (2020) Livestock manure and the impacts on soil health: a review. Soil Syst 4(4):64. https://doi.org/10.3390/soilsystems4040064

Rehman A, Nawaz S, Alghamdi HA, Alrumman S, Yan W, Nawaz MZ (2020) Effects of manure-based biochar on uptake of nutrients and water holding capacity of different types of soils. Case Stud Chem Environ Eng 2:100036. https://doi.org/10.1016/j.cscee.2020.100036

Ro KS, Cantrell KB, Hunt PG (2010) High-temperature pyrolysis of blended animal manures for producing renewable energy and value-added biochar. Ind Eng Chem Res 49(20):10125–10131. https://doi.org/10.1021/ie101155m

Rodriguez Alberto D, Tyle AC, Trabold TA (2021) Phosphate adsorption using biochar derived from solid digestate. Bioresour Technol Rep 16:100864. https://doi.org/10.1016/j.biteb.2021.100864

Saady NMC, Rezaeitavabe F, Ruiz Espinoza JE (2021) Chemical methods for hydrolysing dairy manure fiber: a concise review. Energies 14(19):6159. https://doi.org/10.3390/en14196159

Sanford JR, Johnstone D, Larson RA (2022) Assessing changes to nutrient density and availability following separation, drying, and pyrolysis of manure solids. Bioresour Technol Rep 19:101169. https://doi.org/10.1016/j.biteb.2022.101169

Sapkota S, Ghimire R, Brewer CE, Fernando S (2025) Contrasting effects of plant and animal residue biochars on soil health, carbon stability, and crop yield. J Soils Sediments. https://doi.org/10.1007/s11368-025-03968-1

Schmidt H, Pandit BH, Cornelissen G, Kammann CI (2017) Biochar-based fertilization with liquid nutrient enrichment: 21 field trials covering 13 crop species in Nepal. Land Degrad Dev 28(8):2324–2342. https://doi.org/10.1002/ldr.2761

Schmidt HP, Anca-Couce A, Hagemann N, Werner C, Gerten D, Lucht W, Kammann C (2019) Pyrogenic carbon capture and storage. GCB Bioenergy 11(4):573–591. https://doi.org/10.1111/gcbb.12553

Schmidt HP, Kammann C, Hagemann N, Leifeld J, Bucheli TD, Sánchez Monedero MA, Cayuela ML (2021) Biochar in agriculture-a systematic review of 26 global meta-analyses. GCB Bioenergy 13(11):1708–1730. https://doi.org/10.1111/gcbb.12889

Sharma PB, Kumar P, Tiwari S, Kumari S, Singh M (2022) The potential of cow's dung in sustainable agriculture and environmental health. In: Animal manure: agricultural and biotechnological applications. Springer International Publishing, Cham, pp 107–122

Shin H, Chun D, Cho I, Hanif MA, Kang S, Kwac LK, Kim HG, Kim YS (2024) Systematic characterization of cow manure biochar and its effect on *Salicornia herbacea L.* Growth Sustain 16(8):3396. https://doi.org/10.3390/su16083396

Smith P, Adams J, Beerling DJ, Beringer T, Calvin KV, Fuss S, Griscom B, Hagemann N, Kammann C, Kraxner F, Minx JC, Popp A, Renforth PV, Vicente JLV, Keesstra S (2019) Land-management options for greenhouse gas removal and their impacts on ecosystem services and the sustainable development goals. Annu Rev Environ Resour 44(1):255–286. https://doi.org/10.1146/annurev-environ-101718-033129

Tagele SB, Kim RH, Jeong M, Lim K, Jung DR, Lee D, Shin JH (2023) Soil amendment with cow dung modifies the soil nutrition and microbiota to reduce the ginseng replanting problem. Front Plant Sci 14:1072216. https://doi.org/10.3389/fpls.2023.1072216

Tomczyk A, Sokołowska Z, Boguta P (2020) Biochar physicochemical properties: pyrolysis temperature and feedstock kind effects. Rev Environ Sci Biotechnol 19(1):191–215

Tratzi P, Paolini V, Torre M, Palma A, Petracchini F (2023) Anaerobic digestion manure conversion and recycling. In: Manure technology and sustainable development. Springer Nature Singapore, Singapore, pp 33–63. https://doi.org/10.1007/978-981-19-4120-7_2

Tung NT, Chiemchaisri W, Chiemchaisri C, Sanguanpak S (2021) Removal of greenhouse gas in biofilter using organic and inorganic media. Vietnam J Sci Technol Eng 63(1):83–89. https://doi.org/10.31276/VJSTE

Twagirayezu G, Cheng H, Wu Y, Lu H, Huang S, Fang X, Irumva O (2024) Insights into the influences of biochar on the fate and transport of pesticides in the soil environment: a critical review. Biochar 6(1):9. https://doi.org/10.1007/s42773-024-00301-w

United Nations Framework Convention on Climate Change (UNFCCC) (n.d.). Kyoto protocol–targets for the first commitment period. https://unfccc.int/process-and-meetings/the-kyoto-protocol/what-is-the-kyoto-protocol/kyoto-protocol-targets-for-the-first-commitment-period

Wang L, Ok YS, Tsang DC, Alessi DS, Rinklebe J, Mašek O, Hou D (2022) Biochar composites: emerging trends, field successes and sustainability implications. Soil Use Manag 38(1):14–38. https://doi.org/10.1111/sum.12731

Weimer PJ (2015) Ruminal fermentations to produce liquid and gaseous fuels. In: Rumen microbiology: from evolution to revolution, pp 265–280. https://doi.org/10.1007/978-81-322-2401-3_18

Werner C, Schmidt HP, Gerten D, Lucht W, Kammann C (2018) Biogeochemical potential of biomass pyrolysis systems for limiting global warming to 1.5 °C. Environ Resour Lett 13(4):aabb0c. https://doi.org/10.1088/1748-9326/aabb0c

Werner C, Lucht W, Gerten D, Kammann C (2022) Potential of land-neutral negative emissions through biochar sequestration. Earth's Future 10(7):e2021EF002583. https://doi.org/10.1029/2021EF002583

Williams SRO, Hannah MC, Eckard RJ, Wales WJ, Moate PJ (2020) Supplementing the diet of dairy cows with fat or tannin reduces methane yield, and additively when fed in combination. Animal 14:s464–s472. https://doi.org/10.1017/S1751731120001032

World Bank (2020) Moving towards sustainability: the livestock sector and the world bank. https://www.worldbank.org/en/topic/agriculture/brief/moving-towards-sustainability-the-livestock-sector-and-the-world-banksustainability-the-livestock-sector-and-the-world-bank

Yadav SPS, Bhandari S, Bhatta D, Poudel A, Bhattarai S, Yadav P, Oli B (2023) Biochar application: a sustainable approach to improve soil health. J Agric Food Res 11:100498. https://doi.org/10.1016/j.jafr.2023.100498

Yin J, Zhao L, Xu X, Li D, Qiu H, Cao X (2022) Evaluation of long-term carbon sequestration of biochar in soil with biogeochemical field model. Sci Total Environ 822:153576

Zou X, Debiagi P, Amjed MA, Zhai M, Faravelli T (2024) Impact of high-temperature biomass pyrolysis on biochar formation and composition. J Anal Appl Pyrolysis 179:106463. https://doi.org/10.1016/j.jaap.2024.106463

Chapter 18
Applications of Artificial Intelligence (AI) and Machine Learning (ML) in Biochar Optimization

S. G. Sarowar, Pradarshak Sen, Tripti Pal, Chandra Saha, Debmalya Sarkar, Jyotirmay Roy, Subrata Garain, and Suman Dutta

Abstract The application of Artificial Intelligence (AI) and Machine Learning (ML) in biochar optimization is revolutionizing the way biochar is produced, characterized and applied in environmental and agricultural systems. Traditional biochar optimization methods rely on trial-and-error approaches, which are time-consuming and resource-intensive. AI and ML techniques provide a transformative alternative by enabling data-driven predictions, real-time process optimization and decision-making support across various stages of biochar production and application. This chapter will explore how AI-driven models enhance the efficiency and precision of biochar synthesis by optimizing key parameters such as feedstock selection, pyrolysis temperature, heating rate and residence time. Advanced ML algorithms, including artificial neural networks (ANNs), support vector machines (SVMs), random forests (RF) and deep learning frameworks, have demonstrated exceptional accuracy in predicting biochar yield, surface area, porosity and adsorption efficiency. Reinforcement learning and Bayesian optimization techniques further refine pyrolysis conditions for targeted applications, including carbon sequestration, soil enhancement and contaminant remediation. Moreover, AI-powered image recognition and spectroscopy-based ML models facilitate rapid characterization of biochar properties, reducing reliance on expensive and labour-intensive analytical methods. Predictive

S. G. Sarowar · D. Sarkar · J. Roy
Division of Soil Science and Agricultural Chemistry, ICAR-Indian Agricultural Research Institute, New Delhi, India
e-mail: sarowar13998@gmail.com

P. Sen · T. Pal · C. Saha
ICAR-Indian Institute of Soil Science, Bhopal, Madhya Pradesh, India

S. Garain
Department of Agricultural Economics, Palli Sikshsa Bhavana (Institute of Agriculture), Visva-Bharati (Central University), Sriniketan, West Bengal, India

S. Dutta (✉)
Faculty Centre for Integrated Rural Development and Management, Ramakrishna Mission Vivekananda Educational and Research Institute, Kolkata, West Bengal, India
e-mail: duttasumansiliguri@gmail.com

D. Moulick et al. (eds.), *Biochar: Achieving Agro-Environmental Sustainability*, Sustainability Solutions, https://doi.org/10.1007/978-3-032-19855-6_18

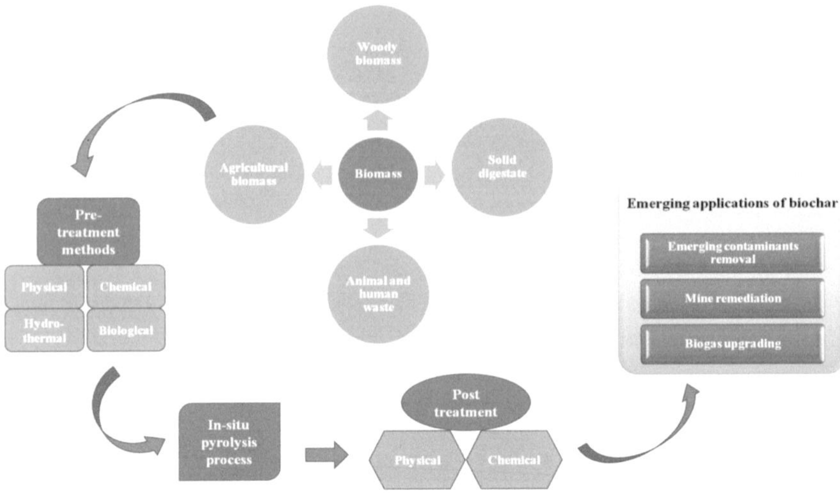

Fig. 18.1 Graphical abstract—biochar production and its application optimization

ML models are also transforming biochar application strategies by assessing its long-term effects on soil health, microbial activity and nutrient retention. Hybrid AI models integrating mechanistic simulations with ML approaches offer high-precision decision-support systems for sustainable biochar utilization in agriculture and environmental management. Despite these advancements, challenges such as data heterogeneity, model interpretability and limited datasets hinder widespread adoption. This chapter will identify key research gaps and proposes innovative strategies, such as AI-integrated bioinformatics, cloud-based modeling and block chain-enabled data-sharing frameworks to drive the next generation of biochar research. By leveraging AI and ML, biochar optimization can achieve unprecedented precision and scalability, offering sustainable solutions for climate resilience, carbon neutrality, and regenerative agriculture.

Keywords Biochar optimization · Artificial intelligence (AI) · Machine learning (ML) · Pyrolysis process modelling · Carbon sequestration

See Fig. 18.1.

18.1 Introduction

The synthesis and use of biochar, a carbon-rich substance obtained from the pyrolysis of biomass, are being optimised by the revolutionary forces of artificial intelligence (AI) and machine learning (ML). The incorporation of these technologies facilitates a more efficient, sustainable, and tailored approach to biochar production and utilization, significantly impacting soil health, agricultural productivity, and carbon sequestration. AI can identify the ideal parameters to maximise the yield of biochar and enhance its advantageous qualities, such as porosity, carbon content, and nutrient availability, by analysing vast datasets about pyrolysis conditions, such as temperature, heating rate, and residence time (Campbell et al. 2018; Ukoba and Jen 2022). Production of biochar, which has historically relied on manual processes, has changed with the integration of AI and ML technologies. This predictive modelling capability enables producers to make dynamic adjustments based on real-time data, ensuring consistent biochar quality suitable for a variety of applications (Ukoba and Jen 2022). The benefits of using AI and ML go beyond just increasing output. With the help of these tools, farmers may make well-informed decisions about the use of biochar by enabling reliable forecasts about how it will affect various soil types and agricultural systems. In order to suggest the best biochar formulations for raising agricultural yields and improving soil quality while reducing environmental impact (Palansooriya et al. 2022a, b; Yadav et al. 2017). By addressing particular issues faced by farmers in various locations, these insights can assist maximise the use of biochar to enhance nutrient cycling and water retention (Nazeer et al. 2022). Furthermore, biochar itself is a practical carbon sequestration method that efficiently absorbs CO_2 from the atmosphere and lowers greenhouse gas emissions. By predicting the long-term stability of biochar in a variety of applications, AI techniques can enhance knowledge of its potential to mitigate climate change (Yadav et al. 2017). Because it can transform biomass into a stable form of carbon that might linger in soils for decades to centuries, the Intergovernmental Panel on Climate Change (IPCC) views it as a negative emissions technology (Zhang et al. 2022; Mohan et al. 2018). Optimising manufacturing techniques to produce biochar with higher recalcitrance can significantly extend the lifespan of carbon storage (Mohan et al. 2018).

Moreover, the efficiencies gained from AI and ML technologies can also contribute to waste management practices. Producers can efficiently reduce waste and produce a valuable soil amendment by turning biomass byproducts such as wood waste and agricultural residues into biochar. This waste valorization process not only deals with excess biomass but also recycles vital nutrients back into the agricultural system (Jatuwong et al. 2025; Nurmalasari et al. 2022). AI-driven analytics assist in identifying the best feedstocks and production protocols for maximizing biochar yield and environmental benefits, supporting the argument for its wider adoption (Ippolito et al. 2020). The economic effects of AI and ML in the manufacture of biochar are significant in addition to the environmental advantages. These technologies streamline procedures and enhance quality control, which lowers production costs and time.

This enables a more customized strategy that satisfies particular agricultural requirements (Ukoba and Jen 2022; Viet et al. 2024). Biochar can be used for purposes other than agriculture. Biochar is a viable remedy for soil remediation since studies show that it may successfully stabilize heavy metals in contaminated soils (Palansooriya et al. 2022a, b; Nazeer et al. 2022). By forecasting biochar's potential for pollutant adsorption based on soil and biochar features, AI-driven evaluations can further increase the efficacy of biochar in environmental applications. According to Canlas et al. (2019), this competence facilitates effective decision-making on the application of biochar in remediation operations, which may eventually lead to healthier ecosystems. Moreover, biochar increases the resilience and fertility of soils by encouraging advantageous microbial activity. A review by Kaur et al. (2024) claims that using biochar improves microbial activity, which in turn improves nutrient retention and reduces leaching. Biochar can reduce the requirement for artificial fertilizers and enhance crop yield in a sustainable way by promoting better microbial communities (Ukoba and Jen 2022). By using AI technologies to examine how microorganisms react to biochar additions, management procedures and dosages can be optimized (Wang 2024).

Despite these important benefits, it is critical to acknowledge the difficulties in using AI-driven technology to biochar research and manufacturing. For instance, the data utilized to train predictive models must be carefully considered because of the heterogeneity in biochar attributes caused by the variety of feedstocks and production techniques. For AI and ML systems to produce reliable insights and results, it will be crucial to comprehend these processes (Al-Wabel et al. 2018). To establish standardized procedures that support the efficient application of these technologies, cooperation between researchers, farmers, and technologists will be essential (Ukoba and Jen 2022). A revolutionary change in the best way to make and use biochar is represented by the incorporation of AI and ML into the technology. Through improved operational efficiency and predictive capabilities, these technologies tackle urgent issues in environmental management and agriculture. Biochar is important in larger attempts to mitigate climate change because of its potential to enhance soil health, trap carbon, and encourage sustainable farming practices. The advancement of biochar applications will be greatly aided by AI and ML as research and technology develop further, providing avenues for healthier ecosystems and more resilient agricultural systems around the world.

There are several key research gaps in the application of AI and ML in biochar optimization. These include the need for better integration of AI with traditional mechanistic models, addressing data scarcity and quality issues, and scaling AI-driven process optimization to industrial levels. Additionally, there is a gap in the use of deep learning frameworks for complex biochar property prediction and predictive modeling for carbon sequestration. The development of AI-powered adaptive biochar strategies for climate resilience and real-time pollutant removal optimization is also underexplored. Furthermore, AI-enhanced environmental impact assessments, biochar modification for heavy metal adsorption, and standardization of AI models remain crucial areas for advancement. Lastly, ethical considerations and policy frameworks for AI in biochar research need to be established to guide its responsible and transparent application. This chapter aims to address the identified research gaps by exploring the potential of AI and ML in biochar optimization, particularly focusing on process modeling, property prediction, and environmental impact assessment. This chapter aims to address research gaps by exploring AI and ML techniques in biochar optimization, focusing on process modeling, property prediction, carbon sequestration, and pollutant removal. It investigates AI-driven methods for optimizing biochar production, enhancing property prediction, and improving real-time monitoring for pollutant removal. By providing solutions for data quality, scalability, and ethical AI use, the chapter seeks to drive more sustainable, efficient, and responsible biochar production and application strategies.

18.2 Applications of AI in Biochar Synthesis and Application Optimization

18.2.1 AI-Based Process Optimization

18.2.1.1 Role of AI in Optimizing Feedstock Selection, Pyrolysis and Yield Prediction

In many industrial processes, particularly waste management and the production of bioenergy, AI is becoming more and more crucial in optimizing feedstock selection, pyrolysis conditions, and yield prediction. AI approaches have the potential to greatly improve the pyrolysis process, which is the heat breakdown of organic molecules without oxygen. By identifying the best feedstocks, precisely controlling the pyrolysis parameters, and accurately predicting product outputs, these techniques can increase the sustainability and efficiency of pyrolysis operations, briefly describes in Table 18.1.

Table 18.1 AI techniques for the aspects below

Aspect	AI techniques used	Purpose	Reference
Feedstock selection	Clustering, classification	Identify suitable feedstocks based on properties	Guo et al. (2021)
Pyrolysis	Neural networks, genetic algorithms	Optimize temperature, pressure, and heating rate	Wang et al. (2024a, b)
Yield prediction	Regression, deep learning models	Predict product yields based on input data	Jaffari et al. (2024)

Feedstock Selection

The choice of feedstock is critical to the performance of the pyrolysis process. Different feedstocks, such as agricultural leftovers, polymers, and biomass, have distinct qualities that impact the efficiency and quality of pyrolysis products. AI models can handle enormous datasets comprising data on the chemical composition, moisture content, and physical qualities of possible feedstocks. Using classification and clustering algorithms, AI can identify the best feedstocks for pyrolysis based on these properties. This lowers the typical trial-and-error method to feedstock selection (Guo et al. 2021).

Pyrolysis

Pyrolysis conditions must be carefully managed to improve product production. AI approaches like as neural networks and genetic algorithms allow for the modeling and optimization of process parameters by assessing historical data and real-time sensor feedback (Liu et al. 2020). These models can forecast the ideal circumstances for maximum bio-oil, gas, or char production. Furthermore, AI-based feedback systems enable continuous monitoring and correction of the pyrolysis process, resulting in improved control and resource efficiency (Han et al. 2022).

Yield Prediction

Process efficiency and economic viability depend on the precise estimation of pyrolysis yields, such as bio-oil, gas, and char. Regression and deep learning networks are two examples of AI algorithms that can forecast yield based on feedstock properties and pyrolysis conditions. AI improves the capacity to predict pyrolysis performance by comprehending the intricate connections among process parameters, feedstock characteristics, and product results (Garg et al. 2022).

18.2.2 Integration of AI with Process Control Systems for Automated Biochar Production

The combination of AI with process control systems for automated biochar manufacturing is increasing the efficiency and uniformity of the process. Biochar, a useful byproduct of pyrolysis, has a variety of uses, including soil improvement and carbon sequestration. ML algorithms improve pyrolysis management by analyzing real-time data from sensors that monitor factors like as temperature, pressure, and feedstock composition. These data-driven AI algorithms can anticipate the best conditions for biochar synthesis, resulting in a high-quality product with low energy usage (Li et al. 2023).

AI may also automate parameter adjustments throughout the pyrolysis process, resulting in a feedback loop that constantly increases biochar output and quality. Reinforcement learning approaches, for example, allow systems to learn from previous operations and automatically fine-tune process parameters (Gao et al. 2024).

This results in a more consistent and energy-efficient production process.

AI integration with process control systems decreases human error, increases operational efficiency, and facilitates scaling in biochar production. As a result, the combination of AI and automation promotes more sustainable and cost-effective biochar manufacturing, making it an appealing solution for sectors focused on sustainable practices (Zhao et al. 2022).

18.2.3 Long-Term Impact Assessment of AI Optimizing Balance Biochar on Soil Fertility, Microbial Activity

AI-driven systems can optimize pyrolysis conditions, producing biochar with specific properties that improve soil structure, nutrient retention, and microbial diversity. Over time, biochar can increase soil organic carbon (SOC), water retention, and enhance nutrient availability, all of which promote soil fertility. The long-term effects of AI in optimizing biochar production on soil fertility and microbial activity hold great promise for improving agricultural sustainability (Ai et al. 2021).

Biochar may be adapted to individual soil types and agricultural demands using artificial intelligence. Based on soil health data, AI models can forecast the most effective biochar properties, which optimize microbial activity and improves nutrient cycling. Furthermore, AI integration allows for real-time monitoring of microbial populations, modifying circumstances for optimal microbial development, and preserving soil ecosystem balance.

The long-term use of AI-optimized biochar has the potential to improve soil health by encouraging beneficial microbial activity, increasing soil resilience, and lowering reliance on chemical fertilizers. AI-driven monitoring and modification will assure biochar's long-term favorable effects throughout numerous growing seasons (Liu et al., 2020).

18.3 Utilization of ML Models for Biochar Property Prediction

18.3.1 Supervised Learning Models for Biochar Property Characterization

Biochar, a carbon-rich material produced from biomass, has a variety of applications in agriculture, environmental remediation, and energy. Accurate prediction of its properties (e.g., surface area, pH, porosity) is essential for optimizing its use, and machine learning provides a robust framework. Supervised learning models, such as Artificial Neural Networks (ANN), Support Vector Machines (SVM), and Random Forest (RF), are increasingly used for biochar property characterization (Table 18.2).

The layers of linked nodes (neurons) that make up artificial neural networks ANNs, which are computer models modeled after the human brain, analyze input data to find patterns and generate predictions. ANNs are well suited for tasks like regression and classification because of their exceptional ability to represent intricate, non-linear connections. Supervised learning models, or SVMs, are employed for problems involving regression and classification. They function by determining the appropriate hyperplane to maximize the margin between data points of various types. In order to improve accuracy and resilience, RF ensemble learning techniques construct many decision trees during training and output the mean prediction (regression) or mode of classes (classification) of the individual trees (Liaw and Wiener 2002).

Figure 18.2 is described the schematic diagram of ANN model in ML.

Table 18.2 ML models encouraging results for forecasting biochar properties

Model	Strengths	Applications	Reference
ANN	Handles complex non-linear data	Predicting surface area, porosity	Zhang et al. (2020)
ANN	Captures intricate patterns	Predicting carbon content, thermal properties	Singh et al. (2022)
SVM	Efficient with high-dimensional data	Predicting adsorption properties	Wang et al. (2021)
SVM	Effective for classification	Classifying biochar properties	Lin et al. (2025)
RF	Robust and interpretable	Estimating biochar yield, surface area	Jia et al. (2021)
RF	Handles large datasets	Characterizing biochar from different feedstocks	Zhou et al. (2022)

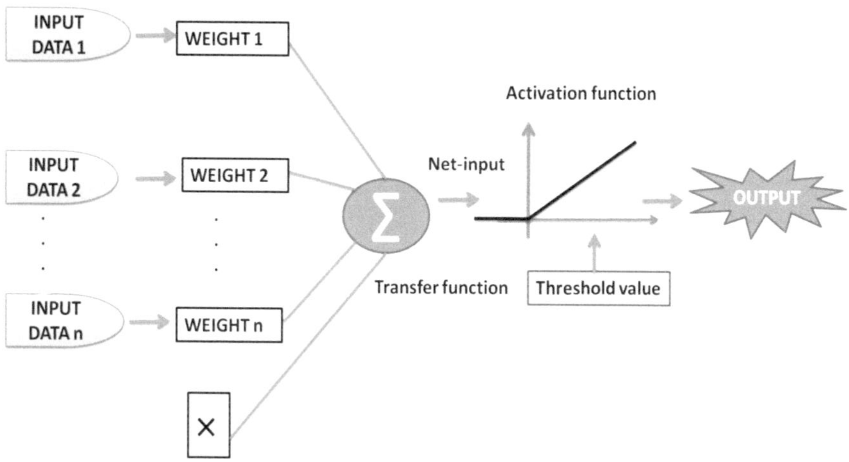

Fig. 18.2 Graphical representation of working principles of Artificial Neural Networks (ANN) model in machine learning

18.3.2 Deep Learning Frameworks for Biochar Predictive Modeling and Complex Feature Extraction

In the complicated feature extraction and predictive modeling of biochar characteristics, deep learning frameworks including Convolutional Neural Networks (CNNs), Recurrent Neural Networks (RNNs), and Auto-encoders have demonstrated considerable promise. Because of its varied physicochemical characteristics and heterogeneous structure, biochar necessitates sophisticated models to precisely forecast characteristics such as porosity, surface area, carbon content, and thermal stability.

CNNs are very good in identifying spatial patterns in photos of biochar, which makes it possible to extract morphological characteristics. CNNs are able to recognize complex structural and textural elements that conventional models can be missed.

Long Short-Term Memory (LSTM) networks and other RNNs are appropriate for analyzing temporal data. They are helpful in accurately representing sequential relationships when modeling the characteristics of biochar over time or under different production settings. Unsupervised models such as autoencoders are excellent at detecting anomalies and reducing features. They help increase the effectiveness of prediction algorithms by revealing hidden patterns in high-dimensional biochar datasets (Table 18.3).

Table 18.3 Deep learning approaches enhance predictive accuracy of diverse biochar sources

Framework	Strengths	Applications	Reference
CNN	Extracts spatial features	Morphological analysis of biochar	Chen et al. (2021)
RNN (LSTM)	Models temporal dependencies	Predicting biochar evolution over time	Li et al. (2025a, b)
Auto-encoder	Dimensionality reduction, anomaly detection	Feature extraction in complex datasets	Dinh et al. (2023)
CNN + Auto-encoder	Hybrid for robust feature learning	Combined structural and chemical analysis	Patel et al. (2022)

18.4 AI and ML in Carbon Sequestration and Climate Change Mitigation

18.4.1 AI-Based Modelling for Carbon Storage Potential

18.4.1.1 Predictive Models for Long-Term Carbon Stability in Biochar

AI and ML have emerged as potent techniques for forecasting biochar's long-term carbon stability, which is critical to its success as a carbon sequestration approach. These new computational tools have transformed our capacity to anticipate biochar performance over long time periods, giving critical insights for climate change mitigation efforts.

Recent research has shown that machine learning approaches may effectively evaluate and forecast the carbon storage capacity of biochar. Several machine learning techniques, including ANN, RF, and NN, have been effectively used to estimate biochar's adsorption efficiency and long-term stability. These models outperformed established experimental approaches in estimating the adsorption efficacy of biochar for heavy metals in water (Chen et al. 2023).

One of the primary benefits of employing AI-based models is their capacity to minimize the number of experimental trials and working hours while retaining excellent prediction accuracy. This efficiency is especially useful in biochar research, where lengthy investigations can be time-consuming and resource-intensive.

A thorough data-driven model has been created to anticipate biochar output and compositions for varied biomass input feedstocks and pyrolysis settings. This approach improved prediction accuracy by 12.7% when compared to current literature (Chen et al. 2023). These advances in predictive modelling are critical for improving biochar production and application tactics for optimum carbon sequestration.

18.4.1.2 AI-Driven Simulations for Biochar's Role in Reducing Greenhouse Gas Emissions

AI-powered simulations have proven useful in understanding and forecasting biochar's function in lowering greenhouse gas emissions. These advanced models can account for the intricate interactions between biochar, soil, and the atmosphere, giving a more complete picture of biochar's contribution to climate change mitigation. Machine learning techniques were used to simulate the impacts of biochar on carbon and nitrogen transformation processes in soils. These models have helped illuminate several methods by which biochar affects greenhouse gas emissions, including:

1. Soil pH changes impact the ratio of N_2O to N_2duringdenitrification.
2. Changes in soil microbial populations, especially those engaged in denitrification.
3. Improved soil capacity to absorb CO_2 or CH_4.
4. Improved soil aeration reduces denitrification rates.

AI-powered simulations have also proven essential in determining biochar's viability as a negative emissions technology. These models can anticipate the carbon removal capability of biochar under various situations, assisting policymakers and researchers to improve methods for large-scale implementation.

18.4.2 ML-Optimized Biochar Strategies for Climate Resilience

18.4.2.1 Adaptive Biochar Application Techniques Based on Climate Data

Machine learning algorithms have enabled the creation of adaptive biochar application strategies that may be customized to individual climatic conditions. By evaluating massive volumes of climatic data, ML models can improve biochar application tactics for optimal performance in a variety of environmental settings. Recent research has focused on applying machine learning to anticipate the nonlinear interactions between variables in complicated biochar-soil systems. These models may use characteristics like soil type, climate, and biochar qualities to predict the most effective application rates and procedures (Fig. 18.3). One interesting option is to apply ensemble ML approaches and explainable AI algorithms to anticipate biochar characteristics and efficiency (Nguyen et al. 2024). These improved models can forecast biochar output and properties from a variety of biomass sources, allowing for more accurate and adaptive application tactics (Table 18.4).

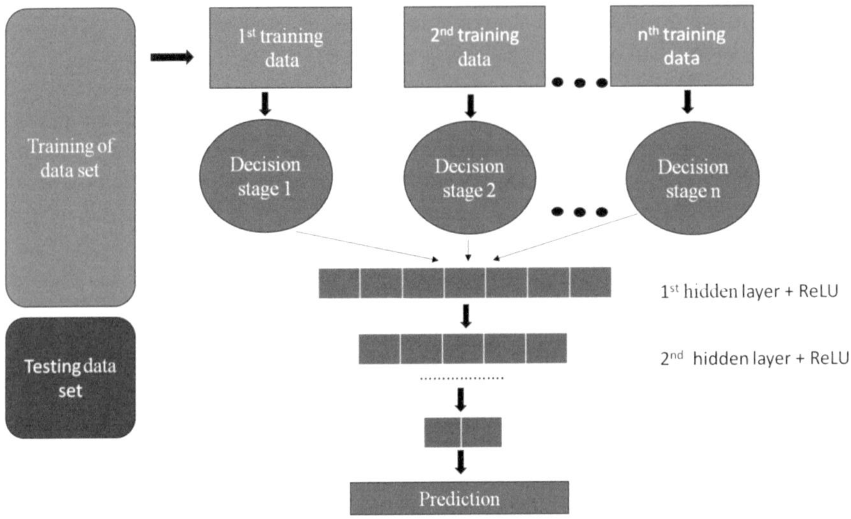

Fig. 18.3 A schematic diagram—working principles of RF model

Table 18.4 Comparison of ML techniques for biochar application optimization

ML technique	Advantages	Limitations	Application in biochar strategies
ANN	High accuracy and complex systems	Requires large datasets	Predicting biochar adsorption efficiency
RF	Handles non-linear relationships well	May overfit with noisy data	Optimizing biochar application rates
SVM	Effective for high-dimensional data	Sensitive to parameter selection	Classifying optimal biochar types for different soils
Gradient Boosting	Robust to outliers	Computationally intensive	Forecasting biochar's long-term carbon stability

18.4.2.2 AI-Powered Decision Support Systems for Large-Scale Implementation

AI-powered decision support systems are becoming more vital for the widespread use of biochar as a climate change mitigation method. These systems combine many data sources and machine learning algorithms to give full advice on biochar production, application, and monitoring.

Machine learning-based predictions, optimization, and feature selection are essential for enhancing biomass management strategies (Nguyen et al. 2024). These sophisticated analytical methods can assist identify key biomass features that impact biochar efficiency, allowing for more focused and successful application tactics.

The creation of hybrid models that integrate ML and mechanism-based analysis is a potential future direction in biochar research (Wang et al. 2024a, b). These models address the "black-box" aspect of some machine learning techniques, resulting in more interpretable and reliable forecasts for large-scale biochar applications.

While substantial progress has been made in applying machine learning to biochar investigations, current research mostly uses lab-scale data for model training (Wang et al. 2024a, b). To fully utilize the promise of machine learning (ML) approaches for biochar field applications, further research is needed to construct models based on pilot or industrial data.

The use of AI and ML into biochar research and implementation has enormous potential for improving carbon sequestration and climate change mitigation efforts. These tools enable more precise forecasts of biochar's long-term stability, optimize application tactics, and provide decision assistance for large-scale deployment. As we perfect these AI-driven techniques, we can anticipate biochar to be used more efficiently and effectively in our worldwide efforts to address climate change.

18.5 Cost–Benefit Analysis Using AI Models

18.5.1 AI-Based Life Cycle Assessment for Biochar Production and Utilization

Life Cycle Assessment (LCA) is a systematic procedure to assess the environmental effects that arise from the interaction between human actions and the environment. Using LCA, it is possible to analyze the potential effects over the whole life cycle of a product. Combining ML methods with LCA can greatly improve the precision and effectiveness of impact analysis, which eventually leads to minimizing environmental loads. The LCA methodology has four main stages: setting the objective and scope of the assessment, developing the Life Cycle Inventory, performing the Life Cycle Impact Assessment and lastly, interpreting the results to inform decision-making and improvement (Guinee 2002). Numerous ML models such as random forest (RF), Gaussian Process Regression (GPR), support vector machines (SVM), extreme Gradient Boosting (XGBoost) and artificial neural networks (ANN) have been created to forecast the production of biochar (Li et al. 2022). The process of pyrolyzing agricultural waste to produce biochar for use in soil is gaining more attention as a method of minimizing carbon emissions. In order to determine the carbon footprint of producing biochar, Li et al. (2025a, b) developed a framework for assessing environmental impact that integrates LCA with ML modelling. Out of the five ML models, the multi-layer perceptron neural network (MLP-NN) and GPR models demonstrated high accuracy in predicting the yield of biochar as well as its carbon and nitrogen contents ($R^2 = 0.97$, RMSE $= 3.5$; $R^2 = 0.92$, RMSE $= 3.2$; $R^2 = 0.94$, RMSE $= 0.36$, respectively). For the greatest reduction in global warming potential, the MLP-NN model determined the ideal pyrolysis conditions

(PT = 400 °C, RT = 40 min, HR = 15 °C/min). When fertilizer replacement and carbon sequestration were considered, the analysis showed that the highest carbon savings for the calcium ammonium nitrate and urea ammonium nitrate cases were − 1355 and − 1323 kg CO_2-eq/t feedstock, respectively.

18.5.2 *Predictive Analytics for Economic Feasibility of Biochar Applications*

For a process to be feasible, apart from the environmental considerations, it is required to evaluate their related costs. The process of determining a process's overall cost is called Life Cycle Cost Assessment or LCCA (Carvalho et al. 2022). By combining ML approaches, LCA and economic analysis, Cheng et al. (2020) compared the energy, climate change and economic performance of slow pyrolysis of various feedstocks under different processing conditions. Monte Carlo simulation was utilized to evaluate the uncertainty related with forecasted energy, climate change and economic indicators. Based on LCA outcomes, slow pyrolysis of lignocellulosic biomass, such as woody wastes and crop residues, results in net positive energy output and net negative greenhouse gas emissions at high temperature. According to economic analysis, the minimum product selling price (MPSP) for biochar rises monotonically with temperature, meaning that lower pyrolysis temperatures are associated with improved economic performance. For all temperatures, the minimum MPSP occurs in biochar from pyrolysis of sludge because it produces more biochar than the other feedstocks that were assessed. The calculated MPSP values in this study varied from $774 to 1256/t based on feedstocks and temperature.

18.6 AI-Assisted Environmental Impact Assessments

18.6.1 *AI-Enhanced Risk Assessment for Soil, Water and Air Pollution*

Pesticides are a new class of water pollutants that have serious impacts on human microbiomes and biodiversity (Finckhet al. 2022). Of all the routes of pesticide contamination, wastewater treatment plant (WWTP) effluents have been recognized as the major contributors to water bodies of the surrounding area and freshwater resources (Qian et al. 2021). Adsorption through biochar has been strongly recommended as an economically sound and ecologically sustainable method of curing pesticide pollution in water bodies (Ponnuchamy et al. 2021). The data-based modeling intervention provides an empirical viewpoint on the optimal use and well-balanced design of biochar for the removal of emerging micro-pollutants from water in agricultural systems, improving soil health through the application of biochar as soil amendments, reducing environmental pollution and aids in soil and water conservation efforts (Table 18.5). Decision-tree-based ensemble ML models, such as CatBoost, Light GBM and RF were employed to predict the effectiveness of adsorptive pesticide sequestration from an aqueous medium by using biochar. These models combine several individual models to improve the accuracy of predictions (Nighojkar et al. 2023a). Nighojkar et al. (2023b) found that the XG Boost-ensemble ML model offers the most accurate prediction for pesticide adsorption on the biochar matrix, yielding good scores for the regularization coefficient ($R^2_{train} = 0.998$, $R^2_{test} = 0.981$). The organophosphorus compound concentration and biochar morphology had a significant impact on the pesticide adsorption behavior. These findings prove the usefulness of employing ensemble learning algorithms in achieving balanced carbon-enriched biomaterial to effectively remove emerging micro-pollutants from water.

Zhu et al. (2019) explored the use of ML techniques to predict metal adsorption efficiency onto biochars. The adsorption of heavy metals (zinc, nickel, arsenic, cadmium, copper and lead) on 44 biochars was predicted with ANNs and RF models. The research found that RF model ($R^2 = 0.973$) was more precise compared to the ANN model. Zhu et al. (2021) utilized ML techniques to create general prediction models for the adsorption of sulfamethoxazole and tetracycline onto carbon-based materials (CBMs) like biochar, activated carbon. The RF models effectively predicted the adsorption capacity of these antibiotics on CBMs outperforming both gradient boosting trees and ANNs models. Zhu et al. (2022) developed interpretable ML models like RF to build and assess predictive models for how pharmaceutical and personal care products adsorption are adsorbed onto biochar and activated carbon based on detailed data regarding the properties of biochar including its surface functionalities and porous structure.

18.6.2 Sustainable Biochar Strategies Through AI-Integrated Environmental Modeling

Effective modeling of intricate relationships and improved comprehension of biochar's performance are made possible by machine learning, which results in more specialized and efficient solutions for a range of agricultural and environmental applications (Supraja et al. 2023). ML algorithms (ANN and RF) have been used to forecast crop yield and organic matter conservation with the addition of biochar (Wang and Yao 2023). ML algorithms such as SVM, XGB, RF, ANN, DT and KNN have also been employed to study biochar adsorption capacity in waste water (Nie et al. 2023). Ding et al. (2018) used boosted regression trees (BRTs) algorithms to detect variables influencing the effect of biochar in soil carbon priming. Incubation conditions (Incubation duration and soil moisture) and biochar characteristics (biochar N content, C/N ratio, pH and pyrolysis duration) are crucial elements, whereas soil characteristics (pH, C/N ratio, silt content and land use type) are less important factors. Liu et al.

Table 18.5 ML applications on biochar utilization for effectively removing different pollutants

Application	Data size	Descriptors	Target output	ML tools	References
Heavy metal removal	476	pH of biochar in water, C %, O/C,(O + N)/C, H/C, SSA, Ash %, CEC, adsorption temperature, initialsolution concentration, pH of the solution, type of heavy metal	Adsorption efficiency %	LR, KNN, SVR, RF, DT, GBDT, XGBoost	Duan et al. (2024)
	162	Biochar pH, pyrolysis temperature, C %, H %, O %, N %, Ash %, SSA, biochar dosage, time, initial heavy metal content, solution pH, Soil EC, electronegativity	Adsorption efficiency %	SVR, NN, RF	Palansooriya et al. (2022a, b)
	30	pH, adsorbent dosage, Contact time, initial lead concentration	Adsorption capacity (mg/g)	DT, RF, GMDH, MLP, RBF, ANFIS, SVM	Talebkeikhah et al. (2022)
	353	Biochar pH in water, pyrolysis temperature, SSA, ash %, C %, CEC, particle size, O/C, H/C, (O + N)/C, solution pH, initial heavy metal concentration, ionic radius, electro negativity, and molecular weight	Adsorption capacity	ANFIS, CSA-LSSVM, PSO-ANFIS, GP	Dashti et al. (2023)
Removal of Remazol Orange Brilliant 3R (RBO-3R)	30	Temperature, persulphate concentration, initial dye concentration, time	Removal efficiency %	RSM, canonical and ridge analysis	John et al. (2024)
Removal of pharmaceuticals	1033	Abraham parameters (A, B, S, E, V), H/C, N/C, O/C, pore volume, pore structure, SSA, initial concentration, biochar dosage	Equilibrium adsorption capacity (mg/g)	Cubist, KNN, RF, NNET, LM, DNN	Nguyen et al. (2022)

Adapted from Oral et al. (2024)

(2019) constructed RF models to comprehend the effects of soil properties, biochar type and addition rate and climate zone on soil N preservation following biochar amendment. Also they employed the predictive model to determine the best biochar application techniques based on worldwide soil conditions.

Shen et al. constructed a Naïve Bayes model using ground penetrating radar signal attributes as inputs to forecast soil organic carbon in soil treated with biochar. Zhang et al. (2023) investigated simultaneous soil N_2O and NO emissions following the application of biochar using the meta analysis approach in conjunction with the BRT model. Generally, the incorporation of biochar significantly reduced concomitant soil emissions of NO (14.7%) and N_2O (16.2%). Twelve ML models were thoroughly assessed to forecast the equilibrium adsorption capacity of NH_4^+-N onto biochar. Tree-based ensamble approaches consistently outperformed deep learning and kernel-based models among these models, with the CatBoost algorithm obtaining the highest accuracy ($R^2 = 0.9329$, RMSE = 0.5378). By combining superior ML methods and interpretability measures, this research increases knowledge on biochar's ammonia adsorption and promotes sustainable methods by reducing nitrogen pollution (Liu et al. 2025).

18.7 Applications of AI in Permuted Biochar for Heavy Metal Adsorption and Pollutant Removal

The rising concern over environmental pollution, particularly heavy metal contamination, has fostered interest in creative and effective remediation technology. Biochar is a carbon-rich substance formed by the combustion of organic substances. It has emerged as a viable candidate for heavy metal adsorption and pollution removal. The use of AI and ML in sophisticated biochar modification processes provides a unique way to optimize biochar for better performance in pollution remediation.

18.7.1 Current Developments in Adsorption and Desorption Research for the Removal of Heavy Metals Using Modified Biochar

18.7.1.1 Development of Advanced Biochar Modification Techniques

The heavy metal removal efficiency of biochar depends on its various characteristics, including pore structure, surface area, pH, stability, and elemental composition. Nowadays, these characteristics are modified using recent advanced techniques such as electrochemical treatment, ball milling, metal impregnation, ultrasonication, composite biochar, plasma-treated biochar, and magnetic biochar. Ultrasonication equally distributes the nanoparticles on the biochar surface and helps in pore formation with no clogging (Sajjadi et al. 2020). On the other hand, ball milling improves the pore distribution and enhances surface complexation, electrostatic interaction, and cation-π interactions that ultimately help increase adsorption. In magnetic biochar, the surface area is increased using particles of iron oxides. Besides, the ratio of biochar and iron also controls the pore size (Lakshmi et al. 2021). Gurav et al. (2020) reported that the paramagnetic properties of iron oxide (Fe_3O_4) significantly increased biochar's surface area, facilitating easy contaminant removal. Recently, metal impregnation in the biochar fabrication technique prevents clogging of pores, increases the reduction, precipitation, and retention time of metals (Lakshmi et al. 2021). Coating of biochar with different substances increases the surface area and facilitates heavy metal removal. For example, coating with magnesium oxide (MgO) increased the surface area of biochar by 300 times, suggesting the importance of coating on the surface area (Wang et al. 2017). Additionally, microwaves provide thermal energy for biochar activation with less electricity as compared to conventional methods. This surface area increase is due to the removal of ash and volatile substances from biochar by microwave radiation (Lakshmi et al. 2021). Lastly, plasma treatment converts gas molecules to ions and radicals, increasing the surface area of biochar and improving adsorption (Niu et al. 2017; Luo et al. 2019).

18.7.1.2 Developments in the Enhancement of the Adsorption Capacity of the Biochar

Current developments in biochar adsorption capability optimization have focused on finding important pyrolysis parameters that are impacted by interrelated feedstock and manufacturing factors (Hodgson et al. 2016). Heating duration, heating temperature, and heating rate are the most essential process parameters, followed by particle size and the gaseous environment employed during preparation (Lakshmi et al. 2021). ANN paired with Taguchi approach and Particle Swarm Optimization improve adsorption efficiency significantly (Hodgson et al. 2016; Balasubramani et al. 2020). Moreover, Response Surface Methodology (RSM) is used to evaluate interactions between these factors in order to obtain maximum performance of biochar (Zhou et al. 2019). Statistical validation using methods such as ANOVA assures that the optimization process consistently supports effective removal of pollutants (Dissanayake Herath et al. 2019). Besides, ML techniques and statistical methods have demonstrated efficacy in tackling the nonlinear trends of multi-component adsorption methods, which are influenced by parameters such as adsorbate quantity and inter-ion interactions. Multicomponent adsorption systems frequently have poorer removal efficiency than single-metal methods due to competing or negative interactions between ions. However, improved biochar has shown potential in addressing these issues. Biochar calibration with ANN models successfully removed Cr^{6+} and Pb^{2+} with 85% efficiency, outperforming standard models by 15%. This is accomplished by balancing pore geometry with oxygen-containing functional groups. Hybrid models that combine ANN with the adsorption isotherms and kinetics improve prediction precision in multicomponent systems, exceeding classic kinetic approaches (Rodríguez-Romero et al. 2020). Furthermore, investigations utilizing new hydrogel adsorbents have identified processes such as hydrogen bonding and channels of transportation that increase adsorption in ternary frameworks (Ma et al. 2019).

18.7.1.3 Improvement of Desorption Kinetics and Regeneration of Modified Biochar

Recent improvements in biochar research have spotlighted the importance of desorption and regeneration, especially for removing heavy metals from the polluted environment (Baltrėnaitė-Gedienė et al., 2020). This approach not only helps to lower costs related to sourcing feedstock but also reduces energy consumption during pyrolysis.Patterns have emerged regarding the effectiveness of various eluents used for desorbing metals from biochar. Strong acids like sulphuric acid (H_2SO_4), nitric acid (HNO_3), and hydrochloric acid (HCl) at concentrations between 0.1 and 1 M have demonstrated remarkable efficiency, achieving over 95% removal rates. Besides, salts formed from acid–base reactions, like sodium carbonate (Na_2CO_3) and sodium nitrate ($NaNO_3$), have shown moderate to high effectiveness in desorbing metals as well. Table 18.6 highlights some desorption investigations that employed different elements at varying concentrations. However, the efficiency of these substances can vary significantly depending on their concentration used and the metal characteristics. This variability underscores the importance of hybrid models and ML algorithms to optimize both adsorption and desorption processes, ensuring that biochar can be effectively utilized in environmental remediation efforts (Lakshmi et al. 2021).

18.8 Biochar Adsorption Modelling for Machine Learning-Based Pollutant Removal

18.8.1 Development of Machine Learning Models for Pollutant Adsorption Prediction

Industrial wastewater treatment needs adaptive solutions owing to variable levels of pollutants and composition. Machine learning handles this by implementing automated adsorption forecasting and process optimization, which reduces the need for resource-intensive trial-and-error procedures. The algorithms used for supervised learning (e.g., random forest, gradient boosting regression) use labelled datasets to predict biochar's adsorption kinetics, efficiency, and thermodynamic characteristics according to the conditions of pyrolysis (feedstock type, temperature) and modification pathways (Bhagat et al. 2020; Zhou et al. 2020). For example, ML models can predict biochar output, surface characteristics, and heavy metal elimination capacity, allowing custom designs for specific contaminants. Classification

Table 18.6 Investigation of heavy metal desorption from biochar using various eluents at varying concentrations (Lakshmi et al. 2021)

S. No.	Biochar type	Removed heavy metal	Efficiency of removal (%)	Desorption eluent	Concentration of eluent (M)
1	Kelp-derived magnetic biochar	Cu(II)	1	–	0.1
		Cu(II)	0.67	–	0.025
2	Gasified dairy biochar manure	Cd(II)	0.99	H_2SO_4	1
			0.97	HCl	1
			0.97	HNO_3	1
		Cu(II)	0.81	HNO_3	1
		Zn(II)	0.57	HCl	1
3	Biochar derived from rice husk	Cd(II)	0.73	$NaNO_3$	0.04
			0.67	$NaNO_3$	0.08
			0.60	$NaNO_3$	0.02
			0.55	$NaNO_3$	0.01
4	Biochar derived from hardwood	Cd(II)	0.61	$NaNO_3$	0.04
			0.57	$NaNO_3$	0.02
			0.55	$NaNO_3$	0.08
			0.52	$NaNO_3$	0.01
5	Sponge gourd biochar activated with phosphoric acid	Pb(II)	0.95	HNO_3	0.15
			0.92	HNO_3	0.2
			0.83	HCl	0.15
			0.81	HCl	0.2
			0.80	HNO_3	0.1
			0.63	H_2SO_4	0.15
			0.63	H_2SO_4	0.2
			0.62	HCl	0.1
			0.50	H_2SO_4	0.1
6	Ball-milled biochar with phytic acid	U(VI)	0.97	Na_2CO_3	0.1
			0.95	HCl	0.6
			–	HCl	1
			–	Acetic acid	0.1
7	Sulfuric acid-treated biochar from posidonia oceanica leaf sheaths	Pb(II)	0.9	HNO_3	0.1
8	Miscanthus biochar	Cd(II)	0.99	HNO_3	0.05
	Wheat biochar	Cd(II)	0.99	HNO_3	0.05

(continued)

Table 18.6 (continued)

S. No.	Biochar type	Removed heavy metal	Efficiency of removal (%)	Desorption eluent	Concentration of eluent (M)
	Oil seed rape biochar	Cd(II)	0.98	HNO_3	0.05
9	Nano zinc oxide-modified biochar derived from bamboo shoot shells	Re(VII)	0.85	KOH	0.5

algorithms help identify pollutant types and make eluent selection decisions (e.g., HCl vs. Na_2CO_3). Hybrid models, such as Kriging-LFER (R^2: 0.940), outperform conventional approaches to predict adsorption behaviour, whereas deep learning techniques use encoders to reduce data dimensionality and enhance resilience for large data sets (Pyo et al. 2020). Dataset quality, overfitting, and validation over conventional models are all challenges, but ML is capable of optimizing biochar alteration (for example, acid treatment intensity) and forecasting regeneration effectiveness, making industrial-scale remediation more efficient. Recent research shows that ML can map the process of heavy metal removal by engineered biochar, favouring cost-effective and eco-friendly options (Bhagat et al. 2020; Li et al. 2019). Future research should address peer acceptability of data formats and guarantee the transfer of technology from lab-scale forecasts to real-world applications. Figure 18.4 represents a schematic illustration of machine learning-driven biochar application for enhanced heavy metal remediation.

18.8.2 *Real-Time Monitoring and Adaptive Control Systems of Biochar Characteristics in Pollutant Removal Optimization*

Adsorption of heavy metals is strongly dependent on biochar properties. Biochar can be changed using a variety of ways, as described in Sect. 5.1.1. Machine learning is essential for anticipating changes in biochar properties as a result of alteration. Predictive ML approaches can be used to improve biochar modification. Wallace et al. (2019) supplemented biochar with polymer compounds and created a model for estimating tensile modulus in biological composites, demonstrating ML's use in biochar design. Microbial engineering, paired with AI, can enhance wastewater treatment by increasing microorganism efficiency. GA-ANN and GA-ANFIS are AI technologies that aid in optimising process parameters, reconstructing metabolic pathways, and predicting microbial populations for metal remediation. Tao et al. (2020) adopted an innovative strategy, combining biochar with living organisms, resulting in enhanced biodegradability over plain biochar. Metal–organic frameworks (MOFs), which are noted to have structural flexibility and customizable pore widths, are increasingly being employed for wastewater treatment. The adsorption

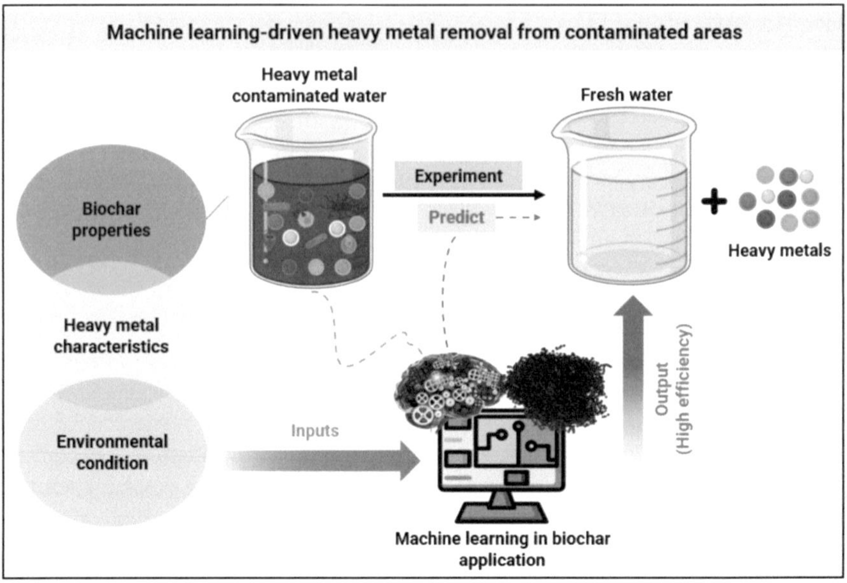

Fig. 18.4 Schematic representation of machine learning-driven biochar application for efficient heavy metal remediation

capacity of MOFs is determined by their structure, functional properties, and surface area. Mohammadi et al. (2019) achieved 99% photocatalytic activity with MOF/activated carbon nano-composites. Machine learning aids MOF design by predicting characteristics and allowing the creation of biochar/MOF composites for various metal adsorption.

18.9 Future Aspects

The production and use of biochar, a carbon-rich byproduct of biomass pyrolysis, is being revolutionised by artificial intelligence (AI) and machine learning (ML), which improve efficiency, sustainability, and customisation. The creation of biochar, which was previously done by hand, now benefits from AI-driven predictive modelling that optimises important variables like temperature, heating rate, and residence time to increase yield and improve the quality of the biochar, including its porosity, carbon content, and nutrient availability. Real-time modifications during production are made possible by AI and ML, guaranteeing constant quality for a variety of applications. Along with lowering environmental effects, these technologies allow farmers choose the best kinds of biochar based on the properties of their soil, enhancing soil fertility, water retention, and nutrient cycling. Large soil records can be analysed using machine learning algorithms to forecast the effects of applying biochar, which helps improve decision-making in agricultural systems. Biochar stabilises carbon in soil for decades and is used as a carbon sequestration strategy outside of agriculture.

Biochar's potential to mitigate climate change is further supported by AI's capacity to predict its long-term stability in various conditions. AI assists in determining the optimal feedstocks for optimal environmental benefit, while biochar helps manage waste by turning wood waste and agricultural wastes into useful soil nutrients. Economically speaking, AI lowers production costs and improves quality control, resulting in scalable, reasonably priced biochar models that are customised to meet particular requirements. AI is used in environmental applications like as soil remediation to forecast how well biochar can absorb contaminants and stabilise heavy metals. Additionally, biochar encourages microbial activity, which enhances soil health and lessens reliance on artificial fertilisers. In order to improve pyrolysis- the thermal breakdown of organic materials without oxygen- which is utilised in waste management and the generation of bioenergy, artificial intelligence is becoming increasingly important. AI improves the process's sustainability and efficiency by selecting feedstocks optimally and managing important variables like temperature and heating rate. Real-time modifications and improved decision-making are made possible by machine learning models' ability to forecast product yields and attributes.

These results in better resource use, less waste, and more consistent outputs. Overall, through more intelligent biomass conversion, AI integration in pyrolysis promotes cleaner, more affordable energy generation and helps create a more sustainable, circular economy. Automated biochar production is becoming more efficient, consistent, and scalable through the integration of AI with process control systems. The important pyrolysis byproduct biochar, which is used to enhance soil and sequester carbon, benefits from real-time sensor data processing and AI-driven monitoring. In order to guarantee high-quality output with the least amount of energy consumption, machine learning algorithms optimise parameters such as temperature and feedstock composition. AI further improves product quality and lowers energy usage by enabling automatic parameter modifications through reinforcement learning and feedback loops. Because of this automation, which lowers operational expenses and human error, the production of biochar is more sustainable and appealing to businesses who prioritise environmentally friendly operations. The complex and variable structure of biochar makes it a promising candidate for prediction using deep learning frameworks such as CNNs, RNNs, and autoencoders. CNNs are excellent at extracting morphological and spatial characteristics from photos of biochar, catching structural details that conventional models frequently overlook. When modelling temporal data, RNNs- particularly LSTM networks- are beneficial for comprehending how the properties of biochar change over time or under different circumstances. In high-dimensional data, autoencoders can uncover hidden patterns through feature reduction and anomaly detection.

These models increase the precision of predictions, effectively handle intricate information, and adjust to various kinds of biochar and manufacturing procedures. For the purpose of carbon sequestration, biochar's long-term carbon stability can be predicted with the help of artificial intelligence (AI) and machine learning (ML).

In comparison to conventional experimental methods, techniques such as Artificial Neural Networks (ANN), Random Forests (RF), and Neural Networks (NN) have demonstrated efficacy in predicting the adsorption effectiveness and stability of biochar. These models preserve great precision while reducing the need for lengthy testing, saving time and money. By optimising biochar synthesis and feedstock selection, a data-driven model increased prediction accuracy by 12.7% when compared to previous experiments. These developments provide more effective, focused approaches to the use of biochar in sustainable resource management and climate change mitigation. Life Cycle Cost Assessment (LCCA) is essential for assessing a process's environmental impact as well as its economic viability.

18.10 Conclusions

In conclusion, the integration of Artificial Intelligence (AI) and Machine Learning (ML) in biochar optimization has the potential to revolutionize biochar production and its application in environmental and agricultural systems. AI and ML technologies facilitate a transition from traditional trial-and-error methods to data-driven approaches, improving the precision and efficiency of biochar synthesis. By leveraging large datasets, AI models can optimize key production parameters such as feedstock selection, pyrolysis conditions (temperature, heating rate, and residence time), and the prediction of biochar yield. These data-driven models enable a more targeted approach to biochar production, ensuring consistent quality and maximizing the desired properties, including surface area, porosity, and carbon content, which are critical for biochar's effectiveness in various applications such as carbon sequestration, soil enhancement, and pollutant removal. Additionally, the use of advanced machine learning algorithms, including supervised models like Artificial Neural Networks (ANN), Support Vector Machines (SVM), and Random Forests (RF), has significantly advanced biochar property prediction. These models have been applied to predict a wide range of biochar characteristics, including its adsorption efficiency for pollutants, nutrient retention capacity, and the potential for improving soil health. Deep learning frameworks, such as Convolutional Neural Networks (CNNs) and Recurrent Neural Networks (RNNs), are particularly effective in extracting complex, high-dimensional features from biochar data, enhancing predictive accuracy for biochar's long-term stability and its impact on soil microbial activity and nutrient cycling. Furthermore, AI-powered predictive models can assess biochar's performance across different soil types and environmental conditions, enabling the optimization of biochar application strategies for specific agricultural or remediation needs.

In the context of climate change mitigation, AI and ML have shown promise in enhancing biochar's role in carbon sequestration. AI-based models have been developed to predict the long-term carbon stability of biochar, helping to refine biochar production processes that maximize carbon retention. These models support the development of biochar as a reliable negative emissions technology, providing critical insights into its potential for reducing atmospheric CO_2 and mitigating global warming. AI-driven simulations also allow for the modeling of biochar's role in altering greenhouse gas emissions, such as nitrogen and methane, in agricultural soils, offering pathways to more climate-resilient farming practices. Moreover, AI and ML are pivotal in advancing biochar's use in environmental remediation, particularly in the adsorption and removal of heavy metals and other pollutants from contaminated environments. AI models can predict biochar's adsorption capacity based on its physicochemical properties and pyrolysis conditions, allowing for the design of customized biochar adsorbents for specific contaminants. This capability enhances biochar's effectiveness in cleaning up polluted water and soil, thereby improving environmental quality and public health. The ability to integrate real-time monitoring systems, powered by AI, into biochar-based remediation systems further optimizes the adsorption process and ensures the regeneration of biochar for repeated use.

However, despite these advancements, there are still challenges that need to be addressed for widespread adoption. One of the major hurdles is the quality and heterogeneity of the data used to train AI and ML models, which can limit the reliability and generalizability of the models. Data scarcity and inconsistencies in biochar production methods across different studies further complicate model development. Additionally, while AI has proven effective in optimizing biochar properties at the lab scale, translating these models to industrial-scale biochar production remains a significant challenge. The scalability of AI-driven process optimization, along with the integration of AI with traditional mechanistic models, needs further exploration to ensure that these techniques can be applied effectively at larger production scales. Furthermore, the interpretability of AI and ML models remains a key concern, particularly when dealing with complex systems such as biochar production and application. Many AI models, particularly deep learning frameworks, often function as "black boxes," making it difficult to understand the underlying mechanisms driving predictions. Increasing the transparency and explainability of these models will be essential to build trust among stakeholders and ensure that the results can be translated into actionable insights for biochar producers and environmental policymakers. Lastly, while AI and ML hold great potential for enhancing biochar optimization, interdisciplinary collaboration between researchers, industry experts, and policymakers is critical to overcoming these challenges. The development of standardized protocols, data-sharing frameworks, and ethical guidelines for AI use in biochar research will be necessary to ensure that these technologies are used responsibly and sustainably. By addressing these gaps and continuing to refine AI and ML models, the biochar industry can leverage these tools to develop more efficient, sustainable, and scalable biochar production and application strategies, ultimately contributing to climate change mitigation, environmental restoration, and sustainable agriculture.

References

Ai J, Lu C, van den Berg FW, Yin W, Strobel BW, Hansen HCB (2021) Biochar catalyzed dechlorination–which biochar properties matter? J Hazard Mater 406:124724. https://doi.org/10.1016/j.jhazmat.2020.124724

Al-Wabel MI, Hussain Q, Usman AR, Ahmad M, Abduljabbar A, Sallam AS, Ok YS (2018) Impact of biochar properties on soil conditions and agricultural sustainability: a review. Land Degrad Dev 29(7):2124–2161. https://doi.org/10.1002/ldr.2829

Balasubramani K, Sivarajasekar N, Muthusaravanan S, Ram K, Naushad M, Ahamad T, Sharma G (2020) Efficient removal of antidepressant Flupentixol using graphene oxide/cellulose nanogel composite: particle swarm algorithm based artificial neural network modelling and optimization. J Mol Liq 319:114371. https://doi.org/10.1016/j.molliq.2020.114371

Baltrėnaitė-Gedienė E, Leonavičienė T, Baltrėnas P (2020) Comparison of CU(II), MN(II) and ZN(II) adsorption on biochar using diagnostic and simulation models. Chemosphere 245:125562. https://doi.org/10.1016/j.chemosphere.2019.125562

Bhagat SK, Tung TM, Yaseen ZM (2020) Development of artificial intelligence for modeling wastewater heavy metal removal: state of the art, application assessment and possible future research. J Clean Prod 250:119473. https://doi.org/10.1016/j.jclepro.2019.119473

Campbell JL, Sessions J, Smith D, Trippe K (2018) Potential carbon storage in biochar made from logging residue: basic principles and Southern Oregon case studies. PLoS ONE 13(9):e0203475. https://doi.org/10.1371/journal.pone.0203475

Canlas JJ, Go JC, Mendoza AC, Dimaano MN (2019) Talisay (Terminalia catappa) seed husk biochar for adsorption of lead (II) ions in artificially contaminated soil. In: MATEC web of conferences, vol 268. EDP Sciences, p 04011. https://doi.org/10.1051/matecconf/201926804011

Carvalho J, Nascimento L, Soares M, Valério N, Ribeiro A, Faria L, Silva A, Pacheco N, Araújo J, Vilarinho C (2022) Life cycle assessment (LCA) of biochar production from a circular economy perspective. Processes 10(12):2684. https://doi.org/10.3390/pr10122684

Chen MW, Chang MS, Mao Y, Hu S, Kung CC (2023) Machine learning in the evaluation and prediction models of biochar application: a review. Sci Prog 106(1):00368504221148842. https://doi.org/10.1177/00368504221148842

Chen CFR, Panda R, Ramakrishnan K, Feris R, Cohn J, Oliva A, Fan Q (2021) Deep analysis of CNN-based spatio-temporal representations for action recognition. In: Proceedings of the IEEE/CVF conference on computer vision and pattern recognition, pp 6165–6175

Cheng F, Luo H, Colosi LM (2020) Slow pyrolysis as a platform for negative emissions technology: an integration of machine learning models, life cycle assessment, and economic analysis. Energy Convers Manage 223:113258. https://doi.org/10.1016/j.enconman.2020.113258

Dashti A, Raji M, Harami HR, Zhou JL, Asghari M (2023) Biochar performance evaluation for heavy metals removal from industrial wastewater based on machine learning: application for environmental protection. Sep Purif Technol 312:123399. https://doi.org/10.1016/j.seppur.2023.123399

Ding F, Van Zwieten L, Zhang W, Weng Z, Shi S, Wang J, Meng J (2018) A meta-analysis and critical evaluation of influencing factors on soil carbon priming following biochar amendment. J Soils Sediments 18:1507–1517

Dinh PV, Nguyen QU, Hoang DT, Nguyen DN, Bao SP, Dutkiewicz E (2023) Constrained twin variational auto-encoder for intrusion detection in IoT systems. IEEE Internet Things J 11(8):14789–14803. https://doi.org/10.1109/JIOT.2023.3344842

Dissanayake Herath GA, Poh LS, Ng WJ (2019) Statistical optimization of glyphosate adsorption by biochar and activated carbon with response surface methodology. Chemosphere 227:533–540. https://doi.org/10.1016/j.chemosphere.2019.04.078

Duan Q, Yan P, Feng Y, Wan Q, Zhu X (2024) Machine learning assisted adsorption performance evaluation of biochar on heavy metal. Front Environ Sci Eng 18(5):55

Finckh S, Beckers LM, Busch W, Carmona E, Dulio V, Kramer L, Krauss M, Posthuma L, Schulze T, Slootweg T, Von der Ohe PC, Brack W (2022) A risk based assessment approach for chemical

mixtures from wastewater treatment plant effluents. Environ Int 164:107234. https://doi.org/10.1016/j.envint.2022.107234
Gao W, Li N, Cheng Z, Yan B, Peng W, Wang S, Chen G (2024) Accelerated screening of active sites on biochar for catalysis and adsorption via multidimensional fingerprint factor descriptors. Biores Technol 408:131156. https://doi.org/10.1016/j.biortech.2024.131156
Garg A, Wani I, Zhu H, Kushvaha V (2022) Exploring efficiency of biochar in enhancing water retention in soils with varying grain size distributions using ANN technique. Acta Geotech 17(4):1315–1326. https://doi.org/10.1007/s11440-021-01411-6
Guinee JB (2002) Handbook on life cycle assessment operational guide to the ISO standards. Int J Life Cycle Assess 7(5):311–313. https://doi.org/10.1007/BF02978897
Guo J, Zheng L, Li Z, Zhou X, Cheng S, Zhang L, Zhang Q (2021) Effects of various pyrolysis conditions and feedstock compositions on the physicochemical characteristics of cow manure-derived biochar. J Clean Prod 311:127458. https://doi.org/10.1016/j.jclepro.2021.127458
Gurav R, Bhatia SK, Choi T-R, Park Y-L, Park JY, Han Y-H, Vyavahare G, Jadhav J, Song H-S, Yang P, Yoon J-J, Bhatnagar A, Choi Y-K, Yang Y-H (2020) Treatment of furazolidone contaminated water using banana pseudostem biochar engineered with facile synthesized magnetic nanocomposites. Biores Technol 297:122472. https://doi.org/10.1016/j.biortech.2019.122472
Han H, Buss W, Zheng Y, Song P, Rafiq MK, Liu P, Mašek O, Li X (2022) Contaminants in biochar and suggested mitigation measures–a review. Chem Eng J 429:132287. https://doi.org/10.1016/j.cej.2021.132287
Hodgson E, Lewys-James A, Rao Ravella S, Thomas-Jones S, Perkins W, Gallagher J (2016) Optimisation of slow-pyrolysis process conditions to maximise char yield and heavy metal adsorption of biochar produced from different feedstocks. Biores Technol 214:574–581. https://doi.org/10.1016/j.biortech.2016.05.009
Ippolito JA, Cui L, Kammann C, Wrage-Mönnig N, Estavillo JM, Fuertes-Mendizabal T, Cayuela ML, Sigua G, Novak J, Spokas K, Borchard N (2020) Feedstock choice, pyrolysis temperature and type influence biochar characteristics: a comprehensive meta-data analysis review. Biochar 2:421–438. https://doi.org/10.1007/s42773-020-00067-x
Jaffari ZH, Abbas A, Kim CM, Shin J, Kwak J, Son C, Lee YG, Kim S, Chon K, Cho KH (2024) Transformer-based deep learning models for adsorption capacity prediction of heavy metal ions toward biochar-based adsorbents. J Hazard Mater 462:132773. https://doi.org/10.1016/j.jhazmat.2023.132773
Jatuwong K, Aiduang W, Kiatsiriroat T, Kamopas W, Lumyong S (2025) A review of biochar from biomass and its interaction with microbes: enhancing soil quality and crop yield in brassica cultivation. Life 15(2):284. https://doi.org/10.3390/life15020284
Jia Y, Jin S, Yan Q, Savi P (2021) The sensitivity analysis on GNSS-R soil moisture retrieval. In: 2021 photonics and electromagnetics research symposium (PIERS). IEEE, pp 2307–2311. https://doi.org/10.1109/PIERS53385.2021.9694804
John J, Gandhimathi R, Sillanpää M, Chellam PV (2024) Fe_3O_4-functionalised biochar for persulphate systems towards the removal of Remazol brilliant orange 3R: machine learning–based approach and toxicity analysis. Biomass Convers Biorefinery 14(9):10319–10334
Kaur D, Singh A, Sarkar S, Thakur S, Kaur J (2024) Biochar: a sustainable tool for soil health, reducing greenhouse gas emissions and mitigating climate change. J Appl Nat Sci 16(3). https://doi.org/10.31018/jans.v16i3.5665
Lakshmi D, Akhil D, Kartik A, Gopinath KP, Arun J, Bhatnagar A, Rinklebe J, Kim W, Muthusamy G (2021) Artificial intelligence (AI) applications in adsorption of heavy metals using modified biochar. Sci Total Environ 801:149623. https://doi.org/10.1016/j.scitotenv.2021.149623
Li M, Wei D, Liu T, Liu Y, Yan L, Wei Q, Du B, Xu W (2019) EDTA functionalized magnetic biochar for Pb(II) removal: adsorption performance, mechanism and SVM model prediction. Sep Purif Technol 227:115696. https://doi.org/10.1016/j.seppur.2019.115696
Li Y, Gupta R, You S (2022) Machine learning assisted prediction of biochar yield and composition via pyrolysis of biomass. Biores Technol 359:127511. https://doi.org/10.1016/j.biortech.2022.127511

Li H, Ai Z, Yang L, Zhang W, Yang Z, Peng H, Leng L (2023) Machine learning assisted predicting and engineering specific surface area and total pore volume of biochar. Biores Technol 369:128417. https://doi.org/10.1016/j.biortech.2022.128417

Li Y, Gupta R, Li W, Fang Y, Toney J, You S (2025a) Machine learning-assisted life cycle assessment of biochar soil application. J Clean Prod 498:145109. https://doi.org/10.1016/j.jclepro.2025.145109

Li Y, Shi K, Zhu M, Li H, Guo Y, Miao S, Zheng Z (2025b) Data-driven models for forecasting algal biomass in a large and deep reservoir. Water Res 270:122832. https://doi.org/10.1016/j.watres.2024.122832

Liaw A, Wiener M (2002) Classification and regression by randomForest. R News 2(3):18–22

Lin Y, Guo Z, Meng Q, Li C, Ma T (2025) Prediction of peak strength under triaxial compression for sandstone based on ABC-SVM algorithm. Expert Syst Appl 264:125923. https://doi.org/10.1016/j.eswa.2024.125923

Liu Q, Liu B, Zhang Y, Hu T, Lin Z, Liu G, Wang X, Ma J, Wang H, Jin H, Ambus P, Amonette JE, Xie Z (2019) Biochar application as a tool to decrease soil nitrogen losses (NH_3 volatilization, N_2O emissions, and N leaching) from croplands: options and mitigation strength in a global perspective. Glob Change Biol 25(6):2077–2093. https://doi.org/10.1111/gcb.14613

Liu J, Huang S, Chen K, Wang T, Mei M, Li J (2020) Preparation of biochar from food waste digestate: pyrolysis behavior and product properties. Biores Technol 302:122841. https://doi.org/10.1016/j.biortech.2020.122841

Liu C, Balasubramanian P, An J, Li F (2025) Machine learning prediction of ammonia nitrogen adsorption on biochar with model evaluation and optimization. NPJ Clean Water 8(1):13

Luo J, Jin M, Ye L, Cao Y, Yan Y, Du R, Yoshiie R, Ueki Y, Naruse I, Lin C, Lee Y (2019) Removal of gaseous elemental mercury by hydrogen chloride non-thermal plasma modified biochar. J Hazard Mater 377:132–141. https://doi.org/10.1016/j.jhazmat.2019.05.045

Ma J, Li T, Liu Y, Cai T, Wei Y, Dong W, Chen H (2019) Rice husk derived double network hydrogel as efficient adsorbent for Pb(II), Cu(II) and Cd(II) removal in individual and multicomponent systems. Biores Technol 290:121793. https://doi.org/10.1016/j.biortech.2019.121793

Mohammadi R, Azadmehr A, Maghsoudi A (2019) Fabrication of the alginate-combusted coal gangue composite for simultaneous and effective adsorption of Zn(II) and Mn(II). J Environ Chem Eng 7(6):103494. https://doi.org/10.1016/j.jece.2019.103494

Mohan D, Abhishek K, Sarswat A, Patel M, Singh P, Pittman CU (2018) Biochar production and applications in soil fertility and carbon sequestration–a sustainable solution to crop-residue burning in India. RSC Adv 8(1):508–520. https://doi.org/10.1039/c7ra10353k

Nazeer M, Khan MJ, Muhammad D, Khan A (2022) Biochar application stabilized the heavy metals in coal mined soil. Can J Soil Sci 103(2):297–304. https://doi.org/10.1139/cjss-2022-0073

Nguyen XC, Ly QV, Nguyen TTH, Ngo HTT, Hu Y, Zhang Z (2022) Potential application of machine learning for exploring adsorption mechanisms of pharmaceuticals onto biochars. Chemosphere 287:132203. https://doi.org/10.1016/j.chemosphere.2021.132203

Nguyen VG, Sharma P, Ağbulut Ü, Le HS, Truong TH, Dzida M, Tran VD (2024) Machine learning for the management of biochar yield and properties of biomass sources for sustainable energy. Biofuels, Bioprod Biorefin 18(2):567–593. https://doi.org/10.1002/bbb.2596

Nie Y, Zhao C, Zhou Z, Kong Y, Ma J (2023) Hydrochloric acid-modified fungi-microalgae biochar for adsorption of tetracycline hydrochloride: performance and mechanism. Biores Technol 383:129224. https://doi.org/10.1016/j.biortech.2023.129224

Nighojkar A, Nagpal J, Soboyejo W, Plappally A, Pandey S (2023a) Prediction of organophosphorus pesticide adsorption by biochar using ensemble learning algorithms. Environ Monit Assess 195(8):984

Nighojkar A, Pandey S, Naebe M, Kandasubramanian B, Soboyejo WW, Plappally A, Wang X (2023b) Using machine learning to predict the efficiency of biochar in pesticide remediation. NPJ Sustain Agric 1(1):1. https://doi.org/10.1038/s44264-023-00001-1

Niu Q, Luo J, Xia Y, Sun S, Chen Q (2017) Surface modification of bio-char by dielectric barrier discharge plasma for Hg^0 removal. Fuel Process Technol 156:310–316. https://doi.org/10.1016/j.fuproc.2016.09.013

Nurmalasari AI, Rahayu M, Owena ENN (2022) Growth of soybean (Glycine max L.) on various types of biochar. In: IOP conference series: earth and environmental science, vol 1. IOP Publishing, p 012005. https://doi.org/10.1088/1755-1315/1016/1/012005

Oral B, Coşgun A, Günay ME, Yıldırım R (2024) Machine learning-based exploration of biochar for environmental management and remediation. J Environ Manage 360:121162. https://doi.org/10.1016/j.jenvman.2024.121162

Palansooriya KN, Li J, Dissanayake PD, Suvarna M, Li L, Yuan X, Sarkar B, Tsang DCW, Rinklebe J, Wang X, Ok YS (2022a) Prediction of soil heavy metal immobilization by biochar using machine learning. Environ Sci Technol 56(7):4187–4198. https://doi.org/10.1021/acs.est.1c08302

Palansooriya KN, Li J, Dissanayake PD, Suvarna M, Li L, Yuan X, Sarkar B, Tsang DCW, Rinklebe J, Wang X, Ok YS (2022b) Prediction of soil heavy metal immobilization by biochar using machine learning. Environ Sci Technol 56(7):4187–4198

Patel N, Patel S, Mankad SH (2022) Impact of autoencoder based compact representation on emotion detection from audio. J Ambient Intell Humaniz Comput 13(2):867–885. https://doi.org/10.1007/s12652-021-02979-3

Ponnuchamy M, Kapoor A, Senthil Kumar P, Vo DVN, Balakrishnan A, Mariam Jacob M, Sivaraman P (2021) Sustainable adsorbents for the removal of pesticides from water: a review. Environ Chem Lett 19:2425–2463

Pyo J, Hong SM, Kwon YS, Kim MS, Cho KH (2020) Estimation of heavy metals using deep neural network with visible and infrared spectroscopy of soil. Sci Total Environ 741:140162. https://doi.org/10.1016/j.scitotenv.2020.140162

Qian Y, Wang X, Wu G, Wang L, Geng J, Yu N, Wei S (2021) Screening priority indicator pollutants in full-scale wastewater treatment plants by non-target analysis. J Hazard Mater 414:125490. https://doi.org/10.1016/j.jhazmat.2021.125490

Rodríguez-Romero JA, Mendoza-Castillo DI, Reynel-Ávila HE, De Haro-Del Rio DA, González-Rodríguez LM, Bonilla-Petriciolet A, Duran-Valle CJ, Camacho-Aguilar KI (2020) Preparation of a new adsorbent for the removal of arsenic and its simulation with artificial neural network-based adsorption models. J Environ Chem Eng 8(4):103928. https://doi.org/10.1016/j.jece.2020.103928

Sajjadi B, Chen W Y, Mattern DL, Hammer N, Dorris A (2020) Low-temperature acoustic-based activation of biochar for enhanced removal of heavy metals. J Water Process Eng 34:101166. https://doi.org/10.1016/j.jwpe.2020.101166

Singh J, Goyal KK, Kumar R, Gupta V (2022) Development of artificial intelligence-based neural network prediction model for responses of additive manufactured polylactic acid parts. Polym Compos 43(8):5623–5639. https://doi.org/10.1002/pc.26876

Supraja KV, Kachroo H, Viswanathan G, Verma VK, Behera B, Doddapaneni TRKC, Kaushal P, Ahammad SZ, Singh V, Awasthi MK, Jain R (2023) Biochar production and its environmental applications: recent developments and machine learning insights. Biores Technol 387:129634

Talebkeikhah F, Rasam S, Talebkeikhah M, Torkashvand M, Salimi A, Moraveji MK (2022) Investigation of effective processes parameters on lead (II) adsorption from wastewater by biochar in mild air oxidation pyrolysis process. Int J Environ Anal Chem 102(16):3975–3995. https://doi.org/10.1080/03067319.2020.1777291

Tao Y, Han S, Zhang Q, Yang Y, Shi H, Akindolie MS, Jiao Y, Qu J, Jiang Z, Han W, Zhang Y (2020) Application of biochar with functional microorganisms for enhanced atrazine removal and phosphorus utilization. J Clean Prod 257:120535. https://doi.org/10.1016/j.jclepro.2020.120535

Ukoba K, Jen TC (2022) Biochar and application of machine learning: a review. IntechOpen. https://doi.org/10.5772/intechopen.108024

Viet NTT, Anh BP, Hanh ND, Huy LM, Duy NT, Lieu PTB (2024) Research on building a biochar production model from agricultural by-products in Gia Lai province. J Forest Sci Technol 9(2):003–010. https://doi.org/10.55250/jo.vnuf.9.2.2024.003-010

Wallace CA, Saha GC, Afzal MT, Lloyd A (2019) Experimental and computational modeling of effective flexural/tensile properties of microwave pyrolysis biochar reinforced GFRP biocomposites. Compos B Eng 175:107180. https://doi.org/10.1016/j.compositesb.2019.107180

Wang HSH, Yao Y (2023) Machine learning for sustainable development and applications of biomass and biomass-derived carbonaceous materials in water and agricultural systems: a review. Resour Conserv Recycl 190:106847. https://doi.org/10.1016/j.resconrec.2022.106847

Wang H, Xia W, Lu P (2017) Study on adsorption characteristics of biochar on heavy metals in soil. Korean J Chem Eng 34(6):1867–1873. https://doi.org/10.1007/s11814-017-0048-7

Wang K, Liu L, Yuan C, Wang Z (2021) Software defect prediction model based on LASSO–SVM. Neural Comput Appl 33:8249–8259. https://doi.org/10.1007/s00521-020-04960-1

Wang W, Chang JS, Lee DJ (2024a) Machine learning applications for biochar studies: a mini-review. Biores Technol 394:130291. https://doi.org/10.1016/j.biortech.2023.130291

Wang Y, Xu L, Li J, Ren Z, Liu W, Ai Y, Zhou Y, Li Q, Zhang B, Guo N, Qu J, Zhang Y (2024b) Multi-output neural network model for predicting biochar yield and composition. Sci Total Environ 945:173942. https://doi.org/10.1016/j.scitotenv.2024.173942

Wang M (2024) Exploration of the application of artificial intelligence in modern agricultural production—take orchard management as an example. Appl Comput Eng 87:127–131. https://doi.org/10.54254/2755-2721/87/20241537

Yadav RK, Yadav MR, Kumar R, Parihar CM, Yadav N, Bajiya R, Ram H, Meena RK, Yadav DK, Yadav B (2017) Role of biochar in mitigation of climate change through carbon sequestration. Int J Curr Microbiol Appl Sci 6(4):859–866. https://doi.org/10.20546/ijcmas.2017.604.107

Zhang Y, Pan G, Chen B, Han J, Zhao Y, Zhang C (2020) Short-term wind speed prediction model based on GA-ANN improved by VMD. Renew Energy 156:1373–1388. https://doi.org/10.1016/j.renene.2019.12.047

Zhang Y, He M, Wang L, Yan J, Ma B, Zhu X, Ok YS, Mechtcherine V, Tsang DCW (2022) Biochar as construction materials for achieving carbon neutrality. Biochar 4(1):59. https://doi.org/10.1007/s42773-022-00182-x

Zhang X, Lv J, Zhang Y, Li S, Chen X, Sha Z (2023) A meta-analysis study on the use of biochar to simultaneously mitigate emissions of reactive nitrogen gases (N_2O and NO) from soils. Sustainability 15(3):2384. https://doi.org/10.3390/su15032384

Zhao Y, Fan D, Li Y, Yang F (2022) Application of machine learning in predicting the adsorption capacity of organic compounds onto biochar and resin. Environ Res 208:112694. https://doi.org/10.1016/j.envres.2022.112694

Zhou R, Zhang M, Zhou J, Wang J (2019) Optimization of biochar preparation from the stem of Eichhornia crassipes using response surface methodology on adsorption of Cd^{2+}. Sci Rep 9(1):17538. https://doi.org/10.1038/s41598-019-54105-1

Zhou X, Sun J, Tian Y, Lu B, Hang Y, Chen Q (2020) Hyperspectral technique combined with deep learning algorithm for detection of compound heavy metals in lettuce. Food Chem 321:126503. https://doi.org/10.1016/j.foodchem.2020.126503

Zhou J, Huang S, Zhou T, Armaghani DJ, Qiu Y (2022) Employing a genetic algorithm and grey wolf optimizer for optimizing RF models to evaluate soil liquefaction potential. Artif Intell Rev 55(7):5673–5705. https://doi.org/10.1007/s10462-022-10140-5

Zhu X, Wang X, Ok YS (2019) The application of machine learning methods for prediction of metal sorption onto biochars. J Hazard Mater 378:120727. https://doi.org/10.1016/j.jhazmat.2019.06.004

Zhu X, Wan Z, Tsang DC, He M, Hou D, Su Z, Shang J (2021) Machine learning for the selection of carbon-based materials for tetracycline and sulfamethoxazole adsorption. Chem Eng J 406:126782. https://doi.org/10.1016/j.cej.2020.126782

Zhu X, He M, Sun Y, Xu Z, Wan Z, Hou D, Alessi DS, Tsang DC (2022) Insights into the adsorption of pharmaceuticals and personal care products (PPCPs) on biochar and activated carbon with the aid of machine learning. J Hazard Mater 423:127060. https://doi.org/10.1016/j.jhazmat.2021.127060

Chapter 19
Optimizing Biochar Production and Substrate Composition Using Artificial Intelligence and Machine Learning: Advances in Process Efficiency and Resource Utilization

Anudeb Ghosh, Binoy Kumar Show, Debasis Maji, and Srinivasan Balachandran

Abstract Biochar production involves complex thermochemical interactions influenced by key process parameters, including feedstock composition, pyrolysis temperature, heating rate, and residence time. Conventional optimization techniques, relying on empirical models and large-scale experimental trials, are often time- and resource-intensive. Recent advancements in Artificial Intelligence (AI) and Machine Learning (ML) have introduced data-driven approaches to predict, optimize, and control biochar production with enhanced accuracy. This work examines the application of AI/ML models, such as artificial neural networks (ANNs), support vector regression (SVR), and deep learning architectures, for predicting biochar yield, carbon content, specific surface area, and pore structure based on experimental datasets. Furthermore, multi-objective optimization methods, including genetic algorithms and Bayesian optimization, have been employed to fine-tune process parameters to achieve desired biochar properties for specific applications, such as soil amendment or wastewater treatment. SHapley Additive exPlanations (SHAP)-based feature importance analysis and sensitivity analysis enhance model interpretability, enabling real-time process control and adaptive decision-making. Integrating AI-based modelswith experimental and pilot-scale manufacturing data reduces reliance on trial-and-error methods, leading to improved resource efficiency, a lower environmental footprint, and optimized biochar functionality. This overview highlights recent developments, addresses challenges related to data heterogeneity and model generalization, and outlines future directions for AI-assisted biochar manufacturing.

A. Ghosh · B. K. Show · S. Balachandran (✉)
Bioenergy Laboratory, Department of Environmental Studies, Siksha Bhavana, Visva-Bharati, Santiniketan, India
e-mail: s.balachandran@visva-bharati.ac.in

D. Maji
Department of Computer and System Sciences, Siksha-Bhavana, Visva-Bharati, Santiniketan, India

D. Moulick et al. (eds.), *Biochar: Achieving Agro-Environmental Sustainability*, Sustainability Solutions, https://doi.org/10.1007/978-3-032-19855-6_19

Keywords Biochar · Process optimization · Machine learning · SHAP · Carbon footprint

19.1 Introduction

The escalating global challenges of climate change, waste management, and resource scarcity necessitate innovative and sustainable solutions. Biochar, a carbon-rich material produced via biomass pyrolysis, has emerged as a particularly promising candidate due to its diverse applications and environmental benefits. However, the complexity of its production, combined with variability in feedstock and desired end-use properties, presents significant challenges for conventional optimization methods.

Biochar's physicochemical properties, determined by feedstock type and process parameters such as heating rate, residence time, and temperature, make it suitable for a wide range of applications (Tomczyk et al. 2020). It is widely used to improve soil fertility, water retention, and microbial activity, thereby supporting sustainable and resilient agriculture (Bulkinside 2025; Li et al. 2021). Biochar can enhance soil pH, nutrient uptake, and nutrient retention while immobilizing hazardous substances in contaminated soils, leading to increased crop yields. Its effectiveness is strongly influenced by application rate and pyrolysis temperature. The IPCC recognizes biochar as a scalable carbon removal strategy and an effective adsorbent for various water contaminants, including heavy metals, organic and inorganic pollutants, and pathogens, while also aiding nutrient recycling through biochar-amended biofilters (Al-Wabel et al. 2024a, b, c; Bulkinside 2025; Mace et al. 2021; Alori et al. 2025; Graham et al. 2021). Moreover, biochar's excellent electrical conductivity, tunable surface functional groups, and porous structure enable its use as a catalyst or catalyst support, and make it promising for energy storage applications such as supercapacitors, batteries, and hydrogen storage (Al-Wabel et al. 2024a, b, c; Liu et al. 2019). It can also be incorporated into concrete and insulation materials to improve strength, durability, thermal performance, and carbon sequestration (Bulkinside 2025).

Biochar production is governed by intricate thermochemical interactions, with parameters such as feedstock composition, pyrolysis temperature, heating rate, and residence time exerting significant influence on yield and final properties (Do and Nguyen 2024). Traditional engineering approaches, including empirical models and large-scale experimental trials, have been used to optimize these characteristics, yet they face notable limitations that hinder efficiency and consistency. Conventional optimization often involves extensive experimentation across the full range of pyrolysis parameters to assess the combined effects of multiple factors (Al-Wabel et al. 2024a, b, c). This trial-and-error approach is inherently labor-intensive and costly (Chen et al. 2023a, b). Furthermore, inadequate temperature and airflow control in traditional kilns can result in substantial variability in biochar quality (Al-Wabel et al. 2024a, b, c), making it difficult to ensure consistent properties—an essential factor for reliable applications such as soil amendment, where inconsistent pH or heavy metal content may be detrimental (Yu et al. 2019). The diversity of biomass feedstocks—ranging from lignocellulosic materials with varying cellulose, hemicellulose, and lignin content to animal litter and wood—further complicates standardization, as feedstock composition directly affects the resulting biochar's properties (Thakur et al. 2025; Gupta et al. 2022). To maximize biochar yield and optimize its properties, production methods must be tailored to specific biomass types and process conditions, a task that remains challenging under traditional approaches (Al-Wabel et al. 2024a, b, c).

While statistical methods such as Design of Experiments (DoE)—including central composite design and Box–Behnken design—are well recognized for optimizing synthesis processes by systematically controlling multiple factors, they still require numerous trials and can be computationally expensive when exploring complex interactions. The highly non-linear relationships between pyrolysis temperature, residence time, gas flow rate, and biochar yield or properties limit the practicality of empirical models (Sharma et al. 2023). These models often rely on simplified assumptions, which can reduce predictive accuracy. Traditional process control systems also tend to lack the adaptability needed for dynamic, non-linear chemical processes such as pyrolysis (Al-Wabel et al. 2024a, b, c). Additionally, the time lag associated with experimental characterization and analysis often prevents real-time adjustments, resulting in suboptimal outcomes or wasted resources (Chen et al. 2023a, b). Effective monitoring and adjustment of parameters such as temperature, heating rate, and residence time to maximize specific biochar properties (e.g., carbon content, porosity, surface area) require precise control that conventional systems may not provide (Zheng et al. 2023).

The multitude of interacting parameters—including feedstock type, pyrolysis temperature, heating rate, residence time, gas flow rate, and post-treatment methods (Mong et al. 2022; Hakeem et al. 2022)—combined with numerous target biochar properties such as yield, carbon content, specific surface area, pore structure, pH, and heavy metal profile, creates a vast and complex experimental space (Liu and Zhang 2022; Wang et al. 2019). As Cedrone et al. (2024) note, conducting experiments across the entire domain of pyrolysis process parameters and examining the simultaneous effects of multiple factors would require an impractically large number of trials, both costly and time-consuming, leading to a combinatorial explosion of possibilities. This complexity renders comprehensive optimization through traditional empirical or statistical methods unmanageable. The large number of variables and their non-linear interactions make exhaustive experimental exploration practically impossible, limiting the ability to fully understand and control the process.

Several examples illustrate the inherent trade-offs in biochar production. For instance, increasing pyrolysis temperature typically decreases biochar yield while increasing synthesis gas production (Al-Wabel et al. 2024a, b, c). Likewise, reaction conditions that improve pellet tensile strength may reduce fixed carbon yield in biochar-enhanced concrete (Castillo et al. 2025). Higher temperatures may enhance specific surface area and carbon content but often reduce overall yield (Al-Wabel et al. 2024a, b, c; Gou et al. 2025). These cases demonstrate that biochar optimization is not about maximizing a single property but instead balancing conflicting objectives.

This "trade-off dilemma" highlights the limitations of conventional single-objective optimization approaches and underscores the need for sophisticated multi-objective optimization frameworks capable of navigating such complexities and identifying solutions that balance performance, economic feasibility, and environmental impact.

Recent advancements in Artificial Intelligence (AI) and Machine Learning (ML) have introduced powerful data-driven methodologies for optimizing complex thermochemical processes, including biochar production. AI/ML approaches can overcome the limitations of traditional empirical and trial-and-error methods by enabling more accurate prediction, precise control, and efficient optimization of biochar properties and production parameters (Balakumar et al. 2025). These algorithms can process large volumes of data to detect patterns and correlations, offering distinct advantages over conventional models that rely on simplified assumptions or human expertise and are capable of capturing the complex, non-linear interactions often missed by traditional approaches (Murphy et al. 2021). This capability is especially relevant given the intricate interplay between feedstock properties and pyrolysis conditions (Al-Wabel et al. 2024a, b, c).

Integrating AI/ML into biochar production promises a paradigm shift toward more efficient, sustainable, and economically viable manufacturing. This transition moves beyond trial-and-error, embracing data-driven optimization for improved resource utilization, reduced environmental impact, and enhanced biochar functionality (Gou et al. 2025; Barkhordari and Qi 2025). This chapter examines the transformative potential of AI and ML in revolutionizing biochar production, focusing on specific models and methodologies, their applications in optimizing production and substrate composition, and their contributions to advancing process efficiency, resource utilization, and the broader realization of biochar's environmental and economic benefits.

19.2 Factors Influencing Biochar Yield and Properties

Optimizing biochar yield and properties is a complex task, as the final characteristics are closely tied to both the feedstock type and the specific pyrolysis conditions employed. Understanding these relationships is essential for tailoring biochar to diverse applications, including soil amendment, wastewater treatment, and energy storage. Biochar production involves the thermal decomposition of biomass in the absence of oxygen, a process known as pyrolysis. Several key parameters significantly influence the yield and physicochemical properties (Fig. 19.1) of the resulting biochar.

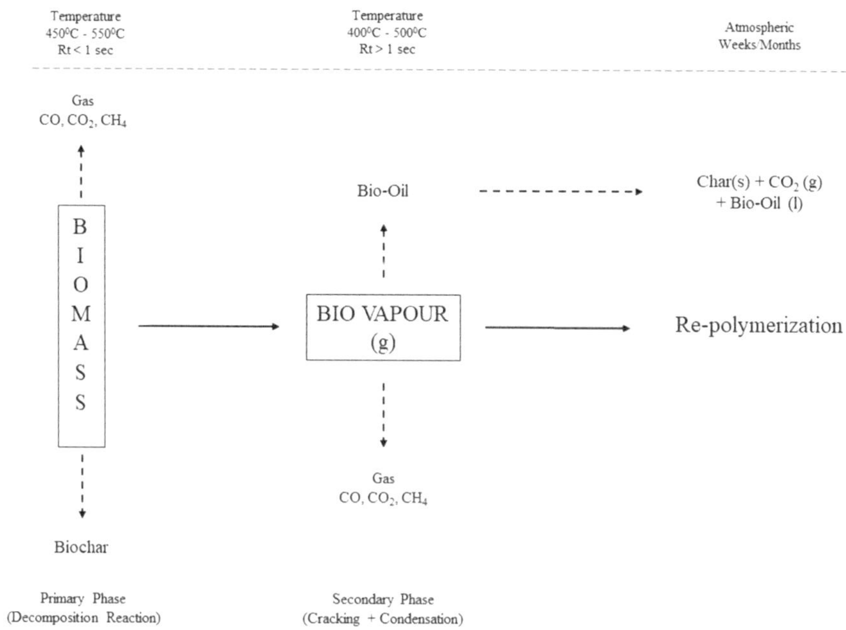

Fig. 19.1 Pyrolysis process

19.2.1 Pyrolysis Temperature

This is one of the most critical factors. In general, increasing pyrolysis temperature decreases biochar yield while enhancing properties such as carbon content, specific surface area (SSA), and porosity. Higher temperatures reduce the oxygen and hydrogen content, thereby increasing carbon concentration (Al-Wabel et al. 2024a, b, c). Elevated pyrolysis temperatures promote greater thermal cracking of the high-molecular-weight hydrocarbons present in biomass, which in turn increases the production of liquid and gaseous products and reduces the overall yield of biochar.

Studies consistently demonstrate an inverse relationship between pyrolysis temperature and biochar yield. For example, when pyrolyzing biomass such as hazelnut shells and sesame stalks within a temperature range of 400 °C to 700 °C, notable yield reductions were observed: 10% for hazelnut shells (Putun et al. 1999) and 17% for sesame stalks (Ates et al. 2004). Similarly, Choi et al. (2012) reported a decrease in biochar yield from 31 to 17% when the temperature increased from 365 °C to 606 °C. Conversely, lower pyrolysis temperatures tend to result in higher biochar yields, likely due to the partial decomposition of biomass, which leaves a larger solid residue compared to the more complete breakdown occurring at higher temperatures.

19.2.2 Heating Rate and Residence Time

These parameters also play a significant role. Slower heating rates are generally favorable for maximizing biochar production. Longer residence times can influence bio-oil and synthesis gas yields, while shorter holding times tend to produce less porous biochar. Precise control of these conditions is essential to achieve the desired biochar characteristics (Fig. 19.2).

Residence time—the duration for which material remains in the pyrolysis reactor—significantly impacts the final product composition. Shorter residence times generally favor the production of liquid and gaseous products, whereas longer residence times tend to increase biochar yield (Sun et al. 2016) by allowing more opportunity for repolymerization, which contributes to char formation. In a study on sugarcane bagasse pyrolysis at 525 °C, Scott et al. (1999) observed a direct correlation between residence time and biochar yield, and an inverse correlation with bio-oil yield: a mere 0.7-s increase in residence time reduced bio-oil yield from 75 to 57%. Similarly, investigations by Kim et al. (2011) into yellow–brown coal and wood biomass found that longer residence times produced greater quantities of biochar. Beyond yield, residence time also influences the composition and quality of pyrolysis products.

The interplay between residence time and temperature is critical, as the effect of residence time is not always straightforward. Some studies suggest that temperature can moderate its influence. For example, high temperatures combined with longer residence times have been shown to increase biochar yield, whereas low temperatures with longer residence times can reduce it. Liang et al. (2016) reported that increasing residence time at lower temperatures (e.g., 300 °C) decreased biochar yield, while at higher temperatures (above 600 °C) residence time had little to no

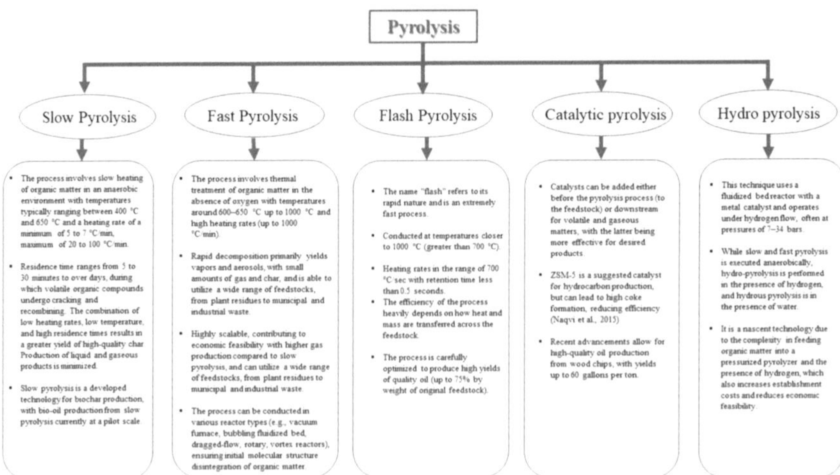

Fig. 19.2 Pyrolysis temperature and heating rates

effect. Identifying the optimal residence time is therefore essential for maximizing biochar yield, particularly when co-pyrolyzing biomass and plastics.

19.2.3 Feedstock Composition

The type of biomass used—such as wood, agricultural residues, or animal waste—profoundly influences biochar properties. Wood-based biochar typically exhibits higher carbon content and specific surface area than that derived from animal sources, whereas poultry litter biochar often has higher pH and extractable phosphorus. Differences in lignin, cellulose, hemicellulose, and moisture content contribute to these variations. Woody biomass generally has high bulk density and calorific value, along with low moisture, ash content, and voidage. In contrast, non-woody biomass (e.g., crop residues, sludge, animal manure) often contains higher moisture and ash levels, lower bulk density, lower calorific value, and greater voidage.

The origin of woody biomass also affects yield; coniferous species tend to produce more biochar than deciduous species. Generally, biomass with higher cellulose and hemicellulose content favors the production of liquids and gases, whereas higher lignin content is more conducive to char formation. Studies show a direct correlation between lignin content and biochar yield, with high-lignin woody biochar particularly suitable for soil applications (Lv et al. 2025). However, there are trade-offs: woody biochar tends to be more acidic than non-woody biochar, largely due to differences in composition and salt content. Non-woody biochar often exhibits greater cation exchange capacity (CEC) and lower surface acidity compared to woody biochar. For example, biochar from non-woody biomass and plastic blends has been reported to be alkaline (pH 10) with high ash content (44%), high water-holding capacity (41%), and a CEC of 37 cmolc kg^{-1} (Rodriguez et al. 2021). Lignin content also influences specific surface area and porosity, with higher lignin correlating positively with both.

Moisture content is another critical factor affecting pyrolysis products. Low moisture levels are preferable for biochar production, whereas high-moisture biomass is more suitable for bio-oil production, as water inhibits char formation and increases energy demand. Plastics, despite having very low moisture content (0.00–0.80%), generally produce less char due to their low fixed carbon (0.00–13.1%) and high volatile matter (86.83–99.88%) (Sharuddin et al. 2016). Product distribution also depends on plastic type: polyethylene (PE), with a melting point of 105–115 °C, typically yields about 1.34 wt% biochar, with the remainder being 69.82% liquid and 28.84% gas (Wang et al. 2015). Similarly, polypropylene (PP) pyrolysis produces approximately 8.4 wt% biochar, 59.7% liquid, and 31.9% gas (Aboulkas et al. 2008).

Blending different feedstocks offers a strategic means of tailoring input composition to control both yield and physicochemical properties of pyrolysis products for specific applications. For instance, increasing the proportion of plastic in a blend is particularly effective for maximizing pyrolysis oil production (Chen et al. 2017).

19.2.4 Gas Flow Rate

Carrier gases play a crucial role in the pyrolysis process by removing vapors produced during thermal decomposition, thereby preventing secondary reactions that could negatively affect the quality and quantity of the final products. Without this purging action, desired products such as biochar may exhibit altered physicochemical properties and reduced yields. Increasing the carrier gas flow rate has been reported to enhance the yield of stable carbon in biochar (Crombie and Mašek 2015). Conversely, other studies indicate an inverse relationship between carrier gas flow rate and biochar yield, suggesting that higher flow rates can reduce overall production. For example, Jamal and Fletcher (2023) reported the highest biochar yield of 25.3% at a maximum flow rate of 250 mL CO_2/min, with yields decreasing as the gas flow rate decreased. Their findings also emphasized the significant influence of process temperature and residence time on biochar yield.

Commonly used carrier gases include argon, nitrogen, carbon dioxide, and steam, with nitrogen being the preferred choice in many commercial pyrolysis operations due to its inert nature, wide availability, and cost-effectiveness (Araki et al. 2021). The effect of different carrier gases on pyrolysis products has also been investigated. For instance, Abbas Abadi et al. (2013) demonstrated that, in the pyrolysis of polypropylene, hydrogen (H_2) was more effective than inert gases in increasing the yield of condensed products.

19.2.5 Reactor Types

The choice of reactor is paramount to successful pyrolysis, as it directly influences the conversion of organic matter into gaseous, liquid, or solid fuels. Reactor design must account for parameters such as heating temperature, pressure, and vapor residence time. Common reactor types include batch, semi-batch, fixed-bed, fluidized-bed, conical sprouted-bed, and microwave-assisted reactors, which can be classified according to feedstock movement and heat transfer mechanisms (Fig. 19.3).

19.3 Integration of AI and ML in Biochar Production

Developing new biochar materials or optimizing existing ones for specific applications often requires extensive preliminary experimental exploration, which slows the process of achieving desired microstructures or properties (Chen et al. 2023a, b). The International Biochar Initiative (IBI) highlights the need for cost-effective and reliable methods for testing biochar stability, underscoring the resource intensity of current characterization practices (International Biochar Initiative 2018).

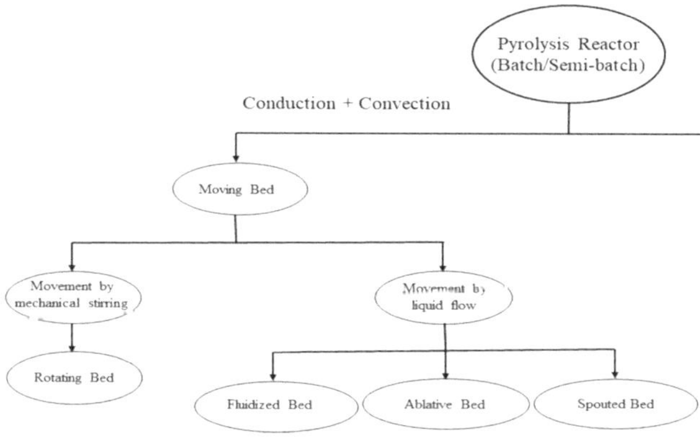

Fig. 19.3 Types of reactors in pyrolysis

To make ML models more transparent and interpretable, several techniques can be employed. One key approach is feature importance analysis, which quantifies the influence of each input factor on the model's output. This can be achieved through methods such as permutation feature importance or feature contribution graphs. By identifying which biomass characteristics are most influential, researchers can prioritize efforts to control these variables, ultimately producing biochar with targeted properties.

Other powerful interpretability tools include LIME (Local Interpretable Model-agnostic Explanations) and SHAP (Shapley Additive Explanations) values (Mangalathu et al. 2020; Man and Chan 2020). LIME generates simplified, interpretable "proxy" models for individual predictions, enabling researchers to understand why a model made a specific prediction for a given sample. SHAP values, in contrast, offer a global perspective, revealing how each feature contributes to predictions across the entire dataset, thus providing a more comprehensive understanding of model behavior. Decision trees and rule-based models, inherently more interpretable, can be used alongside complex models to shed light on post-prediction reasoning.

Comparing the results of simple and complex models helps validate prediction accuracy and reliability. Interpretability is also essential for identifying potential flaws or biases in a model's decision-making process. For example, in biochar production, interpretability analysis could reveal whether the model consistently favors or disregards certain carbon sources (Pathy and Meher 2020), enabling more equitable and informed decision-making.

19.3.1 The Transformative Role of Artificial Intelligence and Machine Learning

ML models can learn from data and improve performance over time without explicit programming, shifting the industry from reactive to proactive process management (Zong and Guan 2024). Their computational efficiency and flexibility enable the handling of massive datasets and the generation of predictions comparable to—or surpassing—those of conventional models (Table 19.1), substantially reducing the need for extensive experimental trials. This leads to significant savings in time, labor, and cost in biochar development.

19.3.2 Multi-Objective Optimization Methods

Multi-objective optimization techniques are essential for addressing conflicting objectives in biochar production via pyrolysis, such as maximizing biochar yield and carbon stability while minimizing energy requirements and emissions. Unlike single-objective models, these methods generate a set of optimal compromise solutions—known as the Pareto front—allowing decision-makers to select operating conditions that simultaneously satisfy multiple criteria.

One widely used approach is the Genetic Algorithm (GA), which applies principles of natural selection and evolutionary biology to iteratively refine solutions. For example, Xie et al. (2025a) used a GA to optimize pyrolysis conditions for soybean straw, balancing surface area and electrical conductivity. Bayesian Optimization (BO) is another powerful technique that constructs a probabilistic model—often using Gaussian processes—to explore the parameter space efficiently with fewer experiments. Nguyen et al. (2024c) applied BO to optimize reactor temperature and feed rate, reducing optimization time by more than 40% compared to conventional grid search methods.

Other algorithms such as Particle Swarm Optimization (PSO) and the Non-dominated Sorting Genetic Algorithm II (NSGA-II) have also gained prominence. Malashin et al. (2026) employed NSGA-II for concurrent optimization of biochar yield, energy efficiency, and CO_2 emissions, producing a suite of Pareto-optimal reactor designs. Similarly, PSO has been used to dynamically adjust heating rates in response to fluctuating lignocellulosic compositions (Aghbashlo et al. 2019).

These multi-objective algorithms not only improve process efficiency but also support informed and sustainable decision-making in pyrolysis systems.

Table 19.1 Key AI/ML approaches for biochar optimization

AI/ML category	Specific AI/ML Models	Primary application in biochar	Key benefit	References
Predictive modeling	ANNs	Predicting biochar yield, carbon content, elemental composition, and physicochemical properties	High accuracy, adept at complex, non-linear relationships, computationally efficient	(Khan et al. 2022)
	Support vector regression (SVR)	Predicting biochar fuel properties, heavy metal adsorption capacity, general characteristics	Strong generalized performance, effective for high-dimensional data	(Al-Wabel et al. 2024a, b, c)
	Deep learning architectures (XGBoost, LightGBM, DfNNs, Random orest)	Predicting biochar yield specific surface area (SSA), pore structure, carbon content, adsorption capacity	Superior performance handles large datasets and complex tasks, robust generalization	(Chen et al. 2023a, b; Gou et al. 2025; Jiang et al. 2025)
Multi-objective optimization	Genetic algorithms (GAs)	Optimizing process parameters for desired properties (e.g.,yield, emissions, specific applications)	Balances conflicting objectives, explores wide design spaces, avoids local minima	(Cheng and Li 1998; Stack Overflow 2021)
	Bayesian optimization	Efficiently finding optimal parameters for specific properties (e.g., SSA, adsorption capacity)	Effective for optimizing expensive-to- evaluate functions, identifies optimal conditions	(Yang et al. 2023)

(continued)

Table 19.1 (continued)

AI/ML category	Specific AI/ML Models	Primary application in biochar	Key benefit	References
Model interpretability	SHAP	Explaining feature contributions to model predictions, identifying influential variables	Provides consistent and objective explanations, builds trust in "black-box" models	(Chen et al. 2023a, b; Nguyen et al. 2024a, b)
	Sensitivity analysis	Determining the impact of input parameters on output properties, identifying influential factors	Quantifies direct and interactive effects, crucial for adaptive decision-making	(Yang et al. 2023; Nguyen et al. 2024a, b)

19.3.3 Data-Driven Prediction of Biochar Properties

AI and ML models—including Artificial Neural Networks (ANNs), Support Vector Regression (SVR), Decision Trees, Gradient Boosting models (e.g., XGBoost, LightGBM), and Deep Learning frameworks—are increasingly used to model and predict key biochar properties from experimental datasets. These models can capture complex nonlinear relationships between pyrolysis conditions, feedstock properties, and resulting biochar characteristics, producing faster and more accurate predictions than empirical methods (Gou et al. 2025; Chen et al. 2023a, b; Jiang et al. 2025; Khan et al. 2022; Wang et al. 2025).

19.3.3.1 Prediction of Biochar Yield

ANNs have shown strong predictive capability for biochar yield, with high coefficients of determination (R^2 values up to 0.988 for training and 0.941 for testing), particularly when inputs include volatile matter, fixed carbon, ash content, elemental composition, and pyrolysis temperature (Khan et al. 2022; Al-Wabel et al. 2024a, b, c). Ensemble models such as XGBoost and Bagging Regressors have also achieved high accuracy ($R^2 > 0.95$) across diverse biomass types and process conditions. Feature importance analyses consistently identify pyrolysis temperature and feedstock type as key controlling variables (Gou et al. 2025; Chen et al. 2023a, b).

19.3.3.2 Carbon Content Prediction

ML models such as XGBoost, Random Forests, and ANNs have been effective in predicting biochar carbon content, with reported R^2 values exceeding 0.96 in some studies (Gou et al. 2025; Al-Wabel et al. 2024a, b, c). Sensitivity analyses identify feedstock carbon content, pyrolysis temperature, and heating rate as dominant variables, in agreement with thermochemical degradation theory (Jiang et al. 2025).

19.3.3.3 SSA and Pore Structure Prediction

Predicting specific surface area (SSA) and pore structure parameters—such as pore volume and mean pore diameter—is critical for adsorption and catalysis applications. SVR, LightGBM, and deep neural networks (DNNs) have demonstrated strong predictive accuracy for these physical properties (Chen et al. 2023a, b; Gou et al. 2025; Jiang et al. 2025). Key input features include pyrolysis temperature, ash content, activation agent ratio, and residence time. Jiang et al. (2025) further demonstrated that incorporating plant organ identity into DNN models enhances SSA prediction, suggesting that anatomical differences between feedstocks significantly influence porosity outcomes.

19.3.3.4 Prediction of Other Biochar Properties

Beyond conventional physicochemical properties, ML models are increasingly applied to predict parameters such as higher heating value (HHV), heavy metal adsorption capacity, and mechanical strength in composite materials containing biochar. For example, Yadav et al. (2025a, b) and Wang et al. (2025) used ensemble ML algorithms to predict the Pb^{2+} and Cd^{2+} adsorption capacities of activated biochars, achieving strong correlations with experimental results. These models typically employ input features such as functional group composition, BET surface area, and elementalratios. Similarly, models trained on experimental mechanical testing data have been used to predict tensile strength and modulus in biochar–polymer composites (Al-Wabel et al. 2024a, b, c; DataCamp n.d.).

Different machine learning regression models are widely employed in biochar production to forecast pyrolysis outcomes based on process parameters such as temperature, feedstock type, and residence time. Each model has its own strengths depending on the complexity, nonlinearity, and interpretability required. Figure 19.4 illustrates the predictive capabilities of different machine learning models in terms of yield, carbon content, and specific surface area (SSA).

Multiple Linear Regression (MLR) is a traditional technique that assumes a linear relationship between the target and independent variables. While MLR is easy to interpret and understand, it generally cannot capture the complex nonlinearities inherent in pyrolysis processes (Li et al. 2022). Decision Trees (DT) provide a non-parametric approach, splitting data according to feature thresholds. They are interpretable and effective at capturing nonlinear relationships but are prone to overfitting (Mienye and Jere 2025). Random Forest (RF) mitigates overfitting by constructing an ensemble of decision trees through bootstrap aggregation (bagging) and random feature selection, thereby improving accuracy and generalizability. RF is widely

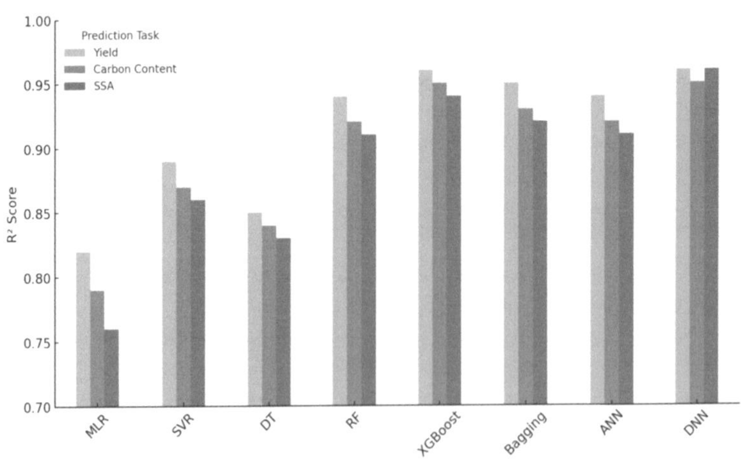

Fig. 19.4 Model prediction comparison (R^2) for biochar property prediction as per various research

applied in predicting biochar properties due to its robustness and ability to handle high-dimensional data (Su and Jiang 2024).

The Bagging Regressor (BR) is an ensemble method similar to RF but capable of using any base regressor. It reduces variance and enhances model stability. BR with DT as the base estimator has demonstrated high predictive accuracy and robustness against noise in biochar modeling (Song et al. 2024). The AdaBoost Regressor (AR) improves predictive performance through sequential fitting of weak learners—typically shallow trees—on the residuals of preceding models. AR performs well with moderately noisy data and has been applied in biochar yield optimization studies (Zhu et al. 2019). Extreme Gradient Boosting (XGBoost or XGB) is an advanced gradient boosting algorithm incorporating regularization, parallel computation, and sophisticated pruning.

It has shown exceptional performance in the simultaneous prediction of multiple biochar properties and has been integrated with SHAP values for model interpretability (Pathy et al. 2020). Table 19.2 presents the equations used in the predictive models.

Each of these models serves a distinct role in modeling pyrolysis processes, with ensemble-based approaches such as RF and XGB proving particularly effective for handling the nonlinear, multivariable, and noisy datasets characteristic of biomass-to-biochar conversion systems. Table 19.3 summarizes the different models employed by various authors in their studies.

19.3.4 Model Interpretability and Real-Time Control

Model interpretability and real-time control are essential for the effective application of AI and machine learning (ML) in optimizing biochar production via pyrolysis. Recent studies have shown that interpretable ML models are critical for improving the transparency and credibility of predictive systems used in pyrolysis. Methods such as SHAP, LIME, and partial dependence plots have been employed to elucidate the roles of variables such as pyrolysis temperature, heating rate, residence time, and feedstock properties in determining biochar yield and quality. For instance, Hooda and Mondal (2024) applied XGBoost combined with SHAP to reveal that pyrolysis temperature and feedstock moisture were major predictors of fixed carbon content and BET surface area. Li et al. (2022) demonstrated that neural networks integrated with feature attention mechanisms could dynamically adjust pyrolysis conditions to achieve optimal carbon stability. Similarly, Nguyen et al. (2024b) used random forest models coupled with LIME to predict biochar pH and cation exchange capacity from a range of lignocellulosic feedstocks. Concurrently, real-time control approaches—such as reinforcement learning (Faridi et al. 2024) and digital twin systems (Song et al. 2024)—are being developed to adapt to variations in feedstock and environmental conditions. These models enable open-loop monitoring and automated reactor parameter adjustment to reduce emissions without compromising product quality.

Table 19.2 Predictive modelling equations

Sl. No	Model	Equation	References
1	Multiple linearregression (MLR)	$Y = \beta_0 + \beta_1 X_1 + \beta_2 X_2 + \cdots + \beta_n X_n + \in$ Where Y is the predicted property (e.g., yield), Xi are inputfeatures(e.g., temperature, ash), β$i$are coefficients, and ϵ is the error term	(Draper and Smith1998)
2	SVR	$f(x) = \sum_{i=1}^{n} (\alpha_i - \alpha_i^*) K(x_i, x) = b$ Where $K(xi,x)$ is the kernel function (e.g., RBF), and αi, α_i^* are Lagrange multipliers	(Smola and Schölkopf 2004)
3	Decision tree (DT)	Recursive binary splits of the form: If Xj < t, then go left; else go right Where Xj is a feature and t is the threshold	(Breiman et al. 2017)
4	Random forest (RF)	$\widehat{Y} = \frac{1}{T} \sum_{t=1}^{T} h_t(X)$ Where h_t is the prediction from the tth decision tree	(Breiman 2001)
5	XGBoost (XGB)	$\widehat{Y} = \sum_{\substack{k=1 \\ f_k \in F}}^{T} f_k(x_1)$ With F being the space of regression trees and objective: $L(\emptyset) = \sum_i l(y_i, \widehat{y}_i) + \sum_k \Omega(f_k)$	(Chen and Guestrin 2016)
6	Bagging regressor (BR)	$\widehat{Y} = \frac{1}{B} \sum_{b=1}^{B} f_b x$ Where each f_b is trained on a bootstrap sample	(Breiman 1996)
7	ANN	$\widehat{Y} = \sigma(W_2 \cdot \sigma(W_1 \cdot X \cdot b_1) + b_2$ Where W and b are weights and biases, and σ is an activation function (e.g., ReLU or sigmoid)	(Haykin 1994)

Together, model interpretability and real-time control are driving the development of smart pyrolysis systems that are reliable, transparent, scalable, and sustainable.

19.3.5 SHAP Values

SHAP values play a key role in enhancing the interpretability and trustworthiness of AI/ML models used to optimize biochar production through pyrolysis. They provide a game-theoretic approach to quantify the contribution of each input variable to a model's output, enabling researchers to understand the influence of pyrolysis conditions—such as temperature, heating rate, residence time, and feedstock properties—on biochar attributes including carbon content, surface area, and volatile matter. For example, Hooda and Mondal (2024) integrated SHAP analysis with XGBoost models

Table 19.3 Key findings from the application of machine learning Models

Feedstock types	Input parameters	ML Models used	Predicted outputs	Key findings/ performance (R^2 Values)	References
Various organic wastes (lignocellulosic biomass)	Feedstock proximate (FC VM, Ash), Ultimate (C, H, O, N) analysis, Pyrolysis temperature (PT), Heating rate (HR), Residence time (RT)	Bagging regressor (BR), Multiple linear regression (MLR), Decision trees (DT), Adaboost regressors (AR)	Biochar yield, Fixed carbon, Volatile matter, Ash, Ultimate composition parameters (C, H, O, N)	Bagging regressor (BR) demonstrated superior efficacy: Biochar Yield: 0.94 and Fixed Carbon: 0.93 and Volatile Matter: 0.93 and Ash: 0.94 and C: 0.95 and H: 0.90 and O: 0.92 and N: 0.96	Gou et al. (2025)
46 different types of agricultural biomass	Proximate (Ash, Fixed carbon, Volatiles), Ultimate (C, H, O, N), Lignin, Cellulose, Hemicellulose content, Pyrolysis Temperature (T), Heating rate (HR), Particle size (PS), Flow rate (Nitrogen)	XGBoost, Random forest, Lasso regression, Tweedie regression, Gradient boosting regression	Biochar Yield, Specific Surface Area (SSA)	XGBoost was superior for biochar yield: and Training R^2: 0.9739 and Test R^2: 0.8875 whereas Random Forest (RF) for SSA: Maximum R^2 value of 0.85	Uppalapati et al. (2025)
Biomass (yellow-brown coal and wood biomass)	Biochar elemental composition, industrial analysis, pyrolysis conditions	XGBoost, Random forest, AdaBoost, GBDT, LightGBM	Biochar yield, Specific surface area (SSA)	XGBoost showed excellent performance: Biochar Yield: R^2 = 0.79 and SSA: R^2 = 0.92	Zhou et al. (2024)

(continued)

Table 19.3 (continued)

Feedstock types	Input parameters	ML Models used	Predicted outputs	Key findings/ performance (R^2 Values)	References
Biomass and plastics (co-pyrolysis)	Feedstock properties, pyrolysis temperature, plastic ratio	XGBoost, Dense neural network	Biochar yield, Bio-oil yield	XGBoost for biochar Yield: $R^2 = 0.96$ and dense neural network for bio-oil yield: $R^2 = 0.96$	Alabdrabalnabi et al. (2022)
Aquatic weed biomass	Not explicitly detailed in search results for ML inputs	Gradient boosted regression trees (BRT), Random Forest (RF), SVM, ANN	Aquatic weed biomass yield (for biochar production)	Gradient boosted regression trees (BRT) demonstrated superior performance: $R^2 = 0.63$	Banda et al. (2025)
Biomass (general)	Not explicitly detailed for input parameters, but typically includes pyrolysis conditions and feedstockproperties	ANN coupled with Rao-2 algorithm (and other metaheuristic-optimized ANNs)	Biochar yield	ANN coupled with Rao-2 algorithm: $R^2 \approx 0.93$	Khan et al. (2022)

(continued)

Table 19.3 (continued)

Feedstock types	Input parameters	ML Models used	Predicted outputs	Key findings/ performance (R^2 Values)	References
Various biomass sources	Temperature (most important), other input features for proximate analysis	BRT, Linear regression, AdaBoost	Biochar Yield, Fixed carbon, Volatile matter, Ash	BRT-based model delivered the best prediction performance: training R^2: 0.982 to 0.999 and Test R^2: 0.968 to 0.988 (range across different outputs)	Le et al. (2023)
Lignocellulosic biomass	Lignin properties, chemical activators (KOH, $ZnCl_2$ H_3PO_4, K_2CO_3, NaOH, Na_2CO_3), pyrolysis conditions	Hybrid ML framework (interpolation model + regressor)	Specific surface area (SSA), Total pore volume (TPV), Microporosity (MP)	High predictive accuracy: SSA: R^2 = 0.82, TPV: R^2 = 0.86 & MP: R^2 = 0.81	Xie et al. (2025a, 2025b)

to identify pyrolysis temperature and biomass moisture content as the most significant factors in determining biochar yield and fixed carbon percentage, allowing for parameter prioritization during process optimization. Likewise, Xiang et al. (2022) applied SHAP analysis to disentangle the nonlinear effects of lignin and cellulose fractions on biochar pH and adsorption capacity, thereby improving feedstock selection strategies. In parallel, sensitivity analysis– particularly variance-based methods such as Sobol indices and Morris screening—has been employed to quantify the variation in model outputs resulting from changes in input parameters. Gao et al. (2021) performed a global sensitivity analysis of pyrolysis conditions, demonstrating that reactor temperature and residence time accounted for more than 70% of the variance in biochar surface area predictions. These analytical tools not only enhance transparency but also support rigorous model validation and experimental design, ensuring that AI-augmented pyrolysis models function as scientifically validated decision-support systems rather than "black boxes."

19.3.6 Real-time Process Control and Adaptive Decision-Making

The integration of AI, ML, and digital twin technologies has made real-time process control and adaptive decision-making increasingly feasible in biochar production via pyrolysis. Such systems are essential to ensure product consistency, maximize carbon yield, and minimize emissions under dynamic operating conditions. Several studies have explored predictive ML algorithms embedded within control loops for continuous pyrolysis reactors. For example, Song et al. (2024) demonstrated the use of a digital twin platform that forecasts pyrolysis dynamics in real time, enabling proactive adjustments to reactor temperature and feedstock flow rate. Reinforcement learning methods, including Deep Q-Networks (DQNs), have also shown promise in adaptive control. Wang et al. (2024) developed a reinforcement learning-based controller that adjusted pyrolysis parameters in response to changes in feedstock moisture content, achieving a 12% increase in carbon stability. Faridi et al. (2024) implemented a model predictive control (MPC) system coupled with artificial neural networks to regulate temperature gradients in a fixed-bed reactor, maintaining steady biochar yields despite environmental disturbances. Additionally, Li et al. (2023) applied an edge computing and sensor fusion hybrid framework for real-time feedback control in distributed biochar production units located in remote regions. Collectively, these studies demonstrate that AI-enabled pyrolysis systems facilitate continuous monitoring, anticipatory control, and adaptive optimization—advancing industrial-scale, intelligent biochar production toward greater resilience, efficiency, and sustainability. Figure 19.5 presents the decision chart for AI- and ML-assisted biochar production.

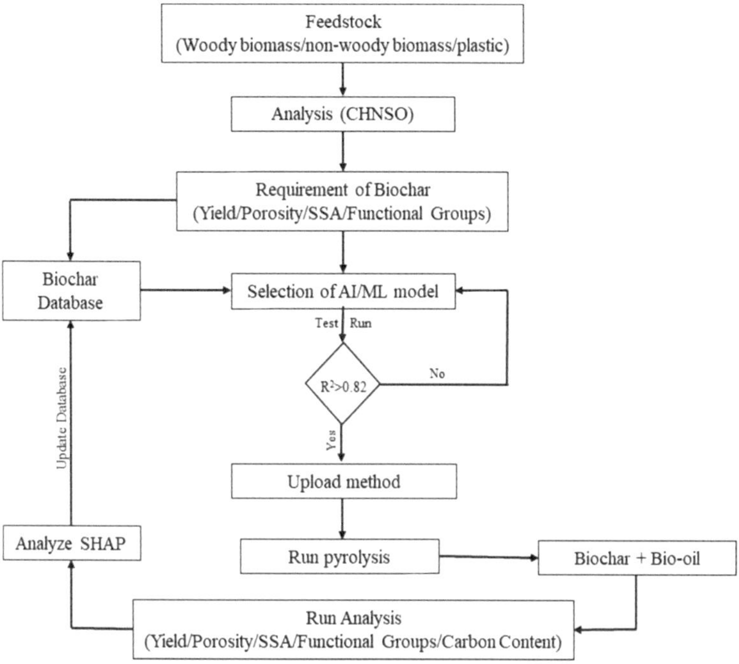

Fig. 19.5 Decision chart for biochar production implementing AI & ML

19.4 Underlying Implications

The characterization of AI and ML models as "black-box" systems presents a significant barrier to their adoption, particularly in safety–critical and industrial contexts such as biochar production. The lack of transparency in model decision-making can undermine trust and limit real-time applicability (Molnar 2022; Ribeiro et al. 2016). However, recent advances in explainable AI (XAI)—particularly the application of SHAP—have alleviated this challenge by providing a structured framework for interpreting model outputs. SHAP-based feature importance and sensitivity analyses decompose model predictions into the contributions of individual input variables, thereby enhancing transparency and interpretability (Nguyen et al. 2024a, b; Lundberg and Lee 2017). Such capabilities are vital for real-time process control and adaptive decision-making in pyrolysis, where understanding the influence of parameters such as temperature, feedstock type, and residence time is essential for optimizing biochar yield and quality (Pathy and Meher 2020).

In this context, explainable AI in biochar manufacturing is not solely concerned with predictive accuracy but also with building operational trust and delivering actionable insights for engineers and process managers. XAI transforms AI from an opaque computational mechanism into a "transparent intelligence," enabling users to debug models, detect systemic bias, and intervene confidently in real time (Tjoa and Guan 2020). This capability is indispensable for scaling AI solutions in industrial pyrolysis systems, where accountability, safety, and regulatory compliance are paramount. Traditional biochar process optimization has relied heavily on trial-and-error experimentation, which is both time- and resource-intensive (SimAnalytics n.d.). By contrast, AI/ML leverages historical and real-time sensor data to support predictive modeling and intelligent control. This transition allows manufacturers to shift from reactive to proactive management, anticipating system bottlenecks, dynamically adjusting parameters, and continuously optimizing performance (Jiang et al. 2025; Chen et al. 2023a, b).

One of the most impactful applications of AI/ML in this domain is predictive maintenance. By continuously monitoring operational signals and forecasting equipment degradation, ML algorithms can prevent unplanned downtime, extend the lifespan of critical pyrolysis components, and substantially reduce maintenance costs (Kane et al. 2022). At a broader level, these technologies facilitate a strategic shift from corrective to preventive action. Rather than responding to suboptimal yields or variable product properties after they occur, AI/ML enables early detection and timely correction. This proactive, data-driven approach improves product quality and consistency, minimizes feedstock losses, optimizes energy efficiency, and enhances the economic and environmental sustainability of the entire production process. Ultimately, AI/ML paves the way toward an intelligent, autonomous, and self-optimizing biochar production ecosystem.

19.5 Challenges and Future Directions

While AI and ML hold transformative potential for optimizing biochar production, large-scale industrial deployment is hindered by several critical challenges. Chief among these is the difficulty of managing data heterogeneity and ensuring model generalizability. Effective ML models require extensive, diverse, and high-quality datasets to capture the complex nonlinearities inherent in pyrolysis processes (Akinpelu et al. 2023). Yet, biochar datasets are often affected by experimental variability, sensor errors, and a lack of standardization—compounded by proprietary restrictions and reliance on pilot-scale setups (Gou et al. 2025; Jiang et al. 2025). As a result, models trained on laboratory-scale data may underperform in pilot or industrial-scale applications, a limitation referred to as poor external validity or transferability (Wang et al. 2025). Low-quality or incomplete datasets—common due to the high cost and complexity of full-factorial pyrolysis experiments—further reduce model accuracy, even when data imputation or augmentation is applied. Robust AI/ML frameworks therefore require not only large datasets but also representative and diverse data covering a wide range of feedstocks, temperatures, heating rates, and reactor configurations (Chen et al. 2023a, b). Emerging initiatives, such as the creation of open-access pyrolysis databases and the integration of real-time sensor data through Industrial Internet of Things (IIoT) systems, offer promising solutions (Zhou et al. 2024; Kondratenko et al. 2022).

Equally important is the issue of explainability. Although black-box models such as deep neural networks and ensemble methods (e.g., Random Forest, XGBoost) often deliver superior predictive performance, their opacity limits trust, accountability, and industrial adoption (Molnar 2022; Ribeiro et al. 2016). While interpretability tools such as SHAP, LIME, and global sensitivity analyses are increasingly used, challenges remain. Some methods yield inconsistent or context- dependent explanations and may overlook multicollinearity or interaction effects, potentially leading to misguided decisions in high-stakes scenarios (Tjoa and Guan 2020; Murdoch et al. 2019). Ultimately, the effectiveness of AI/ML depends on the quality, completeness, and interpretability of the underlying data and models. Bridging the gap between laboratory and industrial scales will require coordinated efforts to standardize data collection, enable cross- institutional data sharing, and improve both generalizability and explainability across diverse pyrolysis systems and biomass feedstocks (Zhou et al. 2024; Nguyen et al. 2024a, b). Addressing these foundational limitations is essential to realizing the full potential of AI/ML for smart, sustainable, and scalable biochar production.

19.6 Conclusion

The synthesis of biochar involves complex, multi-variable process parameters, making it challenging to consistently achieve the desired product quality. The application of AI and ML represents a significant transformation in biochar production, overcoming the limitations of traditional trial-and-error approaches. Conventional optimization, constrained by its empirical nature, is often time-consuming and resource-intensive, struggling to address the intricate, non-linear relationships among diverse feedstocks and pyrolysis process parameters. This study demonstrates that AI/ML offers a robust, data-driven framework to tackle these challenges, enabling a more efficient, predictable, and sustainable approach to producing customized biochar. The results highlight the strong predictive capabilities of models such as artificial neural networks (ANNs), support vector regression (SVR), and ensemble techniques like XGBoost in accurately forecasting key biochar properties, including yield, carbon content, and specific surface area. By leveraging large datasets, these models can identify optimal process conditions, substantially reducing the need for costly and labor-intensive experimental trials. This shift from reactive to proactive management improves resource efficiency, reduces environmental impacts, and ensures consistent production of biochar with tailored functionalities for applications ranging from soil enhancement to advanced materials.

A notable challenge remains the "black-box" nature of many advanced algorithms. Interpretability methods, particularly SHAP, are essential for building trust and fostering industrial adoption. By elucidating the influence of each input variable, these approaches provide transparent and actionable insights, enabling operators to make informed, real-time decisions. Nonetheless, large- scale adoption faces significant obstacles, including data variability and the need for improved model generalizability. Addressing these challenges through standardized, open-access databases and the development of adaptable, transferable models is critical. By confronting these issues, the integration of AI with pyrolysis research can fully realize the potential of biochar production, paving the way for intelligent, automated, and self-optimizing manufacturing systems. Such advancements hold promise for contributing substantially to a sustainable and circular economy.

References

Abbas-Abadi MS, Haghighi MN, Yeganeh H (2013) Evaluation of pyrolysis product of virgin high density polyethylene degradation using different process parameters in a stirred reactor. Fuel Process Technol 109:90–95. https://doi.org/10.1016/j.fuproc.2012.09.042

Aboulkas A, El Harfi K, El Bouadili A (2008) Non-isothermal kinetic studies on co-processing of olive residue and polypropylene. Energy Convers Manage 49(12):3666–3671. https://doi.org/10.1016/j.enconman.2008.06.029

Aghbashlo M, Tabatabaei M, Nadian MH, Davoodnia V, Soltanian S (2019) Prognostication of lignocellulosic biomass pyrolysis behavior using ANFIS model tuned by PSO algorithm. Fuel 253:189–198. https://doi.org/10.1016/j.fuel.2019.04.169

Akinpelu DA, Adekoya OA, Oladoye PO, Ogbaga CC, Okolie JA (2023) Machine learning applications in biomass pyrolysis: from biorefinery to end-of-life product management. Digital Chem Eng 8:100103. https://doi.org/10.1016/j.dche.2023.100103

Alabdrabalnabi A, Gautam R, Sarathy SM (2022) Machine learning to predict biochar and bio-oil yields from co-pyrolysis of biomass and plastics. Fuel 328:125303. https://doi.org/10.1016/j.fuel.2022.125303

Alori ET, Gabasawa AI, Elenwo CE (2025) A systematic review on machine learning-aided design of engineered biochar for soil and water contaminant removal. Front Soil Sci. https://doi.org/10.3389/fsoil.2025.1623083

AlphaNOME (n.d.) Data heterogeneity in AI: a deep dive into the challenges and solutions. https://www.alphanome.ai/post/data-heterogeneity-in-ai-a-deep-dive-into-the-challenges-and-solutions. Assessed on 27 June 2025

Al-Wabel MI, Al-Omran A, El-Naggar AH, Nadeem M (2024) Advancing biochar applications: a review of production. Sustainability 17(6):2685. https://www.mdpi.com/2071-1050/17/6/2685

Al-Wabel MI, Al-Omran A, El-Naggar AH, Nadeem M (2024) AI/ML in microalgae processes: a review. Sustainability 17(6):2685. https://pmc.ncbi.nlm.nih.gov/articles/PMC11592280/

Al-Wabel MI, Al-Omran A, El-Naggar AH, Nadeem M (2024) Bayesian optimization for activated biochar production. Sustainability 17(6):2685. https://www.mdpi.com/2071-1050/17/4/1554

Araki Y, Matsukawa Y, Saito Y, Matsushita Y, Aoki H, Era K, Aoki T (2021) Effects of carrier gas on the properties of soot produced by ethylene pyrolysis. Fuel Process Technol 213:106673. https://doi.org/10.1016/j.fuproc.2020.106673

Ates F, Putun E, Putun AE (2004) Fast pyrolysis of sesame stalk: yields and structural analysis of bio-oil. J Anal Appl Pyrol 71:779–790. https://doi.org/10.1016/j.jaap.2003.11.001

Balakumar S, Mahesh N, Kamaraj M, Aravind J (2025) Harnessing artificial intelligence for sustainable environmental remediation a review. Int J Environ Sci Technol 1–18. https://doi.org/10.1007/s13762-025-06528-9

Banda F, Simukoko L, Kalumba M, Kaoma M (2025) Utilizing machine learning to map aquatic weed yields for biochar production: a case study in the Bangweulu Wetlands, Zambia. Adv Environ Eng Res 6(2):1–22. https://doi.org/10.21926/aeer.2502020

Barkhordari MS, Qi C (2025) Integrating machine learning and reliability analysis: a novel approach to predicting heavy metal removal efficiency using biochar. Ecotoxicol Environ Saf 299:118381. https://doi.org/10.1016/j.ecoenv.2025.118381

Biochar (n.d.) Biochar production. https://biochar.co.uk/. Assessed on 27 June 2025

Breiman L (1996) Bagging predictors. Mach Learn 24:123–140. https://doi.org/10.1007/BF00058655

Breiman L (2001) Random forests. Mach Learn 45:5–32. https://doi.org/10.1023/A:1010933404324

Breiman L, Friedman J, Olshen RA, Stone CJ (2017) In: Classification and regression trees. Routledge.https://doi.org/10.1201/9781315139470

Bulkinside (2025) Innovative uses of biochar across different sectors. https://bulkinside.com/biomass-processing/innovative-uses-of-biochar-across-differentsectors/. Assessed on 27 June 2025

Castillo KAA, Johnson RL, Castillo CJA (2025) Experimental optimization of pellet tensile strength from spruce biochar produced at elevated pressure. Energy Fuels 39(1):123–132. https://pubs.acs.org/doi/abs/10.1021/acs.energyfuels.4c05424

Cedrone G, Bracciale MP, Cafiero L, Langone M, Mattioli D, Scarsella M, Tuffi R (2024) Optimization of pyrolysis parameters by design of experiment for the production of biochar from sewage sludge. Environments 11(10):210. https://doi.org/10.3390/environments11100210

Chen W, Shi S, Chen M, Zhou X (2017) Fast co-pyrolysis of waste newspaper with high-density polyethylene for high yields of alcohols and hydrocarbons. Waste Manage 67:155–162. https://doi.org/10.1016/j.wasman.2017.05.032

Chen C, Wang Z, Ge Y, Liang R, Hou D, Tao J, Yan B, Zheng W, Velichkova R, Chen G (2023a) Characteristics prediction of hydrothermal biochar using data enhanced interpretable machine learning. Biores Technol 377:128893. https://doi.org/10.1016/j.biortech.2023.128893

Chen MW, Chang MS, Mao Y, Hu S, Kung CC (2023b) Machine learning in the evaluation and prediction models of biochar application: a review. Sci Prog 106(1):00368504221148842. https://doi.org/10.1177/00368504221148

Chen T, Guestrin C (2016) Xgboost: a scalable tree boosting system. In: Proceedings of the 22nd ACM SIGKDD international conference on knowledge discovery and data mining, pp 785–794

Cheng FY, Li D (1998) Genetic algorithm development for multiobjective optimization of structures. AIAA J 36(6):1105–1112

Choi HS, Choi YS, Park HC (2012) Fast pyrolysis characteristics of lignocellulosic biomass with varying reaction conditions. Renew Energy 42:131–135. https://doi.org/10.1016/j.renene.2011.08.049

Crombie K, Mašek O (2015) Pyrolysis biochar systems, balance between bioenergy and carbon sequestration. Gcb Bioenergy 7(2):349–361. https://doi.org/10.1111/gcbb.12137

DataCamp (n.d.) Introduction to machine learning with Python. https://www.datacamp.com. Assessed on 27 June 2025

DataCamp (n.d.) An introduction to SHAP values and machine learning. https://www.datacamp.com/tutorial/introduction-to-shap-values-machine-learning-interpretability. Assessed on 27 June 2025

Dhar SA, Sakib TU, Hilary LN (2022) Effects of pyrolysis temperature on production and physicochemical characterization of biochar derived from coconut fiber biomass through slow pyrolysis process. Biomass Convers Biorefinery 12(7):2631–2647. https://doi.org/10.1007/s13399-020-01116-y

Do PT, Nguyen LX (2024) A review of thermochemical decomposition techniques for biochar production. Environ Dev Sustain. https://doi.org/10.1007/s10668-024-05841-6

Draper NR, Smith H (1998) In: Applied regression analysis, vol 326. Wiley

Faridi IK, Tsotsas E, Kharaghani A (2024) Advancing process control in fluidized bed biomass gasification using model-based deep reinforcement learning. Processes 12(2):254. https://doi.org/10.3390/pr12020254

Frontiers in Environmental Science (n.d.) Applications of AI/ML techniques for global optimization of thermal energy conversion systems. https://www.frontiersin.org/research-topics/70818/applications-of-aiml-techniques-for-global-optimization-of-thermal-energy-conversion-systems. Assessed on 27 June 2025

Gao X, Yu J, Lu L, Rogers WA (2021) Coupling particle scale model and SuperDEM-CFD for multiscale simulation of biomass pyrolysis in a packed bed pyrolyzer. AIChE J 67(4):e17139. https://doi.org/10.1002/aic.17139

Gou J, Sajid GH, Sabri MM, El-Meligy M, El Hindi K, Othman NA (2025) Optimizing biochar yield and composition prediction with ensemble machine learning models for sustainable production. Ain Shams Eng J 16(1):103209. https://doi.org/10.1016/j.asej.2024.103209

Graham KE, Anderson CE, Boehm AB (2021) Viral pathogens in urban stormwater runoff: occurrence and removal via vegetated biochar-amended biofilters. Water Res 207:117829. https://doi.org/10.1016/j.watres.2021.117829

Gupta M, Savla N, Pandit C, Pandit S, Gupta PK, Pant M, Thakur VK (2022) Use of biomass-derived biochar in wastewater treatment and power production: a promising solution for a sustainable environment. Sci Total Environ 825:153892. https://doi.org/10.1016/j.scitotenv.2022.153892

Hakeem IG, Halder P, Dike CC, Chiang K, Sharma A, Paz-Ferreiro J, Shah K (2022) Advances in biosolids pyrolysis: roles of pre-treatments, catalysts, and co-feeding on products distribution and high-value chemical production. J Anal Appl Pyrol 166:105608. https://doi.org/10.1016/j.jaap.2022.105608

Haykin S (1994) In: Neural networks: a comprehensive foundation. Prentice Hall PTR

Hooda S, Mondal P (2024) Predictive modeling of plastic pyrolysis process for the evaluation of activation energy: explainable artificial intelligence based comprehensive insights. J Environ Manag 360:121189. https://doi.org/10.1016/j.jenvman.2024.121189

International Biochar Initiative (2018) IBI report: biochar stability test method final. https://biochar-international.org/wp-content/uploads/2018/04/IBI_Report_Biochar_Stability_Test_Method_Final.pdf. Assessed on 27 June 2025

Jamal MU, Fletcher AJ (2023) Design of experiments study on Scottish wood biochars and process parameter influence on final biochar characteristics. Bioenergy Res 16:2342–2355. https://doi.org/10.1007/s12155-023-10595-6

Jiang Z, Xia Q, Lu X, Yue W, Chen A, Liu X, Zhao C (2025) Machine learning prediction of biochar-specific surface area based on plant characterization information. Renew Energy 243:122633. https://doi.org/10.1016/j.renene.2025.122633

Kane AP, Kore AS, Khandale AN, Nigade SS, Joshi PP (2022) Predictive maintenance using machine learning. arXiv preprint arXiv:2205.09402. https://doi.org/10.48550/arXiv.2205.09402

Khan M, Ullah Z, Mašek O, Naqvi SR, Khan MNA (2022) Artificial neural networks for the prediction of biochar yield: a comparative study of metaheuristic algorithms. Biores Technol 355:127215. https://doi.org/10.1016/j.biortech.2022.127215

Kim KH, Eom IY, Lee SM, Choi D, Yeo H, Choi IG, Choi JW (2011) Investigation of physicochemical properties of biooils produced from yellow poplar wood (*Liriodendron tulipifera*) at various temperatures and residence times. J Anal Appl Pyrol 92(1):2–9. https://doi.org/10.1016/j.jaap.2011.04.002

Kondratenko Y, Kozlov O, Topalov A, Korobko O, Gerasin O (2022) Automation of control processes in specialized pyrolysis complexes based on industrial internet of things. In: Dependable IoT for human and industry, River Publishers, pp 367–388. https://doi.org/10.1016/j.jclepro.2022.130403

Le AT, Pandey A, Sirohi R, Sharma P, Chen WH, Pham NDK, Hoang AT (2023) Precise prediction of biochar yield and proximate analysis by modern machine learning and SHapley additive exPlanations. Energy Fuels 37(22):17310–17327. https://doi.org/10.1021/acs.energyfuels.3c02868

Li L, Li J, Li Y, Luo J, Wang Y, Wang X (2023) Perspective of waste to energy and fuel with negative emission potential: smart environmental and techno-economic analysis. Energy & Fuels 37(19):14556–14573. https://doi.org/10.1021/acs.energyfuels.3c02109
Li L, Zhang Y-J, Novak A, Yang Y, Wang J (2021) Role of biochar in improving sandy soil water retention and resilience to drought. Water 13(4):407. https://doi.org/10.3390/w13040407
Li J, Wang Y, Zhang B, Liu Q (2022) Application of interpretable neural networks for optimizing biochar production from agricultural waste. Renew Energy 189:1196–1207. https://doi.org/10.1016/j.renene.2022.03.038
Liang C, Gascó G, Fu S, Méndez A, Paz-Ferreiro J (2016) Biochar from pruning residues as a soil amendment: effects of pyrolysis temperature and particle size. Soil and Tillage Res 164:3–10. https://doi.org/10.1016/j.still.2015.10.002
Liu C, Zhang HX (2022) Modified-biochar adsorbents (MBAs) for heavy-metal ions adsorption: a critical review. J Environ Chem Eng 10(2):107393. https://doi.org/10.1016/j.jece.2022.107393
Liu WJ, Jiang H, Yu HQ (2019) Emerging applications of biochar-based materials for energy storage and conversion. Energy Environ Sci 12(6):1751–1779. https://doi.org/10.1039/C9EE00206E
Lundberg SM, Lee S-I (2017) A unified approach to interpreting model predictions. In: Advances in neural information processing systems, vol 30. https://proceedings.neurips.cc/paper_files/paper/2017/file/8a20a8621978632d76c43dfd28b67767-Paper.pdf. Assessed on 27th June 2025
Lv B, Ma B, Li Y, Wu L, Huang M, He X, Xue J, Yang L (2025) Biochar derived from feedstock with high lignin content leads to better soil improvement performance in red soils: from the perspective of soil microbial regulation and carbon stabilization. Environ Geochem Health 47(6):1–15. https://doi.org/10.1007/s10653-025-02522-9
Mace MJ, Fyson CL, Schaeffer M, Hare WL (2021) Large-scale carbon dioxide removal to meet the 1.5 C limit: key governance gaps, challenges and priority responses. Global Pol 12:67–81. https://doi.org/10.1111/1758-5899.12921
Malashin IP, Martysyuk D, Nelyub V, Borodulin A, Gantimurov A, Tynchenko V (2026) Multi-objective optimisation in biofuel and bioenergy systems: a comprehensive review. Int J Sustain Eng 19(1):2619227. https://doi.org/10.1080/19397038.2026.2619227
Man X, Chan EP (2020) The best way to select features? comparing mda, lime, and shap. J Finan Data Sci. https://doi.org/10.3905/jfds.2020.1.047
Mangalathu S, Hwang SH, Jeon JS (2020) Failure mode and effects analysis of RC members based on machine-learning-based SHapley additive exPlanations (SHAP) approach. Eng Struct 219:110927. https://doi.org/10.1016/j.engstruct.2020.110927
Mienye ID, Jere N (2024) A survey of decision trees: concepts, algorithms, and applications. IEEE Access. https://doi.org/10.1109/ACCESS.2024.3416838
Modern Ag (n.d.) The dark side of biochar. https://modern-ag.net/the-dark-side-of-biochar/. Assessed on 27 June 2025
MolnarC (n.d.) Interpretable machine learning: a guide for making black box models explainable. Christoph Molnar. https://christophm.github.io/interpretable-ml-book/shap.html
Molnar C (2022) In: Interpretable machine learning. 2nd ed. Leanpub. https://christophm.github.io/interpretable-ml-book/
Mong GR, Chong CT, Chong WWF, Ng JH, Ong HC, Ashokkumar V, Yasin MFM (2022) Progress and challenges in sustainable pyrolysis technology: reactors, feedstocks and products. Fuel 324:124777. https://doi.org/10.1016/j.fuel.2022.124777
Murdoch WJ, Singh C, Kumbier K, Abbasi-Asl R, Yu B (2019) Definitions, methods, and applications in interpretable machine learning. Proc Natl Acad Sci 116(44):22071–22080. https://doi.org/10.1073/pnas.1900654116
Murphy R Jr, Estabrooks A, Gauvin J, Gray S, Kroska AC, Wolf N, Harris BP (2021) Using mental models to quantify linear and non-linear relationships in complex fishery systems. Mar Policy 132:104695. https://doi.org/10.1016/j.marpol.2021.104695
Nguyen VG, Sharma P, Ağbulut Ü, Le HS, Cao DN, Dzida M, Tran VD (2024a) Improving the prediction of biochar production from various biomass sources through the implementation of

eXplainable machine learning approaches. Int J Green Energy 21(12):1543–5075. https://www.tandfonline.com/doi/abs/10.1080/15435075.2024.2326076

Nguyen VG, Sharma P, Ağbulut Ü, Le HS, Cao DN, Dzida M, Tran VD (2024b) Improving the prediction of biochar production from various biomass sources through the implementation of eXplainable machine learning approaches. Int J Green Energy 21(12):2771–2798. https://doi.org/10.1080/15435075.2024.2326076

Nguyen VG, Sharma P, Ağbulut Ü, Le HS, Tran VD, Cao DN (2024c) Precise prognostics of biochar yield from various biomass sources by Bayesian approach with supervised machine learning and ensemble methods. Int J Green Energy 21(9):2180–2204. https://doi.org/10.1080/15435075.2023.2297776

Nguyen VG, Sharma P, Ağbulut Ü, Le HS, Truong TH, Dzida M, Tran VD (2024d) Machine learning for the management of biochar yield and properties of biomass sources for sustainable energy. Biofuels, Bioprod Biorefin 18(2):567–593. https://doi.org/10.1002/bbb.2596

Pathy A, Meher S (2020) Predicting algal biochar yield using eXtreme gradient boosting (XGB) algorithm of machine learning methods. Algal Res 50:102006. https://doi.org/10.1016/j.algal.2020.102006

Plastics Engineering (2025) Machine learning for Pyrolysis optimization. https://www.plasticsengineering.org/2025/03/machine-learning-for-pyrolisis-optimization-008234/

Pütün AE, Özcan A, Pütün E (1999) Pyrolysis of hazelnut shells in a fixed-bed tubular reactor: yields and structural analysis of bio-oil. J Anal Appl Pyrol 52(1):33–49. https://doi.org/10.1016/S0165-2370(99)00044-3

Ribeiro MT, Singh S, Guestrin C (2016) Why should I trust you?: explaining the predictions of any classifier. In: Proceedings of the 22nd ACM SIGKDD international conference on knowledge discovery and data mining, pp 1135–1144. https://doi.org/10.1145/2939672.2939778

Rochester Institute of Technology (n.d.) What is biochar and how it's made. https://www.rit.edu/sustainabilityinstitute/blog/what-biochar-and-how-it-made. Assessed on 27 June 2025

Rodriguez JA, Lustosa Filho JF, Melo LCA, de Assis IR, de Oliveira TS (2021) Co-pyrolysis of agricultural and industrial wastes changes the composition and stability of biochars and can improve their agricultural and environmental benefits. J Anal Appl Pyrol 155:105036. https://doi.org/10.1016/j.jaap.2021.105036

Scott DS, Majerski P, Piskorz J, Radlein D (1999) A second look at fast pyrolysis of biomass—The RTI process. J Anal Appl Pyrol 51:23–37. https://doi.org/10.1016/S0165-2370(99)00006-6

Shariff A, Aziz NSM, Saleh NM, Ruzali NSI (2016) The effect of feedstock type and slow pyrolysis temperature on biochar yield from coconut wastes. Int J Chem Mol Eng 10(12):1410–1414. https://www.researchgate.net/profile/Syairah-Aziz/publication/311349228_The_Effect_of_Feedstock_Type_and_Slow_Pyrolysis_Temperature_on_Biochar_Yield_from_Coconut_Wastes/links/584249e208ae61f75dd10802/The-Effect-of-Feedstock-Type-and-Slow-Pyrolysis-Temperature-on-Biochar-Yield-from-Coconut-Wastes.pdf. Assessed on 27 June 2025

Sharma V, Tsai ML, Chen CW, Sun PP, Nargotra P, Dong CD (2023) Advances in machine learning technology for sustainable biofuel production systems in lignocellulosic biorefineries. Sci Total Environ 886:163972. https://doi.org/10.1016/j.scitotenv.2023.163972

Sharuddin SDA, Abnisa F, Daud WMAW, Aroua MK (2016) A review on pyrolysis of plastic wastes. Energy Convers Manage 115:308–326. https://doi.org/10.1016/j.enconman.2016.02.037

Shi L, Gong J, Zhai C (2022) Application of a hybrid PSO-GA optimization algorithm in determining pyrolysis kinetics of biomass. Fuel 323:124344. https://doi.org/10.1016/j.fuel.2022.124344

SimAnalytics (n.d.) AI-Driven process control in pyrolysis systems. Retrieved from https://simalytics.ai/pyrolysis-ai-control. Assessed on 27 June 2025

SimAnalytics (n.d.) Machine learning for chemical production. https://simanalytics.com/insights/machine-learning-for-chemical-production/. Assessed on 27 June 2025

Smola AJ, Schölkopf B (2004) A tutorial on support vector regression. Stat Comput 14:199–222. https://doi.org/10.1023/B:STCO.0000035301.49549.88

Song Y, Huang Z, Jin M, Liu Z, Wang X, Hou C, Zhang Y (2024) Machine learning prediction of biochar physicochemical properties based on biomass characteristics and pyrolysis conditions. J Anal Appl Pyrol 181:106596. https://doi.org/10.1016/j.jaap.2024.106596

Stack Overflow (2021) Multi-objective optimization in genetic algorithm, September. https://stackoverflow.com/questions/66644214/multi-objective-optimization-in-genetic-algorithm. Assessed on 27 June 2025

Su G, Jiang P (2024) Machine learning models for predicting biochar properties from lignocellulosic biomass Torre faction. Biores Technol 399:130519. https://doi.org/10.1016/j.biortech.2024.130519

Sun J, He F, Pan Y, Zhang Z (2016). Effects of pyrolysis temperature and residence time on physicochemical properties of different biochar types. Acta Agriculturae Scandinavica, Section B—Soil Plant Sci 67(1):12–22. https://doi.org/10.1080/09064710.2016.1214745

Suvarna M, Li J, Wang X (2020) Applied machine learning for prediction of biochar properties and applications. In: 2020 Virtual AIChE annual meeting. AIChE, November. https://proceedings.aiche.org/conferences/aiche-annual-meeting/2020/proceeding/paper/583f-applied-machine-learning-prediction-biochar-properties-and-applications. Assessed on 27 June 2025

Thakur A, Kumar A, Somya A (2025) Forestry and agricultural residues-based wastes: fundamentals, classification, properties, and applications. In: Biomass wastes for sustainable industrial applications, CRC Press, pp 95–139

The Royal Horticultural Society (n.d.) Biochar. https://www.rhs.org.uk/soil-composts-mulches/biochar. Assessed on 27 June 2025

Tjoa E, Guan C (2020) A survey on explainable artificial intelligence (Xai): toward medical Xai. IEEE Trans Neural Netw Learn Syst 32(11):4793–4813. https://doi.org/10.1109/TNNLS.2020.3027314

Tomczyk A, Sokołowska Z, Boguta P (2020) Biochar physicochemical properties: pyrolysis temperature and feedstock kind effects. Rev Environ Sci Biotechnol 19:191–215. https://doi.org/10.1007/s11157-020-09523-3

Uppalapati S, Paramasivam P, Kilari N, Chohan JS, Kanti PK, Vemanaboina H, Gupta R (2025) Precision biochar yield forecasting employing random forest and XGBoost with Taylor diagram visualization. Sci Rep 15(1):7105. https://doi.org/10.1038/s41598-025-91450-w

Wakefield BioChar (n.d.) How biochar aids in municipal wastewater remediation. https://wakefieldbiochar.com/learning-center/how-biochar-aids-in-municipal-wastewater-remediation/. Assessed on 27 June 2025

Wang S, Wang Q, Hu YM, Xu SN, He ZX, Ji HS (2015) Study on the synergistic co-pyrolysis behaviors of mixed rice husk and two types of seaweed by a combined TG-FTIR technique. J Anal Appl Pyrol 114:109–118. https://doi.org/10.1016/j.jaap.2015.05.008

Wang Y, Wang HS, Tang CS, Gu K, Shi B (2019) Remediation of heavy-metal-contaminated soils by biochar: a review. Environ Geotech 9(3):135–148. https://doi.org/10.1680/jenge.18.00091

Wang Y, Xu L, Li J, Ren Z, Liu W, Ai Y, Zhou Y, Li Q, Zhang B, Guo N, Qu J, Zhang Y (2024) Multi-output neural network model for predicting biochar yield and composition. Sci Total Environ 945:173942. https://doi.org/10.1016/j.scitotenv.2024.173942

Wang Y, Li Y, Zhang Y (2025) Multi-objective optimization of biochar pyrolysis temperature and application rate for potato cultivation in arid and semi-arid regions. Front Plant Sci 15:1571305. https://www.frontiersin.org/journals/plant-science/articles/10.3389/fpls.2025.1571305/pdf

Xiang L, Harindintwali JD, Wang F, Redmile-Gordon M, Chang SX, Fu Y, Xing B (2022) Integrating biochar, bacteria, and plants for sustainable remediation of soils contaminated with organic pollutants. Environ Sci & Technol 56(23):16546–16566. https://doi.org/10.1021/acs.est.2c02976

Xie H, Zhou X, Zhang Y, Yan W (2025a) Prediction of biochar characteristics and optimization of pyrolysis process by response surface methodology combined with artificial neural network. Biomass Convers Biorefinery 15(3):4745–4757. https://doi.org/10.1007/s13399-023-05194-6

Xie Z, Cao Y, Luo Z (2025b) Machine learning prediction of physical properties of lignin derived porous carbon via catalytic pyrolysis. Green Chem. https://doi.org/10.1039/D4GC05515B

Xu Q, Du L, Deng R (2024) Using machine learning to predict biochar yield and carbon content: enhancing efficiency and sustainability in biomass conversion. J Environ Managem 350:119567. https://www.researchgate.net/publication/382846825_Using_machine_learning_to_predict_biochar_yield_and_carbon_content_Enhancing_efficiency_and_sustainability_in_biomass_conversion. Assessed on 27 June 2025

Yadav S, Rajput P, Balasubramanian P, Liu C, Li F, Zhang P (2025b) Machine learning-driven prediction of biochar adsorption capacity for effective removal of Congo red dye. Carbon Res 4(1):11. https://doi.org/10.1007/s44246-024-00168-3

Yadav S, Rajput P, Balasubramanian P, Liu C, Li F, Zhang P (2025) Machine learning- driven prediction of biochar adsorption capacity for effective removal of Congo red dye. Carbon Res 4(1):11. https://ouci.dntb.gov.ua/en/works/4LYan6Vv/

Yang H, Liu X, Liu Y, Cui J, Xiao Y (2023) Revolutionizing biochar synthesis for enhanced heavy metal adsorption: harnessing machine learning and Bayesian optimization. J Environ Chem Eng 11(5):110593. https://doi.org/10.1016/j.jece.2023.110593

Ye P, Guo B, Qin H, Wang C, Liu Y, Chen Y, Yu Q (2025) The state-of-the-art review on biochar as green additives in cementitious composites: performance, applications, machine learning predictions, and environmental and economic implications. Biochar 7(1):21. https://doi.org/10.1007/s42773-024-00423-1

Yu H, Zou W, Chen J, Chen H, Yu Z, Huang J, Gao B (2019) Biochar amendment improves crop production in problem soils: a review. J Environ Manage 232:8–21. https://doi.org/10.1016/j.jenvman.2018.10.117

Zheng Y, Yu C, Fu L (2023) Biochar-based materials for electroanalytical applications: an overview. Green Anal Chem 7:100081. https://doi.org/10.1016/j.greeac.2023.100081

Zhou X, Liu X, Sun L, Jia X, Tian F, Liu Y, Wu Z (2024) Prediction of biochar yield and specific surface area based on integrated learning algorithm. C, 10(1):10. https://doi.org/10.3390/c10010010

Zhu X, Li Y, Wang X (2019) Machine learning prediction of biochar yield and carbon contents in biochar based on biomass characteristics and pyrolysis conditions. Biores Technol 288:121527. https://doi.org/10.1016/j.biortech.2019.121527

Zong Z, Guan Y (2024) AI-driven intelligent data analytics and predictive analysis in industry 4.0: transforming knowledge, innovation, and efficiency. J Knowled Econ 1–40. https://doi.org/10.1007/s13132-024-02001-z

Index

D. Moulick et al. (eds.), *Biochar: Achieving Agro-Environmental Sustainability*, Sustainability Solutions, https://doi.org/10.1007/978-3-032-19855-6